# BIO-TARGETS AND DRUG DELIVERY APPROACHES

# BIO-TARGETS
# AND DRUG
# DELIVERY
# APPROACHES

EDITED BY
# Sabyasachi Maiti
# Kalyan Kumar Sen

**CRC Press**
Taylor & Francis Group
Boca Raton London New York

CRC Press is an imprint of the
Taylor & Francis Group, an **informa** business

CRC Press
Taylor & Francis Group
6000 Broken Sound Parkway NW, Suite 300
Boca Raton, FL 33487-2742

First issued in paperback 2022

© 2017 by Taylor & Francis Group, LLC
CRC Press is an imprint of Taylor & Francis Group, an Informa business

No claim to original U.S. Government works

ISBN-13: 978-1-498-72999-4 (hbk)
ISBN-13: 978-1-03-233987-0 (pbk)
DOI: 10.1201/9781315370118

### Library of Congress Cataloging-in-Publication Data

Names: Maiti, Sabyasachi, editor. | Sen, Kalyan Kumar, editor.
Title: Bio-targets and drug delivery approaches / editors, Sabyasachi Maiti and Kalyan Kumar Sen.
Other titles: Biotargets and drug delivery approaches
Description: Boca Raton : Taylor & Francis, 2017. | Includes bibliographical references and index.
Identifiers: LCCN 2016012233 | ISBN 9781498729994 (alk. paper)
Subjects: | MESH: Drug Delivery Systems
Classification: LCC RS199.5 | NLM QV 785 | DDC 615/.6--dc23
LC record available at http://lccn.loc.gov/2016012233

**Visit the Taylor & Francis Web site at**
**http://www.taylorandfrancis.com**

**and the CRC Press Web site at**
**http://www.crcpress.com**

# Contents

Contents vii

# Preface

In recent years, the search for biological targets and the consequent development of targeted delivery systems have been an intensive area of pharmaceutical research. Drugs cannot exert their therapeutic effects if they do not reach their target sites in the body at the appropriate concentration and persist for a sufficient length of time. To maximize drug utilization, it is necessary to deliver drugs at the target tissue. Therefore, the design of controlled release or bioresponsive formulation, which can liberate the active constituents over a long period of time, has been proven beneficial in terms of better pharmacological performance at low doses with reduced side effects. New modes of drug administration are also under investigation to overcome the obstacles associated with biodegradation and pharmacokinetics. Most general pharmacy textbooks focus on conventional pharmaceutical formulations such as tablets, capsules, ointments, suppositories, etc.

Although the targeting of therapeutics is an expanding field of research, there is currently a scarcity of books that covers all aspects of novel drug targeting strategies. There are excellent reviews and book chapters that deal with one or more topics, but a book containing comprehensive coverage and up-to-date progress in the area of drug targeting is not available. Our attempt is to bridge this gap by providing a chapterwise discussion on various aspects of drug delivery and targeting in a single comprehensive text. This book emphasizes ongoing worldwide research on biological target identification for a particular disease or disorder and the delivery of bioactive molecules at the molecular, cellular, and higher levels. This book captures current topics of interest and the latest research updates in this field.

The book is divided into 20 chapters to provide a clear overview of each topic. The book starts with a basic understanding of drug targeting (Chapter 1) followed by bio-target identification (Chapter 2). This is subsequently illustrated with an example of therapeutic targets for diabetes in Chapter 3. The prodrug strategy that may augment target specificity and drug development is discussed in Chapter 4. Later on, cell/organ-based approaches such as those for targeting central nervous systems (CNS), colon, lymphatic systems, ocular regions, bone, solid tumors, and mitochondria are covered in Chapters 5 through 11. In addition, different drug targeting devices are included in Chapters 12 through 18 emphasizing antisense oligonucleotide-based targeting strategies, biodegradable polymeric carriers, organic–inorganic composites, carbon nanotubes, functionalized cyclodextrin, nanopolymer scaffolds, and nano-therapeutic systems of polyionic glucan derivatives. Herbal medicines are currently receiving increasing attention due to their near-zero side effects. Therefore, Chapter 19 is devoted to a discussion of different novel carriers for herbal medicines of clinical significance. The toxicity of nanomaterials has been a great concern before their clinical application. Keeping this in mind, Chapter 20 takes into account the toxicity of nanomaterials that may cause harm to various organs of our body.

In this book we have tried our best to cover recent developments in targeted delivery approaches for therapeutic molecules. To make this book reader-friendly and useful, we have provided in-depth literature reports and suitable illustrations. Important references have been included in each chapter for the benefit of readers who wish to pursue any of these topics in greater depth. We hope that this book will meet the demand of a reference book for concerned professionals and researchers who intend to conceptualize, develop, and optimize targeted drug delivery approaches.

This book is primarily intended for undergraduate and postgraduate students undertaking programs in pharmaceutical sciences, medicine, biotechnology, biomedical engineering, and other related subjects. It will also be helpful as a reference for research workers seeking information concerning the design and development of drug targeting systems. We strongly believe that this book will also provide assistance to research workers in developing innovations for humankind. We are most grateful to the authors for their contribution and kind cooperation for the successful completion of this book. Without the skillful sharing of knowledge from a diverse class of expertise, the completion of this book would not have been possible. The contributions of all authors are acknowledged overleaf. We are proud of our family members for their continuous moral support and inspiration during the preparation of this book. We are thankful to CRC Press for their keen interest in and expert assistance with the design of an impressive book cover, and preparation and publication of this book. Constructive comments and suggestions from readers in improving the quality of this book are welcome.

Together with our contributing authors, we will be extremely pleased if our efforts fulfill the needs of pharmaceutical and biomedical students and researchers.

**Dr. Sabyasachi Maiti**
**Dr. Kalyan Kumar Sen**

# Contributors

**Kumar Anand**
Division of Pharmacology
Department of Pharmaceutical
    Technology
Jadavpur University
Kolkata, India

**Subham Banerjee**
Centre for Biodesign and Diagnostics
Translational Health Science and
    Technology Institute
Faridabad, India

**Sugata Banerjee**
Department of Pharmaceutical Sciences
    and Technology
Birla Institute of Technology
Ranchi, India

**Anish Bhattacharya**
Carbon Nanotechnology Laboratory
National Institute of Technology
Durgapur, India

**Sanchari Bhattacharya**
Department of Pharmaceutical
    Technology
Jadavpur University
Kolkata, India

**Sankhadip Bose**
Department of Phytochemistry
Gupta College of Technological
    Sciences
Asansol, India

**Amit K. Chakraborty**
Carbon Nanotechnology Laboratory
National Institute of Technology
Durgapur, India

**Samrat Chakraborty**
Department of Pharmaceutical
    Technology
Jadavpur University
Kolkata, India

**Ankan Choudhury**
Department of Pharmaceutical
    Technology
Jadavpur University
Kolkata, India

**Avik Das**
Department of Pharmacology and
    Toxicology
Gupta College of Technological
    Sciences
Asansol, India

**Nandita G. Das**
College of Pharmacy and Health
    Sciences
Department of Pharmaceutical Sciences
Butler University
Indianapolis, Indiana

**Sudip K. Das**
College of Pharmacy and Health
    Sciences
Department of Pharmaceutical Sciences
Butler University
Indianapolis, Indiana

**Rana Datta**
Department of Pharmacology and
    Toxicology
Gupta College of Technological
    Sciences
Asansol, India

**Manabendra Dhua**
Department of Pharmaceutical
   Chemistry
Gupta College of Technological
   Sciences
Asansol, India

**Arijit Gandhi**
Department of Quality Assurance
Albert David Ltd.
Kolkata, India

**Shovanlal Gayen**
Department of Pharmaceutical Sciences
Dr. Harisingh Gour Central University
Madhya Pradesh, India

**Animesh Ghosh**
Department of Pharmaceutical Sciences
   and Technology
Birla Institute of Technology
Ranchi, India

**Bhargav Hirapara**
College of Pharmacy
University of South Florida Health
Tampa, Florida

**Sougata Jana**
Department of Pharmaceutics
Gupta College of Technological
   Sciences
Asansol, India

**Subrata Jana**
Department of Chemistry
Indira Gandhi National Tribal
   University
Madhya Pradesh, India

**Emily Johnson**
College of Pharmacy
University of South Florida Health
Tampa, Florida

**Sanmoy Karmakar**
Department of Pharmaceutical
   Technology
Jadavpur University
Kolkata, India

**Amit K. Keshari**
Department of Pharmaceutical Sciences
Babasaheb Bhimrao Ambedkar
   University
Lucknow, India

**Venkateshwaran Krishnaswami**
Department of Pharmaceutical
   Technology
Bharathidasan Institute of Technology
Anna University
Tamil Nadu, India

**Raghavendra V. Kulkarni**
Department of Pharmaceutical
   Technology
BLDEA's College of Pharmacy
BLDE University Campus
Karnataka, India

**Vaishnavi Suresh Kumar**
Department of Pharmaceutical
   Technology
Bharathidasan Institute of Technology
Anna University
Tamil Nadu, India

**Priya Singh Kushwaha**
Department of Pharmaceutical Sciences
Babasaheb Bhimrao Ambedkar
   University
Lucknow, India

**Buddhadev Layek**
Department of Pharmaceutics
University of Minnesota
Minneapolis, Minnesota

**Sabyasachi Maiti**
Department of Pharmaceutics
Gupta College of Technological Sciences
Asansol, India

**Siddhartha Maity**
Council of Scientific and Industrial
    Research
Department of Pharmaceutical
    Technology
Jadavpur University
Kolkata, India

**Laboni Mondal**
Department of Pharmaceutical
    Technology
Jadavpur University
Kolkata, India

**Biswajit Mukherjee**
Department of Pharmaceutical
    Technology
Jadavpur University
Kolkata, India

**Subramanian Natesan**
Department of Pharmaceutical
    Technology
Bharathidasan Institute of Technology
Anna University
Tamil Nadu, India

**Amit Kumar Nayak**
Department of Pharmaceutics
Seemanta Institute of Pharmaceutical
    Sciences
Odisha, India

**Jayabalan Nirmal**
School of Materials Science and
    Engineering
Nanyang Technological University
Singapore

**Mintu Pal**
Biotechnology Division
CSIR-North East Institute of Science
    and Technology
Assam, India

**Yashwant Pathak**
College of Pharmacy
University of South Florida Health
Tampa, Florida

**Shubhajit Paul**
Department of Pharmaceutics
University of Minnesota
Minneapolis, Minnesota

**Saranya Radhakrishnan**
Department of Pharmaceutical
    Technology
Bharathidasan Institute of Technology
Anna University
Tamil Nadu, India

**Somasree Ray**
Department of Pharmaceutics
Gupta College of Technological
    Sciences
Asansol, India

**Sanika A. Rege**
College of Pharmacy and Health
    Sciences
Department of Pharmaceutical Sciences
Butler University
Indianapolis, Indiana

**Biswanath Sa**
Department of Pharmaceutical
    Technology
Jadavpur University
Kolkata, India

**Sudipta Saha**
Department of Pharmaceutical Sciences
Babasaheb Bhimrao Ambedkar
    University
Lucknow, India

**Bhabani Sankar Satapathy**
Department of Pharmaceutical
    Technology
Jadavpur University
Kolkata, India

**Kalyan Kumar Sen**
Department of Pharmaceutics
Gupta College of Technological
    Sciences
Asansol, India

**Ashok K. Singh**
Department of Pharmaceutical Sciences
Babasaheb Bhimrao Ambedkar
    University
Lucknow, India

**Jagdish Singh**
Department of Pharmaceutical Sciences
College of Health Professions
North Dakota State University
Fargo, North Dakota

**Aum Solanki**
College of Pharmacy
University of South Florida Health
Tampa, Florida

**Nirmal Sonali**
School of Materials Science and
    Engineering
Nanyang Technological University
Singapore

**Lay Poh Tan**
Division of Materials Technology
School of Materials Science and
    Engineering
Nanyang Technological University
Singapore

# 1 Basic Concepts in Drug Targeting

*Kalyan Kumar Sen and Sabyasachi Maiti*

## CONTENTS

## 1.1 INTRODUCTION

Since the advent of modern methods of treatment, drugs are administered in various dosage forms such as tablets, capsules, pills, creams, ointments, liquids, aerosols, injectables, suppositories, and so on to treat various diseases. Even today, these types of formulations are used as major pharmaceutical products. However, these conventional drug delivery systems (DDSs) may not provide maximum therapeutic responses always (Garg and Kokkoli 2005). The conventional type of dosage forms are required to be administered several times a day to achieve and then to maintain the minimum effective concentration of the drug at the site of action. This results in a fluctuating drug level, adverse drug reaction, prior biodegradation of the drug, occurrence of drug toxicity, and patient noncompliance. Scientists are engaging in continuously developing new drug formulations to minimize those effects as effective tools against diseases (Garg and Kokkoli 2005). New generations of drugs are also developing due to improved knowledge of cellular biology

1

at the molecular level, decoding of the human genome, and a technological break-through in the field of proteomics and deoxyribonucleic acid (DNA) microarrays. These help us to obtain many agents as drug-like proteins, peptides, nucleic acids, and so on. It is now well established that the drug action occurs due to molecular interaction(s) of a drug molecule with a receptor molecule in certain cells; it is therefore easily concluded that it is necessary for the drug to reach somehow the site of action after the administration (oral, intravenous (IV), local, transdermal, etc.) at sufficient concentrations. Therefore, to increase the effectiveness of the drug, it is usually administered in large quantities so that it can reach the site of action in required quantities after distribution in body fluid. But generally, this results in deleterious effects on the healthy organs of the body commonly referred to as *side effects*. In 1981, Gregoriadis introduced the method drug targeting using a novel DDS as old drug in new devices (Gregoriadis 1981). The concept of design-ing a targeted delivery system was originated by Paul Ehrlich, who was a micro-biologist and proposed the idea of drug delivery in the form of a magic bullet. Selective drug targeting yet remains unachieved (Farokhzad and Langer 2009). Targeted drug delivery means the accumulation of pharmacologically active moi-ety at the desired target in therapeutic concentration and at the same time restrict-ing its access to normal cellular lining, thus minimizing the therapeutic index. In site-specific targeted drug delivery, the active drug is delivered to very specific preselected compartments with maximum activity while reducing the concentra-tion of the drug to normal cells. The researchers have targeted the drug through novel technologies to intracellular sites, virus cells, bacteria cell, and parasites, which have been proven highly effective (Forssen and Willis 1998). The minimum distribution of the parent drug to the nontarget cells and with higher and effective concentration at the targeted site certainly maximizes the benefits of targeted drug delivery.

The conventional DDS available for focused oral delivery incorporates those that utilize enteric coatings, prodrugs, osmotic pumps, colloidal bearers, and hydrogels; drug delivery dosage forms varying from implantable electronic gadgets to single polymer chains are required to be perfect with procedures in the body (biocompat-ibility) and with the medication to be delivered. DDS adjusts the biodistribution and pharmacokinetics (PK) of the related medication, that is, the time-dependent rate of the regulated drug release in the diverse organs of the body. Moreover, hindrances emerging from low drug solubility, losses (natural or enzymatic), quick distribution rates, nonparticular toxic quality, failure to cross bioboundaries, and so on must be considered during development of DDS (Paul et al. 2010).

Nanoparticles (NPs) have novel physicochemical properties, for example, ultra-micron size, substantial surface zone-to-mass proportion, and high reactivity that varies from those of mass materials of the same arrangement. Due to reduction in particle size, the drug molecules resides over the surface to an extent greater than its center and render the nanoparticles more reactive. Likewise, the specific surface area of the nanoparticles also increases exponentially. This nanoparticulate drug enhances *in vivo* drug absorption due to increase of dissolution rate and saturation solubility. These properties can be utilized to beat a portion of the constraints experi-enced in the utilization of conventional dosage forms. Liposomes are the most widely utilized NPs as they are made out of the same material as cell films. Different sorts of liposomes as of now are being used as an anticancer delivery system and immunizations. The utilization of NPs (Missirlis et al. 2005) as a cell-targeted delivery system is still in the initial stages but emerging as potential sources of nanomedicine. The first examples of targeted DDS were described in the literature nearly 30 years ago (Heath et al. 1980; Leserman et al. 1980).

Technologies available for preparation of targeted DDSs are highly developed for parenteral formulations. Such technologies are concerned with the delivery of drugs to specific targets in the body. These also prevent drug degradation and premature elimination. These technologies are developed using various carrier systems like soluble carriers (e.g., monoclonal antibodies, dextrans, and soluble synthetic polymers), particulate carriers (liposomes, micro- and nanoparticles, and microspheres), and target-specific recognition moieties (e.g., monoclonal antibodies, carbohydrates, and lectins).

There has been a strong focus toward their development since then, and it depends on four primary parameters described below:

a. *Target site*: For successful targeting, it is important that the target receptor or antigen should have high density on target cells and should have been overexpressed on the diseased tissue and underexpressed on the healthy tissue.

b. *Targeting ligands*: Choice of the targeting ligands is a crucial parameter in determining the ability of the DDS to identify the target molecule of interest. The ligands should bind with affinity and specificity.

c. *Carrier*: There are several choices for drug carriers as discussed earlier. Liposomes are a great choice for a drug carrier.

d. *Drug*: The suitable choice of the drug is another important factor as this ultimately governs the effectiveness of the targeted delivery system. The drug delivery system should be optimal and should present an effective means of controlling drug delivery rate at the target site. In spite of the assurances, one of the several challenges for designing new drug delivery system is to distinguish the target organ and consequently deliver the drug to it. Usually, it is difficult to identify the target of interest. However, at a minimum, a delivery system must have the ability to target the receptors overexpressed on the diseased tissues and specifically liberate its active load when it reaches its therapeutic site of action (Torchilin 2006). The first and foremost concern is the unregulated expression of target molecule in healthy tissues. The second challenge is the target vector that binds to a carrier—stable in the targeted region of undesired extravasation of drugs/carriers in the healthy tissues/organs (e.g., liver, spleen, etc.) that may also occur and can be deleterious when delivering a highly potent drug. After the carriers reach its target site, it should be able to penetrate into specific cells or tissue.

Even if the drug is able to recognize and reach the target and does not accumulate in other tissues, release of drugs from the carrier poses another problem. Last but not the least, control of residence time at the receptor site may be another issue that could ultimately also affect the drug efficacy.

## 1.2  CLASSIFICATION OF TARGETED DDSs

We can categorize the DDSs based on their physical form (Diederichs and Muller 1998; Molema and Meijer 2001) or their functions (Dumitriu 2002; Barratt 2003). Targeted DDSs (TDDSs) are comprehensively separated into particle type, soluble, and cell transporter type on the basis of physical form. Liposomes, microspheres, and NPs are the examples of particle type. Examples of soluble TDDS are plasma proteins, peptides, polysaccharides, and monoclonal antibodies. The cellular TDDS involves entire cells and viruses. The functional characterizations classify the TDDS into first, second, and third generation (Dumitriu 2002; Barratt 2003). The first-generation systems do not have the ability to deliver drugs to their targeted site, and are thus embedded as close as could reasonably be expected to the target. Microspheres, microcapsules, and other delivery systems qualify in this classification. The first generation of anticancer nanomedicine was developed with biomaterials as carrier to target and to treat primary tumors based mainly on the enhanced permeation and retention effect (EPR) (Maeda et al. 2000). The EPR effect is defined as the accumulation of NPs in tumor facilitated by the highly permeable nature of the tumor vasculature and poor lymphatic drainage of the interstitial fluid surrounding a tumor. The second-generation systems are both equipped for delivering (when managed through a general route) and discharging the drug at the required site of activity. These include particulate and soluble carriers (<1 μm diameter) like passively targeted liposomes, nanocapsules, nanospheres, and more advanced TDDS like temperature- or pH-controlled liposomes and nanospheres that discharge their bioactive load in response to a particular signal. Targeted liposomes, polymeric NPs, and other second-generation DDS adjusted with ligands like peptides or antibodies fall under the classification of third generation.

## 1.3  APPROACHES INVOLVED IN DRUG TARGETING

Several methods of drug targeting are available and are used in the experimental and clinical settings (Torchilin 2006).

   a. *Direct application of drugs*: Here the drug is directly applied onto the affected organ or tissue.
   b. *Passive targeting*: Generally, passive targeting is an approach which helps drugs to bypass the different physiological processes like metabolism, excretion, or opsonization followed by phagocytosis (Couvreur and Vauthier 2006; Bae and Park 2011). This method describes the accumulation of the drug or drug carrier system at a particular (like in case of anticancerous drug) site due to physicochemical or pharmacological factors of the disease. Here, the method utilizes the increase in permeability of vascular endothelium in locales of inflammation and tumors. Drug carrier within the size of 10–500 nm can extravasate and aggregate in the interstitial space in

the region of improved permeability, for example, tumors (Chrastina et al. 2011). The permeability of the drug carrier is size limited so the particle size of the carrier may be utilized to control the drug efficacy. Hence, for cancer treatment, the surface properties and size of drug-containing NPs must be controlled, specifically to bypass the uptake by the reticuloendothelial system (RES) to make best use of circulation times, and optimal size should be less than 100 nm in diameter, and the surface should be hydrophilic to avoid clearance by macrophages (large phagocytic cells of the RES) (Gref et al. 1994). Examples of these approaches are targeting of antimalarial drugs for treatment of leishmaniasis, brucellosis, and candidiasis. Tumor cells in the liver can be targeted by the hydrophobic carrier-coated formulation for a drug or DNA vaccine (e.g., an antitransforming growth factor β [TGF-β] DNA vaccine), with optimum size and charge (Schilling et al. 2000; Lam and Dean 2010) targeted to the liver cells, but due to its coating system, the formulation remains intact until its entry to the tumor cells in the liver (Schilling et al. 2000).

c. *Physical targeting*: This type of targeting could be accomplished by both endogenous and exogenous methods. The previous system exploits the differences in the physical environment (e.g., temperature/pH) and in the influenced zone contrasted with the healthy areas, for example, temperature-controlled liposomes and pH-sensitive liposomes. For example, negatively charged liposomal system can be successfully utilized to overcome the multiple drug resistance phenomena. pH-sensitive liposomes, cationic liposomes (lipoplexes), and so on, have shown much potential in gene delivery (Mecke et al. 2006; Ratnam et al. 2006). In the exogenous methodology, use of an outside signal assists the drug carrier to discharge its substance at the target site. For instance, an external signal, such as light, heat, electricity, magnetic field, or ultrasound directed the drug to the target tissue from an outside source (Alexiou et al. 2003; Eljarrat-Binstock and Domb 2006; Bejjani et al. 2007). Structural change of the delivery system caused by the induction of these external signals facilitates drug access to the target site or its release from the formulation. Physical targeting of the drug carriers into their targets is unnecessary if activation of the system occurs only by the external signal in the target zone. In principle, this approach might improve drug delivery at the target site and minimize off-site effects. Drug is administered in the delivery system that is sensitive to a specific external signal. Triggering mechanism on top of physical targeting may further improve the treatment.

d. *Active targeting*: This is a more advanced approach in which the drug delivery transporter has particular affinity for the targeted tissue or organ (Moghimi and Szebeni 2003). The drug could either be coupled to the targeted moiety or could be epitomized in a carrier, which in turn is coupled to the targeting vector where specific receptors are present (Allen 2002; Mohanraj and Chen 2006; Bae and Park 2011; Yoo et al. 2011). Antibodies and antibody fragments (Dinauer et al. 2005), vitamins (Na et al. 2003), peptides (Fahr et al. 2002), folate (Leamon and Reddy 2004), and transferrin

(Sahoo and Labhasetwar 2005) are studied as targeting ligands for cancer therapy. The selection of a specific ligand is based on its specificity, stability, availability, and selective display of its corresponding pair on the target cells, as well as its cost. In addition to the above considerations, conjugation chemistry (Nobs et al. 2004), density, and accessibility of the ligand need to be properly designed for efficient vector targeting.

The active targeting methods can be further categorized into four different levels based on the target site: (1) first-order targeting or organ-level accumulation, for example, the target sites in this case are the capillary bed of organs or tissues like the lymphatic cavity, peritoneal cavity, cerebral ventricles, lungs, eyes, and so on; (2) second-order targeting or cellular-level targeting system where the delivery system targets specific cells like the Kupffer cells in the liver, tumor cells, and so on; (3) third-order targeting or intracellular level of targeting defines targeting intracellular area like the cytoplasm or organelles such as the nucleus, and so on (Charman et al. 1999; Santini Jr. et al. 2000; Kopecek 2003). Presently, to achieve an active third-order targeted drug delivery, the body's immune response is stimulated against cancer cells by the gene therapy (e.g., a DNA cancer vaccine; Bae and Park 2011). Cancer cell death may be induced by introducing cancer cells with genes encoding apoptosis. This most critical approach for intracellular targeting involves four steps: (1) interaction of the active delivery system with the extracellular plasma membrane receptor; (2) entry of the drug in the cell by receptor-mediated endocytosis; (3) fusion of drug with lysosomes; (4) degradation of the homing device and release of the drug or ribonucleic acid (RNA) into the intracellular target (Bae and Park 2011). A *fourth level of targeting* referred to as molecular-level targeting is defined as targeting a specific molecule in the cell. For example, in many gene delivery systems, the target is the DNA present in the nucleus. Active targeting complements passive accumulation into tumors; selectivity and retention are improved as a result of specific interactions with the target cells, at the expense of increased complexity, cost, and risks (e.g., adverse biological reactions to ligand).

Examples of targeting moieties are charged molecules, polysaccharides, antibodies and their fragments (Bendas et al. 1999), lectins, peptides (Garg et al. 2009), lipoproteins (Lundberg et al. 1993), hormones, oligonucleotides (Farokhzad et al. 2004), and proteins.

### 1.3.1 INVERSE TARGETING STRATEGY

The inverse targeting approach refers to the mechanism to escape the body's defense system by an alternative way of passive targeting approaches by blocking the RES. For example, the rapid uptake of a colloidal carrier system by the RES can be escaped by pretreatment of blank colloidal carriers or macromolecules like dextran sulfate to saturate the RES rather than passive targeting approaches that involve an increment in molecular size. This approach leads to an RES blockade and consequent impairment of the host defense system and accumulation of targeting moieties to other organs (Lee et al. 1996).

### 1.3.2 DUAL TARGETING STRATEGY

The dual targeting methodology is a proficient methodology against infection disease and its medication resistance where the transporter system used to stack the antiviral medication synergistically affects the activity of the medication (Kircheis et al. 2001; Martin and Rice 2007; Jain 2008; Bae and Park 2011). In light of this methodology, bioconjugates can be arranged utilizing diverse sorts of normal and manufactured particles, for example, antibodies, immunotoxins, CD4, glycoprotein, and so on. Conjugation can be accomplished utilizing different noncovalent and covalent systems. These methodologies can likewise be utilized for a compound immune enzymatic system (e.g., antibody–enzyme conjugates), vaccine research, and nonviral gene delivery system.

### 1.3.3 DOUBLE TARGETING STRATEGY

This new idea of medication delivery is intended to enhance the drug therapeutic efficacy in terms of selectivity and control release. Usually, the nanoparticulate DDS specifically accumulates in the selective organ or tissue, and then it infiltrates inside the target cells delivering its content (drug or DNA) intracellularly. Aggregation at a selective organ or tissue (tumor, infarct) may be obtained by the passive targeting via the EPR effect (Palmer et al. 1984; Maeda et al. 2000) or by the antibody-driven active targeting (Torchilin 2004; Jaracz et al. 2005), while the intracellular delivery is achieved by certain ligands (folate, transferrin [Tf]) (Widera et al. 2003; Gabizon et al. 2004) or by cell-penetrating peptides (CPPs) those are engulfed by cells (such as trans-activator of transcription (TAT) or poly-Arg; Lochmann et al. 2004; Gupta et al. 2005) Therefore, the system will contain a drug loader that can be controlled spatially, and a linked antibody to govern its selection to specific molecular sites (Sawant et al. 2006).

### 1.3.4 COMBINATION TARGETING STRATEGY

In several cases, the combination targeting approach is investigated. Researchers are facing several challenges to develop brain-targeted delivery. The combination targeting approach is an improved strategy for the delivery of site-specific drug, proteins, and peptides in those cases. Generally, nanoparticulate carriers could provide a direct approach to the target site through different targeting methods (Roth et al. 2008). This approach is widely investigated in case of proteins and peptide delivery techniques due to several factors (Roth et al. 2008; Wei et al. 2014). These are: (1) permeability of large peptides through several membranes; (2) chemical decomposition of targeting carrier; (3) nonspecific drug release due to target tissue heterogeneity; (4) immune response against the system; and so on (Roth et al. 2008). Surface modification of peptides using natural polymers such as polyethylene glycol (PEG) permeability can be increased (Greish 2007). The chemical decomposition of the TDDS may be avoided through the prodrug approach.

## 1.4  ENDOCYTIC PATHWAYS FOR TARGETED DELIVERY TO ENTER CELLS

Endocytosis is the significant course for NPs to transport over the membrane. Actively, three methods of endocytosis can be characterized: fluid phase, adsorptive, and receptor-mediated endocytosis (Amyere et al. 2000). Fluid phase endocytosis alludes to the mass uptake of solutes to the accurate ratio to maintain the concentration in the extracellular liquid. This is a low-proficiency and nonspecific procedure. Interestingly, in adsorptive and receptor-induced endocytosis, macromolecules are attached to the cell surface and accumulated before uptake. In adsorptive endocytosis, particles specially interface with nonspecific complementary binding areas (e.g., by lectin or charged connection). The bound molecules then to a great extent occupy the coated pit region of cell surface membrane and after removal of the binding molecule, the solute recycled in cell membrane. In receptor-interceded endocytosis, certain ligands can bind to receptors on the cell surface and concentrate before entering into the cell. The proficiency of receptor-mediated endocytosis reflects both the affinity of the ligand–receptor binding and the centralization of these complexes in clathrin-covered pits. Liposomes can incorporate lipophilic and hydrophilic drugs. By coating these liposomes with proteins (human serum albumin, Tf, and insulin), a receptor-mediated endocytosis can be achieved. Cationic liposomes can be internalized by adsorptive-mediated endocytosis, and thus overcome the blood–brain barrier (BBB; Schnyder and Huwyler 2005). Endocytosis is by and large classified into phagocytosis and pinocytosis (Conner and Schmid 2003). The endocytosis mechanism is illustrated in Figure 1.1.

Phagocytosis or cell-eating procedure was initially found in macrophages. Pinocytosis or cell-drinking procedure is available in a wide range of cells in four structures, for example, clathrin-subordinate endocytosis, caveolae-subordinate endocytosis, macropinocytosis, and clathrin- and caveolae-free endocytosis (Sawachi et al. 2010; Taniguchi et al. 2010).

Phagocytosis is normally limited to concentrated mammalian cells, whereas pinocytosis happens in all cells (Conner and Schmid 2003). Accordingly, the terms endocytosis and pinocytosis are incidentally thought to be synonymous (Lamaze and Schmid 1995). Targeting of cell-specific receptors by substrates (Tf, insulin, and modified lipoproteins) bound to NP surface leads to an uptake of the particles by receptor-mediated endocytosis, and this is often used for targeted drug delivery by gold NPs. The release of BODIPY, a model drug, which was coupled to functionalized gold NPs, has been demonstrated (Hong et al. 2006). Also the chemotherapeutic drug, paclitaxel, has been conjugated to gold NPs and used for drug delivery (Gibson et al. 2007). Large biomolecules, such as proteins or DNA, are also successfully delivered by gold NPs. Plasmid DNA has been bound to the surface of gold NPs. The particles prevent enzymatic degradation of the DNA which is then released by glutathione treatment (McIntosh et al. 2001; Hong et al. 2006). Thus, gold NPs are also suitable for gene delivery. NPs with positive net charges are known to interact with the negative charges of the extracellular matrix and promote endocytosis (Thorek and Tsourkas 2008). The most important characteristics that influence the uptake of NPs in cells are thought to be size and surface properties (Dobrovolskaia and McNeil 2007) as well as the type of cells. Anionic dendrimers are known to

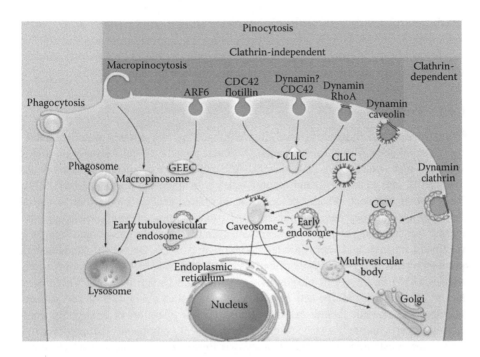

**FIGURE 1.1** Different mechanisms of endocytosis. There are multiple pathways for cellular entry of particles and solutes. In all cases, the initial stage of endocytosis proceeds from the plasma membrane portals of cellular entry and involves engulfment of the cargo into intracellular vesicles. The second stage often involves sorting of the cargo through endosomes. It is followed by the final stage during which the cargo is delivered to its final destination, recycled to extracellular milieu, or delivered across cells (not shown). The figure is a simplified representation of complex trafficking mechanisms and their cross-talks. Abbreviations: CCV, clathrin-coated vesicles, CLIC, clathrin-independent carriers; GEEC, GPI-anchored protein-enriched compartment; GPI, glycophosphatidylinositol, MVB, multivesicular body. (Reprinted from *J Control Release*, 145, Sahay, G., Alakhova, D.Y., and Kabanov, A.V., Endocytosis of nanomedicines, 182–95, Copyright 2010, with permission from Elsevier.)

internalize via caveolin-mediated endocytosis, while mesoporous silica NPs are internalized via a clathrin-mediated pathway (Kunzmann et al. 2011). The interaction of NPs with the plasma membrane and thus the amount of internalization can be tuned by a positive net charge on the NP surface and by the targeting molecules such as ligands for receptors or cell-penetrating peptides (Liang et al. 2009).

Depending on the type of ligand, the NPs are internalized by different endocytotic pathways such as caveolin- or clathrin-mediated pathways and are entrapped in endosomes. In addition to the internalization by classic endocytotic pathways, it has been demonstrated using model lipid membranes that high positively charged gold NPs overcome the lipid cell membrane by penetration followed by diffusion into the cell (Lin et al. 2010). Taylor et al. also demonstrated that certain gold NPs are internalized by passive diffusion instead of active endocytotic pathways (Taylor et al. 2010). Tf receptor (TfR) is also overexpressed in many malignant cells, including breast cancer, pancreatic cancer, prostate cancer, colon cancer, lung cancer, and leukemia cells (Niitsu et al. 1987; Högemann-Savellano et al. 2003; Nakamaki et al. 2004). It is a carrier protein for Tf, imports iron into the cell by internalizing the Tf–iron complex through receptor-mediated endocytosis, and is regulated in response to intracellular iron concentration (Yan et al. 2015).

### 1.4.1 Phagocytosis

Phagocytosis is an endocytosis procedure exhibited by a few types of cells, including epithelial cells, fibroblast, safe cells, particular phagocytic cells (monocyte, macrophages, and neutrophils), cells that produce inflammatory mediators (basophils, eosinophils, and mast cells), and natural killer cells (Aderem and Underhill 1999). Phagocytosis has been broadly described at morphological and molecular levels in the literature (Conner and Schmid 2003). In mammalian creatures, phagocytosis engulfs the crippled particles, senescent cells, and irresistible microorganisms as a reaction of intrinsic and versatile safety (Rabinovitch 1995). One of the fundamental qualities of this exceptional type of endocytosis is the vast size of the endocytosed vesicles >250 nm, known as phagosomes (Aderem and Underhill 1999). The procedure of phagocytosis can be activated either through the cooperation of cell surface receptors with specific ligands exhibited by the outside operators or through the connection of particular cell surface receptors with solvent components that perceive the remote specialists and encourage phagocytosis (opsonization). The solvent elements included in the opsonization procedure incorporate proteins of the supplement framework, antibodies, acetylcholine, laminin, fibronectin, C-receptive protein (CRP), and type I collagen (Owens and Peppas 2006). The most vital receptors that partake in phagocytosis are the Fc receptor family for immunoglobulin G (IgG; FcγRI, FcγRIIA, and FcγRIIA), the complement receptors (CR1, CR3, and CR4), and the α5β1 integrin (Underhill and Goodridge 2012). A lot of research endeavors have been centered on controlling the NPs' internalization by means of phagocytosis. The cell uptake of NPs by means of phagocytosis in macrophages includes attractive forces (i.e., van der Waals, electrostatic, and ionic, hydrophobic/hydrophilic) between the cells and NPs' surfaces. The phagocytosis of NPs can also be activated by the receptor-intervened acknowledgment of opsonins adsorbed on the surface of NPs. NPs should be perceived by the opsonin first, for example, IgG and IgM, supplement segment (C3, C4, and C5), and blood serum proteins. Then the opsonized NPs attach to the cell surface and connect with the receptor, instigating the cup-shaped expansion development. The layer expansions enclose the NPs and after that internalize them, framing the phagosomes which have a diameter of 0.5–10 μm. Finally, the phagosomes move to bond with lysosomes (Yameen et al. 2014). Be that as it may, the carrier contained in the phagosomes will be decimated by fermentation and enzymolysis in the lysosomes. So nanomedicines must sidestep this path to be stable and able to create desired impacts. Because it is primarily performed by specialized cells, phagocytosis is not expected to play a significant role in gene delivery. However, a phagocytosis-like mechanism was proposed for the uptake of large cationic lipid–DNA complexes (lipoplexes) and polyethyleneimine (PEI) polyplexes (Matsui et al. 1997; Kopatz et al. 2004). Mitragotri and colleagues (Champion and Mitragotri 2006; Champion and Mitragotri 2009) have described that the molecule geometry can help in regulating their cell uptake by means of phagocytosis. Different local particle shapes at the purpose of cell adhesion create diverse points between the membrane and particle. This contact point significantly affects the capacity of macrophages to uptake particles through actin-driven development of the macrophage membrane. They observed that elongated particles with higher aspect ratio are less inclined to phagocytosis. Geng et al. (2007) have additionally reported a comparable finding. Interestingly, a higher aspect ratio has likewise as of late been involved with a special restriction into endosomes and lysosomes (Harguindey et al. 2009); thus, care ought to be taken while considering the particle shape for adjusting phagocytosis and intracellular targeting at the same time. Hart et al. (2012) reported that the cross-connecting of macrophage CD44 antibody quickly and particularly increases macrophage phagocytosis of apoptotic neutrophils. They additionally theorized that CD44 cross-connecting supports the adjustment of association of macrophages with apoptotic neutrophil targets that help in the cellular engulfment.

In another study, lectin-mediated endocytosis of sugar-conjugated N-(2-hydroxypropyl)methacrylamide (HPMA) copolymer conjugates in three different human cancer cell lines suggested their potential use for targeted delivery of chemotherapeutics to colon adenocarcinoma (David et al. 2002). However, to compensate for the weak binding affinity of carbohydrates, multiple or multivalent molecules should be conjugated to the surface of NPs to achieve multivalent interactions. In the case of galactosylated liposomal carriers, it was shown that the targeting efficacy depended on the galactose ligand density (Managit et al. 2003).

### 1.4.2 Pinocytosis

Every cell requires performing fundamental functions, such as "eating" and "drinking." Pinocytosis is truly deciphered from the Greek word for "cell drinking." This mechanism is used by cells to uptake fluids. Inside of cells, we have the liquid portion called the cytoplasm, and every cell has an outer layer of a semiporous cell membrane made of lipids. Through pinocytosis, cells will uptake everything that is in the extracellular liquid outside of the cell, including solutes. This is one of the major routes for the cells to uptake liquid, solutes, and dispersions containing small particles. Just little amounts of material will enter the cell by this procedure, and an insignificant measure of adenosine triphosphate (ATP utilized by the cell) is fundamental. At the point when the cell takes in the liquid, it is put away in a little vesicle. A vesicle is membranous sac—a spherical and hollow organelle bounded by a lipid bilayer membrane. Pinocytosis brings fluid containing dissolved substances into a cell through membrane vesicles. The membrane of the vesicle is produced using the same lipid layer that is the cell layer. The procedure of pinocytosis itself is somewhat basic, in that it just truly includes the cell membrane, the vesicle, and a lysosome. Initially, the cell layer will permit the liquid it needs to take into push inward, bringing about a vacuole in the membrane, then the liquid fills the vacuole and the curve gets bigger. In the end, the membrane squeezes it off, catching the liquid inside the vesicle, which is currently inside the cell.

### 1.4.2.1 Clathrin-Dependent Endocytosis

An essential part in cellular entry of drugs is through clathrin-dependent endocytosis (CDE) in every mammalian cell that includes intercellular signaling, layer reusing, and uptake of nutrients (Kirchhausen 2000). After nanomaterials collaborate with receptors on the cytomembrane, a sort of cytosolic protein named clathrin-1 polymerizes on the cytosolic side of the plasma where the carrier is disguised. An extensive protein structure instigates the formation of curvature in the membrane to start vesicle development (e.g., epsin) (Ford et al. 2002) and clathrin assembly lymphoid myeloid leukemia (CALM) protein (Tebar et al. 1999) which are fundamental for the arrangement of the circular clathrin-covered pit. Then, with the help of a small guanosine triphosphatase (GTPase) called dynamin, the invaginated pits are squeezed off into the cytoplasm as clathrin-covered vesicles that accelerate the fission process (Schmid 1997; Mukherjee et al. 1997). With energy supplied by actin, the vesicles move inside the cells, and the course is managed by the cytoskeleton (Doherty and McMahon 2009; Pucadyil and Schmid 2009). The clathrin coat is shed off in the cytosol. At the point when the covered vesicles are discharged, the clathrin pit is dismantled by the activity of auxilin and heat shock cognate 70 (HSC70)-dependent proteins (McMahon and Boucrot 2011). After the uptake through clathrin-mediated endocytosis (CME), the uncoated vesicles are either guided to right on time endosomes or reused to the plasma membrane surface. The vesicles can likewise be developed to late endosomes and later to compartments, for example, lysosomes and multivesicular bodies. Most of the receptor-mediated cellular internalization of NPs takes place through CDE. The receptor that NPs' ligands attach to is associated with the vesicles. For instance, low-density lipoprotein (LDL) particles are uptaken through LDL receptor and exchanged to lysosomes for degradation, while TfRs normally engulf the iron-loaded Tf and are recycled to the cell surface (Benmerah and Lamaze 2007). This course can be hindered by its inhibitors or some different variables, for example, chlorpromazine, a hypertonic medium, or potassium depletion (Richard et al. 2005; Delva et al. 2008). The cationic NPs of around 100 nm size got from the polylactide-co–polyethylene glycol (PLA–PEG) have been found to uptake solely by means of CME. The poly(L-lysine), which is a cationic polymer, functionalized at the surface of poly(lactide-co-glycolide) (PLGA) NPs have additionally been found to fundamentally upgrade the cell uptake through CME (Vasir and Labhasetwar 2008). Another study reported mesoporous silica NPs (~110 nm) labeled with fluorescein isothiocyanate indicating effective internalization into human mesenchymal stem cells (hMSCs) and adipocytes (3T3-L1) prevalently by means of CME (Huang et al. 2005). In spite of the fact that the basic instruments that intercede the disguise of nontargeted NPs are not completely comprehended, it appears that the high rate of cell disguise through CME under ordinary cell action has brought about numerous studies closing the CME as a primary course of internalization for NPs of size ~100 nm. The absolutely charged NPs of ~100 nm distance across have been observed to be internalized dominatingly through CME, which can be a consequence of expanded communication between the emphatically charged NPs and adversely charged cell surface that upgrades the NPs internalized by means of CME basically on the grounds that the CME is the most plausible internalized material for NPs of ~100 nm measurement (Fujimoto et al. 2000). Keeping the fermentation of the endosome restrains their combination and development (Johnson et al. 1993). Regarding quality conveyance, CME can be focused by utilizing certain ligands, for example, transferrin (Tf), which can particularly perceive certain receptors on the cell surface (Stoorvogel et al. 1991). These resulted in the clustering of complexes in the coated pit of plasma membrane and assisted in the engulfing of the particles and offers the likelihood of targeting on particular cells that significantly overexpressed the receptors. On the other hand, qualities that are disguised through CME are generally caught in endosomes after enzymatic corruption in lysosomes, and the last result is that qualities have a little or no entrance to their objective destinations. Really, entanglement and debasement can be viewed as two separate boundaries, in light of the fact that avoiding lysosomal corruption results in a collection of qualities in intracellular vesicles without improving cytosolic discharge. In this manner, to achieve the core, qualities must keep away from corruption in lysosomes and should likewise discharge from intracellular vesicles into the cytosol. A few methodologies have been created to upgrade the cytosolic arrival of endocytosed qualities (Plank et al. 1994; Tachibana et al. 1998; Simões et al. 1999). This includes the consolidation of vesicular dangerous components to the DNA transporter buildings, which annoy the honesty of the vesicular film and permit the cytosolic arrival of their substance, while not harming the DNA. Some cationic polymers, for example, PEI, and a few lipids likewise can upgrade the cytosolic arrival of qualities through diverse instruments as will be accordingly explained.

### 1.4.2.2   Caveolae-Dependent Endocytosis

Classically, caveolae were defined as flask-shaped invaginations of the plasma membrane, but they can also be flat, tubular, or detached vesicles (Smart et al. 1999; Pelkmans et al. 2001). Caveolae are present in many cell types and are especially abundant in endothelial cells (Conner and Schmid 2003) and are rich in cholesterol and glycosphingolipids (Matveev et al. 2001). They are involved in several cellular processes, including cholesterol homeostasis and glycosphingolipid transport (Harris et al. 2002). Caveolae-dependent endocytosis is also a common cellular entry pathway. It could bypass lysosomes, thus many pathogens including viruses and bacteria select this way to avoid lysosomal degradation. For the same reason, this route is believed to be beneficial for the enhancement of concentration of targeting position and improvement of therapeutic effect. Caveolae are also involved in transcytosis and endocytosis of certain viruses such as the simian virus 40 (SV40), as well as some bacteria and bacterial toxins, for example, cholera toxin. Caveolae are characterized by their association with a family of cholesterol-binding proteins called caveolins, which function to create and/or mediate these structures (Lamaze and Schmid 1995). There are three isoforms of caveolin in mammalian cells. Caveolin-3 is muscle specific, while caveolin-1 and 2 are abundant in most nonmuscle cells (such as endothelial cells, fibroblasts, and adipocytes) and absent in neurons and leukocytes. By binding to the receptors on the plasma membrane, NPs or pathogens, like SV40 and cholera toxin, can interact with the receptors to induce the formation of the flask-shaped vesicles, which are cut off from the membrane by dynamin. The mechanisms of caveolar internalization have been elucidated by visualizing the trafficking of the SV40 that uses caveolae to gain entry into the cells (Pelkmans et al. 2001). SV40 initially associates with the cell membrane and then becomes trapped in relatively stationary caveolae. The presence of caveolin in these organelles gave rise to the name *caveosome*. SV40 then segregates from caveolin and is sorted out of caveosomes for delivery to the endoplasmic reticulum (ER). The caveosomes containing nanomedicine move along with microtubules to the ER (Pelkmans et al. 2002; Khalil et al. 2006). It is thought that nanomaterials in ER penetrate into the cytosol, and then enter the nuclear envelope via the nuclear pore complex (Kasamatsu and Nakanishi 1998). Compared to CDE, this pathway takes a longer time and has smaller vesicles in the process (Pelkmans et al. 2001). According to those described above, nanomaterials are taking this way in some cases to avoid a degradative fate and enhance the delivery of the drug to a target organelle (such as ER or nucleus), which is critical for improvement of therapeutic delivery. Another major difference is that the caveolar uptake is a nonacidic and nondigestive route of internalization (Ferrari et al. 2003). Caveolae do not suffer a drop in pH and avoid normal lysosomal degradation. Researchers first described the term *potocytosis* that is usually associated with caveolae to explain the uptake of folic acid. Potocytosis describes the internalization of small molecules without the merging of an endocytic vesicle with endosomes (Anderson et al. 1992). In uptake of folic acid, it is thought that folic acid binds to the folate receptors that are clustered in invaginated caveolae, but the caveolae stay attached to the plasmalemma proper and generate a distinct microenvironment by pinching the neck region closed. The ligand is then released from the receptors, and 5-methyltetrahydrofolic acid moves across the caveolar membrane where it stays in the cytosol after modification with polyglutamate, and the caveolae begin to reopen to the extracellular space to repeat the cycle. Another term associated with caveolae is *lipid rafts*. Markers for lipid rafts are frequently found within caveolae. In general, caveolin-containing rafts are referred to as caveolae, whereas caveolin-devoid rafts are denoted by a variety of names such as glycolipid-enriched membranes and caveolae-like domains (Matveev et al. 2001). Cholesterol is required for caveolar uptake and drugs that specifically bind to cholesterol perturb internalization through the caveolae (Lamaze and Schmid 1995). Caveolae also depend on the actin cytoskeleton and drugs that cause the depolymerization of the actin cytoskeleton such as cytochalasin D can inhibit caveolae uptake without affecting CME (Parton et al. 1994). Genistein, a tyrosine kinase inhibitor, can also inhibit caveolae (Orlandi and Fishman 1998). It is generally believed that caveolar uptake does not lead to lysosomal degradation. Therefore, this pathway seems to be advantageous in terms of DNA delivery. Evidence supporting the existence of a role of caveolae in the uptake of cationic polymer–DNA complexes and the class of protein transduction domains (PTDs), such as the TAT peptide, has appeared (Ferrari et al. 2003; Fittipaldi et al. 2003; Rejman et al. 2005). Another report suggested that large particles (500 nm) are preferentially taken up through caveolae where they do not suffer lysosomal degradation (Rejman et al. 2004). However, caveolae are slowly internalized and small in size, and their fluid phase volume is small. Thus, it is unlikely that they contribute significantly to constitutive endocytosis, although the situation is different in endothelial cells in which caveolae constitute 10%–20% of the cell surface (Conner and Schmid 2003). Caveolae-mediated endocytosis is still a promising strategy for gene delivery especially if the internalization can be increased, possibly through the use of specific receptors for caveolae.

In this pathway, caveolin, a protein that exists in most cells, plays a dominant role. Similar to CDE, caveolar vesicles require actin to move and intact microtubules to traffic within the cell. The caveolae vesicles traffic to fuse with caveosomes or multivesicular bodies (MTV), which have a neutral pH. Different types of cells, such as fibroblasts, endothelial cells, smooth muscle, and adipocytes contain abundant caveolae, but are absent in neurons and leukocytes. It has been observed that adipocyte caveolae can occupy as much as ~50% of plasma membrane, whereas in the endothelial cells in blood capillaries, the percentage of caveolae can be as high as ~70% of plasma membrane (Wang et al. 2011). Three types of caveolin (CAV1, CAV2, and CAV3) are the main protein constituent of caveolae with an estimate of about 140–150 of CAV1 protein molecules per caveolae (Pelkmans and Zerial 2005). Caveolins are known to work together with cavins, the coat proteins (cavin 1–4), to regulate the formation of caveolae, and also potentially participate in the signals that regulate the caveolae fate (Parton and del Pozo 2013). For many years, the intracellular destinations of caveolae have not been clearly understood. However, it has emerged that in endothelial cells, caveolae are able to perform transendothelial transport, which may be utilized for the release of NPs in the subendothelial tissues (Oh et al. 2007). The material that is endocytosed via caveolin-mediated pathway is initially localized in the caveosomes. The neutral pH of caveosomes can be considered as a means to avoid the hydrolytic environment of lysosomes. The sorting of caveosomes cargo to the Golgi apparatus and ER may also be exploited for the targeted delivery of theranostic agents to these subcellular compartments. The negative surface charge has been found to trigger the cellular internalization predominantly via caveolae. Liu et al. (2009) have studied the rabies virus glycoprotein RVG29 (29-amino acid peptide) as a targeting moiety for DNA-conjugated poly(amido amine) (PAMAM) dendrimer, and they revealed significant accumulation of carrier in the brain of mice after intravenously administration. They observed that the process of cellular internalization of PAMAM–RVG29 in the brain capillary endothelial cells occurs through a combination of clathrin and caveolae-mediated energy-dependent endocytosis that involved an interaction with gamma-aminobutyric acid B (GABAB) receptor. Interestingly, previous studies about the mechanism of rabies virus glycoprotein crossing the BBB and cellular internalization have described a specific interaction with the nicotinic acetylcholine receptor (AchR; Lafon 2005).

### 1.4.2.3 Macropinocytosis

Macropinocytosis is commonly defined as a transient, clathrin and caveolin independent, growth factor induced, and a series of events initiated by extensive plasma membrane reorganization or ruffling to form an external macropinocytic structure generated by an actin-driven envagination of the plasma membrane that internalizes the surrounding fluid into large vacuoles (Swanson and Watts 1995; Amyere et al. 2002; Conner and Schmid 2003). The process is constitutive in some organisms and cell types, but in others, it is only pronounced after growth factor stimulation. A ruffle is formed by a linear band of outward-directed actin polymerization near the plasma membrane, which lengthens into a planar extension of the cell surface. After stimulation by any mitogenic factor, the ruffles become longer and broader and frequently close into large macropinosomes (Swanson and Watts 1995). Internalized macropinosomes share many features with phagosomes and both are distinguished from other forms of pinocytic vesicles by their large size, sometimes being as large as 5 μm in diameter, morphological heterogeneity, and lack of coat structures. Macropinosomes do not concentrate receptors. Because they are relatively large, macropinocytosis is an efficient route for the nonselective endocytosis of solute macromolecules (Conner and Schmid 2003). Macropinocytosis fulfills diverse functions, especially when massive fluid-phase endocytosis is necessary. This route facilitates the bulk uptake of soluble antigens by the immature dendritic cells (Conner and Schmid 2003). After the formation of macropinosomes, these vesicles lose their F-actin and their intracellular fate differs, depending on the cell type (Meier and Greber 2003). In macrophages, they move toward the center of the cell, shrink by loss of water, become acidified, and then completely merge into the lysosomal compartment (Meier and Greber 2003). A study illustrates that in human A431 cell line (epidermoid carcinoma), they do not interact with endocytic compartments other than macropinosomes (Swanson and Watts 1995). They constitute a distinct vesicle population, which eventually recycles most of its contents back to the cell surface. Although the pH of macropinosomes decreases, they do not fuse into lysosomes in this case. Macropinosomes are thought to be inherently leaky vesicles compared with other types of endosomes (Wadia et al. 2004). A paucity of information is available on other distinguishing features for macropinocytosis such as specific marker proteins and drugs that interfere with its mechanism over other endocytic processes. This has hampered efforts to characterize the dynamics of this pathway and to identify regulatory proteins that are expressed in order to allow it to proceed. Upon internalization, macropinosomes acquire regulatory proteins common to other endocytic pathways, suggesting that their identities as unique structures are short lived. There is, however, less consensus regarding the overall fate of the macropinosome cargo or its limiting membrane, and processes such as fusion, tubulation, recycling, and regulated exocytosis have all been implicated in shaping the macropinosome and directing cargo traffic. Macropinocytosis has also been implicated in the internalization of cell-penetrating peptides that are of significant interest to researchers aiming to utilize their translocation abilities to deliver therapeutic entities such as genes and proteins into cells. This review focuses on recent findings on the regulation of macropinocytosis, the intracellular fate of the macropinosome, and discusses evidence for the role of this pathway as a mechanism of entry for cell-penetrating peptides. The cargo absorbed through this way is nonspecific. Actually, macropinocytosis can be found in almost all cells with a few exceptions, such as brain microvessel endothelial cells. This pathway is generally started with external stimulations which activate the receptor tyrosine kinases. The activation of receptor mediates a signaling cascade that induces the formation of membrane ruffles. However, according to the form of the ruffles, there are different mechanisms of the macropinosomes pinched off from the membrane. Circular ruffles are cut off by the multifunctional GTPase of dynamin. In contrast, the lamellipodial macropinosomes separated from the membrane is free of dynamin. The macropinosomes with a diameter of 0.5–10 μm are distinct from other vesicles that are formed in other pinocytosis. The surrounding fluid and particles can be internalized into the macropinosomes. In macrophages, after separating from the membrane, macropinosomes move into the cytosol and fuse with lysosomes. In contrast, in human A431 cells, the macropinosomes travel back to the cell surface of the membrane and release the contents to the extracellular space. Therefore, the final fate of macropinosomes depends on the cell type (Mayor and Pagano 2007). Macropinocytosis is a typical route for the uptake of apoptotic cell fragments (Fiorentini et al. 2001) and bacteria (Kolb-Maurer et al. 2002), and contributes substantially to the antigen presentation in major histocompatibility complex (MHC) class II (Sallusto et al. 1995; Steinman and Swanson 1995). Unlike the receptor-mediated endocytosis and phagocytosis, the activation of macropinocytosis is not regulated by the direct action of a receptor or the carrier molecules. In this case, the activation of tyrosine kinase receptor such as the epidermal growth factor and the platelet-derived growth factor receptor leads to an increment of the actin polymerization, actin-mediated ruffling, and macropinosome formation (Kerr and Teasdale 2009). Interestingly, macropinosomes share some proteins (cell division cycle 42 [CDC42], adenosine diphosphate [ADP] ribosylation factor 6 [ARF6], and Rab5, a regulatory guanosine triphosphatase) with other endocytosis processes, suggesting a relationship between the mechanisms of macropinosomes biogenesis and other endocytosis routes (Nobes and Marsh 2000; Schafer et al. 2000). The macropinosomes are sensitive to cytoplasmic pH and undergo acidification and fusion events (West et al. 1989). In macrophages, the macropinosomes present a fate similar to endosomes, and during their maturation, gain and loss markers that are typical for early and late endosomes before their fusion with lysosomes (Racoosin and Swanson 1993). Micron size particles are generally known to internalize the cells via macropinocytosis (Gratton et al. 2008); however, most of the literature reports highlight that the NPs undergo cellular internalization via more than one endocytic pathway. Recently, Zhang et al. (2014) reported lapatinib-loaded NPs formulated with a core of albumin and a lipid corona formed by the egg yolk lecithin. The NPs exhibited ~62 nm and a zeta potential of 22.80 mV and were demonstrated to internalize the breast tumor (BT 474) cells through energy-dependent endocytosis involving clathrin-dependent pinocytosis and macropinocytosis (Zhang et al. 2014).

### 1.4.2.4 Clathrin- and Caveolae-Independent Endocytosis

This is a unique pathway, which relies on cholesterol and requires specific lipid compositions. According to GTPases, which play a role of regulation in the cellular entry pathway, the clathrin-independent endocytosis (CIE) and caveolae-independent endocytosis are classified as ARF6 (dependent, CDC42 dependent and RhoA dependent) (Mayor and Pagano 2007). Dynamins also play a dominant part in these ways, while it is not deeply understood. This field draws more and more attention, but unfortunately, it is still far away from deep understanding and needs further research. The involved endocytic apparatus may contain clathrin-independent carrier (CLIC) or glycophosphatidilinositol (GPI)-anchored protein-enriched early endosomal compartment (GEEC). Furthermore, their later stages are not yet clearly identified.

The CIE pathway was initially described as a mode of entry for a number of bacterial toxins and cell surface proteins, and recently was proposed in the plasma membrane repair, cellular polarization, cellular spreading, and modulation of intercellular signaling (Sandvig et al. 2008). CIE does not require the presence of coat proteins for the vesicle formation and internalization; however, the actin and actin-associated proteins are important players for the vesicle formation during CIE (Robertson et al. 2009). The CIE involves different subtypes of pathways that include the participation of proteins such as ARF6, RhoA, and CDC42 (Sandvig et al. 2011). Studies have shown that ARF6-dependent CIE participates in the endocytosis of the MHC class I (Radhakrishna and Donaldson 1997), the β-integrins (Powelka et al. 2004), the glucose transporter 1 (GLUT1), and other proteins that are involved in amino acid uptake and cell–extracellular matrix interactions (Eyster et al. 2009). In addition, RhoA and CDC42 endocytosis are dependent on the lipid rafts for vesicle formation. In a study, it has been shown that both the *in vivo* membrane targeting and *in vitro* binding to artificial lipid vesicles of RhoA and CDC42 depend upon sphingomyelin. This study also suggests that sphingolipids are differentially required for distinct mechanisms of CIE (Cheng et al. 2006). RhoA is a dynamin-dependent pathway that has been described in the internalization of the β-chain of the interleukin-2 receptor (IL-2R-β) and other proteins in both immune cells and fibroblasts (Lamaze et al. 2001). In contrast, the CDC42 is a dynamin-independent pathway described as a principal route for the uptake of cholera toxin B (CtxB) and the *Helicobacter pylori* vacuolating toxin (VacA) (Llorente et al. 1998; Gauthier et al. 2005). The cargos entering the cell through CIE are usually delivered to the early endosomes, followed by the transfer to late endosomes and lysosomes. In addition, the cargo can be routed to the trans-Golgi network or recycled back to the plasma membrane (Grant and Donaldson 2009). CIE is the internalization route described preferentially for polyplexes of self-branched and trisaccharide-substituted chitosan oligomers' (SBTCOs) NPs for the delivery of DNA (Garaiova et al. 2012), and for cowpea mosaic virus (CPMV), which has been extensively studied in the last years as a strategy for vaccine development, *in vivo* vascular imaging, and tissue-targeted delivery (Plummer and Manchester 2013). Recent studies suggest that CIE is involved in a new mechanism for the uptake of NPs that was described as a type of macropinocytosis. This new mechanism was found to be dependent on the actin filaments and dynamin, and was designated as an *excavator shovel*-like mechanism (Lerch et al. 2013). In another study, Garaiova et al. (2012) showed that the NPs derived from SBTCOs generated a higher uptake and better transfection efficacy than the NPs prepared from a linear chitosan (LCO). SBTCOs were primarily taken up by the cells via CIE, and successfully escaped from the endocytic vesicles. In contrast, LCO suspension in the cell culture medium resulted in the NPs' aggregation and a relatively lower extent of cellular internalization was observed when compared to the SBTCO NPs (Yameen et al. 2014).

## 1.5   MODELING OF PK AND PHARMACODYNAMICS

A large number of probable drug candidates are excluded during drug discovery and advancement process. Recent investigations have demonstrated that this high rate of exclusion to a great extent is due to the absence of efficacy and safety features of new drugs. An imperative inquiry is along these lines how to enhance the expectation of medication efficacy and required safety. Significant difficulties in the advancement of drug dosage forms (DDFs) have been the short half-life, poor bioavailability, lacking aggregation, and infiltration of the DDFs into the tumor tissue. Comprehensive

PK parameters of the DDF are the keys to overcome these difficulties. A useful component of pharmacokinetic–pharmacodynamic (PK–PD) study is that they are expressed to depict, in a quantitative way, the relationship between plasma concentration and effect. To this end, PK–PD modeling uses ideas from physiologically based PK demonstrating, receptor hypothesis, and dynamical systems' investigation and infection systems' examination. PK/PD values have been apparent as key variables in selecting proper dosage regimen and accessing the *in vivo* efficacy of antimicrobial medications showing concentration-dependent bactericidal effect, for example, natamycin (Schentag et al. 2001; Liu et al. 2002). Despite the fact that the prime parameter minimum inhibitory concentration (MIC90) is utilized as a measure of the intensity of an antimicrobial agent, its worth cannot be dependent upon for anticipating *in vivo* effectiveness (Liu et al. 2002). Subsequently, its connection to other PK parameters, for example, $C_{max}$ and area under curve (AUC), is the way to correlate with the efficacy.

Chandasana et al. (2014) compared the PK/PD values of both NPs 5% and NPs 1% with Natamet. They observed that in case of NPs 5%, the parameters such as $C_{max}$/MIC90 (lowest concentration at which 90% of the isolates were inhibited) and AUC (0–10 h)/MIC were more prominent than the standard values indicating the therapeutic effectiveness of the nanoformulation.

The circulation of a medication to the site of activity is a vital determinant in the time course and intensity of medication effects in central nervous system (CNS) drug treatment as described by Syvanen et al. (2009). They described that drug circulation to the cerebrum is portrayed by both detached dissemination and dynamic efflux by transporters present at the luminal surface of the BBB. Along these lines, the drug target at the site of activity may be unique in relation to the fixation in the plasma compartment. It is accounted for that medications that are permeability glycoprotein (P-gp) substrates in rats are prone to likewise be P-gp substrates in higher species (Syvanen et al. 2009).

Nanotechnology has, as of, late been intensively investigated to benefit the medical field because of the possibility to develop smart DDSs, new imaging contrast device, and diagnostics (Bhattacharyya et al. 2011; Kudgus et al. 2011).

Researchers examined that decreasing size of the formulations increases the therapeutic effectiveness in case of emulsions, liposomes, and different metal NPs (Arvizo et al. 2012). The ultimate impact of drug–target interaction and initiation relies on the target affinity and intrinsic affinity and receptor expression at the site (Gill et al. 1996). PK–PD demonstrating methodologies for target binding and initiation depend on the ideas of the receptor hypothesis. Danhof et al. (2007) explained that receptor hypothesis characterizes that receptor affinity and intrinsic efficacy are drug-specific properties, though receptor expression is a biological process.

Liposomal formulations considered as one of the major targeted DDSs being used commonly show nonlinear, saturable PK after IV administration, with moderately short disposal half-lives at low, nonsaturated dosages (Allen et al. 1995). Although sterically balanced out liposomes might demonstrate linier PK over an extensive variation of doses, their PK and biodistributional example is additionally complexed at low dosages and upon repeated administration (Laverman et al. 2000; Zhou et al. 2002; Moghimi and Szebeni 2003; Ishida et al. 2003, 2006a,b).

The PK properties, for example, elimination half-life, biodistribution, penetrability, and drug discharge rate are controlled by the substances present in the lipid bilayer. The outer surface of the liposome can be adjusted in a few ways that can change biodistribution altogether: (i) glycolipids or manufactured hydrophilic polymers, for example, PEG covalently bound to the membrane can create sterically stabilized liposomes (SSLs), which have decreased opsonization and broadened flowing plasma half-life; (ii) the surface can be attached covalently with target ligands that upgrade binding and internalization by cancer cells communicating a receptor for the ligand (Li et al. 2013a,b).

The procedures by which liposomes and different NPs are cleared from the circulation system have been researched in impressive point of interest (Ishida et al. 2002; Moghimi et al. 2012). The general clearance of liposomal formulations is poor upon three variables: (i) the rate of disposal of the liposome bearer itself; (ii) the rate of arrival of the epitomized or layer-consolidated medication from the transporter; and (iii) the rate of elimination and metabolized discharged medication that is no more connected with the transporter.

The principle component for their disposal is through acknowledgment and uptake by macrophages of the RES, which dwell essentially inside of the liver and spleen (Gregoriadis 1976a,b; Weinstein 1984; Senior 1987). Liposome stability may be affected by physicochemical variables, the vesicle size, lipid portion of the film, and discharge rate of liposome substance. Liposomes can likewise connect with plasma constituents, for example, proteins, in this way influencing their destiny *in vivo*, either by influencing their security and/or balancing their ensuing communication with the objective cells (Ishida et al. 2002). Plasma protein connections might separate or trade lipids from the transporter, bargaining its trustworthiness. The efficient uptake of liposomes is mediated via plasma protein opsonins, fibronectin, CRP, the C3b supplement section, $\beta2$-glycoprotein I, or the Fc segment of an IgG (Patel 1992; Devine and Marjan 1997).

## 1.6   REGULATORY CONSIDERATIONS

Targeted DDSs are developed mostly based on nanomaterials. Nanotechnological approach-based DDSs combine advances in the fields of biology, chemistry, engineering, and medicine to offer novel solutions to some of the limitations associated with traditional therapeutic agents. It uses NPs, engineered materials generally in the 1–100 nm dimension range, for the diagnosis and treatment of diseases. These materials can improve the solubility of water-soluble drugs, prolong the drug circulation half-life in the blood by reducing immunogenicity, minimize degradation of the drug after administration, decrease side effects, and increase bioavailability (Zhang et al. 2008).

In general, nanomaterials can be categorized into carbon-based materials, such as fullerenes and carbon nanotubes, and inorganic NPs, including those based on metal oxides (iron oxide, titanium dioxide, silicon dioxide, etc.), metals (gold and silver), and semiconductor NPs or so-called quantum dots (typically, cadmium sulfide and cadmium selenide). Mixtures of different phases are also manufactured. For drug delivery, not only engineered NPs may be used as carriers but also the drug itself may be formulated at the nanoscale, and thus may function as its own carrier

(De Jong and Borm 2008). NPs due to their nanosize range are able to target a specific organ. They can be amorphous or crystalline, or nanoparticulate DDSs, including micelles, microemulsions, liposomes, drug–polymer conjugates, and antibody–drug conjugates. These NPs are either transient or persistent—depending on whether the integrity of their structure and size is maintained until reaching the site of drug action. Examples of several approved drug products are included as pharmaceutical nanoparticulate systems along with a commentary on the current development issues and paradigms for various categories of NPs.

The Food and Drug Administration (FDA) or other regulatory agencies are responsible for protecting the public health by assuring the safety, efficacy, and security of human and veterinary drugs, biological products, medical devices, our nation's food supply, cosmetics, and products that emit radiation. They are also responsible for advancing the public health by helping to speed innovations that make medicines and foods more effective, safer, and more affordable; and helping the public get the accurate, science-based information they need to use medicines and foods to improve their health. The marketing approval of any product is normally given on a product-by-product basis based on premarket approval, market clearance, and postmarket review. The different types of products available nowadays are based on a multicomponent system that may consist of carrier/delivery system (drug or device), therapeutic agent (drug or biologic), imaging agent, targeting agent, implantable microchip-based delivery systems that deliver different drugs under controlled conditions, and injectable delivery systems (transdermal microneedles). Nanomedicines are likely to be three-dimensional constructs of multiple components with preferred spatial arrangements for their functions. As a result, subtle changes in process or composition can adversely affect the complex superposition of the components with negative consequences. A thorough understanding of the components through detailed physicochemical characterization as well as functional tests may be essential in order to support the highly reproducible manufacturing processes for nanomedicines. Multiple hurdles exist before a nanomedicine can reach the clinic, starting with detailed characterization and the successful manufacture of this complex construct. Other than the standard criteria for acceptable safety and efficacy, and desirable pharmaceutical characteristics (e.g., stability, ease of administration, etc.) that are applicable to most drugs, the ideal NP system or nanomedicine to be utilized for therapeutic purpose may embody some additional features. These are: (a) the detailed understanding of critical components and their interactions, (b) the identification of key characteristics and their relation to performance, (c) the ability to replicate key characteristics under manufacturing conditions, (d) easy to produce in a sterile form, (e) the ability to target or accumulate in the desired site of action by overcoming the restrictive biological barriers, and (f) good in-use stability, easy to store and to administer.

But to study the effects on activity and safety, some relevant information is necessary. This is regarding (a) the accessibility of NPs to tissues and cells that normally would be bypassed by larger particles; (b) the time of retention of the particle after entry; (c) clearance from tissues and blood; and (d) the effects of NPs on cellular and tissue functions (transient and/or permanent after their entry).

The major safety considerations that should be addressed by the regulatory bodies are related to (a) route-specific issues (local respiratory toxicity, distribution in respiratory tissues, systemic bioavailability for inhalation; sensitization for subcutaneous; intravitreal retention for ocular; increased bioavailability for oral; increased dermal and systemic bioavailability, follicle retention, distribution to local lymph nodes, phototoxicity for dermal exposure; and hemocompatibility, sterility, different tissue distribution, and half-life of active pharmaceutical ingredient (API) for IV route [with targeted delivery and liposomes]), (b) absorption, distribution, metabolism, and excretion (ADME), and (c) environmental considerations.

An additional issue in the manufacturing of NPs is environmental safety. The handling of dry materials of the nanometer size scale demands special caution as the airborne NPs distribute as aerosols. Lung deposition of such NPs can lead to pulmonary toxicities (Nel et al. 2006; Song et al. 2009).

During dosing solution preparation, aerosolization of solutions needs to be avoided to prevent unintended exposure. Some NPs are capable of penetrating the skin barrier, making dermal exposure a potential risk, so adequate protection of personnel is essential (Nel et al. 2006). In this respect, NPs that are created entirely within a liquid environment may have significantly lower environmental impact, presumably no different from standard manufacturing of liquid pharmaceutical products. So some safety-concerned issues to be considered are (a) release of NPs into the environment following human and animal use, (b) proper methodologies of NP release in the environment, and (c) the environmental impact on other species (animals, fish, plants, and microorganisms).

International standard-setting bodies have recognized this implication and agreed that "as a minimum set of measurements size, zeta potential (surface charge), and solubility" of NPs should be used as predictors of NP toxicity (ISO/IEC/NIST/OECD Workshop: FINAL REPORT June 2008). For example, when inhaled, nanomaterials less than 100 nm can induce pulmonary inflammation and oxidative stress (Nel et al. 2006) and disrupt distal organ functions through mechanisms including hydrophobic interactions, redox cycling, and free radical formation. Unstable NPs may form large aggregates in micrometer size scale, which can be entrapped in the capillary bed of the lungs and pose a serious danger to patients. Notwithstanding these suggestions, it should be recognized that standard toxicology studies required before moving a product into the clinic will more than likely pick up any manifestation of such toxicities due to the extensive histopathology required.

In June 2011, FDA published a draft guidance, *Considering Whether an FDA-Regulated Product Involves the Application of Nanotechnology*, as a starting point for the nanotechnology discussion (US Food and Drug Administration 2007). Based on its current scientific and technical understanding of nanomaterials and their characteristics, the agency is proposing certain points it will use to determine whether an FDA-regulated product contains nanomaterials or otherwise involves the use of nanotechnology. Industry is encouraged to consult with the FDA early in the product development process to address questions related to the regulatory status, safety, effectiveness, or public health impact of products that use nanotechnology. The proposed guidelines are the first step toward developing policies that will guide the regulation of products using nanotechnology. The agency plans to develop additional guidelines for specific products in the future.

# REFERENCES

Aderem, A. and Underhill, D.M. 1999. Mechanisms of phagocytosis in macrophages. *Annu Rev Immunol* 17:593–23.

Alexiou, C., Jurgons, R., Schmid, R.J. et al. 2003. Magnetic drug targeting—Biodistribution of the magnetic carrier and the chemotherapeutic agent mitoxantrone after locoregional cancer treatment. *J Drug Target* 11:139–49.

Allen, T.M. 2002. Ligand-targeted therapeutics in anticancer therapy. *Nat Rev Cancer* 2:750–63.

Allen, T.M., Newman, M.S., Woodle, M.C. et al. 1995. Pharmacokinetics and anti-tumor activity of vincristine encapsulated in sterically stabilized liposomes. *Int J Cancer* 62:199–204.

Amyere, M., Mettlen, M., Van Der Smissen, P. et al. 2002. Origin, originality, functions, subversions and molecular signaling of macropinocytosis. *Int J Med Microbiol* 291:487–94.

Amyere, M., Payrastre, B., Krause, U. et al. 2000. Constitutive macropinocytosis in oncogene-transformed fibroblasts depends on sequential permanent activation of phosphoinositide 3-kinase and phospholipase C. *Mol Biol Cell* 11:3453–67.

Anderson, R.G., Kamen, B.A., Rothberg, K.G. et al. 1992. Potocytosis: Sequestration and transport of small molecules by caveolae. *Science* 255:410–11.

Arvizo, R.R., Bhattacharyya, S., Kudgus, R.A. et al. 2012. Intrinsic therapeutic applications of noble metal nanoparticles: Past, present and future. *Chem Soc Rev* 41:2943–70.

Bae, Y.H. and Park, K. 2011. Targeted drug delivery to tumors: Myths, reality and possibility. *J Control Release* 153:198–205.

Barratt, G. 2003. Colloidal drug carriers: Achievements and perspectives. *Cell Mol Life Sci* 60:21–37.

Bejjani, R.A., Andrieu, C., Bloquel, C. et al. 2007. Electrically assisted ocular gene therapy. *Surv Ophthalmol* 52:196–08.

Bendas, G., Krause, A., Bakowsky, U. et al. 1999. Targetability of novel immunoliposomes prepared by a new antibody conjugation technique. *Int J Pharm* 181:79–93.

Benmerah, A. and Lamaze, C. 2007. Clathrin-covered pits: Vive La Différence? *Traffic* 8:970–82.

Bhattacharyya, S., Kudgus, R.A., Bhattacharya, R. et al. 2011. Inorganic nanoparticles in cancer therapy. *Pharm Res* 28:237–59.

Champion, J.A. and Mitragotri, S. 2006. Role of target geometry in phagocytosis. *Proc Natl Acad Sci U S A* 103:4930–34.

Champion, J.A. and Mitragotri, S. 2009. Shape induced inhibition of phagocytosis of polymer particles. *Pharm Res* 26:244–49.

Chandasana, H., Prasad, Y.D., Chhonker, Y.S. et al. 2014. Corneal targeted nanoparticles for sustained natamycin delivery and their PK/PD indices: An approach to reduce dose and dosing frequency. *Int J Pharm* 477:317–25.

Charman, W.N., Chan, H.K., Finnin, B.C. et al. 1999. Drug delivery: A key factor in realising the full therapeutic potential of drugs. *Drug Dev Res* 46:316–27.

Cheng, Z.J., Singh, R.D., Sharma, D.K. et al. 2006. Distinct mechanisms of clathrin-independent endocytosis have unique sphingolipid requirements. *Mol Biol Cell* 17:3197–210.

Chrastina, A., Massey, K.A., and Schnitzer, J.E. 2011. Overcoming *in vivo* barriers to targeted nanodelivery. *Wiley Interdiscip Rev Nanomed Nanobiotechnol* 3:421–37.

Conner, S.D. and Schmid, S.L. 2003. Regulated gateways of section into the cell. *Nature* 422:37–44.

Couvreur, P. and Vauthier, C. 2006. Nanotechnology: Intelligent design to treat complex disease. *Pharm Res* 23:1417–49.

Danhof, M., de Jongh, J., De Lange, E.C. et al. 2007. Mechanism-based pharmacokinetic-pharmacodynamic modeling: Biophase distribution, receptor theory, and dynamical systems analysis. *Annu Rev Pharmacol Toxicol* 47:357–400.

David, A., Kopeckova, P., Kopecek, J. et al. 2002. The role of galactose, lactose, and galactose valency in the biorecognition of N-(2-hydroxypropyl)methacrylamide copolymers by human colon adenocarcinoma cells. *Pharm Res* 19:1114–22.

De Jong, W.H. and Borm, P.J.A. 2008. Drug delivery and nanoparticles: Applications and hazards. *Int J Nanomed* 3:133–49.

Delva, E., Jennings, J.M., Calkins, C.C. et al. 2008. Pemphigus vulgaris IgG-induced desmoglein-3 endocytosis and desmosomal disassembly are mediated by a clathrin-and dynamin-independent mechanism. *J Biol Chem* 283:18303–313.

Devine, D.V. and Marjan, J.M. 1997. The role of immunoproteins in the survival of liposomes in the circulation. *Crit Rev Ther Drug Carrier Syst* 14:105–31.

Diederichs, J.E. and Muller, R.H. 1998. *Future Strategies for Drug Delivery with Particulate Systems.* Boca Raton, FL: CRC Press.

Dinauer, N., Balthasar, S., Weber, C. et al. 2005. Selective targeting of antibody-conjugated nanoparticles to leukemic cells and primary T-lymphocytes. *Biomaterials* 26:5898–906.

Dobrovolskaia, M.A. and McNeil, S.E. 2007. Immunological properties of engineered nanomaterials. *Nat Nanotechnol* 2: 469–78.

Doherty, G.J. and McMahon H.T. 2009. Mechanisms of endocytosis. *Annu Rev Biochem* 78:857–902.

Dumitriu, S. 2002. *Polymeric Biomaterials Revised and Expanded.* New York: Marcel Dekker Inc.

Eljarrat-Binstock, E. and Domb, A.J. 2006. Iontophoresis: A non-invasive ocular drug delivery. *J Control Release* 110:479–89.

Eyster, C.A., Higginson, J.D., Huebner, R. et al. 2009. Discovery of new cargo proteins that enter cells through clathrin-independent endocytosis. *Traffic* 10:590–99.

Fahr, A., Müller, K., Nahde, T. et al. 2002. A new colloidal lipidic system for gene therapy. *J Liposome Res* 12:37–44.

Farokhzad, O.C., Jon, S., Khademhosseini, A. et al. 2004. Nanoparticle-aptamer bioconjugates: A new approach for targeting prostate cancer cells. *Cancer Res* 64:7668–72.

Farokhzad, O.C. and Langer, R. 2009. Impact of nanotechnology on drug delivery. *ACS Nano* 3:16–20.

Ferrari, A., Pellegrini, V., Arcangeli, C. et al. 2003. Caveolae-mediated internalization of extracellular HIV-1 tat fusion proteins visualized in real time. *Mol Ther* 8:284–94.

Fiorentini, C., Falzano, L., Fabbri, A. et al. 2001. Activation of rho GTPases by cytotoxic necrotizing factor 1 induces macropinocytosis and scavenging activity in epithelial cells. *Mol Biol Cell* 12:2061–73.

Fittipaldi, A., Ferrari, A., Zoppe, M. et al. 2003. Cell membrane lipid rafts mediate caveolar endocytosis of HIV-1 Tat fusion proteins. *J Biol Chem* 278:34141–149.

Ford, M.G., Mills, I.G., Peter, B.J. et al. 2002. Curvature of clathrin-coated pits driven by cpsin. *Nature* 419:361–66.

Forssen, E. and Willis, M. 1998. Ligand-targeted liposomes. *Adv Drug Deliv Rev* 29:249–71.

Fujimoto, L.M., Roth, R., Heuser, J.E. et al. 2000. Actin assembly plays a variable, but not obligatory role in receptor-mediated endocytosis in mammalian cells. *Traffic* 1:161–71.

Gabizon, A., Shmeeda, H., Horowitz, A.T. et al. 2004. Tumor cell targeting of liposome-entrapped drugs with phospholipid-anchored folic acid-PEG conjugates. *Adv Drug Deliv Rev* 56:1177–92.

Garaiova, Z., Strand S.P., Reitan, N.K. et al. 2012. Cellular uptake of DNA–chitosan nanoparticles: The role of clathrin- and caveolae-mediated pathways. *Int J Biol Macromol* 51:1043–51.

Garg, A. and Kokkoli, E. 2005. Characterizing particulate drug-delivery carriers with atomic force microscopy. *IEEE Eng Med Biol Mag* 24:87–95.

Garg, A., Tisdale, A.W., Haidari, E. et al. 2009. Targeting colon cancer cells using PEGylated liposomes modified with a fibronectin-mimetic peptide. *Int J Pharm* 2009. 366:201–10.

Gauthier, N.C., Monzo, P., Kaddai, V. et al. 2005. Helicobacter pylori VacA cytotoxin: A probe for a clathrin-independent and Cdc42-dependent pinocytic pathway routed to late endosomes. *Mol Biol Cell* 16:4852–66.

Geng, Y., Dalhaimer, P., Cai, S. et al. 2007. Shape effects of filaments versus spherical particles in flow and drug delivery. *Nat Nanotechnol* 2:249–55.

Gibson, J.D., Khanal, B.P., and Zubarev, E.R. 2007. Paclitaxel-functionalized gold nanoparticles. *J Am Chem Soc* 129:11653–61.

Gill, P.S., Wernz, J., Scadden, D.T. et al. 1996. Randomized stage III trial of liposomal daunorubicin versus doxorubicin, bleomycin, and vincristine in AIDS-related Kaposi's sarcoma. *J Clin Oncol* 14:2353–64.

Grant, B.D. and Donaldson, J.G. 2009. Pathways and mechanisms of endocytic recycling. *Nat Rev Mol Cell Biol* 10:597–608.

Gratton, S.E., Ropp, P.A., Pohlhaus, P.D. et al. 2008. The effect of particle design on cellular internalization pathways. *Proc Natl Acad Sci U S A* 105:11613–18.

Gref, R., Minamitake, Y., Peracchia, M.T. et al. 1994. Biodegradable long-circulating polymeric nanospheres. *Science* 263:1600–03.

Gregoriadis, G. 1976a. The carrier potential of liposomes in biology and medicine (second of two parts). *N Engl J Med* 295:765–70.

Gregoriadis, G. 1976b. The carrier potential of liposomes in biology and medicine (first of two parts). *N Engl J Med* 295:704–10.

Gregoriadis, G. 1981. Targeting of drugs: Implications in medicine. *Lancet* 2:241–46.

Greish, K. 2007. Enhanced permeability and retention of macromolecular drugs in solid tumors: A royal gate for targeted anticancer nanomedicines. *J Drug Target* 15:457–64.

Gupta, B., Levchenko, T.S., and Torchilin, V.P. 2005. Intracellular delivery of large molecules and small particles by cell-penetrating proteins and peptides. *Adv Drug Deliv Rev* 57:637–51.

Harguindey, S., Arranz, J.L., Wahl, M.L. et al. 2009. Proton transport inhibitors as potentially selective anticancer drugs. *Anticancer Res* 29:2127–36.

Harris, J., Werling, D., Hope, J.C. et al. 2002. Caveolae and caveolin in immune cells: Distribution and functions. *Trends Immunol* 23:158–64.

Hart, S.P., Rossi, A.G., Haslett, C. et al. 2012. Characterization of the Effects of cross-linking of macrophage CD44 associated with increased phagocytosis of apoptotic PMN. *PLoS One* 7:e33142.

Heath, T.D., Fraley, R.T., and Papahadjopoulos, D. 1980. Antibody targeting of liposomes—cell specificity obtained by conjugation of F(Ab')2 to vesicle surface. *Science* 210:539–41.

Högemann-Savellano, D., Bos, E., Blondet, C. et al. 2003. The transferrin receptor: A potential molecular imaging marker for human cancer. *Neoplasia* 5:495–506.

Hong, S., Leroueil, P.R., Janus, E.K. et al. 2006. Interaction of polycationic polymers with supported lipid bilayers and cells: Nanoscale hole formation and enhanced membrane permeability. *Bioconjugate Chem* 17:728–34.

Huang, D.M., Hung, Y., Ko, B.S. et al. 2005. Highly efficient cellular labeling of mesoporous nanoparticles in human mesenchymal stem cells: Implication for stem cell tracking. *FASEB J* 19:2014–16.

Ishida, T., Atobe, K., Wang, X. et al. 2006b. Accelerated blood clearance of PEGylated liposomes upon repeated injections: Effect of doxorubicin-encapsulation and high-dose first injection. *J Control Release* 115:251–58.

Ishida, T., Harashima, H., and Kiwada, H. 2002. Liposome clearance. *Biosci Rep* 22:197–224.

Ishida, T., Ichihara, M., Wang, X. et al. 2006a. Injection of PEGylated liposomes in rats elicits PEG-specific IgM, which is responsible for rapid elimination of a second dose of PEGylated liposomes. *J Control Release* 112:15–25.

Ishida, T., Maeda, R., Ichihara, M. et al. 2003. Accelerated clearance of PEGylated liposomes in rats after repeated injections. *J Control Release* 2003, 88:35–42.

ISO, IEC, NIST, and OECD International workshop on documentary standards for measurement and characterization for nanotechnologies; 2008 02/26-02/28/2008; Gaithersburg, MD.

Jain, K.K. 2008. *The Handbook of Nanomedicine.* Totowa: Humana Press.

Jaracz, S., Chen, J., Kuznetsova, L.V. et al. 2005. Recent advances in tumor-targeting anticancer drug conjugates. *Bioorg Med Chem* 13:5043–54.

Johnson, L.S., Dunn, K.W., Pytowski, B. et al. 1993. Endosome acidification and receptor trafficking: Bafilomycin A1 slows receptor externalization by a mechanism involving the receptor's internalization motif. *Mol Biol Cell* 4:1251–66.

Kasamatsu, H. and Nakanishi, A. 1998. How do animal DNA viruses get to the nucleus? *Annu Rev Microbiol* 52:627–86.

Kerr, M.C. and Teasdale, R.D. 2009. Defining macropinocytosis. *Traffic* 10:364–71.

Khalil, I.A., Kogurem K., Akita, H. et al. 2006. Uptake pathways and subsequent intracellular trafficking in nonviral gene delivery. *Pharmacol Rev* 58:32–45.

Kircheis, R., Wightman, L., and Wagner, E. 2001. Design and gene delivery activity of modified polyethylenimines. *Adv Drug Deliv Rev* 53:341–58.

Kirchhausen T. 2000. Clathrin. *Annu Rev Biochem* 69:699–27.

Kolb-Maurer, A., Wilhelm, M., Weissinger, F. et al. 2002. Interaction of human hematopoietic stem cells with bacterial pathogens. *Blood* 100:3703–09.

Kopatz, I., Remy, J.S., and Behr, J.P. 2004. A model for non-viral gene delivery: Through syndecan adhesion molecules and powered by actin. *J Gene Med* 6:769–76.

Kopecek, J. 2003. Smart and genetically engineered biomaterials and drug delivery systems. *EurJ Pharm Sci* 20:1–16.

Kudgus, R. A., Bhattacharya, R., and Mukherjee, P. 2011. Disease nanotechnology: Rising part of gold nanoconjugates. *Anticancer Agents Med Chem* 11:965–73.

Kunzmann, A., Andersson, B., Thurnherr, T. et al. 2011. Toxicology of engineered nanomaterials: Focus on biocompatibility, biodistribution and biodegradation. *Biochim Biophys Acta* 1810:361–73.

Lafon, M. 2005. Rabies virus receptors. *J Neurovirol* 11:82–87.

Lam, A.P. and Dean, D.A. 2010. Progress and prospects: Nuclear import of nonviral vectors. *Gene Ther* 17:439–47.

Lamaze, C., Dujeancourt, A., Baba, T. et al. 2001. Interleukin 2 receptors and detergent-resistant membrane domains define a clathrin-independent endocytic pathway. *Mol Cell* 7:661–71.

Lamaze, C. and Schmid, S.L. 1995. The rise of clathrin-free pinocytic pathways. *Curr Opin Cell Biol* 7:573–80.

Laverman, P., Brouwers, A.H., Dams, E.T. et al. 2000. Preclinical and clinical evidence for disappearance of long-circulating characteristics of polyethylene glycol liposomes at low lipid dose. *J Pharmacol Exp Ther* 293:996–01.

Leamon, C.P. and Reddy, J.A. 2004. Folate-targeted chemotherapy. *Adv Drug Deliv Rev* 56:1127–41.

Lee, H.J., Ahn, B.N., Paik, W.H. et al. 1996. Inverse targeting of reticuloendothelial system-rich organs after intravenous administration of adriamycin-loaded neutral proliposomes containing poloxamer 407 to rats. *Int J Pharm* 131:91–96.

Lerch, S., Dass, M., Musyanovycha, A. et al. 2013. Polymeric nanoparticles of different sizes overcome the cell membrane barrier. *Eur J Pharm Biopharm* 84:265–74.

Leserman, L.D., Barbet, J., Kourilsky, F. et al. 1980. Targeting to cells of fluorescent liposomes covalently coupled with monoclonal antibody or protein A. *Nature* 288:602–04.

Li, L., ten Hagen, T.L., Bolkestein, M. et al. 2013a. Improved intratumoral nanoparticles extravasation and penetration by mild hyperthermia. *J Control Release* 167:130–37.

Li, L., ten Hagen, T.L., Hossann, M. et al. 2013b. Mild hyperthermia triggered doxorubicin release from optimized stealth thermosensitive liposomes improves intratumoral drug delivery and efficacy. *J Control Release* 168:142–50.

Liang, G., Pu, Y., Yin, L. et al. 2009. Influence of different sizes of titanium dioxide nanoparticles on hepatic and renal functions in rats with correlation to oxidative stress. *J Toxicol Environ Health A* 72:740–45.

Lin, J., Zhang, H., Chen, Z. et al. 2010. Penetration of lipid membranes by gold nanoparticles: Insights into cellular uptake, cytotoxicity, and their relationship. *ACS Nano* 4:5421–29.

Liu, P., Müller, M., and Derendorf, H. 2002. Rational dosing of antibiotics: The use of plasma concentrations versus tissue concentrations. *Int J Antimicrob Agents* 19:285–90.

Liu, Y., Huang, R., Han, L. et al. 2009. Brain-targeting gene delivery and cellular internalization mechanisms for modified rabies virus glycoprotein RVG29 nanoparticles. *Biomaterials* 30:4195–202.

Llorente, A., Rapak, A., Schmid, S.L. et al. 1998. Expression of mutant dynamin inhibits toxicity and transport of endocytosed ricin to the Golgi apparatus. *J Cell Biol* 140:553–63.

Lochmann, D., Jauk, E., and Zimmer, A. 2004. Drug delivery of oligonucleotides by peptides. *Eur J Pharm Biopharm* 58:237–51.

Lundberg, B., Hong, K., and Papahadjopoulos, D. 1993. Conjugation of apolipoprotein B with liposomes and targeting to cells in culture. *Biochimica Biophysica Acta* 1149:305–12.

Maeda, H., Wu, J., Sawa, T. et al. 2000. Tumor vascular permeability and the EPR effect in macromolecular therapeutics: A review. *J Control Release* 65:271–84.

Managit, C., Kawakami, S., Nishikawa, M. et al. 2003. Targeted and sustained drug delivery using PEGylated galactosylated liposomes. *Int J Pharm* 266:77–84.

Martin, M.E. and Rice, K.G. 2007. Peptide-guided gene delivery. *The AAPS Journal* 9:E18-E29.

Matsui, H., Johnson, L.G., Randell, S.H. et al. 1997. Loss of binding and entry of liposome-DNA complexes decreases transfection efficiency in differentiated airway epithelial cells. *J Biol Chem* 272:1117–26.

Matveev, S., Li, X., Everson, W. et al. 2001. The role of caveolae and caveolin in vesicle-dependent and vesicle-independent trafficking. *Adv Drug Deliv Rev* 49:237–50.

Mayor, S. and Pagano, R.E. 2007. Pathways of clathrin-independent endocytosis. *Nat Rev Mol Cell Biol* 8:603–12.

McIntosh, C.M., Esposito, E.A., Boal, A.K. et al. 2001. Inhibition of DNA transcription using cationic mixed monolayer protected gold clusters. *J Am Chem Soc* 123: 7626–29.

McMahon, H.T. and Boucrot, E. 2011. Molecular mechanism and physiological functions of clathrin-mediated endocytosis. *Nat Rev Mol Cell Biol* 12:517–33.

Mecke, A., Dittrich, C., and Meier, W. 2006. Biomimetic membranes designed from amphiphilic block copolymers. *Soft Matter* 2:751–59.

Meier, O. and Greber, U.F. 2003. Adenovirus endocytosis. *J Gene Med* 5:451–62.

Missirlis, D., Tirelli, N., and Hubbell, J.A. 2005. Amphiphilic hydrogel nanoparticles. Preparation, characterization and preliminary assessment as new colloidal drug carriers. *Langmuir* 21:2605–13.

Moghimi, S.M., Hunter, A.C., and Andresen, T.L. 2012. Factors controlling nanoparticle pharmacokinetics: An integrated analysis and perspective. *Annu Rev Pharmacol Toxicol* 52:481–503.

Moghimi, S.M. and Szebeni, J. 2003. Stealth liposomes and long circulating nanoparticles: Critical issues in pharmacokinetics, opsonization and protein-binding properties. *Prog Lipid Res* 42:463–78.

Mohanraj, V.J. and Chen, Y. 2006. Nanoparticles: A review. *Trop J Pharm Res* 5:561–73.

Molema, G. and Meijer, D.K.F. 2001. Drug targeting: Organ-specific strategies. In *Methods and Principles in Medicinal Chemistry*, ed. R. Mannhold, H. Kubinyi, and H. Timmerman, 381–382. New York: Wiley-VCH.

Mukherjee, S., Ghosh, R.N., and Maxfield, F.R. 1997. Endocytosis. *Physiol Rev* 77:759–803.

Na, K., Bum Lee, T., Park, K.H. et al. 2003. Self-assembled nanoparticles of hydrophobically-modified polysaccharide bearing vitamin H as a targeted anti-cancer drug delivery system. *Eur J Pharm Sci* 18:165–73.

Nakamaki, T., Kawabata, H., Saito, B. et al. 2004. Elevated levels of transferrin receptor 2 mRNA, not transferrin receptor 1 mRNA, are associated with increased survival in acute myeloid leukaemia. *BrJ Haematol* 125:42–49.

Nel, A., Xia, T., Madler, L. et al. 2006. Toxic potential of materials at the nanolevel. *Science* 311:622–27.

Song, Y., Li, X., and Du, X. 2009. Exposure to nanoparticles is related to pleural effusion, pulmonary fibrosis and granuloma. *Eur Respir J* 34:559–67.

Niitsu, Y., Kohgo, Y., Nishisato, T. et al. 1987. Transferrin receptors in human cancerous tissues. *Tohoku J Exp Med* 153:239–43.

Nobes, C. and Marsh, M. 2000. Dendritic cells: New roles for Cdc42 and Rac in antigen uptake? *Curr Biol* 10:R739–41.

Nobs, L., Buchegger, F., Gurny, R. et al. 2004. Current methods for attaching targeting ligands to liposomes and nanoparticles. *J Pharm Sci* 93:1980–92.

Oh, P., Borgstrom, P., Witkiewicz, H. et al. 2007. Live dynamic imaging of caveolae pumping targeted antibody rapidly and specifically across endothelium in the lung. *Nat Biotechnol* 25:327–37.

Orlandi, P.A. and Fishman, P.H. 1998. Filipin-dependent inhibition of cholera toxin: Evidence for toxin internalization and activation through caveolae-like domains. *J Cell Biol* 141:905–15.

Owens, D.E. and Peppas, NA. 2006. Opsonization, biodistribution, and pharmacokinetics of polymeric nanoparticles. *Int J Pharm* 307:93–102.

Palmer, T.N., Caride, V.J., Caldecourt, M.A. et al. 1984. The mechanism of liposome accumulation in infarction. *Biochim Biophys Acta* 797:363–68.

Parton, R.G. and del Pozo, M.A. 2013. Caveolae as plasma membrane sensors, protectors and organizers. *Nat Rev Mol cell Biol* 14:98–112.

Parton, R.G., Joggerst. B., and Simons, K. 1994. Regulated internalization of caveolae. *J Cell Biol* 127:1199–215.

Patel, H.M. 1992. Influence of lipid composition on opsonophagocytosis of liposomes. *Res Immunol* 143:242–44.

Paul, S.M., Mytelka, D.S., Dunwiddie, C.T. et al. 2010. How to improve R&D productivity: The pharmaceutical industry's grand challenge. *Nat Rev Drug Discov* 9:203–14.

Pelkmans, L., Kartenbeck, J., and Helenius, A. 2001. Caveolar endocytosis of simian virus 40 reveals a new two-step vesicular-transport pathway to the ER. *Nat Cell Biol* 3:473–83.

Pelkmans, L., Püntener, D., and Helenius, A. 2002. Local actin polymerization and dynamin recruitment in SV40- induced internalization of caveolae. *Science* 296:535–39.

Pelkmans, L. and Zerial, M. 2005. Kinase-regulated quantal assemblies and kiss-and-run recycling of caveolae. *Nature* 436:128–33.

Plank, C., Oberhauser, B., Mechtler, K. et al. 1994. The influence of endosome-disruptive peptides on gene transfer using synthetic virus-like gene transfer systems. *J Biol Chem* 269:12918–924.

Plummer, E.M. and Manchester, M. 2013. Endocytic uptake pathways utilized by CPMV nanoparticles. *Mol Pharm* 10:26–32.

Powelka, A.M., Sun, J., Li, J. et al. 2004. Stimulation-dependent recycling of integrin beta1 regulated by ARF6 and Rab11. *Traffic* 5:20–36.

Pucadyil, J. and Schmid, S.L. 2009. Conserved functions of membrane active GTPases in coated vesicle formation. *Science* 325:1217–20.

Rabinovitch, M. 1995. Professional and non-professional phagocytes: An introduction. *Trends Cell Biol* 5:85–87.

Racoosin, E.L. and Swanson, J.A. 1993. Macropinosome maturation and fusion with tubular lysosomes in macrophages. *J Cell Biol* 121:1011–20.

Radhakrishna, H. and Donaldson, J.G. 1997. ADP-ribosylation factor 6 regulates a novel plasma membrane recycling pathway. *J Cell Biol* 139:49–61.

Ratnam, D.V., Ankola, D.D., Bhardwaj, V. et al. 2006. Role of antioxidants in prophylaxis and therapy: A pharmaceutical perspective. *J Control Release* 113:189–207.

Rejman, J., Bragonzi, A., and Conese, M. 2005. Role of clathrin- and caveolae-mediated endocytosis in gene transfer mediated by lipo- and polyplexes. *Mol Ther* 12:468–74.

Rejman, J., Oberle, V., Zuhorn, I.S. et al. 2004. Size-dependent internalization of particles via the pathways of clathrin- and caveolae-mediated endocytosis. *Biochem J* 377:159–69.

Richard, J.P., Melikov, K., Brooks, H. et al. 2005. Cellular uptake of unconjugated TAT peptide involves clathrin-dependent endocytosis and heparan sulfate receptors. *J Biol Chem* 280:15300–306.

Robertson, A.S., Smythe, E., and Ayscough, K.R. 2009. Functions of actin in endocytosis. *Cell Mol Life Sci* 66:2049–65.

Roth, J.C., Curiel, D.T., and Pereboeva, L. 2008. Cell vehicle targeting strategies. *Gene Ther* 15:716–29.

Sahay, G., Alakhova, D.Y., and Kabanov, A.V. 2010. Endocytosis of nanomedicines. *J Control Release* 145:182–95.

Sahoo S.K. and Labhasetwar, V. 2005. Enhanced antiproliferative activity of transferrin-conjugated paclitaxel-loaded nanoparticles is mediated via sustained intracellular drug retention. *Mol Pharm* 2:373–83.

Sallusto, F., Cella, M., Danieli, C. et al. 1995. Dendritic cells use macropinocytosis and the mannose receptor to concentrate macromolecules in the major histocompatibility complex class II compartment: Downregulation by cytokines and bacterial products. *J Exp Med* 182:389–400.

Sandvig, K., Pust, S., Skotland, T. et al. 2011. Clathrin-independent endocytosis: Mechanisms and function. *Curr Opin Cell Biol* 23:413–20.

Sandvig, K., Torgersen, M.L., Raa, H.A. et al. 2008. Clathrin-independent endocytosis: From nonexisting to an extreme degree of complexity. *Histochem Cell Biol* 129:267–76.

Santini Jr, J.T., Richards, A.C., Scheidt, R. et al. 2000. Microchips as controlled drug-delivery devices. *Angew Chem Int Ed* 39:2396–407.

Sawachi, K., Shimada, Y., Taniguchi, H. et al. 2010. Cytotoxic effects of activated alveolar macrophages on lung carcinoma cells via cell-to-cell contact and nitric oxide. *Anticancer Res* 30:3135–41.

Sawant, R.M., Hurley, J.P., Salmaso, S. et al. 2006. SMART drug delivery systems: Double-targeted pH-responsive pharmaceutical nanocarriers. *Bioconjug Chem* 17:943–49.

Schafer, D.A., D'Souza-Schorey, C., and Cooper, J.A. 2000. Actin assembly at membranes controlled by ARF6. *Traffic* 1:892–903.

Schentag, J.J., Gilliland, K.K., and Paladino, J.A. 2001. What have we learned from pharmacokinetic and pharmacodynamic theories? *Clin Infect Dis* 32:S39–S46.

Schilling, C.L., Schuster, M.J., and Wu, G. 2000. Gene therapy for liver disease. In *An Introduction to Molecular Medicine and Gene Therapy*, ed. T.F. Kresina, 153–182. New Jersey: John Wiley & Sons, Inc.

Schmid, S. L. 1997. Clathrin-coated vesicle formation and protein sorting: An integrated process. *Annu Rev Biochem* 66:511–48.

Schnyder, A. and Huwyler, J. 2005. Drug transport to brain with targeted liposomes. *NeuroRx* 2:99–107.

Senior, J.H. 1987. Fate and behavior of liposomes *in vivo*: A review of controlling factors. *Crit Rev Ther Drug Carrier Syst* 3:123–93.

Simões, S., Slepushkin, V., Pires, P. et al. 1999. Mechanisms of gene transfer mediated by lipoplexes associated with targeting ligands or pH-sensitive peptides. *Gene Ther* 6:1798–807.

Smart, E.J., Graf, G.A., McNiven, M.A. et al. 1999. Caveolins, liquid-ordered domains and signal transduction. *Mol Cell Biol* 19:7289–304.

Steinman, R.M. and Swanson, J. 1995. The endocytic activity of dendritic cells. *J Exp Med* 182:283–88.

Stoorvogel, W., Strous, G.J., Ciechanover, A. et al. 1991. Trafficking of the transferrin receptor. *Targeted Diagn Ther* 4:267–304.

Swanson, J.A. and Watts, C. 1995. Macropinocytosis. *Trends Cell Biol* 5:424–28.

Syvanen, S., Lindhe, O., Palner, M. et al. 2009. Species differences in blood-brain barrier transport of three positron emission tomography radioligands with emphasis on P-glycoprotein transport. *Drug Metab Dispos* 37:635–43.

Tachibana, R., Harashima, H., Shono, M. et al. 1998. Intracellular regulation of macromolecules using pH-sensitive liposomes and nuclear localization signal: Qualitative and quantitative evaluation of intracellular trafficking. *Biochem Biophys Res Commun* 251:538–44.

Taniguchi, H., Shimada, Y., Sawachi, K. et al. 2010. Lipopolysaccharide-activated alveolar macrophages having cytotoxicity toward lung tumor cells through cell-to-cell binding-dependent mechanism. *Anticancer Res* 30:3159–65.

Taylor, U., Klein, S., Petersen, S. et al. 2010. Nonendosomal cellular uptake of ligand-free, positively charged gold nanoparticles. *Cytometry A* 77:439–46.

Tebar, F., Bohlander, S.K., and Sorkin, A. 1999. Clathrin assembly lymphoid myeloid leukemia (CALM) protein: localization in endocytic-coated pits, interactions with clathrin, and the impact of overexpression on clathrin-mediated traffic. *Mol Biol Cell* 10:2687–702.

Thorek, D.L.J. and Tsourkas, A. 2008. Size, charge and concentration dependent uptake of iron oxide particles by non-phagocytic cells. *Biomaterials* 29:3583–90.

Torchilin, V.P. 2004. Targeted polymeric micelles for delivery of poorly soluble drugs. *Cell Mol Life Sci* 61:2549–59.

Torchilin, V.P. 2006. Recent approaches to intracellular delivery of drugs and DNA and organelle targeting. *Annu Rev Biomed Eng* 8:343–75.

Underhill, D.M. and Goodridge, H.S. 2012. Information processing during phagocytosis. *Nat Rev Immunol* 12:492–502.

US Food and Drug Administration. Nanotechnology—A Report of the U.S. Food and Drug administration Nanotechnology Task Force, July 25, 2007. Available at: http://www.fda.gov/downloads/ScienceResearch/SpecialTopics/Nanotechnology/ucm110856.pdf. Accessed October 13, 2011.

Vasir, J.K. and Labhasetwar, V. 2008. Quantification of the force of nanoparticle-cell membrane interactions and its influence on intracellular trafficking of nanoparticles. *Biomaterials* 29:4244–52.

Wadia, J.S., Stan, R.V., and Dowdy, S.F. 2004. Transducible TAT-HA fusogenic peptide enhances escape of TAT-fusion proteins after lipid raft macropinocytosis. *Nat Med* 10:310–15.

Wang, Z., Tiruppathi, C., Cho, J. et al. 2011. Delivery of nanoparticle: Complexed drugs across the vascular endothelial barrier via caveolae. *IUBMB Life* 63:659–67.

Wei, Q., Zhu, H., Qian, X. et al. 2014. Targeted genomic capture and massively parallel sequencing to identify novel variants causing Chinese hereditary hearing loss. *J Transl Med* 12:311–18.

Weinstein, J.N. 1984. Liposomes as drug carriers in cancer therapy. *Cancer Treat Rep* 68:127–35.

West, M.A., Bretscher, M.S., and Watts, C. 1989. Distinct endocytotic pathways in epidermal growth factor-stimulated human carcinoma A431 cells. *J Cell Biol* 109:2731–39.

Widera, A., Norouziyan, F., and Shen, W.C. 2003. Mechanisms of TfR-mediated transcytosis and sorting in epithelial cells and applications toward drug delivery. *Adv Drug Deliv Rev* 55:1439–66.

Yameen, B., Choi, W.I., Vilos, C. et al. 2014. Insight into nanoparticle cellular uptake and intracellular targeting. *J Control Release* 190:485–99.

Yan, J.J., Liao, J.Z., Lin, J.S. et al. 2015. Active radar guides missile to its target: Receptor-based targeted treatment of hepatocellular carcinoma by nanoparticulate systems. *Tumour Biol* 36:55–67.

Yoo, J.W., Irvine, D.J., Discher, D.E. et al. 2011. Bio-inspired, bioengineered and biomimetic drug delivery carriers. *Nat Rev Drug Discov* 10:521–35.

Zhang, L., Zhang, S., Ruan, S.B. et al. 2014. Lapatinib-incorporated lipoprotein-like nanoparticles: Preparation and a proposed breast cancer-targeting mechanism. *Acta Pharmacol Sin* 35:846–52.

Zhang, L., Gu, F.X., Chan, J.M. et al. 2008. Nanoparticles in medicine: Therapeutic applications and developments. *Clin Pharmacol Ther* 83:761–69.

Zhou, R., Mazurchuk, R., and Straubinger, R.M. 2002. Antivasculature effects of doxorubicin-containing liposomes in an intracranial rat brain tumor model. *Cancer Res* 62:2561–66.

# 2 Biological Targets
## *Identification, Selection, and Validation*

Rana Datta and Sugata Banerjee

## CONTENTS

## 2.1 INTRODUCTION

Drug discovery may be defined as the discovery, creation, or design of molecules that shows promise to cure diseases ailing humankind. The process of drug discovery is elaborative, utilizing time and resources, both human and financial. It involves the identification of lead molecules followed by their synthesis, characterization, and screening for the desired therapeutic efficacy. Increasingly, scientists have been using the word *target* in drug designing and development, but the definition of the word *target*, their roles in disease pathways, drug development, and even their number remains controversial. Defining targets can solve, in a way, the puzzle of number of targets (druggable targets). The meaning of druggability is that researchers have the appropriate science and technology in hand to develop antagonists to a particular target. The druggable targets for the small-molecule drugs belong to protein families, which include G-protein-coupled receptors, ion channels, nuclear receptors, proteases, phosphodiesterases, kinases, and other key enzymes. The secreted proteins such as cell surface receptors are the druggable targets for the large-molecule drugs.

A biological target is an entity toward which potential new therapeutics may be targeted rationally. It may be a protein, a nucleic acid, or a gene product. Genetic analysis and biological observation are crucial to the identification of such a target. A drug target is the specific binding site of a drug *in vivo* through which the drug exerts its action. Targets are normally protein in nature, normally a complex protein. Small molecules, either endogenous or extraneous, can interact with such molecular targets. The structure of the biomolecular targets may change/alter when small molecules interact with them, which may be reversible or irreversible. Any changes in the configuration of targets may trigger signal transduction processes, producing various biochemical and physiological responses. The physiological responses that follow alteration in configuration of biomolecules play an important role in treating an underlying pathological process. Increasingly, disease processes have become dynamic and the targets get modified during pathological processes, making the process of target identification and targeting even more challenging. Target-based drug research is advantageous over the classical physiology-based approach. Screening capacity and rationality in the process has evolved enormously (Wang et al. 2004).

Every target has a specified role in the disease process. Human diseases are complicated and many different drug targets may be involved in preventing disease pathogenesis. Though it does not necessarily imply, each target is equally important. Due to evolutionary changes in the disease mechanisms involved, the role of targets may change with time. As insight into the disease mechanisms is deepening day-by-day, already-elucidated targets may lose significance, emphasizing the need of a dynamic drug screening program. A drug may modulate variable targets; similarly, variable drugs may trigger the same target.

Depending on whether drugs are available for a particular molecular target/disease type, the targets may be divided into those that have been already elucidated and established. Moreover, those that show promise are the potential drug targets. The former category refers to the targets that have been well accepted by the scientific community, of which there is a publication history regarding their involvement in disease pathogenesis, and the drugs available that interact with them. While there are also the targets whose roles are not fully understood in disease pathways or those which lack the drugs targeting them. It is estimated that out of the 3000 druggable genes, only ~600–1500 real targets are involved in the pathogenesis of diseases (Du 2004).

Modern drug discovery aims to develop and identify lead molecules that are safe and effective in treating pathological processes. They should be selective to modulate a specific biological target without altering similar gene products (thereby minimizing the side effects and toxicity). Nature since time immemorial has provided humans with potent bioactive molecules. Nevertheless, screening natural biomolecules is a tedious and time-consuming process. In recent decades, targeted drug screening has taken center stage. Although natural products find some space in the modern drug discovery processes, emphasis is more on developing and screening potent selective, synthetic molecules. It is a challenge for scientists to develop and screen for the biological activity from such large libraries of synthetic molecules. Along with scientific progress and advancement in assay technologies, cell- or organism-based phenotypic assays are being adopted. Better *in vitro–in vivo* correlation can hence be attained owing to the preservation and restoration of the cellular and physiological functions of the protein targets.

Any new drug discovery process initiates with target identification. It involves identifying the role of biomolecules that are most likely to allow the development of drugs capable of symptomatic relief or modulation of disease pathogenesis. Once identified, the field of potential targets is narrowed by target validation, a process by which potential targets are tested against factors, such as the extent to which the target is involved in the disease, and the ability to generate compounds to modify the target. Once a biomolecule is identified, it is evaluated at the molecular, cellular, or whole-animal level. This is known as validation of a drug target. The evaluation may be physiological, biochemical, or pharmacological to ascertain the role of the target in disease pathways (Schenone et al. 2013).

Increasingly, molecular scientists are focusing on the target—a view too narrow. They are neglecting the complexity of the physiological processes or the physiological role the target plays in whole animal or living organism. Validation of a target properly is important to avoid these loopholes. Many pharmaceutical companies and research organizations tend to pursue similar targets for different diseases leading to duplication of the research work and collective loss of the resources in case of a failure of the project (Dodd 2005).

## 2.2 EVOLUTION OF DRUG DISCOVERY: SERENDIPITY TO RATIONALITY

Every year pharmaceutical companies worldwide spend huge sums and resources on drug discovery research and development (R&D). In the modern competitive era of product patent and intellectual property rights (IPR), innovation is a key to success for the survival of a pharmaceutical firm. However, the process of drug discovery is elaborative, utilizing time and resources, both human and financial. The output and outcome may not be guaranteed. A new lead molecule can cost 1.2 billion euros and take more than 10 years to develop (Chen and Du 2007). Increasingly, the regulatory aspects and legal obligations are preventing pharmaceutical companies to go all out for innovations. The early bioactive molecules being focused on in drug discovery were from the natural products, their derivatives, and herbal remedies. Since time immemorial, the use of Siddha, Unani, and Ayurvedic systems of drugs or their forms were prevalent worldwide. In ancient times, the bioactive molecules were discovered by mere chance. The information was accumulated and disseminated to others and subsequent generations. Trial-and-error approach or chance discovery were the sole resources available. Serendipity and discovery by chance led to the discovery of many molecules that became commercial successes in due course. Penicillin, sulfa drugs, and chloroform are a few names to mention. Insulin was discovered in one such serendipitous discovery. An observation was made that flies are attracted to sugar-rich urine of the experimental dogs with pancreas removed. This gave scientists the clue that the pancreas may be related to diabetes. Thereafter, extract of the pancreas was tried for use in diabetic patients. Nevertheless, not all succeeded in the attempt, mainly due to the improper extraction techniques adopted.

Quite often, the nonavailability of systemic approaches led to development of toxic congeners emphasizing on the need to adopt a rational approach. With the development of modern analytical, biotechnological tools, molecular biology and high-throughput screening (HTS) techniques, numerous compounds are being developed and screened. In an attempt to achieve a favorable result/outcome, scientists are trying out newer strategies. Traditional drug discovery methodologies relied on experience and chance discoveries. However, these strategies can no longer meet the needs of the current global pharmaceutical industry. In recent times, with new innovations in the field of chemical and biological sciences, rational and semirational approaches are increasingly being adopted. A shift from serendipity to rationality in drug discovery is underway. With an increased knowledge of biological systems, highly selective drugs are being developed. Selective agonists and antagonists are allowing maximization of the efficacy and reduction of the adverse effects. On many occasions, the research molecules are hence progressing to successful drug molecules (Wang et al. 2004).

After the era of chance discoveries came the era of physiology-based drug discovery. The drugs are being characterized on the basis of their physiological and pharmacological effects in complex disease-relevant models and disease pathways (e.g., animal models or isolated organ systems). Knowledge of the disease mechanisms and the underlying physiology helps the scientists develop assays. Even without a prior knowledge about the disease mechanism or etiology of the disease, drugs were developed successfully, but the approach had its own limitations. The throughput was low, and no empirical relationship could be established between the mechanism of action of a drug and its physiological effect. Strengths of the approach include just the requirement of a disease-relevant model to validate a lead molecule. The latest approach has been target-based drug discovery. Humans are seen as a combination of many genes. The physiology and biology is the outcome of signal transduction. The goal of scientists is to develop lead molecules that can selectively target these genes of interest by modulating molecular pathways or one or more gene products, so that it can selectively cure the disease/deficit without altering other physiological processes, thereby minimizing the side effects. The approach consists of five steps: target identification, target validation, assay development, lead identification, and lead optimization. Figure 2.1 outlines the various steps of biological target-based drug discovery. The strengths of target-based drug design are numerous, including high screening capacity and the ability to formulate simple, clear requirements to the drug, which allows the implementation of *rational drug design*. The approach has dissociated the element of physiology from the drug discovery process; thus, the drugs cannot be optimized simultaneously against multiple targets. However, similar to the earlier used physiology-based approach, target-based drug screening also necessitates the validation in animal disease models to prove effectiveness (Williams 2003).

## 2.3   BIOLOGICAL TARGETS: IMPACT ON DRUG
## DISCOVERY AND PHARMACEUTICAL INDUSTRIES

It is a universally accepted, unspoken rule of the pharmaceutical industry that a significant part (almost 50%) of published work in scientific journals by the academic

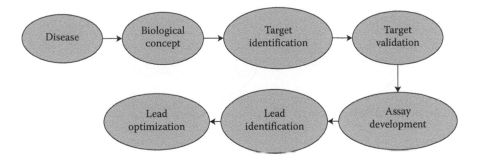

**FIGURE 2.1** Outline of the steps of target-based drug research.

institutions and scientists cannot be reproduced in an industrial setting. Some others believe that this percentage may go up to as high as 65%. In a unique study, researchers at the pharmaceutical company Bayer reported that the in-house experimental data do not match the literature claims in 65% of target-validation projects, which led to the discontinuation of the project. About 20% of the candidates who failed in phase II and phase III of the clinical trials is because of safety issues. It is equally important to study the safety profile of a molecule to enable successful marketing and economic viability to pharmaceutical firms. Selectivity of the drug targets is crucial. Drugs may bind to on and off targets, which may account for both therapeutic and side effects. Evaluation of both efficacy and safety has to be done regarding the characterization and validation of a target (Swinney and Anthony 2011).

Commercial organizations strive for the discovery of new drug molecules to better equip human society to fight various disorders. It must be remembered that the project should be profitable in order to continue the research work (Black 1999), thereby obtaining commercial benefits from marketing of the aforesaid medicines. The organizations may vary in sizes and complexities, and in their ability to take risks. In either case, the project should provide realistic financial return within an acceptable time frame to the drug sponsor. Target-based drug research seems to be their best bet in attainting these goals. Clearly, the drug research and methodologies adopted are strongly influenced by financial considerations (Wang et al. 2013).

Scientists believe that the human genome contains 30,000–35,000 genes. On the basis of sequence data, about 50% of the genes can be assigned a possible physiological function. More than 100,000 proteins can be transcribed by such genes. It is a challenge for the scientists to identify the biological molecules, which can be possible targets for the lead molecules. About 600–1500 to 5000–10,000 targets have been identified which can be targeted by drugs, and whose role in various disease mechanisms have been understood, though not completely. In Figure 2.2, the Venn diagram depicts the number of drug targets, where an intersection between druggable genome and the disease-modifying genes gives the most promising targets for drug development and targeting. Screening such a wide array of biological targets, for possible involvement in disease pathogenesis, can be tedious. Systemic approach may help achieve the goal in a realistic time frame (Hopkins and Groom 2002).

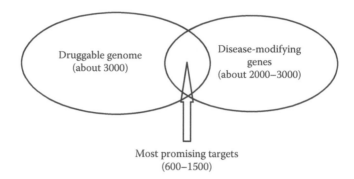

Druggable genome
(about 3000)

Disease-modifying
genes
(about 2000–3000)

Most promising targets
(600–1500)

**FIGURE 2.2** Number of drug targets depicted in Venn diagram. An intersection between druggable genome and the disease-modifying genes gives the most promising targets for drug development.

To attract investors and sponsors, scientists and pharmaceutical companies should be able to convince them about the possible outcome. The business plan presented to them should be comprehensive. A detailed description and justification of the biological target to be selected for targeting should be included. The arguments in favor of the research and the approach may vary, depending on whether the pharmaceutical company is small or large. After the submission of the plan, an external consultant usually evaluates the plan. The sponsor to a great extent is influenced by such opinions. The long-term economic validity of the project is analyzed and opinions are expressed (Knowles and Gromo 2003).

## 2.4  APPROACHES TO TARGET IDENTIFICATION AND SELECTION

Identifying targets that are most likely to allow the development of drugs capable of symptomatic relief or disease modification is one of the most important steps in drug discovery. Target identification develops from great ideas, which can be investigated and supported by an intensive review of literature, evaluation of strength and weaknesses of earlier researchers, experimentation, consideration of the patenting laws and IPR, and assessment of the potential for market entry and commercialization of the project.

A common approach to identifying targets is to assimilate the information and understand the molecular pathways of diseases of interest by collecting, reviewing, and analyzing data across the spectrum of relevant published literature. Performed by a group of scientists, this can be a labor intensive and dedicated exercise, and absolutely necessary to avoid duplication of the earlier work and to ensure novelty of one's work (Feng et al. 2009).

The director heading a research project has to decide between two broad domains while selecting or rejecting a probable target; the likely financial return from investing in the target and the inherent risk associated with it that may lead to the failure of the project. Novelty of work may not be the only criterion that drives a research project.

Comparing the risk/return ratio of new chemical entities (NCEs) marketed between 1975 and 1984, we found that a majority of them had a low risk of failure. At the same time, they provided low financial returns. Risk/return ratio should not be the sole criterion to select a target. It is a challenge to avoid high risk/hard targets with a potentially low financial return. The research coordinator has to take a decision as to whether the project will provide the company and the sponsor with the right drug, at the right time, and at the right price (Knowles and Gromo 2003).

While identifying and selecting a target, it is often assumed that a direct interaction with a single target is responsible for phenotypic observations. Even while validating a target we try to establish the role of the target in the disease pathogenesis, but while developing a lead to interact with the target we often do not account for the interactions with "off-target" proteins, which may lead to multiple therapeutic expressions and side effects. Thus, it is important to identify both the therapeutic targets and off targets. This would optimize the therapeutic effectiveness and minimize the side effects.

Target selection is a dynamic process, where judgment is made about the future taking into account all the uncertainties associated. Predictions and assumptions have to be made about the possible progression of disease pathways, the clinical status of treatment, and the probable financial return. Many modeling techniques and statistical tools are used to predict the future performance of the lead molecule, both clinically and economically.

Many target identification technologies are available, including genomics, gene expression profiling, and proteomics. Biomarkers and metabolomic profiling are increasingly being used to monitor the biological effects of a lead and the modulation of a potential drug target. No single technology can suffice all the needs of modern drug discovery. Several technologies are being assimilated and integrated to achieve the purpose (Schenone et al. 2013).

## 2.4.1 DIRECT BIOCHEMICAL METHODS

Direct methods involve labeling the protein or small molecule of interest, incubation of the two populations and direct detection of binding, usually following some type of wash procedure. Biological targets that interact with lead molecules can easily be detected by this method. These methods provide useful information about the molecular mechanisms involved in both therapeutic activity and toxicity, which is of prime importance to develop a safe and effective novel therapeutic agent. In the method, the enzymatic extracts are exposed to the immobilized columns, followed by elution. Prerequisites to the method are availability of large amounts of extracts and stringent wash conditions (Burdine and Kodadek 2004).

Selection of drug–target interactions having highest affinity is partly biased. It becomes difficult to identify additional or low-affinity targets, even though they have a prominent physiological role. It is challenging to produce immobilized affinity reagents that retain cellular activity, so that the target proteins will interact with the lead molecules even though it is immobilized. Chemical or ultraviolet light-induced cross-linking techniques are increasingly being used. The protein target is covalently modified. This allows scientists to identify biological targets having scarce distribution and those that have a low affinity for various lead molecules (Zheng et al. 2004).

From the perspective of efficiency and sensitivity, affinity chromatography coupled to mass spectroscopy (MS) provides the most unbiased method in identification of the biological targets (Cuatrecasas et al. 1968). Application of quantitative proteomics to study protein–ligand interactions has opened many prospects in the field of drug discovery. Labeling techniques such as stable isotope labeling by amino acids in cell culture (SILAC) (Ong et al. 2002) and isobaric tags for relative and absolute quantification (iTRAQ), (Ross et al. 2004) both used for protein labeling, have been effectively used in quantitative proteomics. Multiple drug targets, both direct and indirect, can be evaluated by these methods. The success of these methods lies in gentle washing and free soluble competitor preincubation of lysates (Oda et al. 2003).

## 2.4.2 GENETIC INTERACTION AND GENOMIC METHODS

Genetic modifiers (enhancers or suppressors) are utilized to identify and select biological targets. Target–ligand interactions at the genetic level help to screen protein targets. This enables performance of large-scale modifications and assessments at molecular levels. Gene knockout organisms, RNA interference (RNAi), and small molecules (ligands) can be used to modify and alter the functions of target genes (Moffat et al. 2006).

Studying the expression levels of many genes can help establish how a compound can affect the cellular and physiological function at the genomic level. This principle is utilized in the DNA microarray technology (Pierce et al. 2006). The idea is to focus on genes with large expression ratios. Expression arrays are used in the technique to identify the role of a specific gene in a particular disease pathway. If a gene function is not allowed to express itself, it normally leads to alteration in the physiological function. It is possible to identify a biological target by treating a deletion mutant defective in the gene encoding the putative target (Drews 1996). With recent advances in molecular biology, it is possible to study gene function and products of genetic expression by techniques such as RNAi. RNAi keeps the endogenous gene intact, thus it outscores the traditional knockout models. RNAi has the advantage of reversibility. A number of inducible RNAi transgenic lines can be developed by this approach. In addition, it helps to identify phenotypes and unknown functions of various genes. A promising genomics-based technique for target identification involves combining results from small-molecule and RNAi perturbations. This approach enables parallel testing of small-molecule and RNAi libraries for induction of the same cellular phenotype. This allows measuring phenotypic effects in various cell lines. Often human cell lines or that of other mammalian species may be used to establish physiological role of RNAi perturbations. Modern scientific community is increasingly concentrating on mammalian cells for genetic target identification (Wang et al. 2008).

## 2.4.3 COMPUTATIONAL INFERENCE METHODS

Protein targets can also be identified by various computational methods. In addition, they provide supplementary evidence for other analytical techniques. Targets screened by proteomic and genomic methods can be cross-validated using computational inference methods. Further, the methods can be applied to find new drug targets, for established therapeutic moieties, or to identify off-target drug effects thereby explaining the molecular mechanisms of side effects (Feng et al. 2009).

Affinity profiling methods are included under the computational inference methods. It can be of predictive value to estimate the extent of target modulation that may be done by a ligand interaction (Wagner and Clemons 2009). The method may also be used to design biologically diverse screening libraries. Profiling by high-throughput microscopy (sometimes called high-content screening) allows phenotypes to be clustered in a manner analogous to the transcriptional profiles. This facilitates the discovery of potential small-molecule targets. Recent profiling methods have taken full advantage of the high-throughput data available in public and proprietary screening databases, for example, by developing HTS fingerprints (HTS-FP) with a goal of facilitating virtual screening and scaffold hopping (Haupt and Schroeder 2011).

Network-based approaches can also be used as a tool to identify potential new drug targets. This allows extension of systems biology to drug–target and ligand–target networks. The approach offers additional advantage over the screening methods as the complex target–ligand interactions can be uniquely interpreted. Extent of bias may vary depending on the method adopted. They provide an accurate tool for target identification as they often rely on the experiments done by others (Koutsoukas et al. 2011).

## 2.5  DYNAMICS OF TARGET VALIDATION

Once a biological target is identified, the field of potential targets is narrowed by target validation, a process by which potential targets are tested against factors, such as extent to which the target is involved in the disease pathogenesis, and the ability to generate compounds to modify the target and the commercial future of such a project. Target validation helps to rule out the less effective targets, leaving those that are most likely to be druggable, reaping maximum commercial benefits within a practical time frame. With disease pathways dynamically evolving, target validation does not have a clear-cut finish line, rather validation is a continuous process, where new data points and external information can change a target's validity. Nevertheless, this brings about a degree of uncertainty in the project (John and Martyn 2003).

A number of experimental and information resources can help validate a target's involvement in disease progression and pathogenesis. Experimental approaches, including cell-based *in vitro* studies to confirm mechanisms of action, knockin, knockout, and knockdown approaches to modulate or silence the level of a particular target for assessment in animal models, and RNAi approaches to interfere with gene expression are increasingly being used along with standard pharmacological tools and approaches to validation. Increasingly *in vitro* models and HTS tools are being adopted to validate and screen multiple targets. With targets confirmed for their involvement in disease progression, attention shifts to the development of the lead compounds having suitable pharmacological, pharmacokinetic properties that can be delivered with acceptable toxicological limits. Approaches may consist of *in vivo* experimentation on animal models complemented with *in silico* computer modeling to assess how a target and potential drugs interact with one another. Both approaches can help in the optimization of ligands with respect to potency and selectivity, while having limitations based on the extent to which the particular disease model parallels the human pathology. One key area is specificity of the ligand with the simulated target. Targets being isolated and characterized may not always represent the *in vivo* conditions 100%. Moreover, the selection of suitable disease models is a challenge. The model should be stable and reproducible under the test conditions (Wang et al. 2004). Table 2.1 shows some of the drugs whose targets were validated by basic biological and medical research.

High-quality targets are those that are expected to give rise to large-selling drugs (Black 1999). Targets that cannot be addressed with currently available medicinal chemistry are called hard targets. Pharmaceutical companies tend to be driven toward the high-quality targets, having better risk/reward ratio and better commercial viability.

During the process validation of a biological target, it is important to suitably convert the target into a bioassay. This allows quantification of the biological activity. The spectrum of potential targets being diverse, the assay methodologies should be in a position to estimate physiological significance of each after ligand–target interaction. Often receptors and enzymes are implicated in complex physiological and pathological mechanisms; thus the task of adopting a suitable bioassay method can be tricky (Lipinski and Hopkins 2004). Directly or indirectly, the assay method developed should be in a position to quantify the physiological changes the lead brings about after interaction with the protein targets (Drews and Ryser 1996). The next rational approach after designing the bioassay method would be the establishment of an HTS method. Sensitivity and reproducibility are the foremost criteria of an HTS assay. The method should be in a position to screen multiple compounds within a rational time frame. Since the outcome of the studies is never guaranteed, pharmaceutical companies to gain a competitive edge try to screen as many compounds as possible in HTS assays. The method should be robust and suitable for screening thousands or even millions of samples (Inglese et al. 2007). For greater precision, the process of target validation should be checked and cross-checked at multiple levels. Usually it

**TABLE 2.1**

**Drugs Whose Targets Were Validated by Basic Biological and Medical Research**

| Drug | Target | Disease |
|------|--------|---------|
| Erlotinib | EGFR | Cancer |
| Rituximab | $CD_{20}$ receptor | Non-Hodgkin's lymphoma |
| Infliximab | $TNF_{\alpha}$ | Inflammatory disease |
| Bevacizumab | VEGF | Colorectal cancer |

*Source:*   Adapted from Haberman, A.B. 2005. *Genetic Engg News* 25:36.

is done at three levels: the molecular level, the cellular level, and the whole-animal model level. It becomes very easy to validate the new drug targets using HTS. Small chemicals generated from HTS can be utilized to validate the targets at molecular levels. The main limitation of such an approach is the occasional reproducibility *in vivo*. Multiple factors can influence the drug target–ligand interactions. The complex *in vivo* mechanisms may not be predicted 100% in cell-free systems. Nevertheless, such studies do give us a trend or possibility of a probable *in vivo* physiological alteration when the ligand interacts with the drug target. Thus, positive results from molecular level studies do help scientists to design further studies. Cell-free systems do not require the ligand to penetrate the cells. In actual physiological systems, the drug may have to penetrate the cells and interact with the intracellular protein targets. Up to some extent, this may be governed by physiochemical properties of the drug, including ionization at physiological pH, and lipophilicity. Thus, cell-free systems do provide an initial insight into target–ligand interaction; it is not conclusive. Results from cell-free systems can easily be validated at the cellular level. Incubation of ligands with cells in HTS modalities, and changes rendered can be assessed using suitable assay screens. Studies in intact organisms provide conclusive evidence to the role of the target in pathological mechanisms. Ligands interacting and modulating targets *in vivo* probably validates the target most. Yet cell-free systems and cellular level studies are necessary for initial screening. They help avoid unnecessary conduct of animal studies in molecules not having sufficient promise to develop as drugs. This saves resources as well as time, and helps screen out nonpromising molecules fast in the drug discovery pathways (Gad 2005). If positive results are obtained in the whole-animal studies, prospects are good that the ligand modulation of the target can be beneficially utilized for treating pathological conditions. The situation may be otherwise if a ligand shows positive results in initial validation studies, and fails in the whole-animal studies. However, often a "hit" may not display any effect in an animal model. The interpretation may become tricky in such a situation. The research team should ensure they do not rule out a promising lead at the first look, while ensuring at the same time that they do not drag a project too long. Selection of a suitable animal model is the key to success. Some of the other parameters that may have a bearing on the final outcome are proper route and dosage regimen, wrong vehicle or formulation of test material (may lead to nonrelease of the drug from the dosage form in a suitable amount and time), and selection of wrong dose levels. Subtherapeutic dose levels may generate false negative data. If the compound does not show therapeutic/toxic effect at the highest dose levels tested, it may be difficult to conclude the outcome. Toxicological data is of prime importance to most regulatory authorities. Nevertheless, there is a point beyond which the dose levels cannot be increased. If the animals start dying, but there is no observable therapeutic effect, the project ends (Kauselmann et al. 2012).

## 2.6   ROLE OF LITERATURE IN TARGET IDENTIFICATION AND VALIDATION

Historically, academic institutions contributed the most in identification and selection of biological targets. A much lesser role was played by commercial organizations. This was probably because of the fact that academic laboratories had the highest degree of understanding about basic biology. Various inter- and intra-cellular signaling cascades that influence the modulation of target by ligands are better appreciated if one has an insight into the fundamentals. Thus, historically many targets those are therapeutically relevant in modern world were discovered in academic institutions (Imming et al. 2006).

Literature analysis is a critical tool in target identification and validation. It show-cases the strengths and weaknesses of the research work carried on so far. Even before venturing into a field, having an idea of the proposed domain willing to work into would be very useful. The researcher can also assess the infrastructure and resources available at hand, thus reducing the chances of stagnation of a project at an advanced stage due to lack of resources. Before even proceeding in a field, one needs to combine all the published literature, past and present. As this is an unrealistic task to perform manually, researchers need alternative means to iden-tify all the information relevant to their research. Careful observation of an article may disclose characteristics of a target that initially or in the first glance may not stand out. Similar observations by different research groups or notable in diverse research articles may help to identify and provide evidence concerning the validity of the target (Cheng et al. 2011). Several approaches may be adopted to review the literature. Manually reviewed databases may have high-quality data, but because they take longer to update, may not be as current as automatic ones nor may they have the same depth of coverage. Often help of experts in the field may be required to disseminate relevant data or information from a pool of data. Automatic index-ing, the method of automatic extraction of indexed information from the published sources, may ensure compilation of current information. Care should be taken not to solely rely on the abstracts (Drews 2000). Automated approaches such as auto-mated co-occurrence indexing and natural language processing can screen the entire database and extract useful information from potential relationships of key terms. These approaches, if run against a large body of the literature, and the full text, can ensure both comprehensiveness and currency. At the same time, algorithms must be effective in indexing accuracy to reduce the potential for either delivering irrelevant results, or missing important information. It is important to identify the relevant information within a comfortable time frame. New tools may be developed that enable users to tune indexing sensitivity by searching, for example, for terms indexed within the boundaries of the same sentence, paragraph, or article, with stronger rela-tionships likely as the indexing boundaries are tightened. Literature analysis tools outscore manual reviewing: it helps to avoid missing crucial information, to generate fresh ideas, and to improve the ability to discover and validate potential targets in drug discovery (Brown and Superti-Furga 2003).

Identification of targets may be done up to a certain extent by aggregating infor-mation relevant to molecular pathogenesis of diseases, and by collecting, assimilat-ing, and analyzing information across the spectrum of relevant published literature. It is a laborious task, requiring skills and guidance, and absolutely necessary if novel modes of action are to be found. Experimental approaches adopted by previous researchers generate abundance of information, and identification of potential drug-gable targets is intimately associated with specific pathologies. Microarray analy-sis, quantitative polymerase chain reaction (PCR), immunohistochemistry, *in situ* hybridization, high-performance liquid chromatography (HPLC), gel electrophoresis, capillary electrophoresis, protein mass spectrometry, and wet chemistry can all be used to identify the targets experimentally (Knight and Shokat 2007). Modern analytical tools generate vast data, and fruitful data mining can be utilized to identify the genes and gene products of interest. Challenges are associated with data and information management, especially to single out relevant information from the crude or raw data. As a consequence, information is often most valuable once peer reviewed and further corroborated with additional findings that may be reported by other researchers. Data sharing among contemporary workers may be useful. However, it rarely occurs keeping in mind the commercial benefits associated therein. More and more companies are turning to automated mining of the published literature to support target identification. Special groups of scientists work as a team assimilating data. Text mining, the process of automatically analyzing the published literature, whether from abstracts or full text, can identify and extract information from published works and can enable cross-literature comparisons (Lachance et al. 2012).

## 2.7 TARGET-BASED CANCER RESEARCH

Cancer-related morbidity and mortality is increasingly becoming common. Along with diabetes, it is fast gripping the human race. Site of cancer and early detection are crucial to the success of the available treatment strategies. Surgery and radiation may not be curative in most cases, if not localized early. Recent advances in chemotherapy show sufficient promise. However, they have their own limitations. In certain cases, they may not be curative. An additional drawback includes high relapse rate. A scientist venturing into chemotherapeutic research should try to develop durable molecules. Along with efficacy, ample consideration should be given to the inherent toxicities associated with chemotherapeutic drugs. Thus, minimization of adverse drug reactions should be equally considered along with the chemotherapeutic efficacy. Inherent genomic instability of tumors may lead to drug resistance, which aggravates the challenges associated with an effective chemotherapeutic drug (Daub et al. 2008). Evaluation of the basic physiological and pathological processes associated with a disease helps to explore novel strategies. It helps researchers to suitably target drugs to the site of anomaly. With knowledge of human genetics, the task of target validation becomes easy. In addition, the chance of developing a "hit" in the whole-animal studies becomes more probable. In most cases of cancer treatment, the target selection is determined by the specific mutations that have taken place in the disease progression. *p53, ras* are a few such points of anomalies. Along with mutations, overexpression of the specific gene products may also lead to cancer. For example, human epidermal growth factor receptor 2 (HER-2) and epidermal growth factor (EGF) may be causative factors in some cancers (Armstrong et al. 2000). Mere target identification is a job half done. Identification of a genetic defect can be therapeutically exploited only if the target can be modulated by small molecules. On many occasions, targets are known, but it is difficult to develop molecules that reach the specific target and/ or modulate it. Table 2.2 shows some of the mechanism-based anticancer drugs that have been developed or are under development. Identification of a druggable

**TABLE 2.2**

**Mechanism-Based Anticancer Drugs Developed/Under Development**

| Drug | Target |
| --- | --- |
| Raloxifene | Estrogen receptor |
| Flavopiridol | Cyclin-dependent kinases |
| Marimastat | Metalloproteins |
| Vitaxin | Integrin |
| Anti-VEGF | VEGF |

*Source:* Adapted from Gibbs, J.B. 2000. *Drug Discov* 287:1969–73.

target is probably the most important factor influencing the chemotherapeutic drug research. With recent advances in molecular biology, many molecular tools have become available for target validation, including antisense oligonucleotides, ribozymes, dominant negative mutants, neutralizing antibodies, and mouse transgenics/knockouts. However, no method is foolproof. Often, scientists check and cross-check outcomes adopting multiple approaches.

Targets can be best exploited if one has a prior knowledge of the basic physiological mechanisms, molecular biology, and the inter- and intracellular signaling cascades. Recently, much work has been done on protein kinases. The choice of bcr-abl as a target is based on the Philadelphia chromosome. It is of great significance in the diagnosis of chronic myelogenous leukemia. It is also believed that EGF receptors are overexpressed in lung and oral cancers. Autocrine activation of the receptor by EGF and transforming growth factor-$\alpha$ (TGF-$\alpha$) is important to the proliferation of the tumor cells. Scientists have also been working on developing inhibitors of farnesyltransferase (FTase) and matrix metalloproteinases (MMPs). Gene therapy has opened new avenues for the development of target-based medicines. A significant number of therapies developed have been based on the p53 glycoprotein (Dancey and Chen 2006).

Cases of drug resistance and cancer relapses have become very common. This may be attributed significantly to tumor instability. Many of the therapies designed or developed may not be effective as planned. Risk/benefit ratio governs the success of a therapeutic outcome to a great extent. Cellular toxicities commonly encountered with anticancer therapy may be mechanism based or due to specific chemical groups present in the molecules. These groups are essential for the chemotherapeutic activity and cannot be removed from the structure of the parent compound. Individualizing therapy may be one of the options to avoid this problem. Priority should be given to compounds that have better therapeutic index than the currently available and marketed cytotoxic compounds (Wolpaw et al. 2011). Though the process of target identification and validation is slow, tedious, and time and resource consuming, it offers sufficient promise to provide better and effective chemotherapeutic agents.

## 2.8 LIMITATIONS OF BIOLOGICAL TARGET-BASED RESEARCH

Quite often, target-based discovery of drugs, especially for genetic targets, will be limited to a specific subgroup of patients. This limits the overall impact of the therapy on disease treatment. Nevertheless, the cost of such a therapy would go up, owing to the small size of these patient populations. It will take quite some time for molecular biology and screening technologies available at hand to expand its base to individualized therapy (Szymkowski 2001). Serious questions pertaining to the overall impact of human genome projects on health-care needs of the society and on the disease treatment may be raised until such futuristic goals are achieved. Pharmaceutical companies and research organizations do not share data and tend to pursue similar druggable targets. Instead of venturing into unknown domains or domains where the risk/reward ratio is high, they prefer targeting well-validated targets. This may be counterproductive at times. If similar projects fail, it leads to wastage of resources as a whole. One of the solutions to this problem may be to adopt physiology-based approaches, where the risk associated may be lower. Most of the companies have different starting points into initiation of drug development programs. Even if compounds are being designed for similar pathological conditions, or are expected to produce similar physiological effects, the research programs are not expected to overlap. Most compounds develop in different directions. On these lines, many new atypical antipsychotics (such as clozapine and risperidone) were developed using the same complex disease models for the evaluation and screening but had substantial variation in their mechanism of action (Dixon and Stockwell 2009). However, there is a certain amount of risk associated with this line of thinking, as data are not shared and reproduction of the work may lead to wastage of resources, in terms of both time and money.

Proper identification and selection of drug targets is one of the first steps to drug discovery. Moreover, a number of issues have to be evaluated to attain the goal successfully. Commercial viability and marketability is probably the most important criterion. Sales forecast is also an important determinant. There is a certain degree of assumption when one predicts sales forecast. Historical experiences may not hold true always, especially in the modern dynamic world. The disease process and expression of various disorders are changing drastically. Inherent unpredictability in such predictions may be a hindrance to successful biological target-oriented drug research (Dodd 2005).

Lack of resources and limitations of experimental procedures to be able to validate a target successfully emphasize the need to test, or confirm, experimental results against the independent and externally peer-reviewed corpus of scientific research. While independent peer-reviewed research will likely not replace experiments in the lab, it will help form the evidence base, giving confidence in a target's validity, and in turn derisking the potential likelihood of downstream failure. Thus, it minimizes the likelihood of financial losses to sponsors and commercial organizations. Otherwise, lack of funding for innovative research projects may emerge as a big constraint to the future of drug discovery projects (Szymkowski 2001).

With a boom in target-oriented drug research, there has been an increased tendency of pharmaceutical organizations to focus solely on the target. By not giving necessary importance to the role of the target in the overall physiological process, scientists are putting research projects at risk. In an attempt to fast-track drug development, commercial organizations are not giving proper time necessary for the validation of biological targets. Research projects are progressing to lead development, without testing and questioning targets sufficiently. Overall output of pharmaceutical organizations is severely jeopardized because of this approach (Drews 2003). Target validation may be assumed to be successful only if a lead is able to modulate the target suitably. This can be best assessed in a clinical setup. However, this is never possible in the initial stages of the drug discovery programs. Thus, on entry into a program, it is obvious that a target is not properly validated, at least with relevance to its clinical usefulness. This much risk is associated with most R&D programs. As research progresses, the level of risk varies. At initial levels of a research program, the risk is usually low; it is the highest at terminal stages. When drug candidates are being tested against disease models in whole animals, any negative outcome can adversely affect the financial state of a commercial organization. To save time and resources, pharmaceutical companies often assume a target to be validated after some preliminary studies only. After advancing much into the research program, maybe after 3–5 years, they carry out additional studies. Since disease process and the role of the target may get modified in due course, the strategy can be risky. Resources are at stake and any failure in terminal stages can prove to be costly. Yet, in an attempt to gain an edge over rival pharmaceutical companies, many do adopt this strategy. Many factors may influence the ultimate outcome, some of which may be beyond the control of scientists. Ultimately when tested in animal models, the ligand should show effectiveness as expected. Since the expression of disease in the human population may vary with time, when ultimately the molecule is launched into the market, it should be able to bear fruit and reap benefits for the drug sponsor (Drews and Ryser 1996).

## 2.9   CONCLUSION

The process of drug discovery is complex. Scientists may try to predict outcomes of the projects, but the success depends on the planning. To manage the uncertainties, all information available at hand should be processed carefully. Analysis and prediction of every possible outcome of every possible decision will determine ultimate success of a research project. Scientists may try to adopt an optimal strategy to achieve their goal but can only hope for success. Target-based drug discovery has a narrow outlook. Here, the interaction between the drug and a biological system is narrowly looked down to that of the drug and the biological target. This dissociates the overall physiological outlook from the drug discovery process. Drug molecules are being developed keeping in mind their ability to modulate targets, rather than to produce physiological and biological responses. However, target-based drug discovery has been able to increase the overall output of the pharmaceutical industry. It has also enhanced the element of rationality in the drug discovery process. Now research projects are more planned and lot of homework and analysis are done before venturing into a field. However, in initial stages, scientists concentrate more on the target, then screen the validity of the targets. This is best done when compounds are screened for biological activity in whole organisms.

With gradual unrevealing of the human genome, therapeutics has gained new dimensions. Be it curative medicine or diagnosis, with the advent of gene therapy, scientists have been able to target therapy much more selectively. Scientists are still working on elucidating functions of the various genes or their products involved in various pathological mechanisms. Out of the various genes identified so far, the biggest challenge for the scientists is to understand their exact role in the disease pathways. Disease may be due to either overexpression or underexpression of one gene or its product. It is equally important to identify the target and target it selectively.

Over years, the scientific community has advanced a lot with regard to its understanding of the human system. It has yet not been able to predict accurately target modulation by ligands on every occasion. Physiological consequences and selectivity of drugs is crucial to its success among experts. This can only be established by reproducible results in test subjects and in the human population as a whole. This explains the importance of target validation. The future of the project and the resources that may be required to achieve the goal depend directly and indirectly on the process of target validation. Drugs may interact with multiple targets; thus, target validation is a dynamic process. The disease pathways and the role of targets may also get modified during the course of the drug development program. All this adds to the uncertainty in the process. Last, for the design of target validation and proof-of-principle studies, it is necessary to understand the biological role of the target, the clinical manifestations of the disease, and the current treatment practices.

# REFERENCES

Armstrong, K.A., Eisen, A., and Webber, B. 2000. Assessing the risk of breast cancer. *N Engl J Med* 342:564–71.

Black, J. 1999. Future perspectives in pharmaceutical research. *Pharm Policy Law* 1:85–92.

Brown, D. and Superti-Furga, G. 2003. Rediscovering the sweet spot in drug discovery. *Drug Discov Today* 8:1067–77.

Burdine, L. and Kodadek, T. 2004. Target identification in chemical genetics: The (often) missing link. *Chem Biol* 11:593–97.

Chen, X.P. and Du, G.H. 2007. Target validation: A door to drug discovery. *Drug Discov Ther* 1:23–29.

Cheng, T., Li, Q., Wang, Y. et al. 2011. Identifying compound-target associations by combining bioactivity profile similarity search and public databases mining. *J Chem Inf Model* 51:2440–48.

Cuatrecasas, P., Wilchek, M., and Anfinsen, C.B. 1968. Selective enzyme purification by affinity chromatography. *Proc Natl Acad Sci USA* 61:636–43.

Dancey, J.E. and Chen, H.X. 2006. Strategies for optimizing combinations of molecularly targeted anticancer agents. *Nat. Rev. Drug Discov* 5:649–59.

Daub, H., Olsen, J.V., Bairlein, M. et al. 2008. Kinase-selective enrichment enables quantitative phosphoproteomics of the kinome across the cell cycle. *Mol Cell* 31:438–48.

Dixon, S.J. and Stockwell, B.R. 2009. Identifying druggable disease-modifying gene products. *Curr Opin Chem Biol* 13:549–55.

Dodd, F.S. 2005. Target-based drug discovery: Is something wrong? *Drug Discov Today* 10:139–47.

Drews, J. 1996. Genomic sciences and the medicines of tomorrow. *Nat Biotechnol* 14:1517–18.

Drews, J. 2000. Drug discovery: A historical perspective. *Science* 287:1960–64.

Drews, J. 2003. Strategic trends in the drug industry. *Drug Discov Today* 8:411–20.

Drews, J. and Ryser, S. 1996. Innovation deficits in the pharmaceutical industry. *Drug Inf J* 30:97–108.

Du, G.H. 2004. Evaluation and validation of drug targets. *Acta Pharmacol Sin* 25:1566–66.

Feng, Y., Mitchison, T.J., Bender, A. et al. 2009. Multi-parameter phenotypic profiling: Using cellular effects to characterize small-molecule compounds. *Nat Rev Drug Discov* 8:567–78.

Gad, S.C. 2005. Introduction: Drug discovery in the 21st century. In *Drug Discovery Handbook*, ed. S.C. Gad, 1–10. New York: Wiley Press.

Gibbs, J.B. 2000. Mechanism-based target identification and drug discovery in cancer research. *Drug Discov* 287:1969–73.

Haberman, A.B. 2005. Strategies to move beyond target validation. *Genetic Engg News* 25:36.

Haupt, V.J. and Schroeder, M. 2011. Old friends in new guise: Repositioning of known drugs with structural bioinformatics. *Brief Bioinform* 12:312–26.

Hopkins, A.L. and Groom, C.R. 2002. The druggable genome. *Nature Rev Drug Discov* 1:727–30.

Inglese, J., Johnson, R.L., Simeonov, A. et al. 2007. High-throughput screening assays for the identification of chemical probes. *Nat Chem Biol* 3:466–79.

Imming, P., Sinning, C., and Meyer, A. 2006. Drugs, their targets and the nature and number of drug targets. *Nature Rev Drug Discov* 5:821–34.

John, G.H. and Martyn, N.B. 2003. High throughput screening for lead discovery. In *Burger's Medicinal Chemistry and Drug Discovery*, ed. D.J. Abraham, 37–69. New York: Wiley.

Kauselmann, G., Dopazo, A., and Link, W. 2012. Identification of disease-relevant genes for molecularly-targeted drug discovery. *Curr Cancer Drug Targets* 12:1–13.

Knight, Z.A. and Shokat, K.M. 2007. Chemical genetics: Where genetics and pharmacology meet. *Cell* 128:425–30.

Knowles, J. and Gromo, G. 2003. Target selection in drug discovery. *Nat Rev Drug Discov* 2:63–9.

Koutsoukas, A., Simms, B., Kirchmair, J. et al. 2011. From *in silico* target prediction to multi-target drug design: Current databases, methods and applications. *J Proteomics* 74:2554–74.

Lachance, H., Wetzel, S., Kumar, K. et al. 2012. Charting, navigating, and populating natural product chemical space for drug discovery. *J Med Chem* 55:5989–01.

Lipinski, C. and Hopkins, A. 2004. Navigating chemical space for biology and medicine. *Nature* 432:855–61.

Moffat, J., Grueneberg, D.A., Yang, X. et al. 2006. A lentiviral RNAi library for human and mouse genes applied to an arrayed viral high-content screen. *Cell* 124:1283–98.

Oda, Y., Owa, T., Sato, T. et al. 2003. Quantitative chemical proteomics for identifying candidate drug targets. *Anal Chem* 75:2159–65.

Ong, S.E., Blagoev, B., Kratchmarova, I. et al. 2002. Stable isotope labeling by amino acids in cell culture, SILAC, as a simple and accurate approach to expression proteomics. *Mol Cell Proteomics* 1:376–86.

Pierce, S.E., Fung, E.L., Jaramillo, D.F. et al. 2006. A unique and universal molecular barcode array. *Nature Methods* 3:601–03.

Ross, P.L., Huang, Y.N., Marchese, J.N. et al. 2004. Multiplexed protein quantitation in *Saccharomyces cerevisiae* using amine-reactive isobaric tagging reagents. *Mol Cell Proteomics* 3:1154–69.

Schenone, M., Dancik, V., Wagner, B.K. et al. 2013. Target identification and mechanism of action in chemical biology and drug discovery. *Nat Chem Biol* 9:232–40.

Swinney, D.C. and Anthony, J. 2011. How were new medicines discovered? *Nat Rev Drug Discov* 10:507–19.

Szymkowski, D.E. 2001. Too many targets, not enough target validation. *Drug Discov Today* 6:398–99.

Wagner, B.K. and Clemons, P.A. 2009. Connecting synthetic chemistry decisions to cell and genome biology using small-molecule phenotypic profiling. *Curr Opin Chem Biol* 13:539–48.

Wang, J., Zhou, X., Bradley, P.L. et al. 2008. Cellular phenotype recognition for high-content RNA interference genome-wide screening. *J Biomol Screen* 13:29–39.

Wang, S., Sim, T.B., Kim, Y.S. et al. 2004. Tools for target identification and validation. *Curr Opin Chem Biol* 8:371–77.

Wang, X., Thijssen, B., and Yu, H. 2013. Target essentiality and centrality characterize drug side effects. *PLoS Comput Biol* 9:e1003119.

Williams, M. 2003. Target validation. *Curr Opin Pharmacol* 3:571–77.

Wolpaw, A.J., Shimada, K., Skouta, R. et al. 2011. Modulatory profiling identifies mechanisms of small molecule-induced cell death. *Proc Natl Acad Sci USA* 108:E771–80.

Zheng, X.S., Chan, T.F., and Zhou, H.H. 2004. Genetic and genomic approaches to identify and study the targets of bioactive small molecules. *Chem Biol* 11:609–18.

# 3 Emerging Therapeutic Targets for Diabetes

*Sudipta Saha, Ashok K. Singh, Amit K. Keshari,*
*Priya Singh Kushwaha, and Siddhartha Maity*

## CONTENTS

## 3.1  INTRODUCTION

The World Health Organization (WHO) determined that today, diabetes is a high-prevalence disease and 346 million people were affected worldwide in 2011 (Nadine 2012). WHO also recognized two types of diabetes, namely, type-1 (insulin-dependent diabetes mellitus, IDDS) and type-2 (non-insulin-dependent diabetes mellitus, NIDDS). IDDS is also known as juvenile onset diabetes, which is caused via an inadequate insulin secretion through β-cells of the islets of Langerhans in the pancreas. A recent survey suggested that only 10% of the world's population is affected by IDDS (Wild et al. 2004). The remaining population is affected by NIDDS, which is not directly related to the insulin hormone. The major causes of NIDDS are reduced insulin sensitivity, insulin resistance, and/or combined with reduced insulin secretion (Mazzone et al. 2008). The factors responsible for NIDDS include a reduced cellular uptake of insulin, increased hepatic gluconeogenesis, impaired β-cell activity, increased glucose absorption from the intestine, and reduced glucose transporter activity (Colberg 2008).

Most of the diabetic patients have secondary complications related to athero-sclerosis, ischemic heart disease, and nephropathy. Obesity, which is a major pub-lic health concern worldwide, may also increase the risk of NIDDS (Zeller et al. 2008). At least one-third of obese individuals exhibit a decrease in β-cell activity caused by β-cell apoptosis, which renders these individuals unable to compensate for their insulin receptor (IR) state and the resulting hyperglycemia (Graves and Kayal 2008).

As has been widely accepted, both IDDS and NIDDS come with enhanced risk factors of macrovascular and microvascular complications. These complications often coexist and can result in hypertension, altered vascular permeability, and ischemia. The common microvasculature defects evident in diabetes include retinop-athy, nephropathy, and peripheral neuropathy, each of which can impart enervative consequences. In addition to defects in the systemic microvasculature, macrovascu-lar complications such as the peripheral vascular disease, coronary heart disease, and stroke are believed to be the primary causes of deaths in diabetic patients (Forbes and Cooper 2013).

For pharmacotherapy, those compounds that increase the sensitivity of muscle and adipose to insulin are the leading choice for the successful treatment of diabetes and its complications. There are mainly five classes of therapeutic agents, having a different mechanism of action and are commonly used to treat hyperglycemia. The agents are α-glucosidase inhibitors, aldose reductase (ALR) inhibitors, protein tyrosine phosphatase-1B (PTP-1B) inhibitors, dipeptidyl peptidase-4 (DPP-4) inhib-itors, and peroxisome proliferator-activated receptor-γ ($PPAR_\gamma$) activators that are commonly used to treat diabetes but they produce various complications (Mahapatra et al. 2015).

Besides these, some emerging therapeutic targets have been discovered for the treatment of diabetes along with its complications. These targets include estrogen-related receptor-α (ERR-α), fibroblast growth factor 21 (FGF21), islet G-protein-coupled receptors (Islet GPCRs), histone deacetylases (HDACs), sodium–glucose cotransporters (SGLTSs), K-cells receptors, T-cells receptors (TCRs), hepatic targets for glycemic control, and microRNAs (miRNAs).

## 3.2  EMERGING THERAPEUTIC TARGETS FOR DIABETES

### 3.2.1  PROTEIN TYROSINE PHOSPHATASE-1B

Insulin signaling starts via the activation of the PTP-1B enzyme at the IR site. This enzyme increases phosphorylation at the tyrosine site of phosphoinositol-3-kinase (PI3-K). PI3-K binds with phosphatidylinositol-3,4,5-trisphosphate (Ptidines (3, 4, 5) $P_3$) and activates protein kinase B (PKB; also known as AKT) and phosphatidylinositol-dependent kinase 1 (PDK 1). This action leads to the activation of glucose transporter-4 (GLUT-4) for more glucose absorption and inactivation of glycogen synthetase kinase-3 (GSK-3) for decreased glycogen production (Figure 3.1). Finally, the glucose and lipid metabolism increase at cellular levels (Zhang and Zhang 2007).

The activation of PTP-1B mediates through IR and IR-related proteins via two-step reactions. First, the sulfur moiety of cysteine residue binds with the tyrosine

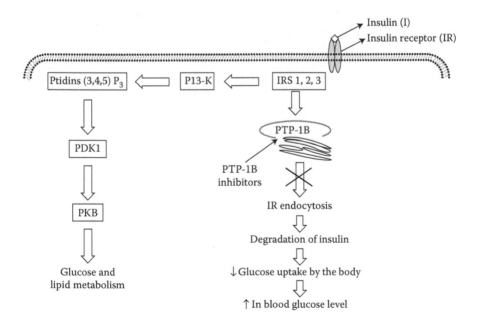

**FIGURE 3.1**  Mechanism of PTP-1B inhibition. (Adapted from Johnson, T.O., Ermolieff, J., and Jirousek, M.R. 2002. *Nat Rev Drug Discov* 1:696–09.)

side of the substrate and this complex is finally attached with Asp181 (aspartic acid) in the PTP-1B enzyme (also called cysteinylphosphate catalytic intermediate). In the second step, hydrolysis of this complex mediates through Gln 262 (glutamine) via the release of one water molecule and one phosphate molecule (Asp181 serves as the general base) (Tonks 2003).

Various molecules are reported to be active against PTP enzymes, which finally increase the glucose and lipid metabolisms. Broadly, these are benzofuran (Dixit et al. 2007), chalcones (Mahapatra et al. 2015), oxazoles (Kumar et al. 2009), pentacyclic triterpenoids (Ramírez-Espinosa et al. 2011), and other non-peptide PTP-1B inhibitors. Among them, chalcones play a major role in the inhibition of PTP-1B, used to manage diabetes and the associated complications. The biochemical processes of glucose utilization and the involvement of receptor PTP-1B have been presented in Figure 3.1. Chen et al. reported 20 novel heterocyclic ring-substituted chalcone derivatives as potent inhibitors of PTP-1B (Chen et al. 2012). PTP-1B inhibitors serve as a valuable tool for a new therapy for diabetes and metabolic disorders (Johnson et al. 2002). More studies would be necessary to know the role of this enzyme in insulin signaling and NIDDS therapy.

### 3.2.2  α-GLUCOSIDASE

α-Glucosidases are mainly responsible for the conversion of carbohydrates into monosaccharides and are located at the epithelial cells of the small intestine. The inhibitors bind with the oligosaccharide site chain of the enzyme and delay the formation of glucose from polysaccharides and starch. Finally, the inhibitors reduce the food digestion, absorption, and blood glucose absorption (Nakai et al. 2005). Acarbose, miglitol, and voglibose are the most widely used drugs for α-glucosidase-induced diabetic therapy (Yina et al. 2014).

α-Glucosidases are exoacting carbohydrases, located at the brush-border epithelium of the small intestine, which catalyze the release of α-D-glucopyranose from the nonreducing ends of substrates by cleaving the terminal non-reducing 1,4-linked α-glucose and play an important role in the processing of glycoproteins and glycolipids (Okuyama et al. 2001). The involvement of the α-glucosidase enzyme, and the biochemical processes of glucose utilization have been depicted in Figure 3.2. A continuous blockade of the enzyme by these "starch blockers" results in a simultaneous reduction in body weight, HbA1c, and serum triglyceride. A wealth of compounds having α-glucosidase inhibitory activity is ubiquitous in medicinal plants. A recent report documented that 411 compounds belonging to different structural frameworks, that is, terpenes, alkaloids, quinines, flavonoids, phenols, phenylpropanoids, sterides, and compounds with other structural and functional motifs isolated from medicinal plants, showed a promising inhibitory activity toward α-glucosidase (Arungarinathan et al. 2011).

### 3.2.3  ALDOSE REDUCTASE

The ALR enzyme is responsible for the polyol pathway. The polyol pathway activates when the normal glycolytic cycle gets saturated. As an alternative route

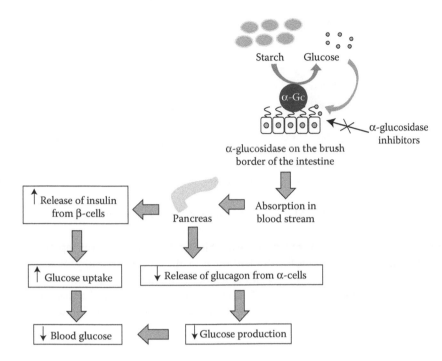

**FIGURE 3.2** Mechanism of α-glucosidase inhibition. (Adapted from Arungarinathan, G., McKay, G.A., Fisher, M. 2011. *Brit J Cardiol* 18:78–81.)

for metabolism, a high concentration of glucose in the nerves, lens, and retina gets metabolized into sorbitol in the absence of insulin, leading to an accumulation in tissues, which precipitates symptoms such as an osmotic imbalance, membrane permeability changes, and oxidative stress causing tissue injury (Muthenna et al. 2009). This condition provokes many diabetic complications such as nephropathy, retinopathy, cataract, neuropathy, etc. Neuropathic pain is a harmful condition characterized by paresthesia, a burning sensation, and central nervous system (CNS) neuronal abnormal function. Glucose or its related compounds mediate renal injury, which is also one of the major diabetic complications that have shown fatal results since the last decade. These conditions also affect the quality of life by precipitating other health issues such as fatigue, depression, and anxiety. Natural products have been recognized as potential inhibitors of ALR and among them, chalcone remained the prime choice among researchers for enzyme inhibition. The biochemical processes of ALR have been presented in Figure 3.3. A series of trisubstituted, heterocyclics and thioglycolic-substituted chalcones were synthesized and investigated for ALR1 and ALR2 inhibitory activity by Severi et al. (1998) where the authors reported that these compounds showed an effective ALR2 inhibition within the range of 35%–40%. Three chalcones, isolated from *Sophora flavescens*, have been found to be active for the inhibition of AR.

To date, several ARIs of three categories, that is, acetic acid compounds (epalrestat), spirohydantoins (sorbinal, fidarestat), and succinimide (AS-3201/ranirestat) have

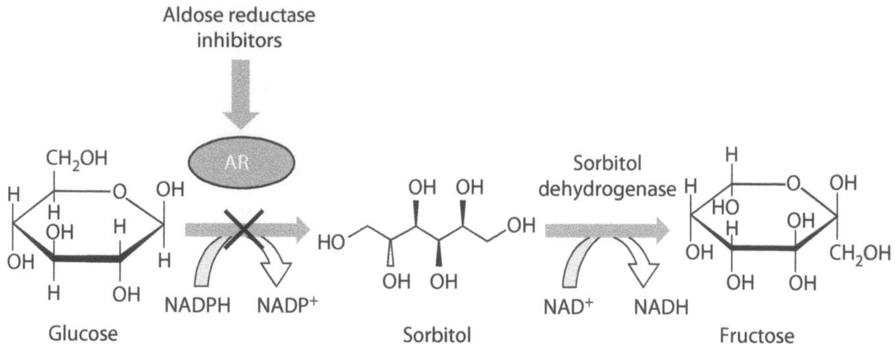

**FIGURE 3.3** Mechanism of ALR inhibition. (Adapted from Zhu, C. 2013. *InTech* 2:17–46.)

been synthesized, none of which are available yet for use. However, epalrestat is currently marketed in Japan for the treatment of diabetic neuropathy. Some well-known medicinal plants such as *Prunus mume, Centella asiatica, Salacia reticulata, Chrysanthemum morifolium, Salacia oblonga, Myrcia multiflora, Salacia chinensis,* and *Chrysanthemum indicum* exhibited potent ALR inhibitory activity (Patel et al. 2012).

## 3.2.4 DIPEPTIDYL PEPTIDASE-4

For the development of NIDDS drugs, another interesting approach is to develop the agonist of glucagon-like peptide-1 (GLP-1). DPP-4 is the enzyme, which acts on one GLP-1 receptor. Therefore, the targets for diabetes treatment are based on peptide based GLP-1 analogs (exenatide and liraglutide), small molecules as nonpeptide based GLP-1 receptor agonist, and DPP-4 inhibitors. DPP-4 inhibitors potentiate the effect of incretin hormones, namely, GLP-1 and gastric-inhibitory peptide (GIP). These hormones are released from the enteroendocrine cells of the gastrointestinal tract (GIT) into the bloodstream in response to food intake. Neuroendocrine K-cells produce GIP and stimulate the glucose-dependent release of insulin by binding to GIP receptors (GIPRs) on pancreatic β-cells. GLP-1 is the most potent incretin hormone and is mainly responsible for a modulatory effect on the pancreatic β-cell functions. The circulating levels of GLP-1 are generally low in the fasting state, which rises quickly to the threshold after taking a meal (Wani et al. 2008). Following the ingestion of food, a neural stimulus and close contact of nutrients with K- and L-cells of the mucosa activates and releases GLP-1 from the GIT. This causes insulin secretion from the pancreatic β-cells, reduces the glucagon secretion of α-cells, increases the growth, survival, proliferation, and differentiation of new β-cells, and slows down gastric emptying along with the increase in gastric acid production (Baggio and Drucker 2007).

Importantly, DPP-4 cleaves incretins at proline and alanine amino acid sites, of which these two amino acids are significant for the biological activity of the incretins; GIP and GLP-1, thus, inactivates by its proteolytic activity nature. Approximately 60% of the postprandial insulin release is promoted by these two hormones only.

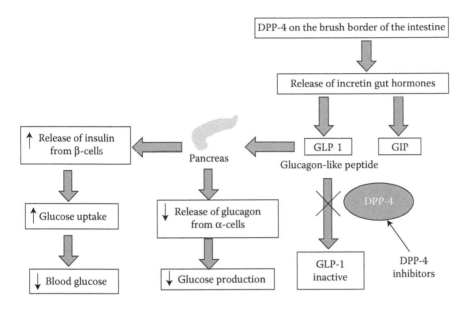

**FIGURE 3.4** Mechanism of DPP-4 inhibition. (Adapted from Lovshin, J.A. and Drucker, D.J. 2009. *Nat Rev Endocrinol* 5:262–69.)

In patients with type-2 diabetes (T2D), a continuous subcutaneous infusion of GLP-1 leads to a reduction in HbA1c levels (Mulvihill and Drucker 2014).

Structurally, two types, namely, peptidomimetic and non-peptidomimetic DPP-4 inhibitors have been reported. The peptidomimetic class are glycine-based (α-series A) and β-alanine-based (β-series B) inhibitors. On the other hand, the non-peptidomimetic inhibitors may be subdivided into fused imidazole, cyclohexylamine/aminopiperidine, quinazolinone/pyrimidinedione/isoquinolone, and fluoroolefin derivatives (Havale and Pal 2009).

Recently, the chalcone inhibitors of *Chana* series have been developed and showed an improved hyperglycemic control in NIDDS patients. The biochemical pathway of DPP-4 has been depicted in Figure 3.4. Bak et al. (2011) synthesized a novel *Chana* series of chalcone derivatives and found them active for its inhibitory activity against α-glucosidase and DPP-4 using C3H10T1/2 cells.

### 3.2.5 Peroxisome Proliferator-Activated Receptor-γ

The nuclear peroxisome proliferator-activated receptors (PPARs) are a group of proteins that play an important role in gene expression related to glucose metabolism. PPARs play a crucial role in cellular differentiation, development, and metabolism of lipids, carbohydrates, and proteins in humans, of which PPAR$_\gamma$ is thought to be specific for glucose homeostasis. It is expressed in the adipose tissue. Thus, it requires binding with retinoid X receptor (RXR) for an optimal deoxyribonucleic acid (DNA) binding and transcriptional activity (Berger and Moller 2002). Then, this complex binds to specific DNA sites of peroxisome proliferator response

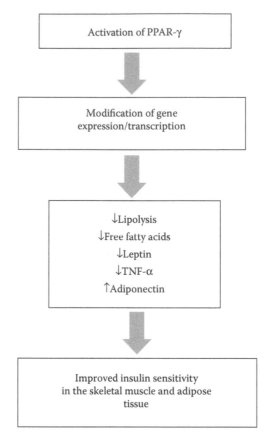

**FIGURE 3.5**   Mechanism of PPAR$_\gamma$ activation. (Adapted from Cheng, A.Y.Y. and Fantus, I.G. 2005. *Can Med Assoc J* 172:213–26.)

elements (PPREs) and regulates the transcription of insulin-responsive genes. PPAR$_\gamma$ is activated in the presence of dietary fatty acids. It performs different physiological functions, including the stimulation of GLUT-4 expression and translocation, which leads to a reduction of blood glucose level, suppression of gluconeogenesis in hepatic tissues, increase in lipid storage, and the entry of glucose in muscles. Thiazolidinediones and chalcones are the most popular ligands but the search is on for newer and better agonists (Jung et al. 2006). The biochemical pathway of PPARγ has been depicted in Figure 3.5.

### 3.2.6   ESTROGEN-RELATED RECEPTOR-α

ERR-α is peroxisome proliferator-activated receptor gamma coactivator 1α (PGC-1α) target gene and regulates mitochondrial oxidative phosphorylation (OXPHOS) and fatty acid oxidation. Genomic studies revealed that OXPHOS genes exhibit a reduced expression of prediabetic and diabetic patients and therefore, regulate the downstream expression of PGC-1α. It is discovered that ERR-α expressed from PGC-1α, enhances the regulation of the OXPHOS transcriptional gene in the diabetic muscle (Schreiber et al. 2004). This transcriptional process provides us a new tool for treating diabetes.

An increased number of ERR-α binding motifs results in a higher activation of ERR-α transcription by PGC-1α and ERR-α. If these two binding sites are inhibited, then NIDDS would be reduced to a higher extent (Wende et al. 2005). Thus, the target of ERR-α with the mitochondrial OXPHOS function is an important target for diabetic patients. Moreover, ERR-α regulates the β-oxidation of a fatty acid and triggers ERR-α for lipid accumulation at the skeletal muscle, which is the major issue in insulin resistance.

Today, the most-promising target for OXPHOS is to interact between PGC-1α and ERR-α. Synthetic inhibitors (toxaphene and chlordane) act via binding with glucocorticoid receptor-interacting protein 1 (GRIP1), and thus reduce diabetic tolerance. XCT790, an inverse agonist of PGC-1α/ERR-α is found to be effective against cellular respiration of the skeletal muscle and OXPHOS genes expression (Willy et al. 2004). This compound enhances the interaction between PGC-1α and ERR-α with the reduction of estrogen production, which indirectly enhances blood glucose concentration.

### 3.2.7 FIBROBLAST GROWTH FACTOR 21

FGF21 is a metabolic hormone, which belongs to the fibroblast growth factor superfamily. The administration of recombinant FGF21 induces favorable metabolic changes in diabetic patients that include the amelioration of hyperglycemia and dyslipidemia, reduction in body weight accompanied by the enhancement of insulin sensitivity and glucose uptake in the peripheral tissue, increase in fat utilization and energy expenditure, and decrease in glucagon production in islet α-cells (Reitman 2013).

Some commonly used antidiabetic drugs may improve metabolic dysregulations in patients with diabetes by acting on the FGF21 signaling pathway. Metformin inhibits respiratory complex I in mitochondria and leads to reduced liver gluconeogenesis in an AMP kinase (AMPK)-dependent pathway. The glucose-lowering effect of metformin also attributes to its counterregulatory effect on glucagon actions in an AMPK-independent pathway. Metformin treatment increases the expression and secretion of FGF21 in a diverse set of cell lines depending on the inhibition of mitochondrial respiratory complex I but independent of AMPK (Kim et al. 2013). These observations suggest that FGF21 may partially mediate the antidiabetic effect of metformin. Fenofibrate is a PPAR-α agonist whereas FGF21 is a downstream target of PPAR-α. In a post hoc analysis, fenofibrate treatment increased the level of circulating FGF21 in patients with NIDDS, suggesting that FGF21 may involve in the cardiovascular-protective actions of fenofibrate in patients with diabetes. FGF21 is a downstream target of $PPAR_\gamma$ whereas it also activates $PPAR_\gamma$ signaling in a feed-forward manner. Moreover, several studies suggested that FGF21 is required to mediate the antidiabetic effects of $PPAR_\gamma$ agonists (Ong et al. 2012). Improving FGF21 sensitivity may be another strategy to enhance FGF21 actions and attain beneficial metabolic effects in patients with diabetes since FGF21 resistance is common in this population. Specifically, the activation of FGF21 signaling by agonistic antibodies led to the reduction in body weight, improvement in dyslipidemia and hepatosteatosis, and amelioration of hyperglycemia and hyperinsulinemia, suggesting that FGFR21 is an interesting target for the treatment of NIDDS associated with dysmetabolism (Foltz et al. 2012).

In a clinical trial study, the administration of LY2405319 (LY), an FGF21 mimetic, to obese patients with NIDDS led to a remarkable improvement in dyslipidemia and a somewhat glucose-lowering trend also. The mechanism underlying the beneficial metabolic effects of FGF21 mimetics is not fully understood. Two independent studies have linked the metabolic-regulating effect of FGF21 with insulin-sensitizing adipokine–adiponectin (Gaich et al. 2013).

### 3.2.8  ISLET G-PROTEIN-COUPLED RECEPTORS

Reduced β-cells of the islets of Langerhans and their dysfunction are the central events underlying the development of IDDS with impaired insulin secretion and decreased glucagon secretion. The main emphasis is to optimize the activity of the islets using proper therapy. Insulin and glucagon secretion is mediated through GPCR, and the antidiabetic drugs act on that target (McKeown et al. 2007).

Free fatty acids (FFAs) serve as energy sources of the human body and act as important constituents of the cell membrane. FFAs also stimulate both insulin and glucagon secretion. FFAs activate the islet GPCR to potentiate the islet function. The function of islet GPCR through FFA was first proposed for GPR40, which stimulates insulin secretion (Itoh et al. 2003). Moreover, FFAs such as GPR119, GPR41, and GPR43 are the activator of GPCR for insulin secretion (Brown et al. 2003; Itoh et al. 2003; Chu et al. 2007). On the other hand, GPCR120 indirectly acts on GPCR for secretion and acts via acting on GLP-1 (Hirasawa et al. 2005).

Incretin hormones are the specific agonist of both GIP and GLP-1 that is the important factor for islet function. These hormones potently enhance glucose-stimulated insulin secretion through increased cyclic adenosine monophosphate (cAMP). The GIPR identified in human-pancreatic islets, which is linked to Gs protein, are predominantly expressed in β-cells. GIP has a prominent role in insulin secretion (glucose stimulated) and inhibits the β-cell function. Since GIP has no important role in insulin secretion, GIP does not appear as a rational treatment for the disease (Drucker 2005).

The expression of the GLP-1 receptor is observed in islet β- and δ-cells. GLP-1 is important in glucose homeostasis and energy metabolism. GLP-1 receptor activation stimulates the secretion of insulin and inhibits the secretion. Further activation substantiates the stimulation of β-cell proliferation and apoptosis. GLP-1 receptor agonists are liraglutide and exenatide. Exenatide also reduces HbA1c concentration in plasma during the combination with sulfonylurea or metformin. A similar action was observed for liraglutide to reduce HbA1c (Drucker and Nauck 2006).

### 3.2.9  HISTONE DEACETYLASES

HDAC causes the acetylation of histone in chromatin remodeling and plays an important role in gene transcription to control cell proliferation, migration, and death. Reports have been documented that HDAC is a promising drug target for a wide range of diseases including cancer, neurodegenerative and psychiatric disorders, autoimmunity, cardiovascular dysfunction, and diabetes mellitus (Wang et al. 2012).

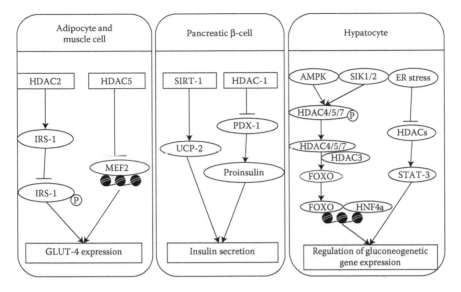

**FIGURE 3.6** Mechanism of HDAC inhibition. (From Crosson, C.E. et al. 2010. *Invest Ophthalmol Vis Sci* 51:3639–45.)

To date, a total of 18 HDAC enzymes have been identified in mammals made up of four classes according to their sequence identity and catalytic activity. Among them, classes I, II, and IV HDACs possess a $Zn^{2+}$-dependent mechanism of action whereas class-III HDACs are the class of nicotinamide adenine dinucleotide (NAD)-dependent deacetylases that comprise a similar-sequence homology of yeast Sir2 protein. Because of this, class-III HDACs are also called "sirtuins" (SIRTs). These $Zn^{2+}$-dependent HDACs are known to act as the targets of small-molecule HDAC inhibitors that have shown potency in patients of diabetes mellitus (Shakespear et al. 2011).

The regulation of glucose homeostasis is associated with epigenetic mechanisms. Modification of histones occurs by phosphorylation and ubiquitination. Methylation and acetylation play important roles in gene transcription. In particular, the acetylation of histone proteins controls gene expression and cellular signaling during diabetes. The HDACs regulate this acetylation of histone and transcriptional factors, which helps in glucose homeostasis and plays a key role in glucose metabolism (Berger 2002).

HDACs regulate glucose metabolism in the following way (Figure 3.6):

1. HDACs downregulate the expression of GLUT-4 and enhance insulin resistance in adipocytes and muscle cells.
2. HDAC1 reduces the expression of PDX1 leading to the decreased expression of insulin while SIRT1 upregulates UCP2 that enhances the secretion of insulin.
3. In hepatocytes, HDACs are involved in the regulation of gluconeogenesis in various ways: class IIa HDACs are translocated and dephosphorylated from the cytoplasm to the nucleus; class IIa HDACs recruit HDAC3 that can

deacetylate FOXO1 and FOXO3, thereby enhancing FOXO DNA binding to promote gluconeogenetic gene expression; and STAT3-dependent suppression inhibits by ER stress during hepatic gluconeogenic enzymes via STAT3 deacetylation (HDAC dependent) (Wang et al. 2012).

Fortunately, besides regulating glucose metabolism by increasing insulin secretion, improving insulin resistance, and controlling gluconeogenesis, HDAC inhibitors also show a promising role in the treatment of most of the diabetic complications, including diabetic retinopathy (DR) and diabetic nephropathy (DN) (Crosson et al. 2010).

### 3.2.10    SODIUM–GLUCOSE COTRANSPORTERS

The kidneys play an important function in glucose homeostasis and have become a target for diabetes treatment recently. The glomerular filtration-related glucose reabsorption from the kidneys is mediated through sodium glucose cotransporter 2 (SGLT2). SGLT2 inhibition promotes urinary glucose excretion (UGE) and reduces plasma glucose levels. SGLT2 inhibitors serve as a novel class of drugs for the treatment of NIDDS. There are two drugs, namely, sergliflozin and dapagliflozin, which are currently available in the market as SGLT2 inhibitors. Sergliflozin (a GLUT-1 inhibitor, analog of phlorizin) is available in a glycated form and is highly specific as the SGLT2 inhibitor. Oral-administered sergliflozin reduces glucose reabsorption and increases glucose excretion from urine (Tahrani et al. 2010).

The second-generation SGLT2 inhibitor, dapagliflozin approved by the Food and Drug Administration (FDA) is used in combination with canagliflozin (HbA1c reducer). Dapagliflozin is another SGLT2 inhibitor, whose efficacy and safety were evaluated in 14 clinical trials as a monotherapy or in combination with other antidiabetes agents. Dapagliflozin causes dose-dependent glycosuria in humans and reduces the HbA1c level in NIDDS patients. Moreover, NIDDS is cured through empagliflozin (BI 10773) that is a potent and selective SGLT2 inhibitor. Empagliflozin has the highest selectivity for SGLT2 over SGLT1 (Grempler et al. 2012). Another drug of this class is ipragliflozin (ASP1941), developed by Astellas and Kotobuki Pharmaceuticals (Schwartz et al. 2011).

Furthermore, LX4211 is an inhibitor of both SGLT1 and SGLT2 transporters, which is useful for NIDDS. It also blocks glucose absorption in the small intestine via inhibiting SGLT1. The combination of LX4211 and DPP-4 inhibitors is promising and has been tested in preclinical and clinical studies. LX4211 with sitagliptin also produced significant results in glucose reduction (Zambrowicz et al. 2013).

### 3.2.11    K-CELLS RECEPTORS

Gut K-cells also serve as another target for insulin therapy for the management of diabetes. K-cells activate glucokinase (GK) (hexokinase enzyme), which have a high affinity to glucose, and act as a glucose sensor for pancreatic β-cells. The secretion of insulinotropic peptide (glucose dependent), namely, GIP is mediated through K-cells and this action enhances insulin secretion (glucose mediated). The expression of

GK-mediated K-cells was limited because their concentration was very low in the liver and pancreas. GLUT-2 is considered the main transporter of glucose in the liver and β-cells and streptozotocin (STZ) reduces its action. STZ reduces the antioxidant enzyme levels and damages the β-cell by inducing DNA damage, possibly due to the accumulation of nitric oxide (NO) (Krishna et al. 2012). Interleukin-1 (IL-1) and interferon-γ inhibit insulin secretion and stimulate inducible nitric oxide synthase (iNOS) to produce NO that leads to the reduction of adenosine triphosphate (ATP) levels and inhibition of insulin secretion.

K-cells serve as an important target for gene therapy-related diabetes as they are the scavengers of free radicals and resist cytokine action. K-cells are also found in Lieberkühn and may also be an excellent cellular target for gene therapy. In the long term, gene-therapeutic approaches that target glucose-responsive K-cells could provide a newer method for the treatment of patients with IDDS (Lortz et al. 2000).

## 3.2.12 T-Cells Receptors

Although IDDS is a common autoimmune disease, the exact mechanisms still remain unknown. Few researchers suggested that CD4+ and CD8+ T-cells have a crucial role in the development of diabetes. T-cells bind with the major histopatibility complex (MHC) that changes the gene expression, especially the complementarities-determining region 3 (CDR3) gene.

According to the components of peptide chains, TCRs are divided into two subpopulations, for example, α–β TCR and γ–δ TCR. There are variable (V), constant (C), transmembrane, and cytoplasm regions on each of the chains. Antigens are recognized in the V region, in which there are three CDRs, and CDR3 is susceptible to the antigens and looked as the element that could reflect the specific changes of TCR (Jianwei et al. 2014).

### 3.2.12.1 TCR Vα

TCR Vα is considered an important gene for IDDS. A recent investigation revealed that Vα5D-4, Vα7, and Vα17 were the active genes, and their expression is closely related to insulin B: 9-23 peptide components. TCR Vα13-1 was also an initiation factor for IDDS, where the clonal changes of TCR Vα just began. However, thorough studies are necessary for Vα gene families and their potential role in IDDS (Nakayama et al. 2012) to be understood.

### 3.2.12.2 TCR Vβ

Codina-Busqueta et al. (2011) identified TCR Vβ in the spleen and peripheral blood of nonobese mice, and observed that Vβ22 was expressed monoclonally. Further evidence revealed that the Vβ22 presents in peripheral blood mononuclear cells (PBMCs). Liu et al. (2012) identified TCR-Vβ13S1A1, an allele of TCR Vβ, which could prevent IDDS. The Vβ7 gene also serves as a key feature of IDDS, and the excitation of the expression of this gene reduces the blood glucose level to some extent (Zhou et al. 2013). Then, the common or individualized character of IDDS and the value of each of the skewed Vβ genes indeed need further investigations.

### 3.2.12.3   TCR γδ

Although the concentration of TCR γδ is 5% in peripheral blood or tissues, research-ers give more attention to its functions. In the IDDS model of study design, the concentration of γδ–T-cells was higher in the peripheral blood during the IDDS con-dition than in the normal condition. These cells regulate the development of IL-17 in IDDS, which increases blood glucose concentration. In another experiment, TCR Vγ converts into γδ–T-cells and this conversion is more pronounced at B: 9-23 pep-tide of insulin. These findings indicated that both TCR Vγ and γδ–T-cells were the sensitive factors of IDDS (Zhang et al. 2010). However, due to the small number of γδ–T-cells, the real values of the γδ TCR skewness in the generation, development, or prevention remain unclear.

### 3.2.13   New Hepatic Targets for Glycemic Control in Diabetes

#### 3.2.13.1   Counteraction of Glucagon Signaling

Glucagon is another counterregulatory hormone, which opposes the action of insu-lin and enhances glucose production in the liver. It inhibits glycogen synthesis, the glycolysis pathway, and stimulates gluconeogenesis. Adenylcyclase activates the glucagon receptor, and leads to the production of cAMP, which initiates vari-ous phosphorylation events. Glucagon secretes from $\alpha$-cells of the pancreas and the secretion is inhibited by insulin, GLP-1, and higher blood glucose level. The action of glucagon is interrupted by various glucagon receptor antagonists (GRAs), downregu-lation of glucagon receptor, and immunodepletion of glucagon (Sorensen et al. 2006).

GRA (Bay 27-9955) is the antagonist of GR from the Merck Company, which reduces glycogenolysis by glucagon and partially emphasizes insulin-mediated glycogen synthesis in human hepatocytes. It has no effect on gluconeogenesis (Qureshi et al. 2004). Another hybrid approach is to use GLP-1 agonist and GRA properties. The selective GLP-1 agonist and one GRA reduce the glycemic condition and activate adenyl cyclase. One such compound is the fungal bisanthroquinone, skyrin (Parker et al. 2000).

#### 3.2.13.2   Inhibition of Glucose 6-Phosphatase

Glucose 6-phosphatase (G6Pase) helps in the formation of hepatic glucose in both glycogenolysis and gluconeogenesis, and it is a potential target for NIDDS. Both insulin deficiency and hyperglycemia induce G6Pase gene expression. As glucagon increased hepatic glucose production, the hyperglycemia can be suppressed through G6Pase inhibition (Dunning and Gerich 2007). However, these inhibitors reduce the action of G6Pase enzyme, which results in the accumulation of glycogen and induction of lipogenic genes in the liver, leading to hepatic steatosis. They indirectly enhance the mRNA expression of various genes, namely, glycogen synthase, T1 transporter, G6Pase catalytic subunit, GLUT-2 transporter, pyruvate kinase, acetyl-coenzyme A carboxylase, and fatty acid synthase (Harndahl et al. 2006). Altogether, G6Pase has two limitations as a therapeutic target for glycemic control in T2D.

Low concentrations of the inhibitor that cause a mild lowering of blood glucose result in a marked perturbation of the expression of various genes involved in lipogenesis, with the consequent development of a fatty liver; and high concentrations of the inhibitor may cause severe hypoglycemia because G6Pase activity is essential for hepatic glucose production by both gluconeogenesis and glycogenolysis.

### 3.2.13.3 Inhibition of Gluconeogenesis by Fructose 1,6-Bisphosphatase Inhibitors

The activities of the gluconeogenic enzymes phosphoenolpyruvate carboxykinase (PEPCK) and fructose 1,6-bisphosphatase (F16BPase) are controlled by glucagon and insulin, and are raised in case of diabetes and insulin resistance. They are considered to be the potential targets for antihyperglycemic therapy. Erion et al. showed that an AMP mimetic (MB06322, CS-917), which inhibits F16BPase synergistically with fructose 2,6-P2, inhibited gluconeogenesis from various substrates in hepatocytes and lowered blood glucose in fasting male Zucker diabetic fatty (ZDF) rats (Erion et al. 2005).

### 3.2.13.4 Inhibition of Glycogenolysis with Glycogen Phosphorylase

In the normal nondiabetic state, the rapid suppression of hepatic glucose production by insulin is largely due to the inhibition of glycogenolysis via glycogen phosphorylase (GP) inactivation. This enzyme exists in two isoforms, that is, GPa (phosphorylated) and GPb (unphosphorylated). Only GPa is physiologically active in the liver. GPb converts into GPa in the presence of glucagon and phosphorylase kinase while the conversion of GPa into GPb takes place in the presence of insulin (Aiston et al. 2003). Thus, GP inhibitors that mimic the action of either glucose or glucose 6-phosphate are expected to increase the conversion of GPa into GPb, and consequently inhibit glycogenolysis and stimulate glycogen synthesis.

### 3.2.13.5 Activation of GK

An alternative strategy to targeting the enzymes involved in gluconeogenesis and glycogenolysis is to target hepatic glucose utilization by the activation of GK or hexokinase IV. GK is expressed in pancreatic β-cells and hepatocytes. In the liver, GK has high control strength on glucose metabolism by glycogen synthesis and glycolysis. Hepatic GK is regulated by binding to a 68-kDa regulatory protein (GKRP) that functions both as a competitive inhibitor for GK with respect to glucose and as a nuclear receptor for the enzyme. Binding of GK to GKRP is dependent on the concentrations of glucose and of fructose 1-phosphate. To date, the data available on the efficacy of GKAs in lowering blood glucose and improving glucose tolerance is very promising (Matschinsky et al. 2006). However, a question is whether GKAs might have undesirable side effects during chronic therapeutic use.

### 3.2.14 MIRNAS AS PHARMACOLOGICAL TARGETS IN DIABETES

Recent reports have suggested that miRNAs have important roles in controlling insulin biosynthesis and release, pancreatic β-cell development and survival, glucose and lipid metabolism, and their involvements in secondary complications associated with diabetes (Broderick and Zamore 2011). The miRNAs are involved in the pathogenesis of diabetes and have become an intriguing target for therapeutic intervention. These may be of the following types:

### 3.2.14.1   miRNAs Regulating Insulin Biosynthesis and Secretion in Pancreatic β-Cells

miRNAs playing a role in regulating insulin biosynthesis, insulin secretion, and β-cell survival in pancreatic β-cells are miR-7, miR-9, miR-21, miR-29a/b/c, miR-30d, miR-124a, miR-338-3p, and miR-375.

### 3.2.14.2   miRNAs Regulating Insulin Sensitivity in Skeletal Muscle and Adipose Tissue

miRNAs involved in regulating insulin sensitivity in the skeletal muscle and adipose tissue are miR-143, miR-29a/b/c, miR-320, and MyomiRs: miR-1, miR-133, and miR-206.

### 3.2.14.3   miRNAs in Regulating Glucose and Lipid Metabolism in the Liver

miRNAs regulating glucose and lipid metabolism in the liver include miR-103/107, miR-122, miR-181a, and miR-802.

### 3.2.14.4   Circulating miRNAs Are a Promising Clinical Biomarker for Diabetes

Circulating miRNAs, as a promising clinical biomarker for diabetes, are miR-375 and miR-126.

The miRNAs have generated an enormous interest as an intriguing therapeutic target in the treatment of complex diseases such as cardiovascular disease, cancer, and diabetes (Broderick and Zamore 2011). Since miRNAs are naturally endogenous regulators of cell processes that are often dysregulated in diabetes, the restoration of any miRNA functions to normal levels will be the ultimate therapeutic goal. Two main therapeutic approaches have been developed: the first is restoring the expression of miRNAs that are downregulated in diabetes using "miRNA mimics" and the second is inhibiting the activity of miRNAs that are significantly above the normal expression using *miRNA inhibitors*.

The miRNA inhibitors are antisense oligonucleotides having the reverse complementary sequence of the target miRNA. Because miRNAs typically act as repressors of the target gene expression, an miRNA inhibitor binds to the mature miRNA, and activates the target gene expression. Locked nucleic acid (LNA) anti-miRs, antagomirs, and morpholinos are the efficient inhibitors with different modifications and have been proven to be effective. Among them, LNA anti-miRs, which have a high affinity with miRNAs, high efficiency, and low toxicity, show a promise for the future development of therapies. LNA anti-miR-122 against miR-122 has successfully resulted in reduced plasma cholesterol with no indication of hepatic toxicity. siRNAs such as miRNA mimics are synthesized RNA duplexes and are designed to mimic the function of an endogenous miRNA (Orom et al. 2006).

Altogether, miRNAs will not only be potential pharmacological targets in treating diabetes but will also be clinical biomarkers for an earlier diagnosis, and thus intervene in the development of diabetes. The potential of miRNA-based therapies will offer an exciting and powerful alternative to attenuate and hopefully will cure diabetes and its complications. Table 3.1 describes the various miRNAs responsible for diabetes along with their targets, expressions, and functions.

**TABLE 3.1**

**Different miRNAs and Their Biological Targets along with Functions in Biosystems**

| Name of miRNA | Expression | Targets | Functions | Reference |
|---|---|---|---|---|
| **miRNAs Regulating Insulin Biosynthesis and Secretion in Pancreatic β-Cells** | | | | |
| miR-7 | Most-abundant endocrine miRNA | Mknk1, Pax6, p70S6 K, eIF4E, Mknk2, and Mapkap1 | Promotes β-cell and β-cell differentiation, inhibits adult β-cell proliferation, and fine-tunes β-cell development and regeneration | Kredo-Russo et al. (2012) |
| miR-9 | Preferentially expressed in the brain and β-cell | Sirt1, Onecut-2 | Regulates glucose-challenged insulin secretion | Ramachandran et al. (2011) |
| miR-21 | Increased in NOD and db/db mice | Pdcd4, Piccolo | Impairs insulin secretion, prevents cytokine-mediated β-cell death, and controls immune rejection in transplanted islets | Ruan et al. (2011) |
| miR-29a/b/c | Increased in NOD mice | Onecut-2, Mcl1, and Mct1 | Maintains glucose-stimulated insulin secretion, promotes cytokine-mediated β-cell apoptosis | Pullen et al. (2011) |
| miR-30d | Decreased in db/db mice | Map4k4 | Stimulates insulin secretion and production, promotes pancreatic islet-derived mesenchymal cell differentiation, and activates MafA expression | Joglekar et al. (2009) |
| miR-124a | Preferentially expressed in the brain and embryonic pancreas, undetectable in mature mouse islets | Foxa2, Rab27a | Inhibits insulin secretion, regulates pancreas development | Lee et al. (2009) |
| miR-338-3p | Decreased during rat pregnancy and young db/db mice | Glp1, Gpr30 | Inhibits β-cell proliferation and survival, reduction is required for β-cell mass expansion during pregnancy | Jacovetti et al. (2012) |

(*Continued*)

**TABLE 3.1 (*Continued*)**

**Different miRNAs and Their Biological Targets along with Functions in Biosystems**

| Name of miRNA | Expression | Targets | Functions | Reference |
|---|---|---|---|---|
| miR-375 | Increased in ob/ob mice and human subjects with T2D | Pdk1, myotrophin | Maintains β-cell mass, proliferation, and regeneration, inhibits insulin secretion and transcription, and promotes embryonic pancreas development | Wei et al. (2013) |
| **miRNAs Regulating Insulin Sensitivity in the Skeletal Muscle and Adipose Tissue** | | | | |
| miR-143 | Overexpression in preadipocytes | (C/EBP)-β, PPAR$_\gamma$, FABP4, and ORP8 | Promotes adipocyte differentiation, impairs insulin sensitivity | Jordan et al. (2011) |
| miR-320 | Highly increased in insulin-resistant adipocytes | PI3-K, AKT, and GLUT-4 | Increases insulin-stimulated glucose uptake | Ling et al. (2009) |
| miR-29a/b/c | Abnormally induced in three insulin-dependent tissues (i.e., skeletal muscle, liver, and fat) in diabetic GK rats, induction of miR-29c in the kidney of db/db mice | p85α, Akt, SPRY1, and FGF | Increases insulin resistance, promotes cell apoptosis | Long et al. (2011) |
| MyomiRs: miR-1, miR-133, and miR-206 | Muscle-specific expressions, downregulated in the skeletal muscle of type-2 diabetic patients | MEF2C, SREBP-1c, and HOMA1 | Improving insulin sensitivity promotes cardiomyocyte apoptosis, promotes brown adipocyte differentiation | Granjon et al. (2009) |
| **miRNAs Regulating Glucose and Lipid Metabolism in the Liver** | | | | |
| miR-103/107 | Mostly upregulated in the livers of both ob/ob mice and diabetic GK rats, increased in liver biopsies from diabetes-associated human patients | Caveolin-1 | Regulating insulin signaling and glucose uptake | Trajkovski et al. (2011) |

*(Continued)*

**TABLE 3.1** (*Continued*)
**Different miRNAs and Their Biological Targets along with Functions in Biosystems**

| Name of miRNA | Expression | Targets | Functions | Reference |
|---|---|---|---|---|
| miR-122 | Mostly abundant miRNA in the liver | FASN, HMG-CoA reductase, SREBP-1c, and SREBP-2 | Improved insulin sensitivity by increasing fatty acid oxidation, promotes cholesterol synthesis and lipoprotein secretion in the liver, and Antagomir-122, a miR-122 inhibitor, reduces the cholesterol level in both the liver and serum | Li et al. (2009) |
| miR-181a | Overexpression in hepatic cells | Sirtuin-1 | Negatively regulates insulin sensitivity in the liver, impairs hepatic insulin signaling, and attenuates insulin sensitivity in hepatic cells | Zhou et al. (2012) |
| miR-802 | Increase in the liver of high-fat diet mice, db/db mice, and obese human subjects | HNF1b, TCF2 | Impaired glucose tolerance and attenuates insulin sensitivity | Kornfeld et al. (2013) |
| Circulating miRNAs, as a Clinical Biomarker for Diabetes | | | | |
| miR-126 | Ubiquitously expressed, and highly enriched in endothelial cells | — | Maintains endothelial homeostasis and vascular integrity A good diagnostic biomarker for the onset of diabetic vascular complications | Zampetaki et al. (2010) |
| miR-375 | Increased in the NOD mouse model of autoimmune diabetes, increased in human individuals with a newly diagnosed T2D | — | A clinical predictor of diabetes | Erener et al. (2013) |

## 3.3 CONCLUSION

The search for newer therapeutic strategies is the foremost need to control the accelerating diabetic population. Although today, many therapies have been successful in treating diabetes and its complications, there is still a lack of a significant increase in the benefit-to-risk ratio. Hence, this chapter mainly gives an overview of the recent targets, receptors, and signaling pathways that may take part in diabetes and its progression. The vast literature on the diversity of emerging therapeutic targets and the deserving candidates modulating them for management of the disease are highlighted in this chapter. By modulating them in different ways, these therapeutic targets have huge future prospects. Still, none of them has gained an adequate attention in modern-day medicine and, therefore, need to be explored thoroughly so that they may be utilized for the better treatment and cure of diabetes in the future. Several natural and/or synthetic compounds targeting these cellular and molecular defects are entering clinical trials, thus holding out the promise that we might be able to stem the tide of a disease that has reached epidemic proportions.

## REFERENCES

Aiston, S., Coghlan, M.P., and Agius, L. 2003. Inactivation of phosphorylase is a major component of the mechanism by which insulin stimulates hepatic glycogen synthesis. *Euro J Biochem* 270:2773–81.

Arungarinathan, G., McKay, G.A., and Fisher, M. 2011. Drugs for diabetes: Part 4 acarbose. *Brit J Cardiol* 18:78–81.

Baggio, L.L. and Drucker, D.J. 2007. Biology of incretins: GLP-1 and GIP. *Gastroenterology* 132:2131–57.

Bak, E.J., Park, H.G., Lee, C.H. et al. 2011. Effects of novel chalcone derivatives on α-glucosidase, dipeptidyl peptidase-4, and adipocyte differentiation *in vitro*. *BMB Rep* 44:410–14.

Berger, J. and Moller, D.E. 2002. The mechanisms of action of PPARs. *Annu Rev Med* 53:409–35.

Berger, S.L. 2002. Histone modifications in transcriptional regulation. *Curr Opin Genet Dev* 12:142–48.

Broderick, J.A. and Zamore, P.D. 2011. MicroRNA therapeutics. *Gene Ther* 18:1104–10.

Brown, A.J., Goldsworthy, S.M., Barnes, A.A. et al. 2003. The orphan G protein-coupled receptors GPR41 and GPR43 are activated by propionate and other short chain carboxylic acids. *J Biol Chem* 278:11312–19.

Chen, Z., Sun, L., Zhang, W. et al. 2012. Synthesis and biological evaluation of heterocyclic ring-substituted chalcone derivatives as novel inhibitors of protein tyrosine phosphatase 1B. *Bull Korean Chem Soc* 33:1505–08.

Cheng, A.Y.Y. and Fantus, I.G. 2005. Oral antihyperglycemic therapy for type 2 diabetes mellitus. *Can Med Assoc J* 172:213–26.

Chu, Z.L., Jones, R.M., He, H. et al. 2007. A role for β-cell-expressed GPR119 in glycemic control by enhancing glucose-dependent insulin release. *Endocrinology* 148:2598–600.

Codina-Busqueta, E., Scholz, E., Mun~oz-Torres, P.M. et al. 2011. TCR bias of *in vivo* expanded T cells in pancreatic islets and spleen at the onset in human type 1 diabetes. *Immunology* 186:787–97.

Colberg, S.R. 2008. Enhancing insulin action with physical activity to prevent and control diabetes. *ACSM's Health Fit J* 12:16–22.

Crosson, C.E., Mani, S.K., Husain, S. et al. 2010. Inhibition of histone deacetylase protects the retina from ischemic injury. *Invest Ophthalmol Vis Sci* 51:3639–45.

Dixit, M., Tripathi, B.K., Tamrakar, A.K. et al. 2007. Synthesis of benzofuran scaffold-based potential PTP-1B inhibitors. *Bioorg Med Chem* 15:727–34.

Drucker, D.J. 2005. Biologic actions and therapeutic potential of the proglucagon-derived peptides. *Nat Clin Pract Endocrinol Metab* 1:22–31.

Drucker, D.J. and Nauck, M.A. 2006. The incretin system: Glucagon-like peptide-1 receptor agonists and dipeptidyl peptidase-4 inhibitors in type 2 diabetes. *Lancet* 368:1696–05.

Dunning, B.E. and Gerich, J.E. 2007. The role of alpha-cell dysregulation in fasting and postprandial hyperglycemia in type-2 diabetes and therapeutic implications. *Endocrinol Rev* 28:253–83.

Erener, S., Mojibian, M., Fox, J.K. et al. 2013. Circulating miR-375 as a biomarker of beta-cell death and diabetes in mice. *Endocrinology* 154:603–08.

Erion, M.D., van Poelje, P.D., Dang, Q. et al. 2005. MB06322 (CS-917): A potent and selective inhibitor of fructose 1,6-bisphosphatase for controlling gluconeogenesis in type-2 diabetes. *Proc Nat Acad Sci USA* 102:7970–75.

Foltz, I.N., Hu, S., King, C. et al. 2012. Treating diabetes and obesity with an FGF21-mimetic antibody activating the βKlotho/FGFR1c receptor complex. *Sci Transl Med* 4:162–53.

Forbes, J.M. and Cooper, M.E. 2013. Mechanism of diabetic complications. *Physiol Rev* 93:137–88.

Gaich, G., Chien, J.Y., Fu, H. et al. 2013. The effects of LY2405319, an FGF21 analog, in obese human subjects with type 2 diabetes. *Cell Metab* 18:333–40.

Granjon, A., Gustin, M.P., Rieusset, J. et al. 2009. The microRNA signature in response to insulin reveals its implication in the transcriptional action of insulin in human skeletal muscle and the role of a sterol regulatory element-binding protein-1c/myocyte enhancer factor 2C pathway. *Diabetes* 58:2555–64.

Graves, D.T. and Kayal, R.A. 2008. Diabetic complications and dysregulated innate immunity. *Front Biosci* 13:1227–39.

Grempler, R., Thomas, L., Eckhardt, M. et al. 2012. Empagliflozin, a novel selective sodium glucose cotransporter-2 (SGLT-2) inhibitor: Characterisation and comparison with other SGLT-2 inhibitors. *Diab Obes Metab* 1:83–90.

Harndahl, L., Schmoll, D., Herling, A.W. et al. 2006. The role of glucose 6-phosphate in mediating the effects of glucokinase overexpression on hepatic glucose metabolism. *Fed Euro Biochem Soc J* 273:336–46.

Havale, S.H. and Pal, M. 2009. Medicinal chemistry approaches to the inhibition of dipeptidyl peptidase-4 for the treatment of type 2 diabetes. *Bioorg Med Chem* 17:1783–02.

Hirasawa, A., Tsumaya, K., Awaji, T. et al. 2005. Free fatty acids regulate gut incretin glucagon-like peptide-1 secretion through GPR120. *Nat Med* 11:90–4.

Itoh, Y., Kawamata, Y., Harada, M. et al. 2003. Free fatty acids regulate insulin secretion from pancreatic beta cells through GPR40. *Nature* 422:173–76.

Jacovetti, C., Abderrahmani, A., Parnaud, G. et al. 2012. MicroRNAs contribute to compensatory beta cell expansion during pregnancy and obesity. *J Clin Inves* 22:3541–51.

Jianwei, Z., Cui, K., and Xinke, C. 2014. The short review on the studies of T cells receptors relate to type 1 diabetes. *Diab Metab Syndr Clin Res Rev* 8:252–54.

Joglekar, M.V., Patil, D., Joglekar, V.M. et al. 2009. The miR-30 family microRNAs confer epithelial phenotype to human pancreatic cells. *Islets* 1:137–47.

Johnson, T.O., Ermolieff, J., and Jirousek, M.R. 2002. Protein tyrosine phosphatase 1B inhibitors for diabetes. *Nat Rev Drug Discov* 1:696–09.

Jordan, S.D., Kruger, M., Willmes, D.M. et al. 2011. Obesity-induced overexpression of miRNA-143 inhibits insulin stimulated AKT activation and impairs glucose metabolism. *Nat Cell Biol* 13:434–46.

Jung, S.H., Park, S.Y., and Kim-Pak, Y. 2006. Synthesis and PPAR$_\gamma$ ligand-binding activity of the new series of 20-hydroxychalcone and thiazolidinedione derivatives. *Chem Pharm Bull* 54:368–71.

Kim, K.H., Jeong, Y.T., Kim, S.H. et al. 2013. Metformin-induced inhibition of the mitochondrial respiratory chain increases FGF21 expression via ATF4 activation. *Biochem Biophys Res Com* 440:76–81.

Kornfeld, J.W., Baitzel, C., Konner, A.C. et al. 2013. Obesity-induced overexpression of miR-802 impairs glucose metabolism through silencing of Hnf1b. *Nature* 494:111–15.

Kredo-Russo, S., Ness, A., Mandelbaum, A.D. et al. 2012. Regulation of pancreatic microRNA-7 expression. *Exp Diab Res* 2012:695214.

Krishna, G.K., Bir, S.C., and Kevil, C.G. 2012. Endothelial dysfunction and diabetes: Effects on angiogenesis, vascular remodeling, and wound healing. *Int J Vasc Med* 2012:918267.

Kumar, A., Ahmad, P., Maurya, R.A. et al. 2009. Novel 2-aryl-naphtho(1,2-D)oxazole derivatives as potential PTP-1B inhibitors showing antihyperglycemic activities. *Eur J Med Chem* 44:109–16.

Lee, J.H., Song, M.Y., Song, E.K. et al. 2009. Over expression of SIRT1 protects pancreatic beta-cells against cytokine toxicity by suppressing the nuclear factor-kappa B signaling pathway. *Diabetes* 58:344–51.

Li, S., Chen, X., Zhang, H. et al. 2009. Differential expression of microRNAs in mouse liver under aberrant energy metabolic status. *J Lipid Res* 50:1756–65.

Ling, H.Y., Ou, H.S., Feng, S.D. et al. 2009. Changes in microRNA (miR) profile and effects of miR-320 in insulin-resistant 3T3- L1 adipocytes. *Clin Exp Pharmacol Physiol* 36:e32–39.

Liu, Z., Cort, L., Eberwine, R. et al. 2012. Prevention of type 1 diabetes in the rat with an allele-specific anti-T-cell receptor antibody: Vβ13 as a therapeutic target and biomarker. *Diabetes* 61:1160–68.

Long, J., Wang, Y., Wang, W. et al. 2011. MicroRNA-29c is a signature microRNA under high glucose conditions that targets sprouty homolog 1, and its *in vivo* knockdown prevents progression of diabetic nephropathy. *J Biol Chem* 286:11837–48.

Lortz, S., Tiedge, M., Nachtwey, T. et al. 2000. Protection of insulin producing RINm5F cells against cytokine mediated antioxidant enzymes. *Diabetes* 49:1123–30.

Lovshin, J.A. and Drucker, D.J. 2009. Incretin-based therapies for type 2 diabetes mellitus. Nat Rev Endocr*inol* 5:262–69.

Mahapatra, D.K., Asati, V., and Bharti, S.K. 2015. Chalcones and their therapeutic targets for the management of diabetes: Structural and pharmacological perspectives. *Eur J Med Chem* 92:839–65.

Matschinsky, F.M., Magnuson, M.A., Zelent, D. et al. 2006. The network of glucokinase-expressing cells in glucose homeostasis and the potential of glucokinase activators for diabetes therapy. *Diabetes* 55:1–12.

Mazzone, T., Chait, A., and Plutzky, J. 2008. Cardiovascular disease risk in type 2 diabetes mellitus: Insight from mechanistic studies. *Lancet* 371:1800–09.

McKeown, S.C., Corbett, D.F., Goetz, A.S. et al. 2007. Solid phase synthesis and SAR of small molecule agonists for the GPR40 receptor. *Bioorg Med Chem Lett* 17:1584–89.

Mulvihill E.E. and Drucker, D.J. 2014. Pharmacology, physiology, and mechanisms of action of dipeptidyl peptidase-4 inhibitors. *Endocrinol Rev* 35:992–1019.

Muthenna, P., Suryanarayan, P., Gunda, S.K. et al. 2009. Inhibition of aldose reductase by dietary antioxidant curcumin: Mechanism of inhibition, significance, specificity and significance. *FEBS Lett* 583:3637–42.

Nadine, J.B. 2012. Mechanisms and techniques of reprogramming—Using PDX-1 homeobox protein as a novel treatment of insulin dependent diabetes mellitus, diabetes and metabolic syndrome. *Clin Res Rev* 6:113–19.

Nakai, H., Okuyama, M., Kim, Y. et al. 2005. Molecular analysis of α-glucosidase belonging to GH-family 31. *Biologia Bratislava* 60:131–35.

Nakayama, M., Castoe, T., Sosinowski, T. et al. 2012. Germline TRAV5D-4 T-cell receptor sequence targets a primary insulin peptide of NOD mice. *Diabetes* 61:857–65.

Okuyama, M., Okuno, A., Shimizu, N. et al. 2001. Carboxyl group of Asp-647 residues as possible proton donor in catalytic reaction of α-glucosidase from *Schizosaccharomyces pombe*. *Eur J Biochem* 268:2270–80.

Ong, K.L., Rye, K.A., O'Connell, R. et al. 2012. Long-term fenofibrate therapy increases fibroblast growth factor 21 and retinol-binding protein 4 in subjects with type 2 diabetes. *J Endocrinol Metab* 97:4701–08.

Orom, U.A., Kauppinen, S., and Lund, A.H. 2006. LNA-modified oligonucleotides mediate specific inhibition of microRNA function. *Gene* 372:137–41.

Parker, J.C., McPherson, R.K., Andrews, K.M. et al. 2000. Effects of skyrin, a receptor-selective glucagon antagonist, in rat and human hepatocytes. *Diabetes* 49:2079–86.

Patel, D.K., Kumar, R., Sairam, K. et al. 2012. Pharmacologically tested aldose reductase inhibitors isolated from plant sources—A concise report. *Chin J Nat Med* 10:388–400.

Pullen, T.J., da Silva Xavier, G., Kelsey, G. et al. 2011. miR-29a and miR-29b contribute to pancreatic beta-cell-specific silencing of monocarboxylate transporter 1 (Mct1). *Mol Cell Biol* 31:3182–94.

Qureshi, S.A., Rios, C.M., Xie, D. et al. 2004. A novel glucagon receptor antagonist inhibits glucagon mediated biological effects. *Diabetes* 53:3267–73.

Ramachandran, D., Roy, U., Garg, S. et al. 2011. Sirt1 and mir-9 expression is regulated during glucose-stimulated insulin secretion in pancreatic beta-islets. *FEBS J* 278:1167–74.

Ramírez-Espinosa, J.J., Rios, M.Y., López-Martínez, S. et al. 2011. Antidiabetic activity of some pentacyclic acid triterpenoids, role of PTPe1B: *In vitro, in silico*, and *in vivo* approaches. *Eur J Med Chem* 46:2243–51.

Reitman, M.L. 2013. FGF21 mimetic shows therapeutic promise. *Cell Metab* 18:307–09.

Ruan, Q., Wang, T., Kameswaran, V. et al. 2011. The microRNA-21-PDCD4 axis prevents type 1 diabetes by blocking pancreatic beta cell death. *Proc Nat Acad Sci USA* 108:12030–35.

Schreiber, S.N., Emter, R., Hock, M.B. et al. 2004. The estrogen-related receptor alpha (ERRα) functions in PPAR$_\gamma$ coactivator 1α (PGC-1α) induced mitochondrial biogenesis. *Proc Nat Acad Sci USA* 101:6472–77.

Schwartz, S.L., Akinlade, B., Klasen, S. et al. 2011. Safety, pharmacokinetic, and pharmacodynamic profiles of ipragliflozin (ASP1941), a novel and selective inhibitor of sodium-dependent glucose co-transporter 2, in patients with type 2 diabetes mellitus. *Diab Tech Ther* 12:1219–27.

Severi, F., Benvenuti, S., Costantino, L. et al. 1998. Synthesis and activity of a new series of chalcones as aldose reductase inhibitors. *J Diab Complications* 33:859–66.

Shakespear, M.R., Halili, M.A., Irvine, K.M. et al. 2011. Histone deacetylases as regulators of inflammation and immunity. *Trends Immunol* 32:335–43.

Sorensen, H., Brand, C.L., Neschen, S. et al. 2006. Immunoneutralization of endogenous glucagon reduces hepatic glucose output and improves long-term glycemic control in diabetic ob/ob mice. *Diabetes* 55:2843–48.

Tahrani, A.A., Piya, M.K., Kennedy, A. et al. 2010. Glycaemic control in type 2 diabetes: Targets and new therapies. *Pharmacol Ther* 125:328–61.

Tonks, N.K. 2003. PTP1B: From the sidelines to the front lines. *FEBS Lett* 546:140–48.

Trajkovski, M., Hausser, J., Soutschek, J. et al. 2011. MicroRNAs 103 and 107 regulate insulin sensitivity. *Nature* 474:649–53.

Wang, X., Wei, X., Pang, Q. et al. 2012. Histone deacetylases and their inhibitors: Molecular mechanisms and therapeutic implications in diabetes mellitus. *Acta Pharm Sinica B* 2:387–95.

Wani, J.H., John-Kalarickal, J., and Fonseca, V.A. 2008. Dipeptidyl peptidase-4 as a new target of action for type 2 diabetes mellitus: A systematic review. *Cardiol Clin* 26:639–48.

Wei, R., Yang, J., Liu, G.Q. et al. 2013. Dynamic expression of microRNAs during the differentiation of human embryonic stem cells into insulin-producing cells. *Gene* 518:246–55.

Wende, A.R., Huss, J.M., Schaeffer, P.J., Giguere, V., and Kelly, D.P. 2005. PGC-1α coacti-
    vates PDK4 gene expression via the orphan nuclear receptor ERRα: A mechanism for
    transcriptional control of muscle glucose metabolism. *Mol Cell Biol* 25:10684–94.

Wild, S., Roglic, G., Green, A., Sicree, R., and King, H. 2004. Global prevalence of diabetes:
    Estimates for the year 2000 and projections for 2030. *Diab Care* 27:1047–53.

Willy, P.J., Murray, I.R., Qian, J. et al. 2004. Regulation of PPAR$_\gamma$ coactivator 1α (PGC-1α)
    signaling by an estrogen-related receptor alpha (ERRα) ligand. *Proc Nat Acad Sci*
    101:8912–17.

Yina, Z., Zhanga, W., Fenga, F. et al. 2014. α-Glucosidase inhibitors isolated from medicinal
    plants. *Food Sci Hum Wellness* 3:136–74.

Zambrowicz, B., Ding, Z.M., Ogbaa, I. et al. 2013. Effects of LX4211, a dual SGLT1/SGLT2
    inhibitor, plus sitagliptin on postprandial active GLP-1 and glycemic control in type 2
    diabetes. *Clin Ther* 3:273–85.

Zampetaki, A., Kiechl, S., Drozdov, I. et al. 2010. Plasma microRNA profiling reveals loss
    of endothelial miR-126 and other microRNAs in type 2 diabetes. *Circ Res* 107:810–17.

Zeller, M., Steg, P.G., Ravisy, J. et al. 2008. Relation between body mass index, waist circum-
    ference, and death after acute myocardial infarction. *Circulation* 118:482–90.

Zhang, L., Jin, N., Nakayama, M. et al. 2010. Gamma delta T cell receptors confer autono-
    mous responsiveness to the insulin-peptide B:9-23. *J Autoimmun* 34:478–84.

Zhang, S. and Zhang, Z.Y. 2007. PTP1B as a drug target: Recent developments in PTP1B
    inhibitor discovery. *Drug Discov Today* 12:373–81.

Zhou, B., Li, C., Qi, W. et al. 2012. Downregulation of miR-181a upregulates sirtuin-1 (SIRT1)
    and improves hepatic insulin sensitivity. *Diabetologia* 55:2032–43.

Zhou, J., Kong, C., Wang, X. et al. 2013. *In silico* analysis of TCR Vβ7 of two patients with
    type 1 diabetes mellitus. *J Lab Phys* 5:79–82.

Zhu, C. 2013. Aldose reductase inhibitors as potential therapeutic drugs of diabetic complica-
    tions. *InTech* 2:17–46.

# 4 Prodrug Strategy
## *An Effective Tool in Drug Delivery and Targeting*

*Buddhadev Layek, Shubhajit Paul,
and Jagdish Singh*

## CONTENTS

## 4.1   INTRODUCTION

During the past few decades, drug development has experienced a major paradigm shift by changing from a trial-and-error method to a systematic hypothesis-driven computational approach. Although the computational strategy results in lead molecule identification with maximized biological activity, there is still a large fraction of lead molecules rejection during the development phase owing to poor pharmacokinetics, toxicity, and the lack of efficacy (Kerns and Di 2003). Today, numerous physicochemical methods are adopted to modulate inherent properties of the drug molecules. Among them, prodrugs are recognized as an effective strategy to conceal the unwanted properties of the prospective drug candidates (Rautio et al. 2008).

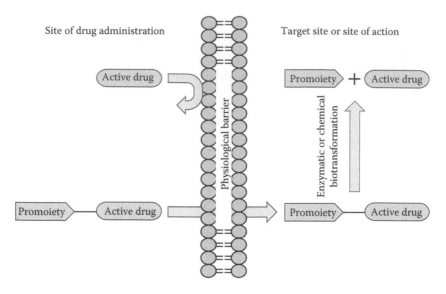

**FIGURE 4.1**   A schematic representation of the prodrug approach.

Prodrugs refer to a class of pharmacologically inactive or less active derivatives of active drugs which must undergo biotransformation *in vivo* to exert their therapeutic effects (Figure 4.1; Rautio et al. 2008). Depending on the objective, prodrugs are designed to release their active moiety before, after, or during their absorption or at the specific target organ or tissue. The main purpose of the prodrug design is to improve bioavailability and reduce adverse effects of the parent molecules (Huttunen et al. 2011). In the past, the prodrug approach has been considered as a last option problem-solving tool after lead optimization when the preferred drug candidate has failed in pharmacokinetics or preclinical studies. Recently, prodrugs are increasingly being integrated into the early stages of drug development, rather than using them as a post hoc approach. An estimated 10% of the globally marketed medicines fall into the prodrug category; in 2008 alone, nearly one-third of all the new approved small-molecule drugs were prodrugs (Huttunen et al. 2011). This fact indicates the immense success of the prodrug strategy in drug delivery.

## 4.2   APPLICATIONS OF PRODRUGS

Prodrugs have emerged as a powerful tool to enhance the drug-like properties of a pharmacologically active agent, and thus increase the outcomes of drug develop-ment. In addition, prodrugs may be employed to increase the target specificity, which not only improves their efficacy but also reduces toxicity. Hence, prodrug strategy could be highly useful for treatments such as chemotherapy, which is often accom-panied by serious undesirable side effects. Development of prodrugs with superior qualities may also provide a life cycle improvement opportunity. Thus, the prodrug approach becomes much more viable, economical, and less tedious to strengthen the druggability of the existing molecules (Karaman et al. 2013). In the following section, we briefly discuss the applications of prodrug strategy in drug development.

## 4.2.1  IMPROVED FORMULATION AND ADMINISTRATION

It has been stated that approximately 40% of the drug candidates derived from combinatorial approach exhibited very poor aqueous solubility (Rautio et al. 2008). Sometimes traditional formulation techniques such as particle size reduction, modification of crystal form, salt formation, solubilizing excipients, complexation agents, and cocrystallization cannot provide adequate solubility. In these circumstances, prodrugs provide an alternative technique to ameliorate the solubility issue of the parent compound by increasing their dissolution rate. This can be accomplished by attaching an ionizable or neutral polar moiety such as phosphate, amino acid, or sugar molecules to the active compound (Fleisher et al. 1996; Stella and Nti-Addae 2007; Muller 2009). Phosphate ester prodrugs are frequently used to improve the aqueous solubility of sparingly soluble drugs that contain a hydroxyl or amine functional group (Stella and Nti-Addae 2007). These prodrugs are rapidly hydrolyzed to their active molecules by the action of endogenous enzymes, for instance alkaline phosphatases, which exist in the intestinal brush border, liver, or plasma (Muller 2009; Testa 2009; Bobeck et al. 2010; Fasinu et al. 2011). Prednisolone sodium phosphate (Figure 4.2a) is an excellent example of the phosphate ester prodrug, which shows 30-fold greater water solubility than prednisolone. The phosphate group is directly attached to the free hydroxyl group of prednisolone. Fosphenytoin sodium is another phosphate prodrug where the phosphate group is connected to an amine group of phenytoin via an oxymethylene linker. It is intended to decrease precipitation of phenytoin and resultant tissue irritation at the site of injection. It exhibited solubility 7000 times higher than phenytoin and undergoes rapid hydrolysis to release free phenytoin (Figure 4.2b; Unadkat and Rowland 1985; Boucher 1996; Browne et al. 1996).

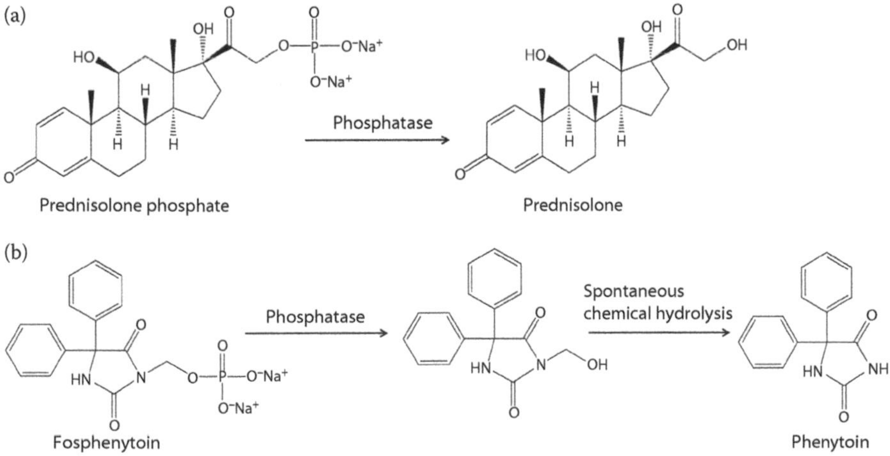

**FIGURE 4.2**  Bioactivation mechanisms of (a) prednisolone phosphate and (b) fosphenytoin.

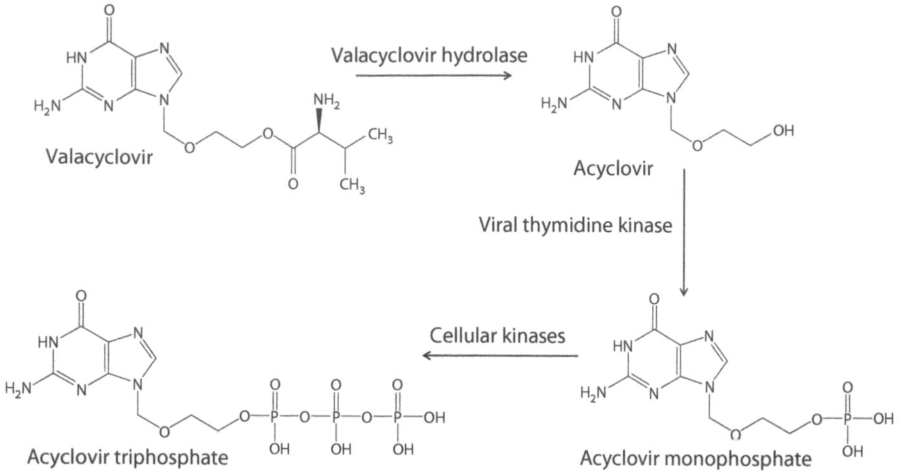

**FIGURE 4.3**   Activation of valacyclovir to its active metabolite, acyclovir triphosphate.

Amino acid prodrugs are extensively used for improving oral bioavailability of drugs with poor water solubility and permeability. The majority of them are either amides or esters in which a carboxyl or amine group of an amino acid is chemically conjugated to amine, carboxyl, hydroxyl, or thiol functionalities of active drug molecules. Various hydrolytic enzymes, including amidases, esterases, and peptidases present in different tissues can readily hydrolyze these prodrugs into their parent molecules (Huttunen et al. 2011). Examples of such prodrugs include valacyclovir and valganciclovir, which are L-valyl esters of the antiviral agent acyclovir and ganciclovir, respectively (Figure 4.3; Soul-Lawton et al. 1995; Pescovitz et al. 2000). Besides having increased water solubility, both of these prodrugs are also substrates of peptide transporters (PEPT1) located in the brush-border membranes of intestinal epithelium that further improved their oral bioavailability (Umapathy et al. 2004; Muller 2009). The amide prodrug midodrine consists of a glycine promoiety linking to the amine functionality of desglymidodrine. Midodrine undergoes enzymatic bioconversion in various tissues, including liver to release the active desglymidodrine molecule (Figure 4.4; Cruz 2000). It is primarily absorbed via PEPT1-mediated transport, which increases its oral bioavailability to 93%, in contrast to 50% for free desglymidodrine (Tsuda et al. 2006). Different sugars such as glucose, galactose, glucuronic acid, and mannose have also been explored to increase solubility of sparingly soluble drugs.

## 4.2.2   Enhanced Permeability, Absorption, and Distribution

A drug has to pass through several biological membranes to reach its target site(s); hence, membrane permeability plays a significant role in its efficacy. After oral and topical administration, most of the drugs are absorbed via passive diffusion. Drug absorption depends on several physicochemical parameters, the two most important of which are solubility and lipophilicity (Bergstrom 2005; Mannhold 2005).

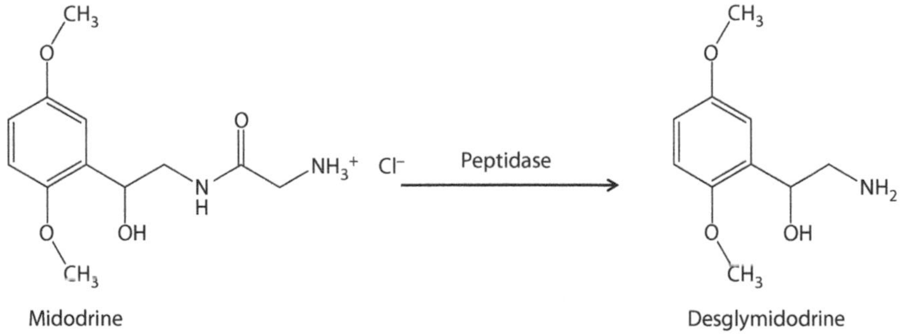

**FIGURE 4.4** Enzymatic hydrolysis of midodrine into its active component, desglymidodrine.

In general, highly lipophilic drugs cross cell membranes more easily through passive transcellular route. The lipophilicity of a drug molecule can be improved by either shielding the polar functional groups or varying the hydrocarbon moieties (Beaumont et al. 2003). One of the most common prodrug approaches used to improve lipophilicity of a hydroxyl, carboxyl, thiol, phosphate, or amine group containing hydrophilic drugs is to convert them into highly lipophilic alkyl or aryl esters (Taylor 1996). These ester prodrugs are rapidly hydrolyzed to their parent drugs by different esterases found all over the body (Taylor 1996; Liederer and Borchardt 2006). Most importantly, lipophilicity of the prodrugs could be precisely controlled by altering the length of alkyl chain.

Oseltamivir is an ethyl ester prodrug used for the prevention and treatment of uncomplicated influenza. Following oral absorption, oseltamivir is rapidly hydrolyzed to its active form oseltamivir carboxylate primarily by carboxylesterase (Figure 4.5a). The oral bioavailability of free carboxylate is only 5%, while that of the oseltamivir is over 80% (McClellan and Perry 2001; Shi et al. 2006). Similarly, adefovir dipivoxil is an orally active phosphoryl diester prodrug indicated for the treatment of chronic hepatitis B virus infection, which demonstrated about four times higher bioavailability as compared to adefovir (Benzaria et al. 1996; Noble and Goa 1999).

## 4.2.3 Prolonged Duration of Drug Action

The use of modified or sustained release formulations offers several advantages, including improved patient compliance due to less frequent dosing and avoidance of toxic exposure to the body by maintaining a constant drug level in the blood. A traditional approach of making sustained release formulation is the synthesis of long-chain aliphatic or polyethylene glycol ester prodrug. The slow hydrolysis of these prodrugs renders them suitable for intramuscular injection (Filpula and Zhao 2008). Sustained release formulations are highly desirable in the treatment of psychoses, as the patients need long-term use of medication that often leads to a high degree of patient noncompliance. An excellent example is haloperidol, which is an orally delivered antipsychotic drug indicated for mental disorders. The peak plasma concentration of this drug is attained within 2–6 h after oral administration. However,

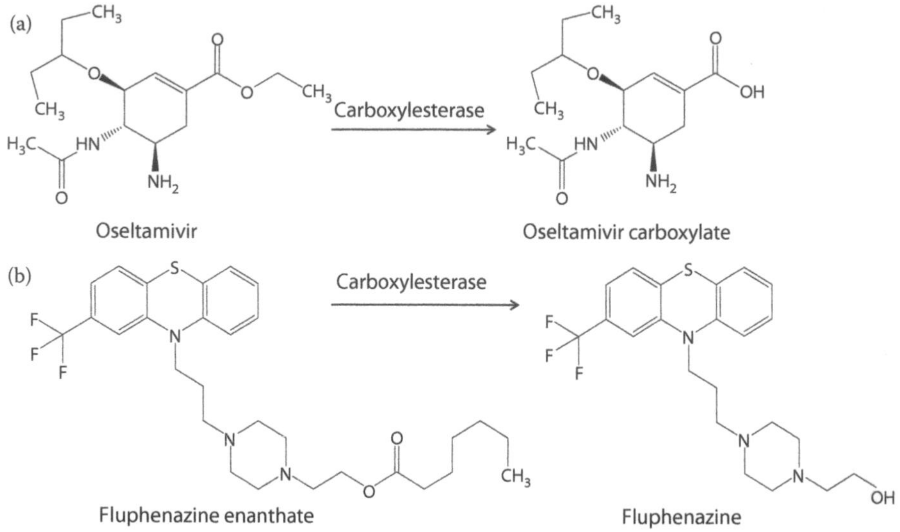

**FIGURE 4.5**  Bioactivation pathways of (a) oseltamivir and (b) fluphenazine enanthate.

the sesame oil formulation of haloperidol decanoate, an ester prodrug, maintains the antipsychotic activity over a month after a single intramuscular injection (Deberdt et al. 1980). A similar approach was implemented for fluphenazine, which itself exhibits a short duration of activity (6–8 h); however, its ester prodrugs fluphenazine enanthate (Figure 4.5b) and fluphenazine decanoate maintain therapeutic activity over a month. Another example of sustained release prodrug is glycine-conjugated anti-inflammatory drug tolmetin sodium, which exhibits duration of action up to 9 h, while the parent drug shows a half-life of about 1 h (Persico et al. 1988).

## 4.2.4  INCREASED STABILITY

Prodrugs can be exploited for improving stability of a particular drug candidate, more frequently to avoid the first-pass effect. In first-pass effect (also known as pre-systemic metabolism), a significant fraction of the drug is metabolized in the liver whereupon a small portion of the drug reaches the systemic circulation to elicit the desired effect. The bioavailability of orally administered propranolol, an antihypertensive drug, is much lower than the intravenous injection due to extensive presystemic metabolism. Cytochrome P450 (CYP) isozymes are involved in its metabolism and the major metabolites are 4-hydroxypropranolol, N-desisopropyl propranolol, and propranolol O-glucuronide. Improved stability was achieved by esterification to hemisuccinate propranolol which hindered the glucuronide formation resulting in eight times increase in plasma concentration than propranolol itself (Garceau et al. 1978).

Naltrexone, an opiate receptor antagonist, is primarily used for the treatment of opioid addiction. Although quickly and well absorbed from the gastrointestinal tract (GIT), naltrexone undergoes extensive first-pass metabolism following oral dosing.

On the contrary, anthranilate and acetylsalicylate esters exhibited significant improvement of naltrexone oral bioavailability (45 times and 28 times greater than naltrexone, respectively; Hussain et al. 1987). Another classic example is the prodrug formulation of penicillin antibiotics. Penicillin G is very unstable in aqueous solution, whereas its procaine and benzathine salts are chemically stable and formulated in suspension form owing to their low aqueous solubility (Stella et al. 1985).

### 4.2.5 MINIMIZED TOXICITY

Active drugs should not present any serious toxicity accompanied with their therapeutic application. Prodrug approaches can be efficiently exploited to minimize toxicity of a specific drug candidate. Epinephrine is used to treat glaucoma, which can cause several ophthalmic and systemic adverse effects. However, its ester prodrug dipivalylepinephrine exhibited enhanced bioavailability and lesser side effects than free epinephrine (Figure 4.6; Mandell et al. 1978). Similar approaches were persuaded for a number of anti-inflammatory drugs where gastric irritation and bleeding are the most serious adverse reactions accompanied with their application. Esterification of aspirin and other nonsteroidal anti-inflammatory agents were found to significantly suppress the ulcerogenic activity (Nielsen and Bundgaard 1989).

### 4.2.6 IMPROVED PATIENT COMPLIANCE

Patient compliance ensures the therapeutic efficacy of a drug molecule by maintaining its intake as guided by the physician. Patient noncompliance generally arises from the painful injection and unpleasant taste or odor, which persuades patients to follow a medication regimen irregularly resulting in therapeutic inefficacy. An excellent example is the antibacterial drug clindamycin, where its phosphate prodrug is generally well absorbed in the body, but it is not well accepted by children due to its bitter taste (DeHaan et al. 1973). This prompted the synthesis of clindamycin palmitate, which is formulated in cherry-flavored syrup for pediatric use (Sinkula et al. 1973). Clindamycin palmitate is tasteless, which is also less water-soluble, resulting in minimal interaction with the saliva and taste receptor. Antibacterial drug sulfisoxazole is also bitter in taste, while its prodrug acetyl sulfisoxazole is tasteless and

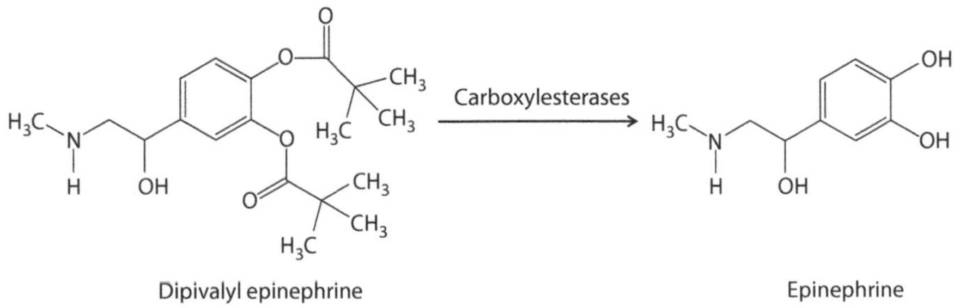

Dipivalyl epinephrine

Epinephrine

**FIGURE 4.6** Structure of dipivalyl epinephrine and its conversion to epinephrine.

commonly used in pediatric medication (Stella et al. 1985). Similar approaches are employed for bitter-tasting drugs chloramphenicol and erythromycin, where their palmitate and succinate salts are tasteless (Glazko et al. 1952; Stella et al. 1985).

### 4.2.7 LIFE CYCLE MANAGEMENT

The term life cycle management (LCM) refers to prolonging the market exclusivity period. The LCM of drugs is highly important for pharmaceutical industries striving for maximizing the revenue from their products. Prodrugs are likely to have the benefit of being considered as new chemical entities and hence securing intellectual property rights against unlawful competition in the course of their development and commercialization. Moreover, prodrug development is much more economical and faster as compared to developing a newly synthesized drug molecule. Therefore, the concept of prodrugs has been utilized successfully for LCM of several approved drug molecules or drugs in the development phase. In 1996, Parke-Davis (Division of Warner-Lambert Company) introduced fosphenytoin as a safer alternative to parenteral phenytoin sodium, thus recapturing the market position that was earlier lost when injectable phenytoin sodium became generic (Boucher 1996). Similarly, GlaxoSmithKline designed phosphate ester prodrug of amprenavir, fosamprenavir, for the treatment of human immunodeficiency virus (HIV; Figure 4.7). This prodrug was originally developed to improve oral bioavailability of amprenavir (Chapman et al. 2004; Wire et al. 2006). Additionally, the prodrug approach helps to extend the patent life of fosamprenavir to 2019, while the patent of amprenavir itself has expired in 2013 (Huttunen et al. 2011).

### 4.2.8 PRODRUGS FOR SITE-SPECIFIC DRUG DELIVERY

The primary goal of the site-specific delivery is to sustain and confine the drug action within the affected tissue. For decades, numerous active and passive targeted delivery systems have been evaluated for improving therapeutic efficacy of drug while reducing its toxicity. These strategies involve antibody or carrier protein-grafted nanocarriers (active targeting), liposomes, nanoparticles, dendrimers, and micelles (passive targeting). However, the overall success rate of these approaches is often questionable. In recent years, the site-specific prodrug design has emerged

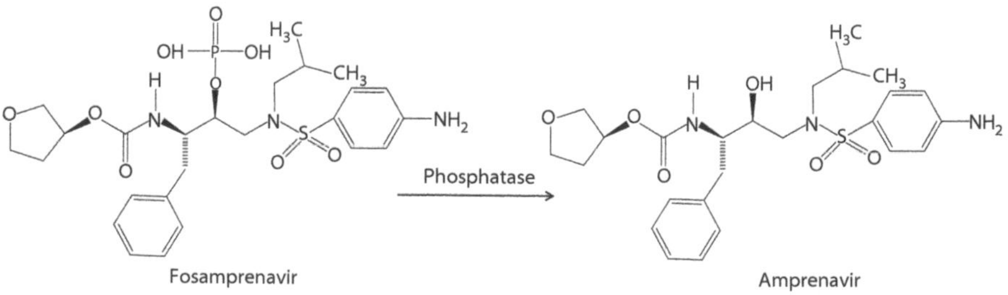

**FIGURE 4.7**   Bioactivation of fosamprenavir to active component, amprenavir.

as a potential targeting strategy. Site selectivity of prodrugs can be attained by targeting tissue-/cell-specific transporters or antigen or by selective metabolic activation through tissue-/cell-specific enzymes. We have briefly discussed a few widely investigated strategies for tumor, liver, kidney, colon, brain, and virus targeting in the section below.

## 4.3  PRODRUGS FOR SITE-SPECIFIC DRUG DELIVERY

### 4.3.1  TUMOR-TARGETED PRODRUGS

Cancer refers to a group of more than 100 diseases involving uncontrolled growth and spread of abnormal cells. It is one of the leading causes of death worldwide and is often treated by surgery, chemotherapy, radiation therapy, or a suitable combination of these approaches. Traditional chemotherapy drugs primarily act by killing cells that are dividing rapidly. Unfortunately, a chemotherapeutic agent does not distinguish between cancer cells and fast growing normal, healthy cells. The healthy cells most likely to be damaged by chemotherapy are blood cells, cells of the GIT, hair follicles, and reproductive tract; resulting in low blood cell counts, mouth soreness, nausea, vomiting, loss of appetite, diarrhea, hair loss, or change of sexuality and fertility. Another major issue is the increasing evidence of drug resistance among various tumor cells. Therefore, the development of tumor-targeted drug delivery systems could be highly valuable to overcome the aforementioned adverse effects while improving the efficacy.

Suitable prodrug design is a promising approach that could enhance the tumor selectivity of chemotherapeutic agents. Prodrugs are intended to target specific antigens, enzymes, receptors, or transporters, which are overexpressed in the tumor cells. This is usually performed by conjugating chemotherapeutic agents with tumor-selective ligands using a cleavable moiety. Among different ligands, monoclonal antibodies (mAbs) are most widely used for tumor targeting due to their high binding affinity for corresponding antigens (Weiner 2007). Several tumor-associated antigens and cell surface receptors such as CD19, CD22, epidermal growth factor receptor, human epidermal growth factor receptor 2, mucin-type glycoproteins 1, prostate-specific antigen, and αv integrin have been widely used for tumor targeting (Mahato et al. 2011). These antibody–drug complexes are mostly inactive and require additional activation in the body. Numerous cytotoxic anticancer drugs, including anthracyclines (Dillman et al. 1988), taxanes (Quiles et al. 2010), vinca alkaloids (Johnson and Laguzza 1987), antifolates (Shen et al. 1986), monomethyl auristatin E (Senter 2009), and calicheamicin (Hamann et al. 2005; Boghaert et al. 2008) have been delivered as antibody conjugates.

Besides antigens, there are several cell surface receptors that provide an effective target for tumor-selective prodrug delivery. Folic acid (FA) is one of the most widely utilized targeting ligands because of its high binding affinity for folate receptors (FR) even after conjugation with drugs or other carrier molecules. Elevated expression of FR has been observed in many tumor cells, including brain, breast, colon, kidney, lung, and ovarian cancer. The expression of FR in healthy tissues is either absent or limited to only some epithelial cell types. Several antitumor drugs such as alkylating agents (Steinberg and Borch 2001), taxol (Lee et al. 2002), camptothecin (Henne et al. 2006), 5-fluorouracil (5-FU) analog (Liu et al. 2001), platinum compounds (Aronov et al. 2003), mitomycin (Reddy et al. 2006), and doxorubicin (Du et al. 2013) have been coupled with FA for efficient targeting of FR-positive tumor cells.

Due to increased proliferation rate, levels of several endogenous enzymes are elevated in tumor cells that can serve as potential targets for tumor-specific prodrug delivery (Rooseboom et al. 2004). Capecitabine is an orally active prodrug of 5-FU with improved tumor selectivity. It is absorbed intact through the intestinal mucosa and bioconverted to the active compound by three sequential enzymatic steps. The bioconversion pathway starts in the liver, where human carboxylesterases 1 and 2 convert it into 5′-deoxy-5-fluorocytidine, then cytidine deaminase further metabolizes it to 5′-deoxy-5-fluorouridine in the liver and tumors, and finally to parent drug 5-FU by tumor-specific thymidine phosphorylase (Miwa et al. 1998). Therefore, an elevated level of 5-FU is produced at the tumor site with minimal exposure of healthy tissues to 5-FU, resulting in reduction in systemic toxicity.

Doxorubicin is an anthracycline antitumor antibiotic effective against a large number of malignant tumors, although the life-threatening cardiac toxicity limits its long-term applications. One potential approach to overcome the cardiac toxicity of doxorubicin is the use of a glucuronide prodrug (HMR 1826) that is cleaved by β-glucuronidase to release doxorubicin in the necrotic regions of tumors (Bosslet et al. 1998). HMR 1826 demonstrated 7-fold higher tumor selectivity index and 100 times less cardiotoxicity than doxorubicin (Murdter et al. 1997; Platel et al. 1999). However, low circulation half-life and bioconversion only at the inflammatory cell infiltration sites have restricted its clinical application (Huang and Oliff 2001).

To expand the scope of enzyme-targeted prodrug therapy, tumor-specific delivery of prodrug-converting exogenous enzymes is accomplished by a number of methods. The most frequently used methods are antibody-directed enzyme prodrug therapy (ADEPT) and gene-directed enzyme prodrug therapy (GDEPT). ADEPT is a two-step treatment strategy where a tumor-specific antibody-conjugated prodrug-converting enzyme is transported first for selective binding to the tumor cells. In a subsequent step, the inactive prodrug is delivered systemically for tumor-selective activation of the cytotoxic drug by the action of localized pretargeted enzyme. The main benefit of ADEPT strategy is that a single antibody–enzyme conjugate molecule can catalyze bioactivation of numerous prodrug molecules. The bystander effect is another potential advantage where the active cytotoxic drug released by the pretargeted enzyme complex would diffuse to kill the adjacent tumor cells that either are devoid of tumor antigen or failed to internalize the enzyme–antibody conjugate. The ADEPT has been extensively used for tumor-specific delivery of various chemotherapy drugs, including doxorubicin, dinitrobenzamides, and paclitaxel. Wang et al. (2001) synthesized a β-lactam prodrug of CC-1065, a potent antitumor agent, linked to a cephalosporin. The prodrug showed 10 times higher $IC_{50}$ value against U937 leukemia cells than the corresponding free drug, indicating better safety profile of the prodrug. The authors anticipated that the prodrug would be activated to its parent drug by the action of β-lactamases localized in tumor tissues using an antibody. As a component of the enzyme/prodrug system, 5-fluorocytosine (5-FC) has been utilized in conjugation with cytosine deaminase to generate biologically relevant concentrations of 5-FU locally. Cytosine deaminase has also been coupled with a single-chain fragment variable (scFv) antibody as a potential candidate for tumor-selective prodrug activation tool (Coelho et al. 2007; Zamboni et al. 2008). The major limitations of ADEPT technology include the enzyme–antibody conjugate-associated immunogenicity, conjugation heterogeneity, and off-target activation of prodrugs by the unbound conjugate (Mahato et al. 2011). The immune responses related to bacterial enzymes and murine antibodies could be eliminated by using human enzymes in combination with humanized antibodies (Alvarez et al. 2002).

GDEPT is a two-step treatment strategy where the gene that encodes a prodrug-activating nonendogenous enzyme is selectively transferred to target tissues. The most commonly used gene delivery vectors are cationic lipids, peptides, and viral vectors (i.e., retroviruses, adenoviruses, and adeno-associated viruses). Subsequently, inactive prodrug is delivered systemically and converted to its active cytotoxic drug by the nonendogenous enzyme expressed in the target tissues. The effectiveness of GDEPT depends on several factors, including the level of enzyme expression at the target site, prodrug activation efficiency, and the capability of the active cytotoxic drug to diffuse into adjacent cells. The most frequently used enzyme/prodrug systems include herpes simplex virus (HSV)-thymidine kinase/ganciclovir, cytosine deaminase/5-FC, carboxypeptidase G2/nitrogen mustard, and nitroreductase/CB1954.

A phase I study for breast cancer treatment revealed that GDEPT approach can be effectively used in humans without any local or systemic adverse effects (Mahato et al. 2011). The gene expression cassette consists of a proximal ERBB2 promoter to specifically express cytosine deaminase enzyme in ERBB2-positive cancer cells, followed by systemic delivery of 5-FC that is bioconverted to release 5-FU in cancer cells. The expression of cytosine deaminase was highly restricted to the cancer cells. Most importantly, patients who received the GDEPT showed substantial tumor regression. In addition, this therapy was devoid of anti-DNA antibody formation, insertional mutagenesis, or tumor nodule ulceration.

The cellular pH difference between normal tissue and most of the solid tumors has been well recognized. A slightly acidic pH is observed in solid tumors primarily due to increased fermentative metabolism and reduced perfusion (Estrella et al. 2013). Hence, formulations sensitive to the acidic pH environment of the tumor tissues can be utilized as a prospective tumor-targeting strategy. Similarly, the delivery systems that are responsive to the acidic pH of endosome or lysosome can also be designed for tumor-selective drug release. Gao et al. (2005) have exploited the acidic pH environment of solid tumor to achieve tumor selective delivery of doxorubicin-loaded polymeric micelles. The *in vivo* pharmacokinetics data demonstrated that the pH responsive polymeric micelles increased both half-life and area under the plasma concentration-time curve (AUC) of doxorubicin approximately 5.2- and 5.8-fold, respectively, as compared to a free doxorubicin solution. Moreover, after an intravenous injection in nude mice, doxorubicin-loaded pH responsive micelles were predominantly accumulated in tumors. Most importantly, the pH-sensitive micelle formulation significantly inhibited the growth of A2780 tumor xenografts in nude mice without serious weight loss.

In a recent study, FA-functionalized bovine serum albumin (BSA)-conjugated, pH-sensitive doxorubicin prodrug, FA-BSA-*cis*-aconitic anhydride (CAD), was prepared for tumor-targeted drug delivery (Du et al. 2013). BSA was used to increase aqueous solubility of the prodrug while CAD served as an acid-cleavable linkage between doxorubicin and BSA. The FA was attached to BSA for improving tumor specificity of the conjugate. The *in vitro* studies demonstrated a pH-dependent release of doxorubicin under different pH conditions with significantly faster release up to pH 5.5. The prodrug conjugate was internalized into cancer cells through FR-mediated endocytosis, and at the acidic pH of endosomes, the prodrug was cleaved to release free doxorubicin. Consequently, the therapeutic efficacy of the FA-BSA-CAD conjugate on FR-positive tumor was 2-fold higher as compared to free doxorubicin.

## 4.3.2 LIVER-TARGETED PRODRUGS

The liver plays a critical role in drug metabolism and detoxification in the body. It is an excellent repository of different carriers, metabolizing enzymes, and transporters that could be exploited for liver-targeted delivery to enhance safety and efficacy of drug products (Worman 2006). Hepatocytes are the structural units of the liver responsible for bioconversion of lipophilic xenobiotics into more hydrophilic metabolites that can be eliminated via urine or bile (Farber 1987). Several acute and chronic liver disorders (e.g., multiple viral infections, overdose of alcohol, drugs, or toxins), metabolic syndromes (e.g., diabetes and hyperlipidemia), and chronic congenital diseases (e.g., Gilbert syndrome, hemochromatosis, and Wilson disease) tremendously affect normal functioning of the liver and translate to high rate of morbidity (Huttunen and Rautio 2011). Therefore, liver-targeted drug delivery systems facilitate complete bioconversion of the drug of interest to ensure higher bioavailability and fewer incidence of adverse side effects.

Several enzymes, transporters, and receptors present on the surface of liver cells have been efficiently exploited for improved bioavailability and attenuation of risk exposure in different disease conditions. The most successful strategy is to target the liver-selective enzymes, for instance, CYP which metabolizes several prodrugs into their active counterparts. The important P450 enzymes include CYP3A4, 3A5, 1A2, 2B6, 2C8, 2C9, 2C19, 2D6, and 2E1. Among them, CYP3A4 is the most predominant isoform, which accounts for oxidation of about two-thirds of the existing drugs (Ortiz de Montellano 2013). The preferential substrate characteristics of the CYP enzymes have been listed in Table 4.1.

The CYPs catalyze the oxidative and reductive metabolism of numerous xenobiotics, including anticancer drugs, antihistamines, anti-inflammatory drugs, cardiovascular drugs, and many more (Evans and Relling 1999). Anticancer drugs, cyclophosphamides, dacarbazine, tegafur, and duomycin; antiplatelet drugs, clopidogrel; antihistamines, loratadine and terfenadine; and anti-inflammatory drug, nabumetone undergo oxidative bioconversion to their active metabolites mainly by CYP1A2, 3A4, and 2D6 enzymes (Ortiz de Montellano 2013). The CYP enzyme-mediated activation of anticancer prodrugs such as cyclophosphamide, ifosfamide, and trofosfamide led to the establishment of novel phosphate and phosphonate prodrugs, also known as HepDirect prodrugs (Erion et al. 2004). HepDirect prodrugs

## TABLE 4.1
## CYP Enzymes and Their Substrate Characteristics

| CYP Enzyme | Substrate Characteristics |
|---|---|
| 1A2 | Planar molecules, neutral or basic in character |
| 2A6 | Diverse, relatively small neutral or basic molecules usually containing one aromatic ring |
| 2B6 | Angular, medium-sized neutral or basic molecules with 1–2 hydrogen bond donor/acceptor atoms |
| 2C8 | Relatively large, elongated molecules, mostly either acidic or neutral in character |
| 2C9 | Medium-sized acidic molecules with 1–2 hydrogen bond acceptors |
| 2C19 | Medium-sized molecules, mostly basic with 2–3 hydrogen bond acceptors |
| 2D6 | Medium-sized basic molecules with protonatable nitrogen 5–7 Å from site of metabolism |
| 2E1 | Relatively small neutral molecules, structurally diverse in character |
| 3A4 | Relatively large, structurally diverse molecules |

*Source:* Reprinted from *Drug Discov Today*, 7, Lewis, D.F. and Dickins, M., Substrate SARs in human P450s, 918–925, Copyright 2002, with permission from Elsevier.

showed excellent stability in plasma and other tissues *in vivo* while rapidly oxidized by liver-specific CYP enzymes such as CYP3A to produce high concentrations of therapeutically active nucleoside triphosphate in the liver. Higher liver nucleoside triphosphate levels are anticipated to enhance efficacy, while lower off-target nucleoside triphosphate levels decrease the risk of dose-dependent toxicity. In another study, Erion et al. (2005) have demonstrated liver-targeting potential of two HepDirect prodrugs like remofovir (prodrug of adefovir) and MB07133 (a prodrug of cytarabine 5′-monophosphate). These prodrugs exhibited higher liver to intestine and liver to kidney ratios in comparison with their active counterparts. Furthermore, pradefovir, a prodrug of adefovir dipivoxil, exhibited marked therapeutic efficiency in hepatitis B patients with lesser extent of nephrotoxicity (Reddy et al. 2008), whereas adefovir dipivoxil was administered at a submaximal dose to reduce nephrotoxicity with compromised therapeutic efficacy. Similarly, MB07133 is expected to undergo bioactivation to cytarabine triphosphate in the hepatocarcinoma cells. *In vivo* experiments in mice had shown a much higher level of cytarabine triphosphate in the liver and a lower concentration in bone marrow, thereby resulting in a decreased toxicity and improved efficacy.

Recently, bile acid transporters have been widely exploited for their liver-targeting potential. Floxuridine is a promising chemotherapeutic agent for colorectal cancer; however, it is spontaneously biotransformed by thymidine phosphorylase to 5-FU, which is 5000 times less potent than floxuridine itself. It also shows low and variable bioavailability (34%–47%) in orally dosed patients. Vivian and Polli (2014) explored the prodrug development possibilities of floxuridine to specifically target Na+/taurocholate cotransporting polypeptide (NTCP) transporter system responsible for hepatocellular uptake of more than 80% of bile acid conjugates. Floxuridine-5′-glutamic acid–chenodeoxycholic acid (CDCA) and floxuridine-3′-glutamic acid–CDCA were synthesized by conjugating CDCA to floxuridine. These prodrugs exhibited excellent stability in rat plasma while releasing floxuridine by esterase degradation in rat liver S9 fraction. Rais et al. (2011) earlier showed a similar approach with gabapentin by targeting to human apical sodium-dependent bile acid transporters (hASBT). The zwitterionic nature of gabapentin exhibits variable absorption with rapid excretion and short half-life, which requires frequent dosing, causing noncompliance in patients. CDCA-based prodrugs of gabapentin were synthesized and cellular uptake potential was evaluated in stably transfected hASBT–Madin–Darby canine kidney (MDCK) epithelial cells. These prodrugs demonstrated increased absorption of gabapentin via a hASBT uptake pathway. The same research group recently reported a similar prodrug strategy to achieve liver-specific delivery of ribavirin (Dong et al. 2015). The therapeutic application of ribavirin is restricted due to occurrence of dose-dependent hemolytic anemia by irreversible phosphorylation of ribavirin inside red blood cells (RBCs), resulting in depletion of adenosine triphosphate (ATP) and oxidative damage to RBCs. Six prodrugs of ribavirin were synthesized using CDCA as a linker which showed strong substrate specificity to NTCP *in vivo*. Among different prodrugs, L-val–glycochenodeoxycholic acid conjugate exhibited comparable ribavirin release to parent drug in mouse liver S9 fraction, while significantly decreasing its accumulation in RBCs indicating its potential as a liver-specific prodrug.

### 4.3.3 KIDNEY-TARGETED PRODRUGS

Renal targeting is highly desired to minimize the extrarenal side effects or to improve intrarenal delivery of the drugs used in renal diseases (Haas et al. 2002). Additionally, renal targeting could be useful for drugs that undergo rapid metabolic inactivation before reaching their target site in kidneys and abolish or reduce the impacts of disease conditions, including proteinuria, on normal renal distribution of the drug (Narang and Mahato 2010). Hence, kidney-targeted prodrugs play a significant role to enhance the therapeutic efficacy and safety of the active drugs. Renal specificity of prodrugs is primarily dependent on the renal-specific uptake and/or metabolism of the promoiety. Proximal tubular cells and glomerular mesangial cells are the primary targets for renal-specific drug development as these cells are involved in the pathophysiology of numerous renal diseases (Haas et al. 2002).

Certain endogenous enzymes such as γ-glutamyl transpeptidase, N-acetyl transferase, β-lyase, or L-amino acid decarboxylase have relatively high concentrations in the kidneys. Therefore, substrates of these enzymes are often chemically attached to parent drug molecules with the hope that the resulting prodrugs would become therapeutically active in the kidney via action of the relevant enzyme. L-γ-glutamyl-L-dopa (gludopa) is a dopamine precursor which is initially metabolized to L-dopa by γ-glutamyl transpeptidase in the kidney, and subsequently to dopamine by aromatic L-amino acid decarboxylase (Figure 4.8; Wilk et al. 1978). Hence, gludopa could be utilized to achieve renal vasodilation without affecting systemic blood pressure. However, the clinical application of gludopa is restricted by its poor oral bioavailability (Lee 1990).

To improve kidney targeting, *p*-nitroaniline, sulfamethizole, and sulfamethoxazole were either conjugated with L-γ-glutamyl or N-acetyl-L-γ-glutamyl promoiety and their pharmacokinetics profiles were investigated in rats after intravenous

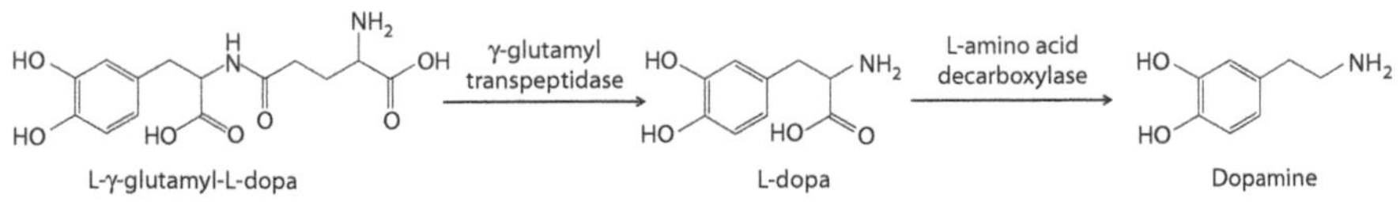

**FIGURE 4.8** Metabolic formation of dopamine from L-γ-glutamyl-L-dopa (gludopa).

administration (Murakami et al. 1998). For all the compounds, N-acetyl-L-γ-glutamyl-based prodrugs displayed superior plasma stability than their L-γ-glutamyl counterparts. The concentrations of parent drugs were higher in the kidney than in the liver and lung for all compounds with significantly increased kidney delivery of N-acetyl-L-γ-glutamyl prodrugs of *p*-nitroaniline and sulfamethoxazole.

Similarly, Su et al. (2003) synthesized an N-acetyl-L-glutamyl derivative of prednisolone and studied its *in vivo* distribution after intravenous administration to Kunming mice. The side effects were assessed by measuring the bone mineral densities of treated animals. The results demonstrated that, compared with parent drug prednisolone, the renal concentration of prodrug was increased markedly and the incidence of prednisolone-induced osteoporosis was significantly reduced.

To improve kidney-targeted delivery of captopril (CAP), a prodrug of CAP was developed by conjugating the drug to G3-C12 (ANTPCGPYTHDCPVKR) carrier peptide (Geng et al. 2012). The G3-C12-CAP conjugate resulted in increased intrarenal drug concentration and renal angiotensin-converting enzyme (ACE) inhibition than free CAP after intravenous bolus administration in mice. These results imply that G3-C12 peptide could serve as a promising carrier moiety for kidney-specific formulation development.

Sugar recognition plays a key role in cell–matrix, cell–cell, and cell–molecule interactions along with receptor-mediated endocytosis (Zhou et al. 2014). This persuaded Suzuki and co-workers to study glycoconjugates as potential renal-specific delivery vectors (Suzuki et al. 1999a, b; Shirota et al. 2001). The model drug arginine vasopressin (AVP) was modified by conjugating it to a variety of glycosylated derivatives via an octamethylene group. Their results suggested that the alkylglucoside moiety was essential for renal specificity and the targeting efficacy was reliant on the properties of the glycosylated moieties such as the length of the alkyl chain, the structure of the peptide, and nature of the linkage. Moreover, the target efficiency of the vector was greatly affected by the size and charge of the ligand molecule with optimal efficacy for small and neutral molecules.

In another study, a series of zidovudine (AZT)–chitosan oligomer (AZT–COS) conjugates was synthesized and evaluated by Liang et al. (2012) for their renal-targeting capacity. *In vitro* release studies had shown that the AZT–COS conjugates successfully released AZT in both mouse plasma and renal homogenate, and also in human plasma. Furthermore, as demonstrated in mouse pharmacokinetics studies, the mean retention time of different AZT–COS conjugates were about 2.5-fold greater than the AZT control group. The biodistribution studies also confirmed an increased accumulation of ATZ–COS conjugates in the kidney than in the lung, liver, heart, and brain following intravenous administration in mice.

To improve kidney-targeting efficacy of prednisolone, a 2-deoxy-2-aminodi-glucose–prednisolone conjugate (DPC) was prepared (Lin et al. 2012). The tissue distribution experiments demonstrated that the DPC possessed an excellent kidney-targeting capacity, which resulted in a 4.9-fold higher prednisolone concentration in the DPC group as compared to the free prednisolone group. These results suggested that 2-deoxy-2-aminodiglucose could serve as a potential renal-targeting drug carrier. In another study, prednisolone carbamate was conjugated with 2-glucosamine and its potential targeting ability was evaluated (PCG; Lin et al. 2013). *In vitro* studies showed that PCG significantly increased the prednisolone uptake into kidney-derived cells. Moreover, following intravenous injection, the PCG group showed 8.1 times higher prednisolone concentration in the kidney than free prednisolone.

### 4.3.4 COLON-TARGETED PRODRUGS

Colon-targeted drug delivery is highly useful for successful treatment of several bowel diseases, including amebiasis, colorectal cancer, Crohn's disease, and ulcerative colitis (Philip et al. 2009). The purpose of colonic delivery is to inhibit the release and absorption of the drug in the upper GIT and facilitate rapid release of the parent drug at the proximal colon (Chourasia and Jain 2003). Several approaches, including pH-, pressure-, redox-, or time-sensitive polymers and enzyme-activated formulations (e.g., coatings, prodrugs, or matrices) have been extensively explored for developing colon-specific delivery systems (Jung et al. 2001; Chourasia and Jain 2003; Friend 2005).

Mesalazine or 5-aminosalicylic acid (5-ASA) is the first-choice anti-inflammatory agent for treatment of ulcerative colitis and Crohn's disease. However, it is primarily absorbed from the upper intestine and only a small portion of the dose reaches to the colon (Crotty and Jewell 1992). Moreover, absorbed 5-ASA can lead to various adverse effects, in particular nephrotoxicity (Novis et al. 1988). Hence, different prodrug approaches have been attempted to deliver 5-ASA selectively to the colon. Salicylazosulfapyridine (sulfasalazine) is an azo prodrug of 5-ASA, which is minimally absorbed from the upper GIT owing to its hydrophilic nature and the unabsorbed portion of the prodrug is cleaved to 5-ASA and sulfapyridine by azoreductase in the colon (Figure 4.9; Van Hees et al. 1981). The toxicity associated with sulfapyridine promoiety has prompted the synthesis of several other prodrugs with safer promoieties such as balsalazide, ipsalazine, and olsalazine (van Hogezand et al. 1985; McIntyre et al. 1988). Dexamethasone 21-β-D-glucoside is another example

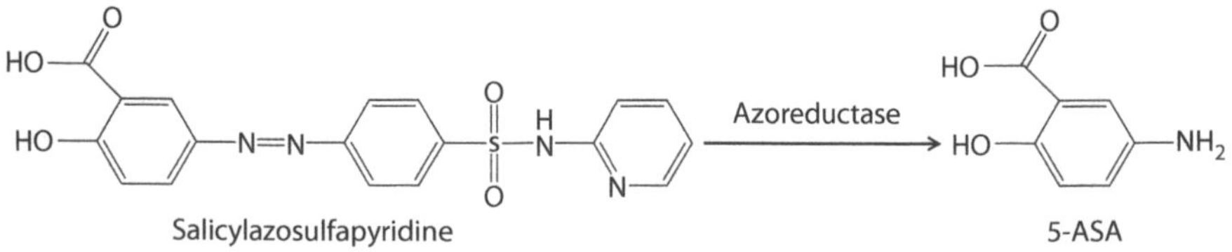

**FIGURE 4.9**    Release of 5-ASA from salicylazosulfapyridine by azoreductase.

of a hydrophilic colon-targeted prodrug, which is bioactivated to its active form by bacterial glucosidases in the colon (Haeberlin et al. 1993).

Several polysaccharide-based prodrugs have also been synthesized for colon-targeted delivery of 5-ASA (Zou et al. 2005). Three pharmacologically inactive polysaccharides: chitosan, hydroxyl propyl cellulose (HPC), and cyclodextrins, were used as drug carriers. The influence of solubility differences of these prodrugs on the release profile of 5-ASA was assessed using rat gastrointestinal contents. Irrespective of gastrointestinal contents, both chitosan-5-ASA and HPC-5-ASA formulations have been found to prevent the release of 5-ASA, while cyclodextrin formulation exhibited greater release of 5-ASA in colonic and cecal contents as compared to stomach and intestinal contents. Furthermore, as the substitution degree increases, the solubility of cyclodextrins-5-ASA decreased significantly, and no detectable amount of 5-ASA was released even after 48 h of incubation with the colonic content. It is suggested that a substitution degree less than 30 could efficiently maintain the balance between colon-specific drug release and 5-ASA loading in the polymer matrix.

A novel 5-aminosalicyltaurine prodrug was developed by Jung et al. (2006) for effective therapy of experimental colitis. Taurine is a free amino acid which can protect colon tissues from damage in the pathophysiology of irritable bowel syndrome (IBD). The prodrug was found to be biochemically stable and enhanced the colon-targeted delivery of 5-ASA. In addition, the prodrug significantly reduced the systemic exposure of 5-ASA. Consequently, the prodrug system enhanced the response of 5-ASA against experimental colitis in rats and had a slightly better efficacy than sulfasalazine.

To improve colon specificity and reduce gastric side effects of carboxylic group containing nonsteroidal anti-inflammatory agents, a glycine-conjugated prodrug of flurbiprofen has been synthesized and evaluated by Philip et al. (2008). Besides acting as a promoiety, glycine is a potent anti-inflammatory, cytoprotective, and immunomodulatory agent, and thus enhances the pharmacological effects of flurbiprofen. The prodrug had higher aqueous solubility than flurbiprofen and showed a tendency to increase with increasing pH. *In vitro* conversion experiments showed that the prodrug remained biochemically stable until colonic pH was achieved, when the amide bond of the prodrug was hydrolyzed by amidase to release free flurbiprofen. Consequently, the prodrug has shown significantly less toxicity and ulcerogenicity in albino rats than the parent drug.

Novel microsphere formulation of pectin–metronidazole prodrug was reported by Vaidya et al. (2015). Pectin is a linear polysaccharide which is resistant to mammalian digestive enzymes, but degraded by bacteria in the colon. Microspheres of pectin-conjugated metronidazole were prepared by water in oil (w/o) emulsification method. The prodrug containing microspheres were small, uniform, and exhibited increased drug loading efficiency (~94%). The *in vitro* release studies demonstrated no detectable amount of metronidazole in the simulated gastric fluid from the prodrug formulation, while approximately 93% of the drug was released when the drug was physically entrapped in pectin microspheres. Recovery analysis in albino rats revealed that about 68% of the drug from prodrug formulation was released over 8 h particularly in the colonic region, suggesting the relevance of this formulation over simple physically entrapped metronidazole in pectin microspheres.

Sharma et al. (2013) reported the synthesis of novel azo prodrugs of gemcitabine, methotrexate, and an analog of oxaliplatin to treat colorectal cancer. These prodrugs remained intact in both acidic (pH 1.2) and slightly basic (pH 7.4) buffers, indicating their stability in the upper GIT conditions. Furthermore, the results of azoreductase assay confirmed that the azoreductase enzyme present in the rat cecal, fecal, or intestinal contents specifically cleaved the azo bond to release active drug molecules. After 24 h of incubation at 37°C, overall 60%–70% of drug release was noticed from these prodrugs. Most importantly, all the prodrugs exhibited comparable efficacy with their respective parent drug on human colorectal cell lines.

### 4.3.5 BRAIN-TARGETED DRUG DELIVERY

The occurrence of several neuronal and cerebrovascular diseases, including amyotrophic lateral sclerosis, Alzheimer's, Huntington's, and Parkinson's diseases have become alarmingly high during the last decades. Effective intervention toward these diseases is an immense challenge attributed to the inherent characteristics of the central nervous system (CNS). The primary hurdle in effective drug delivery to the CNS is the presence of the blood–brain barrier (BBB). Less than 1% of all Food and Drug Administration (FDA)-approved drugs are amenable to cross the intact BBB, among which a majority of them are not the ideal remedies for CNS diseases, substantiating a huge lack of treatment available for these pathological conditions (Dove 2008; Pathan et al. 2009; Gabathuler 2010). The BBB is a permeability barrier composed of densely packed endothelial cells that allows selective delivery of oxygen, glucose, and amino acids into the brain via specific transporters. In addition, capillary endothelial cells express several metabolic enzymes along with multiple potent efflux transporters such as P-glycoprotein and multidrug resistance-associated protein. Conclusively, for a successful delivery to the brain, the drug molecule should be small in size with significant lipid permeability, and not vulnerable to the action of efflux transporters (Garcia-Garcia et al. 2005).

One of the promising strategies to enhance the CNS-specific drug delivery is the design of prodrugs to target-specific endogenous transporters (Stockwell et al. 2014). This can be achieved by either receptor-mediated transport (RMT) or carrier-mediated transport (CMT). In general, proteins and peptides follow RMT pathway using growth factor receptor, insulin receptor, and transferrin receptor, whereas CMT is evident for transport of small drug molecules (MW < 600 Da; Pavan et al. 2008). In the CMT approach, the active drug molecule could be converted to a pseudonutrient prodrug or the drug could conjugate with the nutrient substrate present in the CMT system (Vlieghe and Khrestchatisky 2013). Among various CMT transporters, amino acid and glucose transporters have been extensively studied for CNS-specific drug delivery. L-amino acid transporter 1 (LAT1) is known to enhance the CMT of large neutral amino acid molecules into the brain (Gynther et al. 2008). A typical example of LAT1-targeted prodrug is L-dopa, which is a precursor of dopamine and commonly used for the treatment of Parkinson's disease. Once L-dopa has reached the brain, it is bioconverted to dopamine by the enzyme L-amino acid decarboxylase. Prodrug strategy to the CNS could also be employed by conjugating the active drug molecule with an endogenous transporter substrate. Phenylalanine derivative of valproic acid and L-lysine derivative of ketoprofen were successfully targeted to the brain than their corresponding active counterparts due to efficient LAT1-mediated transport (Gynther et al. 2010; Peura et al. 2011).

Glucose is a crucial energy source for normal functioning of the brain, which utilizes GLUT transporters to cross the BBB. GLUT1 is the primary glucose transporters constituting about 90% of the expressed GLUT transporters. Bilsky et al. (2000) reported the increased transport of opioid agonist peptides to the brain after their conjugation with D-glucose moiety. In another study, Bonina et al. (2003) had developed a glycosyl L-dopa derivative which exhibited better reversing effect against hypolocomotion in reserpine-induced rat model compared to nonglycosylated counterpart.

The transporters for cationic amino acids (cationic amino acid transporter, CAT1), monocarboxylic acids (monocarboxylic acid transporter 1, MCT1), and nucleosides (concentrative nucleoside transporter 2, CNT2) are mainly predominantly located at the BBB, while the transporter for ascorbic acid (sodium-dependent vitamin C transporter 2, SVCT2) is mainly found in the choroid plexus. Ascorbic acid is an essential nutrient for the brain, eyes, and spinal cord, which contributes to the myelin sheath formation and also acts as a neuromodulator of cholinergic, dopaminergic, gamma-amino butyric acid (GABA)-ergic, and glutamatergic transmission. It is not produced in the human body, but taken up by SVCT transporters (Rice 2000). The absorption and recovery of ascorbic acid are facilitated by SVCT1, while its accumulation in the brain and eyes is driven by SVCT2. Ascorbic acid-conjugated diclophenamic, kynurenic, and nipecotic acid demonstrated several-fold higher brain uptake than respective parent drug when used for the treatment of Alzheimer's disease, epilepsy, and Parkinson's disease (Manfredini et al. 2002). Zhao et al. (2014) recently reported the utilization of both GLUT1 and SVCT2 pathways for novel ibuprofen prodrugs. They used thiamine disulfide system (TDS) having a lock-in potential to prevent brain to blood pump out of the drug, thereby maintaining enhanced CNS activity. TDS was reduced by disulfide reductase followed by ring closure to produce thiazolium, which is locked in the brain and cannot be transported across the BBB. Thus, TDS was introduced into the ascorbic acid moiety to form a prodrug of the latter, which was then linked with ibuprofen via esterification. The prodrug was cleaved by esterase and exhibited enhanced pharmacological efficacy.

The BBB possesses a large variety of enzymes such as esterase, adenosine deaminase, and diverse oxidases (CYP enzymes, monoamine oxidase, and xanthine oxidase), which can be exploited as a biotransformation system to transport poorly permeable drugs to the brain. Morgan et al. (1992) showed that adenosine deaminase-activated prodrug 6-chloro-2',3'-dideoxypurine significantly enhanced the CNS delivery of 2',3'-dideoxyinosine due to higher activity of adenosine deaminase in the brain compared to plasma. To improve the brain delivery of 2'-F-ara-2',3'-dideoxyinosine, 2'-F-ara-2',3'-dideoxypurine prodrug was synthesized by Shanmuganathan et al. (1994). The bioconversion of the prodrug was completed by xanthine oxidase which led to significantly higher CNS concentration of 2'-F-ara-2',3'-dideoxyinosine after oral delivery of 2'-F-ara-2',3'-dideoxypurine in experimental mice. Milacemide is a glycine prodrug that readily equilibrates across the BBB. Milacemide that reaches the CNS compartment is rapidly metabolized by monoamine oxidase B to glycinamide, which is subsequently converted to glycine (Semba et al. 1993).

Lipidization followed by enzymatic cleavage via esterases has been recognized as a promising strategy for CNS delivery of opioids. O-methylation or O-acetylation of morphine results in codeine or heroin, respectively, which have been shown to improve the BBB permeability up to 10- to 100-fold (Oldendorf et al. 1972). Both codeine and heroin are bioconverted to morphine before interacting with the opioid receptors in the brain.

CNS delivery of peptides is extremely difficult because of their hydrophilic characteristics and susceptibility toward various peptidases expressed in the capillary endothelium. S-acetylthiorphan and benzyl ester of S-acetylthiorphan (acetorphan) are prodrugs of thiorphan that have higher lipophilicity, hence facilitate their brain delivery. Following CNS delivery, these prodrugs undergo hydrolytic cleavage by esterases to release thiorphan (Lecomte et al. 1986; Lambert et al. 1993). To improve CNS delivery in Alzheimer's disease, Deguchi et al. (2000) synthesized a lipophilic prodrug of ketoprofen (1,3-Diacetyl-2-ketoprofen glyceride) which efficiently crossed the BBB and rapidly hydrolyzed to ketoprofen, exhibiting its potent anti-inflammatory activity.

Prokai-Tatrai et al. (2013) reported the improved pharmacological activity of a novel thyrotropin-releasing hormone analog by the replacement of its N-terminal pyroglutamyl moiety with pyridinium-based moieties. This prodrug was preferentially oxidized to the parent charged molecule in the brain. The oxidized prodrug was expected to be rapidly cleared from the periphery because of its ionic nature, while restricting its efflux from the brain.

### 4.3.6  VIRUS-SPECIFIC DRUG DELIVERY

Design of safe and effective antiviral drugs is challenging because viruses use the host cells to replicate. This makes it hard to find targets for the drugs that would specifically interact with the virus without affecting the host cells. The modern antiviral drug development relies on the identification of viral proteins or segments of proteins that could be inactivated. To avoid the risk of undesired side effects, these *target proteins* should be as different from human proteins (or parts of proteins) as possible. Nowadays, prodrug strategy has been adopted successfully in developing effective viral-specific drug delivery systems.

The discovery of acyclovir, a purine nucleoside analog, opened a new era in antiviral chemotherapy because of its high viral specificity and nontoxicity to host cells (Elion et al. 1977; Schaeffer et al. 1978). It is mainly used for the treatment of HSV and varicella zoster virus (VZV) infection (Nilsen et al. 1982; Serota et al. 1982). The viral-specific thymidine kinase converts acyclovir to its monophosphate, which is subsequently phosphorylated into triphosphate by cellular kinases (Figure 4.3). Cellular kinases can also initiate the phosphorylation of acyclovir but to a negligible extent. Acyclovir triphosphate, in turn, selectively inhibits HSV-specific DNA polymerase (10- to 30-fold more efficiently than cellular DNA polymerase) and prevents viral replication. These features collectively attribute to the acyclovir's selectivity. Consequently, a 3000-fold higher concentration of acyclovir is required to inhibit the replication of uninfected host cells as compared to virus-infected cells (Elion et al. 1977).

Penciclovir is a structural analog of acyclovir, which is also used to treat HSV or VZV infections. Like acyclovir, penciclovir is first converted to its monophosphate derivative by viral thymidine kinase and subsequently to penciclovir triphosphate by cellular kinases (De Clercq and Field 2006). This activated drug inhibits viral DNA polymerase, thus impairing the capability of the virus to replicate within the host cell.

Nitric oxide is an important biologically active molecule involved in various physiological and pathological processes. It also plays a critical role in host immunity against viruses, bacteria, protozoa, and tumor cells (Torre et al. 2002). Sialated diazeniumdiolate is a good example of virus-targeted investigational nitric oxide donor which could be utilized to inhibit replication of influenza viruses (Cai et al. 2004). Influenza viruses have hemagglutinin and neuraminidase in their lipid bilayer that could bind and cut the terminal sialic acid to release nitric oxide in the close proximity of influenza viruses.

## 4.4 RECENT PROGRESS, CHALLENGES, AND FUTURE TRENDS IN PRODRUG DEVELOPMENT

Prodrug strategies have become an emerging trend over the last decade as evidenced by the surge in a number of publications and patents. Most of the newly launched prodrugs were designed to improve their bioavailability. Table 4.2 represents a non-exhaustive list of the recently approved prodrugs.

Tafluprost, abiraterone acetate, azilsartan medoxomil, gabapentin enacarbil, ceftaroline fosamil, dabigatran etexilate, and fingolimod exhibited several-fold higher bioavailability over their corresponding parent drugs (Clas et al. 2014). Identification of specific enzymes that execute the bioconversion of prodrugs such as aminopeptidases, alkaline phosphatases, carboxylesterases, cholinesterases, and retinyl ester hydrolases also facilitated better prodrug design (Imai and Ohura 2010). However, these enzymes often display very high inter- and intraindividual differences in activity, as a result of genetic polymorphisms and environmental factors. Therefore, these parameters can significantly influence the prodrug bioconversion and, subsequently, the efficacy of the prodrug. For example, esterified prodrugs temocapril and oseltamivir exhibited higher plasma concentrations following their bioactivation by liver esterases rather than in the small intestine, highlighting the importance of site of bioconversion. Another classic example is the anticoagulant prodrug dabigatran etexilate, which, unlike ximelagatran is not a CYP substrate. By avoiding CYP-mediated metabolism, dabigatran etexilate has shown lesser extent of side effects compared to ximelagatran (Imai and Hosokawa 2010).

In spite of such intense approaches, the total number of approved prodrugs comprises a small fraction of the marketed drugs, suggesting the challenges faced in understanding, recognition, and translatability of prodrug strategies to address the limitations of a parent drug. A better understanding of these issues requires a definite approach to attain the desired goal with proper experimental strategies such as determining prodrug's stability and solubility in formulation and at absorption site, recognizing the bioactivation site(s), and toxicity evaluation of the breakdown products (Beaumont et al. 2003). The productivity of this huge experimental work could be

# TABLE 4.2
## List of Approved Prodrugs by the US FDA, 2005 Onward

| Prodrug Name (Therapeutic Area) | Prodrug Strategy | Structure | Reference |
|---|---|---|---|
| Abiraterone acetate (hormonal drug for prostate cancer) | • Ester prodrug of abiraterone<br>• Improved bioavailability | | Stappaerts et al. (2015) |
| Azilsartan medoxomil (antihypertensive) | • 4-(hydroxymethyl)-5-methyl-1,3-dioxol-2-one prodrug of azilsartan<br>• Modulated metabolic profile | | Perry (2012) |

*(Continued)*

**TABLE 4.2 (Continued)**
**List of Approved Prodrugs by the US FDA, 2005 Onward**

| Prodrug Name (Therapeutic Area) | Prodrug Strategy | Structure | Reference |
|---|---|---|---|
| Ceftarcline fosamil (broad-spectrum antibiotic) | • Phosphoramidate prodrug of ceftaroline<br>• Improved solubility more than 50-fold relative to parent drug | | Laudano (2011) |
| Dabigatran etexilate (anticcagulant) | • Hexyl carbamate and ethyl ester of dabigatran<br>• Improved oral bioavailability | | Wienen et al. (2007) |

(Continued)

**TABLE 4.2 (Continued)**
**List of Approved Prodrugs by the US FDA, 2005 Onward**

| Prodrug Name (Therapeutic Area) | Prodrug Strategy | Structure | Reference |
|---|---|---|---|
| Fingolimod (immunomodulatory drug used in multiple sclerosis) | • Phosphorylated in vivo to sphingosine-1-phosphate<br>• Able to cross BBB and reduces autoaggresive lymphocyte infiltration in CNS | | Kennedy et al. (2011) |
| Fosamprenavir (anti-HIV) | • Phosphate ester of amprenavir<br>• Improved patient compliance and 10-fold increase in aqueous solubility | | Wire et al. (2006) |
| Fosaprepitant dimeglumine (antiemetic) | • Phosphate-based meglumine (sorbitol-derived amino sugar) salt of aprepitant<br>• Increased aqueous solubility | | Colon-Gonzalez and Kraft (2010) |

(Continued)

## TABLE 4.2 (Continued)
### List of Approved Prodrugs by the US FDA, 2005 Onward

| Prodrug Name (Therapeutic Area) | Prodrug Strategy | Structure | Reference |
|---|---|---|---|
| Gabapentin enacarbil (CNS drug for restless leg syndrome) | • Isobutanoyl oxyethoxy carbamate of gabapentin<br>• Improved bioavailability due to better absorption in the GIT | | Lal et al. (2010) |
| Isavuconazonium sulfate (antifungal) | • Water-soluble azaheterocycle sulfate prodrug of isavuconazole<br>• Exhibiting 98% oral bioavailability and highly effective against invasive fungal diseases | | McCormack (2015) |
| Lisdexamfetamine dimesylate (CNS stimulant used to treat attention-deficit hyperactivity disorder) | • Lysine-conjugated mesylate salt of dextroamphetamine<br>• Reduced abuse potential | | Mattingly (2010) |

*(Continued)*

Bio-Targets and Drug Delivery Approaches

**TABLE 4.2 (*Continued*)**
**List of Approved Prodrugs by the US FDA, 2005 Onward**

| Prodrug Name (Therapeutic Area) | Prodrug Strategy | Structure | Reference |
|---|---|---|---|
| Midodrine (vasopressor) | • Glycyl amide of desglymidodrine<br>• Oral bioavailability enhanced by ~2-fold | | Werling and Chalasani (2011) |
| Phosphonooxymethyl propofol (anesthetic) | • Phosphonooxymethyl ether of propofol<br>• Rapid conversion to propofol by alkaline phosphatase and 3-fold increase in aqueous solubility | | Kumpulainen et al. (2008) |
| Pradefovir mesylate (antiviral) | • 2-(3-chlorophenyl)-[1,2,3] dioxaphosphinane of adefovir<br>• Improved liver targeting ability | | Reddy et al. (2008) |

*(Continued)*

## TABLE 4.2 (*Continued*)
## List of Approved Prodrugs by the US FDA, 2005 Onward

| Prodrug Name (Therapeutic Area) | Prodrug Strategy | Structure | Reference |
|---|---|---|---|
| Tafluprost (antiglaucoma) | • Ester prodrug of tafluprost acid<br>• Increased permeability | | Fukano and Kawazu (2009) |
| Ximelagatran (anticoagulant) | • Hydroxyamidine and ethyl ester of melagatran<br>• Improved bioavailability (4–5-fold greater than melagatran) | | Ho and Brighton (2006) |

*Note:* CNS, central nervous system; CYP, cytochrome P450; GIT, gastrointestinal tract; HIV, human immunodeficiency virus.

enhanced using modern assay techniques, high-throughput screening (HTS), and *in silico* modeling. Lynch highlighted the importance of modern assay techniques and HTS in the selection of optimized lead molecule PF-543, a sphingosine-1-phosphate kinase inhibitor, from a chemical library of more than half a million compounds (Lynch 2012). Another key consideration is the evaluation of safety and tolerability followed by a projection of the human dose. A good understanding of the metabolic products after bioconversion also needs to be accounted for in the design of prodrugs. Encompassing the abovementioned strategies, a good deal of deliberation needs to be put forth by formulation teams and scientists for further productive outcomes.

## REFERENCES

Alvarez, R.D., Huh, W.K., Khazaeli, M.B. et al. 2002. A Phase I study of combined modality (90)Yttrium-CC49 intraperitoneal radioimmunotherapy for ovarian cancer. *Clin Cancer Res* 8:2806–11.

Aronov, O., Horowitz, A. T., Gabizon, A. et al. 2003. Folate-targeted PEG as a potential carrier for carboplatin analogs. Synthesis and *in vitro* studies. *Bioconjug Chem* 14:563–74.

Beaumont, K., Webster, R., Gardner, I. et al. 2003. Design of ester prodrugs to enhance oral absorption of poorly permeable compounds: Challenges to the discovery scientist. *Curr Drug Metab* 4:461–85.

Benzaria, S., Pelicano, H., Johnson, R. et al. 1996. Synthesis, in vitro antiviral evaluation, and stability studies of bis(S-acyl-2-thioethyl) ester derivatives of 9-[2-(phosphonomethoxy)ethyl]adenine (PMEA) as potential PMEA prodrugs with improved oral bioavailability. *J Med Chem* 39:4958–65.

Bergstrom, C.A. 2005. In silico predictions of drug solubility and permeability: Two rate-limiting barriers to oral drug absorption. *Basic Clin Pharmacol Toxicol* 96:156–61.

Bilsky, E.J., Egleton, R.D., Mitchell, S.A. et al. 2000. Enkephalin glycopeptide analogues produce analgesia with reduced dependence liability. *J Med Chem* 43:2586–90.

Bobeck, D.R., Schinazi, R.F., and Coats, S.J. 2010. Advances in nucleoside monophosphate prodrugs as anti-HCV agents. *Antivir Ther* 15:935–50.

Boghaert, E.R., Sridharan, L., Khandke, K.M. et al. 2008. The oncofetal protein, 5T4, is a suitable target for antibody-guided anti-cancer chemotherapy with calicheamicin. *Int J Oncol* 32:221–34.

Bonina, F., Puglia, C., Rimoli, M.G. et al. 2003. Glycosyl derivatives of dopamine and L-dopa as anti-Parkinson prodrugs: Synthesis, pharmacological activity and in vitro stability studies. *J Drug Target* 11:25–36.

Bosslet, K., Straub, R., Blumrich, M. et al. 1998. Elucidation of the mechanism enabling tumor selective prodrug monotherapy. *Cancer Res* 58:1195–201.

Boucher, B.A. 1996. Fosphenytoin: A novel phenytoin prodrug. *Pharmacotherapy* 16:777–91.

Browne, T.R., Kugler, A.R., and Eldon, M.A. 1996. Pharmacology and pharmacokinetics of fosphenytoin. *Neurology* 46:S3–7.

Cai, T.B., Lu, D., Landerholm, M. et al. 2004. Sialated diazeniumdiolate: A new sialidase-activated nitric oxide donor. *Org Lett* 6:4203–5.

Chapman, T.M., Plosker, G.L., and Perry, C.M. 2004. Fosamprenavir: A review of its use in the management of antiretroviral therapy-naive patients with HIV infection. *Drugs* 64:2101–24.

Chourasia, M.K. and Jain, S.K. 2003. Pharmaceutical approaches to colon targeted drug delivery systems. *J Pharm Pharm Sci* 6:33–66.

Clas, S.D., Sanchez, R.I., and Nofsinger, R. 2014. Chemistry-enabled drug delivery (prodrugs): Recent progress and challenges. *Drug Discov Today* 19:79–87.

Coelho, V., Dernedde, J., Petrausch, U. et al. 2007. Design, construction, and in vitro analysis of A33scFv::CDy, a recombinant fusion protein for antibody-directed enzyme prodrug therapy in colon cancer. *Int J Oncol* 31:951–7.

Colon-Gonzalez, F. and Kraft, W.K. 2010. Pharmacokinetic evaluation of fosaprepitant dimeglumine. *Exp Opin Drug Metab Toxicol* 6:1277–86.

Crotty, B., and Jewell, D. P. 1992. Drug therapy of ulcerative colitis. *Br J Clin Pharmacol* 34:189–98.

Cruz, D.N. 2000. Midodrine: A selective alpha-adrenergic agonist for orthostatic hypotension and dialysis hypotension. *Exp Opin Pharmacother* 1:835–40.

De Clercq, E. and Field, H.J. 2006. Antiviral prodrugs—the development of successful prodrug strategies for antiviral chemotherapy. *Br J Pharmacol* 147:1–11.

Deberdt, R., Elens, P., Berghmans, W. et al. 1980. Intramuscular haloperidol decanoate for neuroleptic maintenance therapy. Efficacy, dosage schedule and plasma levels. An open multicenter study. *Acta Psychiatr Scand* 62:356–63.

Deguchi, Y., Hayashi, H., Fujii, S. et al. 2000. Improved brain delivery of a nonsteroidal anti-inflammatory drug with a synthetic glyceride ester: A preliminary attempt at a CNS drug delivery system for the therapy of Alzheimer's disease. *J Drug Target* 8:371–81.

DeHaan, R.M., Metzler, C.M., Schellenberg, D. et al. 1973. Pharmacokinetic studies of clindamycin phosphate. *J Clin Pharmacol* 13:190–209.

Dillman, R.O., Johnson, D.E., Shawler, D.L. et al. 1988. Superiority of an acid-labile daunorubicin-monoclonal antibody immunoconjugate compared to free drug. *Cancer Res* 48:6097–102.

Dong, Z., Li, Q., Guo, D. et al. 2015. Synthesis and evaluation of bile acid-ribavirin conjugates as prodrugs to target the liver. *J Pharm Sci* 104:2864–76.

Dove, A. 2008. Breaching the barrier. *Nat Biotechnol* 26:1213–5.

Du, C., Deng, D., Shan, L. et al. 2013. A pH-sensitive doxorubicin prodrug based on folate-conjugated BSA for tumor-targeted drug delivery. *Biomaterials* 34:3087–97.

Elion, G.B., Furman, P.A., Fyfe, J.A. et al. 1977. Selectivity of action of an antiherpetic agent, 9-(2-hydroxyethoxymethyl) guanine. *Proc Natl Acad Sci U S A* 74:5716–20.

Erion, M.D., Reddy, K.R., Boyer, S.H. et al. 2004. Design, synthesis, and characterization of a series of cytochrome P(450) 3A-activated prodrugs (HepDirect prodrugs) useful for targeting phosph(on)ate-based drugs to the liver. *J Am Chem Soc* 126:5154–63.

Erion, M.D., van Poelje, P.D., Mackenna, D.A. et al. 2005. Liver-targeted drug delivery using HepDirect prodrugs. *J Pharmacol Exp Ther* 312:554–60.

Estrella, V., Chen, T., Lloyd, M. et al. 2013. Acidity generated by the tumor microenvironment drives local invasion. *Cancer Res* 73:1524–35.

Evans, W.E. and Relling, M.V. 1999. Pharmacogenomics: Translating functional genomics into rational therapeutics. *Science* 286:487–91.

Farber, J.L. 1987. Xenobiotics, drug metabolism, and liver injury. *Monogr Pathol* 28:43–53.

Fasinu, P., Pillay, V., Ndesendo, V.M. et al. 2011. Diverse approaches for the enhancement of oral drug bioavailability. *Biopharm Drug Dispos* 32:185–09.

Filpula, D. and Zhao, H. 2008. Releasable PEGylation of proteins with customized linkers. *Adv Drug Deliv Rev* 60:29–49.

Fleisher, D., Bong, R., and Stewart, B.H. 1996. Improved oral drug delivery: Solubility limitations overcome by the use of prodrugs. *Adv Drug Deliv Rev* 19:115–30.

Friend, D.R. 2005. New oral delivery systems for treatment of inflammatory bowel disease. *Adv Drug Deliv Rev* 57:247–65.

Fukano, Y. and Kawazu, K. 2009. Disposition and metabolism of a novel prostanoid anti-glaucoma medication, tafluprost, following ocular administration to rats. *Drug Metab Dispos* 37:1622–34.

Gabathuler, R. 2010. Approaches to transport therapeutic drugs across the blood-brain barrier to treat brain diseases. *Neurobiol Dis* 37:48–57.

Gao, Z.G., Lee, D.H., Kim, D.I. et al. 2005. Doxorubicin loaded pH-sensitive micelle targeting acidic extracellular pH of human ovarian A2780 tumor in mice. *J Drug Target* 13:391–97.

Garceau, Y., Davis, I., and Hasegawa, J. 1978. Plasma propranolol levels in beagle dogs after administration of propranolol hemisuccinate ester. *J Pharm Sci* 67:1360–63.

Garcia-Garcia, E., Andrieux, K., Gil, S. et al. 2005. Colloidal carriers and blood-brain barrier (BBB) translocation: A way to deliver drugs to the brain? *Int J Pharm* 298:274–92.

Geng, Q., Sun, X., Gong, T. et al. 2012. Peptide-drug conjugate linked via a disulfide bond for kidney targeted drug delivery. *Bioconjug Chem* 23:1200–10.

Glazko, A.J., Edgerton, W.H., Dill, W.A. et al. 1952. Chloromycetin palmitate; a synthetic ester of chloromycetin. *Antibiot Chemother (Northfield)* 2:234–42.

Gynther, M., Jalkanen, A., Lehtonen, M. et al. 2010. Brain uptake of ketoprofen-lysine prodrug in rats. *Int J Pharm* 399:121–8.

Gynther, M., Laine, K., Ropponen, J. et al. 2008. Large neutral amino acid transporter enables brain drug delivery via prodrugs. *J Med Chem* 51:932–6.

Haas, M., Moolenaar, F., Meijer, D.K. et al. 2002. Specific drug delivery to the kidney. *Cardiovasc Drugs Ther* 16:489–96.

Haeberlin, B., Rubas, W., Nolen, H.W., 3rd et al. 1993. In vitro evaluation of dexamethasone-beta-D-glucuronide for colon-specific drug delivery. *Pharm Res* 10:1553–62.

Hamann, P.R., Hinman, L.M., Beyer, C.F. et al. 2005. An anti-MUC1 antibody-calicheamicin conjugate for treatment of solid tumors. Choice of linker and overcoming drug resistance. *Bioconjug Chem* 16:346–53.

Henne, W.A., Doorneweerd, D.D., Hilgenbrink, A.R. et al. 2006. Synthesis and activity of a folate peptide camptothecin prodrug. *Bioorg Med Chem Lett* 16:5350–5.

Ho, S.J. and Brighton, T.A. 2006. Ximelagatran: direct thrombin inhibitor. *Vasc Health Risk Manag* 2:49–58.

Huang, P.S. and Oliff, A. 2001. Drug-targeting strategies in cancer therapy. *Curr Opin in Genet Dev* 11:104–10.

Hussain, M.A., Koval, C.A., Myers, M.J. et al. 1987. Improvement of the oral bioavailability of naltrexone in dogs: A prodrug approach. *J Pharm Sci* 76:356–8.

Huttunen, K.M., Raunio, H., and Rautio, J. 2011. Prodrugs—From serendipity to rational design. *Pharmacol Rev* 63:750–71.

Huttunen, K.M. and Rautio, J. 2011. Prodrugs—An efficient way to breach delivery and targeting barriers. *Curr Top Med Chem* 11:2265–87.

Imai, T. and Hosokawa, M. 2010. Prodrug approach using carboxylesterases activity: Catalytic properties and gene regulation of carboxylesterase in mammalian tissue. *J Pestic Sci* 35:229–39.

Imai, T. and Ohura, K. 2010. The role of intestinal carboxylesterase in the oral absorption of prodrugs. *Curr Drug Metab* 11:793–05.

Johnson, D.A. and Laguzza, B.C. 1987. Antitumor xenograft activity with a conjugate of a Vinca derivative and the squamous carcinoma-reactive monoclonal antibody PF1/D. *Cancer Res* 47:3118–22.

Jung, Y., Kim, H.H., Kim, H. et al. 2006. Evaluation of 5-aminosalicyltaurine as a colon-specific prodrug of 5-aminosalicylic acid for treatment of experimental colitis. *Eur J Pharm Sci* 28:26–33.

Jung, Y.J., Lee, J.S., and Kim, Y.M. 2001. Colon-specific prodrugs of 5-aminosalicylic acid: Synthesis and in vitro/in vivo properties of acidic amino acid derivatives of 5-aminosalicylic acid. *J Pharm Sci* 90:1767–75.

Karaman, R., Fattash, B., and Qtait, A. 2013. The future of prodrugs—Design by quantum mechanics methods. *Exp Opin Drug Deliv* 10:713–29.

Kennedy, P.C., Zhu, R., Huang, T. et al. 2011. Characterization of a sphingosine 1-phosphate receptor antagonist prodrug. *J Pharmacol Exp Ther* 338:879–89.

Kerns, E.H. and Di, L. 2003. Pharmaceutical profiling in drug discovery. *Drug Discov Today* 8:316–23.

Kumpulainen, H., Jarvinen, T., Mannila, A. et al. 2008. Synthesis, in vitro and *in vivo* characterization of novel ethyl dioxy phosphate prodrug of propofol. *Eur J Pharm Sci* 34:110–7.

Lal, R., Sukbuntherng, J., Luo, W. et al. 2010. Clinical pharmacokinetic drug interaction studies of gabapentin enacarbil, a novel transported prodrug of gabapentin, with naproxen and cimetidine. *Br J Clin Pharmacol* 69:498–507.

Lambert, D.M., Mergen, F., Poupaert, J.H. et al. 1993. Analgesic potency of S-acetylthiorphan after intravenous administration to mice. *Eur J Pharmacol* 243:129–34.

Laudano, J.B. 2011. Ceftaroline fosamil: A new broad-spectrum cephalosporin. *J Antimicrob Chemother* 66(Suppl 3):iii11–8.

Lecomte, J.M., Costentin, J., Vlaiculescu, A. et al. 1986. Pharmacological properties of acetorphan, a parenterally active "enkephalinase" inhibitor. *J Pharmacol Exp Ther* 237:937–44.

Lee, J.W., Lu, J.Y., Low, P.S. et al. 2002. Synthesis and evaluation of taxol-folic acid conjugates as targeted antineoplastics. *Bioorg Med Chem* 10:2397–14.

Lee, M.R. 1990. Five years' experience with gamma-L-glutamyl-L-dopa: A relatively renally specific dopaminergic prodrug in man. *J Auton Pharmacol* 10(Suppl 1):s103–8.

Lewis, D.F. and Dickins, M. 2002. Substrate SARs in human P450s. *Drug Discov Today* 7:918–25.

Liang, Z., Gong, T., Sun, X. et al. 2012. Chitosan oligomers as drug carriers for renal delivery of zidovudine. *Carbohydr Polym* 87:2284–90.

Liederer, B.M. and Borchardt, R.T. 2006. Enzymes involved in the bioconversion of ester-based prodrugs. *J Pharm Sci* 95:1177–95.

Lin, Y., Li, Y., Wang, X. et al. 2013. Targeted drug delivery to renal proximal tubule epithelial cells mediated by 2-glucosamine. *J Control Release* 167:148–56.

Lin, Y., Sun, X., Gong, T. et al. 2012. Synthesis and *in vivo* distribution of 2-deoxy-2-aminodiglucose–prednisolone conjugate (DPC). *Chinese Chem Lett* 23:557–60.

Liu, J., Kolar, C., Lawson, T.A. et al. 2001. Targeted drug delivery to chemoresistant cells: Folic acid derivatization of FdUMP[10] enhances cytotoxicity toward 5-FU-resistant human colorectal tumor cells. *J Org Chem* 66:5655–63.

Lynch, K.R. 2012. Building a better sphingosine kinase-1 inhibitor. *The Biochem J* 444:e1–2.

Mahato, R., Tai, W., and Cheng, K. 2011. Prodrugs for improving tumor targetability and efficiency. *Adv Drug Deliv Rev* 63:659–70.

Mandell, A.I., Stentz, F., and Kitabchi, A.E. 1978. Dipivalyl epinephrine: A new pro-drug in the treatment of glaucoma. *Ophthalmology* 85:268–75.

Manfredini, S., Pavan, B., Vertuani, S. et al. 2002. Design, synthesis and activity of ascorbic acid prodrugs of nipecotic, kynurenic and diclophenamic acids, liable to increase neurotropic activity. *J Med Chem* 45:559–62.

Mannhold, R. 2005. The impact of lipophilicity in drug research: a case report on beta-blockers. *Mini Rev Med Chem* 5:197–205.

Mattingly, G. 2010. Lisdexamfetamine dimesylate: A prodrug stimulant for the treatment of ADHD in children and adults. *CNS Spectr* 15:315–25.

McClellan, K. and Perry, C. M. 2001. Oseltamivir: A review of its use in influenza. *Drugs* 61:263–83.

McCormack, P.L. 2015. Isavuconazonium: First global approval. *Drugs* 75:817–22.

McIntyre, P.B., Rodrigues, C.A., Lennard-Jones, J.E. et al. 1988. Balsalazide in the maintenance treatment of patients with ulcerative colitis, a double-blind comparison with sulphasalazine. *Aliment Pharmacol Ther* 2:237–43.

Miwa, M., Ura, M., Nishida, M. et al. 1998. Design of a novel oral fluoropyrimidine carbamate, capecitabine, which generates 5-fluorouracil selectively in tumours by enzymes concentrated in human liver and cancer tissue. *Eur J Cancer* 34:1274–81.

Morgan, M. E., Chi, S. C., Murakami, K. et al. 1992. Central nervous system targeting of 2',3'-dideoxyinosine via adenosine deaminase-activated 6-halo-dideoxypurine pro-drugs. *Antimicrob Agents Chemother* 36 :2156–65.

Muller, C.E. 2009. Prodrug approaches for enhancing the bioavailability of drugs with low solubility. *Chem Biodivers* 6:2071–83.

Murakami, T., Kohno, K., Yumoto, R. et al. 1998. N-acetyl-L-gamma-glutamyl derivatives of p-nitroaniline, sulphamethoxazole and sulphamethizole for kidney-specific drug deliv-ery in rats. *J Pharm Pharmacol* 50:459–65.

Murdter, T.E., Sperker, B., Kivisto, K.T. et al. 1997. Enhanced uptake of doxorubicin into bronchial carcinoma: Beta-glucuronidase mediates release of doxorubicin from a gluc-uronide prodrug (HMR 1826) at the tumor site. *Cancer Res* 57:2440–5.

Narang, A.S. and Mahato, R.I. 2010. *Targeted Delivery of Small and Macromolecular Drugs.* Boca Raton, FL: CRC Press.

Nielsen, N.M. and Bundgaard, H. 1989. Evaluation of glycolamide esters and various other esters of aspirin as true aspirin prodrugs. *J Med Chem* 32:727–34.

Nilsen, A.E., Aasen, T., Halsos, A.M. et al. 1982. Efficacy of oral acyclovir in the treatment of initial and recurrent genital herpes. *Lancet* 2:571–73.

Noble, S. and Goa, K.L. 1999. Adefovir dipivoxil. *Drugs* 58:479–87.

Novis, B.H., Korzets, Z., Chen, P. et al. 1988. Nephrotic syndrome after treatment with 5-aminosalicylic acid. *Br Med J (Clin Res Ed)* 296(6634):1442.

Oldendorf, W.H., Hyman, S., Braun, L. et al. 1972. Blood-brain barrier: Penetration of mor-phine, codeine, heroin, and methadone after carotid injection. *Science* 178:984–6.

Ortiz de Montellano, P.R. 2013. Cytochrome P450-activated prodrugs. *Future Med Chem* 5:213–28.

Pathan, S.A., Iqbal, Z., Zaidi, S.M. et al. 2009. CNS drug delivery systems: Novel approaches. *Recent Pat Drug Deliv Formul* 3:71–89.

Pavan, B., Dalpiaz, A., Ciliberti, N. et al. 2008. Progress in drug delivery to the central ner-vous system by the prodrug approach. *Molecules* 13:1035–65.

Perry, C.M. 2012. Azilsartan medoxomil: A review of its use in hypertension. *Clin Drug Investig* 32:621–39.

Persico, F.J., Pritchard, J.F., Fisher, M.C. et al. 1988. Effect of tolmetin glycine amide (McN-4366), a prodrug of tolmetin sodium, on adjuvant arthritis in the rat. *J Pharmacol Exp Ther* 247:889–96.

Pescovitz, M.D., Rabkin, J., Merion, R.M. et al. 2000. Valganciclovir results in improved oral absorption of ganciclovir in liver transplant recipients. *Antimicrob Agents Chemother* 44:2811–5.

Peura, L., Malmioja, K., Laine, K. et al. 2011. Large amino acid transporter 1 (LAT1) pro-drugs of valproic acid: new prodrug design ideas for central nervous system delivery. *Mol Pharm* 8:1857–66.

Philip, A.K., Dabas, S., and Pathak, K. 2009. Optimized prodrug approach: A means for achieving enhanced anti-inflammatory potential in experimentally induced colitis. *J Drug Target* 17:235–41.

Philip, A.K., Dubey, R.K., and Pathak, K. 2008. Optimizing delivery of flurbiprofen to the colon using a targeted prodrug approach. *J Pharm Pharmacol* 60:607–13.

Platel, D., Bonoron-Adele, S., Dix, R.K. et al. 1999. Preclinical evaluation of the cardiac tox-icity of HMR-1826, a novel prodrug of doxorubicin. *Brit J Cancer* 81:24–7.

Prokai-Tatrai, K., Nguyen, V., Szarka, S. et al. 2013. Design and exploratory neuropharma-cological evaluation of novel thyrotropin-releasing hormone analogs and their brain-targeting bioprecursor prodrugs. *Pharmaceutics* 5:318–28.

Quiles, S., Raisch, K.P., Sanford, L.L. et al. 2010. Synthesis and preliminary biological evalu-ation of high-drug-load paclitaxel-antibody conjugates for tumor-targeted chemother-apy. *J Med Chem* 53:586–94.

Rais, R., Fletcher, S., and Polli, J. E. 2011. Synthesis and in vitro evaluation of gabapentin prodrugs that target the human apical sodium-dependent bile acid transporter (hASBT).
*J Pharm Sci* 100:1184–95.

Rautio, J., Kumpulainen, H., Heimbach, T. et al. 2008. Prodrugs: Design and clinical applications. *Nat Rev Drug Discov* 7:255–70.

Reddy, J.A., Westrick, E., Vlahov, I. et al. 2006. Folate receptor specific anti-tumor activity of
folate-mitomycin conjugates. *Cancer Chemother Pharmacol* 58:229–36.

Reddy, K.R., Matelich, M.C., Ugarkar, B.G. et al. 2008. Pradefovir: A prodrug that targets
adefovir to the liver for the treatment of hepatitis B. *J Med Chem* 51:666–76.

Rice, M.E. 2000. Ascorbate regulation and its neuroprotective role in the brain. *Trends
Neurosci* 23:209–16.

Rooseboom, M., Commandeur, J.N., and Vermeulen, N.P. 2004. Enzyme-catalyzed activation of anticancer prodrugs. *Pharmacol Rev* 56:53–102.

Schaeffer, H.J., Beauchamp, L., and de Miranda, P. et al. 1978. 9-(2-hydroxyethoxymethyl)
guanine activity against viruses of the herpes group. *Nature* 272:583–5.

Semba, J., Curzon, G., and Patsalos, P.N. 1993. Antiepileptic drug pharmacokinetics and
neuropharmacokinetics in individual rats by repetitive withdrawal of blood and cerebrospinal fluid: milacemide. *Br J Pharmacol* 108:1117–24.

Senter, P.D. 2009. Potent antibody drug conjugates for cancer therapy. *Curr Opin Chem Biol*
13:235–44.

Serota, F.T., Starr, S.E., Bryan, C.K. et al. 1982. Acyclovir treatment of herpes zoster infections. Use in children undergoing bone marrow transplantation. *JAMA* 247:2132–5.

Shanmuganathan, K., Koudriakova, T., Nampalli, S. et al. 1994. Enhanced brain delivery of
an anti-HIV nucleoside 2′-F-ara-ddI by xanthine oxidase mediated biotransformation.
*J Med Chem* 37:821–7.

Sharma, R., Rawal, R.K., Gaba, T. et al. 2013. Design, synthesis and ex vivo evaluation of colon-
specific azo based prodrugs of anticancer agents. *Bioorg Med Chem Lett* 23:5332–8.

Shen, W.C., Ballou, B., Ryser, H.J. et al. 1986. Targeting, internalization, and cytotoxicity of
methotrexate-monoclonal anti-stage-specific embryonic antigen-1 antibody conjugates
in cultured F-9 teratocarcinoma cells. *Cancer Res* 46:3912–6.

Shi, D., Yang, J., Yang, D. et al. 2006. Anti-influenza prodrug oseltamivir is activated by carboxylesterase human carboxylesterase 1, and the activation is inhibited by antiplatelet
agent clopidogrel. *J Pharmacol Exp Ther* 319:1477–84.

Shirota, K., Kato, Y., Suzuki, K. et al. 2001. Characterization of novel kidney-specific delivery system using an alkylglucoside vector. *J Pharmacol Exp Ther* 299:459–67.

Sinkula, A.A., Morozowich, W., and Rowe, E.L. 1973. Chemical modification of clindamycin: Synthesis and evaluation of selected esters. *J Pharm Sci* 62:1106–11.

Soul-Lawton, J., Seaber, E., On, N. et al. 1995. Absolute bioavailability and metabolic disposition of valaciclovir, the L-valyl ester of acyclovir, following oral administration to
humans. *Antimicrob Agents Chemother* 39:2759–64.

Stappaerts, J., Geboers, S., Snoeys, J. et al. 2015. Rapid conversion of the ester prodrug abiraterone acetate results in intestinal supersaturation and enhanced absorption of abiraterone: In vitro, rat in situ and human *in vivo* studies. *Eur J Pharm Biopharm* 90:1–7.

Steinberg, G. and Borch, R.F. 2001. Synthesis and evaluation of pteroic acid-conjugated nitro-
heterocyclic phosphoramidates as folate receptor-targeted alkylating agents. *J Med
Chem* 44:69–73.

Stella, V.J., Charman, W.N., and Naringrekar, V.H. 1985. Prodrugs. Do they have advantages
in clinical practice? *Drugs* 29:455–73.

Stella, V.J. and Nti-Addae, K.W. 2007. Prodrug strategies to overcome poor water solubility.
*Adv Drug Deliv Rev* 59:677–94.

Stockwell, J., Abdi, N., Lu, X. et al. 2014. Novel central nervous system drug delivery systems. *Chem Biol Drug Des* 83:507–20.

Su, M., He, Q., Zhang, Z. R. et al. 2003. Kidney-targeting characteristics of N-acetyl-L-glutamic prednisolone prodrug. *Acta Pharmaceutica Sinica* 38:627–30.

Suzuki, K., Susaki, H., Okuno, S. et al. 1999a. Renal drug targeting using a vector "alkylglycoside." *J Pharmacol Exp Ther* 288:57–64.

Suzuki, K., Susaki, H., Okuno, S. et al. 1999b. Specific renal delivery of sugar-modified low-molecular-weight peptides. *J Pharmacol Exp Ther* 288:888–97.

Taylor, M.D. 1996. Improved passive oral drug delivery via prodrugs. *Adv Drug Deliv Rev* 19:131–48.

Testa, B. 2009. Prodrugs: Bridging pharmacodynamic/pharmacokinetic gaps. *Curr Opin Chem Biol* 13:338–44.

Torre, D., Pugliese, A., and Speranza, F. 2002. Role of nitric oxide in HIV-1 infection: Friend or foe? *Lancet Infec Dis* 2:273–80.

Tsuda, M., Terada, T., Irie, M. et al. 2006. Transport characteristics of a novel peptide transporter 1 substrate, antihypotensive drug midodrine, and its amino acid derivatives. *J Pharmacol Exp Ther* 318:455–60.

Umapathy, N.S., Ganapathy, V., and Ganapathy, M.E. 2004. Transport of amino acid esters and the amino-acid-based prodrug valganciclovir by the amino acid transporter ATB(0,+). *Pharm Res* 21:1303–10.

Unadkat, J.D. and Rowland, M. 1985. Pharmacokinetics of prednisone and prednisolone at steady state in the rabbit. *Drug Metab Dispos Biol Fate Chem* 13:503–9.

Vaidya, A., Jain, S., Agrawal, R.K. et al. 2015. Pectin–metronidazole prodrug bearing microspheres for colon targeting. *J Saudi Chem Soc* 19:257–64.

Van Hees, P.A., Van Lier, H.J., Van Elteren, P.H. et al. 1981. Effect of sulphasalazine in patients with active Crohn's disease: A controlled double-blind study. *Gut* 22:404–9.

van Hogezand, R.A., van Hees, P.A., Zwanenburg, B. et al. 1985. Disposition of disodium azodisalicylate in healthy subjects. A possible new drug for inflammatory bowel disease. *Gastroenterology* 88:717–22.

Vivian, D. and Polli, J.E. 2014. Synthesis and in vitro evaluation of bile acid prodrugs of floxuridine to target the liver. *Int J Pharm* 475:597–604.

Vlieghe, P. and Khrestchatisky, M. 2013. Medicinal chemistry based approaches and nanotechnology-based systems to improve CNS drug targeting and delivery. *Med Res Rev* 33:457–516.

Wang, Y., Yuan, H., Wright, S.C. et al. 2001. Synthesis and preliminary cytotoxicity study of a cephalosporin-CC-1065 analogue prodrug. *BMC Chem Biol* 1:4–8.

Weiner, L.M. 2007. Building better magic bullets—Improving unconjugated monoclonal antibody therapy for cancer. *Nat Rev Cancer* 7:701–6.

Werling, K. and Chalasani, N. 2011. What is the role of midodrine in patients with decompensated cirrhosis? *Gastroenterol Hepatol (N Y)* 7:134–6.

Wienen, W., Stassen, J. M., Priepke, H. et al. 2007. Effects of the direct thrombin inhibitor dabigatran and its orally active prodrug, dabigatran etexilate, on thrombus formation and bleeding time in rats. *Thromb Haemost* 98:333–8.

Wilk, S., Mizoguchi, H., and Orlowski, M. 1978. Gamma-glutamyl DOPA: A kidney-specific dopamine precursor. *J Pharmacol Exp Ther* 206:227–32.

Wire, M.B., Shelton, M.J., and Studenberg, S. 2006. Fosamprenavir: Clinical pharmacokinetics and drug interactions of the amprenavir prodrug. *Clin Pharmacokinet* 45:137–68.

Worman, H. 2006. *The Liver Disorders and Hepatitis Sourcebook*. New York: McGraw-Hill Companies, Inc.

Zamboni, S., Mallano, A., Flego, M. et al. 2008. Genetic construction, expression, and characterization of a single chain anti-CEA antibody fused to cytosine deaminase from yeast. *Inte J Oncol* 32:1245–51.

Zhao, Y., Qu, B., Wu, X. et al. 2014. Design, synthesis and biological evaluation of brain targeting l-ascorbic acid prodrugs of ibuprofen with "lock-in" function. *Eur J Med Chem* 82:314–23.

Zhou, P., Sun, X., and Zhang, Z. 2014. Kidney-targeted drug delivery systems. *Acta Pharm Sin B* 4:37–42.

Zou, M., Okamoto, H., Cheng, G. et al. 2005. Synthesis and properties of polysaccharide prodrugs of 5-aminosalicylic acid as potential colon-specific delivery systems. *Eur J Pharm Biopharm* 59:155–60.

# 5 Drug-Targeting Strategies to CNS Disorders

*Somasree Ray*

## CONTENTS

## 5.1 INTRODUCTION

The brain is one of the critical organs of our body, which is difficult to access when it comes to the treatment of brain-related ailments. The two main barriers, which act as hindrance, are the blood–brain barrier (BBB) and the blood–cerebrospinal fluid barrier (BCSFB) that prevent the entrance of endogenous substances and xenobiotics. The tightly bound BBB is composed of endothelial cells, astrocytes end feet, and pericytes, and tight junctions are present between the cerebral endothelial cells. This cerebral endothelial cell differs from the rest of the endothelial cells of the body due to the absence of fenestrations with very few pinocytic vesicles and increased numbers of mitochondria. There is also a lack of lymphatic drainage and absence of major histocompatibility complex (MHC) antigen in the central nervous system (CNS) with immune reactivity inducible on temporary demand in order to provide maximum protection to neuronal functions. It may be in trace amounts, but some enzymes also play a part in this BBB such as glutamyltranspeptidase, alkaline phosphatase, esterase, and monoamine oxidase.

The pericytes and the perivascular antigen presenting cells surround the endothelial cells. The astrocytes form a sheath around the vessels. Nevertheless, it does not mean that facilitation of all types of molecules is not possible. Depending upon various characteristics, such as molecular size, the physical nature of access through the tightly bound BBB can be considered. Generally, lipophilic molecules access the capacity of crossing the BBB. Small molecules such as $O_2$ and $CO_2$ diffuse freely across the plasma membrane along the concentration gradients. Nutrients such as glucose and amino acid take the help of a transporter to cross the BBB. Larger molecules such as insulin, leptin, and iron transferrin take the help of receptor-mediated endocytosis.

There are some factors that are involved in drug permeation through BBB and distribution into the brain. Considering the CNS as a separate compartment in the brain and its evolution with time, the plasma concentration will depend on the following factors:

1. The absorption, distribution, metabolism, and excretion characteristics of drugs
2. The degree of plasma–protein binding because only the unbound fraction diffuses across the barrier
3. The metabolic modification by barrier enzymes and the *sink effect* of the continual drainage of CSF
4. The effective permeability across the BBB, which depends on the combination of the passive permeability and the contribution of efflux and influx carrier-mediated transport
5. The nonspecific binding to brain tissues

Pharmacotherapy of CNS diseases such as schizophrenia, meningitis, Parkinson's disease, Alzheimer's disease, epilepsy, and brain cancer is severely constrained by these barriers. The clinical failure to treat these diseases is not due to the lack of potency of the drug but rather no availability of a suitable method by which the drug can be targeted to that region. Consequently, the delivery of hydrophilic drugs and high molecular weight compounds to the proper site of the brain requires extensive research efforts for treating such diseases. The permeability of various molecules to the brain highly depends on the permeability of the molecules through these barriers.

This chapter intends to review different strategies to target drug molecules to the brain. A brief introduction of brain-related diseases and the structure of BBB, blood–tumor barrier (BTB), and BCSFB are also included to illustrate the complexity of the problem.

## 5.2   CNS DISEASES

The concern for CNS diseases has highly increased with an increase in the number of sufferings. Treating CNS disorder is often challenging, particularly due to a wide variety of obstacles, which often hinder the process of targeting. Nowadays, CNS disorders have become the causes of mortality and morbidity worldwide. Some of the CNS disorders are as follows:

1. Bipolar disorder: A disorder of CNS, characterized by depression and elevation of moods and continues for several weeks.
2. Catalepsy: Considered as a symptom of serious CNS diseases. This disease is characterized by immobility and muscular rigidity along with a decreased sensitivity to pain.
3. Epilepsy: Caused by disturbances in electrical activity in the brain.
4. Encephalitis: Inflammation of the brain. Foreign substance or a viral infection is the reason behind its occurrence.
5. Meningitis: An infection that affects the meninges of the brain and spinal cord. Inflammation of meninges is observed.
6. Migraine: An illness of the nervous system associated with recurrent moderate-to-severe headache.
7. Huntington's disease: An inherited neurodegenerative disorder. Degeneration of neuronal cells occurs throughout the brain, especially in the striatum. There is a progressive decline that results in abnormal movements.
8. Alzheimer's disease: A neurodegenerative disease typically found in people over 65 years of age. Short-term memory loss is the early symptom of Alzheimer's disease. However, with the advancement of the disease, degeneration of neuronal cells occurs in the brain, and several new symptoms develop such as mood swings, loss of motivation, worsened ability to take and remember new information, and inability to plan complex or sequential activities.
9. Parkinson's disease: A progressive illness of the nervous system caused by the death of dopaminergic brain cells that affect motor skills and speech.

## 5.3   PHYSIOLOGICAL BARRIERS TO CNS DRUG DELIVERY

The major obstacles in the brain target various barriers having unique composition that prevent the entry of the drug molecules to the desired site. The delivery systems should be designed in such a way that they are able to penetrate these barriers to reach the targeted site.

### 5.3.1   BLOOD–BRAIN BARRIER (BBB)

BBB is one of the strictest barriers of the human body that separates the brain from the circulating blood (Begley 1996). The surface area of human BBB is about 20 m$^2$, which allows the passage of molecules, having molecular weight of less than 400 Da. Capillaries present in the CNS are structurally different from the blood capillaries present in other parts of the body. BBB prevents the entry of hydrophilic molecules, but $O_2$ and $CO_2$ can freely cross the barrier. It is also permeable to nutrients and amino acids via transporters. Three cellular elements of the brain microvasculature compose the BBB: endothelial cells, astrocytes end feet, and pericytes (Ballabh et al. 2004). It acts as a diffusion barrier, which prevents the entry of blood-borne substances to the brain. Astrocytic end feet tightly ensheath the vessel wall and appear to be responsible for the maintenance of the tight junction barrier of the mammalian brain. The presence of tight junctions between the endothelial cells produces transendothelial electrical resistance of 1500–2000 $\Omega cm^2$ compared with 3–33 $\Omega cm^2$ of other tissues (Brightman 1992, Lo et al., 2001). Higher mitochondrial density is observed in the brain microvessels than in other capillaries and that is only due to the small dimensions of the brain microvessels and smaller cytoplasmic area (Misra et al. 2003). Due to the absence of intercellular cleft, the brain is practically inaccessible to lipid-insoluble polar compounds. As the surface of the BCSFB faces the ventricle that is filled with CSF, not the blood (Rip et al. 2009), this in combination with the high turnover rate of CSF leads to continuous flushing of the drug (injected into the ventricles) back to blood (Pathan et al. 2009). Efflux transporters such as p-glycoprotein and multidrug-resistant protein are also present in the luminal membranes of cerebral capillary endothelium, which prevent entry of nonessential compounds to brain parenchyma. Though nonselective inhibition property of efflux transporters provides some therapeutic benefit, it is another obstacle in delivering drugs to the CNS because drugs are substrates of efflux transporters (Fromm 2000).

## 5.3.2 Blood–Cerebrospinal Fluid Barrier (BCSFB)

Four major fluid compartments are present in the brain: the blood flowing through the entire brain, interstitial fluid (ISF) present in neurons and neuroglia, CSF present in brain ventricles and spinal cord, and intercellular fluid present in brain cells. One of the barriers, which a delivery system encounters, is BCSFB. The BCSFB is composed of the choroid plexus epithelial cells, which are joined together by tight junctions and the arachnoid membrane that envelopes the brain. The cells of this membrane are also linked by tight junctions. The CSF is primarily secreted by the choroid plexus. The capillaries in the choroid plexus differ from those of the brain in which there is free movement of molecules across the endothelial cell through fenestrations and intercellular gaps. However, the unique apical tight junctions in between the choroid plexus epithelial cells inhibit the paracellular movement of water-soluble molecules across the barrier. Thus, the choroid plexus capillaries allow free movement of small molecules, whereas the tight junction prevents the passage of macromolecules from the blood to the brain. The arachnoid membrane is generally impermeable to hydrophilic substances, and its role in forming the BCSFB is largely passive. The CSF circulates around brain ventricles and spinal cord, and the intracellular fluid (ICF) circulates within brain cells. There is no major diffusional barrier between the ISF and the CSF. Thus, materials present in either of these two fluid compartments are free to exchange and reach their destination cells. Since the CSF can exchange molecules with the ISF of the brain parenchyma, the passage of blood-borne molecules into the CSF is also carefully regulated by the BCSFB.

Moreover, the BCSFB is reinforced by an active organic acid transporter system in the choroids plexus, which is capable of driving CSF-borne organic acids into the blood. Hence, a variety of therapeutic organic acids such as the antibiotic penicillin, antineoplastic agent methotrexate, and antiviral agent zidovudine are actively removed from the CSF, and therefore inhibited from diffusing into the brain parenchyma. Furthermore, substantial incompatibilities often exist between the composition of the CSF and ISF of the brain parenchyma, rationalizing the presence of what is sometimes called the CSF–brain barrier. This barrier is attributed to the insurmountable diffusion distances required for equilibration between the CSF and the brain ISF. Therefore, entry into the CSF does not guarantee a drug's penetration into the brain.

### 5.3.3 Blood–Tumor Barrier

Drug targeting to tumor is a complicated task due to the complex structure of the tumor. There are three different microvessel populations present in brain tumors. The first one consists of continuous and nonfenestrated capillaries such as those found in normal brain, for example, ethylnitrosourea-induced gliomas in animals (Blasberg et al. 1983). The second type consists of continuous but fenestrated capillaries. These types of microvessels allow the entry of small molecules but not large molecules. The third capillary population has interendothelial gaps, which can measure up to 1 µm, and they do not show any selectivity to large molecules in RG-2 rat gliomas (Schlageter et al. 1999) and D-54 MG human glioma model (Blasberg et al. 1987). Spatial distribution of the target capillaries is another factor, which affects capillary permeability. The brain tumor capillaries may be much more permeable at the tumor site, but the permeability of surrounding brain regions even at a few millimeters periphery of the tumor site is similar to normal brain values. This property of neoplastic vasculature may aid in the process of local invasion, the chief mode of spreading of CNS neoplasia. Individual tumor cells may reside centimeters away from the edge of a tumor, and it may be quite difficult to target them due to differences in spatial permeability. Hence, drug delivery to tumor is a challenge due to this spatial changeability in capillary function.

For better targeting, the drug is administered by intravascular route, and after intravascular administration, the drug mixes with the total body volume of the blood. The human body acts as an enormous sink in which the drug is distributed throughout the body fluid. As the drug is distributed not only to the brain tumor but also in the other tissues, a small portion of available dose is available in the tumor region. Furthermore, the vascular surface area exposed to the tumor gets decreased as the tumor size increases with time, and results in the reduction of transvascular exchange of blood-borne molecules. Simultaneously, the space between two capillaries also increases, leading to increase in the diffusion path length for drug delivery to tumor cells. Hydrostatic pressure in the normal brain parenchyma adjacent to the tumor gets increased due to changes in the intestinal tumor pressure and the associated peritumoral edema. As a result, permeability of the drug across the tumor region decreases. The neoplastic vasculature develops by vasculogenesis rather than angiogenesis, which artifices the structural makeup of vessels in the peritumoral vascular network also poses certain anatomical challenges to tissue-specific drug delivery in CNS neoplasia.

## 5.4  MECHANISM OF SOLUTE TRANSPORT ACROSS BBB

There are several routes through which solute molecules can cross the BBB but to a limited extent. These are as follows:

1. Paracellular pathway of water-soluble agents
2. Lipid-mediated transport
3. Carrier-mediated transport
4. Receptor-mediated transcytosis
5. Adsorptive-mediated transcytosis

Figure 5.1 depicts the different drug transport mechanisms across the BBB. Small water-soluble molecules can cross the tight junction of BBB by simple diffusion technique but not to a great extent. Small lipid-soluble molecules of molecular weight less than 400 Da can cross the barrier by dissolving in the plasma membrane. Small water-soluble nutrients, glucose, and amino acids traverse the BBB via carrier-mediated transport with the expenditure of energy in the form of adenosine triphosphate (ATP). Receptor-mediated transcytosis is responsible for transport of large molecules such as peptides or plasma proteins. Receptors of endothelial cells can uptake many different types of ligands including growth factors, enzymes, and plasma proteins. Adsorptive-mediated transcytosis, also known as pinocytosis, is triggered by an electrostatic interaction between a positively charged substance, usually the charged moiety of a peptide, and the negatively charged plasma membrane surface (heparin sulfate proteoglycans) (Chen and Lihong 2012). Efflux transporters (p-glycoprotein, multidrug-resistant protein) are present in the barrier and prevent entry of nonessential compounds to the brain. While the natural effect of efflux

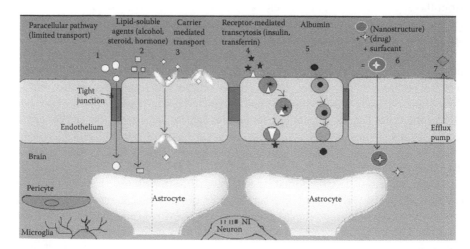

**FIGURE 5.1**   Different mechanisms of solute transport across the blood–brain barrier.

transporters is essential for detoxification, it is another obstacle for delivery of a wide range of biologically active molecules to the brain.

## 5.5 PHYSICOCHEMICAL FACTORS AFFECTING DRUG PERMEATION THROUGH BBB

The rate and extent of availability of drug in brain circulation depends on several factors:

1. The extent of protein binding and plasma concentration of unbound drug as only unbound fraction crosses the BBB
2. The rate of transfer of drug from blood to brain circulation, which depends on the combination of passive permeability and the contribution of efflux and influx carrier-mediated transport
3. The interaction between drug and some receptors in the brain (Feng 2002; Alavijeh et al. 2009), and nonspecific binding to brain tissue

However, it is not easy to identify the exact parameter that predicts good brain permeation, but it can be positively correlated with three major physicochemical parameters: lipophilicity, polar surface area, and molecular weight. Increasing lipophilicity with the intent of improving membrane permeability increases the volume of distribution in a particular plasma protein binding and oxidative metabolism by cytochrome P450 (van De Waterbeemd et al. 2001; Lewis and Dickins 2002). The measure of brain uptake can be expressed by two parameters: (a) ratio of brain to plasma drug concentration at steady state ($K_p$) and (b) BBB permeability quantified as permeability surface area product. Another relevant term, $K_{p,uu}$, is the ratio of unbound concentration of drug in brain interstitial fluid (ISF) over its unbound concentration in plasma.

$$K_{p,uu} = \frac{C_{u,brain\,ISF}}{C_{u,\,plasma}} = \frac{K_p}{f_u \cdot V_{u,\,brain}} \tag{5.1}$$

where $V_{u,brain}$ unbound volume of distribution of drug in brain. It describes the relationship between amount of drug in brain and the unbound drug concentration in brain ISF and $f_u$ is the ratio of unbound drug concentration in brain to that of total drug concentration in brain. The value much higher than the combination of ISF and intracellular fluid volumes (~0.8 mL/gm brain) indicates the nonspecific binding of drug to the brain tissue (Mangas-Sanjuan et al. 2010). The BBB permeability surface area product measures clearance of drug from blood to brain across BBB, and it has units of µl/min/gm.

## 5.6 NOVEL CARRIERS AND NEWER OPTIONS FOR BRAIN TARGETING

BBB provides a formidable obstacle for the delivery of drugs in the CNS, but they are not insurmountable due to the rapid progress in multidiscipline approaches combining biology, nanotechnology, and biophysics (see Figure 5.2).

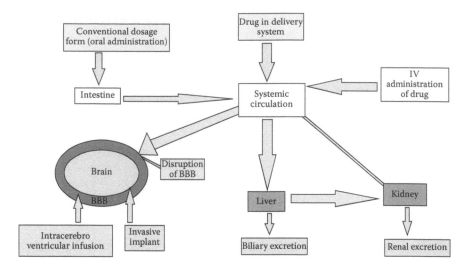

**FIGURE 5.2** Different strategies for drug delivery to the brain.

### 5.6.1 INTRACEREBRO VENTRICULAR INFUSION

In this method, drug is injected directly into the CSF, and after administration, concentration of drug in the brain parenchyma decreases exponentially with each millimeter of distance removed from the ependymal surface of the brain. Concentration of drug in the brain is only 1%–2% of the CSF concentration at just 1–2 mm from the surface. The major drawback of this method is that the diffusion of drug in the brain parenchyma is very low. An effective drug delivery can be obtained only if the target is close to the ventricles.

### 5.6.2 USE OF IMPLANT

Direct implantation into brain parenchyma of a polymer matrix containing nerve growth factor is another approach to circumvent BBB (Krewson et al. 1995). Because of the limited diffusion of the drug in brain, which exponentially decreases with distance, precise mapping of the injection site is required for successful delivery.

### 5.6.3 DRUG TRANSPORT VIA TIGHT JUNCTION OPENING

There is growing evidence that transient opening of the BBB is observed in epileptic seizures and Parkinson's disease due to transient secretion of inflammatory mediators. On the other hand, both primary and secondary brain tumors can increase BBB permeability likely caused by the disturbance of tight junction complex and/or the accumulation of growth factors and proinflammatory cytokines (Sato et al. 1994). After analyzing the factors, it can be concluded that process of inflammation and inflammatory mediators has some effect on the opening and closing of BBB.

## TABLE 5.1
## Effect of Some Biological, Chemical, and Physical Stimuli on Drug Transport through BBB

| Stimuli | Effect on BBB | References |
|---------|---------------|------------|
| | **Biological** | |
| Cereport (RMP-7) | Enhanced delivery of carboplatin, loperamide through BBB opening | Emerich et al. (1999) |
| HIV-1 clade-specific Tat protein B | Increased paracellular transport due to the disruption of blood–brain barrier integrity | Gandhi et al. (2010) |
| | **Chemical** | |
| Cyclodextrin | Facilitated delivery of doxorubicin due to the extraction of cholesterol from brain capillary endothelial cells | Tilloy et al. (2006) |
| Oleic acid | The reversible blood–brain barrier opening to alpha-aminoisobutyric acid after the infusion of oleic acid | Sztriha and Betz (1991) |
| | **Physical** | |
| Ultrasound | Enhanced delivery of chemotherapeutic agents due to transendothelial openings | Hynynen (2008) |

Different chemical and biological substances and physical stimuli such as ultrasound and electromagnetic field can be used as modulators for the reversible opening of tight junctions of the brain (Table 5.1; Hynynen 2008; Deli 2009).

### 5.6.3.1 BBB Disruption by Biological Stimuli

Several modulators have been reported to provide transient BBB opening. Karyekar et al. (2003) reported the ability of *zonula occludens* toxins to reversibly open tight junction in bovine brain microvessel endothelial cells, which increases paracellular transport of sucrose and inulin (permeability markers) in a concentration-dependent manner. These modulations of paracellular transport with *zonula occludans* toxins increase permeation of anticancer drug. Inflammatory stimuli-like histamine and bradykinin increase microvascular permeability in tumor tissues. These substances, in combination with imaging materials, gene, or anticancer drugs, can potentially boost the preferential delivery of these materials to the brain tumor for tumor diagnosis and chemotherapy (Vajkoczy and Menger 2004). Another bradykinin analog, cereport (also known as RMP-7), is highly effective in enhancing the transport of different drugs including carboplatin, loperamide, and acyclovir in different types of diseased animal models (Emerich et al. 1998, 1999). When attached to the surface of liposome, cereport is more effective in transporting different drugs due to the transient opening of BBB (Zhang et al. 2004). However, further research is required to confirm the role of cereport in drug targeting to the brain. Not only the cereport but the virus can also act as stimuli for the disruption of BBB. Dallasta et al. (1999) studied CNS tissue obtained from human immunodeficiency virus (HIV)-1-infected patients both with and without encephalitis, and alterations in BBB integrity were observed via immunohistochemical analysis of the occulin and zonula occludens. Study reveals that BBB is the main route of entry of HIV-1-infected monocyte as significant tight junction disruption was observed in a patient who died of HIV encephalitis.

### 5.6.3.2   BBB Disruption by Chemical Stimuli

During intra-arterial chemotherapy, transient opening of BBB was observed after arterial injection of hyperosmolar solution of mannitol and arabinose (Doolittle et al. 2011). Another chemical that was found to be a modulator of tight junction permeability is the phospholipid lysophosphatidic acid. It causes rapid, reversible, dose-dependent decrease in transcellular electrical resistance in brain endothelial cells (Schulze et al. 1997). Sodium dodecyl sulfate, which is basically an anionic surfactant, can be used to increase the permeability of BBB (Saija et al. 1997). Monnaert et al. (2004) investigated the effects of gamma- and hydroxypropyl-gamma-cyclodextrins on the transport of doxorubicin across an *in vitro* model of BBB. Higher cyclodextrin concentration increased the doxorubicin delivery to the brain, but this effect is due to the loss of BBB integrity.

### 5.6.3.3   Development of Physical Stimuli

Development of acoustic technology has enabled ultrasound a modality with high therapeutic and diagnostic applicability. This method can induce topical biological effects without any surgery. High-intensity focused ultrasound device was used in the treatment of localized prostate cancer (Chapelon et al. 1999), malignant bone tumor (Chen et al. 2002), and localized breast cancer (Wu et al. 2003). The two major mechanisms mainly responsible for these effects are thermal and nonthermal (such as radiation pressure and cavitation). Relatively low temperatures (~43°C) maintained for 30–60 min can be used for hyperthermia to sensitize tumors to chemotherapy, and radiotherapy (Diederich and Hynynen 1999) using higher temperature (>60°C). The focused ultrasound is used as a thermal ablation method to treat tumors in many organs including prostate, liver, kidney, breast, bone, uterus, and pancreas or as a technique of thermal coagulation of blood vessels. After elaborative research, BBB disruption without producing lesion was first demonstrated by Ballantine et al. (1960) using high-power defocused ultrasonic beam of frequency 2.7 MHz, intensity 4000 w/cm², pulse width 0.3 S, pulse period 1.0 s, and pulse number 15. A new approach to open BBB without any damage is the use of focused ultrasound in combination with encapsulated gas-filled microbubbles of less than 5 μm diameter. This method utilizes preformed microbubbles injected into the blood stream before ultrasound burst exposures. These microbubbles act as energy concentrators in the blood. They are made up of albumin and lipids, and contain gas-like air or perfluorocarbon. Microbubbles show cyclic expansion and contraction due to reduction and increase in pressure associated with ultrasound wave. Due to expansion and contraction, large shear force is generated surrounding the bubbles, which causes the surrounding fluids to flow. Due to inertia of the surrounding fluid, bubbles collapse and a shock wave is generated that propagates at a supersonic speed radially from the collapse site (Apfel 1982). If the bubbles collapse to the vessel wall, they can create fluid jets that can puncture the wall (Margulis 1995). As a result, the bubbles absorb and concentrate energy on the ultrasound wave into a microscopic tissue volume reducing

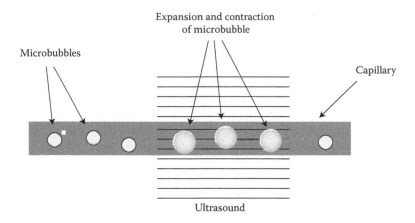

**FIGURE 5.3**  Expansion and contraction of microbubbles due to ultrasound.

the ultrasound power levels by at least two orders of magnitude from that required to induce bioeffects without the bubbles (Hynynen et al. 2001). Though the drug can be transported to the brain using a noninvasive approach such as magnetic resonance imaging scanner and ultrasound, extensive research is required in this field to quantify brain uptake and therapeutic effectiveness (see Figure 5.3).

### 5.6.4 NANOSCALE SYSTEMS FOR BRAIN DRUG DELIVERY

The preexisting approaches such as BBB disruption and superficial and ventricular application of drug to brain parenchyma are invasive methods and require skill, and there are chances of damage to the brain. Considering the drawbacks of preexisting approaches, current areas of research have focused on the development of novel drug delivery systems such as liposome, nanoparticles, supermolecular complexes, endogenous transport systems, and their surface-modified products (Herrero-Vanrell et al. 2005).

Nanoscale systems are colloidal particles of size 10–1000 nm. They are unique in their size and can be easily tailored to produce surface modification; these properties make them a popular alternative for transporting drug across BBB. The following advantages of nanoparticles may be envisaged for CNS-targeted drug delivery:

1. Nontoxic, biodegradable, and biocompatible
2. Due to their small particle size (<100 nm), they can easily penetrate the capillaries
3. Use of biodegradable matrix-forming material allows sustained drug release at the target site after injection
4. They are stable in blood and nonimmunogenic
5. Surface modifications of nanoparticles prevent their rapid clearances by phagocytes after intravascular administration
6. They are useful in targeting small molecules, proteins, peptides, or nucleic acids

### 5.6.4.1  Polymeric Nanoparticles

Modern drug delivery technology has become possible by the use of different polymeric and metal nanoparticles, liposomes, micelles, quantum dots, dendrimers, solid lipid nanoparticles, and many other nanoassemblies, which have been developed mainly to reduce the side effects and increase the therapeutic efficacy of active moiety (Wang et al. 2008). Polymeric nanocarriers have unique properties because of their size and easily tailored structures due to the material used. These systems have certain advantages such as prolonged drug action that enable hydrophobic drug administration, increased stability, and ease of manipulation of polymers for desired characteristics. With the advancement of polymer chemistry, polymer chemists synthesized new polymeric structures of improved drug solubility and stability (Mikhail and Allen 2009). Size of nanoparticles should be less than 1.0 μm and depending on the method of preparation, it may be either nanosphere where the drug is dispersed in the matrix or nanocapsules, which are vesicular systems with drug-containing cavity, surrounded by a polymeric membrane. Solubility of drug in the polymer matrix, composition of polymer, drug–polymer interaction, and the presence of end functional groups are the important determinants of drug loading and drug entrapment efficiency of the nanoparticles. Various polymers such as polylactic acid (PLA), polyglycolic acid (PGA), poly-d, l-lactide-co-glycolide (PLGA), poly(ε-caprolactone) (PCL), poly(alkylcyanoacrylate), and poly(methylmethacrylate) are mainly used for brain-specific delivery of drugs. For active targeting of the drug to the brain, the size of the nanoparticles should be less than 100 nm, as particle size greater than 100 nm is easily captured by Kupffer cells/other phagocytic cells and restricts their biodistribution (Wisse and Leeuw 1984). It prolongs the duration of drug action, which is very much essential for active drug targeting. Table 5.2 enlists some of the remarkable examples of polymeric nanoparticles for brain targeting.

Drug release from the nanoparticles depends on the type and length of the polymer, nanoparticle type, and the structure of nanoparticle.

Drug release from nanoparticle occurs in two steps:

1. Initial rapid release of drug from nanoparticle surface leading to burst effect
2. Slow release due to degradation of nanoparticle matrix or diffusion of drug through the matrix. The slow drug release follows zero-order or first-order kinetics

The transport of nanoparticle from plasma to brain parenchyma occurs by (a) passive diffusion; (b) carrier-mediated transport; and (c) receptor-mediated endocytosis.

- Passive diffusion
  Nanoparticles can be transported across the BBB considering some factors such as lipophilicity of the drugs, charge, concentration gradient, molecular weight, and degree of protein binding, and the transportation is generally governed by passive diffusion. It is characterized by Fick's law of diffusion:

$$-\frac{dc}{dt} = K(C_1 - C_2) \tag{5.2}$$

**TABLE 5.2**

**Examples of Nanoparticles for Brain Targeting**

| Drug Used | Polymer Used | Surfactant Type | Results | References |
|---|---|---|---|---|
| Tacrine | Polysorbate 80-coated poly(n-butlycyanoacrylate) | Polysorbate 80 | Brain concentration of intravenously injected tacrine enhanced by binding to 1% solution of polysorbate 80-coated on poly(n-butlycyanoacrylate) nanoparticles | Wilson et al. (2008) |
| Camphotericin | Soybean lecithin (solid lipid nanoparticles) | Poloxamer 188 | Increased brain AUC | Yang et al. (1999) |
| Dalagrin | Polybutyl cyanoacrylate (nanoparticle) | Polysorbate 80 | Analgesia study: increased latency by 50% | Schroder (1996) |
| Tetanus toxin C fragment | Biotin-binding proteins used as cross-linkers to conjugate protein to nanoparticle | – | *In vitro* neuroblastomacells were selectively targeted with success | Langer et al. (2007) |
| Odorranalectin, urocortin peptide | Odorranalectin was conjugated to poly(ethylene glycol)-poly(lactic-co-glycolic acid) naonoparticles | – | Odorranalectin modification increased brain delivery of nanoparticles and enhanced effects of urocortin-loaded nanoparticles in Parkinson's disease | Lai et al. (2011) |
| Doxorubicin | Non-stealth and stealth solid lipid nanoparticle | Epikuron 200 | In all rat tissues, except the brain, the amount of doxorubicin was always lower after the injection | Fundaro et al. (2000) |
| Tubocurarine | Polybutyl cyanoacrylate (solid nanoparticle) | Polysorbate 80 | Increased transport of tubocurarine | Alyautdin et al. (1998) |
| Dooxorubicin | Polybutylcyanoacrylate (solid nanoparticle) | Polysorbate 80 | 60-fold increase in brain concentration after systemic administration due to coating | Gulyaev et al. (1999) |
| Coumarin 6 | 12-amino acid peptide covalently conjugated onto surface of PEG-PLGA | – | High brain drug targeting index of coumarin 6 incorporated in target nanoparticles | Li et al. (2011) |

where $dc/dt$ is the rate of drug diffusion, $K$ represents first-order diffusion rate constant, and $(C_1 - C_2)$ is the concentration gradient. Fenart et al. (1999) prepared neutral, anionic, and cationic nanoparticles using cross-linked maltodextrin with phosphate (anionic) and quaternary ammonium (cationic) ligands with or without coating with lipid bilayer of dipalmitoylphosphatidylcholine and cholesterol. Rate of permeation through the BBB can be altered by coating with lipid. However, it may also vary depending upon the charge of the particles. Lipid coating over ionically charged nanoparticle increases the permeability three- to fourfold, whereas there is no significant alteration in the permeation character of uncoated particles. They showed that nanoparticle transport was due to transcytosis, not due to altered BBB integrity. Similar results were observed by Schroder et al. (1998). After intravenous administration, the Leu-enkephalin dalargin and the Met-enkephalin kyotorphin normally do not cross the blood–brain barrier. Transport of these substances across the BBB can be increased with the help of nanoparticles. These neuropeptides were adsorbed onto the surface of poly(butylcyanoacrylate) nanoparticles coated with polysorbate 80. Central analgesia was measured by the hot plate test in mice. The antidepressant amitriptyline, which normally penetrates the BBB, was used to examine the versatility of the nanoparticle method. Significant increase in amitriptyline level was observed when these substances are adsorbed on the surface of nanoparticles. The authors hypothesized that greater amount of drug enters through passive diffusion due to larger gradient at the BBB, which is caused by the enhancement of the plasma concentration.

• Carrier-mediated transport
  Drug permeation through the BBB does not only take place by receptor-mediated endocytosis or passive diffusion, it has been found that it can also be mediated by carrier protein at the capillary endothelial cells. Mechanism of drug permeation through BBB can be explained by Michaelis–Menten kinetics, which may be along and against the concentration gradient.
• Receptor-mediated endocytosis
  The high rate of reticuloendothelial system (RES)-mediated detection and clearance of colloidal carriers by liver is the main problem associated with nanoparticles. To prevent the nanoparticles from being taken up by the RES and to increase their availability at the target site, nanoparticles are coated with polysorbate 80. The coated nanoparticles cross BBB by cellular endothelial endocytosis. The nanoparticles are coated with polysorbate 80 and have the ability to adsorb apolipoprotein from the blood stream after injection into the blood stream. Coating of nanoparticle by polysorbate 80 reduces uptake of nanoparticles by liver and other organs of RES which ultimately prolonged the residence time of drug in circulation. Polysorbate 80 plays a vital role in delivering drug to brain. Thus polysorbate coating acts mainly as a helping hand for apolipoprotein over coated nanoparticles and thus would mimic lipoprotein particles and could interact with and then be taken up by the brain capillary endothelial cells via receptor-mediated endocytosis (Kreuter et al. 2002).

### 5.6.4.2 Micelles

Micelles are generally amphiphilic molecules present in cluster that are spherical in nature with a hydrophobic tail toward the surface of the water and the hydrophilic head toward the bulk of the water. These are used for the solubilization of water-insoluble drugs. The amphiphilic molecules in micelles are in constant exchange with those in the bulk solution. Polymeric micelles have the ability to self-assemble as polymer shells which are known as polymersomes such as polyethylene glycol (PEG)–polylactic acid and PEG–poly(ε-caprolactone) (Discher et al. 2006). In block copolymers, the chains are joined covalently forming two or more segments, but they possess some amphiphilic properties as that of lipids. By optimizing the block composition and block length, there is a fair chance of preparing different desirable drug delivery systems. This makes amiphillic copolymers an emerging material in this field.

### 5.6.4.3  Liposomes

Liposomes are phospholipid vesicles having hydrophilic head group and hydrophobic tail group. Liposomes have structural similarity to phospholipids containing both hydrophobic tails packed in the core to avoid water interaction and hydrophilic heads that are readily available to interact with the surrounding aqueous media. Several liposome-based products such as doxorubicin, long-circulating liposomes, are present in the market for cancer therapy. The rationale behind the use of liposomes for cancer therapy is based on slow release of entrapped drug from liposomes, reduced renal clearance, and toxicity profile of the formulation.

Although liposome is effective for cancer therapy, these formulations have not entered in the market in great number. Some of the major problems associated with liposomes are their interaction with body's own lipoproteins, stability problem, poor batch-to-batch reproducibility, and expensive for large-scale production procedure. However, one success story is PEGylation of liposomes, which increases accumulation of liposomes in the ISF at the tumor site and extends their residence time in the body.

### 5.6.4.4  Functionalized Nanocarriers: PEGylation of Nanocarriers

Until today, targeting drugs across the BBB has posed a challenge. Day-by-day advancements are taking place, one of which is the advancement in nanotechnology coupled with smart material planning and designing.

Recently, nanoparticles, liposomes, and polymeric micelles have been explored in detail to increase circulation half life of drug, and to reduce reticuloendothelial system uptake. One of the major problems associated with the hydrophobic particles is that our body treats these particles as protein particles and thus they are rapidly taken up by the mononuclear phagocytic system, finally ending in the liver or spleen. To prevent phagocytosis and increased systemic circulation time, the surface of the particles must be modified, and the goal of their modification technique is to produce particles with hydrophilic nature. PEG is a hydrophilic nonionic polymer that has been used for surface modification of nanoparticles because of its superior biocompatibility. The presence of PEG brush on the surface significantly increases the residence time of nanoparticles *in vivo* as they are not recognized as foreign particles in the body and are therefore not taken up by the phagocytic system. PEGylation produces opsonin resistance property due to its steric repulsion effect (Woodle 1995). Sometimes PEG containing surfactants, poly(oxy-ethylene), poly(oxypropylene) block copolymers (poloxamer 338 = pluronic $F_{108}$ and poloxamine 908 = tetronic®) were also found to be effective in prolonging the blood circulation time even by coating the nanoparticles with the surfactants (Stolnik et al. 1995; Coombes et al. 1994).

Different researchers compared the effectiveness of PEG-coated nanoparticles, and self-assembly from PEGylated polymer was the better choice. Calvo et al. (2001) studied biodistribution profile of polysorbate 80 or poloxamine 908-coated polyhexadecyl cyanoacrylate nanoparticles and uncoated polyhexadecyl cyanoacrylate after intravenous administration. The extent of penetration in the brain can be enhanced by coating the nanoparticle with PEG and thus making them PEGylated, whereas poloxamine 908-coated nanoparticles fail to penetrate BBB due to their inability to interact with cells. Hence, PEGylated nanocarriers hold enormous potential as a promising delivery system for drug targeting.

### 5.6.4.5   Use of Drug Transport via Cell-Penetrating Peptides

In the treatment of any brain disorders, drug should effectively cross the brain barriers such as BBB, BCSFB, and BTB, and in the case of certain diseases, drug may also have to permeate through the cellular particles of the brain such as microglia, astrocytes, neurons, and oligodendrites. Passive diffusion technique is not effective in the transport of large water-soluble and ionic substances across the brain barriers. Cell-penetrating peptides play an important role in this regard. Cell-penetrating peptides are composed of less than 30 amino acids of short peptides. They are positively charged exhibiting amphipathic characteristics. They possess the ability to penetrate the cell membrane and translocate different cargoes into cell (Zorko and Langel 2005).

Cell-penetrating peptides will enter living cells using adsorptive-mediated endocytosis without producing any cytolytic effects. Some cell-penetrating peptides are model amphipathic peptide (MAP), sequence signal-based peptide (SBP), fusion sequence-based peptide (FBP), TAT, and HIV-I transactivating transcriptor, and they differ in length and sequence, amphipathic nature, net positive charge, hydrophobicity, and helical moment. Drin et al. (2003) presented an *in vitro* study of the cellular uptake of peptides, originally derived from protegrin. Protegrin is a synthetic B-type peptide vector that also takes part in enhancing the transport of drug through BBB. An *in vitro* study is performed to evaluate the cellular uptake of one of its derivative peptides, and evaluation of internalization process of two lipid-interacting peptides also takes place.

After the study, it was concluded that adsorptive-mediated endocytosis governed the mechanism of cell penetration for syn B and pAntp peptides in comparison to temperature-independent translocation. Other cell-penetrating peptides such as TAT-fusion proteins are rapidly internalized by lipid raft-dependent macropinocytosis (Wadia et al. 2004).

### 5.6.4.6   Drug Transport via Endogenous Receptor-Mediated Transcytosis

One of the drawbacks associated with adsorptive-mediated endocytosis is its lack of selectivity to the target organ. To overcome this problem, cargo molecules with associated ligands can be transported via receptor-mediated transcytosis. It is also known as a molecular Trojan horse approach, which denotes various brain transport vectors such as monoclonal antibodies, modified proteins, and endogenous peptides. They are useful in transporting various substances such as antisense radiopharmaceuticals (Suzuki et al. 2004), β-galactosidase (Zhang and Pardridge 2005), and brain-derived neurotrophic factor (Wu and Pardridge 1999). Receptor-mediated transcytosis is observed in three steps: (a) endocytosis at the luminal blood side after receptor–ligand binding; (b) transversing the endothelia cytoplasm; and (c) endocytosis of the drug or ligand-attached drug or cargo at the brain side (Pardridge 2002). Sometimes movement through the endothelial cytoplasm is associated with endosomal/liposomal systems, which are responsible for degradation of drug and therapeutic protein, and pH-sensitive liposomes are used to overcome this problem.

## 5.7 CONCLUSION

Brain, a vibrant and extremely regulated organ, is separated from the remaining part of the body by BBB and BCSFB. The normal functioning of the homeostatic mechanism of brain regulates the uptake and accumulation of normal nutrients, xenobiotic, and drugs as well as toxins. In this chapter, the focus is given on the barrier properties of brain from biological and pathological perspectives to provide a better idea on the challenges and opportunities associated with the drug targeting to BBB. In this regard, we need to remind ourselves that during drug targeting, our target organ is the diseased brain not the healthy brain. Thus, based on the disease of the brain, the proper selection and designing of the drug delivery system considering the target ligand should be made. For successful brain targeting, proper understanding of the localization, functional expression, and specificity of the brain receptors is essential to increase the permeability of most of the brain therapeutics through the barriers, which will improve pharmacokinetic and pharmacodynamic profile of CNS-active pharmaceuticals.

## REFERENCES

Alavijeh, M.S., Chishty, M., Qaiser, M.Z., and Palmer, A.M. 2009. Drug metabolism and pharmacokinetics, the blood–brain barrier, and central nervous system drug discovery. *NeuroRx* 2:554–71.

Alyautdin, R.N., Tezikov, B.E., Ramge, P., Kharkevich, D.A., Begley, D.J., and Kreuter, J. 1998. Significant entry of tubocurarine into the brain of rats by adsorption to polysorbate-80 coated polybutylcyanoacrylate nanoparticles: An *in situ* brain perfusion study. *J Microencapsul* 15:67–74.

Apfel, R.E. 1982. Acoustic cavitation: A possible consequence of biomedical use of ultrasound. *Br J Cancer* 5:140–46.

Ballabh, P., Braun, A., and Nedergaard, M. 2004. The blood–brain barrier: An overview Structure, regulation, and clinical implications *Neuroniol Dis* 16:1–13.

Begley, D.J. 1996. The blood-brain barrier: Principles for targeting peptides and drugs to the central nervous system. *J Pharm Pharmacol* 48:136–46.

Blasberg, R.G., Kobayashi, T., Horowitz, M., Rice, J., Groothuis, D.R., Molnar, P., and Fenstermacher, J.D. 1983. Regional blood-to-tissue transport in ethylnitrosourea-induced brain tumors. *Ann Neurol* 14:202–15.

Blasberg, R.G., Nakagawa, H., Bourdon, M.A., Groothuis, D.R., Patlak, C.S., and Bigner, D.D. 1987. Regional localization of a glioma-associated antigen defined by monoclonal antibody 81C6 in vivo: Kinetics and implications for diagnosis and therapy. *Cancer Res* 47:4432–87.

Ballantine, H.T. Jr., Bell, E., and Manlapaz, J. 1960. Progress and problems in the neurological applications of focused ultrasound. *J Neurosurg* 17:858–76.

Brightman, M. 1992. Ultrastructure of brain endothelium. In *Physiology and Pharmacology of the Blood-Brain Barrier: Handbook of Experimental Pharmacology*, ed. M.W.B. Bradbury, 1–22. Berlin: Springer-Verlag.

Calvo, P., Gouritin, B., Chacun, H. et al. 2001. Long-circulating PEGylated polycyanoacrylate nanoparticles as new drug carrier for brain delivery. *Pharm Res* 18:1157–66.

Chapelon, J.Y., Ribault, M., Vernier, F., Souchon, R., and Gelet, A. 1999. Treatment of localised prostate cancer with transrectal high intensity focused ultrasound. *Eur J Ultrasound* 9:31–38.

Chen, W., Wang, Z., Wu, F. et al. 2002. High intensity focused ultrasound in the treatment of primary malignant bone tumor. *Chinese J Oncol* 24:612–15.

Chen, Y. and Lihong, Liu. 2012. Modern methods for delivery of drugs across the blood-brain barrier. *Adv Drug Deliv Rev* 64:640–65.

Coombes, A.G.A., Scholes, P.D., Davies, M.C., Illum, L., and Davis, S.S. 1994. Resorbable polymeric microspheres for drug delivery—production and simultaneous surface modification using PEO-PPO surfactants. *Biomaterials* 15:673–80.

Dallasta, L.M., Pisarov, L.A., Esplen, J.E. et al. 1999. Blood–brain barrier tight junction disruption in human immunodeficiency virus-1 encephalitis. *Am J Pathol* 155:1915–27.

Deli, M.A. 2009. Potential use of tight junction modulators to reversibly open membranous barriers and improve drug delivery. *Biochim Biophys Acta* 1788:892–10.

Diederich, C.J. and Hynynen, K. 1999. Ultrasound technology for hypethermia. *Ultrasound Med Biol* 25:871–87.

Discher, D.E. and Ahmed, F., 2006. Polymersomes. *Annu. Rev. Biomed. Eng* 8:323–341.

Drin, G., Cottin, S., Blanc, E., Rees, A.R., and Temsamani, J. 2003. Studies on the internalization mechanism of cationic cell-penetrating peptides. *J Biol Chem* 278:31192–01.

Doolittle, N.D., Miner, M.E, Hall, W.A. et al. 2011. Transport of drugs across the blood-brain barrier by nanoparticles. *J Control Release* 59:71–81.

Emerich, D.F., Snodgrass, P., Dean, R. et al. 1999. Enhanced delivery of carboplatin into brain tumors with intravenous cereport (RMP-7): Dramatic differences and insight gained from dosing parameters. *Br J Cancer* 80:964–70.

Emerich, D.F., Snodgrass, P., Pink, M., Bloom, F., and Bartus, R.T. 1998. Central analgesic actions of loperamide following transient permeation of the blood brain barrier with Cereport (RMP-7). *Brain Res* 801:259–66.

Fromm, M.F. 2000. P-glycoprotein: A defense mechanism limiting oral bioavailability and CNS accumulation of drugs. *Int J Clin Pharmacol Ther* 38:69–74.

Feng, M.R. 2002. Assessment of blood–brain barrier penetration: *In silico, in vitro* and *in vivo*. *Curr Drug Metab* 3:647–57.

Fundaro, A., Cavalli, R., Bargoni, A., Vighetto, D., Zara, G.P., and Gasco, M.R. 2000. Non-stealth and stealth solid lipid nanoparticles (SLN) carrying doxorubicin: Pharmacokinetics and tissue distribution after i.v. administration to rats. *Pharmacol Res* 42:337–43.

Fenart, L., Casanova, A., and Dehouck, B. 1999. Evaluation of effect of charge and lipid coating on ability of 60 nm nanoparticles to cross an in vitro model of the blood-brain barrier. *J Pharmacol Exp Ther* 291:1017–22.

Gandhi, N., Saiyed, Z.M., Napuri, J. et al. 2010. Interactive role of human immunodeficiency virus type 1(HIV-1) clade-specific Tat protein and cocaine in blood–brain barrier dysfunction: Implications for HIV-1-associated neurocognitive disorder. *J Neurovirol* 16:294–05.

Gulyaev, A.E., Gelperina, S.E., Skidan, I.N., Antropov, A.S., Kivman, G.Y., and Kreuter, J. 1999. Significant transport of doxorubicin into the brain with polysorbate-80 coated nanoparticles. *Pharm Res* 16:1564–69.

Herrero-Vanrell, R., Rincón, A.C., Alonso, M., Reboto, V., Molina-Martinez, I.T., and Rodríguez-Cabello, J.C. 2005. Self-assembled particles of an elastin-like polymer as vehicles for controlled drug release. *J Control Release* 102:113–22.

Hynynen, K. 2008. Ultrasound for drug and gene delivery to the brain. *Adv Drug Deliv Rev* 60:1209–17.

Hynynen, K., McDannold, N., Vykhodtseva, N., and Jolesz, F.A. 2001. Noninvasive MR imaging-guided focal opening of the blood-brain barrier in rabbits. *Radiology* 220:640–46.

Karyekar, C.S., Fasano, A., Raje, S., Lu, R., Dowling, T.C., and Eddington, N.D. 2003. Zonula occludens toxin increases the permeability of molecular weight markers and chemotherapeutic agents across the bovine brain microvessel endothelial cells. *J Pharm Sci* 92:414–423.

Krewson, C.E., Klarman, M.L., and Saltzman, W.M. 1995. Distribution of nerve growth factor following direct delivery to brain interstitum. *Brain Res* 680:196–06.

Kreuter, J., Shamenkov, D., Petrov, V. et al. 2002. Apolipoprotein-mediated transport of nanoparticle-bound drugs across the blood–brain barrier. *J Drug Target* 10:317–25.

Lai, R., Wen, Z., Yan, Z. et al. 2011. Odorranalectin-conjugated nanoparticles: Preparation, brain delivery and pharmacokinetic study on parkinson's disease following intranasal administration. *J Control Release* 151:131–38.

Langer, R., Townsend, S.A., Evrony, GD, Gu, F.X., Schulz, M.P., and Brown, R.H. 2007. Tetanus toxin C fragment-conjugated nanoparticles for targeted drug delivery to neurons. *Biomaterials* 28:5176–84.

Lewis, D.F.V. and Dickins, M. 2002. Substrate SARs in human P450 s. *Drug Discov Today* 7:918–25.

Li, J., Feng, L., Zha, Y. et al. 2011. Targeting the brain with PEG-PLGA nanoparticles modified with phage-displayed peptides. *Biomaterials* 32:4943–50.

Lo, E.H., Singhal, A.B., Torchilin, V.P., and Abbott N.J. 2001. Drug delivery to damaged brain. *Brain Res Rev* 38:140–48.

Mangas-Sanjuan, V., González-Alvarez, M., Gonzalez-Alvarez, I., and Bermejo, M. 2010. Drug penetration across the blood–brain barrier: An overview. *Ther Deliv* 1:535–62.

Margulis, M.A. 1995. *Sonochemistry of Cavitation*. Luxembourg: Gordon & Breach Publishers.

Mikhail, A.S. and Allen C. 2009. Block copolymer micelles for delivery of cancer therapy: Transport at the whole body, tissue and cellular levels. *J Control Release* 138:214–23.

Misra, A., Ganesh, S., Shahiwala, A., and Shah, S.P. 2003. Drug delivery to the central nervous system: A review. *J Pharm Pharmaceut Sci* 6:252–73.

Monnaert, V., Betbeder, D., Fenart, L. et al. 2004. Effects of gamma- and hydroxypropyl-gamma-cyclodextrins on the transport of doxorubicin across an in vitro model of blood–brain barrier. *J Pharmacol Exp Ther* 311:1115–20.

Pardridge W.M. 2002. Drug and gene targeting to the brain with molecular Trojan horses. *Nat Rev Drug Disco* 1:131–39.

Pathan, S.A., Iqbal, Z., Zaidi, S.M. et al. 2009. CNS drug delivery systems: Novel approaches. *Recent Pat Drug Deliv Formul* 3:71–89.

Rip, J., Schenk, G.J., and Boer de, A.G. 2009. Differential receptor-mediated drug targeting to the diseased brain. *Expert Opin Drug Deliv* 6:227–37.

Saija, A., Princi, P., Trombetta, D, Lanza M., and Pasquale De A. 1997. Changes in the permeability of the blood–brain barrier following sodium dodecyl sulphate administration in the rat. *Exp Brain Res* 115:546–51.

Sato, S., Suga, S., Yunoki, K., Mihara, B., 1994. Effect of barrier opening on brain edema in human brain tumors. *Acta Neurochir Suppl. (Wien)* 60, 116–118.

Schlageter, K.E., Molnar, P., Lapin, G.D., and Groothuis, D.R. 1999. Microvessel organization and structure in experimental brain tumors: Microvessel populations with distinctive structural and functional properties. *Microvasc Res* 58:312–28.

Schroder, U. and Sabel, B.A. 1996. Nanoparticles, a drug carrier system to pass the blood-brain barrier, permit central analgesic effects of i.v. dalargin injections. *Brain Res* 710:121–24.

Schroder, U., Sommerfeld, P, Ulrich, S., and Sabel, B.A. 1998. Nanoparticle technology for delivery of drugs across the blood-brain barrier. *J Pharm Sci* 87:1305–07.

Schulze, C., Smales, C., Rubin, L.L., and Staddon, J.M. 1997. Lysophosphatidic acid increases tight junction permeability in cultured brain endothelial cells. *J Neurochem* 68:991–1000.

Stolnik, S., Illum, L., and Davis, S.S. 1995. Long circulating microparticulate drug carriers. *Adv Drug Deliv Rev* 16:195–214.

Suzuki, T., Wu, D., Schlachetzki, F., Li, J.Y., Boado, R.J., and Pardridge, W.M. 2004. Imaging endogenous gene expression in brain cancer in vivo with 111 in peptide nucleic acid antisense radiopharmaceuticals and brain drug targeting technology. *J Nucl Med* 45:1766–75.

Sztriha, L. and Betz, A.L. 1991. Oleic acid reversibly opens the blood–brain barrier. *Brain Res* 550:257–62.

Tilloy, S., Monnaert, V., Fenart, L., Bricout, H., Cecchelli, R., and Monflier, E. 2006. Methylated beta-cyclodextrin as P-gp modulators for deliverance of doxorubicin across an in vitro model of blood–brain barrier. *Bioorg Med Chem Lett* 16:2154–57.

Vajkoczy, P. and Menger, M.D. 2004. Vascular microenvironment in gliomas. *Cancer Treat Res* 117:249–62.

van De Waterbeemd, H., Smith, D.A., Beaumont, K., and Walker, D.K. 2001. Property-based design: Optimization of drug absorption and pharmacokinetics. *J Med Chem* 44:1313–33.

Wadia, J.S., Stan, R.V., and Dowdy, S.F. 2004. Transducible TAT-HA fusogenic peptide enhances escape of TAT-fusion proteins after lipid raft macropinocytosis. *Nat Med* 10:310–15.

Wang, YC., Liu, X.Q., Sun, T.M., Xiong, M.H., and Wang, J. 2008. Functionalized micelles from block copolymer of polyphosphoester and poly ε–caprolactone for receptor mediated drug delivery. *J Control Release* 128:132–40.

Wilson, B., Samanta, M.K., Santhi, K., Kumar, K.P.S., Pramakrishnan, N., and Suresh, B. 2008. Targeted delivery of tacrine into the brain with polysorbate 80-coated poly (N-butylcyanoacrylate) nanoparticles. *Eur J Pharm Biopharm* 70:75–84.

Wisse, E. and Leeuw, A.M. 1984. Structural elements determining transport and exchange process in the liver. In: *Microspheres and Drug Therapy, Pharmaceutical, Immunological and Medical Aspects*, eds. S.S. Daviss, L. Illum, J.G. McVie, and E. Tomlinson, 1–23. Amsterdam: Elsevier.

Woodle, M.C. 1995. Sterically stabilized liposome therapeutics. *Adv Drug Deliv Rev* 16:249–65.

Wu, D. and Pardridge, W.M. 1999. Neuroprotection with noninvasive neurotrophin delivery to the brain. *Proc Nat Acad Sci USA* 96:254–59.

Wu, F., Wang, Z.B., Cao, Y.D., Chen, W.Z., Bai, J., Zou, J.Z., and Zhu, H. 2003. A randomised clinical trial of high intensity focused ultrasound ablation for the treatment of patients with localised breast cancer. *Br J Cancer* 89:2227–33.

Yang, C.S., Lu, F.L., and Cai, Y. 1999. Body distribution in mice of intravenously injected camphotothericin solid lipid nanoparticles and targeting effect on the brain. *J Control Release* 59:299–07.

Zorko, M. and Langel U. 2005. Cell-penetrating peptides: Mechanism and kinetics of cargo delivery. *Adv Drug Deliv Rev* 57:529–45.

Zhang, Y. and Pardridge, W.M. 2005. Delivery of beta –galactosidase to mouse brain via the blood brain barrier transferring receptor. *J Pharmacol Exp Ther* 313:1075–81.

Zhang, X., Xie, Y., Jin, Y., Hou, X., Ye, L., and Lou, J. 2004. The effect of RMP-7 and its derivative on transporting Evans blue liposomes into the brain. *Drug Deliv* 11:301–09.

# 6 Natural Polymers in Colon Targeting
## *Approaches and Future Perspectives*

*Sougata Jana, Arijit Gandhi, and Subrata Jana*

## CONTENTS

## 6.1   INTRODUCTION

The gastrointestinal tract (GIT) is generally divided into three parts: stomach, small intestine, and large intestine. The large intestine extending from the ileocecal junction to the anus is divided into three main parts: the colon, rectum, and anal canal. The important function of the colon is the creation of a suitable environment for the growth of colonic microorganisms, expulsion of the contents of the colon at an appropriate time, and potassium absorption, storage reservoir of fecal contents, and water from the lumen. The colon tissue generally contains the villi, lymph, nerves, muscle, and blood vessels (Sarasija and Hota 2000; Chourasia and Jain 2003; Kumar and Pratibha 2012). The anatomical features of the colon are displayed in Figure 6.1.

The colon is considered an important site of drug targeting for treating local diseases. Targeted drug delivery into the colon is highly essential for the local

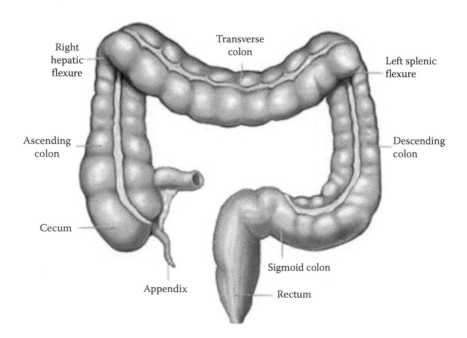

FIGURE 6.1    Anatomy of the colon.

treatment of a variety of bowel diseases such as amebiasis, ulcerative colitis, colonic cancer and Crohn's disease, local treatment of colonic pathologies, and systemic delivery of active drugs. The colon-targeted drug delivery system (CTDDS) should be capable of protecting the drug in route to the colon, that is, the absorption and drug release (DR) should be avoided in the stomach (acidic pH) and the small intestine pH, and neither the bioactive agent should be degraded in either of the dissolution sites, but the drug should be released and absorbed only in the colonic site. The colon is believed to be an ideal absorption site for protein and peptide drugs for the following various factors such as (i) less degradation occurs by the digestive enzymes and (ii) comparative proteolytic activity of colon mucosa is much less than that observed in the small intestine, thus CTDDS protects the peptide drugs from enzymatic degradation and hydrolysis in the duodenum and jejunum, and eventually the drug releases into the ileum or colon which leads to greater systemic bioavailability and, finally, the colon has a long residence time and is highly responsive to absorption enhancers (Sarasija and Hota 2000; Bajpai et al. 2003; Chourasia and Jain 2003; Fatima et al. 2006). The simplest method for targeting of the drugs to the colon is to obtain slower release rates or prolong the periods of drug release by the application of thicker layers of conventional enteric coating or by developing the polymeric matrices. These sustained mechanisms are designed to improve the efficacy of the drug by concentrating the drug molecules, where they are needed most and also minimize the potential side effects and drug instability issues associated with premature drug release in the upper parts of the GIT, namely the stomach and small intestine. CTDDS would ensure lower dosing, direct treatment at the disease site, and less systemic side effects. In addition to restricted therapy, the colon can also be utilized as a portal for the entry of drugs into the systemic circulation (Aurora 2006; Brahma 2007). The use of natural biopolymers holds great applications to achieve targeted drug release to the colon. The class of natural biopolymers plays an important role in drug delivery as it comprises polymers with a large number of functional groups, a wide range of molecular weights, varying chemical compositions, biodegradability, low toxicity, and high stability. The natural polysaccharides represent one of the most important industrial raw materials and have been the subject of intensive research due to their biodegradability, sustainability, and biosafety (Dey et al. 2008; Zhang et al. 2011; Shukla and Tiwari 2012). These natural polymers are degraded by the colonic bacterial enzymes. The important bacteria present in the colon such as *Bifidobacterium, Bacteroides, Clostridium, Eubacterium, Peptococcus,* and *Lactobacillus* secrete a variety of reductive and hydrolytic enzymes such as β-xylosidase, β-glucuronidase, β-galactosidase, nitroreductase, α-arabinosidase, deaminase, azoreductase, and urea hydroxylase. These enzymes are responsible for the degradation of di-, tri-, and polysaccharides; for this reason, natural polymers are gaining much more attention in the development of colon-specific drug delivery systems (Cavalcanti et al. 2002; Sinha and Rachna 2003). Therefore, the objective of the chapter is to provide the scientific information on the natural polymer-based formulation for colon targeting and the treatment of different types of disease associated with the colon.

## 6.2   IMPORTANCE OF THE COLON FOR DRUG TARGETING

The following advantages are obvious with the systems designed for the localized delivery of therapeutic agents to the colon:

a. Targeted drug delivery to the colon would confirm direct treatment at the disease sites, lower dosing, and less systemic side effects.
b. Site-specific or targeted drug delivery system would allow oral administration of peptide and protein therapeutics; colon-specific formulation could also be used to prolong the drug delivery.
c. Colon-specific drug delivery system is important for the treatment of colon diseases.
d. The colon is an important site where local or systemic drug delivery is easily achieved, and thus this approach can be used for the topical treatment of inflammatory bowel disease (IBD), for example, Crohn's disease or ulcerative colitis. Such types of inflammatory diseases are generally treated with glucocorticoids and sulfasalazine (targeted).
e. A number of other serious diseases of the colon (e.g., colorectal cancer) might also be capable of being treated more completely if drugs were targeted to the colon.
f. Pharmaceutical formulations for colonic delivery are also suitable for the delivery of drugs which are polar and/or susceptible to chemical and enzymatic degradation in the upper GI tract and those that are highly affected by hepatic metabolism, in particular, therapeutic proteins and peptides.
g. It has a longer retention time and appears highly responsive to agents that enhance the absorption of poorly absorbed drugs (Halsas et al. 2001; Philip et al. 2008; Kothawade et al. 2009).

## 6.3   FACTORS AFFECTING COLON-TARGETED DRUG DELIVERY

### 6.3.1   pH OF THE COLON

Different pH is observed in the stomach, small intestine and colon, and this depends on various factors such as food intake, diet, intestinal motility, and disease condition. This variability in the gastric pH makes it a more challenging job for the pharmaceutical research scientist to design a delivery system that allows sustained drug release in the desired site. Despite variation in pH throughout the entire length of GIT, the colon drug delivery system allows the targeting of drug in the specific site. The pH of GIT varies from pH 1.2 (stomach) through pH 6.6 (proximal small intestine) to pH 7.5 (distal small intestine). Also distinct pH values are observed at different parts of the colon such as right colon (pH 6.4), mid colon (pH 6.6), and left colon (pH 7.0). The pH of the colon is often lower than the pH of the small intestine, which is as high as 8 or 9. There is a fall in pH on the entry into the colon due to the presence of short-chain fatty acids produced by the bacterial fermentation of polysaccharides. So, the drug targeting to the small intestine is done by the enteric coating of pH sensitive materials. This variation in the pH in different parts of GIT is the basis for the development of CTDDS. Coating with different polymers is done to target the drug to the specific site (Kumar et al. 2009; Challa et al. 2011).

### 6.3.2 COLONIC MICROFLORA AND THEIR ENZYMES

Drug release in various parts of GIT depends on the presence of intestinal enzymes that are derived from the gut microflora residing in high numbers in the colon. The concentration of bacteria in the human colon is around 1000 CFU/mL. The most important anaerobic bacteria are *Bifidobacterium, Bacteroides, Eubacterium, Peptococcus, Propionibacterium, Peptostreptococcus, Ruminococcus,* and *Clostridium*. In the recent years, the GI microflora in the colonic region is important for drug release which is of main interest to researchers. The majority of bacteria in the GIT are present in the distal gut and the remaining are distributed throughout. Endogenous and exogenous substrates, such as carbohydrates and proteins, escape digestion in the upper GIT but are metabolized by the enzymes secreted by colonic bacteria (Yang et al. 2002; Chourasia and Jain 2003).

### 6.3.3 COLON TRANSIT OF THE MATERIAL

One of the major evaluation parts for the absorption of active compounds in the colon is its residence time in any particular segment of the colon. The comparatively constant transit time in the small intestine of approximately 3–5 h is another physiological characteristic that can be taken advantage of to achieve colon specificity (i.e., time-controlled drug delivery). After gastric emptying, a time-controlled drug delivery system is intended to release the drug after a predetermined lag phase. Gastric emptying of dosage forms is highly variable and depends primarily on whether the subject is fed or fasted and on the properties of the dosage form such as density and size (Kothawade et al. 2011; Mehta et al. 2011).

### 6.3.4 DRUG ABSORPTION IN THE COLON

The poor paracellular absorption of many drugs from the colon is due to the fact that epithelial cell junctions are very tight. The slow rate of transit in the colon allows the drug to stay in contact with the mucosa for a prolonged period than in the small intestine, which compensates the much lower surface area. The colonic contents become more viscous with progressive absorption of water as one travels further through the colon. This causes a decrease in dissolution rate, slow diffusion of the dissolved drug through the mucosa (Reddy et al. 2013). Due to high retention time, the colon may allow sufficient time for absorption of less-absorbed drugs and may cause systemic toxicity.

### 6.3.5 OTHER FACTORS

The selection of carrier for CTDDS depends on the nature of the drug and the disease for which the drug is used. The various physicochemical factors that affect the carrier selection for drugs include chemical nature, partition coefficient, stability, functional groups on the drug molecules, and so on (Mahale et al. 2013).

### 6.3.6  General Considerations

In general, three factors are considered for designing colonic formulations:

a. Pathology of the disease, especially the affected parts of lower GIT and pH.
b. Biopharmaceutical and physicochemical properties of the drug, such as solubility, permeability, and stability at the intended site of delivery.
c. The desired drug release data.

The most common physiological factor is pH gradient of the GIT. In normal healthy subjects, there is a progressive increase in luminal pH from the duodenum (pH 6.6) to the end of the ileum (pH 7.5), a decrease in the cecum (pH 6.4), and then a slow rise from the right to the left colon with a final value of 7.0. Some reports revealed that changes in gastrointestinal pH profiles may occur in patients with IBD, which should be considered in the development of prolonged release of the formulations. Formulation of the drugs for colonic delivery also requires careful consideration of drug dissolution or release rate in the colonic medium. The poor dissolution and drug release rate relates to lower systemic availability of drugs. These matters could be more problematic when the drug substances are lipophilic or require higher doses for therapy. Subsequently, such drugs need to be delivered in a presolubilized form or the formulations should be targeted for proximal colon which has more fluid than in the distal colon (Patel et al. 2006; Kumar and Mishra 2008).

## 6.4  NATURAL POLYMERS IN COLON TARGETING

Natural polymers are emerging in the design of novel drug delivery systems (NDDS) such as those that targeted delivery of the drug to a specific area in the GIT or in response to external stimuli to release the drug. This can be made via different mechanisms including coating of tablets with natural polymers having pH-dependent solubility or incorporating nondigestible polymers that are degraded by the bacterial enzymes in the colon. Nonstarch, linear polysaccharides are resistant to the digestive action of the gastrointestinal enzymes and retain their integrity in the upper GIT. Polymeric matrices manufactured from these polysaccharides therefore remain intact in the stomach and small intestine medium, but once they enter into the colon they are degraded by the bacterial polysaccharidases (Cavalcanti et al. 2002; Kaushik et al. 2009). This property makes these natural polysaccharides exceptionally suitable for the formulation of CTDDS. A large number of natural polymers have already been reported for their potential applications as CTDDS (Figure 6.2). Some important polymers are discussed below.

### 6.4.1  Pectin

Pectin is an anionic and nontoxic polysaccharide extracted from the cell walls of most plants. It is also found in the junctional zone between cells within secondary cell walls including fiber cells and xylem in woody tissue. Pectin is composed

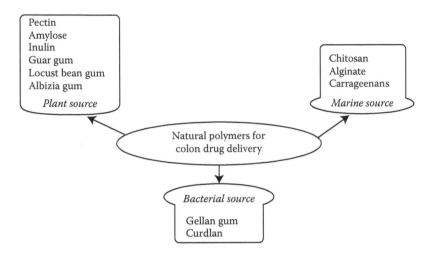

**FIGURE 6.2**   Different natural polymers for colon drug delivery.

of linearly connected α-(1–4)-D-galacturonic acid residues partially esterified with methanol (Figure 6.3a).

The degree of methoxylation (DM) is used to classify the pectins as high methoxyl pectins (HMP) (DM >50) and low methoxyl pectins (LMP) (DM <50) (Sriamornsak et al. 2007; Bhatia et al. 2008). Pectin has been utilized as excipient in many different types of dosage forms such as microparticulate ophthalmic delivery systems and matrix-type transdermal patches, film coating of colon-specific drug delivery systems in combination with ethyl cellulose. It has high capacity as a hydrophilic polymeric compound for controlled release matrix drug delivery systems, but its aqueous solubility causes fast and premature release of the drug from their matrices (Shirwaikar et al. 2008). Nowadays, the high solubility of pectin in aqueous medium is reduced through chemical modification without affecting biodegradability properties. Pectins can be chemically modified by saponification catalyzed by the mineral acids, salts of weak acids, bases, enzymes, primary aliphatic amines, and concentrated ammonium systems. Calcium salts of pectin have been found to decrease the solubility, and matrix tablets prepared with calcium pectinate showed very good capacity to be used in CTDDSs (Liu et al. 2003; Chourasia and Jain 2004a). Pectin is refractory to host gastric and small intestinal enzymes, but is almost completely degraded by the colonic bacterial enzymes to produce a series of soluble oligalacto-runates. Calcium pectinate, the insoluble salt of pectin, was used for colon-targeted drug delivery of indomethacin (Rubinstein et al. 1993). Calcium pectinate nanoparticles were investigated as a potential colonic delivery system by ionotropic gelation method to deliver insulin (Tozaki et al. 1997). Pectin also showed very good capacity in colon-specific drug delivery systems for systemic action or treatment of localized diseases such as Crohn's disease and ulcerative colitis. Gel-forming systems have been widely investigated for sustained drug delivery. Perera et al. (2010) prepared metronidazole-containing microparticles based on a pectin–4-aminothio-phenol conjugate for colon-specific drug delivery (Perera et al. 2010). The pectin-HPMC compression-coated tablets were investigated by Ugurlu et al. (2007) for the

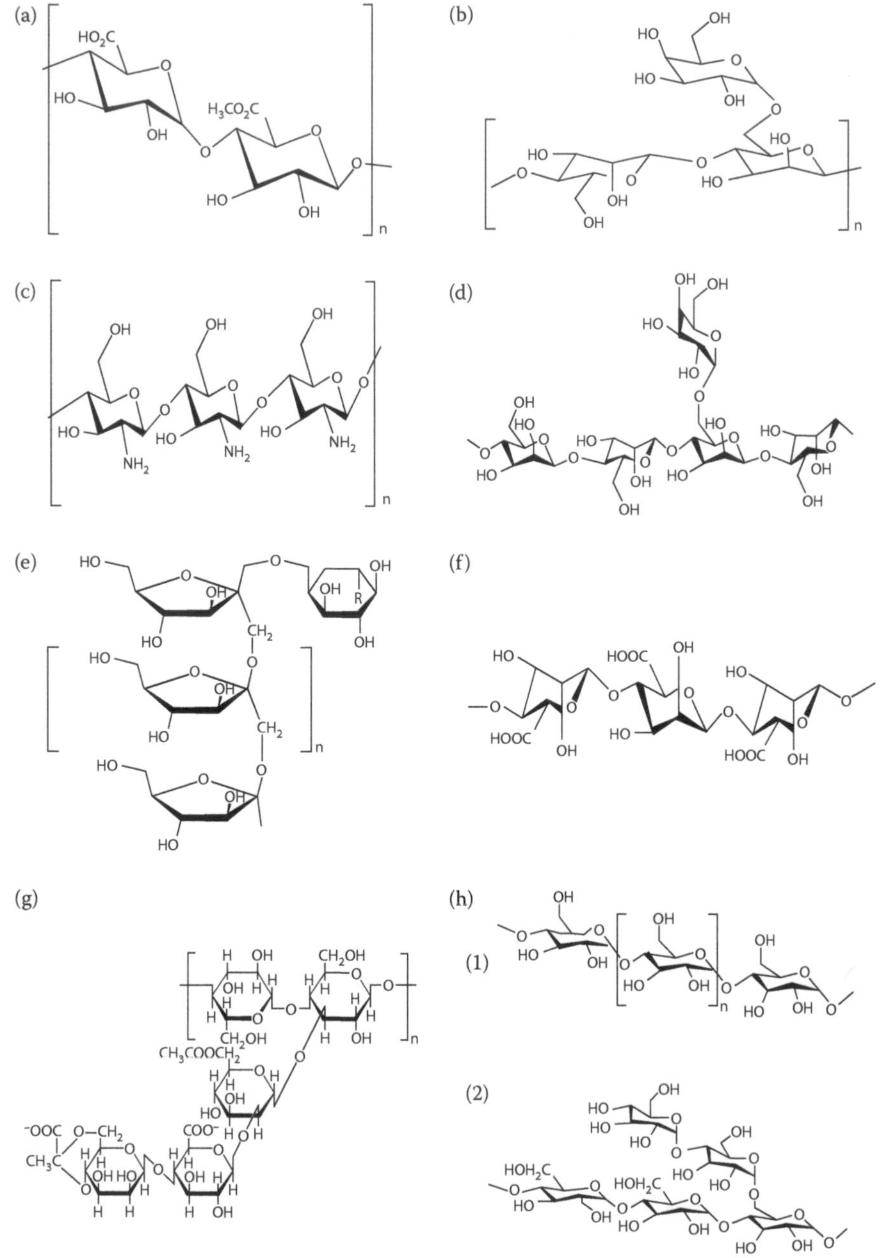

**FIGURE 6.3** Chemical structure of (a) pectin; (b) guar gum; (c) chitosan; (d) locust bean gum; (e) inulin; (f) alginate; (g) xanthan gum; and (h) starch with amylose and amylopectin.

colonic delivery of nisin. Maestrelli et al. (2008) fabricated enteric-coated calcium pectinate microspheres for the same purpose by using theophylline as a model drug (Maestrelli et al. 2008).

## 6.4.2 GUAR GUM

It is a naturally occurring galactomannan polysaccharide; consists of chiefly high-molecular weight hydrocolloidal polysaccharide, chemically composed of galactan and mannan units combined through glycosidic linkages and shows degradation in the large intestine due to the presence of microbial enzymes. Chemically, guar gum (GG) is a galactomannan which occurs as a storage polysaccharide in the seed endosperm of plants in the Fabaceae family. Galactomannans are linear polysaccharides consisting of (1 → 4) diequatorially linked β-D-mannose monomers, some of which are linked to single sugar side chains of α-D-galactose (Figure 6.3b) (Chourasia and Jain 2004b; Gamal-Eldeen et al. 2006; Doyle et al. 2008).

GG is hydrophilic in nature and swells in cold water, forming colloidal viscous dispersions or sols. This gel-like property of the matrix can retard drug release from the dosage form, which mainly degrades in the colon. GG was found to be a colon-specific drug carrier in the form of polymer matrix, compression-coated tablets as well as microspheres. GG is particularly useful for colon delivery because it can be degraded by specific enzymes in this region of the GIT. The gum also protects the drug while in the stomach and small intestine pH and releases the drug to the colon, where it undergoes degradation by the enzymes excreted by these microorganisms or assimilation by specific microorganisms. As hydrogel, GG was not found to be highly suitable for controlled release of water-soluble drugs because of their relatively fast delivery, but is useful for poorly water-soluble drugs (Krishnaiah et al. 2003; Al-Saidan et al. 2005; Coviello et al. 2007). GG-based matrix tablets containing rofecoxib were formulated for their intended use in the chemoprevention of colorectal cancer (Al-Saidan et al. 2005). A colon-specific GG-based tablet composed of 5-fluorouracil has also been reported (Krishnaiah et al. 2003). Salve (2011) prepared sustained release tablets containing ibuprofen for colon targeting. Tablets were also formulated using a combination of guar and xanthan gum. Biodegradability of composite (guar and xanthan gum) was investigated in the presence of 4% w/v rat cecal content (RCC) and galactomannase enzyme (0.1 mg/mL) by viscosity measurement. A significant decrease in viscosity was found with 4% RCC after 24 h incubation. *In vitro* drug release studies revealed that guar and xanthan gum exhibited controlled release for 24 h at a mass ratio of 1:1.5 (Salve 2011). Singh et al. (2012) also developed GG-based colon-targeted delivery carriers for nitrofurantoin. Matrix tablets containing various proportions of GG were prepared by wet granulation technique using starch paste as a binder. The *in vitro* results of release study stated that matrix tablets containing 20% of GG were most likely to provide targeting of nitrofurantoin in the colon (Singh et al. 2012).

## 6.4.3 CHITOSAN

Chitosan, a linear amino polysaccharide composed of randomly distributed (1 → 4)-linked D-glucosamine and N-acetyl-D-glucosamine units is obtained by the deacetylation of chitin, a widespread natural polysaccharide found in the exoskeleton of crustaceans such as crab and shrimp (Figure 6.3c).

The degree of deacetylation has a significant effect on the solubility and rheological properties of the chitosan. Chitosans with a low degree of deacetylation (<40%) are soluble up to a pH of 9, whereas highly deacetylated chitosans (>85%) are soluble only up to pH 6.5. Chitosan is a weak base with a $pK_a$ value in the range 6.2–7.0, depending on the source of the polymer. At low pH (acidic environment), the polymer is soluble with the sol-to-gel transition occurring at ~pH 7.0. Chitosan has favorable biological properties such as biocompatibility, nontoxicity, and biodegradability (Addo et al. 2010; Jana et al. 2013). Chitosan-based delivery systems can protect therapeutic agents from the hostile conditions of the upper GIT and release the entrapped bioactive agents specifically at the colon through degradation of the glycosidic linkages of chitosan by colonic microflora. By using male Wistar rats as an animal model, Tozaki et al. (1999) compared the healing effect of chitosan capsules containing a new thromboxane synthase inhibitor (R68070) on ulcerative colitis induced by 2,4,6-trinitrobenzene sulfonic acid (TNBS) with that of carboxymethylcellulose (CMC) suspension of R68070. Chitosan capsules provided higher concentrations of R68070 in large intestine than the CMC suspension (Tozaki et al. 1999). The same group observed the healing effect of chitosan capsules containing 5-aminosalicylic acid (5-ASA) on TNBS-induced ulcerative colitis. Chitosan capsules loaded with 5-ASA demonstrated higher therapeutic effect than the 5-ASA-CMC suspension (Tozaki et al. 2002). Hydroxypropyl methylcellulose phthalate, an enteric coating material, was used to coat chitosan capsules loaded with insulin. Using male Wistar rats, insulin-containing chitosan capsules were administered orally with a total dose of 20 IU into the stomach with polyethylene tubing. The hypoglycemic effect started 6 h after administration when the capsules were in the colon, and lasted for 24 h (Tozaki et al. 1997). In another experiment, Alaa-Eldeen et al. (2010) revealed that compression-coated 5-fluorouracil tablets with granulated chitosan successfully delivered the drug to the colon. X-ray studies confirmed localization of the system to the colon (Alaa-Eldeen et al. 2010). Eudragit-S100-coated chitosan beads showed a high degree of protection in the upper GIT and released the maximum amount of satranidazole to the colon (Jain et al. 2007). Chitosan microspheres formulated by emulsion cross-linking and coated with Eudragit-S100 by the solvent evaporation technique resulted in successful release of ondansetron in colon (Jose et al. 2010). Microbially triggered colon-targeted osmotic pumps (MTCTOP) were used to deliver budesonide to colon using chitosan as a carrier matrix. Chitosan has gel-forming property at acidic conditions, which was used to prepare drug suspension and produce osmotic pressure; whereas the colonic degradation property was explored to form *in situ* delivery pores for colon-specific drug release. The effects of different formulation variables, including the level of pH-regulating excipient (citric acid) and the amount of chitosan in the core, the weight gain of semipermeable membrane (cellulose acetate) and enteric coating membrane, and the level of pore former (chitosan) in the semipermeable membrane were studied. The study revealed that osmotic technology combined with the microbial degradation attributes of the colon could be effective for colon-specific drug delivery (Liu et al. 2007).

## 6.4.4    LOCUST BEAN GUM

Locust bean gum (LBG) also known as Carob bean gum is derived from the seeds of the leguminous plant *Ceratonia siliqua* Linn. LBG consists mainly of a neutral galactomannan polymer made up of $(1 \rightarrow 4)$-linked D-mannopyranosyl units and every fourth or fifth chain unit is substituted on C6 with a D-galactopyranosyl unit (Figure 6.3d).

The ratio of D-galactose to D-mannose differs depending upon origin of the gum materials and growth conditions of the plant during production. The viscosity and solubility of LBG, a neutral polymer, are little affected by pH changes within the range of 3–11 (Dey et al. 2012). Cross-linked galactomannan leads to water-insoluble film-forming product showing degradation in colonic microflora (Bauer and Kesselhut 1995). Various significant research works have been carried out in combination with other polymers to impart sustained and targeted release properties to the formulation. Raghavan et al. (2002) formulated colon-specific drug delivery of mesalazine using LBG and chitosan in different molar ratios. The dissolution profile and bioavailability of the formulation containing LBG and chitosan in the ratio of 4:1 were favorable for colon drug targeting. LBG coating provided protection to the core tablet of mesalazine in the upper GIT. The studies conducted in healthy human volunteers stated that the drug release was initiated only after 5 h (transit time of stomach and small intestine) and the bioavailability ($AUC_{0-t}$) of the drug was found to be 147.98, 249.57, and 480.07 μg h/ml, respectively. The application of LBG in the form of compression coat was impressive for drug targeting to the colon (Jenita et al. 2010). Syan and Mathur (2011) investigated a successive colon-targeted delivery of compression-coated aceclofenac tablets. The influence of LBG and xanthan gum polymers and their various combinations was studied in *in vitro* drug release profile. Combination of LBG and xanthan gum was found sufficient to prolong the aceclofenac release in colon. A combination of locust bean gum and xanthan gum as a coating material provided protection to the core tablet for immature drug release of aceclofenac (Syan et al. 2011).

## 6.4.5    INULIN

Inulin is a natural plant-derived polysaccharide with a wide range of pharmaceutical and food applications. Inulin belongs to a class of fibers known as fructans. It is used by some plants as a means of storing energy and is typically found in roots and rhizome. Most of the plants that synthesize inulin do not store other forms of carbohydrate such as starch. Plants that contain high content of inulin are dandelion (*Taraxacum officinale*), wild yam (*Dioscorea villosa*), chicory (*Cichorium intybus*), onion (*Allium cepa*), and garlic (*Allium sativum*). Inulin is resistant to digestion in the upper GIT, but it is degraded in the colon by colonic microflora specifically by *Bifidobacterium*. Inulin is chemically composed of a mixture of oligomers and polymers that belong to the group of glucofructans. $\alpha$-*D*-glucopyranosyl-[$\alpha$-*D*-fructofuranosyl] (n-1)-*D*-fructofuranoside is commonly referred to as inulin. The inulin molecules contain 2–60 or more $\beta$-$(2 \rightarrow 1)$-linked D-fructose molecules (Figure 6.3e) (Akhgari et al. 2006; Barday et al. 2010).

Inulin with the combination of eudragit films were prepared in order to resist degradation in the upper GIT, but promote digestion in lower intestine by the action of *Bifidobacterium* and *Bacteroides* (Vervoort and Kinget 1996). Vervoort et al. (1998) investigated methylated inulin hydrogels for water take up and swelling properties. The rate of water transport into inulin was quite high and showed anomalous dynamic swelling behavior. Inulin, derivatized with methacrylic anhydride and succinic anhydride, produced pH-sensitive hydrogel that reduced swelling and low chemical degradation in acidic medium, but had a good swelling and degradation in simulated intestinal fluid in the presence of inulinase enzyme (Castelli et al. 2008). The incorporation of vinyl groups in this sugar polymer by free radical polymerization aided in the formation of hydrogels which were resistant to the upper GIT. So, these chemically modified hydrogels may be used as a carrier for drug delivery to the colon (Vervoort et al. 1999).

### 6.4.6 ALGINATES

Alginates or alginic acids are linear, unbranched polysaccharides found in brown seaweed and marine algae, such as *Laminaria hyperborea*, *Ascophyllum nodosum*, and *Macrocystis pyrifera*. These polymers are composed of two different monomers in varying proportions, namely $\beta$-D-mannuronic acid and $\alpha$-L-guluronic acid linked in $\alpha$- or $\beta$-(1 → 4) glycosidic bonds as blocks of only $\beta$-D-mannuronic acid or $\alpha$-L-guluronic acid in homopolymers or alternating the two in heteropolymeric blocks (Figure 6.3f).

Alginates have molecular weights of 20–600 kDa (Liew et al. 2006; Tuğcu-Demiröz et al. 2007). Alginates have been used as matrix-forming material in the design of various delivery systems for sustained release of drug over a prolonged period due to its hydrogel-forming properties. The gelling properties of alginate's guluronic residues with polyvalent ions, such as calcium or aluminium, allow cross-linking with subsequent formation of gels that can be employed to prepare matrices, films, beads, pellets, microparticles, and nanoparticles (Sarmento et al. 2007; Ching et al. 2008). Alginate gelation takes place when divalent cations (usually $Ca^{2+}$) interact ionically with blocks of guluronic acid residues, resulting in the formation of a three-dimensional (3D) network that is usually described as an "egg-box" model. It is the ion exchange process between $Ca^{2+}$ and $Na^+$ ions that is supposed to be responsible for the swelling and subsequent degradation of sodium alginate in the colon. Singh et al. (2012) formulated a multiparticulate system combining pH-sensitive property and specific biodegradability and evaluated ES-100-coated alginate microspheres for colon targeting of tinidazole. Different ratios of tinidazole and sodium alginate (1:1, 1:2, 1:3, and 1:4) were used for the preparation of hydrogel microspheres. Eudragit coating of tinidazole-alginate microspheres was done at different core:coat ratios (1:4, 1:5, 1:6, and 1:7). The core microspheres sustained the drug release for 8 h in medium mimicking pH conditions of the GIT. In acidic medium, the release rate was much slower; however, it became quicker at pH 7.4 and sustained up to 24 h. It was revealed that eudragit-coated alginate microspheres were promising controlled release carriers for colon-targeted delivery of tinidazole (Singh et al. 2012). Cross-linked alginates had more capacity to entrap the drug. Further, mixing of alginates with other polymers, such as neutral gums, pectins, chitosan, and eudragits can ameliorate the problems of drug leaching. A novel formulation for oral administration using eudragit-S100-coated calcium alginate gel beads-entrapped liposome and prednisolone as drug was tested for colon-specific drug delivery *in vitro*. Drug release studies were done in conditions mimicking stomach to colon transit. Results showed that the drug was protected from being released completely into the physiological environment of stomach and small intestine. The release rate of prednisolone from the eudragit-coated calcium alginate gel beads-entrapped liposome was dependent on the concentration of calcium, prednisolone, and sodium alginate, as well as the coating materials. The results showed that colonic arrival time of the tablets is normally 4–5 h. It was quite evident that the coated calcium alginate gel beads-entrapped liposome is a potential system for colon-specific drug delivery (Goyal et al. 2011).

### 6.4.7 XANTHAN GUM

Xanthan gum is an extracellular bacterial exopolysaccharide synthesized by *Xanthomonas campestris*. It is highly soluble in water, and stable over a wide range of acidic, alkaline, and temperature conditions. It is resistant to enzymatic degradation and also exhibits synergistic interactions with other hydrocolloids. The primary chemical structure of this naturally produced cellulose derivative contains a cellulose backbone (β-D-glucose residues) and a trisaccharide side chain of β-D-mannose-β-D-glucuronic acid–α-D-mannose attached with alternate glucose residues of the main chain. The terminal D-mannose residues may carry a pyruvate function, the distribution of which is dependent on the bacterial strain and the fermentation conditions. The nonterminal D-mannose unit in the side chain contains an acetyl function. The anionic character of this polymer is due to the presence of both glucuronic acid and pyruvic acid groups in the side chain (Figure 6.3g) (Gunasekar et al. 2014).

Xanthan gum is only selectively degraded in colon, not in the stomach and small intestine. This polysaccharide could be a better option for treating colon-associated diseases (Ramasamy et al. 2011). In an experiment related to colon-targeted drug release, matrix tablets of albendazole were prepared using xanthan gum. The ability of the formulated matrices to sustain drug release in the upper GIT and promote enzymatic hydrolysis by the colonic bacteria was evaluated. The *in vitro* drug release studies were carried out in the presence of RCC. Optimum release was observed with albendazole formulation containing 30 and 40% xanthan gum. The rate of albendazole release was found to decrease with increase in the concentration of xanthan gum due to high swelling, which made them more vulnerable to digestion by the microbial enzymes in the colon (Jackson and Ofoefule 2011). Jackson et al. (2011) stated that when xanthan gum and ethyl cellulose were used in matrix tablets for colon drug delivery, higher concentration of the xanthan gum showed more drug retarding capability than the formulation with ethyl cellulose (Jackson et al. 2011). Sinha et al. (2007) prepared compression coated tablets of 5-fluorouracil using with a mixture of xanthan gum and boswellia gum (3:1). Studies also showed that the xanthan gum played an important role in retarding drug release in colon (Sinha et al. 2007). In another experiment, microflora-activated colon drug delivery system was developed by using xanthan gum. Results showed that compression-coated metronidazole tablets with GG and xanthan gum coat showed promise in releasing their content in the colon to treat amebiasis and other local diseases (Niranjan et al. 2013). Sinha et al. (2004) formulated rapidly disintegrating core tablets using a blend of xanthan gum and GG as coating material. It was observed that the xanthan gum:GG (1:2) coat enabled the tablets to deliver 5-fluorouracil in the colon (Sinha et al. 2004).

### 6.4.8  STARCH

Starch is a storage carbohydrate consisting of glucose monomers in plants, such as cereals, root vegetables, and legumes. It is comprised of two polymers, namely amylose (a nonbranching helical polymer consisting of $\alpha$-(1 → 4)-linked D-glucose monomers) and amylopectin (a highly branched polymer consisting of both $\alpha$-(1 → 4)- and $\alpha$-(1 → 6)-linked D-glucose monomers) (Figure 6.3h).

Native starch is not used for designing controlled release systems due to swelling and rapid enzymatic degradation and consequent faster release of drugs. This necessitates the use of different starch derivatives that are more resistant to enzymatic degradation as well as cross-linking and formation of copolymers. Starch acetate was prepared by acetyl esterification, and its ability to prevent enzymatic degradation was investigated for use in CTDDSs (Chen et al. 2007; Thérien-Aubin and Zhu 2009). Cross-linked high amylose starches have been reported as suitable excipients for fabricating controlled release solid oral dosage forms having greater drug loading capacities and quasi-zero-order release kinetics. Other advantages are the absence of erosion, limited swelling, and the ability to control drug release rate by changing the concentration of cross-linker (Lenaerts et al. 1998). Starch has been fabricated for colon-targeted delivery as enteric-coated capsules (Vilivialm et al. 2000). Resistant starch was studied for the improvement of gut microflora and increase of clinical conditions related to IBD, immune-stimulating activities, and protection from colon cancer. Adriano et al. (2008) also investigated starch-modified hydrogel as a potential carrier for the colon-specific drug delivery.

## 6.5  COLON-TARGETING APPROACHES

### 6.5.1  COATING IN pH-SENSITIVE POLYMER

The ideal pH-dependent polymers that are used in colon-specific drug delivery should be able to withstand the lower pH of the stomach and of the proximal part of the small intestine, but able to disintegrate at neutral or shortly alkaline pH of the terminal ileum and preferably at ileocecal junction (Verma et al. 2012). For example, composite film-coated tablets of 5-ASA (5-aminosalicylic acid) were formulated for colon-specific delivery. In this experiment, 5-ASA core tablets were prepared and coated with dispersion of eudragit-RS and de-esterified pectin, polygalacturonic acid, or its potassium (K) and sodium (Na) salts. Less amount of drug release was observed during the first 5 h, where the coated tablets were in the stomach and small intestine. After that, the release of 5-ASA from coated tablets occurred linearly as a function of time due to the action of pectinolytic enzymes (Sriamornsk et al. 2003).

In another study, Jose et al. (2010) combined the pH-dependent solubility of the eudragit-S100 polymers and microbial degradability of chitosan polymers. Chitosan microspheres containing ondansetron were prepared by emulsion cross-linking method. The drug release studies indicated that eudragit-S100 coating around the chitosan microspheres offered a high degree of protection from premature drug release in the stomach and small intestine. Formulation containing core:coat ratio of 1:10 showed less amount of drug release in the upper gastrointestinal conditions and so was selected as the best formulation and then evaluated for *in vitro* drug release studies in the presence of RCCs to reveal biodegradability of chitosan microspheres in the colon. The *in vitro* drug release pattern was evaluated by the various kinetic models. Regression analysis revealed that the drug release pattern followed Peppas model (Jose et al. 2010). Solid unit dosage forms were formulated by Godge and Hiremath (2012) using polysaccharides or synthetic polymer including xanthan gum, chitosan, pectin, and eudragit-E. Meloxicam was used as a model drug. The formulated tablets were enteric coated with eudragit-S100 to give protection in the stomach. The coated tablets were tested *in vitro* for their suitability as colon-specific drug delivery systems. The *in vitro* dissolution data stated that enteric-coated tablets containing 3% chitosan as a binder released only 12.5% drug in the first 5 h, which was the usual upper gastrointestinal transit time, whereas tablets prepared using xanthan gum as binder were unable to protect drug release under similar conditions (Godge et al. 2012).

### 6.5.2   MICROFLORA-ACTIVATED SYSTEM

The basic principle involved in this method is the degradation of polymers coated on the drug delivery system by microflora present in the colon and thereby release of drug load in the colonic region because the bioenvironment inside the human GIT is evaluated by the presence of complex microflora, especially the colon is rich in microorganisms. In this experiment, drugs or dosage forms are coated with the biodegradable polymers to ensure polymer degradation with the influence of colonic microorganisms. When the dosage form passes through the GIT, it remains intact in the stomach and small intestine where very little microbial-degradable activity is present which is insufficient for cleavage of the polymer coating (Sinha and Kumaria 2003). Low swelling GG was prepared by cross-linking with glutaraldehyde (GA) for designing colon-specific drug carrier. Chitosan succinate and chitosan phthalate were synthesized by the chemical reaction of chitosan separately with succinic anhydride and phthalic anhydride, respectively. These semisynthetic polymers produced stable matrices of diclofenac sodium for colon-specific delivery that were more resistant to acidic conditions and improved the drug release profile under basic conditions (Aledeh and Taha 1999). CTDDS for mebendazole was prepared by using GG as a carrier material. In this method, mebendazole matrix tablets containing various proportions of GG were formulated by wet granulation technique using starch paste to act as a binder. The data showed that 20% and 30% GG tablets provided targeted release of mebendazole for local action in the colon (Krishnaiah et al. 2001). Metronidazole tablets were formulated using various polysaccharides, such as GG, xanthan gum, pectin, carrageenan, and β-cyclodextrin for colon-specific drug delivery to prevent amebiasis (Mundargi et al. 2007). 5-Fluorouracil compression-coated tablets were prepared for colonic release of the drug using xanthan gum, boswellia gum, and HPMC as the coating materials (Sinha et al. 2007).

### 6.5.3    BIOADHESIVE POLYMER-BASED DELIVERY

Bioadhesion is the new concept in the design of colon drug delivery system, where the polymer gets adhered to the mucus membrane of the colon. Here the polymer swells and gets adhered; adhesion involves the formation of chemical or physical bonding between the polymer and the surface of mucus membrane. Improvement in both topical and systemic treatment in the colonic inflammatory diseases is achieved by localized drug delivery, thereby improving drug residence time (Varum et al. 2011). A large number of polysaccharides, such as GG, amylase, pectin, chitosan, inulin, chondroitin sulfate, cyclodextrins, dextrans, and locust bean gum (LBG) have been investigated for their use in colon-targeted drug delivery systems. The most important fact in the development of polysaccharide derivatives for colon-targeted drug delivery is the selection of a suitable biodegradable polysaccharide. As these polysaccharides are usually soluble in water, they need to be made water insoluble by cross-linking or hydrophobic derivatization using an optimal proportion of the hydrophobic and hydrophilic parts, respectively, and the number of free hydroxyl groups in the polymeric molecule.

### 6.5.4    EMBEDDED IN BIODEGRADABLE POLYMER MATRIX

These are polysaccharide-based polymers and their monomers remain unaffected and are resistant to digestive enzymes present in the upper GIT; but as they reach the colon, the microflora present in the colon ($\alpha$-L-arabinofunosidase, $\beta$-D-fucosidase, $\beta$-D galactosidase, $\beta$-D-glucosidase and $\beta$-xylosidase) act on the polysaccharide and degrade the tablet matrix and release the drug. A large number of polysaccharides have been observed for their use in CTDDS which include GG, amylase, pectin, inulin, chitosan, chondroitin sulfate, cyclodextrins, dextrans, and locust bean gum; these polysaccharide are inexpensive, nontoxic, and water-soluble and they must make water insoluble by cross-linking or hydrophobic derivatization. An optimal proportion of hydrophilic and hydrophobic part, respectively, and number of free hydroxyl groups in polymeric molecules is very important (Shirwaikar et al. 2008). GG and xanthan gum were used as polymer matrix for the formulation of ibuprofen tablets (Abdullah et al. 2011).

### 6.5.5    TIME-DEPENDENT DELIVERY

Pulsatile release systems are prepared to undergo a lag time of predetermined span of time of no release, followed by a rapid and complete release of the loaded drug(s). The approach is based on the principle of sustaining the time of drug release until the system transits from the mouth to colon. A lag time of 5 h is usually considered sufficient since small intestine transit is about 3–4 h, which is relatively constant and hardly affected by the nature of formulation administered. This system offers many

**TABLE 6.1**

**Approaches for the Development of Natural Polymer-Based Colon-Targeted Drug Delivery**

| Approach | Basic Feature |
|---|---|
| Polymeric prodrugs | The drug is conjugated with the polymer |
| Coating with biodegradable polymer | Drug is released following degradation of the polymer due to the action of colonic bacteria |
| Embedding in biodegradable polysaccharides | The embedded drug in polysaccharide matrices is released by swelling and biodegradable action of polysaccharides |
| Timed release systems | Delaying the time of drug release until the system transits from mouth to colon |
| Bioadhesive system | Drug coated with bioadhesive polymer that selectively provides adhesion to colonic mucosa |
| Coating of microparticles | Drug is released through semipermeable membrane |

advantages over conventional oral drug delivery systems such as reduced dosage and dosage frequency, patient compliance, decrease in side effects, avoidance of peak and valley fluctuation, and constant drug level at the target site (Modasiya and Patel 2011). The various disadvantages of this system include gastric emptying time varies markedly between subjects or in a manner dependent on the type and amount of food intake; gastrointestinal movement, especially peristalsis or contraction in the stomach would result in change in gastrointestinal transit of the drug and accelerated transit through different regions of the colon has been observed in patients with IBD, carcinoid syndrome, and diarrhea and ulcerative colitis. Thus to overcome this problem, certain integration of pH-sensitive and time release functions into a single dosage have been carried out for colon targeting.

A summary of different approaches for natural polymer-based colon-targeted drug delivery is given in Table 6.1.

## 6.6 DIFFERENT CARRIERS FOR NATURAL POLYMER-BASED COLON DRUG DELIVERY

### 6.6.1 NANOPARTICULATE SYSTEMS

Nanoparticle size colloidal carriers composed of natural or synthetic polymers have also been investigated for colon targeting. Orally administered nanoparticles serve as carriers for different types of drugs and have been shown to enhance their solubility, bioavailability, and permeability. Nanoparticles have also been observed for the delivery of protein and peptide drugs. For colonic pathologies, it was shown that nanoparticles tend to accumulate at the site of inflammation in IBD. This is because, in the case of colitis, a strong cellular immune response occurs in the inflamed regions due to increased presence of natural killer cells, neutrophils, macrophages, and so on. It has been observed that microspheres and nanoparticles could be efficiently taken up by these macrophages. This results in accumulation of the particulate carrier system resulting in prolonged residence time in the desired area (Lamprecht et al. 2001). But an important area of concern is to prevent the loss of nanoparticle in the early transit through GIT in order to optimize therapeutic efficacy. Moreover, particle uptake by Payer's patches or enzymatic degradation may cause the release of the entrapped drug leading to systemic drug absorption and side effects. In order to overcome this problem, drug-loaded nanoparticles were entrapped into pH sensitive microspheres which serve to deliver the incorporated nanoparticle to their site of action, thereby preventing a premature drug leakage (Lamprecht et al. 2005). The uses of nanoparticles for bioadhesion purposes have also been investigated. Nanoparticles have a large specific surface, which is indicative of high interactive potential with biological surfaces. Since the interaction is of nonspecific nature, bioadhesion can be induced by binding nanoparticles with different molecules (Krishnaiah et al. 2003).

### 6.6.2 MICROSPHERES

Microspheres are the microparticulate polymer-based carrier systems for drug delivery applications. They offer various advantages such as limited fluctuation of drug–plasma profile within a therapeutic range, reduction in side effects, decreased dosing frequency, and improved patient compliance. Microspheres that are biodegradable can be efficiently taken up by the macrophages. Therefore, the direct uptake of anti-inflammatory agent-loaded microspheres by macrophages would have a superior immunosuppressive effect and can be more useful for treatment of patients with IBD (Ghulam et al. 2009; Mathew et al. 2010). The chitosan-based multiparticulate system was coated with eudragit-L100 or S100 for the colonic delivery of metronidazole for the treatment of amebiasis (Chourasia and Jain 2004c). High amylose cornstarch and pectin blend microparticles containing diclofenac sodium for colon-targeted delivery were prepared by spray drying technique. The blending of high amylose cornstarch with pectin improved the encapsulation efficiency and decreased drug dissolution in the gastric condition from pectin-based microparticles. The drug was released in the colonic region by the action of pectinase from microparticles (Kashappa and Desai 2005). Cross-linked GG microspheres containing methotrexate were prepared and characterized for local release of the drug in the colon for efficient treatment of colorectal cancer. In this preparation, glutaraldehyde (GA) was used as a cross-linking agent and GG microspheres were prepared by the emulsification method. From the results of *in vitro* and *in vivo* studies, the methotrexate-loaded cross-linked GG microspheres delivered most of the drug load (79%) to the colon, whereas plain drug suspension could deliver only 23% of the total dose to the target tissue (Chaurasia et al. 2006). Pandey et al. (2012) formulated a multiparticulate system exploiting pH-sensitive property and specific biodegradability of calcium alginate microbeads for colon-targeted delivery of tinidazole for the treatment of amoebic colitis. Calcium alginate beads containing tinidazole were prepared by ionotropic gelation technique followed by coating with eudragit-S100 using solvent evaporation method to obtain pH-sensitive microbeads. The *in vitro* drug release from eudragit-S100-coated beads was found to be 70.73% in 24 h in the presence of RCC. It is concluded that the calcium alginate microbeads are the potential system for the colon delivery of tinidazole for chemotherapy of amoebic infection (Pandey et al. 2012). Colon-specific microspheres of 5-fluorouracil were prepared and eval-uated for the treatment of colon cancer. In this experiment, core microspheres of alginate were prepared by modified emulsification method in liquid paraffin and by cross-linking with calcium chloride. The core microspheres were coated with eudragit-S100 by the solvent evaporation technique to prevent drug release in the stomach and small intestine. The results showed that this method had great potential in delivery of 5-fluorouracil to the colon region (Rahaman et al. 2006).

### 6.6.3 HYDROGEL BEADS

Hydrogels, the swellable polymeric materials, have been widely investigated as the carrier for drug delivery systems. These biomaterials have gained attention owing to their peculiar characteristics like swelling in aqueous medium, temperature sensi-tivity, and pH or sensitivity toward other stimuli. These swollen polymers are help-ful as targetable carriers for bioactive drugs with tissue specificity (Zhang et al. 2005; Baroli 2007). For example, amidated pectin hydrogel beads containing indo-methacin and sulfamethoxazole for colon-specific delivery (Munjeri et al. 1997). Methacrylated inulin hydrogels were designed for colon targeting the proteins like bovine serum albumin (BSA) or lysozyme. Organic redox-initiated polymerization technique was used to fabricate pH-responsive hydrogels for colon-specific deliv-ery (Emmanuel et al. 2003). GA cross-linked GG hydrogel discs were prepared as vehicles for colon-specific drug delivery of ibuprofen. Percent ibuprofen release increased with glutaraldehyde concentration. Cross-linking decreased the swelling of GG. The formulated hydrogel discs may prove to be beneficial as colon-specific drug delivery vehicles for poorly water-soluble drugs like ibuprofen (Das et al. 2006). Novel hydrogel beads were prepared using pectin and zein for colon-specific drug delivery. Pectin–zein complex hydrogel beads revealed the capability to protect incorporated drugs from premature release into the stomach and small intestine. The inclusion of a small portion of zein (a protein from corn) into the pectin effi-ciently suppressed the swelling behavior of pectin, thus stabilizing the structural property of the pectin networks (Liu and Fishman 2006). In another experiment, hydrogel beads were formed by chitosan and tripolyphosphate (TPP) for the delivery of protein in the colon. TPP was used as a counter ion to positively charged chitosan to form gel beads. The beads were loaded with BSA, a protein that is liable to deg-radation in the upper parts of GIT. The cross-linking of chitosan with TPP resulted in reduced solubility of chitosan, thereby resulting in lesser protein release during the upper GI transit. At the same time, the cross-linking and reduced solubility did not affect the degradability by microbial flora in the colon as shown by the *in vitro* studies where the RCCs were able to attack and degrade the cross-linked chitosan (Healey 1990). The natural polymer-based colon drug delivery systems are sum-marized in Table 6.2.

### 6.6.4 MATRIX TABLETS

Animal models have obvious advantages in assessing colon-specific drug delivery systems. Human subjects are increasingly utilized for the evaluation of this type of

**TABLE 6.2**

**Application Summary of Natural Polymer-Based Colon-Targeted Drug Delivery**

| Natural Polymer(s) | Drug or Active Agent | Formulation | Therapeutic Improvement | References |
|---|---|---|---|---|
| Chitosan/ pectin | Vancomycin | Polyelectrolyte complex | Chitosan–pectin complex (1:9) exhibited pH-sensitive swelling and high mucoadhesive properties, suitable for colon delivery | Bigucci et al. (2008) |
| Guar gum | Ibuprofen | Hydrogel discs | Significant increase in ibuprofen release was observed in release medium containing rat cecal content | Das et al. (2006) |
| Calcium alginate | Tinidazole | Microbeads | Eudragit-S100-coated beads released 70.73% drug in 24 h in the presence of rat cecal content, showed pH sensitivity useful for the treatment of amoebic colitis | Pandey et al. (2012) |
| Guar gum | Methotrexate | Microspheres | The drug release in PBS (pH 7.4) was 91.0% in the medium containing rat cecal content. Guar gum microspheres delivered 79.0% to the colon; whereas pure drug suspensions delivered only 23% of their total dose to the target site | Chaurasia et al. (2006) |
| Pectin | Satranidazole | Compression-coated tablets | HPMC increased the tensile strength of the compression coat and helped in achieving desired lag time for the formulation to reach the colon | Shah et al. (2011) |

(*Continued*)

## TABLE 6.2 (*Continued*)
## Application Summary of Natural Polymer-Based Colon-Targeted Drug Delivery

| Natural Polymer(s) | Drug or Active Agent | Formulation | Therapeutic Improvement | References |
|---|---|---|---|---|
| Pectin | Ketoprofen | Complex as prodrug | Ketoprofen (KP) released from pectin–ketoprofen (PT–KP), prodrug distributed in cecum and colon and KP distributes mainly in the stomach, proximal small intestine, and distal small intestine | Xi et al. (2005) |
| Amylose | 5-Aminosalicylic acid | Films | Under simulated colonic conditions, drug release was more pronounced from coated formulations containing higher proportions of amylose | Siew et al. (2000) |
| Guar gum | Nitrofurantoin | Matrix tablet | In colonic fluid (pH 6.8) containing 4% w/v, rat cecal content liberated drug for 10 h | Singh et al. (2012) |
| Chitosan | Satranidazole | Beads | In the presence of extracellular enzymes, 97.67% drug was released after 24 h; whereas in the presence of 4% cecal content, 64.71% and 96.52% drug was released after 3 and 6 days of enzyme induction, respectively. This indicated the potential of this multiparticulate system to serve as a carrier to deliver macromolecules | Jain et al. (2007) |

(*Continued*)

**TABLE 6.2** (*Continued*)

**Application Summary of Natural Polymer-Based Colon-Targeted Drug Delivery**

| Natural Polymer(s) | Drug or Active Agent | Formulation | Therapeutic Improvement | References |
|---|---|---|---|---|
| Locust bean gum/ xanthan gum | Aceclofenac | Compression-coated tablets | Combination of locust bean gum and xanthan gum was sufficient to sustain the drug release in medium containing 4% rat cecal content for successful colon targeting | Syan et al. (2011) |
| Locust bean gum/guar gum | Mebeverine HCl | Matrix tablet | The optimized tablets provided controlled release over 24 h; whereas the marketed product controlled the drug release over a period of 12 h | Chauhan et al. (2012) |
| Xanthan gum/ boswellia gum | 5-Fluorouracil | Compression-coated tablets | Coating of boswellia gum and HPMC (2:3) released 80.22% drug in 24 h. Rat cecal contents (2% w/v) led to complete release of the drug from the tablets | Sinha et al. (2007) |
| Calcium pectinate | Insulin | Nanoparticles | Hypoglycemic effect started 8 h after the administration of chitosan particles | Tozaki et al. (1997) |
| Pectin | Metronidazole | Microparticles | Without potential colonic release inducers, 34.4-fold more drug is retarded in pectin-ATP microparticles within 6 h. The amount of metronidazole is released rapidly under the influence of pectinolytic enzymes or a reducing agent simulating the colonic environment | Perera et al. (2010) |

delivery system. γ-Scintgraphic studies were conducted in human volunteers with technetium-99m-diethylenetriaminepentaacetic acid (DTPA) as tracers in sodium chloride core tablets, coated with GG. The study showed that compression-coated GG tablets protected the drug (tracer) from being released in the stomach and small intestine. On entering the ascending colon, the tablets commenced to release the tracer indicating the breakdown of gum coat by the enzymatic action of colonic bacteria (Madhu et al. 2011). Technetium-99m-DTPA was used as a tracer for γ-scintigraphy evaluation of colon-specific GG matrix tablets in human volunteers (Krishnaiah et al. 1998). In a study, Krishnaiah et al. (2001) showed the effect of GG on albendazole release from colon-specific matrix tablets. The incubation of RCC with active antimicrobial agents for 7 days decreased the release of albendazole due to the lowering of anaerobic bacteria present in rat (Krishnaiah et al. 2001).

## 6.7 CONCLUSION

A colon drug delivery system requires triggering mechanism for the successful delivery of the drugs to the colon. In the last few decades, considerable research work has been done in the area of colon drug targeting. All the available approaches have their own limitations and advantages and extensive research is being focused for further improvements. Natural polysaccharides are promising biodegradable materials as these can be chemically compatible with the excipients used in drug delivery systems. In addition, natural polysaccharides are freely available, nontoxic, and less expensive compared to their synthetic counterparts. They have a vital role to play in the pharmaceutical industry. The variation in structure and mode of con-jugation/encapsulation could improve site-specific drug release, reducing the need of excessive drug use. More research work must be conducted to improve the specificity of the drug uptake at the colon site using natural polysaccharides, so that they can become clinically available.

## REFERENCES

Abdullah, G.Z., Abdulkarim, M.F., Chitneni, M. et al. 2011. Preparation and in vitro evaluation of mebeverine HCl colon-targeted drug delivery system. *Pharm Dev Technol* 16:331–42.

Addo, R.T., Siddig, A., Siwale, R. et al. 2010. Formulation, characterization and testing of tetracaine hydrochloride-loaded albumin-chitosan microparticles for ocular drug delivery. *J Microencapsul* 27:95–104.

Adriano, V.R., Marcos, R.G., Thais, A.M. et al. 2008. Synthesis and characterization of a starchmodified hydrogel as potential carrier for drug delivery system. *J Polym Sci Part A: Polym Chem* 46:2567–74.

Akhgari, A., Farahmand, F., Garekani, H. et al. 2006. Permeability and swelling studies on free films containing inulin in combination with different polymethacrylates aimed for colonic drug delivery. *Eur J Pharm* 28:307–14.

Alaa-Eldeen, B.Y., Ibrahim, A.A., Fars, K.A. et al. 2010. New targeted-colon delivery system: In vitro and in vivo evaluation using X-ray imaging. *J Drug Target* 18:59–66.

Al-Saidan, S.M., Krishnaiah, Y.S., Satyanarayana, V. et al. 2005. In vitro and in vivo evaluation of guar gum-based matrix tablets of rofecoxib for colonic drug delivery. *Curr Drug Deliv* 2:155–63.

Aledeh, K. and Taha, M.O. 1999. Synthesis of chitosan succinate and chitosan phthalate and their evaluation as suggested matrices in orally administered, colon-specific drug delivery systems. *Arch Pharm Pharm Med Chem* 332:103–07.

Aurora, J. 2006. Colonic drug delivery challenges and opportunities—An overview. *Eur Gastroenterol Rev* 1:1–4.

Bajpai, S.K., Bajpai, M., and Dengree, R. 2003. Chemically treated gelatin capsules for colon-targeted drug delivery: A novel approach. *J Appl Polym Sci* 89:2277–82.

Barday, T., Cooper, M., and Petrosky, N. 2010. Inulin-A versatile polysaccharide with multiple pharmaceutical and food uses. *J Excipients Food Chem* 1:24–42.

Baroli, B. 2007. Hydrogels for tissue engineering and delivery of tissue inducing substances. *J Pharm Sci* 96:2197–23.

Bhatia, M.S., Deshmukh, R., Choudhari, P. et al. 2008. Chemical modifications of pectins, characterization and evaluation for drug delivery. *Sci Pharm* 76:775–84.

Bauer, K.H. and Kesselhut, J.F. 1995. Novel pharmaceutical excipients for colon targeting. *STP Pharm Sci* 5:54–59.

Bigucci, F., Luppi, B., Cerchiara, T. et al. 2008. Chitosan/pectin polyelectrolyte complexes: Selection of suitable preparative conditions for colon-specific delivery of vancomycin. *Eur J Pharm Sci* 35:435–41.

Brahma, N.S. 2007. Modified-release solid formulations for colonic delivery. *Recent Pat Drug Deliv Formul* 1:53–63.

Castelli, F., Sarpietro, M.G., Micieli, D. et al. 2008. Differential scanning calorimetry study on drug release from an inulin-based hydrogel and its interaction with a biomembrane model: pH and loading effect. *Eur J Pharm* 35:76–85.

Cavalcanti, O.A., Van den Mooter, G., Caramico-Soares, I. et al. 2002. Polysaccharides as excipients for colon-specific coatings, permeability and swelling properties of casted films. *Drug Dev Ind Pharm* 28:157–64.

Challa, T., Vynala, V., and Allam, K.V. 2011. Colon specific drug delivery systems: A review on primary and novel approaches. *Int J Pharm Sci Rev Res* 7:171–81.

Chauhan D., Patel, A., and Shah, S. 2012. Influence of selected natural polymers on in-vitro release of colon targeted mebeverine HCl matrix tablet. *Int J Drug Dev Res* 4:247–55.

Chaurasia, M., Chourasia, M.K., Jain, N.K. et al. 2006. Cross-linked guar gum microspheres; A viable approach for improved delivery of anticancer drugs for the treatment of colorectal cancer. *AAPS PharmSciTech* 7:E143–51.

Chen, L., Li, X., Li, L. et al. 2007. Acetylated starch-based biodegradable materials with potential biomedical applications as drug delivery systems. *Curr Appl Phys* 7S1:e90–3.

Ching, A.L., Liew, C.V., Heng, P.W.S. et al. 2008. Impact of cross-linker on alginate matrix integrity and drug release. *Int J Pharm* 355:259–68.

Chourasia, M.K. and Jain, S.K. 2004a. Polysaccharides for colon targeted drug delivery. *Drug Deliv* 11:129–48.

Chourasia, M.K. and Jain, S.K. 2004b. Potential of guar gum microspheres for target specific drug release to colon. *J Drug Target* 2:1–8.

Chourasia, M.K. and Jain, S.K. 2004c. Design and development of multiparticulate system for targeted drug delivery to colon. *Drug Deliv* 11:201–7.

Chourasia, M.K. and Jain, S.K. 2003. Pharmaceutical approaches to colon targeted drug delivery systems. *J Pharm Pharmaceut Sci* 6:33–66.

Coviello, T., Alhaique, F., Dorigo, A. et al. 2007. Two galactomannans and scleroglucan as matrices for drug delivery: Preparation and release studies. *Eur J Pharm Biopharm* 66:200–09.

Das, A., Wadhwa, S., and Srivastava, A.K. 2006. Cross-linked guargum hydrogel discs for colon specific delivery of Ibuprofen: Formulation and in vitro evaluation. *Drug Deliv* 13:139–42.

Dey, P., Maiti, S., and Sa, B. 2012. Locust bean gum and its application in pharmacy and biotechnology: An overview. *Int J Curr Pharm Res.* 4:7–11.

Dey, S.N., Majmudar, S., and Rao, M. 2008. Multiparticulate drug delivery systems for controlled release. *Trop J Pharm Res* 7:1067–75.

Doyle, J.P., Lyons, G., and Morris, E.R. 2008. New proposals on hyperentanglement of galactomannans: Solution viscosity of fenugreek gum under neutral and alkaline conditions. *Food Hydrocolloids* 23:1501–10.

Emmanuel, O., Elekwachi, O., and Chase, V. 2003. Organic Redox-initiated polymerization process for the fabrication of hydrogels for colon-specific drug delivery. *Drug Dev Ind Pharm* 29:375–86.

Fatima, L., Asghar, A., and Chandran, S. 2006. Multiparticulate formulation approach to colon specific drug delivery: Current perspectives. *J Pharm Pharmaceut Sci* 9:327–38.

Gamal-Eldeen, A.M., Amer, H., and Helmy, W.A. 2006. Cancer chemopreventive and anti-inflammatory activities of chemically modified guar gum. *Chem Biol Interact* 161:229–40.

Ghulam, M., Mahmood, A., Naveed, A. et al. 2009. Comparative study of various microencapsulation techniques. Effect of polymer viscosity on microcapsule characteristics. *Pak J Sci* 22:291–300.

Godge, G. and Hiremath, S. 2012. Colonic delivery of film coated meloxicam tablets using natural polysaccharide polymer mixture. *Int Curr Pharm J* 1:264–71.

Goyal, S., Vashist, H., Gupta, A. et al. 2011. Development of alginate gel beads-entrapped liposome for colon specific drug delivery of Prednisolone. *Der Pharmacia Sinica* 2:31–38.

Gunasekar, V., Reshma, K., Treesa, G. et al. 2014. Xanthan from sulphuric acid treated tapioca pulp: Influence of acid concentration on xanthan fermentation. *Carbohydr Polym* 102:669– 73.

Halsas, M., Penttinen, T., Veski, P. et al. 2001. Time controlled release Pseudoephedrine tablets: Bioavailability and in vitro/in vivo correlations. *Pharmazie* 56:718–23.

Healey, J.N. 1990. Gastrointestinal transit and release of mesalazine tablets in patients with inflammatory bowel disease. *Scand J Gastroenterol Suppl* 172:47–51.

Jackson, C. and Ofoefule, S. 2011. Use of xanthan gum and ethylcellulose in formulation of metronidazole for colon delivery. *J Chem Pharm Res* 3:11–20.

Jain, S.K., Jain, A., Gupta, Y. et al. 2007. Design and development of hydrogel beads for targeted drug delivery to the colon. *AAPS PharmSci Tech* 8:E1–E8.

Jana, S., Gandhi, A., Sen, K.K. et al. 2013. Biomedical applications of chitin and chitosan derivatives. In *Chitin and Chitosan Derivatives: Advances in Drug Discovery and Developments*, ed. S.K. Kim, 337–360. Boca Raton, FL: Taylor & Francis Group.

Jenita, J.J.L., Vijaya, K., Suma, R. et al. 2010. Formulation and evaluation of compression coated tablets of mesalazine for colon delivery. *Int J PharmTech Res* 2:535–41.

Jose, S., Dhanya, K., Cinu, T.A. et al. 2010. Multiparticulate system for colon targeted delivery of ondansetron. *Indian J Pharm Sci* 72:58–64.

Kashappa, G. and Desai, H. 2005. Preparation and characteristics of High-amylose corn starch/pectin blend macro particles: A technical note. *AAPS PharmSciTech* 6: E202–08.

Kaushik, D., Sardana, S., and Mishra, D.N. 2009. Implications of biodegradable and bioadhesive systems in colon delivery. *Int J Pharm Sci Drug Res* 1:55–62.

Kothawade K.B., Gattani, S.G., and Amrutkar, J.P. 2009. Colonic delivery of aceclofenac using combination of PH and time dependent polymers. *Indian Drug* 46:67–74.

Kothawade, P.D., Gangurde, H.H., Surawase, R.K. et al. 2011. Conventional and novel approaches for colon specific drug delivery: A review. *e- J Sci Technol* 2:33–56.

Krishnaiah, Y.S.R., Devi, A.S., Rao, L.N. et al. 2001. Guar gum as a carrier for colon specific delivery: Influence of Metronidazole and Tinidazole on in-vitro release of albendazole from guar gum matrix tablets. *J Pharm Pharmaceut Sci* 4:235–43.

Krishnaiah, Y.S.R., Raju, P.V., Kumar, B.D. et al. 2001. Development of colon targeted drug delivery systems for mebendazole. *J Control Release* 77:87–95.

Krishnaiah, Y.S.R., Satyanarayana, V., Kumar, B.D. et al. 2003. In vivo pharmacokinetics in human volunteers: Oral administered guar gum- based colon- targeted 5- fluorouracil tablets. *Eur J Pharm Sci* 19:355–62.

Krishnaiah, Y.S.R., Satyanarayana, V., Rama Prasad, Y.V. et al. 1998. Gamma scintigraphic studies on guar gum matrix tablets for colonic drug delivery in healthy human volunteers. *J Control Release* 55:245–52.

Kumar, P. and Mishra, B. 2008. Colon targeted drug delivery system—An overview. *Curr Drug Deliv* 5:186–98.

Kumar, S.P. and Pratibha, D. 2012. Novel colon specific drug delivery system. *Int J Pharm Pharm Sci* 23:22–29.

Kumar, R., Patil, M.B., Patil, S.R. et al. 2009. Polysaccharides based colon specific drug delivery: A review. *Int J PharmTech Res* 2:334–46.

Lamprecht, A., Yamamoto, H., Takeuchi, H. et al. 2005. A pH-sensitive microsphere system for the colonic delivery of tacrolimus containing nanoparticles. *J Control Release* 104:337–46.

Lamprecht, A., Scaffer, U., and Lehr, C.M. 2001. Size dependent targeting of micro- and nano- particulate carriers to the inflamed colonic mucosa. *Pharm Res* 18:788–93.

Lenaerts, V., Moussa, I., Dumoulin, Y. et al. 1998. Cross-linked high amylase starch for controlled release of drugs: Recent advances. *J Control Release* 53:225–34.

Liew, C.V., Chan, L.W., Ching, A.L. et al. 2006. Evaluation of sodium alginate as drug release modifier in matrix tablets. *Int J Pharm* 309:25–37.

Liu, L.S., and Fishman M.L. 2006. Pectin/zein beads for potential colon-specific drug delivery system: Synthesis and in vitro evaluation. *Drug Deliv* 13:417–23.

Liu, L., Fishman, M.L., Kost, J. et al. 2003. Pectin-based systems for colon-specific drug delivery via oral route. *Biomaterials* 24:3333–43.

Liu, H., Yang, X.G., Nie, S.F. et al. 2007. Chitosan-based controlled porosity osmotic pump for colon-specific delivery system: Screening of formulation variables and in vitro investigation. *Int J Pharm* 332:115–24.

Madhu, E.N., Panaganti, S., Prabakaran, L. et al. 2011. Colon specific drug delivery system: A review. *Int J Pharm Sci Res* 2:2545–61.

Mahale, N.B., Hase, D.P., Bhujbal, S.S. et al. 2013. Colon specific drug delivery system: A review. *Int J Pharm Res Dev* 4:56–64.

Maestrelli, F., Cirri, M., Corti, G. et al. 2008. Development of enteric-coated calcium pectinate microspheres intended for colonic drug delivery. *Eur J Pharm Biopharm* 69:508–18.

Mathew S.T., Gayathri, S.D., Prasanth, V.V. et al. 2010. Suitability of factorial design in determining the processing factors affecting entrapment efficiency of albumin microspheres. *J Pharm Res* 3:1172–77.

Mehta, T.J., Patel, A.D., Patel, M.R. et al. 2011. Need of colon specific drug delivery system: Review on primary and novel approaches. *Int J Pharm Res Dev* 3:134–53.

Modasiya, M.K. and Patel, V.M. 2011. Pulsatile drug delivery system for colon. *Int J Res Pharm Biomed Sci* 2:934–41.

Mundargi, R.C., Patil, S.A., Agnihotri, S.A. et al. 2007. Development of polysaccharide-based colon targeted drug delivery systems for the treatment of ameobiasis. *Drug Dev Ind Pharm* 33:255–64.

Munjeri, O., Collett, J.H., and Fell, J.T. 1997. Hydrogel beads based on amidated pectins for colonspecific drug delivery: The role of chitosan in modifying drug release. *J Control Release* 46:273–78.

Niranjan, K., Shivapooja, A., Muthyala, J. et al. 2013. Effect of guar gum and xanthan gum compression coating on release studies of metronidazole in human fecal media for colon targeted drug delivery systems. *Asian J Pharm Clin Res* 6:315–18.

Pandey, A.K., Choudhary, N., Ra, V.K. et al. 2012. Fabrication and evaluation of tinidazole microbeads for colon targeting. *Asian Pacific J Trop Dis* 2:S197–201.

Patel, G.N., Patel, G.C., and Patel, R.B. 2006. Oral colon-specific drug delivery: An overview. *Drug Deliv Technol* 6:62–71.

Perera, G., Barthelmes, J., and Bernkop-Schnürch, A. 2010. Novel pectin–4-aminothiophenole conjugate microparticles for colon-specific drug delivery. *J Control Release* 145:240–46.

Philip, A.K., Dubey, R.K., and Pathak, K. 2008. Optimizing delivery of flurbiprofen to the colon using a targeted prodrug approach. *J Pharm Pharmacol* 60:607–13.

Raghavan C.V., Muthulingam, C., Jenita, J.A. et al. 2002. An in vitro and in vivo investigation into the suitability of bacterially triggered delivery system for colon targeting. *Chem Pharm Bull* 50:892–95.

Rahaman, Z., Kohli, K., Khar, R.K. et al. 2006. Characterization of 5-fluorouracil microspheres for colonic delivery. *AAPS PharmSciTech* 7: E1–9.

Ramasamy, T., Devi, U., Kandhasami, S. et al. 2011. Formulation and evaluation of xanthan gum based aceclofenac tablets for colon targeted drug delivery. *Brazilian J Pharm Sci* 47:299–11.

Reddy, R.B.D., Malleswari, K., Prasad, G. et al. 2013. Colon targeted drug delivery system: A Review. *Int J Pharm Sci Res* 24:42–54.

Rubinstein, A., Radai, R., Ezra, M. et al. 1993. In vitro evaluation of calcium pectinate: A potential colon-specific drug delivery carrier. *Pharm Res* 10:258–63.

Salve, P.S. 2011. Development and in vitro evaluation colon targeted drug delivery system using natural gums. *Asian J Pharm Res* 1:91–101.

Sarasija, S. and Hota, A. 2000. Colon-specific drug delivery systems. *Indian J Pharm Sci* 62:1–8.

Sarmento, B., Ribeiro, A., Veiga, F. et al. 2007. Alginate/Chitosan nanoparticles are effective for oral insulin delivery. *Pharm Res* 24:2198–06.

Shah, N., Shah, T., and Amin, A. 2011. In-vitro evaluation of pectin as a compression coating material for colon targeted drug delivery. *Int J Pharma Bio Sci* 2:410–18.

Shukla, R.K. and Tiwari, A. 2012. Carbohydrate polymers: Applications and recent advances in delivering drugs to the colon. *Carbohydr Polym* 88:399–16.

Shirwaikar, A., Shirwaikar, A., Prabu, S.L. et al. 2008. Herbal excipients in novel drug delivery systems. *Indian J Pharm Sci* 70:415–22.

Sriamornsak, P., Thirawong, N., Weerapol, Y. et al. 2007. Swelling and erosion of pectin matrix tablets and their impact on drug release behavior. *Eur J Pharm Biopharm* 67:211–19.

Siew L.F., Basit, A.W., and Newton, J.M. 2000. The potential of organic based amylase ethyl-celulose film coatings as oral colon specific drug delivery systems, *AAPS PharmSciTech* 23:E22.

Singh, S.K., Singh, A.K., Mahajan, A.A. et al. 2012. Formulation and evaluation of colon targeted drug delivery of an anti-amoebic drug. *Int J Pharm Innov* 2:138–52.

Singh, R. 2012. Formulation and evaluation of colon targeted drug delivery system. *Int J Pharm Life Sci* 3:2265–68.

Sinha V.R., Mittal, B.R., Bhutani, K.K. et al. 2004. Colonic drug delivery of 5-fluorouracil: An in vitro evaluation. *Int J Pharm* 269:101–08.

Sinha V.R., Singh, A, Singh, S. et al. 2007. Compression coated systems for colonic delivery of 5-fluorouracil. *J Pharm Pharmacol* 59:359–65.

Sinha, V.R. and Rachna, K. 2003. Coating polymers for colon specific drug delivery: A comparative in vitro evaluation. *Acta Pharm* 53:41–47.

Sinha, V.R. and Kumaria, R. 2003. Microbially triggered drug delivery to the colon. *Eur J Pharm Sci* 18:3–18.

Sriamornsk, P., Nuthanid, J., Wan Chana, S. et al. 2003. Composite film-coated tablets intended for colon-specific delivery of 5-amino salicylic acid: Using deesterified pectin. *Pharm Dev Technol* 8:311–18.

Syan, N. and Mathur, P. 2011. Development and evaluation of compression coated colon targeted tablets of aceclofenac by using natural polymers. *Asian J Pharm Clin Res* 4:93–98.

Thérien-Aubin, H. and Zhu, X.X. 2009. NMR spectroscopy and imaging studies of pharmaceutical tablets made of starch. *Carbohydr Polym* 75:369–79.

Tozaki, H., Komoike, J., Tada, C. et al. 1997. Chitosan capsules for colon-specific drug delivery: Improvement of insulin absorption from the rat colon. *J Pharm Sci* 86:1016–21.

Tozaki, H., Fujita, T., Odoriba, T. et al. 1999. Colon, a specific delivery of R68070 new thromboxane synthase inhibitor, using chitosan capsules: Therapeutic effects against 2,4,6-trinitrobenzene sulfonic acid induced ulcerative colitis in rats. *Life Sci* 64:1155–62.

Tozaki, H., Odoriba, T., Okada, N. et al. 2002. Chitosan capsules for colonspecific drug delivery: Enhanced localization of 5-aminosalicylic acid in the large intestine accelerates healing of TNBS-induced colitis in rats. *J Control Release* 82:51–61.

Tuğcu-Demiröz, F., Acartürk, F., Takka, S. et al. 2007. Evaluation of alginate based mesalazine tablets for intestinal drug delivery. *Eur J Pharm Biopharm* 67:491–97.

Ugurlu, T., Turkoglu, M., Gurer, U.S. et al. 2007. Colonic delivery of compression coated nisin tablets using pectin/HPMC polymer mixture. *Eur J Pharm Biopharm* 67:202–10.

Varum, F.J., Veiga, F., Sousa, J.S. et al. 2011. Mucoadhesive platforms for targeted delivery to the colon. *Int J Pharm* 420:11–19.

Verma, S., Kumar, V., and Mishra, D.N. 2012. Colon targeted drug delivery: Current and novel approaches. *Int J Pharm Sci Res* 3:1274–84.

Vervoort, L. and Kinget, R. 1996. In vitro degradation by colonic bacteria of inulin HP incorporated in Eudragit RS films. *Int J Pharm* 129:185–90.

Vervoort, L., Van den Mooter, G., Augustijns, P. et al. 1998. Inulin hydrogels. I. Dynamic and equilibrium swelling properties. *Int J Pharm* 172:127–35.

Vervoort, L., Vinckier, I., Moldenaers, P. et al. 1999. Inulin hydrogels as carriers for colonic drug targeting. Rheological characterization of the hydrogel formation and the hydrogel network. *J Pharm Sci* 88:209–14.

Vilivialm V.D., Illum, L., and Iqbal, K. 2000. Starch capsules, an alternative system for oral drug delivery. *Pharm Sci Technol Today* 3:64–69.

Xi, M.M., Zhang, S.Q., Wang, X.Y. et al. 2005. Study on the characteristics of pectin-ketoprofen for colon targeting in rats. *Int J Pharm* 298:91–97.

Yang, L., Chu, J.S., and Fix, J.A. 2002. Colon-specific drug delivery: New approaches and in vitro/in vivo evaluation. *Int J Pharm* 235:1–15.

Zhang, L., Cao, F., Ding, B. et al. 2011. Eudragit S100 coated calcium pectinate microspheres of curcumin for colon targeting. *J Microencapsul* 28:659–67.

Zhang, X.Z., Lewis, P.J., and Chu, C.C. 2005. Fabrication and characterization of a smart drug delivery systems. *Biomaterials* 26:3299–09.

# 7 Lymphatic Drug Targeting

*Raghavendra V. Kulkarni and Biswanath Sa*

## CONTENTS

## 7.1   INTRODUCTION

The lymphatic system is considered an important route for drug targeting as drugs given through oral route are more effective. Drugs those can absorb from intestine into the systemic circulation possess reduced oral bioavailability due to the decomposition in the gastrointestinal tract before absorption. Usually, pH of stomach varies from 1.0 to 4.0. This highly acidic atmosphere may cause the decomposition of orally given drugs, and also before entering the blood, the drug absorbed into the portal venous system undergoes first-pass metabolism in the liver, and as a result the bioavailability and plasma concentration of the drug lowers. However, highly lipophilic drugs will transfer through the enterocyte and combine with enterocyte lipoproteins to form chylomicrons (Charman and Stella 1986). The chylomicrons along with the drug enter the mesenteric lymph duct, move to the thoracic duct, and then enter the systemic circulation, thus the highly lipophilic drugs bypass hepatic first-pass metabolism. Thus, the lower bioavailability of drugs can be enhanced with the help of the lymphatic system for absorption in the intestine bypassing first-pass metabolism in the liver (Trevaskis et al. 2008).

The lymphatic system also acts as a target route for the treatment of diseases such as acquired immune deficiency syndrome (AIDS), cancer, and so on. In case of AIDS, human immunodeficiency virus replicates vigorously in lymphoid organs; hence, lymphatic-targeted drug delivery is advantageous for the treatment of AIDS (Lalanne et al. 2007). In the case of cancer, the lymphatic system plays an important role in the metastasis, since lymphatic vessels contain pores through which immune cells enter and go out. The tumor cells easily enter the lymphatic vessels through these pores and progress to other organs leading to metastasis. To prevent this, chemotherapeutic agents can be targeted to the lymphatic vessels to control the metastasis of tumors (Garzon-Aburbeh et al. 1983; Arya et al. 2006; Cense et al. 2006).

## 7.2 ANATOMY AND PHYSIOLOGY OF THE LYMPHATIC SYSTEM

In the seventeenth century, Gasparo Aselli revealed the existence of the lymphatic system through his *milky veins* findings in fed dog's mesentery. Lymphatic system comprises lymphatic vessels, lymph nodes (LNs), spleen, thymus, Peyer's patches, and tonsils in a composite arrangement (Figure 7.1). This system plays a prominent role in the surveillance of the immune system and its response (McAllaster and Cohen 2011). The lymphatic system, acting as secondary vascular system in the body of vertebrates, performs its functions, which are interdependent with the cardiovascular system (Maby-El and Petrova 2008; Rinderknecht and Detmar 2008). Lymphatic system without any central driving force exists as a single-way, open-ended transition network (Mohrman and Heller 2006). In dermis, this system originates with the elementary lymphatic vessels, which are blind-end lymphatic capillaries, approximately proportionate to that of blood capillaries size, however, comparatively less in number

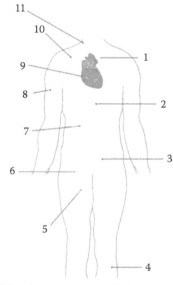

(1) Left lymphatic duct      (2) Thoracic duct
(3) Lumbar lymph nodes      (4) Popliteal lymph nodes
(5) Inguinal lymph nodes      (6) Iliac lymph nodes
(7) Mesenteric lymph nodes    (8) Axillary lymph nodes
(9) Thoracic lymph nodes      (10) Right lymphatic duct
                        (11) Cervical lymph nodes

**FIGURE 7.1** Schematic diagram showing the lymphatic system and LNs.

(Mohrman and Heller 2006). The capillaries of the lymphatic system constitute lymphatic endothelial cells (LECs). The basement membranes of LECs are poorly developed with poor functions and sometimes possess inadequate tightness and adherence junctions. As there exists space between LECs, which measure 30–120 nm approximately in size, lymphatic capillaries are extremely porous and allow the permeation of interstitial fluid, particles, and cells that are larger in size (Baluk et al. 2007). Meanwhile, the space between blood capillaries is less than 10 nm. Numerous hydrophilic passages are present in gaps between the tissues (around 100 nm). The tissue gaps, drug delivery systems (<100 nm) for which physiological aspects were provided, give guidance toward lymphatic system by means of injecting over the tissue space. The void among LECs was increased (300–500 nm) in inflammation or cancer-caused lesions, thereby allowing particles of proper sizes to penetrate the lymphatic system (Cueni and Detmar 2006; Mohrman and Heller 2006).

From various investigations, it was understood that the lymphatic system has a prominent role in the identification of the immune system and its response to an ailment. Spreading of many solid cancers starts from the initial site by means of lymphatic tissue surrounding the tumor before hematological dissemination (Cai et al. 2011). In general, the drugs that are administered by either oral or intravenous route enter into the systemic circulation in the beginning before their distribution in the entire body along with the relevant lesions. When drugs are administered through conventional route, rapid accumulation of drugs occurs in the normal organs and tissues causing unavoidable aftereffects. In such context, administration via the lymphatic system has appreciative benefits. Drug distribution through the lymphatic system is easier in the lymphatic system when compared to blood circulation, and this aids in strengthening the therapeutic benefit and decreasing the aftereffects (Ali Khan et al. 2013). Such mode of administration is effective, especially in lymphatic ailments such as inflammation and tumor metastasis, where the entry of drugs administered intravenously is limited in the lymphatic system.

## 7.3 MECHANISM OF LYMPHATIC-TARGETED DRUG DELIVERY

In tumor metastasis, lymphatic targeting is obtainable by passive targeting and active targeting (Kerjaschki 2014). Passive targeting mechanism is suitable for various drug delivery systems of nanosize, such as nanocapsules and solid lipid nanoparticles (SLNs), which can reach the targeted area. Active targeting can be attained through biochemical interactions such as bonding between the receptors and ligands on tumor cells or between the antigens and respective antibodies on tumor cells (Figure 7.2).

### 7.3.1 PASSIVE LYMPHATIC TARGETING

Passive targeting utilizes the lymphatic system's structural features in drug transportation. For instance, orally administered nano and chyle particles are absorbed through Pye's LNs, and those administered by subcutaneous injection can be absorbed by LNs and the lymphatic system. Targeting of different sizes of particles to different segments in the body can be attained. Emulsion droplets (0.1–0.5 μm in size), which are administered intravenously, are susceptible to macrophage clearance

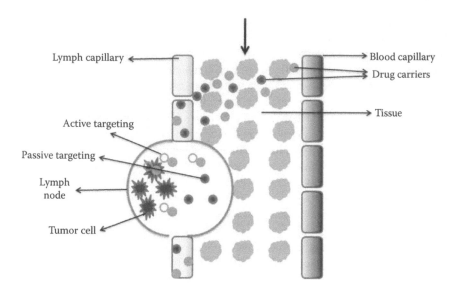

**FIGURE 7.2**  Schematic diagram showing the lymphatic-targeted drug delivery.

in the reticuloendothelial system (RES) of the liver, spleen, lung, and bone marrow (Yao et al. 2014). The entry of fat emulsion droplets into liver circulation cannot be achieved. Many of emulsion droplets agglomerate optionally within the RES macrophages, thereby contributing to the better targeting of drug to lymphatic system for improved anti-inflammatory effects. Mitomycin C that can be delivered in the form of water in oil (W/O) or oil in water (O/W) emulsions through intra-peritoneal or subcutaneous route was the first drug used for lymphatic targeting. From the earlier studies, it was found that targeting of W/O emulsions to the lymphatic system after local injection is more suitable when compared to O/W emulsions. Loading of drug into nanoparticles (NPs) before their dispersion into oil emulsions significantly enhances its targeting to lymphatic system (Flanagan and Singh 2006; Israelachvili 2011). During the process of lymph formation in cancer or inflammation, monolayer endothelial cells form capillary lymphatic vessels with incomplete and highly permeable basement membranes. Also, in contrast to the size of spaces between the blood capillaries, which are lesser than 10 nm in size, the expansion of lymphatic drainage channels occurs, normally from 30–120 nm to 300–500 nm (Cueni and Detmar 2006; Mohrman and Heller 2006). In addition to transportation through the bloodstream, larger molecules and suitable size colloids (<100 nm) are also carried into the lymphatic drainage by diffusing through these lymphatic ducts. Nonetheless, there is a possibility of these macromolecules to be engulfed by the macrophages while flowing through LNs. Hence, lymph node metastasis can be passively targeted.

## 7.3.2  ACTIVE LYMPHATIC TARGETING

Ligand-mediated endocytosis is the means for targeting lymphatic system actively. In the tumor development process, cancer cells secrete several vascular endothelial growth factors (VEGFs), which are capable of binding related receptors on the LECs, inducing tumor lymphatic vessels growth, followed by stimulation of tumor metastasis through lymphatic system (Berggreen and Wiig 2014). He et al. (2008) have synthesized NPs of calcium carbonate ($CaCO_3$) that carry small interfering RNAs (siRNAs) against VEGF-C with the diameter of 58 nm and zeta potential of +28.6 mV. These exhibited high efficiency to transfect in the human gastric cancer cell line SGC-7901 in comparison to $CaCO_3$ NPs comprised of nonspecific siRNA. This was demonstrated by VEGF mRNA and VEGF-C concentration to ~80%. From the experiments carried out on animals, it was found that the inhibition of lymph angiogenesis, metastasis of regional lymph node, and growth of the primary tumor can be attained by $CaCO_3$ NPs siRNA-VEGF-C complexes. When the rate of tumor lymphatic metastasis in the control group was 70%, it was reduced to around 20% in the test group. Cyclic peptide, LyP-1 (Laakkonen et al. 2002) with nine amino acids that can bind specifically to tumor cells, tumor lymphatic cells, and tumor-associated macrophages was recently reported by researchers.

## 7.4 LYMPHATIC-TARGETED DRUG DELIVERY SYSTEMS

Nanosized drug delivery systems play a major role in lymphatic targeting. Nevertheless, these nanoscale drug delivery systems suffer from a few limitations such as placing drugs to a particular site of the body, sustaining desired pharmacological actions, and passing through all the biological barriers. These limitations could be overwhelmed by the surface modification of nanoscale drug delivery systems. The permeation of drug carriers is inversely proportional to the size. The particles of smaller size (<100 nm) can permeate easily through lymphatic metastatic tumors, leading to increased antitumor efficiency. The drugs can be driven into the lymphatic system passively with increased accumulation of nano-carriers in the case of lymphatic metastasis or those can also be surface-modified for lymphatic system specific drug delivery (Yang et al. 2005). The nano-carriers for lymphatic-targeted drug delivery possess the ideal characteristics of high uptake by the lymphatic system, timely drug releases throughout the metastasis with reduced toxicity to normal cells. The various nanosized drug delivery systems have been reported for lymphatic delivery (Table 7.1) and are described as follows.

### 7.4.1 LIPOSOMES

Liposome is a nanosized carrier of drug comprising a lipid bilayer, in which the hydrophilic drugs can be entrapped into an aqueous core and the lipophilic drugs can be entrapped between the phospholipid bilayer (Figure 7.3a). Liposomes possess lymphatic targeting characteristics, which exhibit preferred lymphoid tissue targeting by means of intraperitoneal, subcutaneous, and intramuscular injections. When the liposome is engulfed by tumor cells, the entrapped drugs will be released for longer duration. It is the size of particle, composition, and lipid dose of liposome that affects lymphatic uptake. The temperature/pH-sensitive liposomes can be developed for site-specific controlled release of drugs by the modification of lipids with different fatty acid chain lengths (Singh et al. 2014).

**TABLE 7.1**

**Drug Delivery Systems Reported for Lymphatic Targeting**

| Drug | Delivery Systems | Therapeutic Improvement | References |
|------|------------------|-------------------------|------------|
| Docetaxel | Liposome | Liposomes improved the lymphatic targeting of antineoplastic agents and diagnostic agents | Tiantian et al. (2014) |
| Silymarin | Liposome | Lymphatic targeting was achieved through NLCs | Chaudhary et al. (2015) |
| OVA antigen | Liposome | Lymphatic-targeted vaccine delivery using liposomes effectively improved long-lasting immunological memory | Wang et al. (2014) |
| siRNA | NPs | The NPs were potential for targeting of gene to the LNs after IV injection | Yu-Cheng et al. (2014) |
| PTX | NPs | Delivery of PTX via conjugated NP improved therapeutic approach in ovarian cancer | Lingling et al. (2014) |
| Doxorubicin | NPs | Subcutaneous injection of HA-DOX nanocomplexes in tumor model animals has decreased the tumor volume without any tissue damage or heart toxicity | Cai et al. (2010) |
| Docetaxel | PMs | PMs have improvised the mode of docetaxel delivery in breast cancer | Yunfei et al. (2014) |
| Doxorubicin | PMs | In comparison to unformulated drugs, the drug-loaded PMs have increased cellular uptake of drugs via lymphatic targeting | Qin et al. (2013) |
| Vinorelbine | PMs | Vinorelbine-loaded micelles showed better effectiveness in inhibiting tumor cell growth | Lu et al. (2010) |
| PZQ | Solid lipid NPs | PZQ-loaded solid lipid NPs improved the oral bioavailability through intestinal lymphatic targeting | Mishra et al. (2014) |
| Etoposide | Solid lipid NPs | There was increase in tumor uptake of etoposide from subcutaneous injection of solid lipid NPs | Reddy et al. (2005) |
| MTX | Solid lipid NPs | MTX-loaded SLNs exerted higher bioavailability and lymphatic absorption | Paliwal et al. (2009) |
| Lopinavir | SLNs | Through lymphatic targeting, the lopinavir-loaded SLNs showed increased total dose of drug in lymph than conventional drug solution | Aji Alex et al. (2011) |
| Mitoxantrone | SLNs | The drug concentration of mitoxantrone SLNs in local LNs was much higher than other tissues | Lu et al. (2006) |

*(Continued)*

**TABLE 7.1 (*Continued*)**
**Drug Delivery Systems Reported for Lymphatic Targeting**

| Drug | Delivery Systems | Therapeutic Improvement | References |
|---|---|---|---|
| Propranolol | Dendrimers | Dendrimer-drug prodrugs improved drug solubility and bioavailability of drug | D'Emanuele et al. (2004) |
| Ketoprofen | Dendrimers | The ketoprofen release from drug-dendrimer complex was considerably slower as compared to free ketoprofen. Antinociceptive effect was prolonged | Na et al. (2006) |
| Cisplatin | Carbon nanotubes | The CNTs are retained in the LNs for longer duration | Sahoo et al. (2011), Dhar et al. (2008) |
| Gemcitabine | Carbon nanotubes | The entrapment of camptothecin into multi-walled CNTs was potential in the treatment of breast and skin cancers | Yang et al. (2011) |
| Carboplatin | Carbon nanotubes | The therapeutic benefits of GEM-loaded magnetic multi-walled CNTs had greater antitumor effect | Hampel et al. (2008) |

Tiantian et al. (2014) have formulated hyaluronic acid-modified docetaxel-loaded liposomes for lymphatic targeting. Studies on lymphatic drainage and uptake by lymph node were performed through drug distribution recovery in LNs, site of administration, blood, and pharmacokinetics. Cy7-loaded liposomes lymphatic targeting was examined using near-infrared fluorescence imaging. *In vivo* comparison of unmodified liposomes and hyaluronic acid-modified liposomes determined that hyaluronic acid-modified liposomes have considerably enhanced the docetaxel recovery from LNs and demonstrated increased bioavailability and prolonged retention time. On the contrary, lymphatic drainage was hindered by the free hyaluronic acid, but drug concentration in plasma was increased. Thus, lymphatic drainage and uptake of liposomes by LNs were enriched by the hyaluronic acid-modification. The Cy7-loaded liposomes were remarkably located in the lymphatic system as observed in *in vivo* imaging. This investigation proved that the hyaluronic acid-modified liposomes are promising lymphatic targeting carriers for antineoplastic agents and diagnostic agents (Tiantian et al. 2014).

Chaudhary et al. (2015) have developed nanostructured lipid carriers (NLCs) through emulsification and ultrasonication using a Box–Behnken design. Glyceryl monostearate and oleic acid was found to be the best lipid combination with the particle size of 223.73 nm and drug entrapment efficiency (EE) as high as 78%. The *in vitro* release study indicated that about 84% of drug release was noticed within 24 h. The *in vivo* studies confirmed the drug absorption through lymphatic circulation when only about 5 mg/mL of peak plasma concentration was achieved in blood plasma in the presence of chylomicron inhibitor. They observed that the peak plasma concentration of silymarin loaded NLCs was 25.56 mg/mL; and the relative bioavailability increased by twofold as compared to peak plasma concentration of

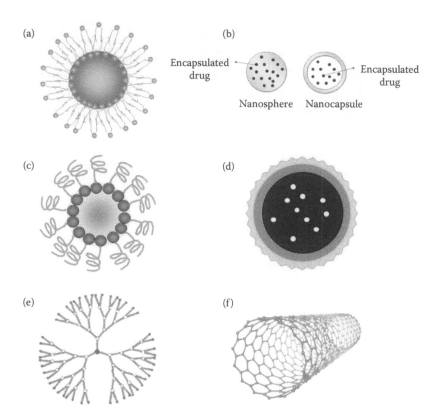

**FIGURE 7.3**   Schematic diagram of liposome (a), drug-loaded nanoparticles (b), PM (c), SLNs (d), dendrimer (e), and carbon nanotube (f).

silymarin suspension (14.05 mg/mL). *In vivo* studies demonstrated that the 19.26 mg of drug has reached liver within 2 h, whereas very low concentration of drug was found in other organs.

Vaccine therapy can be improved in a versatile manner by the utilization of liposomes as carriers in targeting the vaccines to lymphatic system. Wang et al. (2014) encapsulated OVA antigen with 1, 2-dioleoyl-3-trimethylammonium-propane (DOTAP) into cationic liposomes (LP) or DOTAP-PEG-mannose liposomes (LP-Man) to develop lymphatic-targeted liposomal vaccines, respectively. During the investigation it was found that there was accumulation of liposomes around the site of injection, whereas LP-Man were collected LNs, spleen and LP-Man uptake by resident antigen-presenting cells was enhanced. This suggests that lymphatic-targeted vaccine delivery with LP-Man is a potential approach for promoting long-lasting immunological memory.

Another study found the possibility of increase in the size of liposomes due to the interaction of the surface of some liposomes with interstitial proteins, by which lymphatic system cannot absorb liposomes completely; instead, liposomes can be stranded at the site of injection. In comparison with non-modified liposomes, there was no distinct lymphatic absorption improvement with spatially stabilized PEGylated liposomes. This concept can be depicted by the presence of a PEG hydrophilic layer. Liposomes PEGylation can hinder the process of phagocytosis by macrophages and influence their lymph node uptake. Hence, apart from space stability of liposomes and macrophage phagocytosis, lymphatic absorption and liposomes uptake can be affected by other factors as well. With the accumulation of liposomes by mechanical retention or by macrophages of liposomes in normal LNs, regular lymphatic tissue can be damaged (Moghimi and Moghimi 2008; Moghimi et al. 1994).

Generally, it is a passive process by which accumulation of liposomes occurs in LNs. Liposomes cannot discriminate between normal and metastatic LNs. Due to the structural damage in metastatic LNs, the liposomes uptake by metastatic LNs when compared with that of normal LNs is low. This concept was addressed in recent literature by altering the delivery system with the cyclic peptide LyP-133-35, which is composed of nine amino acids and can distinctly have a binding with tumor cells and p32 receptors that are immensely expressed on tumor lymphatic vessels.

The LyP-1-conjugated liposome drug delivery system fabricated by Yan et al. presented sound target ability to lymphatic metastatic tumors after injecting subcutaneously. The definite binding of LyP-1, PEG and liposome with tumor metastases, tumor lymphatics, and tumor-associated macrophages enhances inhibitory effect on lymphatic metastases and improvises the suppressing effect on lymphatic metastasis. Such findings suggested that LyP-1-conjugated liposomes could provide an appropriate platform for active-targeted drug delivery in the treatment of lymphatic metastatic tumors (Yan et al. 2011).

### 7.4.2 NANOPARTICLES

The nanostructures such as NPs, nanospheres, and nanocapsules with a size ranging from 10 to 1000 nm are being developed using natural or synthetic polymers. Nanocapsules comprise polymeric shells containing a liquid core, in which the drug was dissolved, while NPs and nanospheres are carriers containing solid matrix, in which the drug was entrapped. These three are collectively called NPs. Various benefits like retarded drug release, long circulation, better stability, safety, and increased efficiency can be obtained with NPs (Figure 7.3b).

In a recent report, Yu-Cheng et al. (2014) have developed lipid/calcium/phosphate (LCP) NPs with superior siRNA delivery efficiency. They reported the successful incorporation of [111]In into LCP for SPECT/CT imaging. The results of imaging and bio-distribution cited accumulation of polyethylene glycol grafted [111]In-LCP in the LNs at about 70% ID/g in both C57BL/6 and nude mice after adapting improved surface coating method. Accumulation in both the liver and spleen was only about 25% ID/g. Larger LCP (diameter ~67 nm) was less lymphotropic. These results indicated that due to small size, a well-PEGylated lipid surface, and a slightly negative surface charge, 25 nm LCP was able to penetrate into tissues, enter the lymphatic system, and accumulate in the LNs via lymphatic drainage. The ability of [111]In-LCP administered via intravenous route to visualize an enlarged, tumor-loaded lymph node using a 4T1 breast cancer lymph node metastasis model was successfully demonstrated.

A paclitaxel (PTX)-loaded follicle-stimulating hormone (FSHP) NPs (FSHP-NP-PTX) was developed and antitumor effect was carried out using NuTu-19 cells as an example. The FSHP-NP-PTX exhibited considerable stronger anti-cell proliferative and antitumor effects in a dose and time-dependent manner when compared with pristine PTX or pristine PTX loaded NPs (NP-PTX) during *in-vitro* studies. *In-vivo* examinations showed that the size and weight of the LNs were reduced in the FSHP-NP-PTX group. Hence, PTX delivery through conjugated NP (FSHP-NP) might depict a new curative method in ovarian cancer (Lingling et al. 2014).

Eric et al. (2010) have developed Cy5-labeled polylactide (Cy5-PLA) NPs and evaluated for lymphatic bio-distribution pattern. The Cy5-PLA NPs were synthesized through ring-opening polymerization of lactide followed by nano-precipitation. The lymphatic bio-distribution study was carried out in nude mice by fluorescence imaging of whole-body and resected organs. This approach provides inherent optical contrast characteristics and delivering the drug into lymphatic circulation for treating metastatic cancer.

Biodegradable, polymeric NPs were prepared for enhanced lymphatic targeting. Vast experimentation on the preparation of poly (alkylcyanoacrylate) NPs and their curative benefits was carried out by Puisieux and Couvreur research group (Couvreur et al. 1990; Couvreur et al. 1995). Targeting polyhexylcyanoacrylate NPs to lymphatic system was evaluated after administering to rats intraperitoneally and found the competency of NPs in the tumor treatment (Maincent et al. 1991; Maincent et al. 1992). These researchers also revealed that the NPs uptake through Peyer's patches following oral administration, thereby suggested the availability of peptide delivery by peroral means (Damge et al. 1990). Extensive investigations were carried out by Davis and Illum on biodegradable nanospheres with polylactides and poly (lactide-co-glycolide) for attainment of appropriate delivery of drug and diagnostic agent to the lymphatic system. Various mechanisms of surface engineering such as surface coating using poloxamines or poloxamers (Hawley et al. 1997a) and the utilization of poly(ethyleneglycol) (Hawley et al. 1997b) were attempted to improve lymphatic targeting of nanospheres. In addition to drug targeting objective, magnetite-dextran NPs were considered for evaluating diagnostic usage, found to be very effective contrast agents in magnetic resonance imaging (MRI) (Chouly et al. 1996).

Hyaluronic acid (HA) is a natural biodegradable polymer, which was being used for targeting drug delivery to lymphatic system. During their growth for the attainment of P-glycoprotein over-expression cancerous breast cells tend to uptake abundant HA. The pH-sensitive hyaluronic acid-doxorubicin (HA-DOX) complexes were developed with DOX linked to HA by hydrazone bonds that are susceptible for cleavage. Followed by subcutaneous injection in tumor model animals, the comparison of HA-DOX nanocomplexes with the free DOX treatment has revealed that the tumor volume was decreased by 70% after 10 weeks of inoculation. During this period, the transportation from orthotropic tumor to lymphatic metastasis occurs and thus antitumor effect was illustrated. After 18 weeks of inoculation, when death of entire tumor-bearing nude mice of free DOX group was compared with HA-DOX group, survival of a minimum 50% of HA-DOX nude mice group was observed in the entire study duration of 24 weeks. Moreover, there were no observations of tissue damage or heart toxicity in the HA-DOX group (Cai et al. 2010).

### 7.4.3 POLYMER MICELLES

A polymeric micelle (PMs) is comprised of corona developed by means of hydro-philic blocks extending into the aqueous solution and a core formed by the hydropho-bic segments. The intrinsic strength in hydrophobic anticancer drug solubilization with elimination of toxic organic solvents and the use of surfactants was demon-strated in the development of polymeric micelles. It is because of substantial capa-bility in preferential elevation of drug exposure in tumor and improvised anticancer effect, polymer micelles have become true promising carriers in cancer therapy and this still remains to be fully exploited (Figure 7.3c).

There are several inferences available for polymeric micelles PMs to exist as prime carriers in cancer treatment. Primarily, most of the potent anti-cancer medi-cines are insoluble in water and depend on noxious solvents or surfactants for solubi-lization. This results in enhancing dose-limited toxicities of treatments. The ability of PMs to solubilize hydrophobic drugs with no usage of toxic solvents, thereby improvised drug safety was attributed by their specific core-shell structure. Second, it is preferential accumulation of PMs in the tumor through enhanced permeability and retention (EPR) effect. Due to vasculature with leakage and dysfunctional lym-phatic drainage in the tumors, there was extravasation of nanosized carriers through the tumor endothelium and escalation of drug accumulation in tumor tissue. Due to their tightly knitted endothelial lining, the access to normal organs was very limited. At last, PMs are sensitive to modifications in surface with the targeting ligands, which can distinctly identify the receptors that are overexpressed on tumor cells and/ or tumor endothelium surface, leading to intracellular delivery of micellar drugs at higher efficiency.

Poly (D,L-lactide)$_{1300}$-b-(polyethylene glycol-methoxy)$_{2000}$(mPEG$_{2000}$-b-PDLLA$_{1300}$) was used for docetaxel encapsulation to fabricate PMs. Resected and unresected pri-mary tumors were employed in comparison of therapeutic effectiveness of PMs and unformulated docetaxel against breast cancer by means of bioluminescent imaging, lung nodule diagnosis, and histopathological evaluation. The PMs demonstrated iden-tical effectiveness as that of free docetaxel with respect to primary tumor growth suppression; however, superlative chemotherapeutic effect against breast cancer was noticed. Additionally, inflammation of lungs was decreased in the group adminis-tered with PMs, whereas several tumor cells and neutrophils were noticed in the group administered with unformulated docetaxel. Therefore, the PMs can serve as an improvised mode of docetaxel delivery in breast cancer and can depict a valid chemo-therapeutic strategy in breast cancer with resisted primary tumors (Yunfei et al. 2014).

Qin et al. (2013) have systematically prepared doxorubicin-loaded micelles (M-DOX) and vinorelbine-loaded micelles (M-VNR). The *in vivo* antitumor activity of these micelles was evaluated. In comparison with unformulated drugs, the drug-loaded micelles with which membrane fluidity was altered have increased the drug content that has reached the cells; also by adjusting the protein expression control-ling genes, these micelles can enhance cellular uptake of drugs. In 4T1 breast cancer metastasis rat model, both M-DOX and M-VNR micelles suppressed the tumors resulting from lung metastasis and also by increasing the drug distribution in LNs, the systemic and acute toxicity of drugs was reduced as well.

Lu et al. (2010) have reported the mechanism of vinorelbine loaded in glycol-phosphatidylethanolamine (PEG-PE) micelles (M-Vino) on tumor cells. *In-vitro* comparison between free vinorelbine and M-Vino micelles revealed that M-Vino micelles were more effective in inhibiting tumor cells growth by inducing $G_2/M$ phase arrest and apoptosis of tumor cells. M-Vino exhibited rapid passage and more accumulation in 4T1 cells than that of unformulated vinorelbine. Hence, through higher intercellular accumulation of vinorelbine, M-Vino micelles resulted in micro-tubules destabilization, cell death, and enhanced cytotoxic effect.

### 7.4.4  SOLID LIPID NANOPARTICLES

SLNs are drug carriers comprising a solid matrix with a particle size ranging from 50 to 1000 nm. It is comprised of a lipid, which exists in solid form at room tempera-ture and body temperature as well (Figure 7.3d).

Among those introduced in the early 1990s, SLNs are treated as most effective lipid-based carriers. They are more economical due to availability of excipients and production lines at relatively low cost when compared with parenteral emulsions manufacturing (Vaghasiya et al. 2013). Due to the first-pass metabolism by which transportation into the systemic circulation was through the intestinal lymphatics, SLNs could be a good formulation strategy for incorporation of drugs with low oral bioavailability. SLNs comprising triglycerides particularly resemble the chylomi-crons and may alter the absorption behavior of drugs including avoidance of first-pass metabolism. Various anticancer drugs, etoposide, methotrexate (MTX), and idarubicin have been incorporated into SLNs and evaluated for their lymph localiza-tion (Zara et al. 2002).

Praziquantal (PZQ) loaded SLNs (PZQ-SLN) were developed to improve the oral bioavailability by targeting intestinal lymphatic system. PZQ is practically insoluble in water and exhibits extensive hepatic metabolism. Optimization on PZQ-SLNs, which were composed of triglycerides, lecithin, and various aqueous surfactants, was carried out using hot homogenization and ultrasonication method. Particle size of optimized SLNs was $123 \pm 3.41$ nm, and EE is $86.6 \pm 5.72\%$. In PZQ-SLN there was an initial burst release followed by sustained release. Despite the zeta potential existed around $-10$ mV, the optimized SLNs were stable at storage conditions of $25 \pm 2°C/60 \pm 5\%$ RH for 6 months. Transmission electron microscopic study con-firmed almost spherical shape similar to the control formulations. The differential scanning calorimetric analysis and X-ray diffraction analysis confirmed the uniform distribution of PZQ in lipid matrix. The 5.81-fold increase in $AUC_{0\rightarrow\infty}$, after intra-duodenal administration of PZQ-SLNs in rats confirmed its intestinal lymphatic delivery as compared to rats treated with cycloheximide (a blocker of intestinal lym-phatic pathway) (Mishra et al. 2014).

Reddy et al. (2005) investigated the impact of subcutaneous route on bio-distribution and tumor uptake of SLNs. Tripalmitin (ETPL) SLN's radio labeled with 99mTc and loaded with Etoposide were formulated and administered to Dalton's lymphoma tumors bearing mice subcutaneously, intraperitoneally, and intravenously. Bio-distribution and tumor uptake of the ETPL SLNs in the blood and organs was determined by gamma scintigraphy and radioactivity measurements. Remarkable tumor uptake (8- and 59-fold higher than that of the intraperitoneal and intravenous routes, respectively) and reduced build-up in RES organs (i.e., liver, lung, and spleen) after 24 h of ETPL SLNs subcutaneous administration was observed. Concentration of ETPL SLNs in tumor after 1 h of subcutaneous injection was 23.3%. Moderate increase in tumor uptake of drug from subcutaneous injection site after lower initial uptake suggested the use of these SLNs in sustained release of drug. Accordingly, proximal subcutaneous injection is an effective route for lymphatic targeting.

Paliwal et al. (2009) have investigated the effect of lipid on MTX SLNs formulation for oral lymphatic delivery. By means of solvent diffusion method, SLNs were formulated using four different lipids namely stearic acid, monostearin, tristearin, and Compritol 888 ATO. Further, comparative evaluation was carried out for particle shape, size, zeta potential, entrapment efficiency, *in vitro* drug release, and pharmacokinetics. The MTX-loaded SLNs formulated using Compritol 888 ATO generated the uniform SLNs with the smallest size and highest entrapment efficiency. The Compritol 888 ATO-based SLNs (MTX-CA) resulted in 10% higher MTX loading. After intraduodenal administration, bioavailability and intestinal lymphatic uptake of MTX-SLNs were assessed in albino rats. Significant increase in the AUC and lymphatic drug concentration was observed with all MTX SLNs formulations as compared to free drug solution. It is MTX-CA that exerted higher bioavailability and lymphatic absorption. The Compritol 888 ATO formulation was found to be an effective carrier.

Aji Alex et al. (2011) prepared lopinavir loaded SLNs for intestinal lymphatic targeting and reported that total dose of lopinavir secreted into the lymph was increased and it was compared to be 4.91-fold higher than the conventional drug solution; the percentage bioavailability was significantly enhanced. Videira et al. designed technetium-99m-D,L-hexamethylpropyleneamine oxime radiolabeled SLNs. These authors tracked the bio-distribution of SLNs from the lungs and found that 4 h after administration of the radiolabeled SLNs, these were located primarily within inguinal and axillary LNs, suggesting lymphatic uptake and retention of SLNs that had been absorbed from the lungs (Videira et al. 2002). Lu et al. designed and evaluated SLNs of mitoxantrone against breast cancer on mice. The drug concentration of mitoxantrone SLNs in local LNs was much higher and the drug concentrations in other tissues were lower (Lu et al. 2006).

### 7.4.5 DENDRIMERS

Dendrimers are extremely branched well-defined molecules that are synthesized with low polydispersity and definite control on structure (Klajnert and Bryszewska 2001). The word dendrimer is derived from the Greek terms dendron, which means tree or branch, and meros, which means part. Arboroles or cascade polymers are synonyms for dendrimers (Figure 7.3e). In recent times, dendrimers have evolved as dexterous elements in biomedical research applications like imaging and drug delivery (Svenson 2009). Due to precise control in molecular structure, tunable size, availability of number of reactive sites, lesser viscosity as compared to linear polymers of equivalent molecular weight, and interior void space, dendrimers play a major role in developing imaging contrast agents. Various difficulties that are associated with contrast agents of low molecular weight and imprecise synthetic polymers may be addressed by making advantage out of dendrimers regular structure. The *divergent* and *convergent* mechanisms are two main strategies usually involved in dendrimers synthesis. Tomalia (1996) has introduced a divergent method, which starts with multifunctional core and further proceeds with repetitive addition of monomers for increased molecular weight and exponential increase of surface termini. On the contrary, the convergent method was proposed by Hawker and Frechet (1990) that starts from surface and proceeds inward to a multivalent core where the dendrimer segments are joined together.

Dendrimers structure is usually globular, with alterable surface functional groups that are mainly responsible for their environmental interaction with the application in the medical field. This contributes to dendritic units versatile activities, which allows drug encapsulation or attachment to surface or peripheral groups by which solubility and toxicity can be modulated. Dendrimers are repetitively branched macromolecules, which could be an effective system for lymphatic imaging and targeting.

Applicability of dendrimers in the pharmaceutical and medical chemistry fields was expeditive. Dendrimer chemistry and employing dendrimers as drug administration agents has become a more attractive subject for various researchers. For example, D'Emanuele et al. (2004) have synthesized propranolol prodrug with 3G and lauroyl-3G polyamidoamine (PAMAM) dendrimer and the effect of these conjugates on propranolol transport over adenocarcinoma cell line CaCo-2 was determined. From the results, it was found that dendrimer-drug prodrugs can be utilized for drug solubility enhancement and bypassing drug efflux transporters, thereby bioavailability of drug was increased. Subsequently, Na et al. (2006) have made an attempt to prolong ketoprofen delivery employing PAMAM dendrimers. Ketoprofen is less soluble in water and also results in local or systemic gastrointestinal side effects. Therefore, *in vitro* and *in vivo* investigation of ketoprofen loaded PAMAM dendrimers was performed. The *in vitro* release of ketoprofen from the drug-dendrimer complex was significantly slower as compared to free ketoprofen. Anti-nociceptive evaluation of ketoprofen-PAMAM dendrimer complex in mice by means of a writhing model induced by acetic acid has exhibited a prolonged pharmacodynamic behavior.

*In vitro* evaluation of effect of size, charge concentration of PAMAM dendrimer on transition, and uptake across the intestine of an adult rat using the everted rat intestinal sac system was carried out by Wiwattanapatapee et al. (2000). Focus was made on cationic PAMAM dendrimers and anionic PAMAM dendrimers. Findings suggested that I-labeled dendrimers transport over the intestinal membrane depends on charge and increased cationic dendrimers tissue uptake in comparison with serosal transport (Mullin 2003).

### 7.4.6 CARBON NANOTUBES

The major objective in the development of nanocarriers-based drug delivery systems is enhancing curative benefits or reducing therapeutically active substances' toxicity. Carbon nanotubes (CNTs) are typical cylindrical molecules consisting of carbon atoms. CNTs are graphene sheets rolled in the form of open ended or capped cylinder with size range of 1 nm to a few microns (Figure 7.3f). CNTs manufactured from a sole graphene sheet resulted in a single-walled nanotube (SWNT), whereas multi-walled carbon nanotubes (MWNTs) can be obtained from various graphene sheets. Since 1991, that is when they were discovered by Iijima (1991), an immense interest was captured in these carbon allotropes due to their individual physical and chemical properties and important vast applications in various fields, such as electronic devices, sensors to nanocomposite materials of high strength and low weight.

Originally, CNTs were not soluble. The development of an approach to functionalize these molecules containing organic groups and rendering their solubility, bio-applications of CNTs were originated. With the availability of more surface area, they are competent enough to adsorb or conjugate with a large range of therapeutic molecules. Thus, CNTs can be surface engineered (i.e., functionalized) for enhancing their aqueous phase dispersability or to provide appropriate functional groups which can bind to the required curative material or the target tissue to exert therapeutic activity. CNTs may support the adhered therapeutic molecule to permeate the target cell for treating ailments (Pantarotto et al. 2004a, b; Bianco et al. 2005; Chen et al. 2007) and a latest illustration of CNTs with a range of functional groups relevant to cancer treatment (Bhirde et al. 2009) was reported. An overview of CNTs therapeutic applicability with more focus on their use in the treatment of cancer was provided. CNTs were explored to be potential nanocarriers for drug, gene, and protein delivery. Focus in various researches was made on potentiality of CNTs for anticancer drug delivery. This requirement might have ascribed to their unique needle-like shapes that facilitate them to be functionalized for adsorption or covalent linkage with a vast range of therapeutic elements and for their internalization into the target cell.

It is reported that cancer cells overexpress folic acid (FA) receptors. Reports say that the nanocarriers could be designed with surface engineering to which FA derivatives can be tagged. The nonspherical nanocarriers such as CNTs have been found to be retained in the LNs for longer duration of time as compared to spherical nanocarriers (Reddy et al. 2006). Thus, CNTs could be the ideal carriers for lymphatic targeting (Liu et al. 2003; Yang et al. 2008, 2009). Magnetic NPs loaded with anticancer drug cisplatin were encapsulated into folic-acid-functionalized MWNTs. The tumor targeting was done by application of an external magnet to drag the CNTs to the LNs, where the drug release continued for a few days. A recent report describes the entrapment of poorly water-soluble anticancer drug camptothecin into polyvinyl alcohol-functionalized MWNTs, which was found to be potential in treatment of breast and skin cancers (Sahoo et al. 2011).

Dhar et al. (2008) fabricated a "longboat delivery system," where cisplatin complex and FA derivative were adhered to a functionalized SWNT by means of various amide bonds to contain the *longboat* which was reported to be absorbed by endocytosis of cancer cells, followed by drug release and its further interaction with the nuclear DNA. In an *in vitro* study, another platinum anticancer drug, carboplatin, after its incorporation into CNTs exhibited inhibition of urinary bladder cancer cells proliferation. Whereas another study describes that the anticancer effects were dependent on the mode of drug entrapment in CNTs, which highlighted the viable effects of formulation conditions on therapeutic molecules' curative effect related with CNTs (Hampel et al. 2008).

Yang et al. (2011) have developed a novel drug delivery system with magnetic lymphatic targeting, which was based on functionalized carbon nanotubes (fCNTs) with an objective of bettering the outcome of cancer with the involvement of the lymph nodes. The inherent therapeutic benefits of gemcitabine (GEM) loaded magnetic multi-walled carbon nanotubes (mMWNTs) were correlated *in vitro* and *in vivo* with GEM loaded magnetic-activated carbon particles (mACs). mMWNTs-GEM and mACs-GEM both had greater *in vitro* antitumor effects similar to that of free drug. Administering GEM loaded magnetic NPs subcutaneously resulted in efficacious regression of lymph node metastasis under magnetic field, with mMWNTs-GEM superior to mACs-GEM and with greater effectivity in high-dose versus low-dose groups. A noteworthy utilization of chemotherapeutics' intra-lymphatic delivery with mMWNTs features the clinical potential of fCNTs for subsequent cancer metastasis therapy with increased efficacy and low side effects.

## 7.5 CONCLUSION

Though the lymphatic drug targeting is a complex method, it has several advantages. For the development of lymphatic-targeted drug delivery systems, a better knowledge of human physiology and the interface of delivery systems with physiological process is essential. Nanosized drug delivery systems are versatile carriers for the treatment of critical diseases like cancer. The nanosized formulations improve the effectiveness of drug therapy and reduce the drug toxicity because of targeting ability. The poorly soluble and cytotoxic drugs can be successfully targeted to the lymphatic system using nanosized drug delivery systems. While extensive work is being reported by researchers worldwide for lymphatic drug targeting, the major challenging issues like stability, size, commercial productivity, and scalability of nanosized drug delivery systems are to be addressed.

## REFERENCES

Aji Alex, M.R., Chacko, A.J., Jose, S. et al. 2011. Lopinavir loaded solid lipid nanoparticles for intestinal lymphatic targeting. *Eur J Pharm Sci* 42:11–18.

Ali Khan, A., Mudassir, J., Mohtar, N. et al. 2013. Advanced drug delivery to the lymphatic system: Lipid-based nanoformulations. *Int J Nanomedicine* 8:2733–44.

Arya, M., Bott, S.R., Shergill, I.S. et al. 2006. The metastatic cascade in prostate cancer. *Surg Oncol* 15:117–28.

Baluk, P., Fuxe, J., Hashizume, H. et al. 2007. Functionally specialized junctions between endothelial cells of lymphatic vessels. *J Exp Med* 204:2349–62.

Berggreen, E. and Wiig, H. 2014. Lymphatic function and responses in periodontal disease. *Exp Cell Res* 325:130–37.

Bhirde, A.A., Patel, V., Gavard, J. et al. 2009. Targeted killing of cancer cells in vivo and in vitro with EGF-directed carbon nanotube-based drug delivery. *ACS Nano* 3: 307–16.

Bianco, A., Kostarelos, K., and Prato, M. 2005. Applications of carbon nanotubes in drug delivery. *Curr Opinion Chem Biol* 9:674–79.

Cai, S., Thati, S., Bagby, T.R. et al. 2010. Localized doxorubicin chemotherapy with a biopolymeric nanocarrier improves survival and reduces toxicity in xenografts of human breast cancer. *J Control Release* 146:212–18.

Cai, S., Yang, Q., Bagby, T.R. et al. 2011. Lymphatic drug delivery using engineered liposomes and solid lipid nanoparticles. *Adv Drug Deliv Rev* 63:901–08.

Cense, H.A., van Eijck, C.H., and Tilanus, H.W. 2006. New insights in the lymphatic spread of oesophageal cancer and its implications for the extent of surgical resection. *Best Pract Res Clin Gastroenterol* 20:803–906.

Chaudhary, S., Garg, T., Murthy, R.S.R. et al. 2015. Development, optimization and evaluation of long chain nanolipid carrier for hepatic delivery of silymarin through lymphatic transport pathway. *Int J Pharm* 485:108–21.

Charman, W.N. and Stella, V.J. 1986. Estimating the maximum potential for intestinal lymphatic transport of lipophillic drug molecules. *Int J Pharm* 34:175–78.

Chen, X., Kis, A., Zettl, A. et al. 2007. A cell nano injector based on carbon nanotubes. *Proc National Acad Sci USA* 104:8218–22.

Chouly, C., Pouliquen, D., Lucet, I. et al. 1996. Development of super paramagnetic nanoparticles for MRI: Effect of particle size, charge and surface nature on biodistribution. *J Microencapsul* 13:245–55.

Couvreur, P., Dubernet, C., and Puisieux, F. 1995. Controlled drug Enhance delivery with nanoparticles: Current possibilities and future trends. *Eur J Pharm Biopharm* 41:2–13.

Couvreur, P., Roblot-Treupel, L., Poupon, M.F. et al. 1990 Nanoparticles as microcarriers for anticancer drugs. *Adv Drug Del Rev* 5:209–30.

Cueni, L.N. and Detmar, M. 2006. New insights into the molecular control of the lymphatic vascular system and its role in disease. *J Invest Dermatol* 126:2167–77.

D'Emanuele, A., Jevprasesphant, R., Penny, J. et al. 2004. The use of a dendrimer-propranolol prodrug to bypass efflux transporters and enhance oral bioavailability. *J Control Release* 95:447–53.

Damge, C., Mitchel, C., Aprahamian, M. et al. 1990. Nanocapsules as carriers for oral peptide delivery. *J Control Release* 13:233–39.

Dhar, S., Liu, Z., Thomale, J. et al. 2008. Targeted single-wall carbon nanotube-mediated Pt(IV) prodrug delivery using folate as a homing device. *J Am Chem Soc* 130:11467–76.

Eric, J., Chaney, Li., Tang, R. et al. 2010. Lymphatic biodistribution of polylactide nanoparticles. *Molr Imaging* 9:153–62.

Flanagan, J. and Singh, H. 2006. Microemulsions: A potential delivery system for bioactives in food. *Crit Rev Food Sci Nutr* 46:221–37.

Garzon-Aburbeh, A., Poupaert, J.H., Claesen, M. et al. 1983. 1,3-dipalmitoylglycerol ester of chlorambucil as a lymphotropic, orally administrable antineoplastic agent. *J Med Chem* 26:1200–03.

Hampel, S., Kunze, D., Haase, D. et al. 2008. Carbon nanotubes filled with a chemotherapeutic agent: A nanocarrier mediates inhibition of tumor cell growth. *Nanomedicine* 3:175–82.

Hawker, C.J. and Frechet, J.M.J. 1990. Preparation of polymers with controlled molecular architecture. A new convergent approach to dendritic macromolecules. *J Am Chem Soc* 112:7638–47.

Hawley, A.E., Illum, L., and Davis, S.S. 1997a. Lymph node localization of biodegradable nanospheres surface modified with poloxamer and poloxamine block co-polymer. *FEBS Lett* 400:319–23.

Hawley, A.E., Illum, L., and Davis, S.S. 1997b. Preparation of biodegradable, surface engineered PLGA nanospheres with enhanced lymphatic drainage and lymph node uptake. *Pharm Res* 14:657–61.

He, X.W., Liu, T., Chen, Y.X. et al. 2008. Calcium carbonate nanoparticle delivering vascular endothelial growth factor-C siRNA effectively inhibits lymphangiogenesis and growth of gastric cancer in vivo. *Cancer Gene Ther* 15:193–02.

Iijima, S. 1991. Helical microtubules of graphitic carbon. *Nature* 354:56–58.

Israelachvili, J.N. 2011. *Intermolecular and Surface Forces*. 3rd ed., London: Academic Press.

Kerjaschki, D. 2014. The lymphatic vasculature revisited. *J Clin Invest* 124:874–77.

Klajnert, B. and Bryszewska, M. 2001. Dendrimers: Properties and applications. *Acta Biochimica Polonica* 48:199–08.

Laakkonen, P., Porkka, K., Hoffman, J.A. et al. 2002. A tumor homing peptide with a targeting specificity related to lymphatic vessels. *Nat Med* 8:751–55.

Lalanne, M., Paci, A., Andrieux, K. et al. 2007. Synthesis and biological evaluation of two glycerolipidic prodrugs of didanosine for direct lymphatic delivery against HIV. *Bioorg Med Chem Lett* 17:2237–40.

Lingling, F., Jun, C., Xiaoyan, Z. et al. 2014. Follicle-stimulating hormone polypeptide modified nanoparticle drug delivery system in the treatment of lymphatic metastasis during ovarian carcinoma therapy. *Gynecol Oncol* 135:125–32.

Liu, Y., Ng, K.Y., and Lillehei, K.O. 2003. Cell-mediated immunotherapy: A new approach to the treatment of malignant glioma. *Cancer Control* 10:138–47.

Lu, B., Xiong, S.B., Yang, H. et al. 2006. Solid lipid nanoparticles of mitoxantrone for local injection against breast cancer and its lymph node metastases. *Eur J Pharm Sci* 28:86–95.

Lu, X., Fayun, Z., Lei, Q. et al. 2010. Polymeric micelles as a drug delivery system enhance cytotoxicity of vinorelbine through more intercellular accumulation. *Drug Deliv* 17:255–62.

Maby-El, H.H. and Petrova, T.V. 2008. Developmental and pathological lymphangiogenesis: From models to human disease. *Histochem Cell Biol* 130:1063–78.

Maincent, P., Amicabile, C., Thouvenot, P. et al. 1991. Targeting of lymph system with [14]C-polyacrylic nanoparticles. *Arch Pharm* 324:637–48.

Maincent, P., Thouvenot, P., Amicabile, C. et al. 1992. Lymphatic targeting of polymeric nanoparticles after intraperitoneal administration in rats. *Pharm Res* 9:1534–39.

McAllaster, J.D. and Cohen, M.S. 2011. Role of the lymphatics in cancer metastasis and chemotherapy applications. *Adv Drug Deliv Rev* 63:867–75.

Mishra, A., Parameswara Rao, V., and Singh, S. 2014. Intestinal lymphatic delivery of praziquantel by solid lipid nanoparticles: Formulation design, *in vitro* and *in vivo* studies. *J Nanotech* Article ID 351693.

Moghimi, M. and Moghimi, S.M. 2008. Lymphatic targeting of immuno-PEGliposomes: Evaluation of antibody-coupling procedures on lymph node macrophage uptake. *J Drug Target* 16:586–90.

Moghimi, S.M., Hawley, A.E., Christy, N.M. et al. 1994. Surface engineered nanospheres with enhanced drainage into lymphatics and uptake by macrophages of the regional lymph nodes. *FEBS Lett* 344:25–30.

Mohrman, D.E. and Heller, L.J. 2006. *Cardiovascular Physiology*. 6th ed. New York: McGraw-Hill Companies.

Mullin, D. 2003. Prometheus in Gloucestershire. *J Allergy Clin Immunol* 112:810–14.

Na, M., Yiyun, C., Tongwen, X. ct al. 2006. Dendrimers as potential drug carriers. Part II. Prolonged delivery of ketoprofen by in vitro and in vivo studies. *Eur J Med Chem* 41:670–74.

Paliwal R., Rai S., Vaidya B. et al. 2009. Effect of lipid core material on characteristics of solid lipid nanoparticles designed for oral lymphatic delivery. *Nanomedicine* 5:184–91.

Pantarotto, D., Briand, JP., Prato, M. et al. 2004a. Translocation of bioactive peptides across cell membranes by carbon nanotubes. *Chem Comm* 10:16–17.

Pantarotto, D., Singh, R., McCarthy, D. et al. 2004b. Functionalized carbon nanotubes for plasmid DNA gene delivery. *Angew Chem Int Ed Engl* 43:5242–6.

Qin, L., Zhang, F., Lu, X. et al. 2013. Polymeric micelles for enhanced lymphatic drug delivery to treat metastatic tumors. *J Control Release* 171:133–42.

Reddy, L.H., Sharma, R.K., Chuttani K. et al. 2005. Influence of administration route on tumor uptake and biodistribution of etoposide loaded solid lipid nanoparticles in Dalton's lymphoma tumor bearing mice. *J Control Release* 105:185–98.

Reddy, S.T., Rehor, A., Schmoekel, H.G. et al. 2006. In vivo targeting of dendritic cells in lymph nodes with poly(propylene sulfide) nanoparticles. *J Control Release* 112:26–34.

Rinderknecht, M. and Detmar, M. 2008. Tumor lymphangiogenesis and melanoma metastasis. *J Cell Physiol* 216:347–54.

Sahoo, N.G., Bao, H., Pan, Y. et al. 2011. Functionalized carbon nanomaterials as nanocarriers for loading and delivery of a poorly water-soluble anticancer drug: A comparative study. *Chem Commun* 47:5235–37.

Singh, I., Swami, R., Khan, W. et al. 2014. Lymphatic system: A prospective area for advanced targeting of particulate drug carriers. *Expert Opin Drug Deliv* 11:211–29.

Svenson, S. 2009. Dendrimers as versatile platform in drug delivery applications. *Eur J Pharm Biopharm* 71:445–62.

Tiantian, Y.E., Zhang, W., Mingshuang, S. et al. 2014. Study on intralymphatic-targeted hyaluronic acid-modified nanoliposome: Influence of formulation factors on the lymphatic targeting. *Int J Pharm* 471:245–57.

Tomalia, D.A. 1996. Star burst dendrimers-nanoscopic supermolecules according dendritic rules and principles. *Macromolecular Symposia* 101:243–55.

Trevaskis, N.L., Charman, W.N., and Porter, C.J. 2008. Lipid-based delivery systems and intestinal lymphatic drug transport: A mechanistic update. *Adv Drug Deliv Rev* 60:702–16.

Vaghasiya, H., Kumar, A., and Sawant, K. 2013. Development of solid lipid nanoparticles based controlled release system for topical delivery of terbinafine hydrochloride. *Eur J Pharm Sci* 49:311–22.

Videira, M.A., Botelho, M., Santos, A.C. et al. 2002. Lymphatic uptake of pulmonary delivered radiolabelled solid lipid nanoparticles. *J Drug Target* 10:607–13.

Wang, C.E., Liu, P., Zhuang, Y. et al. 2014. Lymphatic-targeted cationic liposomes: A robust vaccine adjuvant for promoting long-term immunological memory. *Vaccine* 32:5475–83.

Wiwattanapatapee, R., Carreno-Gomez, B., Malik, N. et al. 2000. Anionic PAMAM dendrimers rapidly cross adult rat intestine in vitro: A potential oral delivery system? *Pharm Res* 17:991–98.

Yan, Z., Zhan, C., Wen, Z. et al. 2011. LyP-1-conjugated doxorubicin-loaded liposomes suppress lymphatic metastasis by inhibiting lymph node metastases and destroying tumor lymphatics. *Nanotechnology* 22:415103. doi: 10.1088/0957-4484/22/41/415103.

Yang, F., Fu, D.L., Long, J. et al. 2008. Magnetic lymphatic targeting drug delivery system using carbon nanotubes. *Medical Hypotheses* 70:765–67.

Yang, F., Hu, J., Yang, D. et al. 2009. Pilot study of targeting magnetic carbon nanotubes to lymph nodes. *Nanomedicine* 4:317–30.

Yang, F., Jin, D., Yang, D. et al. 2011. Magnetic functionalized carbon nanotubes as drug vehicles for cancer lymph node metastasis treatment. *Eur J Cancer* 47:1873–82.

Yang, L.Z., Sun, L., and Zhang, Y.G. 2005. Distribution, migration and excretion of bio-inert nano-particles in vivo. *Sheng Wu Ji Shu Tong Xun* 16:525–27.

Yao, M., Xiao, H., and McClements, D.J. 2014. Delivery of lipophilic bioactives: Assembly, disassembly, and reassembly of lipid nanoparticles. *Anal Rev Food Sci Technol* 5:53–81.

Yu-Cheng, T., Zhenghong, Xu., Kevin, G. et al. 2014. Lipid/calcium phosphate nanoparticles for delivery to the lymphatic system and SPECT/CT imaging of lymph node metastases. *Biomaterials* 35:4688–98.

Yunfei, Li., Mingji, J., Shuai, S. et al. 2014. Small-sized polymeric micelles incorporating docetaxel suppress distant metastases in the clinically-relevant 4T1 mouse breast cancer model. *BMC Cancer* 14:329–39.

Zara, G.P., Bargoni, A., Cavalli, R. et al. 2002. Pharmacokinetics and tissue distribution of idarubicin-loaded solid lipid nanoparticles after duodenal administration to rats. *J Pharm Sci* 91:1324–33.

# 8 New Drug Targets and Drug Delivery Strategies for Various Ocular Disorders

*Subramanian Natesan,*
*Venkateshwaran Krishnaswami,*
*Saranya Radhakrishnan, Vaishnavi Suresh Kumar,*
*Nirmal Sonali, Jayabalan Nirmal,*
*and Shovanlal Gayen*

## CONTENTS

## 8.1   INTRODUCTION

The eye is a complex organ of the human body with distinguished anatomy and physiology. The optical system is designed with different segments termed the protective segment, anterior segment, posterior segment, and a visual system pathway. The protective segment is composed of the orbit, lids, and sclera. The anterior pole of the eye comprises the cornea, aqueous humor, iris, and ciliary body, whereas the posterior pole is composed of the retina and vitreous humor (Sandeep et al. 2013; Figure 8.1). The visual system pathway constitutes the optic nerve, optic tracts, and visual cortex. The anterior segment of the eye is affected by diseases such as dry eye, allergic conjunctivitis, glaucoma (Rathore and Nema 2009; Thakur and Kashiv 2011,

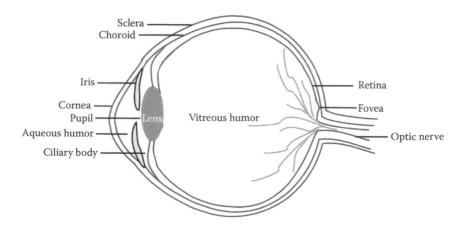

**FIGURE 8.1**   Anatomy and physiology of the human eye.

Venkata et al. 2011), and cataract, whereas the posterior pole is affected by age-related macular degeneration (AMD) and diabetic retinopathy (DR).

The anterior segment of the eye comprises the cornea, conjunctiva, iris, pupil, and aqueous humor. The cornea (10.5–11.5 mm size in adult human eye) comprises five layers, namely, the epithelium, stroma, endothelium, descement's membrane, and Bowman's layer. The main drugs applied topically transport to the anterior segment of the eye through the cornea. The transparent thin mucous membrane lining inside the eyelids and continuous with the cornea, occupying 17 times larger than cornea, is the conjunctiva. The transparent aqueous solution (aqueous humor) secreted by the ciliary body drains the eye through trabecular/uveoscleral routes. The membrane transport process maintains the aqueous humor production and intraocular pressure of the eye. The posterior segment of the eye comprises the lens, sclera, choroid, vitreous, and retina. Vitreous humor (4.0 mL) occupies 80% of a human eye's internal volume. The retina is the transparent tissue of the innermost part of the eye which lines two-thirds of the eye ball and is firmly attached to the ciliary body and to the optic nerve to the other end. Choroid is the highly vascularized tissue present between the retina and the sclera (Colin et al. 2013).

The limited therapeutic success in the anterior and posterior segment diseases due to low ocular bioavailability is associated with conventional formulations. This provides challenging problems to the pharmaceutical scientists to develop effective ocular drug delivery systems. The limitations in the ocular therapeutics like precorneal loss and blood ocular barriers also limit the therapeutic potential. The drug access to the target site gets hampered due to anatomical and physiological ocular barriers. Development of the nanoparticulate drug delivery system paves the way for the ocular therapeutics with improved bioavailability. Still the therapeutic benefit of the active molecule at the desired site of the ocular tissue should be improved to make the treatments more effective. In this regard, the specific targeting to diseased cells using multifunctional/decorated nanoformulations could be exploited as novel drug delivery systems for delivering ocular drugs. Herein, we discussed the pathogenesis, occurrence, and novel ocular drug delivery systems' approaches using nanomedicine-based drug targets (nanoparticles, liposomes, emulsions, and nanodispersion). The major ocular diseases and conditions focused on in this chapter are dry eye syndrome, DR, AMD, and glaucoma.

## 8.2   APPROACHES FOR OCULAR DRUG DELIVERY ASSOCIATED WITH VARIOUS DELIVERY ROUTES

The delivery of drugs to the eye can be achieved by different routes. Route of ocular drug administration depends upon the disease status, patient compliance, and target site of action (Urtti 2006; Dipak et al. 2013). The various routes of ocular drug delivery systems with associated merits and demerits are shown in Table 8.1. Apart from these, considering the different route of administration of the drug delivery to the eye also depends upon the formulation approaches.

In this regard, the various drug delivery systems employed for ocular treatments involve conventional system, retro metabolic system, vascular system, particulate

**TABLE 8.1**
**Various Routes of Ocular Drug Delivery Systems with Associated Merits and Demerits**

| Route of Administration | Merits | Demerits | References |
|---|---|---|---|
| Topical | Ease of delivery | Inefficient to reach posterior segment of the eye. Loss of drug due to nasolacrimal drainage and less residence time | Schopf et al. (2015) |
| Systemic | Delivers large quantity of drugs when compared to topical route | Poor ocular bioavailability and systemic toxicity | Urtti et al. (2006) |
| Intravitreal | Drug is administered specifically into vitreous and retina through injections or implants | Retinal detachment with hemorrhage | Ashaben et al. (2013) |
| Subconjunctival | Delivers drug to the uvea and increases drug penetration across sclera to the choroid | Presence of retinal pigment epithelium makes this route difficult to deliver drugs into the retina | Ashaben et al. (2013) |
| Subretinal | Delivers drug to the retina | Retinal detachment | Ashaben et al. (2013) |

system, controlled drug delivery system, and advanced drug delivery system (Patel and Agrawal 2011). These delivery systems can provide both sustained and controlled release, increase ocular bioavailability, and enhance the therapeutic performance of the drug. Conventional ophthalmic drug delivery systems administer drugs in the form of aqueous solutions (eye drops), suspensions, bioadhesive gels, ointments, Ocuserts, and Lacrisert (Ashaben et al. 2013). The delivery through the vascular system includes liposomes and neosomes. Lipid-based nanoparticulate drug delivery gained importance in recent years as it provides better permeation and prolonged drug release onto the ocular mucosa and allows the drugs to reach the posterior pole of the eye. Whereas topical administration of conventional drugs remains in the anterior pole and fails to reach the posterior pole of the eye (Tangri and Khurana 2011).

The controlled delivery system uses biodegradable and nonbiodegradable implants, hydrogels, dendrimers, iontophoresis, collagen shields, polymeric solution, nanosuspension, microemulsion, and microneedles to deliver drugs. To overcome the problem of rapid elimination in conventional eye drops, nanoparticulate drug delivery has been developed in which the nanoparticulates can be in the form of nanocapsules and nanospheres. Recent report suggests that nanocapsules are more efficient than nanospheres (Valente et al. 2013). The advanced drug delivery includes scleral plugs (Yasukawa et al. 2001), gene therapy (Jacob et al. 1990), small interfering ribonucleic acid (SiRNA), stem cells, cell encapsulation technology (Paswan et al. 2015), electroconvulsive therapy, and laser treatments. Microelectromechanical intraocular drug delivery device is a microelectromechanical system (MEMS) especially for the therapy of chronic and refractory ocular diseases. Drug solution can be refilled in the MEMS device, imparting a long-term drug therapy and avoids repeated surgical treatments.

## 8.3    TREATMENT OPTIONS ADOPTED FOR OCULAR DISORDERS

The various treatment options adopted for ocular disorders involve laser photocoagulation SiRNA, monoclonal antibodies, and surgical interventions. Due to the lack of site specificity associated with the conventional drug delivery systems, the targeted drug delivery approaches have been utilized for ocular disorders such as dry eye syndrome, AMD, DR, and glaucoma (Eugene et al. 2006; Gordon 2006).

### 8.3.1    DRY EYE SYNDROME

The most commonly encountered ocular morbidity, characterized by persistent dryness of the cornea and conjunctiva, is dry eye syndrome or keratoconjunctivitis (Gilbard 1994). Dryness of the eyes is due to reduction of the aqueous phase of the tear film volume (Lubniewski and Nelson 1990). In 2007, International Dry Eye Workshop framed the original definition and classification of dry eye syndrome based on its etiology mechanism and severity of the disease. Dry eye is a multifactorial disease of the tears and ocular surface accompanying indications of discomfort, visual disturbance, tear film instability with potential damage to the ocular surface, and accompanied by increased osmolarity of the tear film and inflammation of the ocular surface. The major classification of dry eye syndrome was tear-deficient eye with evaporative dry eye. The etiopathogenicity of dry eye syndrome is divided into aqueous deficiency (Sjogren's or non-Sjogren's related) and evaporation (resulting from intrinsic or extrinsic causes). The mechanistic causes include tear hyper osmolarity, tear film instability, severity of the disease with respect to visual symptoms, conjuctival injection, conjunctival staining, corneal staining, corneal/tear signs, lid/meibomian glands, tear film breakup, and Schirmer tear (Koray and Dwight 2009).

#### 8.3.1.1    Pathogenesis of Dry Eye Syndrome

The pathogenesis of dry eye is not completely understood and associated with increased osmolarity of the tear film and inflammation, which results in high levels of proinflammatory cytokines such as tumor necrosis factor-$\alpha$ (TNF-$\alpha$), interleukin (IL)-1$\beta$, matrix metalloproteinase, and chemokine macrophage inflammatory protein (MIP)-1$\alpha$, which were found in tear film and ocular epithelial surface (Koray and Dwight 2009). The multiple causes, which lead to dry eye syndrome, were vitamin A deficiency, Sjoeren's syndrome, rheumatoid arthritis, chemical/thermal burns, and drugs such as atenolol, chlorpheniramine, hydrochlorothiazide, isotretinoin, ketorolac, ketotifen, levocabastine, levofloxacin, oxybutynin, and tolterodine (Basilio et al. 2014).

#### 8.3.1.2    Current Treatment Options for Dry Eye Syndrome

The current treatment options adopted for dry eye syndrome involve artificial tear fluids characterized by hypotonic or isotonic buffered solutions with constitutes of electrolytes, surfactants, and other viscosity-building agents. Other treatment options include application of tear retention devices, implants known as punctual plugs, and moisture chamber spectacles (Albietz and Bruce 2001; Yaguchi et al. 2012). Inflammation of eye decreases the tear production and tear turnover which leads to dry eye syndrome, and these complications can be avoided by anti-inflammatory drugs such as corticosteroids and cyclosporine A; the tear film breakup time and ocular protection index were improved by nonsteroidal anti-inflammatory drugs (NSAIDs; Colligris et al. 2014).

### 8.3.1.3 Drug Targets for Novel Drug Delivery Systems to Treat Dry Eye Syndrome

Anti-inflammatory drugs were widely used for the treatment of dry eye syndrome, which works through secretagogue route, and increases the tear production. Anti-inflammatory treatment for the dry eye syndrome may include short-term corticosteroids, oral tetracyclines, oral omega-3 fatty acid supplements, and autologous serum eye drops (Amalia et al. 2010; Isaac et al. 2010). Corticosteroids were not recommended for long-term usage, whereas other NSAIDs at its low dosage are effective to increase tear film breakup time and ocular protection index. The roles of inflammatory cytokines and matrix metalloproteinases in pathogenesis of dry eye are very essential (Koray and Dwight 2009). Cytokines associated with dry eye syndrome include IL-1 and IL-$\beta$. In dry eye conditions, the enhanced levels of proinflammatory forms IL-1 and decreased levels of biologically inactive precursor IL-1$\beta$ were observed in tear fluids. IL-1$\beta$ may increase the production of nitric oxide (NO) through the induction of isoforms of inducible nitric oxide synthase (iNOS) in epithelial cells of the lacrimal gland. The overproduction of this NO may contribute to the pathogenesis in lacrimal gland cell death in case of dry eye syndrome (Beauregard et al. 2003). Thus, the inhibitors of IL-1 and iNOS receptors can be used in the treatment of dry eye disease. It is also documented that treatment of dry eye condition with SP600125 a reversible adenosine triphosphate (ATP)-competitive inhibitor inhibits c-Jun NH$_2$-terminal kinase (JNK), a serine/threonine protein kinase. Thus, it increases the tear production rate by activation upon stress stimuli, as JNK inhibits the lacrimal gland secretion (Bennett et al. 2001). The pathological alterations of wound healing and inflammation to the ocular surface can also be achieved by increased activity of matrix metalloproteins (MMPs), especially MMP-9 (Koray and Dwight 2009). Apart from the cytokine receptor targets, other drug targets involve calcineurin, mechanistic target of rapamycin (mTOR) inhibitors, androgen/estrogen receptor inhibitors, serotonin receptor inhibitors, calcium-activated chloride channel modulators, antilymphangiogenic agents, cycloheptathiphene, zeta chain-associated protein kinase 70 (ZAP-70) inhibitors, and RNA interference (RNAi) were reported (Basilio et al. 2014).

Calceneurin is a ubiquitous immunosuppressant potent enzyme found in cell cytoplasm that prevents the release of proinflammatory cytokines and reversibly inhibits the T-cell proliferation (Figure 8.2). Voclosporin, a novel calceneurin inhibitor, is useful for treating inflammatory ocular surface diseases (Colligris et al. 2014). Due to the decrease in ovarian sex hormones, most post-menopausal women develop

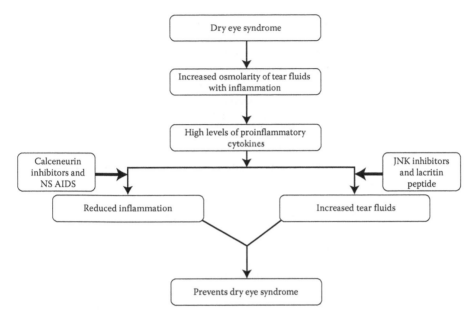

**FIGURE 8.2**    Drug targets in association with dry eye syndrome.

dry eye disease (Mostafa et al. 2012). The mechanism of decrease in sex hormones causes lymphocytic infiltration with apoptosis, and the interaction of lack of sex hormones with the genetic elements was not clear. For the treatment of dry eye condition due to decrease in sex hormone levels, the selective androgen receptor modulators (SARMs) and selective estrogen receptor modulators (SERMs) were recommended (Dalton et al. 2008). Serotonin or 5-hydroxytryptamine (5-HT) is an important neurotransmitter and neuromodulator which plays a significant role in mammalian body, and inhibitors of 5-HT receptors are used to control symptoms of dry eye, as it is found to mediate contractile, secretory, and electrophysiological effects, including vascular/nonvascular smooth muscle contraction and platelet aggregation (Boullin and Glenton 1978).

Recently, the anti-lymphangiogenic agents were used in the treatment of dry eye, as these newly formed corneal lymphatic vessels may function as potential conduits for migration of corneal antigen-presenting cells to lymphoid tissues, where they generate autoreactive TH17 and TH1 cells in dry eye. The antilymphangiogenic agents vascular endothelial growth factor (VEGF)-C and VEGF-D are effective in improving dry eye by inhibiting the binding of VGEF-C/VEGF-D ligand to VEGF receptor (VEGFR)-3 or their stimulatory effect on VEGFR-3, and this inhibition prevents the trafficking of antigen-presenting cells to the drain lymph nodes, and proves to be a potential therapeutic target for the dry eye syndrome (Goyal et al. 2012). Topical application of cycloheptathiophene compounds for the dry eye treatment acts by inhibiting histamine (H1) receptors, stabilizes mast cells, and prevents eosinophil accumulation.

Other drug targets include ZAP-70 which is a protein tyrosine kinase. The inhibitors of ZAP-70 such as pyrimidine derivatives, sulfonamides and sulfamides, or heterocyclylamino pyrimidines were adopted for dry eye syndrome (Alain et al. 2010). In addition to the above treatment options, dry eye disease can also be treated with peptide lacritin which is a prosecretory mitogen that promotes basal tearing as lacritin and lipocalin-1, which are selectively deficient in dry eye (Karnati et al. 2013). Recent development of treatment with RNAi affords high selectivity and specific inhibition of gene expression and reduces cellular processes such as apoptosis of glandular acinar cells (Cha and Pauley 2012). Further understanding is needed to know the immunomodulatory and inflammatory mechanisms of the cornea and conjunctiva. Better understanding of the cellular mechanisms involved in dry eye is needed not only to improve symptoms but also to restore the homeostasis of the ocular surface (Bennett et al. 2001; Colligris et al. 2014).

## 8.3.2 Diabetic Retinopathy

DR is one of the most occurring complications of diabetes, which leads to vision loss and blindness (Crabb et al. 2002; Schwartzman et al. 2010). DR is clinically characterized by microvascular dysfunction, with basement membrane thickening of retinal vessels, loss of pericytes and endothelial cells, blood–retinal barrier (BRB) breakdown, capillary nonperfusion microaneurysms, hemorrhages, lipid exudates, cotton-wool spots, and neovascularization (proliferative DR). Recent findings suggest retinal components such as neurons and glial cells may also get affected (Akiyoshi 2013). The basement membrane thickening, pericute loss, blood and fluid leak from vessels, and accrue in tissues to allow the exudates formation are the remarkable characteristics of nonproliferative DR which was an early stage. The advanced stage, which is proliferative retinopathy, was exemplified by new vessels formation (Osaadon et al. 2014).

### 8.3.2.1 Pathogenesis of DR

The primary pathogenic factors for DR were hyperglycemia (increase in blood glucose level). Receptor of advanced glycation end product (RAGE) plays a pathological role in initiation and progression of DR. The underlying mechanism is the accumulation of AGEs and their interactions with RAGE that increases and adversely affects the retinal microvasculature in patients with diabetes (Leal et al. 2005). Thus, by inhibiting AGE–RAGE interaction using specific inhibitors like aminoguanidine, advanced glycation end-product (AGE) inhibitors, and the putative cross-link breakers N-phenacylthiazolium bromide may be useful for treating the complications of DR (Mohd et al. 2013).

### 8.3.2.2 Current Treatment Options Adopted for DR

The current treatment options adopted for DR include anti-VEGF therapies, focal laser therapy, and steroids. In case of proliferative DR and retinal vein occlusive diseases, laser panretinal photocoagulation is used (Kang et al. 2012). The major biochemical pathways of DR were polyol pathway, protein kinase C (PKC) activation, hexosamine pathway, and oxidative stress.

### 8.3.2.3    Drug Targets for Novel Drug Delivery Systems to Treat DR

PKC is one of the drug targets for DR. The isoforms of PKCs greatly enhance the *de novo* synthesis of diacylglycerol (DAG) that leads to vascular dysfunction. Selective PKC isoform inhibitors like PKC 412 and ruboxistaurin delay the progression of diabetes associated with visual and vascular pathogenesis. Several studies reported that inhibition of aldose reductase enzyme decreases the prevalence of microaneurysms, basement membrane thickness, oxidative stress, VEGF expression, neuronal apoptosis, and gliosis of retina in diabetic animal models. The enzyme activates glucose metabolism to produce sorbitol and as a result, cells are deprived of glutathione and endogenous antioxidants, which leads to increase in oxidative stress. The increased level of oxidative stress stimulates several inflammatory conditions in accordance with DR (Chiara et al. 2009; Geraldes and King 2010; Mohd et al. 2013).

Drugs also target a transcriptional factor (nuclear factor-kappa B [NF-κB]) involved in the pathogenesis of DR. The activation of this transcriptional factor leads to the accumulation of NF-κB subunits of endothelial and glial cells of epiretinal (monocyte chemoattractant protein-1 [MCP-1] and soluble intercellular adhesion molecule-1 [sICAM]) which results in cytokine expression, tumor growth, and angiogenesis. Thus, by suppressing the activation of NF-κB, the complications of DR prevail. In addition, poly(adenosine diphosphate-ribose)polymerase-1 (PARP) is a nuclear enzyme, which causes deoxyribonucleic acid (DNA) damage and increase in oxidative and nitrosative stress on its activation. PARP inhibits the activity of glyceraldehyde 3-phosphate dehydrogenase (GAPDH), inducing the activation of the hexosamine pathway, PKC, and AGE formation, which alters the production of reactive oxygen and nitrogen species. Thus, inhibition of PARP is a promising therapeutic target to inhibit the development of DR (Mohd et al. 2013).

The drug-targeting angiotensin-2 receptor also plays a significant role in treating DR, as hypertension is the major risk factor associated with microvascular complications of DR. Retinal renin angiotensin system (RAS), renin, angiotensin-converting enzyme (ACE), and angiotensin type I and type II have shown increased levels of prorenin, renin, and angiotensin II in vitreous humor of the patients, suggesting the role of RAS in DR. RAS stimulates the expression of VEGF, ICAM-1, MCP-1, and NF-κB by activating the downstream inflammatory responses; thus, by blocking the RAS at the level of ACE inhibition or angiotensin reduces the increase in retinal, and VEGFR-2 is found in diabetic rat models (Leal et al. 2005; Mohd et al. 2013). Also by inhibiting the TGF-β pathway, the inflammatory responses of DR can be controlled and was proven in rat model (Chiara et al. 2009).

In DR, there is an increase in lipid and fatty acid composition, concentration, and tissue distribution contributes to the development and severity of the diseases. The accumulation of high levels of lipid and fatty acids alters the physiological metabolic profile in the retina, and vitreous humor leads to the progression of DR. It also increases oxidative stress and inflammatory responses; hence, the lipid-targeting drugs that lower the lipid levels have a great potential in managing DR. In addition, other drug targets like MK-801 and N-methyl-D-aspartate (NMDA) receptors are responsible for neurodegenerative conditions, and the use of antagonists such as neurotrophins provides protection against the neurodegenerative conditions of DR.

Recently, the therapies targeting stem cells have also been developed as they play a potential role in the formation of normal appearing intraretinal vascularization. In addition to RNAi, the siRNA targets VEGF and inhibits their expression in human retinal pigment epithelium (RPE) cells, and can also be used to treat complications of DR (Gustavsson et al. 2010; Schwartzman et al. 2010). Ranibizumab (monoclonal antibody) binds with the VEGF receptors, which may provide a long acting blockage of VEGF and has been utilized for the treatment of DR and AMD (Prasad et al. 2010).

### 8.3.3    AGE-RELATED MACULAR DEGENERATION

Age macular degeneration (AMD) is a degenerative retinal disease of elderly people, causes progressive loss of central vision, and is characterized by breakdown of macula. Early-onset forms of macular degeneration are caused by genetic mutations and late onset of macular degeneration called *age macular degeneration* is a slow progressive disease with genetic influences and environmental factors. AMD is classified into two types; among the two types of AMD (dry and wet forms), the dry AMD (90%) is more common than the wet AMD (10%). Wet AMD is characterized by aberrant growth of abnormal blood vessels from the choriocapillaris through the RPE, which typically results in hemorrhage, exudation, scarring, and/or serous retinal detachment (Jayakrishna and Fowler 2012). The nonneovascular dry form was characterized by atrophy of RPE and loss of macular photoreceptors. Soft drusen with discrete, round, slightly elevated, variable-sized and sub-RPE deposits in the macula with yellow or white spots in the fundus were the characteristic lesions of dry AMD. The overlying RPE shows thinning, whereas the RPE between drusen shows thickening (Afshari et al. 2010).

#### 8.3.3.1    Pathogenesis of AMD

The pathogenesis of AMD includes aging and oxidative stress, which causes RPE, choriocapillaris injury that further results in chronic inflammatory response in Bruch's membrane, and choroid. The severe inflammation leads to the formation of abnormal extracellular matrix (ECM) and this abnormality leads to altered diffusion of nutrients to the retina and RPE that may further accelerate RPE and retinal damage (Figure 8.3). The altered RPE and choriocapillaris function ultimately leads to atrophy of retina and choriocapillaris to choroidal neovascularization (CNV) growth. The environmental and genetic changes can alter patient's susceptibility in the sequence of pathogenesis toward the disease (Jayakrishna and Fowler 2012).

#### 8.3.3.2    Mechanism of AMD Development

During the ageing process, retinal cells get inflamed due to oxidative stress. The inflamed retinal cell releases cyclooxygenase (COX) and cytokines which upregulate the release of VEGF. VEGF is a naturally secreted protein, which is responsible for the development of new blood vessels in AMD patient. VEGF specifically binds and activates the receptors of vascular endothelial cells inducing CNV and increases vascular permeability and inflammation. The retinal cell layers come in contact with

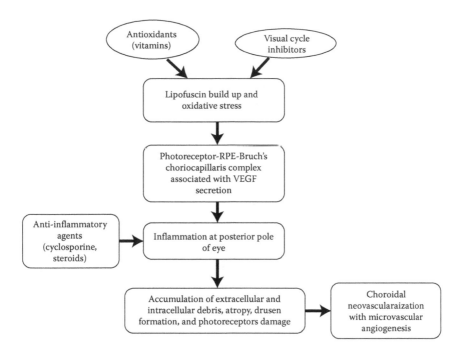

**FIGURE 8.3**   Pathophysiology of age-related macular degeneration.

vitreous cytokines, retinal-pigmented epithelium (RPE) cells proliferate, migrate, and promote angiogenesis. The COX-2 mainly detected in human choroidal neovascular membranes and its evidence that COX pathway is the major pathway for angiogenesis development in macula through the VEGF. The pharmacological inhibition of COX-2 appears to reduce VEGF expression in inflammatory cells and suppresses the formation of new blood vessels (Schoenberger et al. 2012).

Bando et al. (2007) studied the clathrin and adaptin accumulation in drusen, Bruch's membrane, and choroid in AMD-affected eyes. Recently, the clathrin was identified in Bruch's membrane of AMD-affected patients' eyes. The human eyes were collected from AMD and non-AMD donors with different age groups. Bruch's membrane and choroid from the macula of each donor eye were prepared for immunohistochemistry and Western blotting. The greater immune reactivity of clathrin and adaptin were seen in drusen, Bruch's membrane, and choroid of AMD tissues. From the results, they found that the accumulation of clathrin and adaptin in drusen, Bruch's membrane, and choroid might reflect a higher rate of clathrin-mediated endocytosis in AMD tissues. Finally, they came to a conclusion that the accumulation of clathrin and adaptin may reflect a higher susceptibility to oxidative damage of the eye.

### 8.3.3.3   Current Treatment Options Adopted for AMD

The current treatment options adopted for AMD involve laser photocoagulation, photodynamic therapy, anti-VEGF therapy, pharmacological therapy, SiRNA, dual/triple combination therapies, and monoclonal antibodies (Simó and Christian 2009).

### 8.3.3.4   Drug Targets for Novel Drug Delivery Systems to Treat AMD

The current pharmacotherapies for AMD include preocular or intraocular administration of drugs to achieve improved therapeutic concentration at targeted tissue. CNV is a major pathological condition of wet AMD and VEGF-A plays a major role in development of CNV. Thus, the inhibition of VEGF-A acts as the central target for the treatment of wet AMD. The anti-VEGF aptamers selectively binds with VEGF receptors and inhibits the growth of blood vessels and vascular leakage. Autophagy-promoting molecules may also provide therapeutic effect for dry and wet AMD as it inhibits angiogenesis and expression of VEGF-A by RPE cells (John et al. 2007). AMD can also be treated by targeting complement factor in geographic atrophy (GA). GA is a condition of late dry AMD and the complement inhibition of GA suppresses development of CNV. In other cases, the complement inhibition in CNV treatment employs dual function of reduction in secretion of VEGF-A by RPE or it inhibits retinal infiltration of proangiogenic leukocytes. The other targeted treatment approaches for AMD were Toll-like receptor 3 (TLR3), a single nucleotide protein (SNP) in gene coding for the double-stranded RNA (dsRNA) sensor TLR3 in GA and CNV was reported to be associated with prevention in development of GA. Thus, the stimulation of TLR3 causes CNV suppression by protecting against RPE degeneration caused by exogenous dsRNA or by accumulation of all trans-retinaldehyde, which helps in the treatment of dry and wet AMD (David and Fong 2007; Jean et al. 2009).

### 8.3.3.5   Angiogenesis and New Blood Vessel Formation

Angiogenesis is a condition of abnormal growth of new blood vessels from the existing vasculature mediated by two distinct pathways of splitting and sprouting. Angiogenesis is mediated by the formation of tip cells which interact with the angiogenic factor VEGF mediated through VEGF-2 receptor, with the migration and proliferation of endothelial cells mediated through VEGF-A and VEGFR-2, which finally form the matured blood vessels mediated through platelet-derived growth factor β (PDGF-β) and its receptor. VEGF is the potent angiogenic factor with high affinity for VEGFR-1 receptor. Overproduction of VEGF is the prime cause for microvascular diseases like AMD.

### 8.3.3.6   Integrin Receptors

Integrin receptors that were overexpressed on the surface of endothelial cells possess two noncovalently associated transmembranes of $\alpha$ and $\beta$ subunits, around 18 $\alpha$ and 8 $\beta$ subunits form 24 different heterodimers of $\alpha_v\beta_3$ integrin receptors, with most of the integrin receptors bound to the ECM molecules. The ligand specificity of integrins was determined by $\alpha$ and $\beta$ subunits (Ruoslahti 1996). Integrin ligand-binding affinity can be improved by activation with divalent cations and cytoplasmic proteins. Integrin-mediated cell adhesion involves the cascades of cell attachment, cell spreading, organization of actin cytoskeleton, and formation of focal adhesions. The integrin receptors $\alpha_v\beta_3$, $\alpha_5\beta_1$, and $\alpha_{IIb}\beta_3$ were the most predominant receptors.

### 8.3.3.7   RGD Peptides

The RGD peptide is a tripeptide containing arginine, glycine, and aspartate amino acid residues. RGD peptides have high receptor selectivity and affinity with different cell line patterns. In cyclic RGD peptide, the arginine side chain focuses the $\alpha$ sub-unit, whereas the aspartate carboxylate group focuses the metal ion-dependent adhesion site (MIDAS) of the $\beta$ subunit (Ruoslahti 1996). RGD peptides may be linked to biodegradable polymers through covalent amide bonds by reaction of activated surface carboxylic acid group to the peptide (nucleophilic N-terminus).

### 8.3.3.8    Affinity of RGD Motif with Integrin Receptor after Coupling to Activated Polymer

The guanidino group of arginine present in the RGD peptide gets coupled with $\alpha$ subunit of $\alpha_v\beta_3$ integrin receptor, while the N-amino terminus of arginine gets linked with the ($-OH$) hydroxyl group of the copolymer (poly(lactide)-d-$\alpha$-tocopheryl polyethylene glycol succinate). The carboxylate containing aspartate group of RGD peptide gets coupled with the $\beta$ subunit of the integrin receptor by cell adhesion. Thereby, the affinity of RGD peptide with $\alpha_v\beta_3$ integrin receptor will be maintained (Figure 8.4).

### 8.3.3.9    Conjugation of RGD Peptide with Copolymer (Surface Functionalization)

The binding activity of RGD motif-conjugated nanoparticles to $\alpha_v\beta_3$ integrin receptor can be maintained by linking RGD peptide to the free hydroxyl group of random

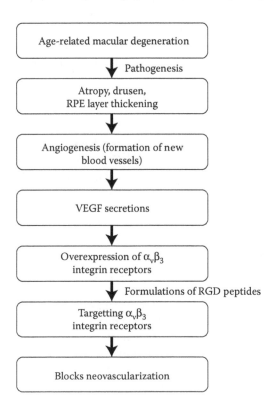

**FIGURE 8.4**    Drug targets in association with age-related macular degeneration.

copolymer. The copolymer will be preactivated with tresyl chloride to form activated material that will react with nucleophilic N-terminus of the arginine side chain of RGD peptide under aqueous conditions (Ruoslahti 1996).

Endothelial cells are crucial in angiogenic development of new blood vessel formation; $\alpha_v\beta_3$ integrin a heterodimeric transmembrane glycoprotein receptor is overexpressed in actively proliferating endothelial cells of AMD patients. Peptides containing RGD motif can bind with the integrin receptors (Singh et al. 2009). Increased iron load in the AMD retina suggests the overexpression of transferrin receptors and the participation of transferrin in the transport of iron (Danhier et al. 2009). Nanoparticles' surface modification with linear RGD peptide or transferrin or both increased the retinal delivery of nanoparticles and subsequently the intraceptor gene expression of the retinal vascular endothelial cells. However, the functionalized nanoparticles show targeted delivery to neovascular eye and not in the control eye, and further significantly reduce the CNV areas in rats compared with vehicle or nonfunctionalized nanoparticle-treated rats (Jayakrishna and Fowler 2012).

PEGylated poly(lactide-co-glycolide) (PLGA)-based nanoparticles grafted with RGD peptide paclitaxel-loaded nanoparticles have been reported for the targeting of tumor endothelium (Anat et al. 2011). Polyglutamic acid-paclitaxel-RGD peptide (polymer-based drug conjugate) targets the tumor tissue which exploits the enhanced permeability and retention effects providing enhanced antitumor activity with decreased toxicity when comparing to free paclitaxel-treated mice (Yujing et al. 2010) with improved antiangiogenic effect. These reports demonstrate the advantage of targeting integrin ($\alpha_v\beta_3$) receptors located on newly formed blood vessels of AMD patients.

## 8.3.4 GLAUCOMA

Glaucoma is a chronic neurodegenerative disease characterized by slow and progressive degeneration of retinal ganglion cells (RCGs). The axons result in progressive vision loss and irreversible blindness. Intraocular pressure (IOP) and mechanical stress on lamina cribrosa or glial cell activation were the factors responsible for development of glaucoma. The changes with visual field lead to loss of nerve fibers within the retina, termed scotomas (Ulrich et al. 2003).

### 8.3.4.1 Pathogenesis of Glaucoma

The pathogenesis of glaucoma involves enlargement and elongation of the optic nerve cup, thinning and eventual notching of the neuroretinal rim, asymmetry in cup size between the two eyes, and disc hemorrhages. In addition to this, the excitotoxic damage causes oxidative damage due to the overproduction of NO and other reactive oxygen species (ROS). Glutamate or glycine released from injured neurons leads to elevated IOP and vascular dysregulation. Aqueous humor production by the ciliary body through the independent pathways of trabecular meshwork and uveoscleral outflow pathway creates the IOP, whereas in patients with open-angle glaucoma, the increased resistance through the trabecular meshwork occurs (Charles 1978).

### 8.3.4.2    Current Treatment Options Adopted for Glaucoma

The treatment options adopted for glaucoma involve IOP-lowering therapy including pharmacological agents and complement pathway inhibitors (Schwartza and Budenzb 2004).

### 8.3.4.3    Drug Targets for Novel Drug Delivery Systems to Treat Glaucoma

IOP is the only proven risk factor for glaucoma (open/closed angle) and its complications occur due to improper drainage of aqueous humor that was considered to be the main cause of retinal ganglion cell (RGC) apoptosis. For centuries, cholinergic agents were used to reduce IOP. By decreasing aqueous inflow rate or by increasing aqueous outflow rate, the pharmacological agents decrease IOP (Kang et al. 2012). Fas ligand binds to the death receptor and promotes apoptosis of RGC cell upon high IOP. The stimulation of apoptosis RGCs is initiated by neurotrophin deprivation, glial activation, excitotoxicity, ischemia, and oxidative stress. Membrane permeability was increased, and release of the mitochondrial apoptosis factor was induced by hypoxia, TNF-$\alpha$, oxidative stress, and elevated levels of intracellular $Ca^{2+}$. Free radicals of nitrous oxide were produced on activation of the mitochondrial apoptosis-inducing factor that triggers caspase-independent apoptotic pathway and stimulates the apoptosis of RGC cells.

VEGF receptors, $\beta$-adrenergic receptors, dopaminergic receptors, cholinergic receptors, prostaglandin receptors, carbonic anhydrase enzyme, and $\alpha$-2 receptors were the common drug targets addressed for glaucoma treatment. VEGF receptor is a common target for most ocular diseases as it promotes proinflammatory cytokines' expression, which ends with inflammation. Retinal vein occlusive disease can be treated by anti-VEGF drugs. Dopaminergic receptors (DA2 and DA3) play a significant role in aqueous humor flow, and activation of these receptors may induce the IOP levels; agonists for these receptors may have a strong influence in IOP levels' modulation, which may be recommended for glaucoma treatment. Activation of $\beta$-adrenergic and cholinergic receptors is also found to increase IOP at diseased state and using antagonist drugs of these receptors and prostaglandin synthase analog lowers the IOP in glaucomatous patients.

Inflammation associated with glaucoma can be targeted by $A_3AR$ receptor, which is a G-protein-coupled receptor highly expressed over inflammatory tissues and activation of these receptors induces apoptosis of inflammatory cells via PKB/NF-$\kappa$B and Wnt signaling pathway. Currently available drugs in the market target these receptors and enzymes for efficient glaucoma treatment.

The change in the actin cytoskeleton and the cellular motility of the outflow pathways were achieved by Rho-associated protein kinase (ROCK) inhibitors (Figure 8.5). ROCK inhibitors are downstream effector of Rho (a small guanine triphosphatase) in the Rho-dependent signal transduction pathway, which induces changes in the actin cytoskeleton, and the cellular motility thereby lowers the IOP (Kang et al. 2012).

## 8.4    NOVEL OCULAR DRUG DELIVERY SYSTEMS

Delivering the drugs to target sites of the eye are posing ever-increasing challenges. The novel drug delivery systems are focused to alleviate the challenges faced by

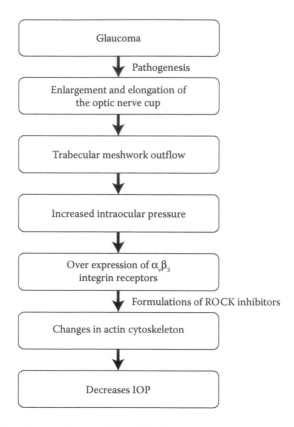

**FIGURE 8.5**  Drug targets in association with glaucoma.

the conventional ophthalmic formulations. The novel drug targets are utilized to treat the disease by designing the novel drug formulations of the drug specific to the target, thereby improving availability of the drug at the site of action, reducing the dosing frequency, limiting the drug dosage, improving the patient compliance, and offering marketing potential. The barriers posed to the conventional formulations are multifold.

The presence of blood aqueous and BRBs limit the entry of topically administered drugs to aqueous humor and vitreous humor of the eye. Natural barriers also limit the passage, therefore, larger amounts of drugs may be needed to achieve an efficient treatment and these may damage the ocular tissues (Silva et al. 2011). Most of the drugs were washed away from ocular surface (less than 30 s) by physiological mechanisms such as lacrimation, tear dilution, and tear turnover. The topically applied drugs find it difficult to cross the cornea due to the lipidic nature of the corneal endothelium; although they cross through sclera/conjunctival pathway, the bioavailability of topically applied drugs was less than 5%. At the posterior segment, additional physical and diffusional barriers reduce the drug concentration at the site. These ultimately lead to difficulty in achieving an effective concentration at the targeted site. Invasive techniques such as intravitreal, periocular, and systemic injections were followed to deliver the drug into the posterior segment. Intravitreal injections may lead to endophthalmitis, hemorrhage, retinal detachment, and poor patient compliance with repeated eye puncture. In order to overcome these difficulties, researchers were attempting to improve the corneal permeability and prolonging the contact time of topically applied drugs into the ocular surface. These attempts were achieved using novel ocular drug delivery systems with the constraints of enhancing the viscosity of the formulation and improving the bioavailability (Rupenthal et al. 2011). Even the problem of poor residence of formulations to the ocular surface can be overcome by *in situ* forming gels (Gratieri et al. 2011). To achieve a prolonged therapeutic drug concentration in ocular target tissues by limiting systemic exposure, side effects, and improving patient adherence, the sustained release drug delivery systems were developed. In order to bring the topically applied dose efficiently to the posterior segment of eye with an improved corneal permeation, Schopf et al. (2015) recently developed loteprednol etabonate with tyrosine kinase inhibitor-loaded nanoparticles using the concept of mucus-penetrating particle technology.

The delivery systems available for the eye were nanoparticles, liposomes, emulsions, nanosuspensions, dendrimers, contact lenses, nanotubes, and quantum dots. In order to achieve a controlled release pattern of the active medicament, organ/site/disease-specific targeting effect, to improve the corneal permeability, to enhance the contact time to the ocular surface, and to avail a convenient administration route, these drug delivery systems were utilized with modulations (Diebold and Calonge 2010).

### 8.4.1  Nanoparticulate-Based Ocular Drug Delivery Systems

The solid, submicron, colloidal particles with a size range of 10–1000 nm, in which drug molecules may be dissolved, entrapped, adsorbed, or covalently attached to form nanoparticles, were used for ocular drug delivery systems. Nanoparticles for ocular drug delivery systems improve the topical passage of large and poorly water-soluble drugs through ocular barrier systems. The challenges of ununiform size distribution and redispersibility of nanoparticles pose a challenge to the ocular drug delivery system. Charged cationic/anionic polymers of nanoparticles interact with the mucin through hydrogen bonding or electrostatic interaction and make them mucoadhesive. Multiple ligands such as peptides, DNA, and antibodies can be loaded into nanoparticles for the transport in the specific cells (Chaurasia et al. 2015). Properly formulated nanoparticles may provide prolonged therapeutic effect with controlled release. The sustained and controlled release of the topically applied nanoparticles was achieved by retaining in the cul-de-sac, and the entrapped drugs have to be released in an appropriate way. The retention time of the topically applied nanoparticles in cul-de-sac were enhanced by the addition of bioadhesive polymer in the nanoparticle formulation. Without the presence of bioadhesive polymer, nanoparticles were quickly eliminated from the eye by precorneal loss. Hence, the corneal residential time of topically applied ophthalmic formulation was increased by the addition of bioadhesive polymers in nanoparticles. The pathways by which nanoparticles penetrate the ocular tissues include endocytosis, cell membrane lysis using metabolic degradation products, and transcellular route pathways. More genetic-based nanomaterials have been developed using nucleic acids (DNA), antisense oligonucleotides, SiRNA, and aptamers. Nanoparticles are exploited to improve the ocular bioavailability and retention of drugs to the ocular tissues. Nanoparticles prepared by using solid lipid nanoparticles were advantageous currently for the development of formulations of both hydrophilic and lipophilic drugs.

Recently, the ocular biodistribution of hypocrellin B-loaded PLGA nanoparticles in different ocular tissues such as retina, aqueous humor, iris, and vitreous body has been checked and found. Hypocrellin B has some exposure to the posterior segment of the eye (Venkateshwaran et al. 2015). Nanoparticle-based drug delivery system has been utilized for the delivery of most lipophilic drugs.

Grama et al. (2013) exploited the efficacy of delaying cataract in diabetic rat model using curcumin nanoparticles. Previously, the authors reported that dietary curcumin delays the cataract in diabetes-induced rats. However, they found difficulties regarding low peroral bioavailability for the utilization of curcumin clinically. Hence, they developed the curcumin-loaded nanoparticles and tested in streptozotocin-induced diabetic cataract rat model. Oral administration of nanocurcumin dose of 2 mg/day was more effective than pure curcumin in diabetic cataracts delaying mechanism. The nanocurcumin significance of the delaying cataract is attributed to its ability to the enhancement of disease progression by intervention of its biochemical pathways. It implies that the performance enhancement of the nanocurcumin achieved by its improved oral bioavailability for the potential in managing diabetic cataract. The gene transfer efficiency of polyethylenimine-conjugated gold nanoparticles was checked both in rabbit and human corneas, and it was found that polyethylenimine-conjugated gold nanoparticles were safe for corneal gene therapy (Sharma et al. 2011).

Nanoparticulate drug delivery system has been also utilized for the efficient treatment of AMD. Efficient delivery of nanoparticle–integrin antagonist C16Y peptide for the treatment of CNV in rats was performed by Kim and Csaky (2010). The aim was to develop a drug delivery system to deliver an integrin antagonist peptide to the subretinal space using poly(lactic acid)/poly(lactic acid)–poly(ethylene oxide) (PLA/PLA–PEO) nanoparticles prepared with water-soluble integrin antagonist peptide, C16Y (C16Y-NP). The particle size of PLA/PLA–PEO nanoparticle-encapsulated C16Y was 302 nm. Intravitreal injection of nanoparticles does not produce any signs of toxicity to retina. The intravitreal administration of coumarin-loaded PLA/PLA–PEO penetrates the retina and localizes in the RPE. These nanoparticles with bio-degradable polymers may have a potential for the intravitreal delivery of drugs for AMD treatment. For the management of inflammation to treat dry eye syndrome using PLGA, Eudragit, and Carbopol, cyclosporine A-loaded nanoparticles have been developed by Aksungur et al. (2011).

In order to achieve a targeted drug delivery at the posterior segment of the eye, the intravenous transferrin/RGD peptide and dual-targeted drug-loaded nanoparticles may be the suitable choice at the diseased site of CNV due to overexpression of VEGF (Singh et al. 2009). Thiolated pectin nanoparticles of timolol maleate prepared by ionotropic gelation technique were developed by Sharma et al. (2012). The results of their *ex vivo* corneal permeation across goat cornea revealed that thiolated pectin nanoparticles of timolol maleate possess significant *ex vivo* corneal permeation than aqueous solution of timolol maleate. Albumin-based aspirin nanoparticles with a size range of 200 nm were developed by Das et al. (2012) for the topical delivery of DR. Sparfloxacin-loaded PLGA-based nanoparticles with a mean particle size of around 180–190 nm at a charge of −22 mV were prepared by Gupta et al. (2010) with an enhanced precorneal residence time and corneal drug permeation. Gatifloxacin-loaded chitosan–sodium alginate nanoparticles with a size range of 205–572 nm and charge of 17.6–47.8 mV were prepared by Motwani et al. (2008) in order to enhance the mucoadhesive property and long-term extraocular drug delivery.

### 8.4.2 LIPOSOMAL-BASED OCULAR DRUG DELIVERY SYSTEMS

Liposomes are lipidic membrane-like vesicles that contain one or more concentric bilayers similar to plasma membrane, promising drug delivery system formulated using cholesterol, and natural nontoxic phospholipids. The properties of liposomes vary accordingly to its size, hydrophilicity, and hydrophobicity. Liposomes are positively, negatively, or neutrally charged vesicles. Liposome improves the efficacy, therapeutic index of drugs, and reduces the toxicity of encapsulated drugs. Commercially, for ocular drug delivery system, liposomes gained specificity due to its biocompatibility, biodegradability, and site-specific targetability. Liposome holds the electrostatic interaction by the positive charge interaction of liposomal membrane to the negatively charged mucin of eye. The usage of bioadhesive polymer in liposome synthesis prolongs the residence time of liposomes to the eye. The structure of liposomes makes a stability issues, which creates a faster drug release in comparison with the nanoparticulate formulations.

Latanoprost drug-loaded liposomes prepared by using cholesterol and phosphatidylcholine were developed for the treatment of glaucoma (Fathalla et al. 2015) and found that latanoprost liposomal gels as an efficient tool to conventional eye drops for the safe and efficient management of glaucoma. As a potential novel system for delivering ocular drugs to the posterior segment of eyes, submicron-sized liposome-based ocular drug delivery system was developed by Hironaka et al. (2009) using coumarin-6. The fluorescence emission of coumarin-6-loaded liposomes was observed in mice after administration of the liposomal eye suspension. However, the similar fluorescence was not observed after administration of either multilamellar vesicles or dimethyl sulfoxide, containing the same amount of coumarin-6. The highest fluorescence of liposomal formulation was observed at 30 min of administration, and fluorescence disappeared after 180 min. This observation suggests that the fluorescence intensity in the retina was related to both liposome rigidity and particle size. The liposomes showed a little cytotoxicity in ocular cells. They suggest that submicron-sized liposomes can be used to deliver drugs to the posterior segment of the eyes.

### 8.4.3 NANODISPERSION/NANOCRYSTALS-BASED OCULAR DRUG DELIVERY SYSTEMS

Nanodispersions hold size with a range of less than 1000 µm in diameter in which the drug is suspended in aqueous medium using surfactants as stabilizers. The nanodispersion preparation methods include bottom-up methods (microprecipitation/

supercritical fluid methods) and top-down methods (wet milling/high-pressure homogenization).

Nanocrystals composed of 100% drugs are devoid of matrix materials with a size range of 200–500 nm. The bottom-up and top-down technologies were widely used in the synthesis of nanocrystals. The techniques used should be monitored to prevent the growth in the microcrystals. The development of drug-loaded nanocrystals based technology has been widely applied to improve the solubility of poorly soluble drugs (budesonide, dexamethasone, hydrocortisone, and prednisolone). NSAIDs were developed recently using nanocrystal formulation approach. Drugs formulated using nanocrystals have been checked for their Oswald ripening. The enhanced bioavailability of drug-loaded nanocrystals was attributed to their enhanced bioadhesion to the surfaces/cell membranes. This may increase the saturation solubility and dissolution velocity, which leads to high drug concentration at the site. Nanodispersion technique apart from improving the bioavailability also elicits a sustained release pattern.

Moxifloxacin-loaded PLGA nanodispersions were developed by Vaidehi et al. (2014) for the enhanced ocular delivery, using high-pressure homogenization of the primary emulsion which was obtained by double emulsification solvent evaporation technique. They found that the particle size was dependent upon PLGA concentration, moxifloxacin:PLGA ratio, concentration of polyvinyl alcohol, and water in oil/phase volume ratio. The encapsulation of moxifloxacin depends upon concentration of PLGA and moxifloxacin:PLGA ratio. Moxifloxacin was encapsulated in the optimized formulation around 62.5% with an average particle size of 118.5 nm suitable for ocular administration. The sustained release profiles of moxifloxacin from nanodispersions were nonirritant and prevent the moxifloxacin ocular drainage with threefold enhanced ocular bioavailability. Ketorolac-loaded alginate/chitosan nanodispersions were prepared by Morsi et al. (2015), using coaservation and ionotropic pregelation techniques in order to provide a sustained drug release pattern and improved transcorneal permeation. The results of their findings indicate that ionotropic pregelation technique provides a sustained release pattern of ketorolac with improved transcorneal permeation.

The brinzolamide-loaded nanocrystal formulations were developed by Tuomelaa et al. (2014) for the treatment of glaucoma. They developed the formulation using different stabilizers for the nanocrystals such as hydroxypropyl methylcellulose (HPMC), poloxamer F127 and F68, polysorbate 80, and found HPMC as the screened stabilizer. The IOP-lowering effect of developed formulation was investigated *in vivo* using a modern rat ocular hypertensive model. They observed an elevated IOP reduction with the developed brinzolamide-loaded nanocrystal formulations.

### 8.4.4 EMULSIONS-BASED OCULAR DRUG DELIVERY SYSTEMS

Emulsion has become one of the promising drug delivery systems due to the unique solubilization property for the delivery of poorly soluble drugs with enhanced bioavailability. Microemulsion/nanoemulsion is characterized as a nonequilibrium, heterogeneous system of two immiscible liquids, where the nanodroplets of oil are dispersed in aqueous continuous phase and stabilized by appropriate surfactants under homogenization. The oil phase in the emulsions serves as the reservoir for the lipophilic drugs, protects the drug molecule from the external physiological environment, offers controlled drug release, and can deliver the drug through various routes of administration such as ocular, oral, topical, pulmonary, and intravenous routes.

Gulsen and Chauhan (2005) developed stabilized microemulsions of poly-2-hydroxyethyl methacrylate and found that the gels released drugs over a period of 8 days and they controlled the delivery rate by modifying the particle size and drug-loading capacity.

The stability of chloramphenicol in microemulsion-based ocular drug delivery system was ascertained by Feng et al. (2005). They developed chloramphenicol microemulsions using span 20/80, tween 20/80, *n*-butanol, and isopropyl palmitate/ isopropyl myristate, and implied that the stability of chloramphenicol microemulsions increased remarkably when compared to commercial eye drops of chloramphenicol. This may be due to the fact that chloramphenicol is trapped into the hydrophilic shells of the microemulsion droplets.

The dorzolamide-loaded nanoemulsion for the treatment of glaucoma was reported by Ammar et al. (2009) using different oils, surfactants, and cosurfactants. They observed a decrease in the number of applications per day providing a sustained release pattern of dorzolamide by utilizing this nanoemulsion, and elicit a better patient compliance compared to marketed formulations. Systematically, they found that this developed dorzolamide-loaded nanoemulsions may enhance the therapeutic efficacy of the drug, increase the membrane permeability due to the presence of surfactants and cosurfactants, act as a penetration enhancer by removing the mucus materials of the eye, and disturb the tight junction layers of the cornea, thereby providing a better therapeutic effect.

## 8.5   CONCLUSION

The use of novel/targeted drug delivery systems like functionalized nanoparticulate/ nanocarrier drug delivery system could be the possible way to overcome the ocular barriers for the transport of drug. Thus, research based on nanoparticles/nanocarriers to overcome ocular barriers provides a clear opportunity to improve the ocular bioavailability. The new drug targets discussed in this chapter can be utilized to develop novel drug delivery systems for various ocular diseases.

## REFERENCES

Afshari, F.T., Kwok, J.C., Andrews, M.R. et al. 2010. Integrin activation or alpha 9 expression allows retinal pigmented epithelial cell adhesion on Bruch's membrane in wet age-related macular degeneration. *Brain* 133:448–64.

Akiyoshi, U. 2013. Identification of novel drug targets for the treatment of diabetic retinopathy. *Diab Metabol* 37:217–24.

Aksungur, P., Demirbilek, M., Denkbas, E.B. et al. 2011. Development and characterization of cyclosporine A loaded nanoparticles for ocular drug delivery: Cellular toxicity, uptake and kinetic studies. *J Control Release* 151:286–94.

Alain, F., Capucine, P., Karine, C. et al. 2010. ZAP70, A master regulator of adaptive immunity. *Semin Immunopathol* 32:107–16.

Albietz, J.M. and Bruce, A.S. 2001. The conjunctival epithelium in dry eye subtypes: Effect of preserved and non-preserved topical treatments. *J Curr Eye Res* 2:8–18.

Amalia, E.S., Evangelina, C., Michael, E.S. et al. 2010. Tear cytokine and chemokine analysis and clinical correlations in evaporative-type dry eye disease. *Mol Vis* 16:862–73.

Ammar, H.O., Salama, H.A., Ghorab, M. et al. 2009. Nanoemulsion as a potential ophthalmic delivery system for dorzolamide hydrochloride. *AAPS PharmSciTech* 10:808–19.

Anat, E.B., Keren, M., Joaquin, S. et al. 2011. Integrin-assisted drug delivery of nano-scaled polymer therapeutics bearing paclitaxel. *Biomater* 32:3862–74.

Ashaben, P., Kishore, C., Vibhuti, A. et al. 2013. Ocular drug delivery systems: an overview. *World J Pharmacol* 2:47–64.

Bando, H., Shadrach, K.G., Rayborn, M.E. et al. 2007. Clathrin and adaptin accumulation in drusen, Bruch's membrane and choroid in AMD and non-AMD donor eyes. *Exp Eye Res* 84:135–42.

Basilio, C., Hanan, A.A., and Jesus, P. 2014. Recent developments on dry eye disease compounds. *Saudi J Ophthalmol* 28:19–30.

Beauregard, C., Brandt, P.C., and Chiou, G.C. 2003. Induction of nitric oxide synthase and over-production of nitric oxide by interleukin-1beta in cultured lacrimal gland acinar cells. *Expl Eye Res* 77:109–14.

Bennett, B. L., Sasaki, D. T., Murray, B. W. et al. 2001. SP600125, An anthrapyrazolone inhibitor of Jun N-terminal *kinase. Proc Natl Acad Sci U S A* 98:13681–86.

Boullin, D.J. and Glenton, P.A. 1978. Characterization of receptors mediating 5-hydroxy-tryptamine- and catecholamine-induced platelet aggregation, assessed by the actions of alpha- and beta-blockers, butyrophenones, 5- HT antagonists and chlorpromazine. *Br J Pharmacol* 62:537–42.

Cha, S. and Pauley, K.M. 2012. Targeted receptor-mediated siRNA. Patent Application No. PCT/US2011/042170.

Charles, D.P. 1978. The pathogenesis of glaucoma in sturge-weber syndrome. *Opthamol* 85:276–86.

Chaurasia, S.S., Lim, R.R., Lakshminarayanan, R. et al. 2015. Nanomedicine approaches for corneal diseases. *J Func Biomater* 6:277–98.

Chiara, G., Zeina, D., Paola, S. et al. 2009. The TGF-β pathway is a common target of drugs that prevent experimental diabetic retinopathy. *Diabetes* 58:1659–67.

Colin, E.W., Diego, P., Stefano, F. et al. 2013. Anatomy and physiology of the human eye: effects of mucopolysaccharidoses disease on structure and function: A review. *J Clin Exp Ophthalmol* 38:2–11.

Colligris, B., Alkozi, H.A., and Pintor, J. 2014. Recent developments on dry eye disease treatment compounds. *Saudi J Ophthalmol* 28:19–30.

Crabb, J.W., Miyagi, M., Gu, X. et al. 2002. Drusen proteome analysis: An approach to the etiology of age-related macular degeneration. *PNAS* 99:14682–87.

Dalton, J.T., Miller, D.D., Rakov, I. et al. 2008. Selective androgen receptor modulators, analogs and derivatives thereof and uses thereof. Patent Application No. PCT/US2007/016311.

Danhier, F., Vroman, B., Lecouturier, N. et al. 2009. Targeting of tumor endothelium by RGD-grafted PLGA-nanoparticles loaded with Paclitaxel. *J Control Release* 140:166–73.

Das, S., Bellare, J.R., and Banerjee, R. 2012. Protein based nanoparticles as platforms for aspirin delivery for ophthalmic applications. *Colloids Surf B Biointerfaces* 93:161–68.

David, G.B. and Fong, Q.L. 2007. Age-related macular degeneration: A target for nanotechnology derived medicines. *Int J Nanomed* 2:65–77.

Diebold, Y. and Calonge, M. 2010. Applications of nanoparticles in ophthalmology. *Prog Retin Eye Res* 29:596–09.

Dipak, R.G., Dinesh, M.S., Pradeep, R.K. et al. 2013. Recent advances in ocular drug delivery systems. *Indo American J Pharm Res* 3:3216–32.

Eugene, W.M.N., Shima, D.T., Calias, P. et al. 2006. Pegaptanib, a targeted anti-VEGF aptamer for ocular vascular disease. *Nat Rev Drug Discov* 5:123.

Fathalla, D., Soliman, G.M., and Fouad, E.A. 2015. Development and in vitro/in vivo evaluation of liposomal gels for the sustained ocular delivery of latanoprost. *Clin Exp Ophthalmol* 6:1–9.

Feng, F., Zheng, Q., Tung, C. et al. 2005. Phase behavior of the microemulsions and the stability of the chloramphenicol in the microemulsion-based ocular drug delivery system. *Int J Pharm* 301:237–246.

Geraldes, P. and King, G.L. 2010. Activation of protein kinase C isoforms and its impact on diabetic complications. *Circ Res* 106:1319–31.

Gilbard, J.P. 1994. Dry eye disorders. In *Principles and Practice of Ophthalmology: Clinical Practice*, ed. D.M. Albert and F.A. Jacobiec, 257–76. Philadelphia: W.B. Saunders Company.

Gordon, L.K. 2006. Orbital inflammatory disease: A diagnostic and therapeutic challenge. *Eye* 20:1196–06.

Goyal, S., Chauhan, S.K., and Dana, R. 2012. Blockade of prolymphangiogenic vascular endothelial growth factor C in dry eye disease. *Arch Ophthalmol* 130:84–9.

Grama, C.N., Suryanarayana, P., Patil, M.A. et al. 2013. Efficacy of biodegradable curcumin nanoparticles in delaying cataract in diabetic rat model. *PLoS One* 8:e78217.

Gratieri, T., Gelfuso, G.M., Freitas, O.D. et al. 2011. Enhancing and sustaining the topical ocular delivery of fluconazole using chitosan solution and poloxamer/chitosan *in situ* forming gel. *Eur J Pharm Biopharm* 79:320–27.

Gulsen, D. and Chauhan, A. 2005. Dispersion of microemulsion in HEMA hydrogel: A potential ophthalmic drug delivery vehicle. *Int J Pharm* 292:95–117.

Gupta, H., Aquil, M., Khar, R.K. et al. 2010. Sparfloxacin loaded PLGA nanoparticles for sustained ocular drug delivery. *Nanomed Nanotechnol Biol Med* 6:324–33.

Gustavsson, C., Agardh, C.D., and Zetterqvist, A.V. 2010. Vascular cellular adhesion molecule-1 (VCAM-1) expression in mice retinal vessels is affected by both hyperglycemia and hyperlipidemia. *PLoS One* 5:e12699.

Hironaka, K., Inokuchi, Y., Fujisawa, T. et al. 2009. Physicochemical properties affecting retinal drug/coumarin-6 delivery from nanocarrier systems via eyedrop administration. *Invest Ophthalmol Vis Sci* 51:3162–3170.

Isaac, A., Hanna, J.G, Irina, S.B. et al. 2010. Treatment of dry eye syndrome with orally administered CF101. *J Ophthamol* 7:1287–93.

Jacob, L., Baure, J.T., and Kaufman, H.E. 1990. Investigation of pilocarpine-loaded polybutyl cyanoacrylate nanocapsules in collagen shields as a drug delivery system. *Invest Opthalmol Vis Sci* 31:485.

Jayakrishna, A. and Fowler, B.J. 2012. Mechanisms of age-related macular degeneration. *Neuron* 75:26–39.

Jean, P.H., Shantan, R., and Steven, D.S. 2009. Age-related macular degeneration: current treatments. *J Clinic Ophthalmol* 3:155–66.

John, R.W.Y., Tiina, S., Baljinder, K.M. et al. 2007. Complement C3 variant and the risk of age-related macular degeneration, for the genetic factors in AMD study group. *N Eng J Med* 357:553–61.

Kang, Z., Liangfang, Z., and Robert, N.W. 2012. Ophthalmic drug discovery: Novel targets and mechanisms for retinal diseases and glaucoma. *Nat Rev Drug Discov* 11:541–59.

Karnati, R., Laurie, D.E., and Laurie, G.W. 2013. Lacritin and the tear proteome as natural replacement therapy for dry eye. *Exp Eye Res* 12:39–52.

Kim, H. and Csaky, K.G. 2010. Nanoparticle integrin antagonist C16Y peptide treatment of choroidal neovascularization in rats. *J Control Release* 142:286–93.

Koray, G. and Dwight, H.C. 2009. The role of inflammation and anti-inflammation therapies in kerato conjunctivitis sicca. *J Clin Ophthalmol* 3:57–67.

Leal, E.C., Santiago, A.R., and Ambrosio, A.F. 2005. Old and new drug targets in diabetic retinopathy: From biochemical changes to inflammation and neurodegeneration. *Curr Drug Targets CNS Neuro Disord* 4:421–34.

Lubniewski, A.J. and Nelson, J.D. 1990. Diagnosis and management of dry eye and ocular surface disorders. *J Clin Ophthalmol* 3575–94.

Mohd, I.N., Marwan, A., Haseeb A.K. et al. 2013. Novel drugs and their targets in the potential treatment of diabetic retinopathy. *Med Sci Monit* 19:300–08.

Morsi, N., Ghorab, D., Refai, H. et al. 2015. Preparation and evaluation of alginate/chitosan nanodispersions for ocular delivery. *Int J Pharm Pharm Sci* 7:234–40.

Mostafa, S., Seamon, V., and Azzarolo, A.M. 2012. Influence of sex hormones and genetic predisposition in Sjogren's syndrome: A new clue to the immunopathogenesis of dry eye disease. *Exp Eye Res* 96:88–97.

Motwani, S.K., Chopra, S., Talegaonkar, S. et al. 2008. Chitosan sodium alginate nanoparticles as submicroscopic reservoirs for ocular delivery: Formulation, optimization and in vitro characterization. *Eur J Pharm Biopharm* 68:513–25.

Osaadon, P., Fagan, X.J., Lifshitz, T. et al. 2014. A review of anti-VEGF agents for proliferative diabetic retinopathy. *Eye* 28:510–20.

Paswan, K., Verma, P., Yadav, M.S. et al. 2015. Review-advanced technique in ocular drug delivery system. *World J Pharm Pharm Sci* 4:346–65.

Patel, V. and Agrawal, Y.K. 2011. Current status and advanced approaches in ocular drug delivery system. *J Global Trends Pharm Sci* 2:131–48.

Prasad, P.S., Schwartz, S.D., and Hubschman, J.P. 2010. Age related macular degeneration: Current and novel therapies. *Maturitas* 66:46–50.

Rathore, K.S. and Nema, R.K. 2009. An insight into ophthalmic drug delivery system. *Int J Pharm Sci Drug Res* 1:1–5.

Ruoslahti. E. 1996. RGD and other recognition sequences for integrins. *Annu Rev Cell Dev Biol* 12:697–15.

Rupenthal, I.D., Green, C.R., and Alany, R.G. 2011. Comparison of ion activated *in situ* gelling systems for ocular drug delivery. Part 1 physicochemical characterization and in vitro release. *Int J Pharm* 411:69–77.

Sandeep, C.A., Nishan, N.B., Vikrant P.W. et al. 2013. Current trends towards an ocular drug delivery system: Review. *Int J Pharm Pharm Sci Res* 3:28–34.

Schoenberger, S.D., Kim, S.J., Sheng, J. et al. 2012. Increased prostaglandin E2 (PGE2) levels in proliferative diabetic retinopathy, and correlation with VEGF and inflammatory cytokines. *Inves Ophthalmol Vis Sci* 53:5906–11.

Schopf, L.R., Popov, A.M., Enlow, E.M. et al. 2015. Topical ocular drug delivery to the back of the eye by mucus-penetrating particles. *Transl Vis Sci Technol* 4:1–11.

Schwartza, K. and Budenzb, D. 2004. Current management of glaucoma. *Curr Opin Ophthalmol* 15:119–26.

Schwartzman, M.L., Iserovich, P., and Gotlinger, K. 2010. Profile of lipid and protein autacoids in diabetic vitreous correlates with the progression of diabetic retinopathy. *J Diab* 59:1780–88.

Sharma, A., Tandon, A., Tovey, J.C. et al. 2011. Polyethylenimine conjugated gold nanoparticles gene transfer potential and low toxicity in the cornea. *Nanomedicine* 7:505–15.

Sharma, R., Ahuja, M., and Kaur, H. 2012. Thiolated pectin nanoparticles preparation, characterization and ex vivo corneal permeation study. *Carbohydr Polym* 87:1606–10.

Silva, G.R.D., Cunha, A.D.S., Cohen, F.B. et al. 2011. Biodegradable polyurethane nanocomposites containing dexamethasone for ocular route. *Mat Sci Eng C* 31:414–22.

Simó, R.O. and Christian H.A. 2009. Advances in the medical treatment of diabetic retinopathy. *Diabetes Care* 32:1556–62.

Singh, S.R., Grossniklaus, H.E., Kang, S.J. et al. 2009. Intravenous transferrin, RGD peptide and dual targeted nanoparticles enhance anti-VEGF intraceptor gene delivery to laser induced CNV. *Gene Ther* 16:645–59.

Tangri, P. and Khurana, S. 2011. Basics of ocular drug delivery systems: Review. *Int J Res Pharm Biomed Sci* 2:1541–52.

Thakur, R.R. and Kashiv, M. 2011. Modern delivery system for ocular drug formulations: A comparative overview with respective to conventional dosage form. *Int J Res Pharm Biomed Sci* 2:8–18.

Tuomelaa, A., Liua, P., Puranenb, J. et al. 2014. Brinzolamide nanocrystal formulations for ophthalmic delivery: Reduction of elevated intraocular pressure in vivo. *Int J Pharm* 467:34–41.

Ulrich, H., Claudia, D., and Hort, K. 2003. RGD modified polymers biomaterials for stimulated cell adhesion and beyond. *Biomater* 24:4385–15.

Urtti, A. 2006. Challenges and obstacles of ocular pharmacokinetics and drug delivery. *Adv Drug Del Rev* 58:1131–35.

Vaidehi, G., Gaurav, K.J, Jayabalan, N. et al. 2014. Development of poly lactide-co-glycolide nanodispersions for enhanced ocular delivery of moxifloxacin. *Sci Advan Mater* 6:990–99.

Valente, I., del Valle, L.J., Casas, M.T. et al. 2013. Nanospheres and nanocapsules of amphiphilic copolymers constituted by methoxypolyethylene glycol cyanoacrylate and hexadecyl cyanoacrylate units. *Polymer Lett* 7:2–20.

Venkata, R.G., Madhavi, S., and Rajesh, P. 2011. Ocular drug delivery: An update review. *Int J Pharm Bio Sci* 4:437–46.

Venkateshwaran, K., Chandrasekar, P., Senthilkumar, S. et al. 2015. Quantification of hypocrellin B in rabbit ocular tissues/plasma by spectrofluorimeter: Application to biodistribution study. *Curr Bioactive Comp* 10:245–53.

Yaguchi, S., Ogawa, Y., and Kamoi, M. 2012. Surgical management of lacrimal punctal cauterization in chronic GVHD-related dry eye with recurrent punctal plug extrusion. *J Bone Marrow Transpl* 47:1465–69.

Yasukawa, T., Kimura, H., Tabata, Y. et al. 2001. Biodegradable scleral plugs for vitreoretinal drug delivery. *Adv Drug Del Rev* 52:25–36.

Yujing, B., ZhiHua, S., Yehong, Z. et al. 2010. In glaucoma the upregulated Truncated TrkC. T1 receptor isoform in Glia causes increased TNF Production, leading to retinal ganglion cell death. *Invest Opthalmol Vis Sci* 51:6639–51.

# 9 Bone-Targeted Drug Delivery Systems

*Amit Kumar Nayak and Kalyan Kumar Sen*

## CONTENTS

## 9.1 INTRODUCTION

Like other organs, bones are susceptible to a range of diseases. Commonly occurring bone-related diseases are bone metastasis, osteomyelitis, osteoporosis, arthrosis, and so on (Odgreen and Martin 2000). These diseases increase the economic toll on the health systems worldwide and also decrease the quality of life for millions of people each year. Since the last few decades, an impressive progress has been recorded in terms of developing drug delivery approaches and devices to treat these bone-related diseases (Pham et al. 2002; Takahashi-Nishioka et al. 2008; Nayak et al. 2010, 2011a). Targeting bone tissues to treat these diseases is an important and effective approach in the field of drug delivery technology.

Bone is a highly specified connective tissue, which consists of 50%–70% minerals, 20%–40% organic matrix, 5%–10% water, and 1%–5% lipids. Generally, bone is mainly composed of three types of cells: osteoblasts, osteoclasts, and osteocytes (Marks and Odgren 2002). Osteoblasts are known as bone-forming cells, which originate from local osteogenitor cells. Osteoclasts are nonnucleated bone resorbing cells originating from various hemopoietic tissues. Generally, osteoblast and osteoclast cells are present only on the bone surface; while osteocytes permeate the mineralized bone matrix. Osteocytes are mature osteoblasts located in the bone matrix and can also resorb to form the bone matrix (Takahashi et al. 2002). The extracellular matrix is composed of organic materials, which are highly mineralized with calcium phosphate in the form of hydroxyapatite (HAp) $[Ca_{10}(PO_4)_6(OH)_2]$ (Marks et al. 2002). HAp provides rigidity, strength, and load-bearing capacity to bones. It consists of very small crystals of about 200 Å in their largest dimension (Shea and Miller 2005). HAp crystals are broken down and replaced during bone turn over. Primarily, monocytes receive several signals pushing them to differentiate into osteoclasts. Osteoblasts present receptor activator of nuclear factor-kappa-B (kB) ligand (RANKL) to the receptor activator of nuclear factor-kB surface receptor triggering the tumor necrosis factor (TNF) receptor-associated factor 6 (TRAF6) cascade in monocytes, ensuring the development of osteoclast (Yasuda et al. 1998). Mature osteoclasts then perform the catabolic process of the bone. Osteoblasts create a sealing zone over the lacunae (receptor sites) and proceed to release cathepsin K to degrade the organic matrix and HCl to dissolve the HAp. These mechanisms in combination with reactive oxygen species (ROS) produced by tartrate-resistant acid phosphatase (TRAP) degrade the bone preparatory for osteoblasts to lay down the organic matrix called osteoid. Osteoids primarily consist of type-I collagen and it is then calcified to become a new bone (Väänänen et al. 2000; Harada 2003; Silverthorn et al. 2004). In this way, the bone drug delivery for the treatment of bone diseases offers both challenges and opportunities for improving the delivery of drug molecules. Moreover, the poor vascularization of the bone presents a unique challenge for the delivery of drug molecules to the effective sites of the diseased bones. A number of drug molecules are effective for the treatment of bone diseases, but their systemic delivery is inevitably related with significant side effects as well as lack of targeting (Wang et al. 2005). Moreover, the targeting of drug molecules to the diseased sites of the bones provides improved efficacy and reduction of the required dose of drugs. The current chapter deals with comprehensive and useful discussions on various important aspects of already reported bone-targeting drug delivery systems. These include relevance to bone targeting, different bone-targeting moieties (e.g., tetracyclines, bisphosphonates, acidic oligopeptides, estradiol analogs, etc.), linkers for bone targeting (i.e., reduced pH or acid-cleavable linkers and enzymatic-cleavable linkers), polymeric carriers for bone-targeting drug delivery, and bone-targeting therapies against common bone disorders like osteoporosis, osteoarthritis, osteomyelitis, and bone metastases.

## 9.2   RELEVANCE TO BONE TARGETING AND BONE-TARGETING MOIETIES

Based on the understanding of the bone biology, bone-targeting therapeutics has been employed in recent years to improve the treatment of various bone diseases including osteosarcoma, bone metastasis, osteomyelitis, osteoporosis, arthrosis, and so on. HAp crystal exposure and bone marrow are mostly the disease-specific sites, and as such modification of the targeting ligand itself or the density of the ligand may provide specific targeting opportunities. A bone-targeting drug delivery system should consist of several essential elements. Every bone disease state has similarities in that each causes local inflammation and/or results in the exposure of HAp to blood. These two important characteristics can be employed in order to deliver high drug loads to the diseased sites specifically. The delivery of drugs to bone depends on the extravasation through vessels in or near the bone (Silverthorn et al. 2004). Tetracyclines, bisphosphonates, acidic oligopeptides, chelating compounds, salivary proteins, and so on, have been employed to target various bone diseases (Wang et al. 2005). Generally, these compounds target HAp crystals of the bone and have specificity for a certain size of HAp crystals. As the crystallinity of the newly deposited bone in a tumor is different from the crystallinity of HAp exposed during bone fracture, the targeting molecules can be selected for specific diseased states (Hirabayashi and Fujisaki 2003). Moreover, bone is turning over continually to synthesize new bone to present the opportunities for the incorporation of drug molecules to these infected sites of bones. The effective targeting of the drug molecules to the infected bone sites provides the desired treatment of various bone diseases (Hirabayashi and Fujisaki 2003; Wang et al. 2005). However, the active targeting should be, in some cases, more useful to reach effective drug concentrations in specific tissues and organs. Also, approaches in which drug molecules are covalently linked to a targeted moiety are able to recognize the bone tissues selectively, and are also employed to convey the whole compound (through prodrug approaches or polymeric conjugates) to the infected bone sites (Hirabayashi and Fujisaki 2003; Torchilin 2010). The binding efficiency of the targeting molecules to the apatite surfaces is governed by several important issues such as blood supply, local extravasation rate of targeting molecules into the interstitial fluid (where bone surface is exposed), the molecular structure of the apatite surface, and the effective binding surface area (Williams 1984). The effective binding surface area is determined by the amount of apatite, the average crystal size, the surface morphology of the apatite crystals, and the organic/inorganic ratio in the bone matrix. The selection of bone-targeting moiety can be done on the following criteria (Wang et al. 2005):

1. Bone-targeting moiety must possess a strong affinity for HAp.
2. These must be chemically modifiable to permit conjugation to the polymer backbone.
3. These must not render the delivery system toxic.
4. Their biological effects should be negligible.

Important bone-targeting moieties include tetracyclines, bisphosphonates, acidic oligopeptides, estradiol analogs, and so on. These are described below.

## 9.2.1 TETRACYCLINES

Tetracyclines represent a family of compounds of polycyclic naphthacene carboxamides (Figure 9.1). These are amphoteric substances and recognized as a category

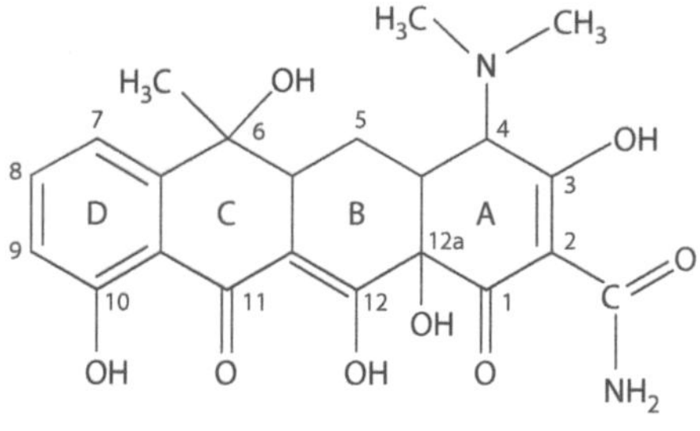

**FIGURE 9.1**  The structure of tetracycline.

of broad-spectrum polyketide antibiotics, which are derived from the metabolites of the actinomycete *Streptomyces rimosus* with a wide range of antibiotic activities (Chopra and Roberts, 2001). These inhibit the synthesis of proteins (protein elongation) by blocking the attachment of charged aminocyl-tRNA to the A-site on the ribosome. Tetracyclines bind to the 30S subunit of microbial ribosomes and prevent introduction of new amino acids (AA) to the nascent peptide chains (Semenkov et al. 1982; van der Bijl and Pitigoi-Aron 1995). Thus, these affect the bacterial metabolism and act as antibiotics. Tetracyclines act against a wide range of gram-positive and gram-negative bacteria. The administration of tetracyclines during odontogenesis is generally not suggested because of the potential discoloration of primary as well as permanent teeth (Albert and Rees 1956). The side effects of tetracycline-associated discoloration of teeth are mainly pronounced in younger age due to their rapid growth rate of bone. Tetracyclines have well-defined metal complexing abilities and the characteristics of their binding to the bone appetite are probably by chelation with surface calcium ions. The model of metal ion binding to the HAp surface was suggested by Perrin (1965). Perrin oriented the tetracycline molecule to the putative three surface calcium ions where tetracyclines replace two $PO_4^{3-}$ ions. The oxygen atom of tetracyclines at $C_2$, $C_{10}$, and $C_{12}$ (from carbomoyl) is able to fit three calcium ion positions of HAp with mineral distortion while maintaining electrostatic neutrality of the complex. It was also suggested by Myers et al. (1983) that the chelation of tetracyclines with the calcium ions on the HAp surface may engage the oxygen atoms of the two sites at $C_2$ (of carbomoyl moiety):$C_1$ (or $C_3$) and $C_{11}$:$C_{12}$ (Myers et al. 1983). Due to the osteotropicity characters, tetracycline molecules have been conjugated directly with some therapeutic molecules used for the effective treatment of bone diseases (Misra 1991; Gittens et al. 2005; Low and Kopeček 2012; Luhmann et al. 2012).

    In an investigation by Pierce and Waite (1984), conjugation of tetracycline molecule to acetazolamide (an example of carbonic anhydrase inhibitor with 1, 6-hexandioyl-dichloride) was studied (Pierce and Waite 1984). The *in vitro* results of this study demonstrated that the product showed a high degree of HAp-binding affinity. After hydrolysis, it was found to release free acetazolamide molecules. This phenomenon indicated the carbonic anhydrase inhibitory actions. Yet, the chemical structural features of the final conjugate were not explained by the work done by Pierce and Waite (1984). In another study, Orme and Labroo (1994) adopted a similar approach to conjugate β-estradiol with tetracycline. After careful characterization, they found a strong binding to the bone. However, no *in vivo* data was supplied for the justification of free estradiol release. The similar approach was also adopted by a group of researchers in a group of investigations. These researchers synthesized 2-[3-estrone-N-ethyl-piperazine-methyl] tetracycline and systemically evaluated the biological effects of the synthesized molecule (Zheng et al. 2000; Weng et al. 2002). They found a high binding affinity of the synthesized molecule to HAp. However, they found that the synthesized molecule had a much lesser degree of estrogen receptor binding in comparison with estrone and estradiol. The vaginal smear assay of oophorectomized mice demonstrated that the estronic activity of the synthesized molecule is lower than that of estradiol and estrone. The result indicated that the compound might enhance endochondral bone formation by the up-regulation of the expression of some important proteins (like C-fos, C-jun, and C-myc proteins) in chondrocytes. From the results obtained from the organ, they also inferred that the estrogenic activity of the synthesized molecules was found to be greater than that of estrone. Neale et al. (2009) attempted to reduce the potential side effects associated with the biological activity of tetracyclines through the mineralization of the tetracycline structure. They found that there was absence of any biological activity, which is capable to bind HAp. They also found that 3-amino-2, 6-dihydroxy benzamide retains 50% of the capability to bind the HAp in comparison with that of native tetracyclines. To achieve the bone anabolism, the new targeting ligand was bound to estradiol via a succinate linker. Following the conjugation process, the compound showed 105% binding affinity over tetracycline alone. The enhanced binding activity may be attributed to the addition of a succinate linker, as estradiol alone did not bind to HAp. The complexity in the chemical structure and the poor solubility of tetracyclines during modification seems to hinder further utilization of tetracycline as a bone-targeting moiety. But its strong bone-binding mode and capacity have inspired the search for smaller molecules with bone-binding capacities such as tetracycline molecules (Wang et al. 2005). A series of smaller heterocyclic molecules, which are associated to thiazole with affinity to HAp were identified by Willson et al. (1996a, b). They also found that this affinity to HAp was comparable to tetracycline.

## 9.2.2 Bisphosphonates

Bisphosphonates are the most studied bone-targeting molecules able to bind HAp strongly, and also retain much of the binding affinity after conjugation to other molecules or carriers (Fleisch et al. 2002; Kavangh et al. 2006). Bisphosphonates are derived from pyrophosphate with a carbon substitution of pyrophosphate's central oxygen. From the central carbon, most bisphosphonates have been functionalized with -OH group and "R"-group by which they are classified as "R"-group-containing nitrogen and those that do not (Figure 9.2). Bisphosphonates have a general structure of P—C—P (Low and Kopeček 2012). The P—C—P bond is resistant to most chemical reagents and inert to enzymatic degradation. The non-nitrogen-containing bisphosphonates are metabolized into methylene-containing ATP analogs in the osteoclasts, which leads to apoptosis and reduced bone turn over. The P—C—P structure of bisphosphonates cannot be further hydrolyzed and as a result of this, these

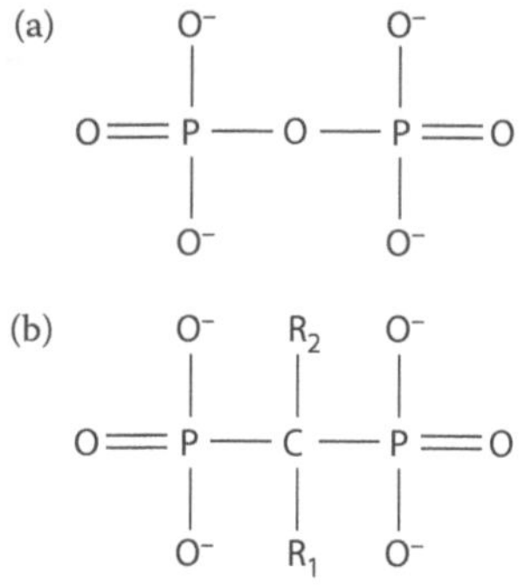

(a)

$$O = P(-O^-)(-O^-) - O - P(=O)(-O^-)(-O^-)$$

(b)

$$O = P(-O^-)(-O^-) - C(R_2)(R_1) - P(=O)(-O^-)(-O^-)$$

**FIGURE 9.2**   The structures of (a) pyrophosphate and (b) germinal bisphosphates.

compounds accumulate in the cells showing substantial cytotoxicity (Fleisch et al. 2002). The nitrogen-containing bisphosphonates are much more potent than the non-nitrogen-containing bisphosphonates. The nitrogen-containing bisphosphonates also inhibit farnesyl pyrophosphate synthase activity and thus, disrupt the mevalonic acid pathway (Wang et al. 2005). Without the mevalonic acid pathway, the protein pre-nylation is inhibited and osteoclast cells lose their functionality (Fleisch et al. 2002). The inhibition of protein prenylation may not completely arrest turn over as meva-lonic acid pathway inhibitors increase the production of bone morphogenic protein (BMP) (Ohnaka et al. 2001; Ayukawa et al. 2009).

At the cellular level, bisphosphonates exert their influences on the bone cells especially on the osteoclasts and these influences are manifested in four different ways: (a) osteoclast recruitment inhibition, (b) inhibition of osteoclast adhesion to the mineral matrix, (c) osteoclast life span shortening, and (d) direct inhibition of osteoclast activity (Fleisch et al. 2002). Bisphosphonates have poor oral bioavail-ability, though a higher percentage of bioavailable drugs deposit to the skeleton. This deposition is heterogenous and is concentrated in the deposition sites with a high turnover rate (Lin 1996). Currently, many efforts have been attempted to conjugate nonspecific bone therapeutic agents with bisphosphonates in order to obtain osteo-tropicity. These include small and large drug molecules (Fujisaki et al. 1995; Niemi et al. 1999; Hirabayashi et al. 2001), proteins and peptides (Uludağ and Yang 2002; Wright et al. 2006), radiopharmaceuticals (Fancis and Fogelman 1987; Lamb and Faulds 1997; Wang et al. 2003), and so on. Among them, the only approved mol-ecules for clinical use are radiopharmaceuticals (Wang et al. 2005). Bisphosphonate-coupled radiopharmaceuticals are extensively investigated for use in clinical settings for imaging purposes and also for pain palliation of the skeleton (Wang et al. 2003). The poor oral bioavailability of the bisphosphonate conjugates may exert problems tackling the future use of these molecules. In addition, retaining the bioactivity of these molecules after conjugation with bisphosphonates is one of the central con-cerns in the development of effective bisphosphonate conjugates with these mol-ecules (Wang et al. 2005).

In recent years, alendronate is successfully conjugated with bisphosphonates and delivered through polymeric systems (Wang et al. 2003). Both polyethylene glycol (PEG) and hydroxypropyl methacrylate (HPMA) copolymers were modified for this purpose. The modified copolymers showed high binding with HAp as compared to unmodified copolymers. When alendronate formulation was injected into balb/c mice, it showed very strong binding capacity to HAp of the bone. Proteins and peptides have also been proposed for the conjugation with bisphosphonates to enhance the bone specificity (Gittens et al. 2005). The conjugated bisphosphonates may be capable of inducing apoptosis in osteoclasts inhibiting migration with the reduction of angiogenic sprouts of endothelial tissue making. It is considered a good antineoplastic agent. However, it is reported to reduce bone healing (Ziebart et al. 2009). Furthermore, the bisphosphonates such as alendronate in the bone are more than 10 years (Marx et al. 2005). The inhibition of osteoclasts for extended periods can have serious detrimental effects on the bone turn over as the osteoclasts are reported to secrete insulin-like growth factors and bone morphogenic proteins to promote the maturation of osteoblasts. To develop a controlled release system, caution must be taken since free bisphosphonates may counter the activity of the payload drug. Avoidance of using bisphosphonates as a bone-targeting moiety is suggested if an osteoclast-mediated releasing mechanism is used (Wang et al. 2005).

In an investigation, bisphosphonate conjugate of methotrexate was formulated using a linkage of formamide bond with $\gamma$-COOH of Glu in methotrexate (Hosain et al. 1996). The rationale for the development of this methotrexate-bearing bone-targeting drug delivery system was to enhance the activity against metastatic tumor and to reduce side effects. Though the antitumor activity of this bone-targeting drug delivery system was unclear, the bisphosphonate conjugate of methotrexate was reported to be visualized with $99_mT_c$-label, *in vivo*. A prostaglandin E2 (a most potent bone-anabolic agent)-bisphosphonate conjugate was also investigated (Gil et al. 1999). The linkage between prostaglandin $E_2$ (drug) and the bisphosphonates was $BrCH_2COBr$ between 15-OH of prostaglandin $E_2$ and -SH of the bisphosphonate. The prostaglandin $E_2$ released from the conjugate followed a nonspecific hydrolysis. This prostaglandin $E_2$–bisphosphonate conjugate minimized the side effects of prostaglandin $E_2$. Diclofenac sodium, a phenyl acetic acid derivative nonsteroidal anti-inflammatory drug (NSAID), was conjugated with amino methylene bisphosphonate using bromoacetic acid as a linkage between diclofenac sodium and the bone-targeting moiety (amino methylene bisphosphonate) to avoid the side effects of NSAID (Hirabayashi et al. 2001). The mechanism of diclofenac sodium release from the conjugate was a nonspecific hydrolysis. Salmon calcitonin was also conjugated with thiol-modified bisphosphonate using sulfosuccinimidyl-4 [N-Maleimidomethyl] cyclohexane-1 carboxylate (Bhandari et al. 2010). Osteoprotegerin was conjugated with thiol-modified bisphosphonate alendronate and investigated in *in vivo* models using rats and mice (Pan et al. 2008; Doschek et al. 2009). Osteoprotegerin is a naturally occurring molecule produced by osteoblasts Wnt/$\beta$-catenin signaling pathway and it is a potent RANKL inhibitor preventing osteoclastogenesis. When osteoprotegerin-thiol-modified bisphosphonate conjugate was administered intravenously to rats, the overall bone deposition of bisphosphonate-targeted osteoprotegerin was formed twice that of free osteoprotegerin. In a model of osteoarthritis rats, targeted osteoprotegerin was accumulated in the matrix and femur 6× that of free osteoprotegerin, which demonstrated bisphosphonates bone-targeting abilities (Doschek et al. 2009). Small molecules like TNP470 (used in the cancer therapy) have been conjugated to alendronate (Segal et al. 2009, 2011). TNP470 conjugated with alendronate through an enzymatic-cleavable linker has already shown its efficacy in the reduction of the tumor size in a human xenogeneic osteosarcoma mouse model (Segal et al. 2011).

### 9.2.3 ACIDIC OLIGOPEPTIDES

Recent research on various acidic oligopeptides demonstrate playing of significant roles in the growth and dissolution of bone appetite (Wasserman et al. 1977). Osteocalcin (Hoang et al. 2003), sialoprotein (Midura et al. 2004), osteopontin (Steitz et al. 2002), dentin-matrix protein (Tartaix et al. 2004), and so on, are the examples of proteins which are able to bind strongly with the calcium ions and the mineral faces of the bone and teeth. These acidic oligipeptides are reported for selective bone-targeting drug delivery (Ishizaki 2009).

Osteocalcin is one of the most abundant noncollagenous calcium-binding proteins in bone which is highly conserved among vertebrates. It consists of three vitamin K-dependent γ-carboxylated glutamic acid residues at positions 17, 21, and 24 (Hoang et al. 2003). The serum concentration of osteocalcin is related to bone metabolism rate. Conventionally, osteocalcin is used as bone formation marker instead (Ivaska et al. 2004). The porcine osteocalcin crystal structure was reported to have a negatively charged protein surface and possess three glutamic acid residues (Hoang et al. 2003). These glutamic acid residues and aspartic acid residues were shown to coordinate five calcium ions in a spatial orientation, which is complementary to the calcium ions in HAp crystal lattice. Similar results were also evidenced in the research with bovine osteocalcin (Dowd et al. 2003).

Bone sialoprotein is a highly modified anionic phosphoprotein that exhibits strong affinity to HAp (Tye et al. 2003). Bone sialoprotein is expressed exclusively in mineralized connective tissues. It has several strings of AA. Modeled after sialoprotein, glutamic acid and aspartic acid oligopeptides 4–10 AA long facilitate a more biocompatible option, when adequate bone turn over is needed (Ishizaki 2009). Like bisphosphonates, the HAp-binding capabilities of sialoprotein are retained after conjugation to a nanocarrier via the peptide's α-amino group, and the nucleation site is the point of binding between bone sialoprotein and HAp surface as the crystal form (Ouyang et al. 2009).

Osteopontin is also a multifunctional protein that consists of several structured domains including an integrin-binding adhesive domain and aspartic acid rich calcium-binding domains (Steitz et al. 2002). It can efficiently bind to the HAp surface. A novel bone-targeting acidic oligopeptide moiety was also developed on the hexapeptide of the aspartic acid (Kasugai et al. 2000). Through the solid-phase peptide synthesis, fluorescein was conjugated to hexapeptide of aspartic acid. The biodistribution as well as histomorphometry analysis of this novel bone-targeting delivery system supported the osteotropicity of the systems. Only 2% of the conjugate was found to deposit in bone.

### 9.2.4 ESTRADIOL ANALOGS

Estradiol analogs are used as bone-targeting moieties when estrogen replacement therapy is required. It is mainly required for the treatment of osteoporosis related to low estrogen levels (Yokogawa et al. 2001). Estradiol analogs were studied to localize in bone tissues. However, these were found for lacking sufficient estrogenic properties. Various bone-targeting systems are also being developed by attaching calcium chelators to estradiol moieties through succinyl or carboxyethyl linker (Neale et al. 2009). For the further improvement in bone targeting, a series of phosphate esters of the carboxyethyl-linker-containing estradiol analog, estratriene, was developed and found to improve bone-targeting potential like tetracyclines (Nasim et al. 2010a, b). Recently, the small molecular anticancer drug, 5-flurouracil, was conjugated to oligopeptides of aspartate and applied systemically in mice (Ouyang et al. 2011). The tested prodrug was shown to search the femur, and derivatives composed of hexa-aspartates sequences were observed to bind more effectively to HAp than tetracyclines in *in vitro* study.

### 9.2.5 MISCELLANEOUS

Polymalonic acid is used as a bone-targeting moiety as it has the ability of string affinity to bind with bone apatite for the therapy of osteoporosis, malignant hypercalcemia, Paget's disease, and so on (Thompson et al. 1989). Estradiol conjugated with polymalonic acid through the urea bond between 17-α OH of estradiol and primary amine of polymalonic acid. The conjugate showed some antiosteoporosis effect with reduced estradiol side effects.

N-methacrylomido salicylic acid and N-acryloyl aspartic acid have also been demonstrated to have some extent of affinity to HAp (Elvia and Roman 1997; Hou et al. 2000). However, their binding capabilities were rather found to be low as compared to hexapeptide of aspartic acid or bisphosphonates. Some important examples of bone-targeting drug delivery systems are summarized in Table 9.1.

## 9.3 LINKERS FOR BONE TARGETING

The release of the drug molecules from the drug delivery systems containing bone-targeting moiety is important in terms of the bioactivity of the drug molecules. However, only very few drug delivery systems to the skeletal systems include a release system, which is processed or cleaved at the site of action. As the drug molecules/bone-targeting moieties conjugate might remain on the bone surface after adsorption for an uncertain time, which can change from minutes to several months, the bone-targeting linkers must be stable against unspecific hydrolysis and should not interfere with the mechanical characteristics of the bone (Luhmann et al. 2012). It can be assumed that the bone-targeting drug delivery systems reach to the hard tissue and it probably remains dissolved in the bone extracellular fluid or binds to the bone surface. The release of drug molecules from the bone-targeting moieties-based systems is dependent on either reduced pH and acid-cleavable linkers or enzymatic-cleavable linkers (Luhmann et al. 2012).

**TABLE 9.1**
**Some Examples of Bone-Targeting Drug Delivery Systems**

| Drug | Bone-Targeting Moieties | Drug/Bone-Targeting Moieties Linkages | Remarks | References |
|---|---|---|---|---|
| Acetazolamide | Tetracycline | Hexanedioic acid | Purification and characterization of the bone-targeting system was not performed. Ubiquitous distribution of drug was evident *in vivo*. | Pierce Jr. and Waite (1984) |
| 3-Estrone | Tetracycline | Piperazine derivative | Piperazine attached to the amide of tetracycline with formaldehyde. Ubiquitous distribution of estrogen was evident *in vivo*. | Zheng and Weng (1997) |
| β-Estradiol-3-benzoate | Tetracycline | Succinic acid | The structure of the prepared drug/bone-targeting moiety conjugate was well characterized. *In vivo* results were absent. Ubiquitous distribution of estrogen was evidenced *in vivo*. Some side effects were seen. | Orme and Labroo (1994) |
| Methotrexate | Bisphosphonate | Amide bond with γ-COOH of Glu in methotrexate | Increased activity against metastatic tumor to reduce side effects. Antitumor activity was not clear. | Hosain et al. (1996) |
| Diclofenac sodium | Amino methylene bisphosphonate | Bromoacetic acid | Attempted to limit the side effects of diclofenac sodium. | Hirabayashi et al. (2001) |
| Prostaglandin E2 (PGE2) | Bisphosphonate | BrCH2COBr between 15-OH of PGE2 and -SH of bisphosphonate | Side effects of PGE2 were seen *in vivo*; effect was marginal. | Gil et al. (1999) |
| TNP40 | Alendronate | N-(2-hydroxypropyl) methacrylamide | *In vivo* evaluation was done after intraperitoneally xenograft-SCID mice model (osteosarcoma). The drug release was controlled by Cathepsin K-cleavable tetrapeptide. | Segal et al. (2011) |

*(Continued)*

**TABLE 9.1 (Continued)**
**Some Examples of Bone-Targeting Drug Delivery Systems**

| Drug | Bone-Targeting Moieties | Drug/Bone-Targeting Moieties Linkages | Remarks | References |
|------|------------------------|--------------------------------------|---------|------------|
| Salmon calcitonin | Thio-modified bisphosphonate | Sulfosuccinimidyl-4 [N-malein imidomethyl]-cyclohexane-1 carboxylate | *In vitro* evaluation was performed. | Bhandari et al. (2010) |
| Osteoprotegerin | Thio-modified bisphosphonate | Thiol-linker, disulfide-linked | *In vivo* evaluation was performed in rat model of osteoarthritis. | Doschek et al. (2009) |
| Osteoprotegerin | Alendronate | N-(2-hydroxypropyl) methacrylamide | Biodistribution and pharmacokinetic parameters were computed. Cathepsin K-cleavable tetrapeptide-linked drug release was experienced. | Pan et al. (2008a, 2008b) |
| Bovine serum albumin | Thio-modified bisphosphonate | N-succinimidyl-3(2-pyridyldithio) propionate, disulfide linked | Stability differences *in vitro* were not observed *in vivo* rat model. | Wright et al. (2006) |
| Estradiol | Polymalonic acid | Urea bond between 17α-OH of estradiol and primary amine of polymalonic acid | No biological effect was explored, ubiquitous distribution of estrogen was noticed. | Thompson et al. (1989) |
| Estradiol | Hexaapeptide of L-aspartic acid | Succinate between 17β-OH and amine of peptide | Some antiosteoporosis effect and limited estradiol side effects were seen *in vivo* with ovariectomized mice. | Yokogawa et al. (2001) |
| 5-Flurouracil | L-aspartate 6 | Carbamate, succinate | *In vivo* evaluation was done and profound systemic effect was observed. | Ouyang et al. (2011) |
| Prostaglandin $E_1$ ($PGE_1$) | D-aspartate 8-FITC | N-(2-hydroxypropyl) methacrylamide | Plasma stability was observed in rat, mouse, and human. The drug release occurred via Cathepsin K-cleavable tetrapeptide linkers. | Pan et al. (2007) |

### 9.3.1 REDUCED pH AND ACID-CLEAVABLE LINKERS

The bone mineral is mainly composed of crystalline HAp and the only process which has been proven to solubilize the inorganic material in the biological environment is acidic dissolution (Wang et al. 2005). Furthermore, reduced pH or acidic environment is correlated with the osteoclast activity and this may be deployed for the release of drug molecules that are immobilized by the pH linkers. This type of drug release mechanism is not specific to bone. This has been investigated for the tumor therapy, as the tumoral extracellular milieu usually exhibits acidic conditions with pH values down to 6.5. Moreover, the extracellular milieu in inflammatory tissue is acidic (Luhmann et al. 2012). Thus, the pH responsiveness is one of the major mechanisms for bone-targeting drug delivery, as endosomes display a pH between 5.5 and 6 rapidly after endocytosis with further acidification on the lysosomal pathway. Several acid-cleavable linkers have been well investigated in the development of bone-targeted drug delivery systems (Figure 9.3) (Kratz et al. 1999).

Hydrazones are one of the most extensively employed classes of acid-cleavable linkers, which are formed by the reactions between hydrazine and ketone in the presence of a catalyst (Luhmann et al. 2012). These generally degrade within minutes at moderate pH values of 5 in solutions. Hydrazones have been employed for the linking of doxorubicin (an anticancer drug) to monoclonal antibodies (Greenfield et al. 1990). In this approach, the 13-keto position of the anthracycline was used as the site of attachment and this type of design was found to allow the release of unmodified doxorubicin in a pH range from 4.5 to 6.5. In addition to this immune conjugate, the hydrazone bond has been utilized in the synthetic polymer-based drug delivery systems. Doxorubicin was conjugated to HPMA copolymer backbone via a low pH-cleavable hydrazone bond, and it was evaluated for the *in vitro* drug release and *in vivo* antitumor activity (Etrych et al. 2001). A majority of the conjugated doxorubicin was released within 48 hours at pH 5. The *in vitro* doxorubicin release correlated well with the *in vivo* inhibition of tumor growth. This strategy showed

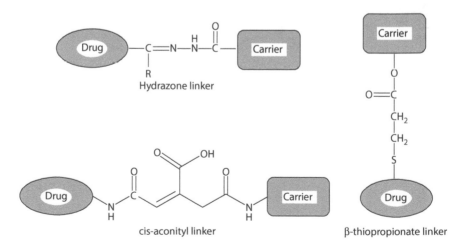

**FIGURE 9.3** Structures of some reduced pH and acid-cleavable linkers.

the applicability of hydrazones to bone targeting. The release pattern and kinetics do alter when the drug delivery systems are adsorbed on the inorganic substances in comparison to solution.

Esters are also investigated to undergo acid-cleavable hydrolysis both in basic as well as acidic solutions. The kinetics of ester hydrolysis is strongly dependent on the chemical environment (Shoenmakers et al. 2004). Examples of esters which have been investigated as acid-cleavable linkers are β-thiopropionate, orthoesters, diorthoesters, and so on (Luhmann et al. 2012). β-Thiopropionate is recognized as the most prominent example of acid-cleavable linkers in the development of bone-targeting drug delivery, which generally results from the Michael addition of thiols to acrylic acid esters. In an investigation, oligodeoxynucleotides coupled to PEG via β-thiopropionate was investigated (Oishi et al. 2005). It exhibited stability at pH 7.4 and 97% release at pH 5.5 after 24 h. The most promising drug delivery approaches are orthoesters and diorthoesters that are well known for the most acid-labile characteristics available to date. Moreover, their pH sensitivity (i.e., rate and extent of hydrolysis) can be tuned by the modification of the adjacent moieties. Five-membered orthoesters are found as promising linkers inhibiting complete stability at neutral pH. However, they complete hydrolysis at pH 4.5–5.5 within 100 minutes (Bruyere et al. 2010). The diorthoesters were also investigated for the bone-targeting drug delivery. An acid-labile conjugate of PEG and disteroyl glycerol via a diorthoester linkage was investigated within the sheddable coating of liposomes (Guo and Szoka Ja 2001). At neutral pH, this conjugate was found stable for 3 h at 37°C and the complete hydrolysis occurred within 1 h at pH 4.0–5.0. The *in vivo* evaluation on mice demonstrated liposome's half-life of 194 minutes. Cis-aconityl linkage is also used in the development of polymer–drug conjugates for bone-targeting drug delivery (Shen and Ryser 1981). These types of linkages can be formed in two steps. First, the drug is reacted with cis-aconityl anhydride to form a β-unsaturated amide bond. Second, the free γ-carboxylate group of cis-aconityl is coupled to a pendant amino group on the polymer backbone. Inspired by the neighbor group effect of hydrolysis in phthalamic acid, also cis-aconityl anhydrides (D'Souza and Topp 2004) as well as carboxylated dimethyl maleic anhydride (Rozema et al. 2007) were employed for linkage of drug molecules to polymers. Cis-aconityl linkers are cleaved at moderate pH values (Frachet and Gillies 2004); but these do not present complete drug release, most probably because of cis–trans isomerization about the double bond between the hydrolytically susceptible amide bond and the carboxylate bond (Zloh et al. 2003). In an investigation, the cis-aconityl linkage between the HPMA polymer backbone and doxorubicin was shown to be cleavable under the acidic pH at a slower rate than that of the hydrazone-containing conjugate (Ulbrich et al. 2003). In most of the cases of reduced pH-cleavable linkage system-based deliveries, susceptibility to hydrolysis of the linkers can be significantly influenced by the local chemical environment of the acid-labile group. In addition, the targeting approach may influence the drug release patterns as demonstrated for the drug release from the HAp-adsorbed polymer–drug conjugates. Therefore, the combination of the targeting strategy, pharmacology of the drug of interest, availability of functional groups for chemical bonding, and practicability of the coupling protocol are eventually decisive factors for selecting suitable linker chemistry from case-to-case.

### 9.3.2 Enzymatic-Cleavable Linkers

#### 9.3.2.1 Cathepsin K-Specific Peptide Linkage

Cathepsin K, a 24 kDa cysteine protease of the papain superfamily, is extensively generated in the bone-resorbing osteoclast (Zhao et al. 2009). It is expressed as an inactive preproenzyme, which is transported to the lysosomal compartments of the osteoclasts. As its levels of expression are high in the osteoclasts and low in other organs and tissues, location of it might be another important issue for the bone-targeting sites with high turnover rate (Luhmann et al. 2012). Enhanced cathepsin K expression is relevant in pathological situations, such as osteoporosis, as cathepsin K-cleavage systems are recognized as sensitive serum markers of bone resorption in osteoporotic patients (Costa et al. 2011). Deficiency of cathepsin K activity in osteo-clasts causes psychodynamic, which is recognized as a rarely seen inherited disease with an osteopetrotic phenotype (Hou et al. 1999). Therefore, it can be said that cathepsin K is the major enzyme candidate which is responsible for the breakdown of the organic matrix of bone.

A drug conjugate based on HPMA copolymers, harboring a bisphosphonate-based bone-targeting moiety was designed, where this was combined covalently via a cathepsin K-cleavable tetrapeptide motif to the antitumor drug, TNP 470 (Segal et al. 2011). In another study, a prostaglandin $E_1$ ($PGE_1$)-HPMA copolymer conjugate was developed (Pan et al. 2006). The bone-targeting moiety was attached covalently via a cathepsin K-responsive spacer. The result of $PGE_1$ indicated the stability of the corresponding polymer. In a study, biodegradable hydrogels were developed by incorporating short peptide fragments of type I collagen ($\alpha$-1) with the sequence Gly-Pro-Ser-Gly, which was specifically cleavable between serine and glycine by recombinant cathepsin K and by osteoblast cell lines, *in vitro* (Hsu et al. 2011).

#### 9.3.2.2 Matrix Metalloprotinases (MMPs)-Specific Peptide Linkage

MMPs are a family of 24 human zinc-containing peptidases that play an important role in bone resorption. They are recognized as integral to the bone remodeling process (Anderson et al. 2004). MMPs are produced as latent zymogens and must be proteolytically processed in the tissues to be enzymatically active. These are responsible for degrading all compounds of the extracellular matrix and other proteins at neutral pH. MMPs are expressed by a variety of cells either as soluble and/or as membrane-bound form. These are the products which are achieved through the degradation of collagen type I helices (Krane and Inada 2008). MMPs participate in an important function during bone remodeling, as these release an endogenous pool of growth factors being stored within the organic phase of mineralized tissues and by the degradation of binding proteins. These also play a pivotal role in osteoclast motility. The MMP-13 acts as the limiting factor for bone resorption (Anderson et al. 2004). Expression of protein mRNA levels of MMP-13 was found to enhance in ovariectomized rat osteoblasts and was down-regulated after the estrogen treatment (Li et al. 2004). The membrane-bound MMP-14 may be an interesting target for the design of responsive linkers used in drug delivery to the bone as migration of active osteoclasts, as prerequisite of osteolysis, might be deployed in that way where antiresorptive drugs could be targeted and released at the same time (Luhmann et al. 2012).

## 9.4 POLYMERIC CARRIERS FOR BONE-TARGETING DRUG DELIVERY

Currently, the conjugation of various bone-targeting ligands to polymeric backbone is considered an obvious advantage over small molecular drug delivery carriers (Duncan 2003). Polymeric bone-targeting drug delivery systems are dependent on the ultimate fate of the polymeric carriers. As the drug delivery systems arrive at bone-targeting sites, both the targeting moiety and the loaded drug may be detached from the polymeric systems to allow their clearance from the body (Kopeček et al. 2000). Moreover, the modification of polymer size has a profound influence on the blood circulation time and also on the biodistribution in the whole body. The bone-targeting polymeric drug delivery carrier systems possess several drug release modes and these range from passive diffusion by various mechanisms like acidic cleavage or enzymatic cleavage of drug-polymer linkers (Low and Kopeček 2012). The molecular weight of polymers and the dimension of bone-targeting polymeric carriers have a profound effect on the efficacy of the bone-targeting drug delivery systems. The higher-molecular weight polymers possess a larger circulating half-life and higher dispositions in the diseased bone sites. The careful selection of molecular weights with appropriate targeting ligand concentration is important and essential to ensure the biocompatibility as well as a high level of disposition in the affected area of the diseased bone (Low and Kopeček 2012). Several polymers are investigated for the development of bone-targeting polymeric systems.

Polyethyleneimine (PEI) and poly (L-lysine) (PLL) are the cationic polymers widely used in the development of bone-targeting drug delivery systems. These two polymers were investigated to incorporate 2-(3-mercaptoproyl sulfanyl)-ethyl-1, 1-bisphosphonic acid using hetero-bifunctional reagents (Zhang et al. 2001). The *in vitro* binding to HAp was found similar or lower after conjugation with bisphosphonates. Both PEI and PLL were found capable of binding almost 100% to HAp. The *in vivo* results in rat model also exhibited significant difference in the bone affinity of modified and unmodified polymers (Zhang et al. 2001). However, PEI and other cationic polymers are associated with toxicity and this was also examined in *in vivo* experiments (Aravindan et al. 2009; Wen et al. 2009).

Poly HPMA is one of the extensively studied polymer candidates used in the development of bone-targeting drug delivery systems. The biodistribution of poly HPMA and its bone-targeting capacity has already been reported (Wang et al. 2006; Pan et al. 2008a, b). The important advantages of poly HPMA include a low toxicity profile and the ability to control the molecular weight and biodistribution through reversible addition fragmentation chain transfer (RAFT) polymerization method (Luo et al. 2011; Yang et al. 2011). In addition, the RAFT polymerization method permits the incorporation of methacrylated drug derivatives into the polymer through copolymerization as well as eliminating the opportunity of having residual (unreacted) groups remaining on the polymer backbone after the postpolymerization modification (Pan et al. 2011; Yang et al. 2011). Poly HPMA also has been investigated to design bone-targeting micelles (Talelli et al. 2010; Krimmer et al. 2011) and dendrimers (Wang et al. 2000).

Poly (lactic-co-glycolic acid) is well studied as bone-target drug delivery carriers. PEG-PLGA block copolymers were intermixed with alendronate-functionalized PLGA to prepare surface-modified nanoparticles (Choi and Kim 2007). The adsorption on HAp revealed decreasing adsorption correlated with a decreasing concentration of alendronate at the surface. The increase in molecular weight of PEG (such as 550, 750, and 2000) resulted in decreased adsorption apparently due to the shielding of alendronate moieties by lower PEG chains. Alendronate-functionalized nanospheres demonstrated *in vitro* blood compatibility and the absence of toxicity (Cenni et al. 2008).

Currently, several liposome-based polymeric bone-targeting drug delivery systems are also being investigated by various research groups. The incorporation of bisphosphonate-derived liposomes into collagen/HAp scaffolds decreased the rate of model drug release from these scaffold systems (Wang et al. 2011). This system also exhibited sustained drug release pattern for the bone regeneration. Various liposomal formulations were also formulated for the treatment of rheumatoid arthritis (van der Hoven et al. 2011).

Bovine serum albumin nanoparticles were also investigated for the bone-targeting delivery of bone morphogenic protein-2 (BMP-2) (Wang et al. 2010). These nanoparticles were surface coated with PEI-PEG-thiol bisphosphonate. These nanoparticles were not much effective for bone targeting after intravenous administration. However, these were found to be useful in the localized bone delivery of BMP-2 for bone repair and bone regeneration.

A PEG-based dendrimer from hetero-functional PEG was formulated and this system was decorated with $H_2N$-PEG-$\beta$-Glu-$(\beta$-Glu$)_2$-$(COOH)_4$ exposing four carboxylic groups for the attachment of alendronate and/or paclitaxel (Clementi et al. 2011). This system was able to achieve a very high percentage of HAp binding following conjugation to paclitaxel. This conjugate was designed for the treatment of bone neoplasms and employed pH-sensitive linker for the rapid release of the drug. The *in vitro* release demonstrated that the drug release was highly effective resulting in conjugated paclitaxel having a similar $IC_{50}$ as paclitaxel alone, apparently due to faster drug release at physiological pH.

## 9.5  BONE-TARGETING THERAPY

### 9.5.1  OSTEOPOROSIS AND OTHER LOW BONE MINERAL DENSITY-RELATED DISEASES

Currently, osteoporosis and other low bone mineral density-related diseases represent major public health problems, especially considering the old age population worldwide and among the most common degeneration diseases (Handa et al. 2008; Luhmann et al. 2012). Osteoporosis is a disease pattern within which the decline in bone mass is beyond what is normal as a function of sex, race, and height (Golob and Laya 2015). It typically leads to reduced bone strength and in turn a higher probability to fracture (Cummings and Melton 2002). The symptoms of this type of fracture include pain, and particular complications may arise for the fractures of spine and hip, mainly. The osteoporosis treatment targets at decreasing the fracture rate through increasing bone strength by enhancing the mineral bone density and bone quality (Handa et al. 2008; Golob and Laya, 2015). The current therapeutic intervention in osteoporosis aims at the resorptive events including bisphosphonates. Cathepsin K inhibitors, RANKL inhibitors, strontium ranelate calcitonin or selective estrogen receptor modulators (SERM) and anabolic agents, such as parathormone fragments, sclerostin inhibitors (in clinical phase) are used in the treatment of osteoporosis and other low bone mineral density-related diseases (Luhmann et al. 2012).

In antiresponsive therapy, bisphosphonates are generally charged in spite of their complicated dosing regimens challenging the patient compliance and from a safety perspective by gastric side effects. This can be avoided by using injectable bisphosphonates. Bisphosphonates generally reduce the osteoclastic activity through the inhibition of farnesyl diphosphate synthase and this leads to a loss in guanosine triphosphate (GTP) binding proteins, which are key factors to osteoclastic activity interference within mevalonate pathway to halt osteoclastic activity and therefore bone resorption (Gittens et al. 2005). Osteoclasts interfere very strongly with osteoblasts and the apoptosis of osteoblast as a result of bisphosphonate exposure ultimately shows osteoblastic activity and therefore bone formation. Several molecules of bisphosphonates such as alendronate, zoledronate, ibandronate, risedronic acid, and so on, are available in the market. SERMs such as raloxifene, bazedoxifene, and so on are also available in the market and are used in the treatment of osteoporosis and other low bone mineral density-related diseases.

Various anabolic agents such as prostaglandins, statins, parathormone, and so on are also employed in the treatment of osteoporosis and other low bone mineral density-related diseases. Several clinical studies in animals and humans after systemic and local administration of prostaglandins (mainly $PGE_1$ and $PGE_2$) caused substantial bone formation (Wang et al. 2005; Luhmann et al. 2012). Statins affect the bone turn over by interacting with the mevalonic acid pathway and down-regulating protein prenylation. These also induce vascular endothelial growth factor (VEGF) that contributes to osteoblast differentiations (Wong and Rabie 2005; Uzzan et al. 2007). The systemic delivery of bone-targeting statins may increase local bone concentration enough for bone mineralization and bone turn over (Gittens et al. 2005; Masuzaki et al. 2010).

### 9.5.2 Osteomyelitis

Osteomyelitis is a bone disease caused by bacterial infection of the bone medullary cavity, cortex, and/or periosteum leading to bone loss, which could cause serious impact on the health and working capacity of the patients (Nayak et al. 2013). Most common bacterial isolates in patients with chronic osteomyelitis are *Staphylococcus aureus, Pseudomonas aeruginosa, Staphylococcus epidermidis,* and *Proteus mirabilis* (Castro et al. 2005; Nayak et al. 2013). The treatment of osteomyelitis usually requires antimicrobial therapy. Therefore, local delivery of antimicrobial therapy for the surgical debridement of the infected bone site is recommended (Nayak et al. 2011b). The conventional prolonged systemic antibiotic therapy is insufficient for achieving high local tissue concentration of antibiotics at the diseased bone site of the infected bone as bones are a poorly perfused organ (Nayak et al. 2013). An alternative approach for the treatment of osteomyelitis would be to develop bone-targeting systems releasing antimicrobial agents that can be delivered systemically (Karau et al. 2013). The chemical conjugation of antimicrobial agents to bone-targeting moieties has been researched to yield bone-targeting delivery systems of antimicrobial agents and this approach can be useful in the treatment of osteomyelitis. Pradama Inc. has developed a bone-targeting system of vancomycin (BT2-peg2-vancomycin) for the treatment of osteomyelitis (Pierce et al. 2013). In an investigation, Karau et al. (2013) have investigated *in vitro* and *in vivo* activity of BT2-peg2-vancomycin with that of conventional vancomycin in the treatment of methicillin-resistant *Staphylococcus aureus* osteomyelitis (Karau et al. 2013). They have studied bone affinity using a HAp-binding assay, assessed the *in vitro* antimicrobial susceptibility of 30 methicillin-resistant *Staphylococcus aureus* isolates, vancomycin, and BT2-peg2-vancomycin in a rat experimental osteomyelitis model. They also found that BT2-peg2-vancomycin exhibited its efficacy in the treatment of methicillin-resistant *Staphylococcus aureus* osteomyelitis. In an investigation, the micelles self-assembled from poly (lactic acid-co-glycolic acid)-block-poly (ethylene glycol) alendronate copolymers for bone-targeted delivery of vancomycin were developed to treat osteomyelitis (Cong et al. 2015). Vancomycin release patterns from these micelles revealed that the conjugation of alendronate to the surface of micelles did not cause any adverse effect on the drug-loading capacity of the bone-targeting system. The cytotoxicities of these vancomycin-releasing bone-targeted delivery micelles and the blank micelles was assessed via 3-(4, 5-dimethylthiazol-2-yl)-2, 5-diphenyltetrazolium bromide assay toward the rat bone marrow stromal cells (rBMSCs) and human embryonic hepatocytes (L02 cells). The results demonstrated that the vancomycin-releasing bone-targeted delivery micelles is safe for the potential treatment of osteomyelitis. The *in vitro* affinity of these micelles to the HAp was also observed in this study. The antibacterial effect of vancomycin-releasing bone-targeted delivery micelles was evaluated against *Staphylococcus aureus* (the main pathogenic bacteria in osteomyelitis). The results showed that the vancomycin-releasing micelles were able to effectively inhibit the growth of *Staphylococcus aureus*, indicating the potential use of poly (lactic acid-co-glycolic acid)-block-poly (ethylene glycol) alendronate micelles for the bone-targeted delivery of vancomycin. In a study, quinolones such as levofloxacin and norfloxacin were conjugated with acidic oligopeptides, L-Asp hexapeptides and investigated as bone-targeting drug delivery carriers after systemic administration for the treatment of osteomyelitis (Takahashi et al. 2008). In this work, levofloxacin was conjugated with L-Asp hexapeptide via a glycolyl ester, and norfloxacin was conjugated via a glycolyl amide and subsequent succinate ester. The bone-targeting properties and therapeutic effectiveness of these acidic oligopeptide-conjugated levofloxacin and norfloxacin were evaluated in a mouse model of osteomyelitis created by inoculating *Staphylococcus aureus* into the tibia of mice. It was found that the conjugated drugs selectively distributed to the bone after intravenous injection with reaching concentrations up to 100-fold those of nonconjugated drugs. The antimicrobial effect of conjugated levofloxacin was observed to persist for at least 6 days after injection, whereas the effect of nonconjugated levofloxacin was found to be temporary. On the other hand, neither conjugated nor nonconjugated norfloxacin was found effective against *Staphylococcus aureus*.

### 9.5.3 Bone Metastases

Bone metastases are manifestations of severe cancers representing a largely unresolved problem on oncology. Paget's 1889 "seed and soil" theory remains the generally accepted theory of bone metastases and this theory simply reasons that as metastases cells (the seeds) enter into the blood stream, these will be carried everywhere. However, these will only attach and grow where the conditions are right (the soil) (Clines and Guise 2005). The treatment of bone metastases often involves small molecular therapeutics that targets various osteoclast mechanisms. Several osteoclast-targeting small molecules have been investigated for the treatment of bone metastases. Among them, reveromycin A, avb₃, C-Src inhibitors, cathepsin K inhibitors, methyl gerfelin, integrin inhibitors, bisphosphonates, and so on, have demonstrated the reduction of bone pain and some degree of bone metastases (Kavatani 2009).

## 9.6 CONCLUSION

Though the bone-targeting drug delivery systems have been investigated from 1984, its development is still in an immature state. Currently, bone-targeting approach is recognized as an emerging strategy for the delivery of a wide range of drug molecules to the bone to treat various bone diseases such as osteoporosis, osteoarthritis, osteomyelitis, bone cancer, and so on. The designing of novel bone-targeting moieties, the drug-releasing mechanisms by the bone-targeting systems, the choice of the suitable bone-targeting therapeutic molecules (drugs) and also optimizing the performances of bone targeting by the use of carriers are the most important concerns in the bone-targeting drug delivery system research and development. In recent years, many new drug molecules used in the treatment of various bone diseases have been synthesized. However, their clinical use has been complicated because of several toxicity hazards. Therefore, more work on the development, designing, and evaluation of suitability of newer and effective bone-targeting drug delivery systems for the treatment of bone diseases like osteoporosis, osteoarthritis, osteomyelitis, bone cancer, and so on, are necessary to advance this approach into effective and safest clinical applications. The successful advancement in this growing approach can yield chemically relevant bone-targeting drug delivery systems, which can be able to reduce pain and improve the quality of life for millions of people.

## REFERENCES

Albert, A. and Rees, C.W. 1956. Avidity of the tetracyclines for the cations of metals. *Nature* 177:433–34.

Anderson, T.L., del Carmen Ovejero, M., Kirkegaard, T. et al. 2004. A scrutiny of matrix metalloproteinases in osteoclasts: Evidence for heterogenecity and for the presence of MMPs synthesized by other cells. *Bone* 35:1107–19.

Aravindan, L., Bicknell, K.A., Brooks, G. et al. 2009. Effect of acyl chain length on transfection efficiency and toxicity of polyethylenimine. *Int J Pharm* 378:201–10.

Ayukawa, Y., Yasukawa, E., Moriyama, Y. et al. 2009. Local application of statin promotes bone repair through the suppression of osteoclasts and enhancement of osteoblasts at bone-healing sites in rats. *Oral Surg Oral Med Oral Pathol Oral Radiol Endod* 107:336–42.

Bhandari, K.H., Newa, M., Uludağ, H. et al. 2010. Synthesis, characterization and in vitro evaluation of a bone targeting delivery system for salmon calcitonin. *Int J Pharm* 394:26–34.

Bruyere, H., Westwell, A.D., and Jones, A.T. 2010. Tuning the pH-sensitives of orthoester based compounds for drug delivery applications by simple chemical modification. *Bioorg Med Chem Lett* 20:2200–30.

Castro, C., Evora, C., Baro, M. et al. 2005. Twomonth ciprofloxacin implants for multibacterial bone infections. *Eur J Pharm Biopharm* 60:401–06.

Cenni, E., Granchi, D., Avnet, S. et al. 2008. Biocompatibility of poly (D,L-lactide-co-glycolide) nanoparticles conjugated with alendronate. *Biomaterials* 29:1400–11.

Choi, S.-W. and Kim, J-H. 2007. Design of surface modified poly (D,L-lactide-co-glycolide) nanoparticles for targeted delivery to bone. *J Control Release* 122:24–40.

Chopra, I. and Roberts, M. 2001. Tetracycline antibiotics: Mode of action, applications, molecular biology, and epidemiology of bacterial resistance. *Microbiol Mol Biol Rev* 65:232–60.

Clementi, C., Miller, K., Mero, A. et al. 2011. Dendritic poly(ethylene glycol) bearing paclitaxel and alendronate for targeting bone neoplasms. *Mol Pharm* 8:1063–72.

Clines, G.A. and Guise, T.A. 2005. Hypercalcemia of malignancy and basic research on mechanisms responsible for osteolytic and osteoblastic metastases to bone. *Endocr Relat Cancer* 12:549–83.

Cong, Y., Quan, C., Liu, M. et al. 2015. Alendronate-decorated biodegradable polymeric micelles for potential bone-targeted delivery of vancomycin. *J Biomater Sci Polym Ed* 26:629–43.

Costa, A.G., Cusano, N.E., Silva, B.C. et al. 2011. Cathepsin K: Its skeletal actions and role as a therapeutic target in osteoporosis. *Nat Rev Rheumatol* 7:447–56.

Cummings, S.R. and Melton, L.J. 2002. Epidemiology and outcomes of osteoporotic fractures. *Lancet* 359:1761–67.

D'Souza, A.J. and Topp, E.M. 2004. Release from polymeric prodrugs: Linkages and their degradation. *J Pharm Sci* 93:1962–79.

Doschek, M.R., Kucharski, C.M., Wright, J.E. et al. 2009. Improved bone delivery of osteoprotogenin by bisphosphonate conjugation in a rat model of osteoarthritis. *Mol Pharm* 6:634–40.

Dowd, T.L., Rosen, J.F., Li, L. et al. 2003. The three dimensional structure of bovine calcium ion-bound osteocalcin using $^1$H NMR spectroscopy. *Biochemistry* 42:7769–79.

Duncan, R. 2003. The dawning era of polymer therapeutics. *Nat Rev Drug Discov* 2:347–60.

Elvia, C. and Roman, J.S. 1997. Synthesis and stereochemistry of isomeric methacrylic polymers derived from 4- and 5- amino salicylic acids. *Polymers* 38:4743–50.

Etrych, T., Jelinkova, M., Rihova, B. et al. 2001. New HPMA copolymer containing doxorubicin bound via pH-sensitive linkage: Synthesis and preliminary in vitro and in vivo biological properties. *J Control Release* 73:89–102.

Fancis, M.D. and Fogelman, I. 1987. 99mTc diphosphonate uptake mechanism on bone. In *Bone Scanning in Clinical Practice*, ed, I. Fogelman, 7–17. New York: Springer-Verlag.

Fleisch, H., Raszeka, A., Rodan, G.A. et al. 2002. Bisphosphonates-mechanism of action. In *Principles of Bone Biology*, ed, J.P. Bilezikian., L.G. Raisz, G.A. Rodan, 1361–1385. San Diego: Academic Press.

Frachet, J.M.J. and Gillies, E.R. 2004. Development of acid-sensitive copolymer micelles for drug delivery. *Pure Appl Chem* 76:1295–07.

Fujisaki, J., Tokunaga, Y., Takahashi, T. et al. 1995. Osteotropic drug delivery system (ODDS) based on bisphosphonic prodrug. I: Synthesis and in vivo characterization of osteotropic carboxyfluorescein. *J Drug Target* 3:273–82.

Gil, L., Han, Y., Opas, E. et al. 1999. Prostaglandin $E_2$-bisphosphonate conjugates: Potential agents for treatment of osteoporosis. *Bioorganic Med Chem* 7:901–19.

Gittens, S.A., Bansal, G., Zernicke, R.F. et al. 2005. Designing proteins for bone targeting. *Adv Drug Deliv Rev* 57:1011–36.

Golob, A.L. and Laya, M.B. 2015. Osteoporosis: Screening, prevention, and management. *Med Clin North Am* 99:587–606.

Greenfield, R.S., Kaneko, T., Daues, A. et al. 1990. Evaluation in vitro of adriamycin immunoconjugates synthesized using an acid-sensitive hydrazone linkers. *Cancer Res* 50:6600–07.

Guo, X. and Szoka Ja, F.C. 2001. Steric stabilization of fusogenic liposomes by a low pH-sensitive PEG-diortho ester-lipid conjugate. *Bioconjug Chem* 12:241–300.

Handa, R., Ali Kalla, A., and Maalouf, G. 2008. Osteoporosis in developing countries. *Best Pract Res Clin Rheumatol* 22:693–08.

Harada, S. 2003. Control of osteoblast function and regulation of bone mass. *Nature* 423:349–55.

Hirabayashi, H. and Fujisaki, J. 2003. Bone-specific drug delivery systems: Approaches via chemical modification of bone-seeking agents. *Clin Pharmacokinet* 42:1319–30.

Hirabayashi, H., Sawamoto, T., Fujisaki, J. et al. 2001. Relationship between physicochemical and osteotropic properties of bisphosphonic derivatives: Rational design for osteotropic drug delivery system (ODDS). *Pharm Res* 18:646–51.

Hirabayashi, H., Takahashi, T., Fujisaki, J. et al. 2001. Bone-specific delivery and sustained release of diclofenac, a non-steroidal anti-inflammatory drug, via bisphosphonic prodrug based on osteotropic drug delivery system (ODDS). *J Control Release* 70:183–91.

Hoang, Q.Q., Sicheri, F., Howard, A.J. et al. 2003. Bone recognition mechanism of procine osteocalcin from crystal structure. *Nature* 425:977–80.

Hosain, F., Spencer, R.P., Couthon, H.M. et al. 1996. Targeted delivery of antineoplastic agent to bone: Biodistribution studies of technetium-99m-labelled gem-bisphosphonate conjugate of methotrexate. *J Nuclear Med* 37:105–07.

Hou, K., Torii, Y., Nishitani, Y. et al. 2000. Effect of self-etching primers containing N-acryloyl aspartic acid on dentin adhesion. *J Biomed Mater Res* 51:496–74.

Hou, W.S., Brömme, D., Zhao, Y. et al. 1999. Characterization of novel Cathepsin K mutations in the pro and mature polypeptide regions causing pycnodysostosis. *J Clin Invest* 103:731–38.

Hsu, C.W., Olabisi, R.M., Olmsted-Davis, E.A. et al. 2011. Cathepsin K-sensitive poly (ethylene glycol) hydrogels for degradation in response to bone resorption. *J Biomed Mater Res A* 98:53–62.

Ishizaki, J. 2009. Selective drug delivery to bone using acidic oligipeptides. *J Bone Mineral Metab* 27:1–8.

Ivaska, K.K., Hentunen, T.A., Vaaraniemi, J. et al. 2004. Release of intact and fragmented osteocalcin molecules from bone matrix during bone resorption in vitro. *J Biol Chem* 279:18361–69.

Karau, M.J., Schmidt-Malan, S.M., Greenwood-Quaintance, K.E. et al. 2013. Treatment of methicillin-resistant Staphylococcus aureus experimental osteomyelitis with bone-targeted vancomycin. *Springerplus* 2:329.

Kasugai, S., Fujisawa, R., Waki, Y. et al. 2000. Selective drug delivery system to bone: Small peptide (Asp)$_6$ conjugation. *J Bone Mineral Res* 15:936–43.

Kavangh, K.L., Dunford, J.E., Wu, X. et al. 2006. The molecular mechanism of nitrogen containing bisphosphonates as osteoporosis drug. *Proc Natl Acad Sci U S A* 103:7829–34.

Kavatani, K. 2009. Osteoclast targeting small molecules for the treatment of neoplastic bone metastases. *Cancer Sci* 100:1999–05.

Kopeček, J., Kopečková, P., Minko, T. et al. 2000. HPMA copolymer-anticancer drug conjugates: Design, activity and mechanism of action. *Eur J Pharm Biopharm* 50:61–81.

Krane, S.M. and Inada, M. 2008. Matrix metalloprotinases and bone. *Bone* 43:7–18.

Kratz, F., Beyer, U., and Schutte, M.T. 1999. Drug-polymer conjugates containing acid-cleavable bonds. *Crit Rev Ther Drug Carr Syst* 16:245–88.

Krimmer, S., Pan, H., Liu, J. et al. 2011. Synthesis and characterization of poly (ε-caprolactone)–block-poly [N-(2-hydroxypropyl) methacrylamide] micelles for drug delivery. *Macromol Biosci* 11:1041–51.

Lamb, H.M. and Faulds, D. 1997. Samarium 153Sm lexidronam. *Drugs Aging* 11:413–18.

Li., J., Liao, E.Y., Dai, R.C. et al. 2004. Effects of 17-beta-estradiol on the expression of interstitial collagenases-8 and -13 (MMP-8 and MMP-13) and tissue inhibitor of petalloprotease-1 (TIMP-1) in overactimized rat osteoblastic cells. *J Mol Histology* 35:723–31.

Lin, J.H. 1996. Bisphosphonates: A review of the pharmacokinetic properties. *Bone* 18:75–85.

Low, A.S. and Kopeček, J. 2012. Targeting polymer therapeutics to bone. *Adv Drug Deliv Rev* 64:1189–204.

Luhmann, T., Germershaus, O., Groll, J. et al. 2012. Bone targeting for the treatment of osteoporosis. *J Control Release* 161:198–13.

Luo, K., Yang, J., Kopečková, P. et al. 2011. Biodegradable multiblock N-(2-hydroxypropyl) methacrylamide copolymers via reversible addition–fragmentation chain transfer polymerization and click chemistry. *Macromolecules* 44:2481–88.

Marks, Jr. S.C. and Odgren, P.R. 2002. Structure and development of the skeleton. In *Principles of Bone Biology*, eds. J.P. Bilezikian., L.G. Raisz, G.A. Rodan, 3–15. San Diego: Academic Press.

Marx, R.E., Sawatari, Y., Fortin, M. et al. 2005. Bisphosphonate-induced exposed bone (osteonecrosis/osteopetrosis) of the jaws: Risk factors, recognition, prevention, and treatment. *J Oral Maxillofacial Surgery* 63:1567–75.

Masuzaki, T., Ayukawa, Y., Moriyama, Y. et al. 2010. The effect of a single remote injection of statin-impregnated poly (lactic-co-glicolic acid) microspheres on osteogenesis around titanium implants in rat tibia. *Biomaterials* 31:3327–34.

Midura, R.J., Wang, A., Lovitch, D. et al. 2004. Bone acidic glycoprotein-75 delineates the extracellular sites of future bone sialoprotein accumulation and apatite nucleation in osteoblastic cultures. *J Biol Chem* 279:25464–73.

Misra, D.N. 1991. Adsorption and orientation of tetracycline on hydroxyapatite. *Calcif Tissue Int* 48:362–67.

Myers, H.M., Tochon-Danguy, H.J., and Baud, C.A. 1983. IR absorption spectrophotometric analysis of the complex formed by tetracycline and synthetic hydroxyapatite. *Calcif Tissue Int* 35:745–49.

Nasim, S., Vartak, A., Pierce Jr. W.M. et al. 2010a. Improved and scalable synthetic route to the synthon 17-C2-carboxymethyl)-1,3,5 (10)-estratriene: An important intermediate in the synthesis of bone-targeting estrogenes. *Synthetic Commun* 40:772–81.

Nasim, S., Vartak, A., Pierce, Jr. W.M. et al. 2010b. 3-O-phosphate ester conjugates of 17-β-o-{1-[2-carboxy-(2-hydroxy-4-methoxy-3-carboxamido)anilido] ethyl} 1,3,5 (10)-estratriene as novel bone-targeting agents. *Bioorg Med Chem Lett* 20:7450–53.

Nayak, A.K., Bhattacharya, A., and Sen, K.K. 2010. Hydroxyapatite-antibiotic implantable minipellets for bacterial bone infections using precipitation technique: Preparation, characterization and in-vitro antibiotic release studies. *J Pharm Res* 3:53–59.

Nayak, A.K., Bhattacharyya, A., and Sen, K.K. 2011b. In vivo ciprofloxacin release from hydroxyapatite-based bone implants in rabbit tibia: A preliminary study. *ISRN Orthopedics* 2011:1–4.

Nayak, A.K., Hasnain, M.S., and Malakar, J. 2013. Development and optimization of hydroxyapatite-ofloxacin implants for possible bone delivery in osteomyelitis treatment. *Curr Drug Deliv* 10:241–50.

Nayak, A.K., Laha, B., and Sen, K.K. 2011a. Development of hydroxyapatite-ciprofloxacin bone-implants using Quality by Design. *Acta Pharm* 61:25–36.

Neale, J.R., Ritcher, N.B., Mertin, K.E. et al. 2009. Bone selective effect of an estradiol conjugate with a novel tetracycline-derived bone-targeting agent. *Bioorg Med Chem Lett* 19:680–83.

Niemi, R., Vepsalainen, J., Taipale, H. et al. 1999. Bisphosphonate prodrugs: Synthesis and in vitro evaluation of novel acyloxyalkyl esters of clodronic acid. *J Med Chem* 2:5053–58.

Odgreen, P.R. and Martin, T.J. 2000. Therapeutic approaches to bone diseases. *Science* 289:1508–14.

Ohnaka, K., Shimoda, S., Nawata, H. et al. 2001. Pitavastatin enhanced BMP-2 and osteocalcin expression by inhibition of Rho-associated kinase in human osteoblasts. *Biochem Biophys Res Commun* 287:337–42.

Oishi, M., Nagasaki, Y., Haka, K. et al. 2005. Lactosylated poly (-ethylene glycol)-siRNA conjugate through acid-labile beta-propiothinate linkage to construct pH-sensitive polyion complex micelles achieving enhanced gene silencing in hepatoma cells. *J Am Chem Soc* 127:1624–65.

Orme, M.W. and Labroo, V.M. 1994. Synthesis of β-estradiol-3-benzoate-17-(succinyl-12a-tetracycline): A potential bone-seeking estrogen. *Bioorg Med Chem Lett* 4:1375–80.

Ouyang, L., He, D., Zhang, J. et al. 2011. Selective bone targeting 5-flurouracil prodrugs: Synthesis and preliminary biological evaluation. *Bioorganic Med Chem* 19:3750–56.

Ouyang, L., Huang, W., He, G. et al. 2009. Bone-targeting prodrugs based on peptide drendrimers, synthesis and hydroxyapatite binding, in vitro. *Org Chem* 6:272–77.

Pan, H., Kopečková, P., Liu, J. et al. 2007. Stability in plasmas of various species of HPMA copolymer-PGE1 conjugates. *Pharm Res* 24:2270–80.

Pan, H., Kopečková, P., Wang, D. et al. 2006. Water-soluble HPMA copolymer-prostaglandin E1 conjugates containing Cathepsin K sensitive specer. *J Drug Target* 94:425–35.

Pan, H., Liu, J., Dong, Y. et al. 2008b. Release of prostaglandin E1 from N-(2-Hydroxypropyl) methacrylamide copolymer conjugates by bone cells. *Macromol Biosci* 8:599–05.

Pan, H., Sima, M., Kopečková, P. et al. 2008a. Biodistribution and pharmacokinetic studies of bone-targeting N-(2-hydroxy propyl) methacrylamide copolymer-alendronate conjugates. *Mol Pharm* 25:2889–95.

Pan, H., Yang, J., Kopečková, P. et al. 2011. Backbone degradable multiblock HPMA copolymer conjugates via RAFT polymerization and thiol-ene coupling reaction. *Biomacromolecules* 12:247–52.

Perrin, D.D. 1965. Biding of tetracyclines to bone. *Nature* 208:787–88.

Pham, H.H., Luo, P., Genin, F. et al. 2002. Synthesis and characterization of hydroxyapatite-ciprofloxacin delivery systems by precipitation and spray drying technique. *AAPS PharmSciTech* 3:1–9.

Pierce, Jr, W.M., Taylor, K.G., Waite, L.C. et al. 2013. Methods and compounds for the targeted delivery of agents to bone for interaction therewith. US Patent US20130029901A1.

Pierce, Jr. W.M. and Waite, L.C. 1984. Bone-targeted carbonic anhydrase inhibitors: Effect of a proinhibitor on bone resorption in vitro. *Proc Soc Exp Biol Med* 186:96–102.

Rozema, D.B., Lewis, D.L., Wakefield, D.H. et al. 2007. Dynamic poly-conjugates for targeted in vivo delivery of siRNA to hepatocytes. *Proc Natl Acad Sci U S A* 104:12982–87.

Segal, E., Pan, H., Benayoum, L. et al. 2011. Enhanced anti-tumor activity and safety profile of targeted nano-scaled HPMA copolymer-alendronate-TNP-470 conjugate in the treatment of bone malignancies. *Biomaterials* 32:4450–63.

Segal, E., Pan, H., Ofek, P. et al. 2009. Targeting angiogenesis-dependent calcified neoplasms using combined polymer therapeutics. *PLoS One* 4:e5233.

Semenkov, Y.P., Makarov, E.M., Makhno, V.I. et al. 1982. Kinetic aspects of tetracycline action on the acceptor (A) site of *Escherichia coli* ribosomes. *FEBS Letter* 144:125–29.

Shea, J.E. and Miller, S.C. 2005. Skeletal function and structure: Implications for tissue-targeted therapeutics. *Adv Drug Deliv Rev* 57:945–57.

Shen, W.C. and Ryser, H.J.P. 1981. Cis-aconityl spacer between daunomycin and macromo-lecular carriers- a model of pH-sensitive linkage releaseing drug from a lysosomotropic conjugate. *Biochem Biophys Res* 102:1048–54.

Shoenmakers, R.G., van de Watering, P., Elbert, D.L. et al. 2004. Effect of the linker on the hydrolysis rate of drug-linker ester bond. *J Control Release* 95:291–300.

Silverthorn, D.U., Ober, W.C., Garrison, C.W. et al. 2004. *Human Physiology: An Integrated Approach*. San Francisco, CA: Pearson/Benjamin Cummings.

Steitz, S.A., Speer, M.Y., McKee, M.D. et al. 2002. Osteopontin inhibits mineral deposition and promotes regression of ectopic calcification. *Am J Pathol* 161:2035–46.

Takahashi, N., Udagawa, N., Takami, M. et al. 2002. Cells of bone: Osteoclast generation. In *Principles of Bone Biology*, eds. J.P. Bilezikian, L.G. Raisz, G.A. Rodan, 109–126. San Diego: Academic Press.

Takahashi, T., Yokogawa, K., Sakura, N. et al. 2008. Bone-targeting of quinolones conjugated with an acidic oligopeptide. *Pharm Res* 25:2881–88.

Takahashi-Nishioka, T., Yokogawa, K., Tomatsu, S. et al. 2008. Targeted drug delivery to bone: Pharmacokinetic and pharmacological properties of acidic oligopeptide-tagged drugs. *Curr Drug Discov Technol* 5:39–48.

Talelli, M., Rijcken, C.J.F., van Nostrum, C.F. et al. 2010. Micelles based on HPMA copoly-mers. *Adv Drug Deliv Rev* 62:231–39.

Tartaix, P.H., Doulaverakis, M., George, A. et al. 2004. In vitro effects of dentin matrix pro-tein-1 on hydroxyapatite formation provide insights into in vivo functions. *J Biol Chem* 279:18115–20.

Thompson, W.J., Thompson, D.D., Anderson, P.S. et al. 1989. Polymalonic acids as bone-affinity agents. EP 034961.

Torchilin, V.P. 2010. Passive and active drug targeting: Drug delivery to tumours as an exam-ple. *Handb ExpPharmacol* 197:3–53.

Tye, C.E., Rattray, K.R., Warner, K.J. et al. 2003. Delineation of the hydroxyapatite- nucleat-ing domains of bone sialoprotein. *J Biol Chem* 278:7949–55.

Ulbrich, K., Etrych, T., Chytil, P. et al. 2003. HPMA copolymers with pH-controlled release of doxorubicin: In vitro cytotoxicity and in vivo antitumor activity. *J Control Release* 87:33–47.

Uludağ, H. and Yang, J. 2002. Targeting systemically administered proteins to bone by bisphosphonate conjugation. *Biotechnol Prog* 18:604–11.

Uzzan, B., Cohen, R., Nicolas, P. et al. 2007. Effect of statins on bone-mineral density: A meta-analysis of clinical studies. *Bone* 40:1581–87.

Väänänen, H., Zhao, H., Mulari, M. et al. 2000. The cell biology of osteoclast function. *J Cell Sci* 113:377–81.

van der Bijl, P. and Pitigoi-Aron, G. 1995. Tetracyclines and calcified tissues. *Ann Dent* 54:69–72.

van der Hoven, J.M., Van Tomme, S.R., Metseler, J.M. et al. 2011. Liposomal drug formula-tions in the treatment of rheumatoid arthritis. *Mol Pharm* 8:1002–05.

Wang, D., Kopečková, P., Minko, T. et al. 2000. Synthesis of star like N-(2-hydroxypropyl) methacrylamide copolymers: Potential drug carriers. *Biomacromolecules* 1:313–19.

Wang, D., Miller, S.C., Kopeckova, P. et al. 2005. Bone-targeting macromolecular therapeu-tics. *Adv Drug Deliv Rev* 57:1049–76.

Wang, D., Miller, S.C., Sima, M. et al. 2003. Synthesis of polymeric bone-targeting drug delivery systems. *Bioconjug Chem* 14:853–59.

Wang, D., Sima, M., Mosley, R.L. et al. 2006. Pharmacokinetic and biodistribution studies of bone-targeting drug delivery system based on N-(2-hydroxy propyl) methacrylamide copolymers. *Mol Pharm* 3:717–25.

Wang, G., Babadagh, M.E., and Uludağ, H. 2011. Bisphosphonate-derivatized liposomes to control drug release from collagen hydroxyapatite scaffolds. *Mol Pharm* 8:1025–34.

Wang, G., Kucharski, C., Lin, X. et al. 2010. Bisphosphonate-coated BSA nanoparticles lack bone targeting after systemic administration. *J Drug Target* 18:611–26.

Wasserman, R.H., Corradino, R.A., Carafoli, E. et al. 1977. *Calcium-Binding Proteins and Calcium Function*. New York: North Holland.

Wen, Y., Pan, S., Luo, X. et al. 2009. A biodegradable low molecular weight polyethylenimine derivative as low toxicity and efficient gene vector. *Bioconjug Chem* 20:322–32.

Weng, L., Li, L., Zhang, Y. et al. 2002. C-nyc protein expression upregulated by 2-[3-estrone-N-ethyl-piperazine-methyl] tetracycline in bone. *Yaoxue Xuebao* 37:771–74.

Williams, C. 1984. Radiopharmaceuticals in bone scanning. In *Bone Scintigraphy*, ed, E.B. Silberstein, 13–38. New York: Futura Publishing Company.

Willson, T.M., Charifson, P.S., Baxter, A.D. et al. 1996a. Bone-targeted drugs: 1. Identification of heterocycles with hydroxyapatite affinity. *Bioorg Med Chem Lett* 6:1043–46.

Willson, T.M., Kenke, B.R., Momtahen, T.M. et al. 1996b. Bone targeted drugs: 2. Synthesis of estrogens with hydroxyapatite affinity. *Bioorg Med Chem Lett* 6:1047–50.

Wong, R.W.K. and Rabie, A.B.M. 2005. Early healing pattern of statin-induced osteogenesis. *Br J Oral Maxillofac Surg* 43:46–50.

Wright, J.E., Gittens, S.A., Bansal, G. et al. 2006. A comparison of mineral affinity of bisphosphonate-protein conjugates constructed with disulfide and thioether linkages. *Biomaterials* 27:769–84.

Yang, J., Luo, K., Pan, H. et al. 2011. Synthesis of biodegradable multiblock copolymers by click coupling of RAFT-generated heterotelechelic polyHPMA conjugates. *React Funct Polym* 71:294–302.

Yasuda, H., Shima, N., Nakagawa, N. et al. 1998. Osteoclast differentiation factor is a ligand for osteoprotegerin/osteoclastogenesis-inhibitory factor and is identical to TRANCE/RANKL. *Proc Natl Acad Sci U S A* 95:3597–602.

Yokogawa, K., Miya, K., Sekido, T. et al. 2001. Selective delivery of estradiol to bone by aspartic acid oligopeptide and its effect on overactized mice. *Endocrinology* 142:1228–33.

Zhang, S., Wright, J.E.I., Ozber, N. et al. 2001. The interaction of cationic polymers and their bisphosphonate derivatives with hydroxyapatite. *Macromol Biosci* 7:656–70.

Zhao, Q., Jia, Y., and Xiao, Y. 2009. Cathepsin K: A therapeutic target for bone disease. *Biochem Biophys Res* 380:721–23.

Zheng, H. and Weng, L.L. 1997. Bone resorption inhibition/osteogenesis promotion pharmaceutical composition. US Patent 5,698,542.

Zheng, H., Li, L., and Weng, L. 2000. Effects of 2-[3-estrone-N-ethyl-piperazine-methyl] tetracycline on levels of C-fos and C-jun mRNAs and their product proteins in epiphyseal plate. *Yaoxue Xuebao* 35:249–52.

Ziebart, T., Pabst, A., Klein, M.O. et al. 2009. Bisphosphonates: Restrictions for vasculogenesis: Inhibition of cell function of endothelial progenitor cells and mature endothelial cells in vitro. *Clin Oral Invest* 15:105–11.

Zloh, M., Dinand, E., and Brochini, S. 2003. Aconityl-derived polymers for biomedical applications. Modeling study of cis-trans isomerization. *Theor Chem Acc* 109:206–12.

# 10 Targeted Drug Delivery in Solid Tumors

## An Overview and Novel Approaches for Therapy

*Mintu Pal and Lay Poh Tan*

## CONTENTS

## 10.1   INTRODUCTION

Solid tumors are highly heterogeneous and comprise cancer cells, and a mixture of non-cancer stromal cells including endothelial cells, pericytes, fibroblasts, myofibroblasts, macrophages, inflammatory cells, dendritic cells, and mast cells. Solid tumors are characterized by their abnormal cellular growth in solid organs such as brain, breast, prostate, colorectum, kidney, and as opposed to leukemia, a cancer affecting the blood, without forming a mass. Abnormal growth of cells can then invade adjoining parts of the body and spread to other distant organs, which is called metastasis, and is responsible for major cause of death in cancer (Egeblad et al. 2010). Cancers are classified in two ways: by their primary site of origin or by their histological grade/tissue types. There are hundreds of different solid tumors. Based on the histological standpoint, these are grouped into five major categories:

a. Carcinoma (malignant neoplasm of epithelial origin found throughout the body or cancer of the internal or external lining of the body that accounts for 80%–90% of all cancer cases)
b. Sarcoma (originates in supportive and connective tissues such as bones, tendons, muscle, and fat)
c. Myeloma (originates in plasma cells of the bone marrow)
d. Lymphoma (develops in the glands or nodes of the lymphatic system, specifically the spleen, tonsils, and thymus), and
e. Mixed types (components may be within one category or from different categories such as adenosquamous carcinoma, mixed mesodermal tumor, carcinosarcoma, and teratocarcinoma)

Solid tumors can also be classified into benign (not cancerous form), and malignant (cancer) (http://www.ijpsr.info/docs/IJPSR11-02-01-01.pdf) as shown in Table 10.1.

Solid tumors appear in over 85% of human cancers. Cancers are the leading cause of death worldwide. According to World Health Organization, in 2012, solid tumors accounted for 8.2 million deaths (14.6% of all human deaths) from about 14.1 million new cases occurred globally. Across the world, cancer incidence rates increased

**TABLE 10.1**

**Cancer Can Also Be Classified into Benign and Malignant Types of Cancers**

| Localized solid tumor (lump or growth of tissue made up from abnormal cells) | Benign tumors (non-cancerous) grow slowly, and do not spread or invade other tissues, not usually life-threatening | Papilloma squamous, choroids plexus, laryngeal papilloma, warts genital, plantar, epidermodysplasia verruciformis, and malignant warts |
|---|---|---|
| | Malignant tumors (tend to grow quite quickly, and invade into nearby tissues and organs, which can cause damage) | Bladder, cervical, colorectal, esophageal, prostate, breast, lung, skin, pancreatic, liver, kidney, bone, ovary, brain, head and neck, melanoma, multiple myeloma, lymphoma, sarcoma, nerve, and so on |

sharply in 2014, and the WHO has predicted "a substantive increase," with new cases predicted to rise to 19.3 million by 2025 (http://www.iarc.fr/en/media-centre/pr/2013/pdfs/pr223_E.pdf). Among men, lung, prostate, colorectum, stomach, and liver cancers are the five most common sites of cancer diagnosed in 2012, whereas among women, the five most common sites diagnosed were breast, colorectum, lung, cervix, and stomach cancer. It is reported that around one third of cancer-related deaths are due to the abnormalities in the genetic material, which may occur due to the effects of carcinogens such as tobacco, smoke, radiation, chemicals, or infectious agents including specific genetic background, long-term exposure to various environmental stresses, and bias diet habit (Boffetta and Nyberg 2003). Despite individual variations such as age, gender, and genetic differences in patients, solid tumors are developed mainly due to the modulation of proto-oncogenes and tumor suppressor genes expression, which ultimately enhances unrestricted growth to develop malignant cancer (Krausova and Korinek 2014). Proto-oncogenes and tumor suppressor genes act like the accelerator and breaks of a car, respectively (Kumar and Manjunatha 2013). If those genes are mutated, they can no longer stop abnormal cell division, which is much like a car with broken brakes, and therefore cells undergo continuous abnormal cell division with loss of normal functions such as accurate DNA replication, orientation, and adhesion within tissues, and interaction with protective cells of the immune system, and as a result form tumor (Furthauer and Gonzalez-Gaitan 2009).

The current conventional treatment strategies are chemotherapy, surgery, radiation, targeted, immunotherapy, and/or the combination therapies. Although, despite advances in surgical and radiation treatments, chemotherapy is one of the best approaches to eradicate cancer, especially in the primary, advanced, and metastatic tumors (Nersesyan and Slavin 2007). The success of chemotherapy depends on the selection of optimum carrier system including liposomes, nanoparticles, solid lipid nanoparticles, microspheres, polymersomes, polymer–drug conjugates, polymeric nanoparticles, nanocrystals, peptide nanoparticles, micelles, nanospheres, and so on (Fernandez-Fernandez et al. 2011).

## 10.2 BASIC KNOWLEDGE ON THE TUMOR MICROENVIRONMENT OF SOLID TUMORS

Current drug development strategies are targeting several key molecules including mutated proteins, which have been recognized as the important contributors to carcinogenesis. Although this common strategy is useful at the earlier stage, drug resistance usually develops eventually, with faster relapse of the disease. This clearly implies some missing links between the actual underlying carcinogenic mechanisms and the current drug development strategies (Galluzzi et al. 2014). Tumor microenvironment (TME) may play a crucial role in these missing links. Recently, TME has been gradually recognized as a key contributor for cancer progression and metastasis. Recently, based on various molecular studies, the concept of cancer stem cell (CSC) models has been proposed as drivers of intratumoral heterogeneity and cancer metastasis modulated by TME. This concept was originally anticipated in the late 1970s and was first described in hematologic malignancies in 1994 (Tan et al. 2006; Visvader and Lindeman 2008). This concept has evolved to better model the complex and highly dynamic processes of tumorigenesis, tumor relapse, and metastasis. Another more current concept suggests that solid tumors are no longer an isolated cellular population of malignant cells; instead, they are the consequence of collaboration of extracellular matrix (ECM) proteins and many other non-malignant cell types such as macrophages, neutrophils, mast cells, and lymphocytes (Diaz-Cano 2012). In fact, in the early 1880s, Steven Paget proposed the *seed and soil* hypothesis suggesting that a fertile *soil* (the microenvironment) is essential for the *seed* (the tumor cells) to grow the tumor cells (Langley and Fidler 2011). As the microenvironment is quite complicated, we would like to focus on the role of dysregulated immune responses and the various components in the microenvironment involved in tumor progression, invasiveness, and development of drug resistance.

### 10.2.1 TUMOR MICROENVIRONMENT IN CANCER INITIATION AND PROGRESSION

Bidirectional communication and interactions between cells and their microenvironment are essential for normal tissue homeostasis and tumor growth (Quail and Joyce 2013). Tumors consist of stromal cells as well as non-cellular components, in addition to neoplastic cells. Non-neoplastic stromal cells function to preserve normal tissue structure but when tissue architecture is disbalanced, it supports cancer initiation and progression (Hansen and Bissell 2000). Besides the cellular components, ECM and secreted extracellular molecules perform in autocrine and/or paracrine manners to support/sustain tumor development and progression (Adjei and Blanka 2015). In their dysfunctional state, fibroblast and immune cells produce many biomolecules such as chemokines and growth factors that stimulate cancer cell growth, proliferation, and invasion, and can recruit other cells including mesenchymal stem cells (MSCs) that replenish cells in the tumor (Erez et al. 2010). In 1971, Judah Folkman communicated a revolutionary article proposing that all vascularized tumor induces host microvessels to undergo angiogenesis, which initiated a model shift throughout the cancer research community despite the initial resistance (Ribatti, 2008). Therefore, understanding the activities of different components of the tumor stroma

**TABLE 10.2**

**Cellular Components of Tumor Stroma**

| Lineage | Role in Tumorigenesis | References |
|---|---|---|
| Tumor-associated macrophages (TAM) | Activated M1 macrophages are pro-inflammatory and anti-tumorigenic and secrete TH1 cytokines, whereas activated M2 macrophages are anti-inflammatory and pro-tumorigenic and secrete TH2 cytokines. They promote cancer metastasis by secreting proteases, such as matrix metalloproteinases (MMPs), urokinase-type plasminogen activator, and cathepsin B. | Saif (2013) |
| Neutrophils | Similar to TAMs, neutrophils have been shown to have opposing functions in regulating cancer progression and metastasis, indicating that they have context-dependent roles within the TME | Sajja et al. (2009) |
| Treg cells | Suppressing tumor-specific immunity, secrete IL-10, and TGFβ, which are expected to be the major obstacles to CD8$^+$ T-cell mediated tumor lysis | Sarris et al. (2000); Schraa et al. (2002) |
| Th cells | CD4+ TH cells have Th1 and Th2 lineages. Th1 cells, characterized by the secretion of IFN-gamma and TNF-alpha, are responsible for activating and regulating the development and persistence of cytotoxic T cell (CTL). Th1 cells secrete anti-inflammatory cytokines and can be protumorigenic. Th2 cells favor a predominantly humoral response, secrete anti-inflammatory cytokines, and can be pro-tumorigenic. The ratio of Th1 to Th2 cells in cancer correlates with tumor stage and grade. | See et al. (1974) |
| B cells | Important mediators of humoral immunity, secrete interleukin-10 (IL-10), and transform growth factor beta (TGF-β) | Semeraro, and Colucci (1997) |
| Mesenchymal stem cells | One of the key components of tumor microenvironment to play critical roles in cancer progression and metastasis | Seymour et al. (2009) |
| Tumor-associated fibroblasts | Key regulators of tumorigenesis, contribute to the integrity, and remodeling of the ECM by secreting proteases such as matrix metalloproteinase (MMPs) including MMP1, MMP3, MMP9, and so on | Shea et al. (2011) |

and their relationships with cancer cells involved in tumor progression, metastasis and drug resistance are very important in order to develop better strategies to treat cancer. Here, we highlight various roles of tumor stromal cells, predominantly those that have been exploited for cancer therapies as shown in Table 10.2.

### 10.2.2 TUMOR MICROENVIRONMENT IN CANCER INVASION AND METASTASIS

Once the primary tumor gains the capacity to escape host immune defenses and cancer cells enter into the circulation, metastatic dissemination is in progress. Metastasis is a very much inefficient process; only 0.01% of cells that intravasate into circulation are accomplished of forming detectable metastases (Valastyan and Weinberg 2011). Hence, the critical question arises on how the TME supports cancer cells in leaving the primary tumor site and then seeding effectively in secondary organs for metastasis. Increasing evidences suggest that epithelial-mesenchymal transition (EMT) is one of the initiating steps of primary tumor invasion during which the tumor cells lose highly polarized epithelial characteristics and gain mesenchymal markers that confer stem-like properties and a migratory phenotype (Maier et al. 2010) as shown in Figure 10.1. In addition to the phenotypic changes of cells during EMT, some other important changes also take place at the cellular and the molecular level (Martin et al. 2013). These include upregulated transcriptional repressors of E-cadherin (including Snail, Slug, Twist, and Zeb), E-cadherin degradation, and replacement of epithelial proteins (such as cytokeratins, apical actin-binding transmembrane protein-1, and zonula occludens-1) with mesenchymal proteins (such as vimentin, N-cadherin, and fibronectin), and so on. A variety of inflammatory mediators from cancer and tumor-infiltrating cells, such as IL-1, IL-6, and IL-8, are secreted and facilitate the development of TME in favor of tumor cell proliferation, motility, invasion, and therefore increase their metastatic ability (Tsai et al. 2014). The infiltrated immune cells can produce a series of EMT-favorable cytokines such as transforming growth factor beta (TGF-β), TNF-α, and so on (Gao et al. 2012; Heinrich et al. 2012; Thuault et al. 2006). The key regulatory role of TGF-β for EMT has been recognized in various models, including alveolar epithelial cells, making them transform to fibroblasts/myofibroblasts (Kim et al. 2006; Willis and Borok 2007). TNF-α alone also mediates

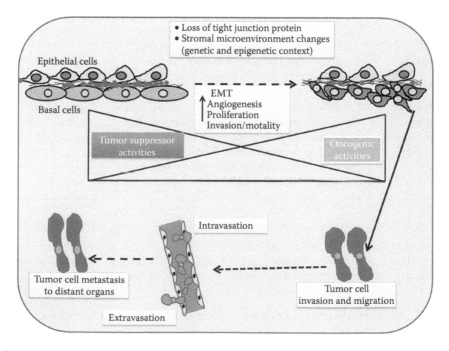

**FIGURE 10.1**   Various factors (genetic and epigenetic) that induce tumor cells epithelial-mesenchymal transition (EMT) and tumor metastasis in tumor microenvironment.

EMT through promoting E-cadherin degradation, mainly via strengthening Snail stability in an NF-κB-dependent manner (Wu et al. 2009). Besides intrinsic factors, matrix stiffness is an important microenvironmental cue that plays a significant role in determining cancer cell invasion and metastasis. Recent studies have developed a 3D biomimetic model of tissue stiffness interface for cancer drug testing, which allows a standardized comparison of spheroid invasion under a 3D stiffness gradient influence. Therefore, reduced invasion of cancer cells accredited to increased tissue stiffness barriers may favor their reduced apoptotic susceptibility to chemotherapeutic treatment (Lam et al. 2014).

### 10.2.3 ROLE OF TUMOR MICROENVIRONMENT IN DEVELOPMENT OF DRUG RESISTANCE

Substantial evidence suggests that TME may contribute to the failure of standard chemotherapy to eradicate the entire tumor population and may facilitate the emergence of acquired drug resistance (Khawar et al. 2015). To avoid tissue microenvironment-mediated drug resistance, the multi-targeting drug or the "cocktail" drug application strategy may give a more favorable long-term outcome. However, the current combination of chemotherapy approaches still fails in many cases because of multi-drug resistance (MDR), which is linked with alterations in apoptotic signaling, and an overexpression of drug-efflux pumps (Kapse-Mistry et al. 2014). The initial MDR was described in the late 1960s and early 1970s. Over-expression of phosphor-glycoprotein (P-gp) is a common feature of drug-resistant mammalian cell lines when compared with the drug-sensitive parent cell lines (See et al. 1974). P-gp is a crucial and the best known membrane transporter including *MDR1* and *MDR2* or *MDR3* in humans. P-gp is present in many types of cancers such as gastrointestinal, liver, pancreatic, and ovarian cancers (Gottesman 2002), and therefore, overexpression of multidrug transporters such as P-gp enables a quick efflux of anticancer drugs out of the tumor cells. So, from the drug delivery viewpoint, drug resistance can be minimized by improving drug delivery to the tumor sites and reducing MDR-based drug efflux. Many anticancer drugs, including doxorubicin and paclitaxel used for the treatment of solid tumors, develop resistance due to poor penetration or altered pharmacokinetics due to poorly perfused vasculature or extensive necrosis of parts of the tumor (Minchinton and Tannock 2006; Khawar et al. 2015). Therefore, understanding in detail, knowledge on mechanisms modulating TME may facilitate the design and greater success of a novel anticancer therapy.

## 10.3 CHALLENGES IN DELIVERING DRUGS TO SOLID TUMORS

Many cancer patients have encouraging quick response to first line chemotherapies but end with cancer recurrence that requires further treatment. Unfortunately, the response to consequent chemotherapies with various drugs usually drops significantly due to the dreadful cancer chemoresistance (Hu and Zhang 2009). A number of mechanisms have been suggested to account for cancer chemoresistance or poor response to chemotherapy. These therapies are restricted by solubility and pharmacokinetic factors on account of their physiochemical properties such as low pH, low oxygenation, and high interstitial fluid pressure as well as some other barriers (Sriraman et al. 2014) as

**TABLE 10.3**

**Barriers Limiting Conventional Systems Bearing Anticancer Agents**

| Biological Barriers | Anatomical Barriers | Physiological Barriers | Chemical Barriers | Clinical Barriers |
|---|---|---|---|---|
| The mononuclear phagocyte system, tumor microenvironment, extracellular matrix, matrix metalloproteinase, hypoxic core, and extracellular pH of tumors | Vesicular and organellar barriers, cellular membrane, blood-brain barrier (BBB), nuclear membrane, drug efflux transporters | Renal filtration, drug efflux pumps, high interstitial fluid pressure, high tumor cell density, hepatic degradation | Low solubility, large volume of distribution, low stability, charge interactions, low molecular weight | Need for hospitalization, low efficacy, high toxicity, low cost effectiveness |

*Source:* Adapted from Jain, R.K., Martin, J.D., and Stylianopoulos, T. 2014. *Annu Rev Biomed Eng* 16:321–46; Lammers, T. 2010. *Adv Drug Deliv Rev* 62:203–30.

shown in Table 10.3. Recently, most of the researchers have concentrated on molecular mechanisms of resistance to chemotherapy, whereas the role of TME has been less focused. One of the biggest challenges is to cross the tumor interstitium or ECM microenvironment, which consists of a cross-linked network and heterogeneous distribution throughout the microspace of collagen and elastin fibers, proteoglycans, and hyaluronic acid. It helps to transport vital nutrients as well as oxygen to support the cell growth and proliferation, despite the structural integrity (Kuppen et al. 2001). Hence, the abnormal TME has recently been suggested to be a promising target for the improvement of therapeutic efficacy against cancer. Strategies to modulate the abnormal TME, referred to as *solid tumor priming,* have shown promising approaches in the improvement of drug delivery and anticancer efficacy (Lu et al. 2007; Khawar et al. 2015).

## 10.4 STRATEGIES FOR DRUG DELIVERY IN SOLID TUMOR

An exciting new generation of clinical trials is now harnessing drug-containing thermosensitive liposomes and other nanoparticle drug carriers that release contained chemotherapy agents upon heating above ~40°C. Combined with localized heating methods as described above, this allows for targeted chemotherapy delivery to the tumors. Thermal ablation or hyperthermia can be combined with heat-activated drug carriers to selectively deposit chemotherapy in the heated area. Initial clinical trial results suggest patient benefits from this combination, and thus, there is considerable excitement among members of our society in this approach.

### 10.4.1 Systemic Delivery Targeted to Solid Tumors

Although numerous anticancer therapies have been developed to treat cancer, a key challenge has been to generate a systemic delivery that can both selectively and efficiently target tumor tissue. Transcriptional targeting was first reported in 1997. Rodriguez and group (Rodriguez et al. 1997) constrained transgene expression in prostate-specific antigen (PSA) producing prostate cells by applying the PSA promoter into adenovirus type 5 DNA to drive transgene expression called *attenuated replication-competent adenovirus*. Until now, two major approaches have been used for tumor targeting: (i) transcriptional targeting, which uses promoters (such as cytomegalovirus (CMV) promoter in adenoviral vectors-based gene therapy) that are only active in the target tumors, and remain at a low basal expression level in normal tissues and (ii) ligand-targeting of vectors to specific receptors expressed within the tumor tissue. Each of these approaches has shown promising results for vector delivery and transgene expression in preclinical tumor models (Pranjol and Hajitou 2015). Here, we have highlighted the recent trends in prodrug and conjugate rationale as well as other pre-targeting approaches for cancer treatment.

### 10.4.1.1  Tumor-Activated Prodrug Therapy

The basic idea of the targeted delivery approach is that conjugation of drug to a tumor-specific molecule renders the drug inactive until it reaches the target site. Once the drug reaches the tumor site, it binds to the surface of tumor cells and is processed (internalized, released from the carrier molecule) to bring back its original anticancer activity. Therefore, drug conjugates can be considered tumor-activated prodrugs (TAPs) (Sajja et al. 2009; Guarnieri et al. 2015). Most conventional prodrugs are converted to active drugs by different mechanisms such as chemical or enzymatic hydrolysis, and restored the activity of TAPs. Finally, activated prodrugs interact with antigens or receptors specifically found on the surface of tumor cells without binding to non-target tissues and as a result will be non-toxic while in circulation *in vivo*.

### 10.4.1.2  Two-Step Targeting Using Bi-Specific Antibody

Many recent researches have revealed preclinical and clinical advances in the use of pre-targeting methods for the radio-immunodetection and radio-immunotherapy of cancer. Over the past 30 years, directly labeled antibodies, fragments, and sub-fragments such as mini-bodies and other constructs have been used for both imaging and therapeutic purposes. However, their clinical acceptance has not satisfied the expectations due to either poor image resolution or insufficient radiation doses delivered selectively to tumors for therapy (Goldenberg et al. 2007). Therefore, bispecific monoclonal antibodies (bsMAs) followed by radiolabeled peptide haptens have a great importance as a new modality of selective delivery of radionuclides for the imaging and therapy of cancer. Antibodies are naturally monospecific, and therefore bsMAbs reactive with a tumor antigen and the effector hapten need to be prepared for pretargeting in one of the three ways. Firstly, the first generation of bsMAbs was produced chemically by generating mostly Fab' fragments, which naturally contain at least one free sulfhydryl group that could be derivatized to make a stable thio-ether linkage, when added to another Fab'-SH. Secondly, bsMAbs can be produced in prokaryotic or eukaryotic expression vectors by the recombinant DNA technology. Thirdly, bsMAbs can be produced by the fusion of two hybridoma cell lines producing the two parental monoclonal antibodies. The resulting quadroma cell line expresses the heavy and light chains of both parental MAbs and these heavy and light chains assemble randomly; theoretically ensuing in 12 different "immunoglobulin-like molecules," of which only 1 is the functional bsMAb (Goldenberg et al. 2007).

### 10.4.1.3 Site-Specific Delivery and Light Activation of Anticancer Proteins

Strained ruthenium (Ru) complexes have been characterized as novel agents for photodynamic therapy (PDT). The complexes are inert until triggered by visible light (Howerton et al. 2012). An increase in cytotoxicity of several folds against 3D tumor spheroids as compared to cisplatin is observed with light activation in cancer cells. While unreactive in the dark, Ru (II) polypyridyl complexes are transformed upon light activation into potent cytotoxic species (Howerton et al. 2012).

### 10.4.2 DRUG DELIVERY TARGETED TO BLOOD VESSELS OF TUMORS

Angiogenesis plays a central role in the growth and spread of tumors. A blood supply is required for tumors to grow ahead of a few millimeters in size. Tumors can also stimulate nearby normal cells to produce angiogenesis signaling molecules. The resulting new blood vessels "feed" growing tumors with oxygen and nutrients, allowing the cancer cells to invade nearby tissue and to form metastases. Tumors cannot grow beyond a certain size or spread without a blood supply. Therefore, researchers are trying to find ways to prevent or slow tumor angiogenesis by developing natural and synthetic angiogenesis inhibitors.

### 10.4.2.1 Anti-Angiogenesis Therapy

The concept of anti-angiogenic therapy of cancer is based on several elegant ideas in tumor biology. First, angiogenesis is enormously required for tumor growth (Folkman 1971). Therefore, the inhibition of angiogenesis could be an important therapeutic target to suppress tumor growth and metastasis. Second, due to high proliferative activity of cancer cells, tumor-derived endothelial cells can be targeted selectively. Third, due to low mutation rate, endothelial cells will be unlikely to gain drug resistance associated with an adaptive mutation (Kerbel 1997). Several endogenous angiogenesis activators have been identified such as fibroblast growth factors, placental growth factor, angiogenin, interleukin-8, hepatocyte growth factor, vascular endothelial growth factor, granulocyte colony-stimulating factor (G-CSF), and platelet-derived endothelial cell growth factor involved in the growth and proliferation of new blood vessels (Bull et al. 1993). Angiogenesis inhibitors such as thrombospondin-1, angiostatin, metallo-proteinase inhibitors, platelet factor 4, interferon alpha, prolactin 16-kd fragment are also an attractive therapeutic strategy for eradicating cancer. Bevacizumab (Avastin) is a humanized anti-VEGF monoclonal antibody (mAb), which is applied for the treatment of patients with metastatic colorectal cancer, non-small cell lung cancer (NSCLC), ovarian cancer, and other types of cancer (Saif 2013).

### 10.4.2.2 Drugs to Induce Clotting in Blood Vessels of Tumors

One of the most important roles of vascular endothelial cells (ECs) in the thrombotic cascade is the induction of tissue factor (TF) (Semeraro and Colucci 1997), a transmembrane protein that serves as a high-affinity receptor for factor VIIa, which initiates the extrinsic blood coagulation series of events (Nemerson, 1988). Drugs that affect TF induction have a well-known effect on thrombosis, for example, rapamycin inhibits tumor growth by an anti-angiogenic mechanism that involves blockage of vascular-endothelial-growth-factor signaling by ECs and upregulates TF expression in ECs (Guba et al. 2002). Expression of TF is further upregulated by the stimulation of TF-mediated cell signaling pathways by vascular-endothelial-growth factor and rapamycin (Guba et al. 2005) leading to thrombosis in tumor-containing vessels (Ramot and Nyska 2007).

### 10.4.2.3 Vascular-Targeting Agents

Vascular-targeting agents (VTAs), for the treatment of cancer, are designed to cause a rapid and selective shutdown of the blood vessels of tumors. Distinct anti-angiogenic drugs that inhibit the formation of new vessels, VTAs occlude the pre-existing blood vessels of tumors to cause tumor cell death from ischemia and extensive hemorrhagic necrosis. Overall, VTAs can kill the tumor cells that are resistant to conventional anti-proliferative cancer therapies or the cells in the areas distant from blood vessels where drug penetration is poor, and hypoxia can lead to radiation and drug resistance (Thorpe 2004).

### 10.4.3 SPECIAL FORMULATIONS AND CARRIERS OF ANTITUMOR DRUGS

Recent evidences suggest that using nanotechnology in cancer chemotherapy has great importance for better efficacy to treat cancer. These carriers include liposomes, nanoparticles, microspheres, mAb, nanobodies, and so on (Fay and Scott 2011) as shown in Table 10.4. Although there are several efficient and validated chemical methods for functionalizing nanoparticles with a wide range of targeting ligands, a number of important issues, including the best possible interplay of physicochemical features, need to be addressed before translation from preclinical to clinical development. Here, we have abridged different formulations and promising drug carriers targeting tumor outlined in detail below (Peer et al. 2007).

### 10.4.3.1 Albumin-Based Drug Carriers

Albumin is playing an increasing role as a drug carrier in the clinical setting. Primarily, three drug delivery technologies are distinguished mainly in context to coupling of low-molecular weight drugs to exogenous or endogenous albumin, conjugation with bioactive proteins, and encapsulation of drugs into albumin nanoparticles. An alternative strategy is to bind a therapeutic peptide or protein covalently or physically to albumin to improve its stability and half-life (Kratz 2008).

### 10.4.3.2 PEGylated Liposomes

Liposomes are versatile, self-assembling, carrier materials that contain unilamellar or multi-lamellar microscopic particles composed of membrane-like lipid layers, often phospholipids and cholesterol, surrounding aqueous compartments (Akbarzadeh et al. 2013). Liposome size is usually limited to 50–150 nm when used for drug delivery purposes (Kraft et al. 2014). Liposomes are now achieving clinical acceptance as intracellular delivery systems for anti-sense molecules, proteins/peptides, and DNA to disease locations, with sustained release functions (Akbarzadeh et al. 2013). Antibody targeting is the most commonly used and extensively studied. One of the ways is through the liposomes with encapsulated chemotherapeutic drugs, such as trastuzumab (Herceptin), targeting HER2 receptors overexpressed

**TABLE 10.4**

**Nanocarriers as an Emerging Platform for Cancer Therapy**

| Drug | Commercial Name | Nanocarriers | Indications | References |
|---|---|---|---|---|
| Styrene maleic anhydride-neocarzinostatin (SMANCS) | Zinostatin/Stimalmer | Polymer-protein conjugate | Hepatocellular carcinoma | Abe and Otsuki (2002) |
| PEG-L-asparaginase | Oncaspar | Polymer-protein conjugate | Acute lymphoblastic leukemia | Sikorska-Fic et al. (1998) |
| PEG-granulocyte colony-stimulating factor (G-CSF) | Neulasta/PEGfilgrastim | Polymer-protein conjugate | Prevention of chemotherapy-associated neutropenia | Frampton and Keating (2005) |
| IL2 fused to diphtheria toxin | Ontak (Denileukin diftitox) | Immunotoxin (fusion protein) | Cutaneous T-cell lymphoma | Kaminetzky and Hymes (2008) |
| Anti-CD20 conjugated to yttrium-90 or indium-111 | Zevalin | Radio-immunoconjugate | Relapsed or refractory, low-grade, follicular, or transformed non-Hodgkin's lymphoma | Iwamoto et al. (2007) |
| Anti-CD20 conjugated to iodine-131 | Bexxar | Radio-immunoconjugate | Relapsed or refractory, low-grade, follicular, or transformed non-Hodgkin's lymphoma | Andemariam and Leonard (2007) |
| Daunorubicin | DaunoXome | Liposomes | Kaposi's sarcoma | Petre and Dittmer (2007) |
| Doxorubicin | Myocet | Liposomes | Combinational therapy of recurrent breast cancer, ovarian cancer, and Kaposi's sarcoma | Rivera (2003); Lao et al. (2013) |
| Doxorubicin | Doxil/Caelyx | PEG-liposomes | Refractory Kaposi's sarcoma, recurrent breast cancer, and ovarian cancer | Duggan and Keating (2011) |
| Vincristine | Onco TCS | Liposomes | Relapsed aggressive non-Hodgkin's lymphoma | Sarris et al. (2000) |
| Paclitaxel | Abraxane | Albumin-bound paclitaxel nanoparticles | Metastatic breast cancer | Gradishar et al. (2005) |
| Paclitaxel | Genexol®-PM | Polymeric micelles | Breast, lung, pancreatic cancer | Shea et al. (2011) |
| Doxorubicin | NK911 | Nanoparticles | Recurrent breast cancer | Uwatoku et al. (2003) |
| Doxorubicin | Transdrug® | Nanoparticles | Various cancers | Reddy and Couvreur (2011) |
| Paclitaxel | Nanoxel® | Nanoparticles | Hepatocarcinoma | Cucinotto et al. (2013) |
| Paclitaxel | Xyotax® (CT-2103) | Polymer–drug conjugates | Advanced breast cancer, Breast, ovarian cancer II, Advanced lung cancer | Melancon and Li (2011) |
| Doxorubicin | PK1 | Polymer–drug conjugates | Breast, lung, colon | Seymour et al. (2009) |
| Paclitaxel | Taxoprexin® | Polymer–drug conjugates | Various | Yared and Tkaczuk (2012) |

on some types of breast cancer and bevacizumab specific for the vascular endothelial growth factor receptor (VEGFR) (Sullivan and Brekken 2010). For targeted
liposomal drug delivery in cancer, two main approaches can be followed that are
distinguished based on the target cell type: (i) vascular targeting and (ii) tumor cell
targeting. Recently, a new gene therapy approach has utilized ultrasound-sensitive
liposomes, commonly referred to as bubble liposomes, to transfer gene therapeutics
in a mouse tumor model to deliver interleukin 12 (IL-12) where a dramatic suppression of tumor growth was observed (Babu et al. 2014).

### 10.4.3.3 Tumor-Targeted Nanoparticles

Nanoparticles loaded with anticancer drugs or therapeutic genes can be targeted to
tumor sites using targeting moieties. These ligands should be specific toward receptors overexpressed by cancer cells (Babu et al. 2014). Targeted drug delivery systems are known to enhance the therapeutic efficacy due to site-specific delivery and
subsequent increase in tumor uptake of the therapeutic drugs compared with nontargeted conventional drug delivery systems (Hughes et al. 1989; Kukowska-Latallo
et al. 2005). Affibodies are small alpha-helical polypeptide ligands that function as
inert antibody mimetics that have gained attention for their use in targeted delivery
of nanoparticles (Babu et al. 2014). Conjugation of different peptides to nanoparticle
surfaces increased the selective intratumoral delivery of the nanoparticle payload as
shown in Table 10.5.

### 10.4.3.4 Carbohydrate-Enhanced Chemotherapy

In addition to being a valuable source of energy, carbohydrates are integral parts
of the cell. Several synthetic and natural origin carbohydrate molecules are widely
used clinically to treat different ailments. Due to their structural diversity in terms of
functional groups, ring size, and linkages, they are used as scaffolds in drug discovery processes (Tiwari et al. 2012). Due to the hydrophilic nature of monosaccharides
they offer good water solubility, optimum pharmacokinetics, and decreased toxicity,

**TABLE 10.5**
**Currently Used Targeting Moieties with Examples**

| Targeting Ligands | Example of Tumor Target | Target | References |
|---|---|---|---|
| RGD, Arg-Gly-Asp | Survival and vasculature endothelial cells in solid tumors | Transmembrane protein $\alpha v \beta 3$-integrin | Schraa et al. (2002); Yu et al. (2012) |
| NGR, Asn-Gly–Arg | Neovasculature endothelial cells in solid tumors | Aminopeptidase N (CD13) | Curnis et al. (2010) |
| Anti-tenascin | Breast cancer | An extracellular matrix protein, which is abundant in the stroma of several solid tumors | Zhang and Liu (2013) |
| Anti-VEGFR | Play a central role in angiogenesis and often are highly expressed in solid tumors | Vasculature endothelial growth factor receptor (VEGFR) | Zhao and Adjei (2015) |

and therefore, they have been extensively used to access a diverse library of compounds with great chemotherapeutic importance.

### 10.4.3.5 Microspheres

The use of microspheres as a vaccine delivery system is being increasingly explored to elicit a comprehensive immune response. The major problem is to get effective encapsulation and stabilization of hydrophilic antigens in hydrophobic matrices of microspheres. To address this issue, an approach of combining hydrophobic ion pairing (HIP) with O/W single emulsion microsphere formulation, such as orntide poly(d,l-lactide-co-glycolide) and poly(d,l-lactide) microspheres, a biodegradable poly(lactic acid) microsphere formulation for *in vivo* cytokine immunotherapy of cancer has been carried out. Polylactic co-glycolic acid microspheres in nanofibrous scaffolds have been shown to control the release of PDGF-BB (platelet-derived growth factor) *in vitro* (Rajput and Agrawal 2010). Drug-eluting microsphere transarterial chemoembolization (DEM-TACE) is a new delivery system to administrate drugs in a controlled manner. DEM-TACE is focused on obtaining higher concentrations of the drug to the tumor with lower systemic concentrations than traditional cancer chemotherapy (Sottani et al. 2009), particularly useful in cancer metastases.

### 10.4.3.6 Monoclonal Antibodies

The use of MAbs has revolutionized both the cancer therapy and the imaging. Hughes et al. 1989 (Hughes et al. 1989) demonstrated the use of the first mAb-targeted liposomes for organ-specific delivery in lung tumor-bearing mouse models. An attractive alternative for MAbs are nanobodies or VHHs, which were first shown to be able to bind to specific tumor antigens (Majidi et al. 2009). The emergence of recombinant technologies has revolutionized the selection and production of MAbs, allowing the design of fully human antibodies of any specificity and for diverse purposes, such as antigen-binding affinity, molecular architecture, and dimerization state, and fused with a vast array of effector moieties to enhance their cell-targeting ability and potency (Oliveira et al. 2013).

### 10.4.3.7 Therapeutic Peptides

Therapeutic peptides (TPs) are peptide-based agents capable of eliciting a therapeutic response by modulation of targets within or on the cell's surface. TPs have a great potential for cancer therapy as they are amenable to rational design, high specificity for their targets, and can be made to target almost any protein of interest including proteins for which we have no small-molecule drugs (Bidwell 2012). However, since TPs have short half-lives in systemic circulation, they are easily degraded by proteases in plasma and target cells, are often cleared by the reticuloendothelial system, and can be immunogenic, for which many obstacles must be overcome for the potential to be realized.

### 10.4.3.8 Nanoparticles

The combination of nanoparticles and biologically active components is of intense interest because of the synergistic properties being exploited by such novel technology for therapeutic and diagnostic purposes. Nanoparticle modifications such as conjugation with polyethylene glycol have been used to increase the sustainable effect of drugs in the body and reduce renal clearance rates (Babu et al. 2014). Moreover, the development of targeted nanoparticles-mediated drug delivery systems has considerably contributed to the therapeutic efficacy of anticancer drugs and cancer gene therapies compared with non-targeted conventional delivery systems.

### 10.4.3.9 PEG Technology

Among the approaches in the field of drug delivery, PEGylation has so far been one of the best choices for protein delivery. Initially, this technology was mainly applied with macromolecular drugs, such as proteins and enzymes, with 10 PEGylated biomacromolecules for the treatment of related diseases. Then, this technology is successfully used with small molecule drugs to overcome the possible limitations of slow solubility, high toxicity, and nonspecific bio-distribution profiles (Pisal et al. 2010; Petros and DeSimone 2010). NKTR-102 (PEGirinotecan), a PEGylated form of the topoisomerase I inhibitor irinotecan, is now under phase III/II clinical trials for the treatment of solid tumors (Kang et al. 2009).

### 10.4.3.10 Single-Chain Antigen-Binding Technology

Single-chain variable domain (Fv) fragments (scFv) are powerful tools in research and clinical settings owing to their better pharmacokinetic properties as compared to the parent MAbs, and low cost. scFv can be administered by systemic injection for diagnostic and therapeutic purposes (Chames et al. 2009), and is expressed *in vivo* through viral vectors in instances where large infection rates and sustenance of high levels of the antibody is required. For example, the CC chemokine receptor 4 (CCR4) ligands are highly expressed in breast cancer, ovarian cancer, and cutaneous T-cell lymphoma (Olkhanud et al. 2009; Han et al. 2012), and therefore can be targeted as antitumor therapy.

### 10.4.3.11 Nanobodies

Molecular imaging involves the noninvasive investigation of biological processes *in vivo* at the cellular and molecular levels, which can play diverse roles in better understanding and treatment of various diseases such as cancer. Recently, single-domain antigen-binding fragments known as "nanobodies" are being bioengineered and tested for several molecular imaging applications. Small molecular size (~15 kDa) of nanobodies offers many desirable features suitable for imaging applications, such as rapid targeting and fast blood clearance, high solubility and stability, and the capability of binding to cavities (Chakravarty et al. 2014). Antibody size reduction into a nanobody provides many advantages over conventional antibodies and their recombinant fragments. First, nanobodies are weakly immunogenic in humans as the genes encoding them share a high degree of similarity with the human type 3 VH domain (VH3) (Smolarek et al. 2012). Second, nanobodies can be easily cloned as they consist of only one domain and can therefore be expressed with high yield (Harmsen and De Haard 2007). Third, nanobodies rapidly and specifically bind tumor antigens, whereas the unbound ones are rapidly cleared from the blood circulation typically by renal elimination (Huang et al. 2008; Xavier et al. 2013). Lastly, the high variability of length and sequence of VHHs and the small size allows nanobodies to efficiently enter into tissues followed by binding to epitopes that typically cannot be reached by conventional antibodies (Hassanzadeh-Ghassabeh et al. 2013).

### 10.4.4 TRANSMEMBRANE DRUG DELIVERY TO INTRACELLULAR TARGETS

The use of transmembrane proteins as target for intracellular drug delivery is an exciting possibility based on the role of cancer-specific growth factor receptor domains in receptor-mediated endocytosis (Popov-Celeketic and van Bergen En Henegouwen 2014). Tetraspanins belong to the superfamily of four transmembrane proteins that interact with other membrane proteins and intracellular proteins forming the so-called *tetraspanin microdomains* (Tarrant et al. 2003). Recent studies have exhibited that tetraspanins not only organize proteins in the plasma membrane but they can also directly regulate the signaling pathways (Levy and Shoham 2005; Lapalombella et al. 2012) involved in tumor progression and metastasis. Remarkably, the fusion of protein transduction domain peptide sequences comprising cell membrane-penetrating properties with heterologous proteins is adequate to cause their rapid transduction in a receptor-independent fashion into a variety of different tumor cells. Overall, this technique including transduction of proteins/peptides and receptor-mediated endocytosis may represent the next paradigm to modulate cell function and offer a unique avenue for the treatment of cancer metastasis.

#### 10.4.4.1 Transduction of Proteins and Peptides

The plasma membrane protects the cell from its environment. Although normally permeable only by lipids and small nonpolar molecules, different chemical or mechanical methods will permit transport across the membrane including liposomes and virus particles to transfect cells with plasmids, genes, micro-RNA, oligonucleotides, and mimics of nucleic acids (Jarver et al. 2012). Internalization of peptides/proteins into live cells is an essential prerequisite for the treatment of certain microbial diseases and for the cancer treatment (Monteiro et al. 2014). Cell-penetrating peptides (CPPs) facilitate the transport of cargo-proteins through the cell membrane into the live cells (Reissmann 2014).

#### 10.4.4.2 Receptor-Mediated Endocytosis

The targeted delivery of a therapeutic molecule aims to enhance its circulation and cellular uptake, decrease systemic toxicity, and improve the therapeutic efficacy with disease specificity. The transferrin peptide, its receptor with their biological significance, has been widely characterized and applied to targeting strategies (DeLaBarre et al. 2014). Owing to the increased expression of the transferrin receptor in brain glioma, the successful delivery of anticancer compounds to the tumor site and the ability to cross the blood-brain barrier (BBB) has shown to be an important breakthrough. Its implication in the development of cancer-specific therapies is revealed to be important by direct conjugation and immunotoxin studies, which use transferrin and anti-transferrin receptor antibodies as the targeting moiety (Tortorella and Karagiannis 2014).

Such conjugates have increased selective cytotoxicity in a number of cancer cell lines and tumor xenograft animal models.

### 10.4.5 Biological Therapies

Biological therapy involves the use of living organisms, substances derived from living organisms, or laboratory-produced versions of such substances to treat disease. Antibodies or fragments of genetic materials such as RNA or DNA are used to target cancer cells directly. Vaccines or genetically modified bacteria do not target cancer cells directly but stimulate the body's immune system to act against cancer cells. Recent advances in clinical and preclinical studies have demonstrated the promise of RNAi therapeutics in offering a safe, effective, and more robust approach for the treatment of cancer. However, there are challenges that must be overcome in order for RNAi therapeutics to reach their clinical potential with the refinement of strategies for delivery and to reduce the risk of mutational escape. Here, we provide an overview of RNAi-based therapies, and discuss other approaches for *ex vivo* delivery and *in vivo* delivery to treat cancer.

#### 10.4.5.1 RNA Interference: Small Interfering RNA (siRNA) and MicroRNA (miRNA) as Promising Biomarkers

MicroRNAs (miRNAs) are highly conserved single-stranded small non-coding RNA (ncRNA) molecules (~19–22 nucleotides long) that play a key role in the post-transcriptional gene regulation in cancer biology. These small RNA molecules bind the 3'UTR region of their messenger RNA (mRNA) targets, inducing posttranscriptional gene regulation by either inhibition of translation or mRNA degradation (Macfarlane and Murphy, 2010; Pal and Pal, 2013). miRNA signatures from normal cancer tissues and metastases have been used to classify different types of cancers and to represent potential biomarkers for diagnosis, prognosis, and therapy (Lan et al. 2015). ncRNAs are divided into a number of different categories according to the size and function; however, one common feature is that they are not translated into proteins. The two main types are: (i) microRNAs, which negatively regulate gene expression either by translational repression or target mRNA degradation, and (ii) small interfering RNAs (siRNAs), which are designed to specifically suppress expression of proteins that are traditionally considered nondruggable (Dogini et al. 2014). The currently developed siRNA delivery systems for cancer therapy can be divided into the following categories: chemical modifications, lipid-based nanovectors, polymer-mediated delivery systems, conjugate delivery systems, and others (exosomes, RNAi-microsponges, oligonucleotide nanoparticles) (Kim et al. 2015). Since nanoparticles are potential vehicles for gene delivery, siRNA formulated in nanoparticles has been extensively studied in clinical trials of cancer therapy.

The discovery of RNA interference has opened the door for the development of a new class of cancer therapeutics (Bumcrot et al. 2006). However, a major obstacle of these molecules to use in the clinic is the absence of safe and reliable means for their specific delivery to target cells. In this regard, a highly promising class of molecules is represented by nucleic acid aptamers, which are short, single-stranded RNAs, or DNAs oligonucleotides. These aptamers bind with high specificity and affinity to target molecules, represent a powerful tool for the selective delivery of therapeutic cargos, including toxins, chemotherapeutics, and nanoparticles to cancer cells or tissues, thus it potentially improve the therapeutic efficacy with lower toxicity (Esposito et al. 2014). Other areas of siRNA therapeutics have shown great promising approach in sensitization to chemotherapy, since overexpression of multidrug-resistant genes has been attributed to chemo resistance.

In the last few years, there has been increasing attention in circulating miRNAs as cancer biomarkers, due to their high stability, putative capability to be more informative than mRNA, and the noninvasiveness of their detection (Creemers et al. 2012). Since their findings in body fluids, considerable effort has been directed to explore the relevance of these small RNAs in different diseases such as cancer (Tiberio et al. 2015). The first study that identified specific circulating miRNAs associated to cancer was by Lawrie et al. (2008). They found high levels of miR-155, miR-210, and miR-21 in patients with diffuse large B-cell lymphoma and demonstrated a significant correlation between high levels of miR-21 and relapse-free survival.

## 10.4.5.2   Other Emerging Biological Approaches

### 10.4.5.2.1   Genetically Modified Bacteria

Genetically engineered bacteria localize and grow exclusively in cancerous tissue. This means that they can be used to visualize tumors using bioluminescence or diagnostic imaging technologies and also to treat tumors through direct cell killing, alternative gene therapy, and a mechanism called bactofection, in which bacteria are used as a vector to deliver genes (Baban et al. 2010). Due to the positive outcomes seen *in vitro* and in animal models, many of these genetically modified bacteria are at present being taken to clinical trials (Panteli et al. 2015).

### 10.4.5.2.2   Oncolytic Viruses

Targeted therapy of cancer using oncolytic viruses has generated much attention over the past few years in the light of limited efficacy and side effects of the standard cancer therapeutics for advanced diseases (Wong et al. 2010). In 2006, the world witnessed the first government approved oncolytic virus for the treatment of head and neck cancer. Even though encouraging results have been demonstrated *in vitro* and in animal models, most oncolytic viruses have been unsuccessful to impress in the clinical setting (Wong et al. 2010).

### 10.4.5.2.3   Cell Therapy

Recently, the cell transfer therapy for cancer has made a rapid progress and the immunotherapy has been documented as the fourth anticancer modality after surgery, chemotherapy, and radiotherapy. Lymphocytes used for cell transfer therapy include dendritic cells, natural killer (NK) cells, and T lymphocytes such as tumor-infiltrating lymphocytes (TILs) and cytotoxic T lymphocytes (CTLs) (Qian et al. 2014). Recently, the great successes in adoptive cell transfer therapy (ACT) and the development of anticancer antibodies such as ipilimumab rekindled the interest of the scientific community in the anticancer immunotherapy (Rosenberg et al. 2008).

### 10.4.5.2.4    Gene Therapy

Gene transfer is a new treatment modality that commences new genes into a cancerous cell or the surrounding tissue to cause cell death or slow the growth of the cancer. This therapy assures a number of innovative treatments that are likely to become important in preventing deaths from cancer. Immunotherapy uses genetically modified cells and viral particles to stimulate the immune system of our body to destroy cancer cells. Recent clinical trials of second and third generation vaccines have demonstrated encouraging results with a wide range of cancers including lung, pancreatic, prostate, and malignant melanoma (Cross and Burmester 2006; Thundimadathil 2012).

### 10.4.5.2.5    Antisense Therapy

Over the last few years, antisense technology has come forward as an exciting and promising strategy in the fight against cancer. The antisense concept is to selectively bind short, modified DNA or RNA molecules to messenger RNA in cells and prevent the synthesis of the encoded protein by targeting mRNA with sequence-specific antisense molecules (Kushner and Silverman 2000; Rubenstein et al. 2006). Although different mechanisms for antisense activity have been exploited, the most widely used methods cause the degradation of the targeted RNA.

## 10.5    CONCLUSION AND FUTURE DIRECTIONS

Over the last decade, significant progress has been made in the field of transition of targeted drug therapy for cancer shifting from broad-spectrum cytotoxic agents to highly targeted therapies. The clinical use of conventional chemotherapies is restricted for some harmful side effects on healthy cells and tissues due to nonspecific targeting, and inability to enter the core or the tumors resulting in impaired treatment with reduced dose and with low survival rate. Therefore, ideal therapy is expected to be fast, effective, relatively less toxic, and inexpensive along with higher cure rates, and may even prevent cancer. Among the different drug delivery approaches, nanotechnology has provided the opportunity to get direct access to cancerous cells selectively with increased drug localization and cellular uptake as discussed before. Gene or cell therapies have emerged as realistic prospects for the treatment of cancer. However, there is still much research needed to be done before an efficient and safe cell/gene therapy is achieved. Furthermore, with the advancement in biological research, much cheaper gene vectors may become commercially available in the near future, which will make gene therapy readily available to the majority of cancer patients. This will transform the future of cancer therapy, from generalized cancer treatment strategies, based on tumor size, nature, and location, to an individualized cancer therapy based on patient-specific genomic constituents, host immune status, and genetic profile. Gene transfer with synthetic viruses and non-viral methods, as well as the success in using autologous and allogenic chimeric antigen receptor integrated T-lymphocytes will improve the effectiveness and safety profile of gene therapy. The microspheres are used not only for controlled release but also for targeted delivery of the drugs to specific sites in the body. The development of tumor-targeted therapies using MAbs has been successful during the last

30 years. Nevertheless, the efficacy of antibody-based therapy is still limited and further improvements are eagerly awaited. One of the promising novel developments that may overcome the drawbacks of mAb-based therapies is the use of nanobodies functioning as receptor antagonists, targeting moieties of effector domains, or targeting molecules on the surface of nanoparticles. The last two decades have revealed intense scientific interest in non-coding RNAs. However, the molecular machinery of RNAome, including RNAi, still needs to be exemplified to answer the question of how the same enzymes of RNAi pathway can sometimes switch the genes on and off. Hence, a thorough understanding of the molecular machinery of RNAi is required before the RNAi strategies can be adapted from bench to bedside. Additionally, synthetic and natural biocompatible, biodegradable polymeric biomaterials play a major role in drug delivery systems to destroy cancer cells or reprogram the microenvironment for tumor inhibition and metastasis. Therefore, the impact of *in vitro* 3D tumor models is extensively increasing for understanding the progression of tumor and preclinical anticancer drug screening.

Tumors possess distinct physiological features, which allow them to resist traditional treatment approaches. Therefore, currently there are a few key challenges that need to be addressed for ideal cancer therapy that would eradicate tumor cells selectively with minimum side effects on normal tissues. First, it is needed to treat the tumor as a whole including the *seed* TME rather than just try to eliminate the cancer cells. Second, further investigations are needed to explain the mechanism for regulation of ECM and cancer cell function in human malignancies, and this may help for the design of novel adjunctive cancer therapy in the near future. Third, current challenges over the upcoming decade will be overcoming the resistance to the existing agents with next-generation drugs and through combinatorial therapeutic approaches, so that the use of tissue-specific therapies might delay or prevent disease recurrence. In the near future, certain cytotoxic, or "cell killing" therapies such as chemotherapy and radiation, used in strategic ways, can synergize with immunotherapies to strengthen or expand the anti-tumor immune response. Fourth, another major challenge is the development of cancer-specific new biomarkers or highly sensitive modern devices for early detection of the cancer, since many cancers have a high chance of cure if detected early and treated effectively. Finally, understanding and targeting cancer stem cells is a new paradigm for the combination therapy in solid cancers.

## REFERENCES

Abe, S. and Otsuki, M. 2002. Styrene maleic acid neocarzinostatin treatment for hepatocellular carcinoma. *Curr Med Chem Anticancer Agents* 2:715–26.

Adjei, I.M. and Blanka, S. 2015. Modulation of the tumor microenvironment for cancer treatment: A biomaterials approach. *J Funct Biomater* 6:81–103.

Akbarzadeh, A., Rezaei-Sadabady, R., Davaran, S. et al. 2013. Liposome: Classification, preparation, and applications. *Nanoscale Res Lett* 8:102.

Andemariam, B. and Leonard, J.P. 2007. Radioimmunotherapy with tositumomab and iodine-131 tositumomab for non-Hodgkin's lymphoma. *Biologics* 1:113–20.

Baban, C.K., Cronin, M., O'Hanlon, D. et al. 2010. Bacteria as vectors for gene therapy of cancer. *Bioeng Bugs* 1:385–94.

Babu, A., Templeton, A.K., Munshi, A. et al. 2014. Nanodrug delivery systems: A promising technology for detection, diagnosis, and treatment of cancer. *AAPS PharmSciTech* 15:709–21.

Bidwell, G.L. 2012. Peptides for cancer therapy: A drug-development opportunity and a drug-delivery challenge. *Ther Deliv* 3:609–21.

Boffetta, P. and Nyberg, F. 2003. Contribution of environmental factors to cancer risk. *Br Med Bull* 68:71–94.

Bull, D.A., Seftor, E.A., Hendrix, M.J. et al. 1993. Putative vascular endothelial cell chemotactic factors: Comparison in a standardized migration assay. *J Surg Res* 55:473–79.

Bumcrot, D., Manoharan, M., Koteliansky, V. et al. 2006. RNAi therapeutics: A potential new class of pharmaceutical drugs. *Nat Chem Biol* 2:711–19.

Chakravarty, R., Goel, S., and Cai, W. 2014. Nanobody: The "magic bullet" for molecular imaging? *Theranostics* 4:386–98.

Chames, P., Van Regenmortel, M., Weiss, E. et al. 2009. Therapeutic antibodies: Successes, limitations and hopes for the future. *Br J Pharmacol* 157:220–33.

Creemers, E.E., Tijsen, A.J., and Pinto, Y.M. 2012. Circulating microRNAs: Novel biomarkers and extracellular communicators in cardiovascular disease? *Circ Res* 110:483–95.

Cross, D., and Burmester, J.K. 2006. Gene therapy for cancer treatment: Past, present and future. *Clin Med Res* 4:218–27.

Cucinotto, I., Fiorillo, L., Gualtieri, S. et al. 2013. Nanoparticle albumin bound Paclitaxel in the treatment of human cancer: Nanodelivery reaches prime-time? *J Drug Deliv* 2013:905091.

Curnis, F., Cattaneo, A., Longhi, R. et al. 2010. Critical role of flanking residues in NGR-to-isoDGR transition and CD13/integrin receptor switching. *J Biol Chem* 285:9114–23.

DeLaBarre, B., Hurov, J, Cianchetta, G. et al. 2014. Action at a distance: Allostery and the development of drugs to target cancer cell metabolism. *Chem Biol* 21:1143–61.

Diaz-Cano, S.J. 2012. Tumor heterogeneity: Mechanisms and bases for a reliable application of molecular marker design. *Int J Mol Sci* 13:1951–11.

Dogini, D.B., Pascoal, V.D.B., Avansini, S.H. et al. 2014. The new world of RNAs. *Genet Mol Biol* 37:285–293.

Duggan, S.T. and Keating, G.M. 2011. Pegylated liposomal doxorubicin: A review of its use in metastatic breast cancer, ovarian cancer, multiple myeloma and AIDS-related Kaposi's sarcoma. *Drugs* 71:2531–58.

Egeblad, M., Nakasone, E.S., and Werb, Z. 2010. Tumors as organs: Complex tissues that interface with the entire organism. *Dev Cell* 18:884–01.

Erez, N., Truitt, M., Olson, P. et al. 2010. Cancer-associated fibroblasts are activated in incipient neoplasia to orchestrate tumor-promoting inflammation in an NF-kappaB -dependent manner. *Cancer Cell* 17:135–47.

Esposito, C.L., Catuogno, S., and de Franciscis, V. 2014. Aptamer-mediated selective delivery of short RNA therapeutics in cancer cells. *J RNAi Gene Silencing* 10:500–06.

Fay, F. and Scott, C.J. 2011. Antibody-targeted nanoparticles for cancer therapy. *Immunotherapy* 3:381–94.

Fernandez-Fernandez, A., Manchanda, R., and McGoron, A.J. 2011. Theranostic applications of nanomaterials in cancer: Drug delivery, image-guided therapy, and multifunctional platforms. *Appl Biochem Biotechnol* 165:1628–51.

Folkman, J. 1971. Tumor angiogenesis: Therapeutic implications. *N Engl J Med* 285:1182–86.

Frampton, J.E. and Keating, G.M. 2005. Spotlight on pegfilgrastim in chemotherapy-induced neutropenia. *BioDrugs* 19:405–07.

Furthauer, M. and Gonzalez-Gaitan, M. 2009. Endocytosis, asymmetric cell division, stem cells and cancer: Unus pro omnibus, omnes pro uno. *Mol Oncol* 3:339–53.

Galluzzi, L., Vitale, I., Michels, J. et al. 2014. Systems biology of cisplatin resistance: Past, present and future. *Cell Death Dis* 5:e1257.

Gao, D., Vahdat, L.T., Wong, S. et al. 2012. Microenvironmental regulation of epithelial-mesenchymal transitions in cancer. *Cancer Res* 72:4883–89.

Goldenberg, D.M., Chatal, J.F., Barbet, J. et al. 2007. Cancer imaging and therapy with bispecific antibody pretargeting. *Update Cancer Ther* 2:19–31.

Gottesman, M.M. 2002. Mechanisms of cancer drug resistance. *Annu Rev Med* 53:615–27.

Gradishar, W.J., Tjulandin, S., Davidson, N. et al. 2005. Phase III trial of nanoparticle albumin-bound paclitaxel compared with polyethylated castor oil-based paclitaxel in women with breast cancer. *J Clin Oncol* 23:7794–03.

Guarnieri, D., Biondi, M., Yu, H. et al. 2015. Tumor-activated prodrug (TAP)-conjugated nanoparticles with cleavable domains for safe doxorubicin delivery. *Biotechnol Bioeng* 112:601–11.

Guba, M., von Breitenbuch, P., Steinbauer, M. et al. 2002. Rapamycin inhibits primary and metastatic tumor growth by antiangiogenesis: Involvement of vascular endothelial growth factor. *Nat Med* 8:128–35.

Guba, M., Yezhelyev, M., Eichhorn, M.E. et al. 2005. Rapamycin induces tumor-specific thrombosis via tissue factor in the presence of VEGF. *Blood* 105:4463–69.

Han, T., Abdel-Motal, U.M., Chang, D.K. et al. 2012. Human anti-CCR4 minibody gene transfer for the treatment of cutaneous T-cell lymphoma. *PLoS One* 7:e44455.

Hansen, R.K. and Bissell, M.J. 2000. Tissue architecture and breast cancer: The role of extracellular matrix and steroid hormones. *Endocr Relat Cancer* 7:95–13.

Harmsen, M.M. and De Haard, H.J. 2007. Properties, production, and applications of camelid single-domain antibody fragments. *Appl Microbiol Biotechnol* 77:13–22.

Hassanzadeh-Ghassabeh, G., Devoogdt, N., De Pauw, P. et al. 2013. Nanobodies and their potential applications. *Nanomedicine (Lond)* 8:1013–26.

Heinrich, E.L., Walser, T.C., Krysan, K. et al. 2012. The inflammatory tumor microenvironment, epithelial mesenchymal transition and lung carcinogenesis. *Cancer Microenviron* 5:5–18.

Howerton, B.S., Heidary, D.K., and Glazer, E.C. 2012. Strained ruthenium complexes are potent light-activated anticancer agents. *J Am Chem Soc* 134:8324–27.

Hu, C.M. and Zhang, L. 2009. Therapeutic nanoparticles to combat cancer drug resistance. *Curr Drug Metab* 10:836–41.

Huang, L., Gainkam, L.O., Caveliers, V. et al. 2008. SPECT imaging with 99mTc-labeled EGFR-specific nanobody for in vivo monitoring of EGFR expression. *Mol Imaging Biol* 10:167–75.

Hughes, B.J., Kennel, S., Lee, R. et al. 1989. Monoclonal antibody targeting of liposomes to mouse lung in vivo. *Cancer Res* 49:6214–20.

Iwamoto, F.M., Schwartz, J., Pandit-Taskar, N. et al. 2007. Study of radiolabeled indium-111 and yttrium-90 ibritumomab tiuxetan in primary central nervous system lymphoma. *Cancer* 110:2528–34.

Jain, R.K., Martin, J.D., and Stylianopoulos, T. 2014. The role of mechanical forces in tumor growth and therapy. *Annu Rev Biomed Eng* 16:321–46.

Jarver, P., Coursindel, T., Andaloussi, S.E. et al. 2012. Peptide-mediated Cell and In Vivo Delivery of Antisense Oligonucleotides and siRNA. *Mol Ther Nucleic Acids* 1:e27.

Kaminetzky, D. and Hymes, K.B. 2008. Denileukin diftitox for the treatment of cutaneous T-cell lymphoma. *Biologics* 2:717–24.

Kang, J.S., Deluca, P.P., and Lee, K.C. 2009. Emerging PEGylated drugs. *Expert Opin Emerg Drugs* 14:363–80.

Kapse-Mistry, S., Govender, T., Srivastava, R. et al. 2014. Nanodrug delivery in reversing multidrug resistance in cancer cells. *Front Pharmacol* 5:159.

Kerbel, R.S. 1997. A cancer therapy resistant to resistance. *Nature* 390:335–36.

Khawar, I.A., Kim, J.H., and Kuh, H.J. 2015. Improving drug delivery to solid tumors: Priming the tumor microenvironment. *J Control Release* 201:78–89.

Khazaie, K., Bonertz, A., and Beckhove, P. 2009. Current developments with peptide-based human tumor vaccines. *Curr Opin Oncol* 21:524–30.

Kim, K.K., Kugler, M.C., Wolters, P.J. et al. 2006. Alveolar epithelial cell mesenchymal transition develops in vivo during pulmonary fibrosis and is regulated by the extracellular matrix. *Proc Natl Acad Sci U S A* 103:13180–85.

Kim, Y.D., Park, T.E., Singh, B. et al. 2015. Nanoparticle-mediated delivery of siRNA for effective lung cancer therapy. *Nanomedicine (Lond)* 10:1165–88.

Kraft, J.C., Freeling, J.P., Wang, Z. et al. 2014. Emerging research and clinical development trends of liposome and lipid nanoparticle drug delivery systems. *J Pharm Sci* 103:29–52.

Kratz, F. 2008. Albumin as a drug carrier: Design of prodrugs, drug conjugates and nanoparticles. *J Control Release* 132:171–83.

Krausova, M. and Korinek, V. 2014. Wnt signaling in adult intestinal stem cells and cancer. *Cell Signal* 26:570–79.

Kukowska-Latallo, J.F., Candido, K.A., Cao, Z. et al. 2005. Nanoparticle targeting of anticancer drug improves therapeutic response in animal model of human epithelial cancer. *Cancer Res* 65:5317–24.

Kumar, G. and Manjunatha, B. 2013. Metastatic tumors to the jaws and oral cavity. *J Oral Maxillofac Pathol* 17:71–75.

Kuppen, P.J., van der Eb, M.M., Jonges, L.E. et al. 2001. Tumor structure and extracellular matrix as a possible barrier for therapeutic approaches using immune cells or adenoviruses in colorectal cancer. *Histochem Cell Biol* 115:67–72.

Kushner, D.M. and Silverman, R.H. 2000. Antisense cancer therapy: The state of the science. *Curr Oncol Rep* 2:23–30.

Lam, C.R., Wong, H.K., Nai, S. et al. 2014. A 3D biomimetic model of tissue stiffness interface for cancer drug testing. *Mol Pharm* 11:2016–21.

Lammers, T. 2010. Improving the efficacy of combined modality anticancer therapy using HPMA copolymer-based nanomedicine formulations. *Adv Drug Deliv Rev* 62:203–30.

Lan, H., Lu, H., Wang, X. et al. 2015. MicroRNAs as potential biomarkers in cancer: Opportunities and challenges. *Biomed Res Int* 2015:125094.

Langley, R.R. and Fidler, I.J. 2011. The seed and soil hypothesis revisited—the role of tumor-stroma interactions in metastasis to different organs. *Int J Cancer* 128:2527–35.

Lao, J., Madani, J., Puertolas, T. et al. 2013. Liposomal Doxorubicin in the treatment of breast cancer patients: A review. *J Drug Deliv* 2013:456409.

Lapalombella, R., Yeh, Y.Y., Wang, L. et al. 2012. Tetraspanin CD37 directly mediates transduction of survival and apoptotic signals. *Cancer Cell* 21:694–08.

Lawrie, C.H., Gal, S., Dunlop, H.M. et al. 2008. Detection of elevated levels of tumour-associated microRNAs in serum of patients with diffuse large B-cell lymphoma. *Br J Haematol* 141:672–75.

Levy, S. and Shoham, T. 2005. The tetraspanin web modulates immune-signalling complexes. *Nat Rev Immunol* 5:136–48.

Lu, D., Wientjes, M.G., Lu, Z. et al. 2007. Tumor priming enhances delivery and efficacy of nanomedicines. *J Pharmacol Exp Ther* 322:80–88.

Macfarlane, L.A. and Murphy, P.R. 2010. MicroRNA: Biogenesis, function and role in cancer. *Curr Genomics* 11:537–61.

Maier, H.J., Schmidt-Strassburger, U., Huber, M.A. et al. 2010. NF-kappaB promotes epithelial-mesenchymal transition, migration and invasion of pancreatic carcinoma cells. *Cancer Lett* 295:214–28.

Majidi, J., Barar, J., Baradaran, B. et al. 2009. Target therapy of cancer: Implementation of monoclonal antibodies and nanobodies. *Hum Antibodies* 18:81–00.

Martin, T.A., Ye, L., Sanders, A.J. et al. 2013. Cancer invasion and metastasis: Molecular and cellular perspective. In *Metastatic Cancer: Clinical and Biological Perspectives*, R. Jandial, (Ed.). Austin, Texas: Landes Bioscience.

Melancon, M.P. and Li, C. 2011. Multifunctional synthetic poly(L-glutamic acid)-based cancer therapeutic and imaging agents. *Mol Imaging* 10:28–42.

Minchinton, A.I. and Tannock, I.F. 2006. Drug penetration in solid tumours. *Nat Rev Cancer* 6:583–92.

Monteiro, N., Martins, A., Reis, R.L. et al. 2014. Liposomes in tissue engineering and regenerative medicine. *J R Soc Interface* 11:20140459.

Nemerson, Y. 1988. Tissue factor and hemostasis. *Blood* 71:1–8.

Nersesyan, H. and Slavin, K.V. 2007. Current aproach to cancer pain management: Availability and implications of different treatment options. *Ther Clin Risk Manag* 3:381–00.

Oliveira, S., Heukers, R., Sornkom, J. et al. 2013. Targeting tumors with nanobodies for cancer imaging and therapy. *J Control Release* 172:607–17.

Olkhanud, P.B., Baatar, D., Bodogai, M. et al. 2009. Breast cancer lung metastasis requires expression of chemokine receptor CCR4 and regulatory T cells. *Cancer Res* 69:5996–04.

Pal, M. and Pal, P. 2013. BRCA1 and miRNAs: An emerging therapeutic target and intervention tool in breast cancer. *J PharmaSciTech* 3:9–19.

Panteli, J.T., Forkus, B.A., Van Dessel, N. et al. 2015. Genetically modified bacteria as a tool to detect microscopic solid tumor masses with triggered release of a recombinant biomarker. *Integr Biol (Camb)* 7:423–34.

Peer, D., Karp, J.M., Hong, S. et al. 2007. Nanocarriers as an emerging platform for cancer therapy. *Nat Nanotechnol.* 2:751–60.

Petre, C.E. and Dittmer, D.P. 2007. Liposomal daunorubicin as treatment for Kaposi's sarcoma. *Int J Nanomedicine* 2:277–88.

Petros, R.A. and DeSimone, J.M. 2010. Strategies in the design of nanoparticles for therapeutic applications. *Nat Rev Drug Discov* 9:615–27.

Pisal, D.S., Kosloski, M.P., Balu-Iyer, S.V. 2010. Delivery of therapeutic proteins. *J Pharm Sci* 99:2557–75.

Popov-Celeketic, D. and van Bergen En Henegouwen, P.M. 2014. Membrane domain formation-a key factor for targeted intracellular drug delivery. *Front Physiol* 5:462.

Pranjol, M.Z. and Hajitou, A. 2015. Bacteriophage-derived vectors for targeted cancer gene therapy. *Viruses* 7:268–84.

Qian, X., Wang, X., Jin, H. 2014. Cell transfer therapy for cancer: Past, present, and future. *J Immunol Res* 2014:525913.

Quail, D.F. and Joyce, J.A. 2013. Microenvironmental regulation of tumor progression and metastasis. *Nat Med* 19:1423–37.

Rajput, M.S. and Agrawal, P. 2010. Microspheres in cancer therapy. *Indian J Cancer* 47:458–68.

Ramot, Y. and Nyska, A. 2007. Drug-induced thrombosis—experimental, clinical, and mechanistic considerations. *Toxicol Pathol* 35:208–25.

Reddy, L.H. and Couvreur, P. 2011. Nanotechnology for therapy and imaging of liver diseases. *J Hepatol* 55:1461–66.

Reissmann, S. 2014. Cell penetration: Scope and limitations by the application of cell-penetrating peptides. *J Pept Sci* 20:760–84.

Ribatti, D. 2008. Judah Folkman, a pioneer in the study of angiogenesis. *Angiogenesis* 11:3–10.

Rivera, E. 2003. Liposomal anthracyclines in metastatic breast cancer: Clinical update. *Oncologist* 8:3–9.

Rodriguez, R., Schuur, E.R., Lim, H.Y. et al. 1997. Prostate attenuated replication competent adenovirus (ARCA) CN706: A selective cytotoxic for prostate-specific antigen-positive prostate cancer cells. *Cancer Res* 57:2559–63.

Rosenberg, S.A., Restifo, N.P., Yang, J.C. et al. 2008. Adoptive cell transfer: A clinical path to effective cancer immunotherapy. *Nat Rev Cancer* 8:299–08.

Rubenstein, M., Tsui, P., and Guinan, P. 2006. Bispecific antisense oligonucleotides with multiple binding sites for the treatment of prostate tumors and their applicability to combination therapy. *Methods Find Exp Clin Pharmacol* 28:515–18.

Saif, M.W. 2013. Anti-VEGF agents in metastatic colorectal cancer (mCRC): Are they all alike? *Cancer Manag Res* 5:103–15.

Sajja, H.K., East, M.P., Mao, H. et al. 2009. Development of multifunctional nanoparticles for targeted drug delivery and noninvasive imaging of therapeutic effect. *Curr Drug Discov Technol* 6:43–51.

Sarris, A.H., Hagemeister, F., Romaguera, J. et al. 2000. Liposomal vincristine in relapsed non-Hodgkin's lymphomas: Early results of an ongoing phase II trial. *Ann Oncol* 11:69–72.

Schraa, A.J., Kok, R.J., Moorlag, H.E. et al. 2002. Targeting of RGD-modified proteins to tumor vasculature: A pharmacokinetic and cellular distribution study. *Int J Cancer* 102:469–75.

See, Y.P., Carlsen, S.A., Till, J.E. et al. 1974. Increased drug permeability in Chinese hamster ovary cells in the presence of cyanide. *Biochim Biophys Acta* 373:242–52.

Semeraro, N. and Colucci, M. 1997. Tissue factor in health and disease. *Thromb Haemost* 78:759–64.

Seymour, L.W., Ferry, D.R., Kerr, D.J. et al. 2009. Phase II studies of polymer-doxorubicin (PK1, FCE28068) in the treatment of breast, lung and colorectal cancer. *Int J Oncol* 34:1629–36.

Shea, J.E., Nam, K.H., Rapoport, N. et al. 2011. Genexol inhibits primary tumour growth and metastases in gemcitabine-resistant pancreatic ductal adenocarcinoma. *HPB (Oxford)* 13:153–57.

Sikorska-Fic, B., Makowska, K., and Rokicka-Milewska, R. 1998. New possibilities of treatment with PEG-L-asparaginase in patients with acute lymphoblastic leukemia sensitized to l-asparaginase E.coli and erwinase. *Wiad Lek* 51:233–36.

Smolarek, D., Bertrand, O., and Czerwinski, M. 2012. Variable fragments of heavy chain antibodies (VHHs): A new magic bullet molecule of medicine? *Postepy Hig Med Dosw (Online)* 66:348–58.

Sottani, C., Leoni, E., Porro, B. et al. 2009. Validation of an LC-MS/MS method for the determination of epirubicin in human serum of patients undergoing drug eluting microsphere-transarterial chemoembolization (DEM-TACE). *J Chromatogr B Analyt Technol Biomed Life Sci* 877:3543–48.

Sriraman, S.K., Aryasomayajula, B., and Torchilin, V.P. 2014. Barriers to drug delivery in solid tumors. *Tissue Barriers* 2:e29528.

Sullivan, L.A. and Brekken R.A. 2010. The VEGF family in cancer and antibody-based strategies for their inhibition. *MAbs* 2:165–75.

Tan, B.T., Park, C.Y., Ailles, L.E. et al. 2006. The cancer stem cell hypothesis: A work in progress. *Lab Invest* 86:1203–07.

Tarrant, J.M., Robb, L., van Spriel, A.B. et al. 2003. Tetraspanins: Molecular organisers of the leukocyte surface. *Trends Immunol* 24:610–17.

Thorpe, P.E. 2004. Vascular targeting agents as cancer therapeutics. *Clin Cancer Res* 10:415–27.

Thuault, S., Valcourt, U., Petersen, M. et al. 2006. Transforming growth factor-beta employs HMGA2 to elicit epithelial-mesenchymal transition. *J Cell Biol* 174:175–83.

Thundimadathil, J. 2012. Cancer treatment using peptides: Current therapies and future prospects. *J Amino Acids* 2012:967347.

Tiberio, P., Callari, M., Angeloni, V. et al. 2015. Challenges in using circulating miRNAs as cancer biomarkers. *Biomed Res Int* 2015:731479.

Tiwari, V.K., Mishra, R.C., Sharma, A. et al. 2012. Carbohydrate based potential chemotherapeutic agents: Recent developments and their scope in future drug discovery. *Mini Rev Med Chem* 12:1497–19.

Tortorella, S. and Karagiannis, T.C. 2014. Transferrin receptor-mediated endocytosis: A useful target for cancer therapy. *J Membr Biol* 247:291–07.

Tsai, M.J., Chang, W.A., Huang, M.S. et al. 2014. Tumor microenvironment: A new treatment target for cancer. *ISRN Biochem* 2014:351959.

Uwatoku, T., Shimokawa, H., Abe, K. et al. 2003. Application of nanoparticle technology for the prevention of restenosis after balloon injury in rats. *Circ Res* 92:e62–69.

Valastyan, S. and Weinberg, R.A. 2011. Tumor metastasis: Molecular insights and evolving paradigms. *Cell* 147:275–92.

Visvader, J.E. and Lindeman, G.J. 2008. Cancer stem cells in solid tumours: Accumulating evidence and unresolved questions. *Nat Rev Cancer* 8:755–68.

Willis, B.C. and Borok, Z. 2007. TGF-beta-induced EMT: Mechanisms and implications for fibrotic lung disease. *Am J Physiol Lung Cell Mol Physiol* 293:L525–34.

Wong, H.H., Lemoine, N.R., and Wang, Y. 2010. Oncolytic viruses for cancer therapy: Overcoming the obstacles. *Viruses* 2:78–106.

Wu, Y., Deng, J., Rychahou, P.G. et al. 2009. Stabilization of snail by NF-kappaB is required for inflammation-induced cell migration and invasion. *Cancer Cell* 15:416–28.

Xavier, C., Vaneycken, I., D'Huyvetter, M. et al. 2013. Synthesis, preclinical validation, dosimetry, and toxicity of 68Ga-NOTA-anti-HER2 Nanobodies for iPET imaging of HER2 receptor expression in cancer. *J Nucl Med* 54:776–84.

Yared, J.A. and Tkaczuk, K.H. 2012. Update on taxane development: New analogs and new formulations. *Drug Des Devel Ther* 6:371–84.

Yu, M.K., Park, J., and Jon, S. 2012. Targeting strategies for multifunctional nanoparticles in cancer imaging and therapy. *Theranostics* 2:3–44.

Zhang, J. and Liu, J. 2013. Tumor stroma as targets for cancer therapy. *Pharmacol Ther* 137:200–15.

Zhao, Y. and Adjei, A.A. 2015. Targeting angiogenesis in cancer therapy: Moving beyond vascular endothelial growth factor. *Oncologist.* 20:660–73.

# 11 Mitochondria as an Emerging Target for the Delivery of Small Therapeutic Molecules

*Avik Das, Sabyasachi Maiti, and Kalyan Kumar Sen*

## CONTENTS

## 11.1   INTRODUCTION

The subcellular organelles mitochondria are the prime targets for pharmacological intervention due to their pivotal role in various fundamental metabolic pathways. The malfunctioning of mitochondria may cause or contribute to a large number of human diseases, which include but are not limited to cancer, diabetes, infertility, kidney and liver diseases, and stroke (Modica-Napolitano and Singh 2002; Murphy and Smith 2007; Skulachev et al. 2009; Cunha et al. 2011). The reactive oxygen species (ROS) produced in mitochondria induces mutations and/or defects in mitochondrial DNA (mtDNA) as well as nuclear DNA (Spencer and Sorger 2011). Further, a number of xenobiotics and pharmaceuticals interfere with mitochondrial functions and manifest their toxicity (Wallace and Starkov 2000).

Several research groups have identified different molecular targets for bioactive molecules in mitochondria (Weissig 2003, 2005; Dias and Bailly 2005; Hail 2005; Petit et al. 2005). The dysfunction of energy transfer system, loss of apoptosis (programmed cell death) regulation, and mtDNA mutations are responsible for mitochondrial diseases (Holt et al. 1988; Green and Reed 1998; Wallace 2005). The mitochondrial permeability transition pore complex, potassium channels, and mitochondria-associated anti- and proapoptotic factors could also be the molecular targets (Szewczyk and Wojtczak 2002; Bouchier-Hayes et al. 2005; Dias and Bailly 2005). Therefore, mitochondria have become promising targets for drug delivery.

Despite a number of mechanism-based studies, the development of adequate drug delivery systems limits an effective mitochondrial therapy. Biological barriers and toxicity further complicate the development of mitochondria-targeted therapeutics. On reaching the target cell and entering the cytoplasm, the drug has to overcome additional hurdles such as intracellular diffusion/transport to the mitochondria and outer and inner mitochondrial membranes (IMM). Membrane barriers and mitochondrial toxicity are among other concerns toward the development of an effective mitochondrial medicine (Durazo and Kompella 2011). The design of delivery systems must consider the drug encapsulation method, selection of specific cell-targeting ligands, and regulation of intracellular trafficking in the event of drug-carrier release from the endosome to the cytosol and consequent delivery to the mitochondria (Yamada et al. 2007).

Various molecules such as vitamins, coenzyme Q, and proteins remain functional in the generation of adenosine triphosphate (ATP) by the mitochondria (Saraste 1999).

The dysfunction of the respiratory chain generates mitochondrial myopathy, encephalopathy, lactic acidosis, stroke-like episodes (MELAS), and ophthalmoplegia (Chinnery and Turnbull 2000).

The oral administration of exogenous coenzymes such as vitamin B1 or B2, succinic acid, coenzyme Q, and ATP itself can compensate for their intrinsic deficits in the mitochondria. Despite some extent of therapeutic benefits, the complete recovery of impaired function is not accomplished (Nishikawa et al. 1989; Suzuki et al. 1995; Tanaka et al. 1997). A good understanding of the molecular pathology of mitochondrial diseases has prompted pharmacological and pharmaceutical scientists to find effective therapies for mitochondrial disorders.

The selective delivery of drugs could overcome many problems associated with the random body distribution of the drug. Moreover, a therapeutically effective concentration at the target site is achievable at a reduced dose. This would in turn contribute to the reduction of nonspecific toxic effects of the drug (Weissig et al. 2004). The active control of tissue distribution, cellular uptake, and intracellular trafficking is only possible with the use of a suitable mitochondria-targeting carrier–ligand system.

However, specific strategies have to be developed for selective release of mitochondriotropic substances at the target site of action (Frantz and Wipf 2010; Fulda et al. 2010; Smith and Murphy 2011). Improving the degree of selective accumulation may reduce the required dose and hence should be the major focus of all targeting approaches. Sometimes, the physicochemical properties of a drug substance are not favorable in overcoming biochemical, anatomical, or immunological barriers to reach its cellular or subcellular target (Moghimi et al. 2012). A molecule may require very specific physicochemical properties to pass through the different barriers to find its final destination. However, chemical modifications of drug substances in order to achieve a higher accumulation at their targeted site may also cause a loss of intended biological activity (Ringsdorf 1975). Thus, a specific, careful physicochemical tailoring of drug molecules may be another approach to overcome these barriers. Modern delivery strategies and delivery technologies could further help in better distribution of drugs to their site of action (D'Souza et al. 2011).

Very often, drug targeting is the ability of the drug molecules to interact only with the target. The subcellular target might be a cytosolic molecule or a molecule that is on or inside a membrane-bound organelle. In the latter case, the drug must also be able to enter the organelle and then find its molecular target.

The drug targeting at a subcellular level has not been widely pursued possibly due to technological limitations or the argument that once a drug gets inside a cell, it will eventually find its way to the subcellular target. As the potential bioactive molecules are not adequately mitochondria-specific, multidisciplinary effort can exploit mitochondrial targets for disease therapy.

Mitochondria-targeted drug delivery systems have been the subject of several reviews published in the last few years in highly regarded journals (Murphy and Smith 2000; Armstrong 2007; Yamada et al. 2007; Pathania et al. 2009; Fulda et al. 2010; Durazo and Kompella 2011; Heller et al. 2012; Biswas and Torchilin 2014; Scarpelli et al. 2014). This chapter discusses mitochondrial structures, their physiological role, therapeutic relevance of intracellular drug targeting, small therapeutic compounds suitable for mitochondria targeting, and recent advances in their delivery to the mitochondria.

## 11.2  MITOCHONDRIAL STRUCTURE

Mitochondria, the double-membrane-enclosed capsule-like organelles, are about 0.5–1 μm in diameter and up to 7.0 μm long and, therefore, are not descriptively visible under a light microscope. Nevertheless, the shape and number of mitochondria per cell vary in different tissues. Mitochondria may be filamentous or granular in shape and structural changes occur depending upon the physiological condition of the cells (McBride and Soubannier 2010). Sometimes, they may be clubbed, racket, vesicular, ring, or round shaped. In the primary spermatocyte of rats, they are granular in shape, and are club shaped in liver cells. Two membranes bind mitochondria, and phospholipids are the architectural material of the membrane (Chen and Chan 2010). Each of the mitochondrial membrane is 6 nm in thickness and is fluid mosaic in ultrastructure. The outer membrane is moderately smooth in nature and is composed of different phospholipids, with a protein-to-lipid ratio of about 50:50, and it has a number of copies of transport protein, namely, *porin*, which creates aqueous channels through the phospholipid bilayer. The molecular weight of all molecules is of 10 kDa or less. Ions and small proteins are sifted through this membrane. An 8–10 nm-wide space is present in between the inner side of the outer membrane and the outer side of the inner membrane. This membrane separates mitochondria into two subcompartments, that is, *internal matrix space* and *intermembrane space*. The rough inner membrane is impermeable, and protein and lipid (ratio 80:20) are the structural material (Chen and Chan 2010). The structural features of mitochondria are given in Figure 11.1.

The internal matrix space is filled up with dense, homogeneous, gel-like protein-aceous materials, namely, *mitochondrial matrix*. Lipids, proteins, circular DNA molecules, 55S ribosomes, and certain granules are present in the mitochondrial matrix. Owing to its responsibility of ATP production, the matrix is considered a highly

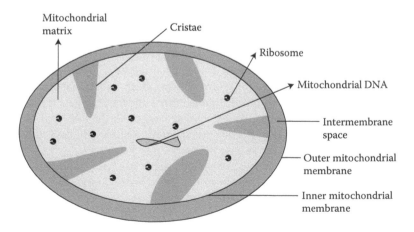

**FIGURE 11.1**   The structural features of a mitochondrion.

functional zone of the mitochondria. The narrow cleft in between the membranes or perimitochondrial space is continuous with the core of cristae and its contents are of low density. The inner membrane has an outer cytosol or C-fact and an inner matrix or M-fact toward the perimitochondrial space and mitochondrial matrix, respectively. A series of folding known as cristae are present in the inner membrane.

Cristae are an important part of mitochondria. In plant mitochondria, cristae are tubular in shape while those of animal are lamellar or plate-like in structure. The packed tubular cristae are present in many protozoa and some steroid-synthesizing tissues, most likely adrenal cortex and corpus luteum (McBride and Soubannier 2010).

Oxysome is another valuable subunit of the inner membrane, attached to the M-fact of the IMM, also called elementary particles. We identify them as $F_1$ particles or $F_0$-$F_1$ particles. Intervals between two $F_1$ particles are 10 nm. According to some estimates, there are 104–105 elementary particles per mitochondrion. Sometimes, submitochondrial vesicles are generated after the disruption of mitochondrial cristae by sonic vibration and $F_0$-$F_1$ particles are attached to their outer surface (Gnaiger 2009).

## 11.3  PHYSIOLOGICAL ROLE OF MITOCHONDRIA

The creation of chemical energy sources (ATP) through oxidative phosphorylation (Saraste 1999) and providing this energy for the biosynthesis and motor activities of the cell are the main physiological functions of mitochondria. Therefore, we call them the cell's "power house." The most important biochemical activities that are accomplished in mitochondria include oxidation, dehydrogenation, oxidative phosphorylation, and respiratory chain of the cell, and their different activities based on their structure and enzymatic system. Among the membranes and matrix, the IMM contains some principal enzymes, that is, cytochromes $b$, $c$, $c_1$, $a$, and $a_3$, for oxidative phosphorylation. ATPase is used along with certain dehydrogenases (Hansen et al. 2000). Monoamine oxidase, a flavoprotein that catalyzes the oxidation of various monoamines, is used as a biochemical marker for the identification of the outer membrane (de la Asunción et al. 1998). The markers used for the matrix system are the enzymes of the tricarboxylic acid (TCA) cycle, namely, malate and glutamate dehydrogenases. Carbohydrates, fatty acids, and amino acids from foodstuffs are oxidized in the mitochondria to carbon dioxide and water, and the released free energy is used for the conversion of adenosine diphosphate (ADP) and inorganic phosphate to ATP. Therefore, they are the chief respiratory organs of the cells. In animal cells, 95% of ATP molecules are produced by mitochondria and the balance 5% are produced during anaerobic respiration outside the mitochondria. In plant cells, chloroplasts produce ATP.

### 11.3.1  OXIDATION OF CARBOHYDRATES

Carbohydrates enter the cells only in the form of monosaccharides such as glucose or glycogen. These hexose sugars are first broken down to pyruvic acid by a set of chemical reactions. After that, pyruvic acid enters into mitochondria for its complete oxidation into carbon dioxide and water. This metabolic pathway can be divided into the following steps:

1. Glycolysis or Embden–Meyerhof pathway (EMP) or Embden–Meyerhof–Parnas pathway (EMPP)
2. Oxidative decarboxylation
3. Krebs cycle or citric acid cycle or TCA cycle
4. Respiratory chain or oxidative phosphorylation

### 11.3.1.1   Glycolysis

In glycolysis, one molecule of glucose degrades into two molecules of the three-carbon compound pyruvate after a series of enzyme-catalyzed reactions. Some of the free energy released in the process is preserved in the form of ATP and NADH. A series of enzymes, located in the cytoplasmic matrix, participate in glycolysis. The total glycolysis pathway can be divided into several steps depending on the enzymatic activity. Three main steps are (a) stage I or activation; (b) stage II or cleavage; and (c) stage III or oxidation.

Reactions 1 through 3 are under the activation stage where the glucose is converted into fructose-1-6-diphosphate and two molecules of ATP are used. Some enzymes such as hexokinase, phosphoglucose (phosphohexose) isomerase, and phosphofructokinase are used in sequence during conversions.

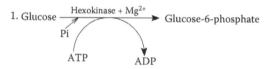

2. Glucose-6-phosphate $\xrightarrow{\text{Phosphohexoisomerase}}$ Fructose-6-phosphate

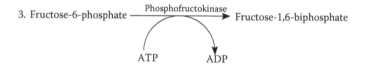

Stage II or cleavage deals with reactions 4 and 5. Here, fructose-1-6-diphosphate breaks into two molecules of glyceraldehyde-3-phosphate and dihydroxyacetone phosphate by the use of two enzymes, fructoaldolase (aldolase) and trios isomerase.

4. Fructose-1,6-biphosphate $\xrightarrow{\text{Fructoaldolase}}$ Glyceraldehyde-3-phosphate (2 molecules)
                                                                        +
                                                                Dihydroxyacetone phosphate

5. Dihydroxyacetone phosphate $\xrightarrow{\text{Triose isomerase}}$ Glyceraldehyde-3-phosphate

The rest of the reaction falls under the oxidation stage. Two molecules of glyceraldehyde-3-phosphate undergo oxidation, and finally convert into pyruvic acid. In this stage, substrate-level phosphorylation produces four molecules of ATP by the use of enzymes such as glyceraldehyde-3-phosphate dehydrogenase, phosphoglycerate kinase, phosphoglycerate mutase, enolase, and pyruvate kinase.

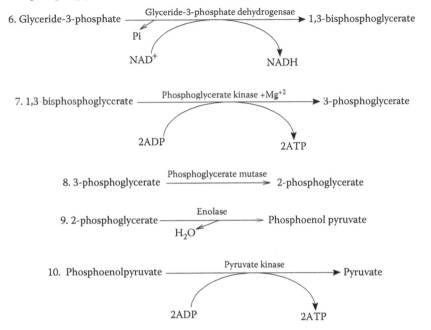

6. Glyceride-3-phosphate $\xrightarrow{\text{Glyceride-3-phosphate dehydrogensae}}$ 1,3-bisphosphoglycerate

Pi

NAD$^+$                                  NADH

7. 1,3-bisphosphoglycerate $\xrightarrow{\text{Phosphoglycerate kinase +Mg}^{+2}}$ 3-phosphoglycerate

2ADP                                  2ATP

8. 3-phosphoglycerate $\xrightarrow{\text{Phosphoglycerate mutase}}$ 2-phosphoglycerate

9. 2-phosphoglycerate $\xrightarrow{\text{Enolase}}$ Phosphoenol pyruvate

H$_2$O

10. Phosphoenolpyruvate $\xrightarrow{\text{Pyruvate kinase}}$ Pyruvate

2ADP                                  2ATP

The net energy yield of these serial reactions of glycolysis pathway is the production of two ATP molecules from one molecule of glucose (Chen and Chan 2010).

### 11.3.1.2 Oxidative Decarboxylation

Pyruvic acid is the end product of glycolysis. The pyruvic acid produced by oxidative decarboxylation undergoes degenerative reactions such as oxidative decarboxylation, Krebs cycle, and oxidative phosphorylation. Oxidative decarboxylation is the process of removal of the carboxyl group from pyruvic acid to convert it to acetyl-CoA with the help of enzyme pyruvic acid dehydrogenase. This process is carried in the mitochondrial matrix utilizing one molecule of ATP for each NADH. The reaction consumes two molecules of ATP for two NADH molecules (McBride and Soubannier 2010).

### 11.3.1.3 Krebs Cycle

The Krebs cycle is also known as the TCA cycle or citric acid cycle. Two acetyl-CoA molecules, which are produced as intermediates in a number of catabolic pathways such as glycolysis and fatty acid oxidation, undergo a set of reactions in the Krebs cycle to produce carbon dioxide, water, and electrons. The enzymes and coenzymes of this cycle are primarily located in the mitochondrial matrix (de la Asunción et al. 1998). Some of them are fused to the M-face of the IMM, the configurable part of complex II of the electron transport chain (ETC). The main function of the citric acid cycle is to oxidize the acetyl component of acetyl-CoA to two molecules of carbon dioxide (Gnaiger 2009). At each cycle, four pairs of hydrogen atoms are liberated as free energy from substrate intermediates by dehydrogenation and two molecules of $CO_2$ are released. This free energy enters into the respiratory chain, and is received by $NAD^+$ and FAD, reducing them to NADH and $FADH_2$, respectively. During the Krebs cycle, two molecules of ATP are also produced via the guanosine triphosphate (GTP) molecule. Most of the cellular needs for the energy carrier ATP are then met by mitochondrial oxidative phosphorylation, for which NADH and $FADH_2$ are the substrates.

### 11.3.1.4  Oxidative Phosphorylation

In mammalian tissues, 80%–90% of ATP molecules are produced in mitochondria by the process of oxidative phosphorylation. It relies on a set of respiratory complexes called ETC, which are located in the IMM. Two molecules of $FADH_2$ and six molecules of NADH produced in the Krebs cycle are oxidized by molecular $O_2$ with a group of enzymes and coenzymes (low-molecular-weight redox intermediates) in the respiratory chain or ETC (McBride and Soubannier 2010). Electron complexes transport hydrogen atoms or just their electrons from respiratory substrates to molecular oxygen, down the redox potential. Thermodynamic calculation shows that for the synthesis of several moles of ATP, oxidation of $NADH_2$ acts as the driving force. In the electron transport system, electron acceptors are at lower energy levels. At each stage, the released energy, the so-called potential energy, is used to form ATP.

In the last four decades, a number of papers have tried to throw light on various concepts of the generation of free energy through the oxidation of substrates stored by the cell and the simultaneous coupling of electron transport and oxidative phosphorylation. Of these hypotheses, Edward Slater's chemical coupling hypothesis and Paul Boyer's conformational coupling hypothesis gained the attention of the scientific community, although Peter Mitchell's chemiosmotic theory, proposed in 1967, is the mostly accepted one until now. The chemiosmotic theory seems to be consistent with scientific experimental information. It suggests the coupling of respiration and ATP synthesis (Duchen 2004). The proton gradient, built up by the enzymatic complexes of the ETC, conserves the free energy of electron transport. These complexes pump protons from the matrix to the intermembrane space and create an electrochemical gradient across the inner membrane. This gradient then enables the ATPase to synthesize ATP.

> *Complex I:* Complex I or NADH-coenzyme Q reductase moves electrons from NADH to CoQ with the help of the iron-sulfur protein, which is used as an electron carrier of mitochondria. This protein transfers one electron at a time. Two coenzymes, flavin mononucleotide (FMN) and $CoQ_{10}$ (a form of ubiquinones) of complex I, have considerable proficiency to accommodate up to two electrons each in stable conformation and donate one or two electrons to the cytochromes of complex III. It has protein-pumping activities; pumps four protons for each pair of electrons.
>
> *Complex II:* Complex II or succinate-coenzyme Q reductase contains succinate dehydrogenase (SDH) and three small hydrophobic subunits. It moves electrons from succinate by using coenzyme (FAD), three iron-sulfur proteins, and cytochrome $b_{560}$, but has no protein-pumping activity.

*Complex III:* Complex III or coenzyme Q-cytochrome *c* reductase trans-
fers electrons from lower CoQ to cytochrome *c*. Two cytochrome *b*, one
cytochrome *c*, and an iron-sulfur cluster are present in this complex. Here,
proton-pumping activity is in active mode and two protons are pumped for
each pair of electrons. In complex III, electron transport and proton translocation are facilitated by the Q-cycle.

*Complex IV:* Complex IV or cytochrome *c* oxidase catalyzes the last stage of
the electron transfer by the reduction of oxygen to water. Four protons for
each pair of electrons are translocated in complex IV.

Proton translocation is an energy-requiring process because it takes place against
an electrochemical gradient. The electrochemical gradient or proton motive force
($\Delta P$) is the result of proton-translocating activities of complex I, II, and IV. It is dependent on two components, electrical potential ($\Delta\Psi_m$) and pH as given in Equation 11.1.

$$\Delta P = \Delta\Psi_m + \Delta pH \tag{11.1}$$

### 11.3.2 ATP SYNTHASE

The ATP synthase or $F_0$-$F_1$ ATPase is one of the principal proteins in the IMM
(Gnaiger 2009). It utilizes proton motive force ($\Delta P$) to convert ADP and phosphate
to ATP and simultaneously couple electron transport and proton pumping to ATP
synthesis (Duchen 2004). Structurally, it is a multi-subunit coupling factor and has
two principal components: (a) $F_0$-complex, the fundamental membrane complex,
composed of hydrophobic proteins and one proteolipid. $F_0$-complex possesses the
proton-transporting mechanism. (b) $F_1$-complex, a complex of five distinct polypep-
tides, namely, alpha ($\alpha$), beta ($\beta$), gamma ($\gamma$), delta ($\delta$), and epsilon ($\epsilon$), with the
probable composition of $\alpha_3 \beta_3 \gamma \delta \epsilon$. $F_1$ forms the knob or "tadpole" that protrudes on
the matrix side of the IMM. $F_1$ particle is capable only of catalyzing the hydrolysis
of ATP into ADP and phosphate when physically separated from the IMM. During
ATP synthesis, protons are translocated into the matrix and can be transported to the
cytosol by the adenine nucleotide translocase.

## 11.4 MITOCHONDRIAL DYSFUNCTION: PRIMARY
## AND SECONDARY

Mitochondrial involvement in the pathology of various diseases is quite well established. The disruption of normal mitochondrial activity is grouped under two main
types, namely, primary and secondary. The etiology behind the development of primary mitochondrial dysfunction is organelle specific. It can be due to some mitochondrial toxin or a mitochondria-specific protein encoded by the nucleus. The
etiology may also include the deterrent expression of a gene encoded by the mtDNA,
which is mutated. MELAS is a classic example of such primary dysfunction that
takes place due to the mutation at np 3243 in the mitochondrial tRNA[leu(UUR)] gene
that leads to the defective assembly of oxidative phosphorylation complexes and the
consequent defects in neuromuscular systems. The genes encoding for mitochondrial

proteins may also suffer mutation, thus giving rise to energy metabolism defects in neonates (Mitchell et al. 2011). These defects are commonly associated with the mutation in the NDUFAF3 gene that encodes for an assembly factor of complex I of the respiratory chain. Nuclear-encoded mitochondrial genes are also associated with a plethora of functional anomalies of mitochondria other than oxidative phosphorylation, which includes assembly, dynamics, and metabolic function (Prime et al. 2009). Although it was earlier thought that primary mitochondrial pathology occupied a minor place in the map of clinical epidemiology, the idea however was altered in view of the large number of clinical cases linked primarily with mitochondrial pathology. This may be partly due to the diversity in the clinical presentation of such cases (Wallace and Starkov 2000). In comparison, the chances of occurrence of secondary mitochondrial pathology are much more common, and are generally precipitated by extra mitochondrial pathologies. One such classical example is ischemia reperfusion injury (I/R injury). I/R injury occurs due to the sudden interruption of blood flow to a region followed by reestablishment of perfusion, which is generally associated with extensive secondary mitochondrial disruption and consequent tissue damage. The underlying etiology may be the consequence of the failure of cascade reduction of oxygen through the respiratory chain. Other disorders in which secondary mitochondrial damage plays a significant role include sepsis, neurodegeneration, metabolic syndrome, organ transplantation, cancer, autoimmune diseases, and diabetes (Smith et al. 2012).

Limitations in the scope of specific targeting of mitochondria restrict the development of newer therapies aimed at the treatment of primary mitochondrial disorders. Moreover, since it is still hard to target mtDNA with the available methods to date, the pharmacological agents used for managing mitochondrial pathology mainly ameliorate rather than address the etiology underpinning it, with a curative approach. On the other hand, the clinical scenario of secondary mitochondrial pathologies is somewhat different where treatment approaches are mainly targeted against the primary disease itself rather than directly targeting mitochondria. Since the development of both primary and secondary mitochondrial pathologies follows nearly a similar pattern (Smith et al. 2012), a good number of drugs available with us in our armament can be used against a wide array of both primary and secondary mitochondrial pathologies. However, specific subcellular targeting indeed has certain therapeutic benefits.

## 11.5 THERAPEUTIC RELEVANCE OF TARGETING SPECIFIC SUBCELLULAR SITES

The distribution of the administered drug to the site of pathology is an important determinant of the agent's therapeutic efficacy. The concept has its basis in the "magic bullet" approach proposed by Ehrlich in the early nineteenth century. Although a plethora of methods has been developed to date for *in vivo* drug targeting (Torchilin 2008), there are still certain limitations in the subcellular targeting of drugs. To date, nucleus and mitochondria are two important sites, which have been used for subcellular targeting with a certain degree of therapeutic relevance (Saito et al. 1998).

The desired effects of therapy directed toward mitochondria are the antioxidant or proapoptotic ones. In fact, the activation of the proapoptotic Bcl-2 family of proteins has been proved to be useful for stimulating mitochondria-based apoptotic pathways in treating various forms of cancers (Sheu et al. 2006). The same approach may be therapeutically relevant in treating diseases, which has a mitochondrial component underpinning their pathology such as neurodegenerative diseases, I/R injury, diabetes, etc. Rapid scientific advances in the field of cytology and biochemistry have provided insight into the structural and functional properties of subcellular locations such as mitochondria, though much is still unknown about mitochondrial physiology (Armstrong 2007). Nevertheless, high-voltage electron tomographic analysis of mitochondrial structure has boosted our efforts to elucidate the full functional potential of mitochondria, which call for a greater scientific effort in the field to project mitochondria as an effective target for a plethora of complicated diseases.

## 11.6 ROLE OF MITOCHONDRIA IN TRIGGERING CELLULAR TOXICITY

Mitochondria play a key role in the maintenance of a variety of cellular functions and act as the powerhouse of the cell. As such, mitochondrial toxicity has been implicated in the pathology of various diseases (Murphy 2009). The majority of cellular processes such as biosynthesis of endogenous compounds, cellular transportation, generation of kinetic energy, etc. require the involvement of ATP, and maintenance of a physiological level of ATP concentration is crucial to the smooth running of cellular activities (Sheu et al. 2006). The production of ATP from ADP and inorganic phosphate ($P_i$) in an ATP synthetase-mediated pathway occurs through a complex cascade mechanism and any perturbation in this process ultimately results in the inhibition of ATP synthesis, thus giving rise to cellular energy crisis.

Three modes of damage generally contribute to the development of primary and secondary mitochondrial pathologies (Mitchell et al. 2011). They include oxidative damage, deregulation of intracellular calcium homeostasis, and disruption to ATP biosynthesis. The inner membrane of mitochondria, which is stacked with unsaturated fatty acids and densely packed proteins, makes it sensitive to oxidative damage (Murphy 2009). The presence of several iron sulfur centers in this organelle adds up to this vulnerability. Damage to the respiratory chain can also contribute to disruption of mitochondrial ATP production that may precipitate cellular death and dysfunction. These anomalies underlie the pathologies of various diseases such as I/R injuries, diabetes, etc. Loss of normal cellular homeostasis of calcium ions can be triggered by disruption in ATP synthesis (Smith et al. 2012) due to altered calcium ATPase activity in the sarcoplasmic /endoplasmic reticulum. Although the increased cytosolic calcium concentration sensitizes the calcium uniporter in the mitochondrial membrane and the calcium load starts building in mitochondria, but with time the rising level of intramitochondrial calcium may prove to be fatal and give rise to toxicity (Murphy and Smith 2007). All of these culminate into the activation of *cyclophilin D*, a matrix *cis-trans* isomerase that aids in the formation of mitochondrial permeability transition (mPT) (Mammucari et al. 2011).

The possibility of mitochondria-specific toxicity is directly connected to the mitochondrial number in a particular tissue that again can be directly linked to its energy demand according to the physiological function (Prime et al. 2009). Thus, the chances of precipitation of mitochondrial toxicity are expected to be high in metabolically active tissues such as muscles, liver, brain, cardiac tissues, etc.

## 11.7 XENOBIOTIC-INDUCED INTERFERENCE IN MITOCHONDRIAL FUNCTIONALITY

Three anatomical features of IMM play a pivotal role in mitochondrial toxicity. This part that is chiefly made up of cardiolipin, a form of lipid, serves as an attractant for a number of xenobiotics because of their high affinity toward cardiolipin. The second feature, that is, the high negative transmembrane potential serves as the motive force for the intramitochondrial transfer of small cationic xenobiotics, which may act as a trigger for various toxicological processes. The third, that is, the mitochondrial permeability transition pore, a megachannel spanning the mitochondrial membrane, serves as a channel for the transfer of different xenobiotic moieties into the mitochondrial compartment. Disarray in the opening and closure of this pore results in the abolishment of the membrane potential and triggers the release of signaling molecules such as cytochrome $c$ into the cytosol, thus precipitating toxicity.

The xenobiotic-induced functional damage of mitochondria can most often occur by any of these four mechanistic pathways:

1. Protonophoretic and uncoupling pathway
2. Disruption of mitochondrial fatty acyl beta-oxidation
3. Disruption of ETC
4. Opening of mitochondrial membrane permeability transition pore (MPTP)

### 11.7.1 PROTONOPHORETIC AND UNCOUPLING PATHWAY

Lipophilic and weakly acidic xenobiotics can act as protonophores. They are generally dissociated in the physiological pH, and pass through the outer layer of the mitochondrial membrane to reach the intermembrane space because of their high lipophilicity. This space consists of high proton population, necessary for driving the ATP synthetase, and so protonates the xenobiotic moiety thus neutralizing it (Saraste 1999). The uncharged lipophilic moieties readily cross the inner membrane and reach the inner mitochondrial matrix where there is a surplus of negative charges, which again make the xenobiotic moiety negatively charged. This negatively charged moiety again gets recycled back into the intermembrane space and this cycle continues. In this process, the protons, which ought to be there in the intermembrane space, get transported to the matrix by this xenobiotic, and thus act as protonophores (Fromenty and Pessayre 1995). This causes the subsequent uncoupling of oxidative phosphorylation. A classic example of a xenobiotic inducing mitochondrial toxicity through this pathway is a widely used fungicide and wood-impregnating agent known as pentachlorophenol (PCP). The hazards associated with its use include increased body temperature, polydipsia, respiratory problems, muscle fatigue, and dizziness. All of these symptoms are indicative of energy crisis, thus signifying mitochondrial toxicity.

The consequences of uncoupling of oxidative phosphorylation are not always detrimental. Since the uncoupling of oxidative phosphorylation results in increased oxygen consumption and greater heat production, this mechanism is often used by many hibernating animals for maintaining body temperature in the absence of ATP. Specific mitochondrial proteins called uncoupling proteins (UCPs) have also been identified in the cells of the brown adipose tissues of such animals, which aid in the process of uncoupling.

## 11.7.2 Disruption of Mitochondrial Fatty Acyl Beta Oxidation

The large fatty acid molecules are progressively curtailed to acetyl-CoA products for their subsequent reduction into ketone bodies or entrance into the citric acid cycle for NADH production. NADH acts as the major reducing equivalent in mitochondria and provides electrons for driving proton pumps, which subsequently maintain cellular energy homeostasis. The process of reducing fatty acids into shorter moieties is called beta-oxidation. The enzymes required for the process are located in the mitochondria. So naturally, for the initiation of the beta-oxidation process, the fatty acids have to first pass through the IMM.

Although the short-chain and medium-chain fatty acids can pass through the membrane readily, long-chain fatty acids require a carrier for the transmembrane passage. For that, they are first activated by coenzyme A to the CoA thioester and are subsequently coupled to a transmembrane shuttle molecule called carnitine.

In the matrix, the fatty acyl-CoA is abridged to acetyl-CoA that subsequently enters the citric acid cycle for NADH generation. The xenobiotics can disrupt any step of this multistage process, which can precipitate toxicity.

The hallmark drug precipitating toxicity through this pathway is valproic acid. A widely used antiepileptic drug, it has a more or less good safety profile, except in reported cases of liver injury associated with microvesicular steatosis. Two main mechanisms have been caught up in the tissue-specific toxicity associated with this drug (Mammucari et al. 2011). First, the valproic acid metabolites cause a decrease in the cellular level of free CoA, which is the main factor underpinning the inhibition of beta-oxidation pathway associated with them. Valproic acid gets coupled to CoA by the medium-chain acyl-CoA that is located chiefly in the mitochondrial matrix. Thus, it can be inferred that the intramitochondrial pool of CoA is depleted by the chronic use of this drug (McLaughlin et al. 2000). Second, the metabolites of valproic acid such as $\Delta^4$ valproic acid, which is activated inside the mitochondrial matrix by CoA can cause direct inhibition of enzymes involved in beta-oxidation. Sustained inhibition of beta-oxidation can lead to an energy crisis that can damage tissues such as the liver to a great extent because of their high metabolic demand. Inhibition of the beta-oxidation process is also implicated in the development of microvesicular steatosis that occurs as a consequence of the accumulation of fatty acids. Thus, an imbalance in mitochondria-specific physiological processes can lead to severe toxicity, which may be triggered by xenobiotics.

### 11.7.3 ELECTRON TRANSPORT CHAIN DISRUPTION AND INCREASED ROS GENERATION

Mitochondria-dependent electron transfer cascade is the main pathway for the generation of ATP in the cell. Inhibition of this cascade can lead to the deregulation of cellular energy homeostasis.

Two types of inhibitors can disrupt ETC. One type blocks the chain by binding to one of the components of the cascade. The second type includes compounds, which stimulate the electron flow initially, but diverts the same from their physiological path by accepting the electrons themselves. Xenobiotics such as doxorubicin (DOX) precipitate toxicity by this second mechanism (Singal et al. 1992). This antineoplastic agent belongs to the class of anthracyclines and is used for the treatment of various solid cancers and lymphomas, but despite having a good efficacy, its use in the treatment of cancer is limited because of the dose-dependent irreversible cardiomyopathy associated with its use that precipitates congestive heart failure in a large proportion of patients. DOX that possesses a quinine moiety acts as a good electron acceptor and as such diverts electron from complex I of the respiratory chain, reducing quinine to its semiquinone radical. The semiquinone so formed is highly unstable and auto-oxidizes readily to the quinone moiety (Zhou et al. 2001). Because it has a high redox potential, the quinine moiety reduces molecular oxygen to superoxide anion radical and initiates a redox cycle. Because of this oxidation, injury sets in. Now, the heart, which is in need of continuous supply of ATP, becomes starved of energy, and because of its relatively low antioxidant reserve, succumbs easily to the oxidative damage.

DOX also affects the mitochondrial DNA by oxidizing it. The oxidized DNA bases gradually accumulate in the mitochondria, which leads to chronic cardiomyopathy, and may be the underpinning mechanism behind the delayed toxicity associated with DOX use. The other type of blockers generally bind to a component of the respiratory chain. A typical example of such a toxicant is cyanide (Mitchell et al. 2011). Cyanide binds avidly to the ferric form of iron, which is found profusely in heme-containing proteins undergoing redox cycling generally to the cytochrome $a_3$ moiety of complex IV (Smith et al. 2012). Blockage of this complex leads to the severe depletion of cellular ATP content resulting in cell death due to energy impairment.

### 11.7.4 OPENING OF MITOCHONDRIAL MEMBRANE PERMEABILITY TRANSITION PORE

The mitochondrial membrane is a sentinel membrane regulating the entry of various solutes into the mitochondrial matrix. Recently, a physiological pore has been identified in the mitochondrial membrane, which has been found to provide ready access to a wide range of solutes across the membrane thus contributing to the mechanism of cell death by the disruption of mitochondrial permeability (Armstrong 2007). This pore is a membrane-bound protein complex consisting of a number of proteins, including voltage-dependent anion channel (VDAC) located in the outer membrane, cyclophilin D of the matrix, and ADP/ATP exchanger of the inner membrane

(Fromenty and Pessayre 1995). The opening of this channel allows the entry of molecules with a molecular weight of about 1500 Da to pass across the channel (Wallace and Starkov 2000). The rapid opening of this channel causes a matrix-bound, inward surge of solutes because of which the mitochondrial proton gradient collapses, leading to the uncoupling of oxidative phosphorylation. In addition, the intramitochondrial solutes are released. Finally, because of the increased osmolarity of the matrix compared to that of the cytosol, mitochondria swell up due to water influx, ultimately leading to the disruption of the mitochondrial membrane (Prime et al. 2009). Oxidative stress and toxic hydrophobic bile acids that accumulate in the liver due to cholestasis are well-established openers of the MPTP.

## 11.8 POTENTIAL THERAPEUTIC COMPOUNDS AFFECTING MITOCHONDRIA: A MECHANISTIC AND DESCRIPTIVE APPROACH

Mitochondria are macromolecular cellular organelles that drive the majority of energy production by producing 80%–90% of the total ATP production of our body. The complexes embedded in the IMM carry out the majority of mitochondrial ATP production through a process known as oxidative phosphorylation. Besides energy production, mitochondria are responsible for other vital functions of cells such as induction of apoptosis, phase II metabolism of xenobiotics, etc. From recent studies, cardiac mitochondria have also been found to play an important role in cardioprotection.

Mitochondria can be affected by a plethora of xenobiotics through a variety of mechanisms that range from the uncoupling of oxidative phosphorylation to dissipation of mitochondrial membrane potential and inhibition of membrane potential (Frøyland et al. 1997). For some xenobiotics, the mitochondrion acts as a primary site of action as intended, while for others, it is a secondary site of action even when it is not intended to do so. The function of mitochondria is also interrupted by various toxins, such as cyanides (blocks cytochrome oxidase) and oligomycin (blocks ATP synthase), etc. As mitochondria are covered by two phospholipid bilayer membranes, both lipophilic and amphiphilic drugs get access across the mitochondrial membrane.

## 11.9 CERTAIN CLASSES OF DRUGS AFFECTING MITOCHONDRIAL FUNCTION

### 11.9.1 POTASSIUM CHANNEL OPENERS

Like the plasma membrane, mitochondria have small conductance potassium channels, which allows $K^+$ ion migration to the inside of mitochondria. Typically, high ATP concentration blocks these channels, and GTP opens these channels (Epps et al. 1982). Two types of potassium channels are mainly present. One is ATP regulated and the other calcium activated. Potassium channel openers (KCOs), which are still considered a cornerstone in cardiovascular therapy, are particularly vulnerable to affect the potassium channels present on the mitochondrial membrane since their cardioprotective effects are thought to involve mitochondrial pathways. The ATP-controlled potassium channels discussed earlier have been discovered in the studies carried out on rat liver and beef heart (Aleksandrowicz et al. 1973). The blocking of the channels by ATP and also by antidiabetic drugs like sulfo-nylureas hinted of the possibilities of their structural and functional homology with the channels present on the cell membranes. Thus, the chances of involve-ment of KCOs in the modulation of mitochondrial functions are bright (Chance et al. 1979). In fact, some KCOs such as cromakaline, nicorandil, and diazoxide activate the potassium channels of mitochondria. These KCOs also have targets in other organs, such as plasma membrane ($K^+$ channels), sarcoplasmic reticulum (Kourie 1998), zymogen granules (Thévenod et al. 1992), etc. KCOs that are still in the pipeline have also been shown to increase potassium influx and subsequent depolarization in liver mitochondria. KCOs such as diazoxide have been shown to affect the mitochondrial energy metabolism by inhibiting complex II of the respiratory chain.

The KCOs have been shown to be effective in providing protection against car-diac ischemia by the mechanism of ischemic preconditioning of the cardiac tissue. The mitochondrial ATP channels conventionally designated as $mitoK_{ATP}$ perhaps are the prime mediator of such action serving as the main targets for molecules such as diazoxide. This has found basis in the fact that diazoxide is capable of induc-ing oxidation of mitochondrial flavoproteins mediated by $mitoK_{ATP}$. The $mitoK_{ATP}$-mediated cardioprotection has three mechanistic facets (Aleksandrowicz et al. 1973). First, the opening of the mitochondrial $K_{ATP}$ channels cause mitochondrial swelling, which improves mitochondrial ATP production. This is thought to be the mecha-nism behind the protective effect of diazoxide in the ischemic heart. Second, the opening of mitochondrial ATP can reduce the intramitochondrial calcium overload-ing, which also contributes to the protective effect of KCOs. Third, the opening of the mitochondrial $K_{ATP}$ channels can increase the production of ROS-mediated protein kinase C, a mechanism pivotal in cardioprotection (Ashcroft and Ashcroft 1992). Thus, the KCOs that are instrumental in giving cardioprotection exert many of their effects mediated through mitochondria.

### 11.9.2 Sulfonylurea Derivatives

There are two classes of sulfonylureas used to treat diabetes and cancer, respectively. Sulfonylurea receptors (SURs) are present on the plasma membranes of various cell types, including pancreatic beta cells where they cause exocytotic release of insu-lin by closure of ATP-sensitive potassium channels (Szewczyk and Wojtczak 2002). Besides its intended action, antidiabetic sulfonyl ureas such as glibenclamide also block ATP-sensitive $K^+$ pump in the mitochondrial inner membrane, which has been previously activated by $Mg^{2+}$ or physiological activators such as GTP (Jabůrek et al. 1998). Glibenclamide increases the proton conductance of mitochondria due to the hydrophobic characteristics of its protonated form (Szewczyk 1997). Glibenclamide and tolbutamide inhibit carnitine palmitoyl transferase and thus affect fatty acid oxi-dation (Patel 1986). The activity of pyruvate carboxylase is blocked as well (White et al. 1988).

A group of antineoplastic sulfonylureas–diarylsulfonylureas uncouples the oxidative phosphorylation in mitochondria and lowers cellular ATP. Antineoplastic agents such as arylsulfonylureas, including 4-chlorophenylureas and *N*-4 methyl analogs, also disrupt oxidative phosphorylation by uncoupling the mitochondrial complexes (Thakar et al. 1991). This uncoupling depends on the dissociation of a single proton but a direct correlation of the antineoplastic activity and uncoupling is yet to be proved.

### 11.9.3 LIGANDS OF PERIPHERAL BENZODIAZEPINE RECEPTORS

Benzodiazepines have a myriad of pharmacological effects, mediated through gamma-aminobutyric (GABA) receptors. Although the GABA receptors are most widely distributed in the central nervous system, their presence in the peripheral nervous system PNS is also abundant and they are designated as PBRs. They are located on the mitochondrial membrane. They typically consist of a pore protein (porin), isoquinoline binding protein, and adenine nucleotide translocase (Epps et al. 1982). They are almost similar structurally and functionally to the mitochondrial PTP. An 18 kDa peptide has been found to activate this receptor and allows the transfer of up to 1.5 kDa molecules from cytosol. This is also known as mitochondrial mega pore and its activation is related to the modulation of apoptosis. The physiological ligands of this receptor are protoporphyrin IV and a neuropeptide known as diazepam binding inhibitor consisting of 104 amino acids. Although the physiological functions of PBR are unclear, it is imperative that many of the mitochondrial functions can be attributed to those receptors (Hansen et al. 2000). The PBRs in steroid synthesizing cells play a key role in the transport of cholesterol across the mitochondrial membrane. This finds support in the action of PBR agonist RO5-4864, which has shown protective activity against TNF-$\alpha$-induced apoptosis through mitochondria-mediated pathway.

### 11.9.4 IMMUNOSUPPRESSIVE DRUGS

Many potent immunosuppressive agents such as cyclosporin A block the transcription of cytokine genes in activated T cells by inhibiting the calcineurin/nuclear factor of activated T cells (NFAT) pathway. Apart from immunosuppressant action, cyclosporin A also blocks mitochondrial PTP by forming complex with cyclophylin D, an effect independent of the immunopsuppressive one because one of its analogs, *N*-methyl-val-4-cyclosporin, that lacks the immunosuppressive function, shows PTP blocking (Zamzami et al. 1996). This beneficial action provides relief from ischemia and traumatic brain injury in animal studies (Osborne et al. 1996).

Traumatic brain injury is often associated with the impairment of long-term potentiation of synaptic transmission, which is ameliorated greatly with agents such as cyclosporin A in a mitochondria-mediated pathway. The inhibitory action of cyclosporin A on PTP is also thought to be associated with its protective action on cardiomyocytes against reperfusion injury. However, not all immunosuppressants show mitochondrial effect as that of cyclosporin A. Tacrolimus that acts through the same signaling pathway as that of cyclosporin A does not show such activity.

## 11.9.5 ANTIVIRAL DRUGS

Many nucleoside analogs are antiviral drugs that often target organelles such as mitochondria. Drugs such as AZT that are used to treat HIV infection can affect mitochondria via two mechanisms. The short-term mechanism involves a detrimental effect on mitochondrial enzymes, disrupting the respiratory chain, while the long-term mechanism damages the mitochondrial DNA, thus impairing mitochondrial protein synthesis (Frøyland et al. 1997). Drugs such as AZT in both muscular cells and hepatic tissue have shown organ-specific mitochondrial damage. It has been reported that AZT increases ROS and peroxynitrite formation in cardiac tissue that causes oxidative damage to mitochondrial DNA. AZT also inhibits the polymerase responsible for mitochondrial DNA replication (Chance et al. 1979). Antiviral drugs such as zidovudine block HIV DNA replication. Besides, they also interact with mitochondria. Their short-term use causes damage to the respiratory proteins of mitochondria, resulting in cardiomyopathy. They also elevate the ROS level and oxidation of cellular proteins. They rapidly decrease ATP production and cause lethal effects to the cell (Szabados et al. 1999). Long-term exposure can also alter mitochondrial circular DNA by causing it oxidative damage (de la Asunción et al. 1998). They can also block complex III production by blocking the mitochondrial DNA-coded mRNA responsible for its synthesis. They also increase the β-myosin heavy-chain protein mRNA content and change the contractility of cardiac muscle (McCurdy and Kennedy 1998).

Antiviral drugs such as dideoxycytidine have significantly increased the formation of ROS in heart and skeletal muscles and caused malfunctions in the cardiac tissue of growing animals (McCurdy and Kennedy 1998). However, the mechanism of damage is independent of the drug's effect on mitochondrial DNA. Thus, it can be inferred that selected antivirals, though not all, have their actions largely mediated through mitochondria-dependent pathways.

## 11.9.6 NSAIDs

Nonsteroidal anti-inflammatory drugs, which have been so effective in the treatment of pain and inflammation associated with arthritis, have also shown promise against colorectal cancers (Hostetler and van den Bosch 1972). In spite of being effective healers, their use is associated with a plethora of damaging effects such as ulceration, perforation, and bleeding; the mechanistic basis being the mitochondrial-mediated pathways. This group of chemicals uncouples the mitochondrial respiratory chain enzymes and alters the mitochondrial transmembrane electrical potential (Mingatto et al. 2000). NSAIDs such as aminosalicylic acid undergo metabolism within the body into their active form, that is, salicylic acid that induces mitochondrial dysfunction *in vivo*. It also facilitates the opening of PTP and elevates the level of $Ca^{2+}$ ion (Al-Nasser 2000). Various NSAIDs have shown an inhibitory effect on the mitochondrial oxidation of NADP-linked substrates. NSAIDs also inhibit β-oxidation of fatty acids in mitochondria. Drugs such as R-ibuprofen show an inhibitory effect on β-oxidation in a stereoselective manner. Both enantiomers of the drug have shown a moderate inhibition of mitochondrial enzymes, thus affecting the respiratory chain.

## 11.9.7 Local Anesthetics

Local anesthetics (LAs) block nervous conductance and cause temporary paralysis (Aleksandrowicz et al. 1973). LAs uncouple oxidative phosphorylation probably as they are tertiary amines with $pK_a$ values ranging from 7 to 9 and are lipophilic in nature, which provides ready access to the mitochondrial membrane (Hostetler and van den Bosch 1972). Although mitochondria are not their primary site of action, the main side effects originate there. LAs inhibit mitochondrial ATPase and other respiratory chain enzymes. However, different LAs have shown different actions in different concentrations. At a dose of 10–50 µM, dibucaine stimulates phospholipase A2 and blocks it (Waite and Sisson 1972) at 200–300 µM concentration. Phospholipase A2 has shown a detrimental effect on isolated mitochondria. Thus, the dose-dependent inhibitory effect of dibucaine on this enzyme can be the reflection of protective mechanism against mitochondrial damage. LAs such as bupivacaine inhibit mitochondrial respiration in a stage-dependent manner. The lipophilic anion tetraphenylboron can precipitate the forgoing effects of LAs (Lorenz et al. 1998). Mitochondria-mediated effects of LAs can play a role in complex processes such as apoptosis. LAs such as dibucane did not only show a decrease in $Ca^{2+}$ level, mitochondrial swelling, and ROS generation, but also arrested the growth of some tumors such as premyelocytic leukemia.

## 11.10 STRATEGIES FOR THE DELIVERY OF DRUGS TO MITOCHONDRIA

In the field of pharmaceutical research, mitochondria hold promise as a target for drug delivery. This section describes the current strategies for the delivery of low-molecular-weight drugs to the organelle. The drug delivery process is illustrated in Figure 11.2.

### 11.10.1 Triphenylphosphonium–Drug Complex

Owing to larger lipophilic surface and delocalized positive charge, the delocalized lipophilic cations (DLCs) can readily permeate plasma and mitochondrial

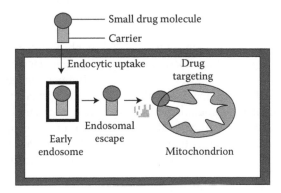

**FIGURE 11.2** The process of cellular uptake and targeting of lipophilic complex of small drug molecules to mitochondria.

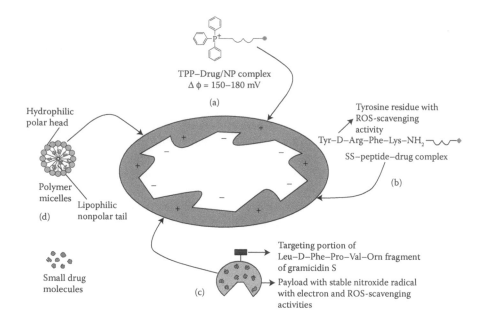

TPP–Drug/NP complex
Δ φ = 150–180 mV

(a)

Hydrophilic
polar head

Tyrosine residue with
ROS-scavenging
activity
Tyr–D–Arg–Phe–Lys–NH$_2$

SS–peptide–drug complex

Polymer
micelles
Lipophilic
(d)          nonpolar tail

(b)

Small drug
molecules

(c)

Targeting portion of
Leu–D–Phe–Pro–Val–Orn fragment
of gramicidin S

Payload with stable nitroxide radical
with electron and ROS-scavenging
activities

**FIGURE 11.3**  Recent mitochondria-targeting strategies: (a) triphenylphosphonium (TPP)–small molecule/nanocarrier conjugates; (b) SS peptide–drug conjugates; (c) dual targeting and scavenging approach; (d) drug incorporated into polymeric micelles.

membranes (Murphy 2008). The lipophilic and positively charged DLCs are rapidly driven by negative membrane potential through the bilayer membrane and cause 100–1000-fold uptake inside the negatively charged cytosol and mitochondrial compartments. Studies have shown that the size, lipophilicity, binding properties, and charge delocalization are crucial parameters that determine selective accumulation in mitochondria (Kurtoglu and Lampidis 2009).

Because of their relatively higher degree of accumulation in cancerous cells, DLCs are now under investigation for targeting drugs to the mitochondria of tumor cells. A more negative plasma and membrane potential in tumor cells could be responsible for higher accumulation. Thus, they are capable of preferential delivery of anticancer drugs to the target cells, with reduced toxicities. Lipophilic TPP is the most widely studied cation for targeting mitochondria with small drug molecules (Figure 11.3a).

Preclinical and clinical studies showed only limited anticancer effects of curcumin, presumably due to low bioavailability in both plasma and tissues, resulting in poor intracellular accumulation. The conjugation of curcumin to TPP cations can promote its delivery to mitochondria. The uptake of the conjugates in mitochondria was significantly higher than curcumin in MCF-7 breast cancer cells. The conjugates induced significant ROS generation, cell-cycle arrest, and apoptosis (Reddy et al. 2014).

The covalent linking of small lipophilic agents such as antioxidants and coenzyme $Q_{10}$ with lipophilic TPP cations leads to a positively charged complex. The concentration of this complex builds up in mitochondria up to 1000-fold. This is associated with large negative membrane potential (−180 to −200 mV) across the mitochondrial membrane. Such a high negative membrane potential is not present in any other cellular organelle, which offers a unique chemical opportunity for selective accumulation of lipophilic cations to mitochondria. An electric potential of 30–60 mV at the plasma membrane can enhance the concentration of charged compounds and facilitates the selective delivery of small molecules in intact cells. A report indicated that TPP–vitamin E conjugate could cross the lipid bilayer membranes and accumulate appreciably in the mitochondrial matrix compared to native vitamin E (Smith et al. 1999; Kelso et al. 2001; James et al. 2005, 2007; Cocheme et al. 2007).

The reports on selective uptake of TPP–ubiquinone conjugates in mitochondria also exist (James et al. 2005, 2007). In mitochondria, the respiratory chain reduces the conjugate to its active ubiquinol form. The ubiquinol derivative prevents lipid peroxidation and oxidative damage to mitochondria and, therefore, can act as an effective antioxidant. The detoxification of ROS by the respiratory chain regenerates the ubiquinol moieties, and henceforth its antioxidant activity is recycled. Thus, this conjugate can provide protection to mitochondria against oxidative damage following oral administration.

Mitoquinone (MitoQ), the TPP-bound ubiquinol, is the widely investigated antioxidant for mitochondria targeting (Smith et al. 1999; Murphy and Smith 2000, 2007). MitoQ uptake primarily occurs at the matrix-facing surface of the IMM, where complex II of the ETC recycles it into the active ubiquinol form (MitoQH$_2$), which remains stable during long-term incubation (Murphy and Smith 2000). The reduction is crucial concerning MitoQ effects because the accumulation of the ubiquinol form can contribute to antioxidant effects. This form is highly effective as an antioxidant by reacting with ROS and inhibiting peroxynitrite (ONOO$^-$) formation. The ubiquinol form of MitoQ is oxidized to the ubiquinone form, which is then rapidly re-reduced by complex II, and regains its antioxidant efficacy. Generally, MitoQ adsorbs onto the IMM, and its associated lipophilic chain enables the penetration of ubiquinol deep into the membrane core and thus behaves as an effective reducing agent against mitochondrial lipid peroxidation (Kelso et al. 2001; Smith et al. 2008).

The partition coefficient (log P) plays an important role in the selective uptake of DLCs such as TPPs into tumor cells. Below the range (−2 to 2), the absorption of compounds is kinetically restricted. The compounds with higher log P values are absorbed to cytosolic and mitochondrial lipids and therefore, lack compartmental selectivity (Trapp and Horobin 2005). The hydrophobicity of TPP compounds, imparted by conjugated alkyl chains of varying length, greatly influences their rate of mitochondrial uptake in cells. The conjugation of hydrophobic moieties results in increased uptake of TPPs in cellular mitochondria compared to more hydrophilic TPPs due to the rapid passage through the plasma membrane lipid layers (Asin-Cayuela et al. 2004).

Once in the cytosol, the TPP-containing molecule adsorbs to the outer surface of the mitochondrial inner membrane, permeates the lipid bilayer, binds to the surface of the inner membrane, and finally desorbs into the mitochondrial matrix (Ross et al. 2005). The hydrophobicity ultimately impairs the function of therapeutic molecules via sequestration, leading to reduced potency and drug-like properties such as solubility and bioavailability.

A wide range of antioxidants, including vitamin E, ebselen, lipoic acid, plasto-quinone, nitroxides, nitrones, and Mito Q have been targeted to mitochondria in the form of TPP conjugates (Smith and Murphy 2011). The Nernst equation can describe the accumulation of lipophilic cations, driven by the mitochondrial membrane potential as follows:

$$\text{Membrane potential (mV)} = 61.5 \log_{10} \left\{ \frac{[\text{cation}]_{in}}{[\text{cation}]_{out}} \right\}$$

As a consequence, every 61.5 mV increase in membrane potential causes a 10-fold increase in the accumulation of lipophilic cations. Therefore, the matrix concentration will be 100–1000-fold higher than that in the cytoplasm. Furthermore, the plasma membrane potential (typically 30–60 mV, negative inside) also drives the accumulation of cations into the cell and subsequently into the mitochondria with about 90%–95% of intracellular cation localization (Burns and Murphy 1997).

The phospholipid vesicles (liposomes) offer the greatest translational potential for mitochondrial delivery because of their biocompatible and nontoxic attributes. Weissig et al. largely contributed to the advancement of mitochondria-targeted liposomal drug delivery systems (Boddapati et al. 2005, 2008, 2010; D'Souza et al. 2005, 2008; Elbayoumi and Weissig 2009; Patel et al. 2010).

The drug-loaded liposomes were surface-functionalized with TPP for effective delivery of therapeutic cargo to mitochondria (Boddapati et al. 2008; Biswas et al. 2012a). TPP cations were conjugated to the lipophilic stearyl moiety to form stearyl triphenylphosphonium (STPP), which was then incorporated into the liposomal lipid bilayer, thereby anchoring the TPP group on the liposomal surface (Boddapati et al. 2008). They fabricated STPP-functionalized liposomes for mitochondria-specific targeting of ceramide. STPP was selected as a targeting molecule for mitochondria due to its cationic and lipophilic properties. Ceramide is a known anticancer agent that targets cytochrome $c$ release, ROS production, and apoptosis. Dioleoylphosphatidylcholine–cholesterol–STPP (83.5:15:1.5 mol%) liposomes (55 nm diameter) are localized within mitochondria of 4T1 breast cancer cells. Besides, the vesicles induced more apoptosis in the cancer cells than nontargeted ones. The drug efficacy was found to be much higher in mitochondria. Thus, the conjugation of a cationic, lipophilic ligand to liposomes could promote targeting drugs to mitochondria.

STPP liposomes also enhanced the targeting efficacy of sclareol, a colon cancer and leukemia drug to mitochondria compared to the free drug (Patel et al. 2010). Sclareol-loaded STPP liposomes (105 nm) exhibited 160% increase in apoptotic events in COLO 205 cells. The same was only 30% for nonmodified liposomes. Further, the functionalized liposomes increased caspase-8 activity and caspase-9 activity by 200% and 300%, respectively. Relative to this, the nonfunctionalized liposomes showed only 75% and 125% increase in caspase-8 and caspase-9 activity, respectively. Hence, the STPP liposomes can reduce the dose for an effective response ($EC_{50}$) compared to nonfunctionalized liposomes.

The toxicity associated with STPP could be overcome by conjugating the TPP group to the amphiphilic poly(ethylene glycol)–phosphatidylethanolamine moiety (Biswas et al. 2012a). The amphiphilic TPP–PEG–PE was incorporated into the liposomes. The lipid group was embedded in the liposomal lipid bilayer and the TPP group was anchored at the surface PEG layers. These mitochondriotropic liposomes effectively delivered a proapoptotic chemotherapeutic paclitaxel to mitochondria in comparison to the non-mitochondria-targeted liposomes and enhanced the antitumor effect *in vitro* and *in vivo*.

In a related study, surface conjugation of TPP to a macromolecule dendrimer resulted in efficient mitochondrial targeting (Biswas et al. 2012b). *N*-(2-Hydroxypropyl) methacrylamide (HPMA) copolymer-based mitochondriotropic drug delivery system was developed by conjugating a TPP group and a photosensitizer mesochlorine e6 (Mce6) on the polymer backbone (Cuchelkar et al. 2008). This mitochondria-targeted HPMA–copolymer–drug conjugate enhanced *in vitro* cytotoxicity compared to nontargeted HPMA–Mce6 conjugates. Rhodamine 123 is another lipophilic cation utilized for drug delivery to mitochondria. A complex of tetrachloroplatinate and rhodamine-123 generated cytotoxic and antitumor effects. They modified paclitaxel liposomes with rhodamine 123-conjugated PEG–PE polymer. This drug delivery system produced efficient mitochondrial targeting and increased cytotoxicity of loaded paclitaxel compared to unmodified liposomes (Biswas et al. 2011).

Cationic DOPE (1,2-dioleoyl-*sn*-glycero-3-phosphoethanolamine)/DOTAP (dioleoyl-1,2-diacyl-3-trimethylammoniumpropane) liposomes delivered the proapoptotic peptide D-(KLAKLAK)$^2$ together with an antisense oligonucleotide into the mitochondria of cancer cells (Ko et al. 2009). The liposomal surface modified with octa-arginine can stimulate cell internalization via macropinocytosis rather than clathrin endocytosis. This MITO-Porter promotes fusion with mitochondrial membranes and selectively delivers cargos to mitochondria (Yamada et al. 2008). This approach may find its application in intramitochondrial delivery of the antioxidants such as plastoquinol (Antonenko et al. 2008), quercetin (Biasutto et al. 2010), and imidazole-substituted fatty acids for the inhibition of peroxidase activity (Jiang et al. 2014).

## 11.10.2 Szeto–Schiller Peptides

Mitochondria-penetrating peptides or Szeto–Schiller (SS) peptides possess a structurally similar aromatic cationic motif in which the aromatic group, either tyrosine or dimethyltyrosine, alternates with a basic amino acid (Zhao et al. 2004; Szeto 2006a). The cells rapidly take up SS peptides, which then target mitochondria by an energy-independent mechanism (Szeto 2006a, b). The SS peptides localize to the IMM, where they serve as antioxidants, scavenge hydrogen peroxide and peroxynitrite, and inhibit lipid peroxidation (Zhao et al. 2004). They can protect ischemia–reperfusion-mediated cell injury (Cho et al. 2007).

Studies with isolated cells demonstrated rapid uptake of SS peptides through the plasma membrane to mitochondria, where they bind to the inner membrane. Despite positive charges at the physiological pH, the mitochondrial uptake of SS peptides does not seem to occur in response to the membrane potential (Figure 11.3b). However, the underlying mechanism of their selective uptake by mitochondria remains unclear.

The cell-permeable SS peptide antioxidants can selectively target the IMM. SS peptides can scavenge ROS generated within cells and inhibit lipid peroxidation due to the presence of tyrosine and dimethyltyrosine residues. The peptides are accumulated in the IMM up to 1000-fold, independent of the membrane potential. These peptide antioxidants can significantly reduce intracellular ROS and cell death caused by $t$-butylhydroperoxide in neuronal $N_2A$ cells even at nanomolar concentration. SS peptide antioxidants also decrease mitochondrial ROS production, inhibit permeability transition (PT) and swelling, and prevent cytochrome $c$ release induced by $Ca^{2+}$ in the isolated mitochondria. In the isolated mitochondria, SS peptides inhibit 3-nitropropionic acid (3NP)-induced PT and prevent mitochondrial depolarization in cells. Zhao et al. (2004) stated that these peptide antioxidants could significantly improve contractile force in an $ex$ $vivo$ heart model because ROS and PT have been implicated in myocardial stunning. They had further speculated that IMM-targeted antioxidants might be effective for treating aging and diseases associated with oxidative stress.

Highly effective anticancer drugs disrupt the synthesis of nucleic acid and become selectively toxic to cancer cells. However, the cellular effects due to the interference of anticancer drugs with nucleic acid synthesis in mitochondria are much less known. In a recent report by Chamberlain et al. (2013), mitochondrial targeting of DOX inhibited DNA topoisomerase II enzyme in both mitochondria and nuclei of human cells. The DOX–peptide conjugates exhibited significant toxicity and thus provided a way to limit drug efflux.

Yousif et al. (2009) indicated that MPPs are able to deliver water-soluble analogs of vitamin E, biotin, and trolox into mitochondria. A mitochondrial-targeting nanosystem, MITO-Porter, developed by Harashima et al. holds promise as an efficacious system for the delivery of both large and small bioactive molecules into mitochondria (Yasuzaki et al. 2010). MITO-Porter is a liposome-based nanocarrier (NC) composed of DOPE/sphingomyelin/Stearyl-R8 (9:2:1) that delivers its macromolecular cargo to the mitochondrial interior via membrane fusion. The liposomal surface modification with high-density R8 was responsible for its efficient cytosolic delivery after cell uptake by macropinocytosis. Figure 11.4 illustrates the mitochondrial delivery of the MITO-Porter contents.

Upon cytosolic release from macropinosomes, MITO-Porters translocate into the mitochondrial membrane via electrostatic interaction between the MITO-Porters and mitochondria and induce fusion (Yamada et al. 2008). The lipid composition of the MITO-Porter promotes both fusion with the mitochondrial membrane and releases its cargo to the intramitochondrial compartment. MITO-Porter encapsulating propidiumiodide (PI), a fluorescent dye commonly used to stain nuclei, successfully delivered PI as cargo to the mitochondrial matrix (Yasuzaki et al. 2010). Yamada et al. (2011) developed a MITO-Porter for both cytoplasmic and mitochondrial macromolecule delivery, wherein MITO-Porter liposomes were surface functionalized

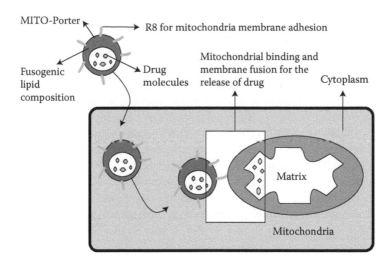

**FIGURE 11.4** Schematic representation of mitochondrial delivery of MITO-Porter-encapsulated small drug molecules—entry of MITO-Porters via pinocytosis, disruption of vesicles and translocation of liberated MITO-Porter to mitochondria via electrostatic interaction, and fusion of mitochondrial membrane with MITO-Porter octa-arginine (R8) component.

with R8. The nanopreparation showed efficient cytoplasmic delivery through the cell membrane due to R8 modification followed by mitochondrial delivery through the mitochondria membrane via membrane fusion.

Oligo-arginine cell-penetrating peptides have gained a great deal of attention as a cell membrane penetration enhancer. R8 modifications succeeded the intracellular delivery of several nanoparticles. Releasable, octa-arginine (R8)-conjugated paclitaxel overcame multidrug resistance (MDR) elicited by Taxol® (Dubikovskaya et al. 2008). Octa-arginine-linked NCs were efficiently internalized and demonstrated improved *in vivo* efficacy (Biswas et al. 2013; Toriyabe et al. 2013). The bleomycin-loaded octa-arginine-modified liposomes resulted in improved tumor growth inhibition (Koshkaryev et al. 2013). Surface modification with octa-arginine-linked PEG–PE polymer enhanced anticancer activity of DOX-loaded liposomes (Doxil® or Lipodox®), *in vitro* and *in vivo*, compared to the unmodified liposomes (Biswas et al. 2013). The confocal laser scanning micrograph revealed that modified liposomes delivered the payloads more efficiently to the cytosol, perhaps via endosome disruption, and resulted in enhanced DOX accumulation in the nucleus.

### 11.10.3 TARGETING–SCAVENGING APPROACH

The mitochondria-targeted electron and ROS scavenger consists of two parts (Figure 11.3c): a payload portion of stable nitroxide radical with electron- and ROS-scavenging activity and a targeting portion, which promotes selective accumulation in mitochondria (Wipf et al. 2005). By accepting one electron, nitroxide radicals are converted to effective ROS scavengers—hydroxylamines, and renewed back into nitroxides in the process (Zhang et al. 1999). Due to superoxide dismutase mimetic activity, nitroxide radicals inhibit the formation of highly toxic species, ONOO$^-$ by preventing reaction with $O_2$. with nitric oxide.

The targeting portion of the coupling system consists of the Leu–D–Phe–Pro–Val–Orn fragment of gramicidin S, which is a membrane-active cyclopeptide antibiotic (Kondejewski et al. 1996). This type of antibiotic has a high affinity for bacterial membranes (Sholtz et al. 1975). Owing to a close relationship between bacteria and mitochondria (Berry 2003), this fragment effectively targets the coupled system to mitochondria. This system ameliorates the peroxidation of mitochondrial phospholipid, cardiolipin, in rat ileal mucosal samples. Based on clinical observation on lethal hemorrhagic shock-induced rat ileal mucosal models, they concluded that mitochondria-targeted electron acceptors and superoxide dismutase mimics are valuable therapeutics for treating serious acute conditions such as lethal hemorrhagic shock, associated with marked tissue ischemia (Berry 2003).

## 11.10.4   pH-Responsive Micelles

Like low-molecular-weight surfactants, the amphiphilic copolymers self-organized in water to form nanoscale polymeric micelles (Kwon and Okano 1996). The micelles consist of a hydrophobic core stabilized by a layer of hydrophilic polymer chains, extended to the aqueous environment (Kataoka et al. 2001).

Some compounds exhibit undesired pharmacokinetics, and low stability in the physiological environment. The hydrophobic core of micellar structures can accommodate these types of poorly soluble compounds and improve their properties (Francis et al. 2004).

The occurrence of intrinsic or acquired MDR of cancer cells hinders the success of chemotherapy. Yu et al. (2015) prepared pH-responsive polymeric micelles of poly(ethylene glycol)-$b$–poly(2-(diisopropylamino) ethyl methacrylate) (PEGb–PDPA), and D-α-tocopheryl polyethylene glycol 1000 succinate (TPGS) to combat DOX resistance in breast cancer cells. Following cellular uptake, the micelles dissociated in early endosomes, triggered by acidic pH and released the DOX. TPGS improved the cytotoxicity of DOX by targeting mitochondrial organelles and reducing mitochondrial transmembrane potential. The $IC_{50}$ of DOX on DOX-resistant MCF-7/ADR cells reduced by sixfold following incorporation into PDPA/TPGS micelles. DOX-loaded micelles also inhibited tumor growth more efficiently than free DOX in a nude mouse model bearing orthotropic MCF-7/ADR tumor.

Recently, Hao et al. (2015) investigated the ability of DOX-loaded D-α-tocopheryl polyethylene glycol 2000 succinate (TPGS 2K) micelles to overcome MDR in breast cancer treatment. The DOX-loaded TPGS 2K micelles (~23 nm) did not reduce mitochondrial membrane potential significantly and deplete intracellular ATP level of MCF-7/ADR cells but showed an effect on the inhibition of verapamil-induced P-gp ATPase activity. The DOX-loaded TPGS 2K micelles showed higher cellular uptake and significant cytotoxic effect against MCF-7/MDR cells than the free DOX solution. The DOX-loaded TPGS 2K micelles displayed significantly higher antitumor activity in MCF-7/ADR tumor-bearing nude mice compared to free DOX solution at the same DOX dosage with less toxicity. Therefore, this type of micellar system can deliver DOX selectively and reverse MDR.

Sharma et al. (2012) reported a nanoscale delivery system that can sufficiently deliver coenzyme Q10 (CoQ10) in mitochondria. Multifunctional poly(ethylene glycol)–polycaprolactone (PCL)–TPP arm polymer self-assembled into nanosized micelles (25–60 nm). The CoQ10 loading capacity was excellent in the micelles due to good compatibility between CoQ10 and PCL. The data suggested that this new carrier did not cause any loss of drug effectiveness in mitochondria. The targeting of micellar drugs to mitochondria is illustrated in Figure 11.3d.

## 11.10.5 DQAsomes

DQAsomes (dequalinium-based liposome-like vesicles) are considered the prototype for all mitochondria-targeted nanovesicular carriers. Their ability to deliver DNA and low-molecular-weight molecules successfully to mitochondria of living mammalian cells has been explored (Weissig 2015).

The vesicle-forming ability of dequalinium chloride (DQA) opened the door for developing a mitochondria-targeted vesicular system (Weissig et al. 1998; Weissig 2011). Dequalinium (1,1′-decamethylene bis (4-aminoquinaldinium chloride)) is a cationic bolaamphiphile composed of two quinaldinium rings linked by 10 methylene groups (Figure 11.5). Dequalinium is a dicationic amphiphilic compound that self-assembles and forms vesicle-like aggregates referred to as dequalinium liposomes or DQAsomes (Weissig et al. 2000; Weissig and Torchilin 2000, 2001).

For the preparation of DQAsomes, dequalinium chloride is dissolved in methanol and then the solvent is evaporated by a rotary evaporator. Salt-free HEPES buffer is added and the suspension is probe-sonicated until a clear opalescent solution is obtained. To prevent overheating of the sonicated material, the round-bottom flask is cooled with ice water. The colloidal solution containing DQAsomes

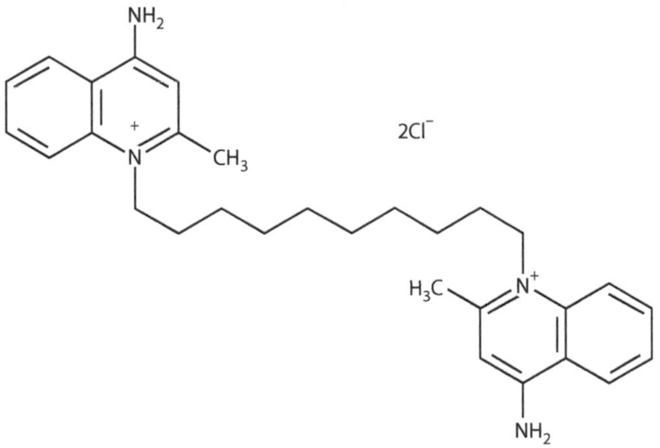

**FIGURE 11.5**  Chemical structure of dequalinium chloride used for the preparation of DQAsomes.

requires centrifugation to remove any undissolved solid material. The supernatant is collected carefully and stored at 4°C.

DQAsomes have been extensively explored for mitochondria-targeted delivery of low-molecular-weight compounds. For example, paclitaxel-loaded DQAsomes showed rod-shaped structures of ~400 nm length under TEM examination. This was quite surprising considering the spherical nature of the empty DQAsomes (Cheng et al. 2005). The paclitaxel formulation with DQA increased the solubility of the drug by a factor of roughly 3000. Thus, DQAsomes can be looked upon as a better alternative to Cremophor-based formulations for the insoluble paclitaxel. Studies demonstrated that DQAsomal paclitaxel preparations triggered apoptosis by directly acting on mitochondria and increased the drug's efficiency. In contrast to random cytosolic diffusion of the free drug, its DQAsomal formulation colocalized with mitochondria. DQAsomal paclitaxel triggered cell death via apoptosis in cancer cells (D'Souza et al. 2008). Moreover, DQAsomal paclitaxel inhibited the tumor growth in nude mice bearing human colon carcinoma cells by about 50%, while the free drug did not have any impact on tumor growth at the concentration used (Cheng et al. 2005). An independent study by Vaidya et al. (2009) also confirmed the rod-like structure of paclitaxel-loaded DQAsomes. Their colocalization with mitochondria as well as the enhanced apoptotic activity of paclitaxel were reported. To increase apoptotic activity of DQAsomal paclitaxel, Vaidya et al. successfully conjugated folic acid to the surface of DQAsomes rendering them more specific for tumor cells overexpressing the folate receptor. Most recently, Zupancic et al. (2014) described the formulation of curcumin-loaded DQAsomes for pulmonary delivery. Curcumin is a potent antioxidant with anti-inflammatory properties, which however has a very low oral bioavailability. Curcumin-loaded DQAsomes (170–200 nm) had an encapsulation efficiency of 90% and displayed better antioxidant activity than free curcumin. Thus, curcumin-loaded DQAsomes proved their promise as inhalation formulation with mitochondrial targeting ability and, consequently, opened a new research avenue for efficient curcumin delivery in acute lung injury.

These vesicles translocate into cells via endocytosis and fusion with the mitochondrial outer membrane. DQAsomes were investigated for their potential utility to serve as a mitochondriotropic, nonviral transfection vector for delivery of mitochondrial DNA to the mitochondrial membrane (Weissig et al. 2001; D'Souza et al. 2003). The results demonstrated that DQAsomes could transport the pDNA to mitochondria by escaping from endosomes without losing their pDNA load.

## 11.10.6 Nanoparticulate Approach

Polymeric and metallic nanoparticles (5–260 nm) can exert mitochondrial effects. The metallic nanoparticles possess unique antioxidant properties, especially for gold and platinum elements. Further, their small size and ease of attachment of ligands helps in targeting mitochondria. However, their safety and efficacy are questionable among scientists. The literature reports that the application of gold nanoparticles as mitochondrial delivery devices are limited.

Salnikov et al. (2007) investigated the effect of gold nanoparticles on the integrity of outer mitochondrial membrane and cytochrome $c$ release from the ETC.

The gold particles (6 nm) did not permeate the outer mitochondrial membrane, but the particles of 3 nm diameter permeated the outer mitochondrial membrane of heart mitochondria. The mitochondrial entry of smaller size gold particles was VDAC-dependent since 15–50 µg/mL VDAC inhibitors inhibited their entry. This entry mechanism may be of therapeutic value for induction of cellular apoptosis. However, other therapeutic mechanisms may require that gold nanoparticles avoid mitochondrial entry by this mechanism. The surface charge and concentration of gold nanoparticles mediates their cellular toxicity (Goodman et al. 2004). The cationic gold nanoparticles were responsible for 10-fold more lysis of anionic liposomes than anionic gold nanoparticles in 5 min. The extent of disruption of 1-stearoyl-2-oleoylphosphatidylcholine (SOPC) neutral liposomes (about 15%) was nearly equivalent to anionic stearoyl–oleoyl–phosphatidylserine (SOPS) liposomes (about 20%). However, anionic gold nanoparticles lysed more neutral liposomes (about 5%) than anionic liposomes (about 2%). The induction of vesicular lysis by cationic nanoparticles was exponential above 175 nM concentration with time. The charge and concentration of the gold nanoparticles also contribute to the cellular toxicity. In addition to their toxic profile, they also exhibit therapeutic effects. For instance, chitosan-modified gold nanoparticles (6–16 nm) were 80 times more efficient at eliminating ROS in an $H_2O_2/FeSO_4$ system than ascorbic acid (Esumi et al. 2003). The antioxidant activity was independent of nanoparticle size, but increased with an increase in chitosan concentration. This indicated that chitosan might be responsible for the antioxidant activity after gold complexation. However, the literature reports did not reveal any antioxidant effects of chitosan; rather cytotoxic effects are known (Qi et al. 2005). The same group demonstrated that gold nanoparticles functionalized with polyamidoamine (PAMAM) dendrimers might behave as antioxidants by reducing ROS to water and oxygen (Esumi et al. 2004). The dendrimer-decorated nanoparticles could be more promising than the chitosan–gold nanoparticles because of their smaller size and albeit higher efficiency at eliminating ROS. The dendrimer-functionalized gold nanoparticles could be suitable for mitochondria targeting. It has recently been reported that gold nanorods functionalized with cationic cetyltrimethylammonium bromide (CTAB) preferentially induced cell death in cancer cells and accumulated in mitochondria of cancer cells more than in normal cells (Wang et al. 2011). CTAB may be partially responsible for cell death caused by gold nanorods due to its well-known toxicity in both animal and cell models (Isomaa et al. 1976). The particles were of ~55 nm long and 13 nm wide. The authors demonstrated that normal cells eliminated and removed gold nanoparticles through lysosomal cleanup. In cancer cells, gold nanoparticles were able to escape the lysosome due to differences in lysosomal protein content between cancerous and normal cells. The addition of CTAB alone to cancer cells significantly reduced the integrity of the lysosomal membrane without any effect on normal cell membrane integrity. Thus, the effects were determined to be a property of CTAB. After 6 h incubation, the gold nanorods entered the cytoplasm and localized within the mitochondria of cancer cells. The authors hypothesized that an extreme negative membrane potential across the IMM was responsible for the preferential uptake of the cationic CTAB-functionalized gold nanorods and caused cancer cell death. The conjugates of TPP–gold nanoparticles are under investigation to reveal their suitability in targeted drug delivery (D'Souza and Weissig 2009).

Platinum nanoparticles also exhibit unique antioxidant properties (Aiuchi et al. 2004). However, their safety profiles are supported by limited studies in cell-free and cell-based assays (Elder et al. 2007). Platinum nanoparticles of varying size (~20–30 nm) and shapes (nanoflowers, spheres, and multipods) produce neither ROS in a cell-free system nor inflammatory responses (interleukin-6 and tumor necrosis factor $\alpha$) in the human umbilical vein endothelial cell (HUVEC) system up to 50 μg dose. This confirms the noncytotoxic nature of the nanoparticles. In addition to antioxidant properties, platinum nanoparticles exhibit therapeutic effects in mitochondria. Hikosaka et al. (2008) demonstrated that pectin-functionalized platinum particles were capable of oxidizing NADH to $NAD^+$. This property could normalize the redox potential by regenerating $NAD^+$ species for glycolysis and other cellular pathways to function properly. The same group demonstrated that platinum nanoparticles (~5 nm) were capable of quenching superoxide anion radical ($O_2^-$) and hydrogen peroxide ($H_2O_2$) (Kajita et al. 2007). Since pectin-functionalized platinum nanoparticles oxidized NADH to $NAD^+$ and reduced ubiquinone (CoQ) to ubiquinol ($CoQH_2$), they may mimic the functions of complex I. The result was astonishing in that if the pectin functionalization can replace the functions of complex I, these nanoparticles may become suitable for treating diseases with complex I deficiencies such as Alzheimer's disease.

Considering the crucial role of mitochondria in cellular energetics, metabolism, and signaling, the targeting of mitochondria with small molecules would lead to severe side effects in cancer patients. Moreover, mitochondrial functions are highly dependent on other cellular organelles such as nucleus. Hence, simultaneous targeting of mitochondria and nucleus could lead to more effective anticancer strategy. To meet this goal, Mallick et al. (2015) developed nanoparticles (<200 nm) from dual drug conjugates. They tethered mitochondria-damaging drug ($\alpha$-tocopheryl succinate) and nucleus-damaging drugs (cisplatin, DOX, and paclitaxel). These dual drug conjugated nanoparticles internalized into acidic lysosomal compartments of the HeLa cervical cancer cells through endocytosis and induced apoptosis through cell cycle arrest. The nanoparticles damaged mitochondrial morphology and triggered the release of cytochrome $c$. Furthermore, the particles targeted the nucleus to induce DNA damage, fragment nuclear morphology, and damage the cytoskeletal protein tubulin. Therefore, such systems could be successful for the simultaneous targeting of multiple subcellular organelles in cancer cells to improve the therapeutic efficacy of free drugs.

Marrache et al. (2015) synthesized mitochondria-targeted poly(DL-lactide-co-glycolide)-b–polyethylene glycol–TPP and a nontargeted poly(DL-lactide-co-glycolide)-b–polyethylene glycol NPs. They reported that the size and charge of NPs were suitable for efficient mitochondrial trafficking.

Cationic lipid-based NCs such as lipidic micelles and nanoemulsions could serve as mitochondria-specific therapeutic and targeting moieties. Genistein (Gen) exhibits proapoptotic anticancer effects after induction of mitochondrial damage. Pham et al. (2015) entrapped this soy isoflavone in NCs. Following incorporation into NCs, a significant cytotoxic effect of Gen-NCs was observed at 5–10-fold lower $EC_{50}$ values compared to all drug controls, in hepatic and colon carcinomas. The mitochondria-specific accumulation of various Gen-NCs correlated well with marked mitochondrial depolarization effects. Therefore, the mitochondriotropic activity of Gen-NCs can improve the anticancer efficacy of Gen.

## 11.11 CONCLUSION

Mitochondria play a vital role in the pathology of various degenerative and metabolic disorders and thus create differences in structural and functional properties from other cellular compartments. As several diseases are associated with mitochondrial dysfunction, mitochondria can be the potential targets for small and macromolecular drugs. The targeted delivery can reduce the dose-related toxicity and lead to manifold increase in therapeutic efficacy. Currently, a number of strategies are being built up to improve cellular uptake and delivery of therapeutic molecules to mitochondria. Once mitochondria-specific delivery is achieved, a dramatic improvement in the interaction between bioactive molecules with target organelles occurs. However, the drug molecules have to traverse some physiological and biochemical barriers before achieving target selectivity and accumulation in mitochondria. The most challenging job is the translation of small-scale findings into the commercial scale for safe and effective use of delivery systems in patients. The scarcity of data on metabolism and elimination of drugs following their transport into mitochondria also pose other hurdles to its commercialization. Thus, future works must focus on preclinical trials that would provide useful information regarding safety and bioactivity prior to final trials in humans. Furthermore, an in-depth understanding of the intracellular trafficking of mitochondria-targeted nanosystems as well as the loaded drugs, time-dependent adverse events, and drug release inside the organelles is desirable. We hope that the ongoing efforts toward the development of effective targeted delivery systems will never be in vain and will find the marketplace in the future for the benefit of many patients suffering from mitochondrial diseases.

## REFERENCES

Aiuchi, T., Nakajo, S., and Nakaya, K. 2004. Reducing activity of colloidal platinum nanoparticles for hydrogen peroxide, 2,2-diphenyl-1-picrylhydrazyl radical and 2,6-dichlorophenol indophenol. *Biol Pharm Bull* 27:736–38.

Aleksandrowicz, Z., Świerczyński, J., and Wrzołkowa, T. 1973. Protective effect of nupercaine on mitochondrial structure. *Biochim Biophys Acta Bioenerg* 305:59–66.

Al-Nasser, I.A. 2000. Ibuprofen-induced liver mitochondrial permeability transition. *Toxicol Lett* 111:213–18.

Antonenko, Y.N., Avetisyan, A.V., and Bakeeva, L.E. 2008. Mitochondria-targeted plastoquinone derivatives as tools to interrupt execution of the aging program. 1. Cationic plastoquinone derivatives: Synthesis and in vitro studies. *Biochem (Mosc)* 73:1273–87.

Armstrong, J.S. 2007. Mitochondrial medicine: Pharmacological targeting of mitochondria in disease. *Br J Pharmacol* 151:1154–56.

Ashcroft, S.J. and Ashcroft, F.M. 1992. The sulfonylurea receptor. *Biochim Biophys Acta* 1175:45–49.

Asin-Cayuela, J., Manas, A., James, A., Smith, R., and Murphy, M. 2004. Fine-tuning the hydrophobicity of a mitochondria-targeted antioxidant. *FEBS Lett* 571:9–16.

Berry, S. 2003. Endosymbiosis and the design of eukaryotic electron transport. *Biochim Biophys Acta* 1606:57–72.

Biasutto, L., Sassi, N., Mattarei, A. et al. 2010. Impact of mitochondriotropic quercetin derivatives on mitochondria. *Biochim Biophys Acta* 1797:189–96.

Biswas, S., Dodwadkar, N.S., Deshpande, P.P., Parab, S., and Torchilin, V.P. 2013. Surface 1Q2584 functionalization of doxorubicin-loaded liposomes with octa-arginine for 1285 enhanced anticancer activity. *Eur J Pharm Biopharm* 84:517–25.

Biswas, S., Dodwadkar, N.S., Deshpande, P.P., and Torchilin, V.P. 2012a. Liposomes loaded with paclitaxel and modified with novel triphenylphosphonium–PEG–PE conjugate possess low toxicity, target mitochondria and demonstrate enhanced antitumor effects in vitro and in vivo. *J Control Release* 159:393–402.

Biswas, S., Dodwadkar, N.S., Piroyan, A., and Torchilin, V.P. 2012b. Surface conjugation of triphenylphosphonium to target poly(amidoamine) dendrimers to mitochondria. *Biomaterials* 33:4773–82.

Biswas, S., Dodwadkar, N.S., Sawant, R.R., Koshkaryev, A., and Torchilin, V.P. 2011. Surface modification of liposomes with rhodamine-123-conjugated polymer results in enhanced mitochondrial targeting. *J Drug Target* 19:552–61.

Biswas, S. and Torchilin, V.P. 2014. Nanopreparations for organelle-specific delivery in cancer. *Adv Drug Deliv Rev* 66:26–41.

Boddapati, S.V., D'Souza, G.G., Erdogan, S., Torchilin, V.P., and Weissig, V. 2008. Organelle-targeted nanocarriers: Specific delivery of liposomal ceramide to mitochondria enhances its cytotoxicity in vitro and in vivo. *Nano Lett* 8:2559–63.

Boddapati, S.V., D'Souza, G.G., and Weissig, V. 2010. Liposomes for drug delivery to mitochondria. *Methods Mol Biol* 605:295–303.

Boddapati, S.V., Tongcharoensirikul, P., Hanson, R.N., D'Souza, G.G., Torchilin, V.P., and Weissig, V. 2005. Mitochondriotropic liposomes. *J Liposome Res* 15:49–58.

Bouchier-Hayes, L., Lartigue, L., and Newmeyer, D.D. 2005. Mitochondria: Pharmacological manipulation of cell death. *J Clin Invest* 115:2640–47.

Burns, R.J. and Murphy, M.P. 1997. Labeling of mitochondrial proteins in living cells by the thiol probe thiobutyltriphenylphosphonium bromide. *Arch Biochem Biophys* 339:33–39.

Chamberlain, G.R., Tulumello, D.V., and Kelley, S.O. 2013. Targeted delivery of doxorubicin to mitochondria. *ACS Chem Biol* 8:1389–95.

Chance, B., Sies, H., and Boveris, A. 1979. Hydroperoxide metabolism in mammalian organs. *Physiol Rev* 59:527–605.

Chen, H. and Chan, D.C. 2010. Physiological functions of mitochondrial fusion. *Ann N Y Acad Sci* 1201:21–25.

Cheng, S.M., Pabba, S., and Torchilin, V.P. 2005. Towards mitochondria-specific delivery of apoptosis-inducing agents: DQAsomal incorporated paclitaxel. *J Drug Deliv Sci Technol* 15:81–86.

Chinnery, P.F. and Turnbull, D.M. 2000. Mitochondrial DNA mutations in the pathogenesis of human disease. *Mol Med Today* 6:425–32.

Cho, S., Szeto, H.H., Kim, E., Kim, H., Tolhurst, A.T., and Pinto, J.T. 2007. A novel cell-permeable antioxidant peptide, SS31, attenuates ischemic brain injury by down-regulating CD36. *J Biol Chem* 282:4634–42.

Cocheme, H.M., Kelso, G.F., James, A.M. et al. 2007. Mitochondrial targeting of quinones: Therapeutic implications. *Mitochondrion* 7:S94–S102.

Cuchelkar, V., Kopeckova, P., and Kopecek, J. 2008. Novel HPMA copolymer-bound constructs for combined tumor and mitochondrial targeting. *Mol Pharm* 5:776–86.

Cunha, F.M., Caldeira da Silva, C.C., Cerqueira, F.M., and Kowaltowski, A.J. 2011. Mild mitochondrial uncoupling as a therapeutic strategy, *Curr Drug Targets* 12:783–89.

D'Souza, G.G.M., Boddapati, S.V., and Weissig, V. 2005. Mitochondrial leader sequence—Plasmid DNA conjugates delivered into mammalian cells by DQAsomes co-localize with mitochondria. *Mitochondrion* 5:352–58.

D'Souza, G.G.M., Cheng, S.M., Boddapati, S.V., Horobin, R.W., and Weissig, V. 2008. Nanocarrier assisted sub-cellular targeting to the site of mitochondria improves the pro-apoptotic activity of paclitaxel. *J Drug Target* 16:578–85.

D'Souza, G.G.M., Rammohan, R., Cheng, S.M., Torchilin, V.P., and Weissig, V. 2003. DQAsome mediated delivery of plasmid DNA toward mitochondria in living cells. *J Control Release* 92:189–97.

D'Souza, G.G.M., Wagle, M.A., Saxena, V., and Shah, A. 2011. Approaches for targeting mitochondria in cancer therapy. *Biochim Biophys Acta Bioenerg* 1807:689–96.

D'Souza, G.G.M. and Weissig, V. 2009. Subcellular targeting: A new frontier for drugloaded pharmaceutical nanocarriers and the concept of the magic bullet. *Expert Opin Drug Deliv* 6:1135–48.

de la Asunción J.G., del Olmo, M.L., Sastre, J. et al. 1998. AZT treatment induces molecular and ultrastructural oxidative damage to muscle mitochondria. Prevention by antioxidant vitamins. *J Clin Invest* 102:4–9.

Dias, N. and Bailly, C. 2005. Drugs targeting mitochondrial functions to control tumor cell growth. *Biochem Pharmacol* 70:1–12.

Dubikovskaya, E.A., Thorne, S.H., Pillow, T.H., Contag, C.H., and Wender, P.A. 2008. Overcoming multidrug resistance of small-molecule therapeutics through conjugation with releasable octaarginine transporters. *Proc Natl Acad Sci U S A* 105:12128–33.

Duchen, M.R. 2004. Mitochondria in health and disease: Perspectives on a new mitochondrial biology. *Mol Aspects Med* 25:365–51.

Durazo, S.A. and Kompella, U.B. 2011. Functionalized nanosystems for targeted mitochondrial delivery. *Mitochondrion* 12:190–201.

Elbayoumi, T. and Weissig, V. 2009. Implications of intracellular distribution of nanovesicles for bioimaging studies. *J Biomed Nanotechnol* 5:620–33.

Elder, A., Yang, H., and Gwiazda, R. 2007. Testing nanomaterials of unknown toxicity: An example based on platinum nanoparticles of different shapes. *Adv Mater* 19:3124–29.

Epps, D.E., Palmer, J.W., Schmid, H.H., and Pfeiffer, D.R. 1982. Inhibition of permeability-dependent $Ca^{2+}$ release from mitochondria by $N$-acylethanolamines, a class of lipids synthesized in ischemic heart tissue. *J Biol Chem* 257:1383–91.

Esumi, K., Houdatsu, H., and Yoshimura, T. 2004. Antioxidant action by gold-PAMAM dendrimer nanocomposites. *Langmuir* 20:2536–38.

Esumi, K., Takei, N., and Yoshimura, T. 2003. Antioxidant-potentiality of gold-chitosan nanocomposites. *Colloids Surf B Biointerfaces* 32:117–23.

Francis, M.F., Cristea, M., and Winnik, F.M. 2004. Polymeric micelles for oral drug delivery: Why and how? *Pure Appl Chem* 76:1321–35.

Frantz, M.-C. and Wipf, P. 2010. Mitochondria as a target in treatment. *Environ Mol Mutagen* 51:462–75.

Fromenty, B. and Pessayre, D. 1995. Impaired mitochondrial function in microvesicular steatosis. *J Hepatol* 2:101–54.

Frøyland, L., Madsen, L., Vaagenes, H. et al. 1997. Mitochondrion is the principal target for nutritional and pharmacological control of triglyceride metabolism. *J Lipid Res* 38:1851–58.

Fulda, S., Galluzzi, L., and Kroemer, G. 2010. Targeting mitochondria for cancer therapy. *Nat Rev Drug Discov* 9:447–64.

Gnaiger, E. 2009. Capacity of oxidative phosphorylation in human skeletal muscle: New perspectives of mitochondrial physiology. *Int J Biochem Cell Biol* 41:1837–45.

Goodman, C.M., McCusker, C.D., Yilmaz, T., and Rotello, V.M. 2004. Toxicity of gold nanoparticles functionalized with cationic and anionic side chains. *Bioconjug Chem* 15:897–900.

Green, D.R. and Reed, J.C. 1998. Mitochondria and apoptosis. *Science* 281:1309–12.

Hail, N. Jr. 2005. Mitochondria: A novel target for the chemoprevention of cancer. *Apoptosis* 10:687–705.

Hansen, H.S., Moesgaard, B., Hansen, H.H., and Petersen, G. 2000. *N*-Acylethanolamines and precursor phospholipids—Relation to cell injury. *Chem Phys Lipids* 108:135–50.

Hao, T., Chen, D., Liu, K. et al. 2015. Micelles of d-α-tocopheryl polyethylene glycol 2000 succinate (TPGS 2K) for doxorubicin delivery with reversal of multidrug resistance. *ACS Appl Mater Interfaces* 7:18064–75.

Heller, A., Brockhoff, G., and Goepferich, A. 2012. Targeting drugs to mitochondria. *Eur J Pharm Biopharm* 82:1–18.

Hikosaka, K., Kim, J., Kajita, M., Kanayama, A., and Miyamoto, Y. 2008. Platinum nanoparticles have an activity similar to mitochondrial NADH: Ubiquinone oxidoreductase. *Colloids Surf B Biointerfaces* 66:195–200.

Holt, I.J., Harding, A.E., and Morgan-Hughes, J.A. 1988. Deletions of muscle mitochondrial DNA in patients with mitochondrial myopathies. *Nature* 331:717–19.

Hostetler, K.Y. and van den Bosch, H. 1972. Subcellular and submitochondrial localization of the biosynthesis of cardiolipin and related phospholipids in rat liver. *Biochim Biophys Acta* 260:380–86.

Isomaa, B., Reuter, J., and Djupsund, B.M. 1976. The subacute and chronic toxicity of cetyltrimethylammonium bromide (CTAB), a cationic surfactant, in the rat. *Arch Toxicol* 35:91–96.

Jabůrek, M., Yarov-Yarovoy, V., Paucek, P., and Garlid, K.D. 1998. State-dependent inhibition of the mitochondrial KATP channel by glyburide and 5-hydroxydecanoate. *J Biol Chem* 273:13578–82.

James, A.M., Cocheme, H.M., Smith, R.A., and Murphy, M.P. 2005. Interactions of mitochondria-targeted and untargeted ubiquinones with the mitochondrial respiratory chain and reactive oxygen species. Implications for the use of exogenous ubiquinones as therapies and experimental tools. *J Biol Chem* 280:21295–12.

James, A.M., Sharpley, M.S., Manas, A.R. et al. 2007. Interaction of the mitochondria-targeted antioxidant MitoQ with phospholipid bilayers and ubiquinone oxidoreductases. *J Biol Chem* 282:14708–18.

Jiang, J.F., Bakan, A., and Kapralov, A.A. 2014. Designing inhibitors of cytochrome *c*/cardiolipin peroxidase complexes: Mitochondria-targeted imidazole-substituted fatty acids. *Free Radic Bio Med* 71:221–30.

Kajita, M., Hikosaka, K., Iitsuka, M., Kanayama, A., Toshima, N., and Miyamoto, Y. 2007. Platinum nanoparticle is a useful scavenger of superoxide anion and hydrogen peroxide. *Free Radic Res* 41:615–26.

Kataoka, K., Harada, A., and Nagasaki, Y. 2001. Block copolymer micelles for drug delivery: Design, characterization and biological significance *Adv Drug Deliv Rev* 47:113–31.

Kelso, G.F., Porteous, C.M., Coulter, C.V. et al. 2001. Selective targeting of a redox-active ubiquinone to mitochondria within cells: Antioxidant and antiapoptotic properties. *J Biol Chem* 276:4588–96.

Ko, Y.T., Falcao, C., and Torchilin, V.P. 2009. Cationic liposomes loaded with proapoptotic peptide D-(KLAKLAK)(2) and Bcl-2 antisense oligodeoxynucleotide G3139 for enhanced anticancer therapy. *Mol Pharm* 6:971–77.

Kondejewski, L.H., Farmer, S.W., Wishart, D.S., Hancock, R.E., and Hodges, R.S. 1996. Gramicidin S is active against both gram-positive and gram-negative bacteria. *Int J Pept Protein Res* 47:460–66.

Koshkaryev, A., Piroyan, A., and Torchilin, V.P. 2013. Bleomycin in octaarginine-modified fusogenic liposomes results in improved tumor growth inhibition. *Cancer Lett* 334:293–301.

Kourie, J.I. 1998. Effects of ATP-sensitive potassium channel regulators on chloridechannels in the sarcoplasmic reticulum vesicles from rabbit skeletal muscle. *J Membr Biol* 164:47–58.

Kurtoglu, M. and Lampidis, T. 2009. From delocalized lipophilic cations to hypoxia: Blocking tumor cell mitochondrial function leads to therapeutic gain with glycolytic inhibitors. *Mol Nutr Food Res* 53:68–75.

Kwon, G.S. and Okano, T. 1996. Polymeric micelles as new drug carriers. *Adv Drug Deliv Rev* 21:107–16.

Lorenz, B., Schlüter, T., Bohnensack, R., Pergande, G., and Müller, W.E. 1998. Effect of flupirtine on cell death of human umbilical vein endothelial cells induced by reactive oxygen species. *Biochem Pharmacol* 56:1615–24.

Mallick, A., More, P., Ghosh, S. et al. 2015. Dual drug conjugated nanoparticle for simultaneous targeting of mitochondria and nucleus in cancer cells. *ACS Appl Mater Interfaces* 27:7584–98.

Mammucari, C., Patron, M., Granatiero, V., and Rizzuto, R. 2011. Molecules and roles of mitochondrial calcium signaling. *BioFactors* 37:219–27.

Marrache, S., Pathak, R.K., and Dhar, S. 2015. Formulation and optimization of mitochondria-targeted polymeric nanoparticles. In *Mitochondrial Medicine: Methods in Molecular Biology*, ed. V. Weissig and M. Edeas, 103–112. New York: Springer Science.

McBride, H. and Soubannier, V. 2010. Mitochondrial function: OMA1 and OPA1, the grandmasters of mitochondrial health. *Curr Biol* 20:274–76.

McCurdy, D.T. 3rd. and Kennedy, J.M. 1998. AZT decreases rat myocardial cytochrome oxidase activity and increases beta-myosin heavy chain content. *J Mol Cell Cardiol* 30:1979–89.

McLaughlin, D.B., Eadie, M.J., Parker-Scott, S.L. et al. 2000. Valproate metabolism during valproate-associated hepatotoxicity in surviving adult patient. *Epilepsy Res* 41:259–68.

Mingatto, F.E., dos Santos, A.C., Rodrigues, T., Pigoso, A.A., Uyemura, S.A., and Curti, C. 2000. Effects of nimesulide and its reduced metabolite on mitochondria. *Br J Pharmacol* 131:1154–60.

Mitchell, T., Rotaru, D., Saba, H., Smith, R.A.J., Murphy, M.P., and MacMillan-Crow, L.A. 2011. The mitochondria-targeted antioxidantmitoquinone protects against cold storage injury of renal tubular cells and rat kidneys. *J Pharmacol Exp Ther* 336:682–92.

Modica-Napolitano, J.S. and Singh, K.K. 2002. Mitochondria as targets for detection and treatment of cancer. *Expert Rev Mol Med* 4:1–19.

Moghimi, S., Hunter, A., and Andresen, T. 2012. Factors controlling nanoparticles pharmacokinetics: An integrated analysis and perspective. *Annu Rev Pharmacol Toxicol* 52:481–503.

Murphy, M. 2008. Targeting lipophilic cations to mitochondria. *Biochim Biophys Acta* 1777:1028–31.

Murphy, M.P. 2009. How mitochondria produce reactive oxygen species. *Biochem J* 417:1–13.

Murphy, M.P. and Smith, R.A. 2000. Drug delivery to mitochondria: The key to mitochondrial medicine. *Adv Drug Deliv Rev* 41:235–50.

Murphy, M.P. and Smith, R.A. 2007. Targeting antioxidants to mitochondria by conjugation to lipophilic cations. *Annu Rev Pharmacol Toxicol* 47:629–56.

Nishikawa, Y., Takahashi, M., Yorifuji, S. et al. 1989. Long-term coenzyme Q10 therapy for a mitochondrial encephalomyopathy with cytochrome *c* oxidase deficiency: A $^{31}$P NMR study. *Neurology* 39:399–403.

Osborne, N.N., Schwarz, M., and Pergande, G. 1996. Protection of rabbit retina from ischemic injury by flupirtine. *Invest Ophthalmol Vis Sci* 37:274–80.

Patel, N.R., Hatziantoniou, S., Georgopoulos, A. et al. 2010. Mitochondria-targeted liposomes improve the apoptotic and cytotoxic action of sclareol. *J Liposome Res* 20:244–49.

Patel, T.B. 1986. Effect of sulfonylureas on hepatic fatty acid oxidation. *Am J Physiol* 251:E241–E246.

Pathania, D., Millard, M., and Neamati, N. 2009. Opportunities in discovery and delivery of anticancer drugs targeting mitochondria and cancer cell metabolism. *Adv Drug Deliv Rev* 61:1250–75.

Petit, F., Fromenty, B., Owen, A., and Estaquier, J. 2005. Mitochondria are sensors for HIV drugs. *Trends Pharmacol Sci* 26:258–64.

Pham, J., Grundmann, O., and Elbayoumi, T. 2015. Mitochondriotropic nanoemulsified genistein-loaded vehicles for cancer therapy. In *Mitochondrial Medicine: Methods in Molecular Biology*, ed. V. Weissig and M. Edeas, 85–101. New York: Springer Science.

Prime, T.A., Blaikie, F.H., Evans, C. et al. 2009. A mitochondria-targeted S-nitrosothiol modulates respiration, nitrosates thiols, and protects against ischemia-reperfusion injury. *Proc Natl Acad Sci U S A* 106:10764–69.

Qi, L.F., Xu, Z.R., Li, Y., Jiang, X., and Han, X.Y. 2005. In vitro effects of chitosan nanoparticles on proliferation of human gastric carcinoma cell line MGC803 cells. *World J Gastroenterol* 11:5136–41.

Reddy, C.A., Somepalli, V., Golakoti, T. et al. 2014. Mitochondrial-targeted curcuminoids: A strategy to enhance bioavailability and anticancer efficacy of curcumin. *PLoS One* 9:e89351.

Ringsdorf, H. 1975. Structure and properties of pharmacologically active polymers. *J Polym Sci C Polym Symp* 51:135–53.

Ross, M., Kelso, G., Blaikie, F. et al. 2005. Lipophilic triphenylphosphonium cations as tools in mitochondrial bioenergetics and free radical biology. *Biochem (Mosc)* 70:222–30.

Saito, K., Yoshioka, H., and Cutler, R.G. 1998. A spin trap in mitochondria saves life in mice. *Biosci Biotechnol Biochem* 62:792–94.

Salnikov, V., Lukyanenko, Y.O., Frederick, C.A., Lederer, W.J., and Lukyanenko, V. 2007. Probing the outer mitochondrial membrane in cardiac mitochondria with nanoparticles. *Biophys J* 92:1058–71.

Saraste, M. 1999. Oxidative phosphorylation at the fin de siecle. *Science* 283:1488–93.

Scarpelli, M., Todeschini, A., Rinaldi, F., Rota, S., Padovani, A., and Filosto, M. 2014. Strategies for treating mitochondrial disorders: An update. *Mol Genet Metabol* 113:253–60.

Sharma, A., Soliman, G.M., Al-Hajaj, N., Sharma, R., Maysinger, D., and Kakkar, A. 2012. Design and evaluation of multifunctional nanocarriers for selective delivery of coenzyme Q10 to mitochondria. *Biomacromolecules* 13:239–52.

Sheu, S.S., Nauduri, D., and Anders, M.W. 2006. Targeting antioxidants to mitochondria: A new therapeutic direction. *Biochim Biophys Acta* 176:256–65.

Sholtz, K.F., Solovjeva, N.A., Kotelnikova, A.V., Snezhkova, L.G., and Miroshnikov, A.I. 1975. Effect of gramicidin S and its derivatives on the mitochondrial membrane. *FEBS Lett* 58:140–44.

Singal, P.K., Iliskovic, N., Li, T., and Kumar, D. 1992. Adriamycin cardiopmyopathy: Pathophysiology and prevention. *FASEB* 11:931–36.

Skulachev, V.P., Anisimov, V.N., and Antonenko, Y.N. 2009. An attempt to prevent senescence: A mitochondrial approach. *Biochim Biophys Acta* 1787:437–61.

Smith, R.A., Adlam, V.J., Blaikie, F.H. et al. 2008. Mitochondria-targeted antioxidants in the treatment of disease. *Ann N Y Acad Sci* 1147:105–11.

Smith, R.A., Hartley, R.C., Cochemé, H.M., and Murphy, M.P. 2012. Mitochondrial Pharmacology. *Trends Pharmacol Sci* 33:341–52.

Smith, R.A. and Murphy, M.P. 2011. Mitochondria-targeted antioxidants as therapies, *Discov Med* 11:106–14.

Smith, R.A., Porteous, C.M., Coulter, C.M., and Murphy, M.P. 1999. Selective targeting of an antioxidant to mitochondria. *Eur J Biochem* 263:709–16.

Spencer, S.L. and Sorger, P.K. 2011. Measuring and modeling apoptosis in single cells. *Cell* 144:926–39.

Suzuki, Y., Kadowaki, H., Atsumi, Y. et al. 1995. A case of diabetic amyotrophy associated with 3243 mitochondrial tRNA (leu; UUR) mutation and successful therapy with coenzyme Q10. *Endocr J* 42:141–45.

Szabados, E., Fischer, G.M., Toth, K. et al. 1999. Role of reactive oxygen species and poly-ADP-ribose polymerase in the development of AZT-induced cardiomyopathy in rat. *Free Radic Biol Med* 26:309–17.

Szeto, H.H. 2006a. Cell-permeable, mitochondrial-targeted, peptide antioxidants. *AAPS J* 8:E277–83.

Szeto, H.H. 2006b. Mitochondria-targeted peptide antioxidants: Novel neuroprotective agents. *AAPS J* 8:E521–31.

Szewczyk, A. 1997. Intracellular targets for antidiabetic sulfonylureas and potassium channel openers. *Biochem Pharmacol* 54:961–65.

Szewczyk, A. and Wojtczak, L. 2002. Mitochondria as a pharmacological target. *Pharmacol Rev* 54:101–27.

Tanaka, J., Nagai, T., Arai, H. et al. 1997. Treatment of mitochondrial encephalomyopathy with a combination of cytochrome *c* and vitamins B1 and B2. *Brain Dev* 19:262–67.

Thakar, J.H., Chapin, C., Berg, R.H., Ashmun, R.A., and Houghton, P.J. 1991. Effect of antitumor diarylsulfonylureas on in vivo and in vitro mitochondrial structure and functions. *Cancer Res* 51:6286–91.

Thévenod, F., Chathadi, K.V., Jiang, B., and Hopfer, U. 1992. ATP sensitive K$^+$ conductance in pancreatic zymogen granules: Block by glyburide and activation by diazoxide. *J Membr Biol* 129:253–66.

Torchilin, V.P. 2008. Recent approaches to intracellular delivery of drugs and DNA and organelle targeting. *Annu Rev Biomed Eng* 8:343–75.

Toriyabe, N., Hayashi, Y., and Harashima, H. 2013. The transfection activity of R8-modified nanoparticles and siRNA condensation using pH sensitive stearylated-octahistidine. *Biomaterials* 34:1337–43.

Trapp, S. and Horobin, R. 2005. A predictive model for the selective accumulation of chemicals in tumor cells. *Eur Biophys J* 34:959–66.

Vaidya, B., Paliwal, R., Rai, S. et al. 2009. Cell-selective mitochondrial targeting: A new approach for cancer therapy. *Cancer Ther* 7:141–48.

Waite, M. and Sisson, P. 1972. Effect of local anesthetics on phospholipase from mitochondria and lysosomes. Probe into the role of the calcium ion phospholipid hydrolysis. *Biochemistry* 11:3098–105.

Wallace, D.C. 2005. A mitochondrial paradigm of metabolic and degenerative diseases, aging, and cancer: A dawn for evolutionary medicine. *Annu Rev Genet* 39: 359–407.

Wallace, K.B. and Starkov, A.A. 2000. Mitochondrial targets of drug toxicity. *Annu Rev Pharmacol Toxicol* 40:353–83.

Wang, L., Liu, Y., Li, W. et al. 2011. Selective targeting of gold nanorods at the mitochondria of cancer cells: Implications for cancer therapy. *Nano Lett* 11:772–80.

Weissig, V. 2003. Mitochondrial-targeted drug and DNA delivery. *Crit Rev Ther Drug Carrier Syst* 20:1–62.

Weissig, V. 2005. Targeted drug delivery to mammalian mitochondria in living cells. *Expert Opin Drug Deliv* 2:89–102.

Weissig, V. 2011. From serendipity to mitochondria-targeted nanocarriers. *Pharm Res* 28:2657–68.

Weissig, V. 2015. DQAsomes as the prototypes of mitochondria-targeted pharmaceutical nanocarriers: Preparation, characterization, and use. *Methods Mol Biol* 1265:1–11.

Weissig, V., Cheng, S.M., and D'Souza, G.G. 2004. Mitochondrial pharmaceutics. *Mitochondrion* 3:229–44.

Weissig, V., D'Souza, G.G., and Torchilin, V.P. 2001. DQAsome/DNA complexes release DNA upon contact with isolated mouse liver mitochondria. *J Control Release* 75:401–08.

Weissig, V., Lasch, J., Erdos, G., Meyer, H.W., Rowe, T.C., and Hughes, J. 1998. DQAsomes: A novel potential drug and gene delivery system made from dequalinium. *Pharm Res* 15:334–37.

Weissig, V., Lizano, C., and Torchilin, V.P. 2000. Selective DNA release from DQAsome/DNA complexes at mitochondria-like membranes. *Drug Deliv* 7:1–5.

Weissig, V. and Torchilin, V.P. 2000. Mitochondriotropic cationic vesicles: A strategy towards mitochondrial gene therapy. *Curr Pharm Biotechnol* 1:325–46.

Weissig, V. and Torchilin, V.P. 2001. Cationic bolasomes with delocalized charge centers as mitochondria-specific DNA delivery systems. *Adv Drug Deliv Rev* 49:127–49.

White, C.W., Rashed, H.M., and Patel, T.B. 1988. Sulfonylureas inhibit metabolic flux through rat liver pyruvate carboxylase reaction. *J Pharmacol Exp Ther* 246:971–74.

Wipf, P., Xiao, J., Jiang, J. et al. 2005. Mitochondrial targeting of selective electron scavengers: Synthesis and biological analysis of hemigramicidin-TEMPO conjugates. *J Am Chem Soc* 127:12460–61.

Yamada, Y., Akita, H., and Kamiya, H. 2008. MITO-Porter: A liposome-based carrier system for delivery of macromolecules into mitochondria via membrane fusion. *Biochim Biophys Acta (BBA) Biomembr* 1778:423–32.

Yamada, Y., Akita, H., Kogure, K., Kamiya, H., and Harashima, H. 2007. Mitochondrial drug delivery and mitochondrial disease therapy—An approach to liposome-based delivery targeted to mitochondria. *Mitochondrion* 7:63–71.

Yamada, Y., Furukawa, R., Yasuzaki, Y., and Harashima, H. 2011. Dual function MITO-Porter, a nano carrier integrating both efficient cytoplasmic delivery and mitochondrial macromolecule delivery. *Mol Ther* 19:1449–56.

Yasuzaki, Y., Yamada, Y., and Harashima, H. 2010. Mitochondrial matrix delivery using MITO-Porter, a liposome-based carrier that specifies fusion with mitochondrial membranes. *Biochem Biophys Res Commun* 397:181–86.

Yousif, L.F., Stewart, K.M., Horton, K.L., and Kelley, S.O. 2009. Mitochondria-penetrating peptides: Sequence effects and model cargo transport. *Chembiochem* 10:2081–88.

Yu, P., Yu, H., Guo C. et al. 2015. Reversal of doxorubicin resistance in breast cancer by mitochondria-targeted pH-responsive micelles. *Acta Biomater* 14:115–24.

Zamzami, N., Marchetti, P., Castedo, M. et al. 1996. Inhibitors of permeability transition interfere with the disruption of themitochondrial transmembrane potential during apoptosis. *FEBS Lett* 384:53–57.

Zhang, R., Goldstein, S., and Samuni, A. 1999. Kinetics of superoxide-induced exchange among nitroxide antioxidants and their oxidized and reduced forms. *Free Radic Biol Med* 26:1245–52.

Zhao, K., Zhao, G.M., Wu, D. et al. 2004. Cell-permeable peptide antioxidants targeted to inner mitochondrial membrane inhibit mitochondrial swelling, oxidative cell death, and reperfusion injury. *J Biol Chem* 279:34682–90.

Zhou, S., Palmeira, C.M., and Wallace, K.B. 2001. Doxorubicin induced persistent oxidative stress to cardiac myocytes. *Toxicol Lett* 121:151–57.

Zupancic, S., Kocbek, P., Zariwala, M.G. et al. 2014. Design and development of novel mitochondrial targeted nanocarriers, DQAsomes for curcumin inhalation. *Mol Pharm* 11:2334–45.

# 12 Antisense Oligonucleotide-Mediated Target-Specific Gene Silencing

## Design, Delivery Strategies, and Therapeutic Applications

*Biswajit Mukherjee, Samrat Chakraborty,
Laboni Mondal, Ankan Choudhury,
Bhabani Sankar Satapathy, and
Sanchari Bhattacharya*

## CONTENTS

## 12.1  INTRODUCTION

Antisense oligonucleotide (ASO), popularly known as oligo, is a single stranded deoxyribonucleotide/ribonucleotide sequence (<50 bp in length) that has the capability to hybridize with a target gene/mRNA through Watson-Crick base pairing, resulting in an arrested or reduced expression of that gene. The potential of ASO in therapeutics was first identified by Zemencnik and Stephenson (Fattal and Bochot 2008) when they observed that oligonucleotide complementary to 3′ end of Rous sarcoma virus could block the viral replications in chicken fibroblast. The exquisite specificity of ASO relies on the concept that a particular sequence of 17 bases in DNA occurs only once in the human genome (Martimprey et al. 2009). Since its advent, ASO has gained a lot of attention among researchers as smarter therapeutic alternatives against some deadly diseases such as cancer and other genetic disorders (Fattal and Bochot 2008).

Oligonucleotide technology experienced a remarkable revolution after the finding of small interfering RNAs (siRNAs) (21-23-mer double stranded RNA molecules) that effectively silence the gene expression. Their mechanism of action is completely elucidated due to the proper understanding of the function of Argonaut protein family. The occurrence of RNA interference pathway is common to all eukaryotes including animals and the process is initiated by the enzyme dicer (endoribonuclease dicer or helicase with RNase motif). The cleavage of long double stranded RNA molecules by dicer results in the generation of short double stranded fragments of around 20 nucleotide siRNAs and finally it separates into two single stranded RNAs (ssRNA)—the passenger strand and the guide strand. Among these two strands, the passenger strand is degraded while guided strand has been incorporated into the RNA-induced silencing complex (RISC). The Argonaut protein representing the active part of RISC separates the guided strand from passenger strand and the separated guided strand binds with target mRNA. Finally, the endonuclease activity of the RISC hydrolyzes the target RNA at the site where it binds with the antisense strand. The single stranded RNA molecule binds with the target RNA strand by Watson-Crick base pairing resulting in recruitment of ribonuclease as well as cleavage of the strand, therefore RNA interference is also recognized as antisense mechanism. In certain cases, oligonucleotides have been designed in such a way that they can chemically cleave the target RNA after hybridization without recruiting the cellular nucleases. Such types of oligonucleotides are known as ribozymes and DNAzyme, which have gained the significant interest of researchers. ASOs that do not directly initiate the degradation of mRNA have also been reported in the literature (Dias and Stein 2002). Instead, they cause translational arrest through blocking of mRNA sterically. For such types of nucleotides, 3'- or 5'-untranslated region appears to be a more appropriate target as compared to translation initiation coding region. Most mammalian RNAs undergo post-transcriptional processing steps in the cell nuclei and they include the addition of 5'-cap structure, splicing and polyadenylation. Therefore, the regulation of RNA processing is considered another efficient approach of utilizing oligonucleotides to regulate gene expression. Numerous reports in the literature have demonstrated the occurrence of alternatively spliced transcripts encoding functionally antagonist protein and data of these reports clearly illustrate the viability of this approach to reversibly "switch" the protein function. For example, the most commonly expressed splice variant of apoptosis-regulating gene (bcl-x) is bcl-xL, which is antiapoptotic whereas its variant (bcl-xS) lacking a portion of exon II is proapoptotic. Therefore, by designing the oligonucleotide specific to the region surrounding the appropriate splice site, the switch over from bcl-xL to bcl-xS is possible resulting in sensitization of cells to chemotherapeutic agents. Moreover, oligonucleotides can be designed to promote the binding with pre-mRNA to mask the polyadenylation sites forcing the cells to exploit alternative poly A sites.

A comparison of different antisense mechanisms in cell-based assays revealed that once optimized, oligonucleotides exploiting any of the above-mentioned mechanisms to exert its function could be potent inhibitors of gene expression. More importantly, no mechanism becomes vastly superior to other and designing of antisense nucleotide solely depends on its biological applications.

To perform the downregulation of gene expression, antisense oligonucleotide should penetrate into the target cells. To date, the precise mechanisms responsible for oligonucleotide penetration have not yet been explored. Fluid-phase pinocytosis and adsorptive endocytosis are the two principal mechanisms together with other factors such as temperature, structure and concentration of oligonucleotide and the cell line-mediated control of the cellular uptake. At a relatively low concentration, oligonucleotides are internalized through the interaction with membrane-bound receptors and this concept has been supported by the study conducted by De Diesbach et al. (2000) who have purified and characterized one of those receptors. Data of numerous studies (Dias and Stein 2002) demonstrated that the ASOs are poorly internalized by the cells, and preferentially accumulate in liver and kidneys and to a lesser extent in the spleen, lymph nodes, and bone marrow. Liver, which serves as a vital location for oligonucleotide accumulation, has gained significant attention due to its ability to control the clearance of foreign matters through the reticuloendothelial system (RES). This is possible because of the presence of phagocytic Kupffer cells, high blood flow, and the existence of fenestrated vasculature of 100–200 nm along the endothelial lining. Therefore, the development of effective delivery strategies is prerequisite to exploit the potential of ASOs as therapeutics.

Despite their immense potential as therapeutics against an array of genetic disorders, there are various key issues, which should be addressed properly in order to develop an efficient antisense therapeutic. Although the research on oligonucleotides had started its journey before three decades; but until today only two FDA-approved products are available (Juliano et al. 2008). This clearly gives evidence in support of hurdles, which hinder the translation of research from the laboratory to clinic. In the present chapter, we will focus on the design of antisense oligonucleotides to promote the site-specific delivery. Apart from that, its potential in various disorders will also be discussed along with recent findings.

## 12.2   DESIGNING ANTISENSE OLIGONUCLEOTIDES

The progress in the area of Antisense development has experienced a slow but steady pace due to the obstacles included as follows:

1. They lack stability against intra and extracellular degradation.
2. The inherent ionic charge hinders their transport across the plasma membrane to reach the target cells or tissues.
3. They sometimes have poor affinity toward intended target sequence.
4. Specific targeting strategy should be adopted for them in order to avoid off-target toxicity.
5. They may have potential to stimulate immunological consequences.

Numerous strategies have been adopted in order to make chemical modifications to native DNA or RNA in order to overcome the aforementioned drawbacks. ASOs act mainly by hybridizing the target site of nucleic acids (gene/DNA/mRNA) with complementary sequence through Crick-Watson base pairing (Chan et al. 2006), which leads to inhibition of gene expression or cleavage of the mother sequence through the induction of RNase L and RNase H endonuclease enzymes (Pirollo et al. 2003; Caffo et al. 2013). The primary mechanism involves the endogenous RNase H enzymes (Dias and Stein 2002; Dean and Bennett 2003; Caffo et al. 2013) which specifically cleave the heteroduplex-like formation of the two different nucleic acids, that is, mRNA and antisense DNA (Chan et al. 2006; Caffo et al. 2013). The pathway that involves binding and subsequent heteroduplex formation, leads to RNase H-mediated mRNA degradation while releasing the oligonucleotide intact and capable of targeting new mRNAs and thus, it inhibits translation and expression of the protein encoded in that mRNA (Behlke et al. 2005; Chan et al. 2006; Caffo et al. 2013). This may be the primary mechanism exploited by most of the researchers working on antisense therapy. Nevertheless, there are other mechanisms independent of RNase H endonuclease activation. They involve sterically blocking of the initiation site of a particular protein translation in an mRNA, thus sterically hindering the ribosomal activity and interfering with mRNA maturation by disrupting splicing which subsequently destabilizes the pre-mRNA in the nucleus (Behlke et al. 2005; Chan et al. 2006; Caffo et al. 2013). These mechanisms also induce downregulation of the target proteins but are around 10–100-fold less potent than the RNase H-dependent oligonucleotides of homologous sequences (Behlke et al. 2005), which are efficient enough to achieve 85%–90% downregulation of the targeted proteins (Dias and Stein 2002). The RNase H-dependent oligonucleotides can function by inhibiting protein expression even if they are targeted to any region of the mRNA since it cleaves the whole system. But for silencing mRNA translation and protein expression, most steric-blocker species have to be specifically oriented to complement and block the 5′- or AUG initiation codon region (Larrouy et al. 1992; Dean et al. 1994; Dias and Stein 2002). This superiority is, however, not general and has been developed over years through the arrival of the different generations of ASOs with varying degrees of structural and functional efficiency (Dias and Stein 2002; Behlke et al. 2005; Chan et al. 2006).

## 12.2.1 Factors Influencing Design of Antisense Oligonucleotide

### 12.2.1.1 Functional Factors

#### 12.2.1.1.1 *RNase H-Mediated Antisense Activity*

RNase H-dependent ASOs are the most widely investigated for their potential in completely destabilizing aberrant mRNAs. However, they require stable and efficient binding to its target sequence for desired action. The RNase H class of endonucleases acts primarily in the nucleus with some activity detectable in the cytoplasm as well (Cazenave et al. 1994). The inlaid mechanism is that the ASOs bind specifically to target mRNA, which initiates RNase H-mediated degradation of the double stranded antisense probe–mRNA heteroduplex (Dagle et al. 1991). The mechanism has also been depicted in Figure 12.1.

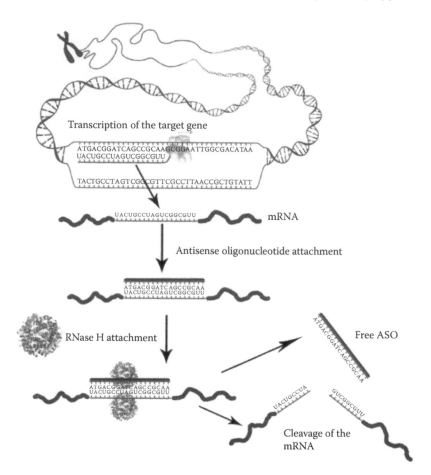

**FIGURE 12.1** Suggested mechanism behind the RNase H-mediated cleavage of target mRNA through attachment of antisense oligonucleotide.

The precise chronology of actions of this is not elaborately documented, but it has been proposed to be similar to that of DNase I (Ishikawa et al. 1993). Although ASO–mRNA heteroduplex stimulates RNase H activity, but since the RNase H activity itself is sequence independent (Ho et al. 1996), the ASO should be designed only to complement the sequence of the target RNA. The presence of certain nucleotides or sequences of nucleotide favor a successful RNase-mediated cleavage such as the content of GC (Ho et al. 1996). The thermodynamic stability of the ASO–mRNA duplexes is correlated with the number of GC residues. A 20 bp ASO containing 9 or less G or C residues will show weaker inhibition of target whereas 11 or more such residues bring about strong ASO effects (Ho et al. 1996).

Most DNA ASOs that have been structurally modified to be nuclease resistant do not form a cleavable heteroduplex structure with the phosphorothioate-modified oligonucleotides being an exception. Besides phosphorothioates, other RNase H-competent backbones include 2′-fluorooligodeoxy nucleotides (Wilds and Damha

2000; Damha et al. 2001). Modifications such as in methylphosphonates, 2'-O-methyl derivatives, peptide nucleic acids, and morpholino oligonucleotides do not preferably induce RNase H activity (Dias and Stein 2002). The exact pathway by which these duplexes are recognized by RNase H is not well understood. This phenomenon was investigated using chimeric oligonucleotides consisting of a central chain of phosphorothioate nucleotides, flanked by 2'-O-methyl phosphorothioates at the 3' and 5' termini (Monia et al. 1993). The study has established that a small stretch of homology (~5 bp) is sufficient to induce RNase H activity in even nontargeted mRNAs besides the original target against which it is designed (Monia et al. 1993). However, this should be investigated whether this phenomenon is equally responsive in living human cells and/or cell lines. Isis 3521, a 20-mer long phosphorothioate oligonucleotide targeted to the protein kinase C (PKC) α-mRNA downregulated PKC-ζ, another protein whose mRNA shares 11 bases of contiguous homology with the sequence of the former mRNA (Benimetskaya et al. 1998). This may be a major concern if the design of the ASO is to be clinically validated because of the wanton cleavage of nontargeted mRNAs. Partial hybridization is the problem that can only be feebly subsidized using chimeric oligonucleotides (Monia et al. 1993; Giles et al. 1998). Human RNase H enzymes degrade the RNA in heteroduplex form completely. However, enzyme levels are limiting certain sites to be cleaved preferentially over the others (Eder and Walder 1991; Eder et al. 1993; Frank et al. 1994; Toulme 1996). The cleavage rates of a single ribonucleotide residue hybridized with single strand (SS) DNA maintains the following manner of $A > U > C > G$, with the rate of cleavage at A being over four times faster than at G as documented previously (Eder et al. 1991). Furthermore, the secondary and tertiary structures also contain certain sites that are topologically favored by the endonucleases. Thus, designing an ASO for the abovementioned pathway must be done on lines, which ensure the accessibility of the target sequences within the folded RNA structure.

### 12.2.1.1.2   Non-RNase H-Mediated Antisense Activity

Non-RNase H-dependent antisense activity can also efficiently inhibit mRNA translation namely through the disruption of ribosomes or by sterically blocking the initiation codon or inhibiting elongation steps of protein translation (Baker et al. 1997; Dias and Stein 2002) such as arresting the elongation of the polypeptide chain. For an example, using a peptide nucleic acid oligonucleotide acts against the H-ras mRNA which yields a truncated peptide of the same size of a truncated product produced by the RNase H-dependent phosphodiester oligonucleotide of same sequence (Dias et al. 1999). There are sequences (present in RNA) that are to regularly interact with proteins, ribosomes, spliceosomes, and other large molecules and so they are also likely to be accessible to oligonucleotides, assuming no unwinding activity is required. These sequences present potent target candidates for many such ASOs (Behlke et al. 2005). Initial investigations had chosen the terminal cap, initiation codon, and 3'-end as targets (Goodchild et al. 1988). Despite the frequent (>70%) (Dias and Stein 2002) use of the strategy involving oligonucleotides to target the translation initiation codon AUG, the site may not be always optimum for targeting purpose. The 3'- or 5'-untranslated region in some mRNAs also represents feasible targets even more than the initiation codon region (Skorski et al. 1997;

Mologni et al. 1998). Sterically blocking of the main coding sequence is even less fruitful evident from the volume of works which have established the ability of ASOs to block the core coding region compared to the 3′- and 5′-untranslated or initiation codon regions (Monia et al. 1993; Gryaznov et al. 1996; Tyler et al. 1999; Dias and Stein 2002). This may be due to the ability of the ribosomal machinery to unwind the oligonucleotide blockage from its targeted mRNA before translation (Dias and Stein 2002).

ASOs can also interfere with the splicing process, resulting in nonfunctional mRNAs that are slowly decayed in a nonsense-mediated RNA decay pathway. The latter approach can also modulate alternative splicing or block aberrant and/or disease-causing splice sites or single strands (SSs) (Kole et al. 2004; Aartsma-Rus et al. 2005). For successful targeting of splice sites, the oligonucleotide must gain access to the nucleus where the pre-mRNA resides whereas inhibition of translation can also be accomplished in the cytoplasm. Target sites of splice-modulating ASOs, involved in exon recognition and inclusion, consist mainly of both the donor and acceptor SSs and branch point sequences. However, they also include exon-internal sequences (exonic splicing enhancers [ESEs] sites) that induce splicing by binding of Ser-Arg-rich (SR) proteins, which in turn recruit the splicing factors U1 small nuclear riboprotein (U1 snRNP) and U2 small nuclear RNA auxiliary factor (U2AF) to the donor and acceptor SSs, respectively (Cartegni et al. 2002). The strategies to target aberrant SSs and ESEs have been shown to be efficient modulators of splicing (Aartsma-Rus et al. 2005; Wilton et al. 2007). Since ESEs are weakly defined motifs and they do not fit the purview of consensus sequences, that is, devoid of frequently repeated nucleotides, they present a wide range of target choices, which act as an advantage of exon-internal ASOs over SS ASOs (Aartsma-Rus et al. 2005). These abovementioned mechanisms have been described in Figure 12.2.

Moreover, the binding energy and stability of the ASOs depend on factors such as length, sequence, and the free energy of local structures (Chan et al. 2006). For efficient association with a target sequence, the free energy of the ASO–target heteroduplex must be higher than the summation of respective free energies of the target complex and the ASO individually (Mathews et al. 2004). Efficient association not only demands the presence of single stranded motifs on the target mRNA but on the ASOs as well. Since ASOs are generally 15–25-nucleotides long, frequency to form a stable secondary structure is very low. However, most ASOs can form ASO–ASO homologous complexes with other similar ASOs. Computational methods also provide the free energy of ASO–ASO complexes and ASO–target complexes, along with the free energy of individual ASOs and the target sequences (Mathews et al. 2004). The requirements of ASOs following the translation blocking design can be incorporated in many computational methods, which require the targeting of the AUG initiation codon of the mRNA, as mentioned before. Software programs can be used for assistance in designing ASOs; however, pitfall exists and a generalized trial and error procedure is required to identify potential of ASOs (Aartsma-Rus et al. 2005). Backbone modifications like 2′-O-alkyl, peptide nucleic acids, and morpholinos may facilitate splicing modification with their ability to inhibit intron and exon excision (Kole and Sazani 2001). These morpholino analogues or 2′-O-alkyl derivative ASOs are widely used despite failing to induce cleavage because of

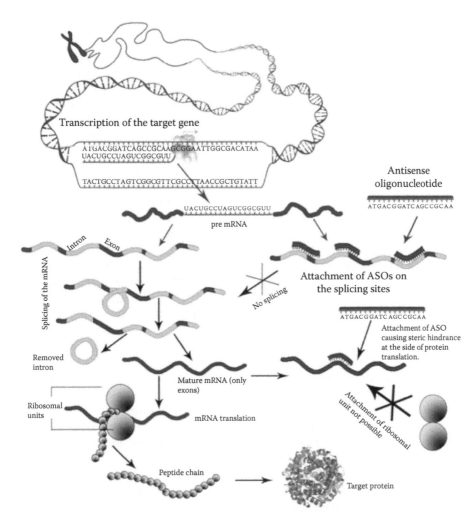

**FIGURE 12.2**   Non-RNase-mediated mechanisms of antisense oligonucleotides including the blocking of the splicing sites on the pre-mRNA (nuclear) and the sites for attachment of ribosomal units for mRNA translation (nonnuclear).

their tendency to hybridize so strongly and leaving the RNA sequence incapacitated for further actions like splicing or translation (Arora et al. 2000; Iversen 2001; Mercatante et al. 2001).

### 12.2.1.2   Conformational Factors

The secondary and tertiary folding of the ribonucleotide chain can leave portions of the mRNA inaccessible to an oligonucleotide. Accurate prediction of the secondary structure of the target mRNA is prime important in order to designing ASO effectively (Andronescu et al. 2005; Chan et al. 2006) as an ASO should bind at the locations where mRNA is prone to hybridization (Ding and Lawrence 2001;

Aartsma-Rus et al. 2005). Since the design of ASOs allows them to be attached preferably on SS bearing parts of the mRNAs (Aartsma-Rus et al. 2008), several software programs developed to predict the secondary structure of RNA which provides a parameter called SS count for the target sequence. The SS count indicates the tendency of a target ribonucleotide to remain single stranded amongst a number of predicted secondary structures (Chan et al. 2006; Aartsma-Rus et al. 2008) of that RNA. The SS availability became one of the obvious determinants of accessibility of all forms of ASOs. This approach closely simulates the actual *in vivo* efficacy than the model, which assesses the most energetically stable structure (Aartsma-Rus et al. 2008). There were studies done in the past (Matveeva et al. 1997) which pronounced the difficulties of predicting accurate and accessible binding sites by comparing experimentally determined RNase H-accessible sites in four RNA species with those gathered from known or predicted secondary or tertiary structure of those RNAs. Older software such as Foldsplit was used to screen and yield more effective ASO than those obtained through random selection, which returned about one good target site per 1000 bases in RNA on average (Patzel et al. 1999; Behlke et al. 2005). Even though the universality and reproducibility along with utmost reliability is still an issue with computational models for truly predicting secondary fold structures of mRNA, they are frequently used for preliminary designing processes. One such widely used program (Mfold program) is available in the public domain (http://www.bioinfo.rpi.edu/applications/mfold). This provides a clue regarding the possible optimal and suboptimal structures of a particular sequence of mRNA and also overall minimum free energy ($\Delta G$) of different possible foldings (Zuker 2003; Chan et al. 2006; Aartsma-Rus et al. 2008). Another popular computer algorithm is the Sfold program (http://sfold.wardsworth. org/index.pl), which returns only the best secondary structure of the target mRNA sequence (Ding and Lawrence 2001). For the best result, and to surmise a comprehensive picture on the design of ASO, one may use both the algorithms in a combination.

The occurrence of mRNA cleavage may be skewed toward single strand sites. In those cases, only some single stranded sites were adjudged as good targets, although many attractive sites were found to be located in double stranded regions (Behlke et al. 2005). The attachment or association of ASOs with RNA may be primarily governed by simple rules of Watson-Crick base pairings but there are additional subtleties that govern the hybridization of oligonucleotides to RNA. Even there are possibilities of sequences that are apparently accessible but may already be hindered through intramolecular hydrogen bonding, stacking interactions, or in solvation. These hindrances are often disrupted by ASO hybridization, but any hybridization-induced rearrangement of the existing RNA structure can compromise the thermodynamics of the system and ultimately be penalizing (Behlke et al. 2005). Small alterations in the length of ASO or a shift in the binding site of one or two nucleotides can skew the kinetics of hybrid formation (Rittner et al. 1993; Kronenwett et al. 1996; Sczakiel 2000). On the other hand, some single stranded sequences within the RNA may also be preordered by stacking into helical conformations, which can turn out to be favorable for hybridization. Local structures easily accessed and recognized by the ASOs for hybridization are those, which are usually located at the terminal end, internal loops, joint sequences, hairpins, and bulges of 10 or more consecutive nucleotides (Kretschmer-Kaxemi et al. 2001). The hybrids formed between the loops of two hairpins (kissing interactions) are markedly stable and often influence the association of natural antisense RNAs with their targets (Behlke et al. 2005). The behavior of oligonucleotides is also dependent on the nature of the terminal nucleotides. The prevalence of preferential hybridization of oligonucleotides to 5′ side of such loop rather than 3′ side can be added to the list conformational influences (Bruice and Lima 1997; Puri and Chattopadhyaya 1999). A new software Target-Finder (http://www.bioit.org.cn/ao/targetfinder.com) has been developed for scouting potential ASO target sites based on the method of mRNA accessible site tagging (MAST) (Bo and Wang 2005). Highly conserved local sequence motifs have been explored and validated as potent ASO target sites (Yang et al. 2003). In order to increase the "hit rate" of the ASO candidate design, one also needs to look for locally conserved structures among various optimal and suboptimal mRNA-predicted secondary structures (Chan et al. 2006).

### 12.2.1.3 Compositional Factors

#### 12.2.1.3.1 The Length of Antisense Oligonucleotides

The stability of ASO–mRNA hybrids can also be influenced by length. This effect is more acute in shorter oligonucleotides since longer oligonucleotides are less likely to encounter a complementary sequence other than its targeted sequence. But on the other hand, longer oligonucleotides have more tendency to bind even to partially complementary sequences in a nontarget mRNA and cleaving through RNase H activation as it requires just around five complementary base pairs (as previously stated) to form a heteroduplex substrate for RNase action (Behlke et al. 2005). Thus, the optimum length for ASOs is fixed around 20 bases. This is because oligonucleotides of this dimension are convenient for synthesis and also that statistically each unique sequence in the human genome is at least 19 bp long (Jiménez and Durbin 2006) and so it is less likely for a 20-mer long ASO to find a duplicate complement site where it will bind instead of the target site. There are certain structural features in the target RNA that may allow the use of shorter oligomers in some instances (Wagner et al. 1996; Disney et al. 2000). However, in most studies, it was seen that the activity of ASOs were either reduced or lost on truncating them to around 10 bases (Cowsert et al. 1993).

#### 12.2.1.3.2 Constituents of Antisense Oligonucleotides

Optimization of length alone will not do the trick if a check on the sequence of ASO is not done for screening and manipulating features that could affect its activity. For instance, the sequences that complement nontarget RNAs or well-defined motifs can enhance or reduce the quality of ASOs or even self-complementing sequences that can interfere with the hybridization process. Certain features or motifs serving biological purposes may be totally detached from the antisense activity and can even be potentially harmful (Matveeva et al. 2000; Behlke et al. 2005; Chan et al. 2006). Certain motifs such as the CCAC, TCCC, ACTC, GCCA, and CTCT have a positive correlation with the ASO activity whereas prevalence of GGGG (G-quartets formation), ACTG, AAA, and TAA motifs reduces the same

(Matveeva et al. 2000). One of the most important features is the abundance of CG sequences in ASOs. Since unmethylated CG sequences appear more frequently in bacterial DNA than in any other eukaryotic genome, human immune system may use their presence as a signal of bacterial infection (Sun et al. 1998; Krieg et al. 1999). Oligonucleotides with frequent CG sequences are possibly recognized by toll-like receptor-9 in immune cells, resulting in the release of cytokines (IL-6, IL-12, and interferon-γ), B lymphocyte proliferation, antibody production and activation of T lymphocyte, macrophages, dendritic cells and natural killer cells (Lipford et al. 1997; Bendigs et al. 1999; Jakob et al. 1999; Krieg et al. 1999; Vollmer et al. 2004). These CG-mediated immune effects vary with the sequences flanking with the CG dimer. The strongest effect is seen when the purine–purine–C–G–pyrimidine–pyrimidine motif is present (Krieg and Stein 1995). These CG effects may be responsible for some of the unwanted activities of oligonucle-otides reported *in vivo*. Few such examples are splenomegaly, lymphoid hyper-plasia, diffused multiorgan mixed mononuclear cell infiltrates, etc. (Levin 1999). Alternatives for avoiding this effect are choosing oligonucleotides devoid of CG sequences, especially those with immune stimulatory flanking sequences and also to replace the C in CG sequences with 5-methylcytidine which, albeit being expen-sive, prevent immune stimulation without affecting hybridization (Behlke et al. 2005). In addition to these, formation of 3D tetraplexes (comprising three, four, or more intrinsic strands in the oligonucleotide, outside Watson-Crick base pairing) may portend some undesirable biological activity to the ASO (Benimetskaya et al. 1997; Wang et al. 1998).

Frequently occurring tetraplexes are formed in oligonucleotides contain-ing multiple adjacent guanine residues occurring in a quartet of G residues (GGGG), or in repeated dimeric (GG) or trimeric (GGG) motifs placed in vicinity (Williamson et al. 1989; Schultze et al. 1994). When they do not form tetraplexes, these G-rich sequences with multiple GG dimers may form other unusual struc-tures depending on the other proximal sequences (Chou et al. 1994). Tetraplex-forming lines of Gs show varying affinity for different proteins which produce a multitude of biological effects when included in ASO sequences. For example, ASOs those bind to proteins like thrombin (Griffin et al. 1993), HIV envelope protein (Wyatt et al. 1994), transcription factors (Mann and Dzau 2000), or sun-dry antiproliferative agents (Bates et al. 1999). By replacing guanosine residues with 7-deazaG (Murchie and Lilley 1994) or 6-thioG (Olivas and Maher 1995) one can reduce the tendency of forming tetraplexes. A less common occurrence of tetraplex is found inphosphorothioate oligonucleotide containing only C residues, which had similar activity to the commonly found G-tetraplex (Wang et al. 1998). Thus, every oligonucleotide should be carefully screened for sequences very rich in a particular nucleoside, especially if they have repeated sequences or have multiple occurrences of two or more adjacent identical bases. They run the risk of forming tetraplexes and other non-Watson-Crick structures. Although not all such features will necessarily lead to formation of these higher-order structures in the ASOs, particularly in physiological conditions, but these patterns of sequences should be treated as warning flags while designing new ASOs and they should be avoided or mitigated at the best.

### 12.2.1.4 Thermodynamic Considerations

The thermodynamic energy of oligonucleotide mechanisms must be taken into consideration before designing the ASO. Software such as OligoWalk from the package RNAstructure 3.5 (http://rna.urmc.rochester.edu/cgi-bin/server_exe/oligowalk/oligowalk_form.cgi) are available for calculating thermodynamic properties between the ASO and mRNA target sequence and also between two ASO molecules (Mathews et al. 1999, 2004). A potent ASO should have a higher binding energy ($\Delta G_{37}^0 \geq -8\,$kcal/mol) between the ASO and mRNA than the binding energy ($\Delta G_{37}^0 \geq -1.1\,$kcal/mol) between two ASOs (Matveeva et al. 2000). If this criterion is met, the hit rate of developing a potent and active ASO is 3–6-fold higher, as supported by the antisense databases from ISIS Pharmaceuticals (Carlsbad, CA, USA) (Matveeva et al. 2003). This algorithm can be reflected in the designing of ASO for the downregulation of vascular endothelial growth factor protein expression with a success rate higher than 85% (Fei and Zhang 2005). But a contrasting picture was seen in the analysis of ASOs for targeting exon splicing sites in Duchenne muscular dystrophy (DMD) gene transcript where the ASO-ASO complexes of efficient ASO species had a higher free energy (Aartsma-Rus et al. 2008). This discrepancy can be explained from the fact that the splice modulation occurs in the nucleus and targets pre-mRNA and not the mature mRNAs in the cytoplasm as in RNase H-mediated system. Thus, stable ASO–ASO complexes may increase the chance of ASO reaching the nucleus intact and prevent the ASO from binding nonspecifically to nontarget RNAs (Aartsma-Rus et al. 2008). To mitigate such convolutions in the design approach, a single enveloping algorithm (http://www.genscript.com/siRNA_target_finder.html) has been developed for ASO prediction, based on a neural network, perusing a wider range of parameters, which includes sequence composition, RNA–ASO binding energy, RNA–ASO terminal properties, ASO–ASO binding properties and several verified sequence motifs correlated with efficacy and RNase H accessibilities (Chan et al. 2006). This model was able to predict ASOs with greater than 50% gene expression inhibition with a success rate of 92%, but it is far from perfect since it had screened out some experimentally effective sequences due to stringent selective criteria set by the program and also because it fails to consider thermodynamics of dimer energy properly. So for efficient ASO scouting purposes, a combination of theoretical algorithm and the "trial and error" linear screening method should be used.

### 12.2.2 OLIGONUCLEOTIDE CHEMISTRY

Though it is not difficult to synthesize phosphodiester oligonucleotides in laboratory conditions, their use is limited due to their rapid degeneration by intracellular endonucleases and exonucleases, usually via $3' \rightarrow 5'$ order, found in biological fluid and charge present in them prevents them from penetrating through the cell membrane (Dias and Stein 2002; Chan et al. 2006). Furthermore, the dNMP2 mononucleotides, the byproducts of phosphodiester oligonucleotide degradation, may have cytotoxic and antiproliferative effects (Vacrman et al. 1997) through mononucleotide dephosphorylation by the cell surface enzyme ecto-5'-nucleotidase which leads to inhibition of critical proteins like thymidine kinase (Koziolkiewicz et al. 2001). Modifications can enhance the intracellular half-life of oligos in the order of days and influence the ability of an oligo to trigger RNase H-mediated degradation of RNA following hybridization (Behlke et al. 2005). The various alterations done to the oligonucleotides through time have a deep impact in ASO designs. All the commonly encountered structural modifications of the nucleotide backbone have been given in the Figure 12.3.

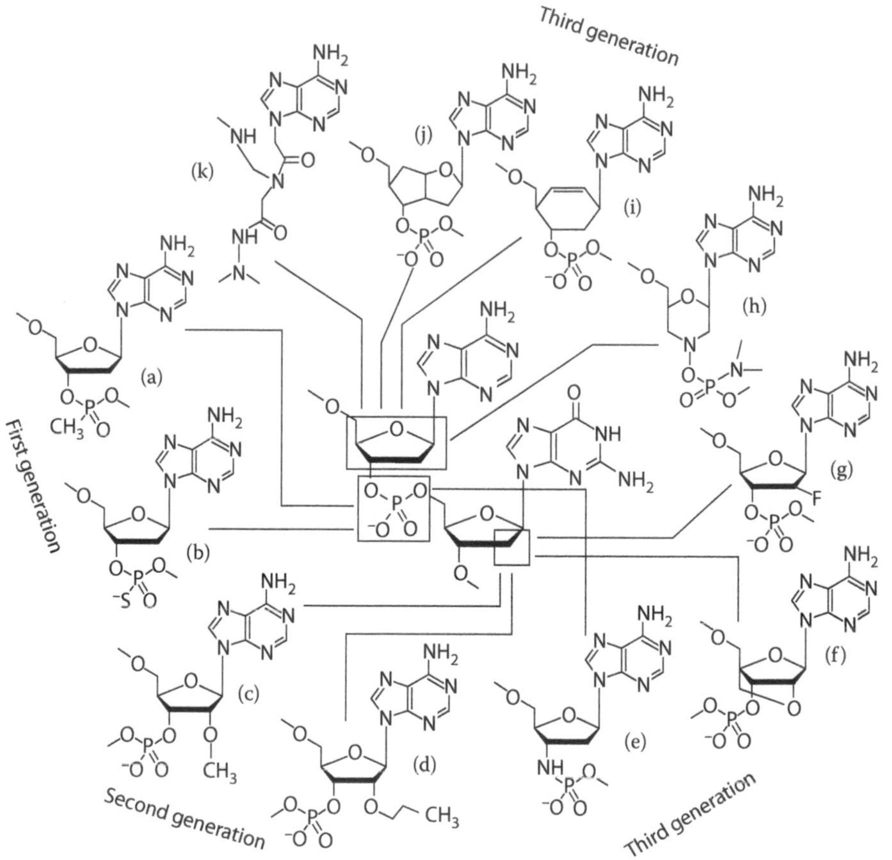

**FIGURE 12.3** First-generation oligonucleotides include (a) methylphosphonates, and (b) phosphorothioates, which modified the phosphate molecule of the normal phosphodiester linkage (central figure). The second-generation oligonucleotides involved 2′ substitution at the ribose sugar, which include (c) 2′-O-methyl, and (d) 2′-O-methoxyethyl derivatives. The third-generation oligonucleotides had a wide variety of modifications which included modification in the phosphate linkage as in (e) phosphoramidate derivatives; substitution at the 2′ of ribose as in (f) locked nucleic acid, and (g) fluoro arabino nucleic acid derivatives and modification/substitution of the ribose molecule as in (h) phosphoramidate morpholino, (i) cyclohexene nucleic acid, (j) tricycle nucleic acid, and (k) peptide nucleic acid derivatives.

## 12.2.2.1 First-Generation Oligonucleotides: Methylphosphonates and Phosphorothioates

Though the methylphosphonates preceded the phosphorothioates as the first serious attempt on chemically modifying the phosphodiester backbone, the phosphorothioates have always been the major player in the scene of oligonucleotide chemistry (Dias and Stein 2002). Methylphosphonate oligonucleotides are synthesized by replacing the nonbridging oxygen atom with a methyl group at each phosphate molecule in the oligonucleotide chain. Despite having excellent biological stability (Miller et al. 1979), these species of oligonucleotide were discarded as ASO because of the total absence of charge that reduces their solubility and cellular uptake (Blake et al. 1985) since it occurs predominately via endocytosis and not via transmembrane diffusion (Tonkinson and Stein 1994). Moreover, methylphosphonate linkages destabilized the natural helix and failed to activate RNase H activity. These restricted their use and led to the emergence of phosphorothioates. Phosphorothioates are synthesized by replacing one of the nonbridging oxygens with sulfur atom at each phosphate molecule in the oligonucleotide chain (Dias and Stein 2002; Behlke et al. 2005; Chan et al. 2006; Caffo et al. 2013). Further, they have fairly stable resistance against nuclease (Dias and Stein 2002; Dean and Bennett 2003; Behlke et al. 2005), and hence provide greater bioavailability. Since the introduction of sulfur atom also introduces chirality to the oligonucleotide molecule, the two resultant diastereomers have varying biological properties with the Sp (or exo) diastereomer being nuclease resistant and Rp (or endo) diastereomer being prone to degradation (Dias and Stein 2002). The Sp species has a tendency to sterically destabilize the helix which may slightly reduce the melting temperature of the ASO–mRNA heteroduplex, by approximately 0.5°C, per nucleotide, compared to phosphodiester linkage; decreasing the affinity of the ASO for its mRNA slightly (Crooke 2000; Dias and Stein 2002; Chan et al. 2006). On the other hand, Rp diastereomers have a higher binding affinity, higher melting temperature, and slightly higher RNase H-activating potential (Koziolkiewicz et al. 1995). However, these oligonucleotides have excellent solubility and antisense activity due to their capability of inducing RNase H activity (Dias and Stein 2002). However, many of the antisense activities observed by initial investigations may be due to the nonspecific, sequence-independent actions (Krieg and Stein 1995) induced by them through the interactions with intracellular and cell surface proteins (Crooke 2000; Dias and Stein 2002; Chan et al. 2006). These include heparin-binding growth factors, vascular endothelial growth factor, and some heparin-binding molecules such as laminin, fibronectin, and Mac-1 (Dias and Stein 2002).

Phosphorothioate ASO candidates that showed promise were G3139 (Oblimersen), an 18-mer long oligo, targeting the initiation codons of the bcl-2 mRNA, and Isis 3521, a 20-mer long oligo, targeting the 3′ untranslated region of the protein kinase C-α isoform; both of which were used in combination with the chemotherapeutics against melanoma and small cell lung cancer, respectively (Geiger et al. 1998; Jansen et al. 2000). The phosphorothioate Fomivirsen (formerly Isis 2922) which is a 21-mer oligo inhibiting the human cytomegalovirus (CMV) immediate-early 2 (IE2) mRNA (Mulamba et al. 1998) is currently the only ASO cleared for clinical use for the treatment of human CMV retinitis (Chan et al. 2006). Other phosphorothioates under clinical trial include GTI-2040 (Phase II over), 20-mer ASO, targeting the coding region of human ribonucleotide reductase (RNR) R2 subunit component mRNA for controlling the drug- resistant growth in cancer cells via c-myc and H-ras oncogenes (Lee et al. 2003). LErafAON (Phase-I over) might inhibit the deregulated expression of raf-1 proto-oncogene but became associated with a hypersensitivity reaction and dose-dependent thrombocytopenia (Rudin et al. 2004).

Removal of the problems associated with phosphorothioates can be done using the strategy of chimeric oligonucleotides having modified 3'- and 5'-ends and a core of phosphorothioate in order to lower the total thiol-modified content of the oligo and decrease the associated side-effects and/or toxicity (Behlke et al. 2005). Early chimeric models with phosphorothioates at the termini and phosphodiester at the core induced optimal RNase H activity when the distance between phosphorothioate-linked bases was at least 6–8 residues with phosphodiester linkage (Eder and Walder 1991). Other modifications like the addition of C-5 propyne pyrimidines into phosphorothioate oligos can increase their binding affinity, compensate for the decrease in melting temperature, increase potency of ASOs at even 100 nm range and allow for shorter oligos (~11 bp) to have antisense effects (Moulds et al. 1995; Flanagan et al. 1996; Wagner et al. 1996). C-7 modified purines (7-deaza-2'-deoxyguanosine and 7-deaza-2'-deoxyadenosine) (Buhr et al. 1996), combined 2'-O-modification plus thiolation (McKay et al. 1996) have been investigated (Behlke et al. 2005) and also they have shown improvement in melting temperature ($T_m$) of the oligonucleotide and enhanced nucleus stability.

## 12.2.2.2 Second-Generation Oligonucleotides: 2' Substitutions

The second-generation ASOs were the result of investigation for mitigating the non-specific actions of phosphorothioate oligonucleotides, increasing the affinity of the oligos toward their targets and enhancing their nuclease stability. However, the resultant species of oligos did not induce antisense effects through RNase H-mediated cleavage and employed different sterical or splice-modulating mechanisms as discussed in the previous sections. The most important and common 2' substitutions are those made by O-alkyl groups with 2'-O-methyl (2'-OMe) and 2'-O-methoxyethyl (2'-MOE) (Chan et al. 2006). They showed stronger affinity, formed heteroduplex with higher melting temperature (Monia et al. 1993) and had even greater resistance to nuclease-mediated degradation. However, they failed to induce RNase H activity (McKay et al. 1999). Their strong binding affinity toward target mRNA made them unambiguously capable of arresting RNA translation by sterical blocking (Johansson et al. 1994). ISIS11158 and ISIS 11159 were two initial phase ASOs with 2'-MOE backbone that were designed to target the 5' cap region of the human intercellular adhesion molecule-1 (ICAM-1) mRNA. They were shown to arrest the translation probably through interference with the initiation codon and sterically hindering ribosomal assembly (Baker et al. 1997). A newer candidate of 2-O alkyl derivatives in clinical trials is OGX-011 (Phase-I over), a 2'-MOE "gapmer" ASO, designed to target the initiation site of human clusterin mRNA that helps in arresting clusterin-mediated tumorigenesis and disease progression in localized prostate cancer. ISIS 104383 is another example of a 2'-MOE gapmer targeting the translation of tumor necrosis factor-α (TNFα) mRNA (Chi et al. 2005; Chan et al. 2006). A "gapmer" is a chimeric oligonucleotide with a phosphorothioate sequence as its core or "gap" in between a pair of 2'-OMe or 2'-MOE sequences (McKay et al. 1999). The phosphorothioate core helps in stimulating the usual RNase cleavage mechanism while the 2' O-alkyl peripheries help in strengthening the binding capacity and increasing nuclease resistance of the ASO (Yu et al. 2004), resulting in a 5–10-fold more activity (in tissue culture) than an identical homogeneous phosphorothioate ASO (Griffey et al. 1996). Further modification in the backbone produced varying binding affinity and sterically dependent antisense activity. A modified C-5 propyne pyrimidine 2'-o-allyl oligo had a melting temperature of greater than 90°C for its heteroduplex with an $IC_{50}$ of 5.0 μM. On contrary, C-5 propyne phosphorothioate oligo of identical sequence had a melting temperature of only 79°C, but showed a 20-fold higher antisense potency with an $IC_{50}$ of 0.25 μM due to a bonus RNase H-activating capability (Moulds et al. 1995). There seems to be a correlation between the addition of bulkier groups at the 2'-position and increasing nuclease resistance in the order of pentoxy > propoxy > methoxy > fluoro = deoxy. Conversely, these bulkier groups also hamper the binding affinities of the oligonucleotides in the following manner: fluoro > methoxy > propoxy > pentoxy = deoxy (Monia et al. 1996). Placing a self-complementary hairpin onto the 3'-end can also increase nuclease resistance (Kuwasaki et al. 1996).

### 12.2.2.3   Third-Generation Oligonucleotides

The third-generation of oligonucleotides was further developed to enhance target affinity, nuclease resistance, stability and pharmacokinetics of the ASOs. These species of ASOs modified or sometimes discarded the natural phosphate-ribose backbone for their purpose (Dias and Stein 2002). The oligonucleotides with N3'-P5' phosphoramidate (PN), locked nucleic acid (LNA), and 2'-fluoro-arabino nucleic acid (FANA) have unmodified furanose ring whereas the phosphoramidate morpholino oligomer (PMO or morpholino), peptide nucleic acid (PNA), cyclohexenenucleic acid (CNA), and tricycle DNA (tcDNA) are devoid of the furanose ring completely (Chan et al. 2006). Among them, the morpholino, PNA, phosphoramidate, and LNA are extensively investigated and used (Dias and Stein 2002; Gleave and Monia 2005).

PNAs are synthetic DNA mimic analogues with the phosphodiester backbone replaced by an uncharged, flexible, polyamide backbone comprised of pseudopeptide polymer (N-(2-aminoethyl) glycine) to which the nucleobases are attached via methylene carbonyl linkers (Nielsen et al. 1991; Egholm 1993; Nielsen 2004). PNA ASOs can form stable hybrids of triplex, triplex invasion, duplex invasion or double duplex invasion configuration with both single and double strand DNA and duplex hybrids with RNA of complementary sequence (Nielsen et al. 1991; Koppelhus and Nielsen 2003). They have higher affinity and specificity than the unmodified DNA–DNA and DNA–RNA duplexes due to the absence of electrostatic repulsion. In addition, PNA demonstrates high resistance to both nucleases and peptidases (Chan et al. 2006). PNAs do not elicit RNase H-based antisense effect but cause sequence-specific steric hindrance of translational machinery or perform transcription arrest by hybridizing with double stranded DNA, both of which lead to protein downregulation (Dias and Stein 2002; Koppelhus and Nielsen 2003; Paulasova and Pellestor 2004). Besides, they can inhibit RNA polymerase initiation and elongation

(Hanvey et al. 1992; Boffa et al. 1996; Cutrona et al. 2000), bind with transcription factors such as nuclear factor-κB (Vickers et al. 1995), and inhibit splicing (Karras et al. 2001) or translation initiation and elongation of mRNA (Dias and Stein 2002).

Phosphoroamidate morpholino oligomer is a noncharged ASO agent where the deoxyribose moiety is replaced by a six-membered morpholine ring, and the charged phosphodiester bond is replaced by an uncharged phosphorodiamidate linkage (Summerton and Weller 1997; Amantan and Iversen 2005). They are biologically stable against nuclease and protease attack (Hudziak et al. 1996) but like PNA, they do not support RNase H activity (Chan et al. 2006). Hence, their primary mechanism behind antisense action is through the steric interference of ribosomal assembly resulting in translational arrest. Despite having fewer nonspecific properties than phosphorothioate, they do not form complexes with cationic lipids or other commonly used cationic delivery reagents for being uncharged (Dias and Stein 2002). Morpholino oligomers do not readily enter mammalian cells but a facilitated entry can be done using an arginine-rich peptide (ARP) conjugation which markedly enhanced its cellular entry and antisense potency by increasing the thermal stability of the ARP–PMO–mRNA heteroduplex (Nelson et al. 2005). Previously, a similar strategy was used for streptolysin O for the cellular delivery of the ASOs (Giles et al. 1999).

Another example of the third generation of oligonucleotide is the N3′ → P5′ phosphoramidate (PN) oligonucleotides, which are formed by replacing the oxygen at the 3′ position on the furanose ring with an amine group (Dias and Stein 2002; Chan et al. 2006). They can form very stable hybridized complexes, with RNA and single/double stranded DNA (Chen et al. 1995; Gryaznov et al. 1995). They can exhibit highly selective and specific antisense activity both *in vitro* and *in vivo* as is seen with an 11-mer PN ASO complementary to the bcr-abl mRNA causing a significant downregulation of the protein in the treated BV173 cells (Gryaznov et al. 1996). This inhibition was not due to RNase H-mediated degradation but rather due to the hithertho-unknown enzyme, which was able to cleave the heteroduplex formed by the PN and the mRNA. Further modifications were performed on the phosphoramidates like the oligo-2′-fluoro-2′-deoxynucleotide phosphoramidites, which apparently increased the melting temperature of the heteroduplex the most (Schultz and Gryaznov 1996).

Locked nucleic acid is a conformationally restricted nucleotide, which has a 2′-O,4′-C-methylene bridge in the β-d-ribofuranosyl configuration. This modification greatly potentiates its hybridization affinity toward target mRNA and DNA, creates a significantly high thermal stability of the duplexes (Vester and Wengel 2004), enhances resistance toward nuclease degradation at the cost of RNase H dependence such as any other 2′-O ribose modification. Despite that, LNA monomers can be freely incorporated into chimeric oligonucleotides along with RNase H-inducing oligonucleotides (Chan et al. 2006). A "gapmer" with a core of 7–10 phosphorothioate flanked by 3–4 LNA oligomers on both the ends provides highly efficient RNase H-mediated mRNA cleavage, high ASO potency, target accessibility, and nuclease resistance (Kurreck et al. 2002). There are nine members of the LNA molecular family among which the stereoisomer α-L-LNA has demonstrated the highest efficacy in mRNA knockdown both *in vitro* and *in vivo* (Fluiter et al. 2005).

## 12.3 DIFFERENT THERAPEUTIC APPROACHES INVOLVING ANTISENSE OLIGONUCLEOTIDES

### 12.3.1 THERAPEUTIC APPROACHES AGAINST CANCER

#### 12.3.1.1 Lung Cancer

Lung cancer has become one of the leading causes of cancerous death due to its higher tendency for early metastasization and development of resistance to a wide range of anticancer drugs. Survivin, an apoptosis inhibitor, is now in the lime light of selective lung cancer targeting due to its ability to be expressed in different tumors while lacking the expression in differentiated adult tissues. Survivin inhibits processing of downstream effector caspase-3 and -7 in cells receiving an apoptotic stimulus (Tamm et al. 1998). Olie et al. (2000) have developed 20-mer phosphorothioate ASO that can target different regions of survivin mRNA. They have studied the downregulation of survivin mRNA, thus overcome the resistance to cancer therapy by facilitating apoptosis in the lung cancer cell line (A549). Another very important apoptosis inhibitor named X-linked inhibitor of apoptosis (XIAP) can be targeted in the lung cancer therapy, as this most potent IAP gene family member can directly bind to caspase-3, -7, and -9 and can inhibit their actions which ultimately block apoptosis (Deveraux et al. 1997; Datta et al. 2000). Hu et al. (2003) studied the downregulation of XIAP mRNA in H460 cells (*in vitro*) and *in vivo* tumor xenograft mice (immunodeficient mice bearing H460 tumors) by using ASO for improved therapeutic treatment in human non-small cell lung cancer. This study of Hu et al. established that the antisense therapy to lung cancer have downregulated the XIAP mRNA by 55% and protein levels up to 60% that may play a crucial role of antisense therapy in the regulation of apoptosis as well as cancer.

Another potent target for cancer therapeutics is a cyclin D1-dependent cell cycle progression in mammalian cells as the overexpression of cyclin D1 occurs in a number of different cancers (Yasui et al. 2006). Saini and Klein (2011) reported the CD1 ASO-associated growth inhibition of cancerous cells particularly targeting pleural mesothelioma cells along with non-small lung cancer cells. Complimentary ASO against the CD1 translation-starting site of cDNA was synthesized and sense oligomer was used as control. The ASO of CD1 showed remarkable growth inhibition (43%–95%) of the experimental cells, which established the efficacy of the technology in future cancer therapy.

In early and advanced stages of small cell lung cancer, c-Met (receptor tyrosine kinase) is considered as a useful prognostic indicator whose activation with hepatocyte growth factor (HGF) causes tumorigenicity in a variety of tumors. c-Met and/or HGF overexpression denote(s) the higher grade of tumors in human (Ichimura et al. 1996; Qian et al. 2002). In the study of Stabile et al. (2004), inhibition of lung tumor cell growth by suppressing c-Met protein expression was investigated *in vitro* (nonsmall cell lung cancer cell lines) and *in vivo* (tumor xenograft model). They used U6 expression plasmid constructs with sense or antisense sequences complementary to the sequence of the translation start site of c-Met gene for blocking of target m RNA translation. Inhibition (50%) in tumor xenograft volume in lung cancer models was concluded as a consequence of downregulation of c-Met and phospho-MAP kinase protein expression. This antitumor effect was also associated with increased apoptosis, which showed the potential value of the method for the future therapy of lung cancer.

### 12.3.1.2 Liver Cancer

Das et al. (2010) have determined the anticancer activity of c-raf 1 oligomer during hepatocarcinogenesis by inhibiting the specific c-raf 1 mRNA overexpression by ASO. The particular ASO showed 68% *in vitro* cell proliferation inhibition. The microscopic study of liver exhibited a predominant inhibition of preneoplastic lesions and hyperplastic nodules in the antisense oligomer-treated group as compared to the carcinogen control group. Cytochrome P-450 content was enhanced and glutathione S-transferase, UDP-glucuronosyltransferase activity, MDA concentration were reduced in the antisense-treated animals as compared to the carcinogen-treated animals. In another study by Mukherjee et al. (2005), the role of insulin-like-growth factor II (IGF-II) gene was investigated in a defined model of hepatocarcinogenesis (during different cancer development stages, i.e., from preneoplastic lesions to hepatocellular carcinoma, HCC) by using the IGF-II ASO. The study concluded that IGF-II expression in liver could be used as a suitable marker for very early detection of the cancer that can save a number of future cancer victims by very early detection of this disease. Further, the study showed the IGF-II ASO effectively controlled the progression of HCC (Mukherjee et al. 2005).

Cai et al. (1998) have examined the fact whether the type of methionine adenosyltransferase (MAT) protein expressed (liver specific and non-liver specific, are products of two genes, MATI A and MAT2A, respectively) in liver cells have any influential role in the cellular growth, by using specific ASO against the expressed MAT gene.

Li et al. (2006) have studied the role and therapeutic potential of inhibition of signal transducer and activator of transcription 3 (STAT3) expression hepatocellular carcinoma by selectively blocking the particular signaling through 2′-O-methoxyethylribose-modified phosphorothioate ASO. ASO-specific knock down STAT3 expression in different human HCC cell lines effectively reduced the DNA-binding ability, hampered the expression of vascular endothelial growth factors such as survivin, matrix metalloproteinases 2 and 9. The antisense treatment induced *in vitro* apoptosis and inhibited angiogenesis (intradermal) as well as tumorigenesis (subcutaneous) when injected in mice, which concluded the utility of STAT-targeted therapy in liver cancer.

In a similar study by Dai et al. (2005), tumor growth reduction by downregulation of survivin expression with specific antisense compound, Lipofectamine™2000 (LiP), in Hep G2 cells was reported a bright prospect for the liver cancer treatment. This study demonstrated the role of survivin protein as a key molecule, for tumor cell proliferation.

Ebinuma et al. (2003) investigated the effect of blocking of c-myc expression by c-myc antisense oligonucleotide in liver cancer, to establish whether c-myc blocking induced apoptosis or not. They have done the experiment by using the liver cancer-derived cell lines, HCC-T, HepG2, and PLC/PRF/5 and c-myc-specific antisense oligos. Morphological detection of apoptosis was performed and examination of c-myc and bcl-2 expressions was done by western blotting. Massive reduction in c-myc mRNA and partial decrease in bcl-2 enhanced apoptosis and cell death, respectively.

Wu et al. (2004) explored the blocking effect of the function of vascular endothelial growth factor (VEGF) expression in cultured Walker-256 cells with the VEGF ASO and also investigated the antitumor effect of the VEGF in combination with lipiodol after intra-arterial infusion. The result of this study claimed that the antiangiogenesis therapy of the ASO in combination with lipiodol showed better inhibitory effect on the liver cancer growth, VEGF expression, and microvessel density than lipiodol alone.

### 12.3.1.3   Colorectal Cancer

Ciardiello et al. (2000) have studied the combination effect of anti-EGFR mAb along with VEGF antisense nucleotides on VEGF production, in the GEO colon cancer, demonstrating *in vitro* and *in vivo* antiangiogenic and antitumor activity of the ASOs. Nakano et al. (2001) studied the effect of antisense gene along with the sense one in the colorectal cancer *in vitro* cell lines as well as *in vivo* animal model. They specifically transduced the cancerous cells with the recombinant adenovirus vector expressing an antisense or sense K-ras gene fragment (AxCA-AS-K-ras or AxCA-S-K-ras). K-ras-specific p21 protein was reduced 25% in the antisense-treated cells showing the potential of using K-ras antisense construct as a useful strategy in the treatment of colorectal cancer. Another study by Tian et al. (2000) showed the G1 arrest, as well as induction of apoptosis in the colon cancer cells *in vitro* and *in vivo* after the treatment with antisense p21 oligodeoxynucleotide. The human P21-targeted inhibition by ASO specifically sensitizes human colon cancer cells to radiation, resulting in conversion of p53-mediated G1 arrest into apoptosis. This may be a rational approach to improve conventional radiotherapy outcomes.

### 12.3.1.4   Pancreatic Cancer

Hotz et al. (2004) studied the effect of VEGF ASO on the angiogenesis and growth of human pancreatic cancer in nude mouse model as neo-angiogenesis is a very important incident that occurs during formation of neoplasm in the pancreatic cancer. Targeting of the VEGF receptor expression can halt the neoplasm expansion. They showed that blocking of the VEGF with the ASOs significantly reduced angiogenesis as evidenced from the reduced microvessel density in the ASO-treated animal group.

Cheng et al. (1996) have studied the effect of antisense RNA on the tumorigenicity in pancreatic cancer. This study demonstrated that the pancreatic cancer cell line, PANC1 cells (overexpressing the AKT2 oncogene) showed a marked reduction in tumorigenicity, when treated with AKT2 RNA antisense.

Denham et al. (1998) have shown the inverse correlation of PKC-α (alpha isoform of protein kinase C) gene expression with the pathologic differentiation of human pancreatic cancers. They compared the survival rate of parent HPAC pancreatic cancer cells with the PKC-α expressing HPAC cells, following the treatment with the PKC-α antisense both in *in vitro* and *in vivo* models. They concluded that the downregulation of PKC-α gene in the ASOs-treated subjects is responsible for the greater survival rate of the animals than the control group treated with normal saline. The study supports the potential of antisense therapy to combat pancreatic cancer.

Aoki et al. (1995) reported that growth of pancreatic cancer cells is hampered significantly by using antisense K-ras construct. In this study, they transduced the sense and plasmid expressing antisense oligonucleotides to model AsPC-1, MIAPaCa-2, and BxPC-3 human pancreatic cell lines, by using liposomal delivery system.

### 12.3.1.5  Ovarian Cancer

RIα, one of the four isoforms of subunit R of cAMP-dependent protein kinases (PKAI), is an ontogenic growth-inducing protein and its overexpression can cause malignancy by disrupting natural ontogenesis (Alper et al. 1999). In the study, they have studied the effect of RIα antisense on the growth of the ovarian cancer cells by using the mixed backbone (hybrid) oligonucleotides consisting of PS-2′-O-methyl-oligo RNA. The mixed ASOs possess better antisense properties than PS-oligonucleotides (Metelev et al. 1994; Cho-Chung et al. 1997). They have examined the effect of one-time treatment (at 5 h after seeding) of RIα ASO on the ovarian cancer (OVCAR-8) cells and the result showed a time- and concentration-dependent inhibition of ovarian cell proliferation (75% inhibition at 200 nM by day 5). They have demonstrated the reduction in expression of RIα at the mRNA and protein level which may be the cause of cellular growth inhibition, morphological changes and also apoptosis induction in the ovarian cancer cells (Alper et al. 1999).

Ling et al. (1998) established the induction of apoptosis in ovarian cancer cells by evaluation of the antiproliferative effects of ASO to insulin-like growth factor II (IGFII). The antiproliferative effect was examined by 3H-thymidine incorporation. The ASO treatment to the ovarian cells inhibited the cell proliferation and induced apoptosis also to the AO cells. The result of this study established the potential of IGFII ASO as a target in the treatment of ovarian cancer to be considered as a useful therapeutic approach.

Phillips et al. (2001) have investigated the therapeutic role of c-raf ASO (both 1st and 2nd generation) on impairment of growth of the ovarian cancer cells. They investigated the potential of the ASO by evaluating the extent of growth inhibition in the ovarian cancer cells as well *in vitro* as *in vivo* xenograft animal model. In this study, they have established that the ASO treatment to ovarian cells may be a novel strategy to treat ovarian cancer provided a high level of c-raf expression.

Popadiuk et al. (2006) have investigated the role of a critical elemental pygopus protein (hPygo2) in the deadliest gynecologic malignancy, namely ovarian cancer. They assayed the amount of expression and growth requirement of this protein by using specific ASO. After the transfection of the phosphorothioated ASO to the cancerous cell lines, the growth assay was done. The results showed in the impairment of growth in the treated cells in comparison to the normal cells. This study showed the application of cell growth inhibition in ovarian cancer by the hindrance of the particular protein expression.

In a study by Sasaki et al. (2000), the effect of downregulation of X-linked inhibitor of apoptosis protein (Xiap) on apoptosis in cisplatin-resistant human ovarian surface epithelial (HOSE) cancer cells was studied by using adenoviral antisense Xiap cDNA. The treatment with ASO induced the apoptosis to the HOSE cancer cells by decreasing the particular Xiap protein in the cancerous cells. The ASO enhanced the cisplatin-induced cell death in p53 wild-type excluding the mutated resistant cells. In this study, the author demonstrated that caspase-3-mediated MDM2 cleavage and increased p53 content ultimately caused the downregulation of Xiap by ASO.

### 12.3.1.6   Prostate Cancer

Iversen et al. (2003) investigated the safety and efficacy of a PMO antisense for c-myc gene, that is, AVI-4126 in the treatment of prostate cancer. Treatment with c-myc-specific ASO significantly hampered the growth (about 75%–80% growth reduction in tumor growth in comparison to the control groups) and caused apoptosis in the experimental prostate cancer cells (PC-3 androgen-independent cells) as well as in tumor xenograft mice.

Balaji et al. (1997) explored use of ASO therapy for prostate cancer by investigating the effect of c-myc-ASO in three types of human prostate cancer cell lines (LNCaP, PC3, and DU145). After the ASO treatment DNA synthesis and cell viability were decreased in all the three cell lines which suggest that the cancerous cell proliferation and growth were decreased due to the reduction of cell viability.

Chi et al. (2001) evaluated the efficacy of combined treatment of antisense therapy and standard anticancer drug for combating the hormone-refractory prostate cancer (HRPC). They have used the phosphorothioate ASO complimentary to bcl-2 mRNA (Genasense) and mitoxantrone, a standard chemotherapy for the treatment of prostate cancer. Effective delay in androgen independence and improvement in sensitivity to other cancer due to the inhibition of bcl-2 expression showed the effectiveness of the therapy against prostate cancer.

A similar study was performed by Miyake et al. (2000) for the treatment of the same type of prostate cancer. They have evaluated the potential of ASO to target testosterone-repressed prostate message-2 (TRPM-2) protein that is upregulated after androgen withdrawal and during androgen-independent progression in prostate cancer. This study showed that ASO enhanced sensitivity toward chemotherapy.

### 12.3.1.7   Breast Cancer

The type I insulin-like growth factor receptor (IGF-IR) has a major effect on the growth and transformation of breast cancer cells. The study was carried out (Chernicky et al. 2000) with an antisense IGF-IR construct on human breast cancer cell line (estrogen receptor negative), MDA-MB-435s. Decreased cell proliferation and reduced IGF-IR expression were observed. In an immune-deficient or nude mice model, a delayed tumor formation and reduction in tumor size occured when the animals were injected with the particular ASO. This has established the fact that IGF-IR has a significant role in breast cancer progression that can be utilized in future for the treatment of breast cancer (Chernicky et al. 2000).

Inhibitor of DNA-binding (ID)-1protein, expressed by mammary epithelial cells, is an important regulator of breast cancer cells. The protein is overexpressed in aggressive and metastatic breast cancer cells. Fong et al. (2003) demonstrated that when the particular protein ID-1 was downregulated by using specific ID-1 ASO, human metastatic breast cancer cells showed less invasiveness *in vitro* and less metastasis *in vivo*.

Stat 3 proteins serve an important role in tumor cell survival and are persistently activated in breast cancer. Gritsko et al. (2006) evaluated the role of stat 3 genes in causing resistance to apoptosis by using the Stat 3 antisense. The signals were downregulated in human MDA-MB-435s breast cancer cells. Induction in apoptosis was noticed as blocking of Stat3 signal directly inhibited the survivin protein expression.

The human $\alpha$ isoform folate receptor ($\alpha$hFR) has a very high affinity for folic acid, and is overexpressed in different cancers. Jhaveri et al. (2004) studied the effect of this receptor to be used as a targeted site for treatment of breast cancer using ASO in combination with doxorubicin. They found that ASO treatment effectively decreased $\alpha$hFR levels, which lead to the reduction in cell survival. The ASO treatment to MDA-MD-435 breast cancer cells made them more sensitized to doxorubicin, which in turn shows the potential of this type of combined therapy to combat the disease like breast cancer.

Simões-Wüst et al. (2002) demonstrated the use of a bispecific ASO (4625) to sensitize breast cancer cells to some standard anticancer drugs. They have used the bcl-2/bcl-xL bispecific antisense oligomer that downregulates the proteins, bcl-2 and bcl-xL. They are two antiapoptotic proteins that play a crucial role in tumor progression and development of drug resistance in breast cancer cells. This study ensures the potentialrole of the ASO as an adjuvant with the chemotherapy in breast cancer therapy.

Tanabe et al. (2003) reported an enhanced sensitivity of the anticancer drugs in breast cancer after ASO administration. They examined the effects of antisense bcl-2 and HER-2 oligonucleotides (ODN) which effectively blocked the related proteins, and this leads ultimately to an induced response of the cancer cells toward the anticancer drugs.

### 12.3.1.8 Other Cancers

Jansen et al. (2000) have investigated the effect of combined therapy of a standard anticancer drug along with the bcl-2-specific ASO (augmerosen, Genasense, G3139) in the treatment of advanced malignant melanoma expressing bcl-2. This therapy effectively reduced the protein amount, induced cellular apoptosis of the tumor in the mouse xenotransplantation model. The phase I–II clinical study of the same combination in patients also showed promising results for the therapy.

Bcl-2 overexpression in non-Hodgkin lymphoma was downregulated in the study of Webb et al. (1997) who examined the effect of ASO in normal functioning of bcl-2 mRNA, that is, apoptosis resistance and tumorigenesis. They have first reported the study in human as the laboratory animals showed toxicity. About nine bcl-2-positive relapsed non-Hodgkin lymphoma patients were administered with the ASO for 2 weeks and both the tumor response and the toxicity were assessed. The reduction of sign and symptoms of the disease condition was demonstrated along with the downregulation of the protein observed in the patients.

### 12.3.2 THERAPEUTIC APPROACHES FOR OTHER DISEASES

#### 12.3.2.1 Diabetes Mellitus

Various types of ASOs have been investigated in the last decades against diabetes mellitus. Zemany et al. (2015) reported that antisense oligonucleotide of circulating transthyretin (TTR-ASO) improves glucose metabolism and sensitivity to insulin by reducing circulating TTR (80%–95%) which controls retinol-binding protein (RBP4), a very crucial determinant for maintaining glucose homeostasis. Another study revealed that antisense drug mipomersen (ISIS 113715) was used for the treatment of type 2 diabetes by inhibiting protein-tyrosine phosphatase-1β and sensitized insulin. It also reduced LDL-cholesterol and provided antiobesity and lipid-lowering potential (Saonere 2011). Phase-III clinical trials of the abovementioned ASO has been done till date in four different groups and it was proved to be effective, safe, and well tolerated by the patients. It provides a promise for the management of familial hypercholesterolemia and severe hypercholesterolemia (Toth 2013).

### 12.3.2.2  Diabetic Retinopathy
Diabetic retinopathy (DR), which is one of the major causes of blindness globally, is characterized by microvascular changes in the retina and a retinal thickening resulting from the accumulation of fluid, known as macular edema. For the treatment of this disease, a second generation ASO iCo-007 has been investigated and it is in the Phase-I clinical trial. The main advantages of this molecule over the conventional therapies (focal photocoagulation, vitrectomy, etc.) are the specificity of target, extended half-life of the drug leading to less frequent dosing, low degradation, and a safe profile (Hnik et al. 2009).

### 12.3.2.3  Cytomegalovirus Retinitis-Associated Acquired Immunodeficiency Syndrome
Cytomegalovirus (CMV) retinitisan inflammation of the retina of the eye caused by human cytomegalovirus that can lead to blindness. People with a compromised immune system are generally affected by this virus. For the treatment of this disease, an ASO, fomivirsen, was invented, and it is capable of inhibiting CMV retinitis when injected into a human eye. After clearing the preclinical tests, it showed positive result in a clinical trial of patients with diagnosed CMV retinitis (de Smet et al. 1999). ISIS 2922 (Vitravene) is another ASO, developed by ISIS pharmaceutical company, is complementary to the human CMV RNA. It showed effective results in *in vitro* trials (Alama et al. 1997; Rayburn and Zhang 2008). The first novel therapeutic ASO, which has been approved for marketing in USA, is vitravene. It is used in the local treatment of cytomegalovirus (CMV) retinitis in patients with acquired immunodeficiency syndrome (AIDS) who are intolerant to or have a contraindication to other treatments for CMV retinitis or who were insufficiently responsive to previous treatments (Galderisi et al. 1999).

### 12.3.2.4  Human Papilloma Virus
Human papilloma virus (HPV) are small double stranded viruses, which are the main cause of cervical cancer, and to date, a suitable treatment of this disease is unavailable (Alam et al. 2005). According to Alam et al. ORI-1001, an antisense drug may target HPV31b genes of HPV and it may serve as a future treatment against the virus-related cancer risks in women. Different molecular targeting of HPV gene expression using RNA interference technology has also been developed. The various

examples are: (i) *in vitro* and *in vivo* study of ASOs on HPV16, HPV18 E6 and E7 genes; (ii) *in vitro* use of ribozymes (Rb) on HPV16-E6/E7 HP ribozyme (R434) and HPV6b and HPV11 E1 genes; (iii) targeting of short interfering RNAs (siRNA) on HPV16, HPV18 E6 and E7 genes (*in vitro*) and HPV16 E6 and E7 genes (*in vivo* in NOD/SCID nude mice); and (iv) use of short hairpin RNA (sh-RNA) on HPV16 and HPV18 E6 and E7 genes (*in vitro*) and HPV18 E6 and E7 genes (*in vivo* in NOD/ SCID nude mice) (Bharti et al. 2009).

### 12.3.2.5   Viral Hemorrhagic Fever

Viral hemorrhagic fever (VHF) is a virus-borne disease, which is characterized by high fever and abnormal bleeding. One of the major causes of this disease is infection of RNA viruses from the family Filoviridae include *Ebola* virus and *Marburg* virus which may lead to 50%–90% fatal case rate in human beings (Warfield et al. 2006). Antisense PMOs were developed by Bavari et al. and the ASO interfered with translation and amplification of mRNA and thus, might provide a successful treatment against those viruses. They have also administered the Ebola virus (EBOV) specific PMOs on VP24 and VP35 genes and as a result, the viral polymerase L protected rhesus macaques from lethal EBOV infection. It was the first successful ASO strategy against nonprimates treating this disease. In late 2008, two products AVI-6002 and AVI-6003 have been developed against *Ebola* virus and *Marburg* virus, respectively, by AVI BioPharma, a U.S. biotechnology firm (Saonere 2011).

### 12.3.2.6   Hepatitis B Virus

Hepatitis B virus (HBV) is the main cause of liver-related diseases worldwide and hepatitis B infection may lead to HCC also. Moriya and Matsukura (1996) showed that intraperitoneal injection of specified ASOs against these HBV genes prevented preneoplastic lesions of the liver, with the minimum toxic side effects in an animal model (Alama et al. 1997). The main target regions for the ASOs for HBV are cap site/SPII and initiator/gene S, 5′ region of pre-S gene, pre-S1 open reading frame, S gene, C gene, S and C gene, pre-S region, upstream sequence of the encapsidation site, pol gene, pregenomic RNA, core promoter, etc. (Bai et al. 2013).

### 12.3.2.7   Herpes Simplex Virus

Herpes simplex virus (HSV) are the most common cause of genital herpes, cold sores, and corneal keratitis and may end up in more severe disease in immunocompromised individuals. ASO technology has been used to target and specifically treat the responsible genes for this viral disease. Two reported ASOs are ISIS 1082 (Alama et al. 1997) and ISIS 5652 (Shogan et al. 2006). Some of the targeted regions of HSV gene for this disease are IE 4 and 5 pre-mRNA, TIS of IE 1 mRNA, IE 1 and 3 mRNA, splicing acceptor site of IE 5 mRNA, TIS of HSV-1 UL30 and UL39, vIL6 mRNA, DNA polymerase, etc. (Bai et al. 2013).

### 12.3.2.8   Neurodegenerative Disorders

ASOs have been tried against common neurological disorders and various clinical and preclinical studies are being carried out. A list of ASOs against specific diseases and their current status has been summarized in Table 12.1.

**TABLE 12.1**
**Applications of ASOs in Different Neurodegenerative Disorders**

| Disease | Name of Drug | Oligo Type (Chemistry) | Mechanism of Action | Target | Clinical Phase/ Status | References |
|---|---|---|---|---|---|---|
| Alzheimer disease (AD) | N/A | 2-O'-methoxyethyl (MOE) | Translation repression | Amyloid precursor protein (APP) | Preclinical (rodents) | Southwell et al. (2012), Wolfe (2014) |
| Frontotemporal dementia with Parkinsonism linked to chromosome 17 (FTDP-17) | N/A | 2-O-Me PS ASO | Exon exclusion | Exon 10 | NA | Southwell et al. (2012), Siva et al. (2014) |
| Duchenne muscular dystrophy (DMD) | Eteplirsen (AVI-4658) NCT01396239 | Phosphoramidate morpholino (PMO) | Exon skipping | Exon 51 | Phase-II completed | Lee and Yokota (2013) |
| DMD | Drisapersen (PRO051/G K2402968) NCT01803412 | 2'OMePS | Exon skipping | Exon 51 | Phase-III recruiting | Lee and Yokota (2013) |
| DMD | NS-065/ NCNP-01 | PMO | Exon skipping | Exon 53 | Phase-I recruiting | Lee and Yokota (2013) |
| Fukuyama congenital muscular dystrophy (FCMD). | N/A | vPMOs | N/A | Intronic or exonic splicing enhancer sites | N/A | Lee and Yokota (2013) |

*(Continued)*

**TABLE 12.1 (Continued)**
**Applications of ASOs in Different Neurodegenerative Disorders**

| Disease | Name of Drug | Oligo Type (Chemistry) | Mechanism of Action | Target | Clinical Phase/ Status | References |
|---|---|---|---|---|---|---|
| Myotonic dystrophy (DM) | PRO135 | N/A | CUG expansion | N/A | Preclinical in progress | Lee and Yokota (2013) |
| Spinal muscular atrophy (SMA) | ISIS-SMNRx NCT01839656 | 2'-MOE | Exon inclusion | Exon 7 | Phase-II recruiting | Lee and Yokota (2013) |
| SMA | N/A | PMO | Splice modification | SMN2 | Preclinical rodents | Southwell et al. (2012) |
| Dysferlinopathy | N/A | N/A | Exon skipping | Exon 32 | NA | Lee and Yokota (2013) |
| Amyotrophic lateral sclerosis (ALS) | ISIS-SOD1Rx/ ISIS 333611 NCT01041222 | 2'-MOE | Gapmer | Exon 1 | Phase-I completed | Lee and Yokota (2013) |
| ALS | NA | MOE | RNase H induction | SOD1 | Phase-I clinical | Southwell et al. (2012) |
| Huntington's disease (HD) | PRO289 | NA | NA | CAG expansion | Preclinical in progress | Lee and Yokota (2013) |
| HD | NA | MOE cEt | RNase H induction | muHTT SNPs | Preclinical rodents | Southwell et al. (2012) |

### 12.3.2.9   Asthma

Asthma is a chronic respiratory disorder characterized by coughing, wheezing, shortness of breath, etc. In the near past, a few antisense technologies were used to target and treat this disease. The most successful ASO was EPI-2010 against adenosine A (1) receptor. Phase-I clinical trial was been done for this drug and results showed that this drug is effective, safe, and well tolerated (Ball et al. 2004). Chang et al. reported that in inhalational therapy for asthma, cytokine receptors were targeted by TPI-ASM8, an ASO. Besides these, some other strategies have been tried, such as RNase P external guide sequence (EGS) delivered into pulmonary tissues, ribozyme strategies, siRNA targeting, delivery of synthetic analogs of microRNA (miRNA) on the specified site, etc. (Saonere 2011).

### 12.3.2.10   Crohn's Disease

It is a form of irritable bowel syndrome, which shows the symptom of diarrhea, abdominal pain, weight loss and fever. Patients with active CD show high expression of Smad7 gene, which increases inflammatory signals by the immune-suppressive activity of transforming growth factor (TGF)-β1. GED 0301, a Smad7 ASO, has successfully undergone the Phase-I clinical trial and showed a reduction in the percentage of inflammatory cytokine in the blood. It was proved to be safe and well tolerated in patients with active CD (Monteleone et al. 2012).

## 12.4   PHARMACOKINETICS OF ANTISENSE OLIGONUCLEOTIDES

ASOs have a wide range of RNA targets both in the nucleus (pre-RNA, mRNA, noncoding RNA, toxic nuclear-localized RNA, etc.) and cytoplasm and thus they are required to traverse different biological membranes and various biofluid systems for executing the desired action. Once bound to the target RNA, ASOs can manipulate the target RNA through various mechanisms, including degradation of the RNA-ASO duplex via the RNase H pathway, sterical hindrance of RNA translation, or modulation of splicing of mRNAs. The pharmacokinetics and biodistribution of ASOs are therefore very important to investigate before fixing a strategy for developing the antisense therapy.

### 12.4.1   Biodistribution of Antisense Oligonucleotides Depending on Routes of Administration

#### 12.4.1.1   Intravenous/Subcutaneous Administration

The primary route of administration for ASOs for the systemic applications is either by intravenous (IV) infusion or by subcutaneous (SC) injection. Following systemic administration, phosphorothioate ASOs rapidly transfer from the blood into tissues within hours. Intracellular entry is predominantly done by endocytotic uptake and after that ASOs remain there for a long period of time (half-life 2–4 weeks) (Levin et al. 2007; Geary et al. 2009; Crooke and Geary 2013; Yu and Xu 2015). SC administration allowed rapid absorption of the ASOs into the blood circulation with the peak plasma concentration reaching in 4–5 h, and animal trials involving small mammals (monkeys) showed near absolute bioavailability (Levin et al. 2007; Geary et al. 2009). The peak level subsidizes in a multiexponential manner, characterized by an initial rapid distribution phase where the molecules migrate from vasculature to other tissues within few minutes to few hours, followed by a much slower terminal elimination phase. The phase itself lingers in a span of several weeks, indicating the establishment of equilibrium between postdistribution phase plasma concentrations and tissue concentrations (Geary et al. 2009). During this phase, the partition ratio between liver and plasma concentrations of ASO was approximately 5000:1 for 2'-MOE ASOs, and it was consistent for a varying species of animals. This study indicates the theoretical applicability of these values for human tissue distribution models as well (Geary et al. 2009). Phosphorothioate ASOs have a tendency of nonspecific interaction with cell surface and intracellular proteins, and hence are subjected to extensive binding with plasma proteins (>80%) (Yu et al. 2007), serum albumin being the prominent one among them. This protein binding is below clinically relevant doses for protein saturation and of very low affinity, and does not constitute the danger of excess renal clearance of the ASO (Kd~150 $\mu$M) (Levin et al. 2007). However, uncharged ASOs like Morpholino and PNA experience a rapid elimination through renal clearance and metabolism in the blood (McMahon et al. 2002; Amantan and Iversen 2005) and thus exhibit a poor tissue distribution. Thus, a balanced plasma protein binding is required for proper tissue distribution of ASOs. Nevertheless, the animal studies revealed that almost all ASOs, especially the second-generation 2'-substituted oligonucleotides, are broadly distributed into a wide range of tissues and organs including liver, kidney, bone marrow, adipocytes, and lymph nodes. They show consistent antisense activity (Altmann et al. 1996; Zhang et al. 2000; Levin et al. 2007; Geary et al. 2009; Yu et al. 2009; Crooke and Geary 2013; Hung et al. 2013; Yu et al. 2013). This phenomenon of ASOs has great potential for being exploited as a means of wider distribution for such molecules during therapy. Slow infusion results in substantially greater uptake of the ASOs in the liver, as seen in *in vivo* experiments in mice compared with the bolus administration. This indicates that the uptake process is a saturable mechanism in which the long duration of infusion allows the liver cells to take up a larger amount of oligonucleotides resulting in low ASO plasma concentrations (Geary et al. 2009). No significant antisense activity was observed in the liver cells where the concentration of the oligonucleotides was high. This also suggests that the uptake pathways in the hepatocytes are largely nonproductive which only endocytose ASOs when presented slowly and at low concentration. After bolus injection, however, the uptake was observed to be productive, that is, with observable antisense effect, which implicates that the bolus administration may overcome the saturable nonproductive pathways. Similarly, ASO activity was seen when the nonproductive pathways of the liver cells were saturated competitively with bulk nonsense RNA sequences followed by the administration of the correct ASO. Later studies showed that the ASOs had a great propensity for distribution in muscular tissues, and this investigation may lead to the development of antisense therapies against disorders such as DMD (Deutekom et al. 2007; Mendell et al. 2013). ASO employs exon splicing modulation (exon skipping) in the dystrophin protein mRNA to remove the mutations in the dystrophin gene (Aartsma-Rus et al. 2008). Spinal muscular dystrophy also employs ASOs to alter the exon splicing sequence in the survival of motor neuron 2 (SMN2) gene to mimic the inactive SMN1 gene and to restore the production of SMN protein (Passini et al. 2011; Rigo et al. 2012). However, for the suboptimal pharmacokinetic profile of the muscle-targeted ASOs, the routes of administration and the delivery mechanisms still require intelligent manipulation.

### 12.4.1.2   Intrathecal Administration

The brain and remaining parts of the central nervous system (CNS) have always remained an elusive target for most of the drugs due to the enveloping by blood–brain barrier (BBB) and the case for ASOs in this regard is not entirely different (Caffo et al. 2013). But it has been observed in various preclinical studies that administration of the oligonucleotides intrathecally into the cerebrospinal fluid (CSF) results in better distribution profile in those organs (Smith et al. 2006; Passini et al. 2011; Kordasiewicz et al. 2012; Rigo et al. 2012). Distribution kinetics from the CSF into CNS is rapid (half-life <1 h) due to the combined effect of uptake into CNS tissues and limited transfer to systemic circulation (Miller et al. 2013). Following the dosing, the CSF concentrations of the molecules exhibit the same biphasic kinetics ending with a long terminal distribution phase with very low CSF concentrations, apparently in equilibrium with the CNS. Intrathecal infusion of 2′-MOE ASO targeting super oxidase dismutase1 gene in amyotrophic lateral sclerosis conditions showed a peak CSF concentration at the end of the 12 h infusion. The plasma concentration was lower than the CSF concentration when the dose was given iv/sc, indicating only a limited transmission of the oligonucleotides into the systemic circulation from the CSF (Rigo et al. 2012; Miller et al. 2013; Rigo et al. 2014). Subsequent studies have also suggested that the pharmacokinetic profile improves if the dose is given in a bolus form than the spinal infusion (Miller et al. 2013). A broader distribution throughout the CNS, invading the entire spinal cord and brain structures with an active uptake into neurons broadly showed efficient antisense activity in rodents and smaller primates. Other methods of improving CNS delivery of ASOs include modification of the carrier (Caffo et al. 2013). Nanocarriers encapsulating the ASOs can penetrate the BBB into the CNS and brain. They can be used to maintain a sustained release of the oligonucleotides through slow degradation of the scaffold or they can be engineered to release the oligos on being triggered by a stimulus unique to the delivery site or they can enhance the uptake of the molecules through a ligand-mediated endocytosis. Nanocarriers such as nanoparticles, liposomes, dendrimers, or micelles can be engineered for brain-specific delivery by the use of neutral lipids such as 1,2-dioleoyl-sn-glycero-3-phosphocholine (DOPC) and 1,2-dioleoyl-sn-glycero-3-phosphoethanolamine (DOPE) that pass through the lipoidal membranes of the BBB. Antibodies against surface antigens of the brain cells and immunoliposomes engineered with polyethylene glycol (PEG) may also be used for this purpose (Yang 2010; Martin-Banderas et al. 2011; Xin et al. 2012).

## 12.5   CELLULAR UPTAKE OF ANTISENSE OLIGONUCLEOTIDES

Unmodified (naked) ASOs have a net negative charge and thus are unable to penetrate the plasma membrane. Cellular uptake and internalization of ASO occur primarily through adsorptive endocytosis process and fluid-phase pinocytosis (Dias and Stein 2002; Chan et al. 2006). This transport of oligonucleotide depends upon the temperature, the structure and the concentration of the oligonucleotide (Dias and Stein 2002). In the previous sections, we discussed how the cellular uptake both *in vitro* and *in vivo* follows two distinct pathways with one resulting in productive antisense activity and the other resulting in nonproductive uptake or sequestration of the ASO (Geary et al. 2009). But even productive internalization depends on the relative concentration of ASOs with a low concentration of oligonucleotides resulting in the membrane-bound receptor-mediated absorption; and a higher concentration of them causes pinocytosis (Loke et al. 1989; Dias and Stein 2002). Despite shedding the negative charge, naked oligonucleotides are poorly internalized and tend to localize in the endosomes/lysosomes, where they are unavailable to exhibit antisense actions (Gray et al. 1997). This phenomenon can be seen in poor adsorptive endocytosis of peptide nucleic acid and PMO in spite of being noncharged because they do not interact well with cell surface proteins (Chan et al. 2006). To improve internalization and subsequent nuclear localization, a range of techniques has been employed including the vectors, which increase the stability against nuclease and allows the use of far lesser concentrations of oligonucleotides for similar response (Dias and Stein 2002; Lysik and Wu-Pong 2003). First-generation phosphorothioates not only enhance nuclease resistance, but also promote interaction of ASOs with the cell surface proteins, resulting in higher internalization through receptor-mediated pathways (Chan et al. 2006; Liang et al. 2014). These pathways were dissimilar to the common transfection mechanisms employed by cells to internalize nucleic acids. A mouse hepatocyte cell line (MHT) was identified which internalized ASOs productively in the absence of transfection agents (Koller et al. 2011). This cell line and other *in vivo* works further identified the protein AP2M1, a clathrin-associated adaptor protein. The protein enables productive uptake of ASOs, which functions even in the absence of clathrin (Koller et al. 2011). This may be due to a yet-to-be-discovered clathrin independent but AP2M1-dependent mechanism (Koller et al. 2011). Furthermore, this investigation also highlighted the importance of both the nucleotide bases and the deoxyribose sugar to ensure the optimal uptake of the ASOs. For other ASOs, vector-mediated delivery has been more extensively investigated. Prominent among the vectors is liposome (Dias and Stein 2002). Liposomes are engineered to carry a positive charge to induce high affinity for the negatively charged cell membranes (under physiological condition) (Dias and Stein 2002). To dope liposomes for better internalization of ASO, cationic lipid carriers such as N-(1-(2,3-dioleoyloxy)propyl)-N,N,N-trimethylammonium chloride (DOTMA) and N-(1-(2,3-dioleoyloxy)propyl)-N,N,N-trimethylammonium methyl sulphate (DOTAP) are widely used (Chan et al. 2006). Certain "helper" molecules such as 1,2-dioleoyl-sn-glycero-3-phosphatidylethanolamine (DOPE) and chloroquine are also added to the liposomes to assist the oligonucleotides for escaping from endosomes by inducing endosomal membrane (Bennett et al. 1992; Felgner et al. 1994; Farhood et al. 1995; Ma and Wei 1996; Lysik and Wu-Pong 2003). The pH-sensitive fusogenic amphipathic lipids cause ASO escape. An example of such lipid is cholesteyl hemisuccinate.

This lipid molecule helps liposome to undergo fusion with endosomal membrane below the pH value 5.5 and thereby it influences ASO escape (Fattal et al. 2004). Other cationic polymers that have been employed for ASO delivery include polyalkylcyanoacrylate (nanoparticles), poly-L-lysine, poly (amidoamine) (dendrimers), and polyethyleneimine (Dias and Stein 2002). Bypassing cationic endocytosis-mediated uptake, one can also use cell-penetrating peptides (CPP) to transport ASOs via an energy-dependent pathway by modulating the permeability of the cellular and nuclear membrane and causing rapid translocation of oligonucleotides (Dias and Stein 2002; Chan et al. 2006). CPPs are short peptide chain (<30 peptides) with net positive charge which can be linked covalently via a disulphide bridge with ASOs. A few examples are penetration (RQIKIWFQNRRMKWKK), HIV TAT peptide 48–60 (GRKKRRQRRRPPQ), and transportan (GWTLNSAGYLLGKINLKALAALAKKIL-amide) (Jarver and Langel 2004; Nelson et al. 2005). Other ligands (e.g., Triantennary GalNAc) that are conjugated with ASOs specifically increases the distribution of the ASOs in the hepatocytes to about 10-fold than the unconjugated ASOs with a corresponding decrease in the oligonucleotide levels in other organs (Prakash et al. 2014).

## 12.6 DIFFERENT DELIVERY APPROACHES TAKEN FOR ANTISENSE THERAPY

It is very important to keep in mind that the advancement in designing of ASOs and siRNA should be complemented by the development of appropriate delivery systems. The developed delivery systems must meet all the critical parameters as described below

i. They should have optimum stability and activity in biological system.
ii. They should increase the biological half-life of antisense drugs.
iii. They should potentiate the penetration of antisense drugs into the cellular compartments.
iv. Finally, they should release the oligonucleotides in therapeutically effective concentration in therapeutically concentrations.

Earlier, the focus of the research on antisense oligonucleotides was in their proper design in order to prevent their degradation, recently this focus has been shifted toward the development of the ideal carrier system to deliver it specifically to the target site. Delivery of nucleic acids specifically to liver and muscle is possible through physical methods such as hydrodynamic injections or electroporation. But these physical methods are inefficient for systemic delivery of gene. In this context, lipid-based systems and polymer-based strategies are most widely exploited as delivery systems. The carrier formulations that are most commonly exploited for the delivery of ASOs include cationic lipid-based systems and polymeric carriers derived from natural and synthetic lipids. ASO vectors also incorporate certain *helper* molecules, which cause destabilization of the endosomal membrane allowing oligonucleotides to be actively transported to the nucleus in high concentration. However, the probability of cationic particles to form aggregates is high after an interaction with the plasma membrane. The aggregates so formed are either readily taken up by the mononuclear phagocytic system (MPS) or pulmonary capillary bed resulting in its clearance. To overcome it, polyethylene glycol (PEG) is widely used. It forms a shield over the surface of particles (stealth or long circulating) resulting in reduced uptake. There are various ways to deliver antisense therapeutics using polymeric nanoparticles. They include encapsulation of nucleotides within the core of nanoparticles, adsorption on the nanoparticle surface, complexation with cationic molecules, etc. However, polymers used must be biocompatible as they are exogenous to the organisms. The size of the carrier system encapsulating DNA is very vital as small particles (with a diameter less than 500 nm) can exploit the phenomenon known as enhanced permeability and retention effect (EPR). The EPR effect drastically decreases if the size of the particles is bigger than 500 nm. Moreover, the extracellular matrix forms a highly hydrated gel-like fibroblast-embedded network of polysaccharides and proteins over the cells that produce them. It acts as a scaffold and allows the faster diffusion of small molecules and proteins as compared to large particles. These polymeric delivery systems for therapeutically active oligos offer significant potential advantages as described below.

1. They provide significantly prolonged circulation time.
2. They are able to interact with cell membrane to promote the uptake.
3. They protect degradation of nucleic acid in the physiological environment and carry the nucleic acids safely to the target cells.
4. They promote escape from the endocytotic vehicles.
5. They dissociate from the active nucleic acids allowing them to perform their function.

In this section, we have discussed in details the formulation aspects of antisense oligonucleotides.

## 12.6.1 LIPID-BASED CARRIER SYSTEMS

Among the various carrier systems, cationic liposomes are heavily exploited for the systemic delivery of oligonucleotides. DNA can be condensed into the cationic particles when two components are mixed together and the complex is known as cationic lipid/DNA complex (lipoplex). They protect the DNA from degradation and deliver the DNA to cells by interacting with negatively charged cell membrane (Martimprey et al. 2009). They have potential for gene delivery. Although plethora of cationic lipids have been synthesized, but none of them produces positive impact in clinical trial due to their low efficiency and toxicity associated with inflammation and complement activation. The exact origin of toxicity has not yet been explored and number of factors such as the combination of unmethylated DNA and cationic molecules; and even larger diameter of lipoplexes would responsible for imposing toxicity. Systemically administered lipoplexes are rapidly distributed in lungs. When pulmonary endothelium cells were transfected, due to quick association of lipoplex with the capillary endothelium of lungs, less than 2% of lipoplex were left within 10 min of administration by tail vein. However, the liver serves as a major organ for the elimination of lipoplexes as they were redistributed from lungs to the liver after injection within 60 min and most of them were taken up by Kuffer cells rather than hepatocytes. As mentioned before, cationic surface mediates strong interaction with plasma proteins as well as with the glycocalyx of many tissues resulting in a rapid elimination of lipoplex (half-life <5 min) by RES. This in turn induces inflammatory adverse effects and limits access to the target tissue. To overcome this drawback, PEG-shielded cationic liposomal bilayer is prepared and it offers numerous advantages such as formation of a bilayer shell around the nucleic acids, smaller particle diameter (~100 nm), increased half-life (1–10 h), better transfection efficiency toward the target cells, etc. The lipoidal carriers are constructed from a combination of following elements such as lipids, PEG-lipids, lipopolymers, pH or reduction sensitive components, targeting ligands, etc (Li and Szoka 2007).

### 12.6.1.1   Cationic Lipid

A multivalent cation is considered a vital component of lipid-based carrier systems as they condense with negatively charged nucleic acid by charge–charge interaction into small particles. Moreover, cationic lipid can bind to the negatively charged cell membranes promoting the uptake of nucleic acids into the cells. However, cationic lipids possess some serious drawbacks such as immune reactivity, aggregation with blood components and enhanced uptake, which make the delivery vehicles inefficient for *in vivo* applications. They can be overcome by adopting certain approaches such as the use of titrable cationic lipids, that are charged at acidic pH but uncharged at pH 7.4, covalent modification of cationic group to an anionic or neutral group, use of disulfide-linked cationic lipid surface of which can be exchanged to charge neutral or negative to encapsulate DNA into small particles (Fattal and Barratt 2009).

### 12.6.1.2   Neutral Lipid/Helper Lipid

Neutral fusogenic lipid 1,2-dioleoyl-sn-glycero-3-phosphoethanolamine and cholesterol are used as helper lipids which may increase the transfection efficiency by membrane fusion and disruption, thus promote the escape of DNA from the endosomal membrane (Li and Szoka 2007).

### 12.6.1.3   Anionic Lipid

The toxicity and *in vivo* rapid clearance of cationic lipoplex can be tackled by using anionic lipidic particles, which reduce nonspecific interaction with negatively charged serum proteins such as albumin and extracellular matrix. In this approach, a cationic lipid such as polyethyleneimine (PEI), protamine sulfate, or polylysine (PLL) is used to form a nanometric cationic core. After that, anionic lipids are added to form a surface coat resulting in the formation of anionic or neutral lipid particle (Li and Szoka 2007).

### 12.6.1.4   PEG-Lipids

It is a biocompatible and inert polymer resulting in lesser interaction with particle and cell surface. Apart from that, they substantially improve the efficiency of gene transfer and safety of the delivery system by prolonging the circulation time, reducing toxicity, and stabilizing the particle *in vivo* (Juliano et al. 2009).

### 12.6.1.5    Targeting Ligand

Poor uptake of PEG-stabilized lipoidal carriers can be improved through the decoration of their surface with targeting ligands. Therefore, significant efforts have been made in order to develop a suitable ligand. Different chemical entities can be used as a ligand such as peptide containing RGD motif (arginine-glycine-aspartic acid), several member of integrin receptor on the cell surface, folic acid, antibody, transferrin, etc. Targeted liposomes are found to have more transfection efficiency as compared to nontargeted liposome as active targeting potentiate internalization of particles resulting in subsequent improvement in uptake of particles in the target cells (Fattal and Bochot 2008).

### 12.6.1.6    Lipid–Polymer Conjugate

Water soluble lipopolymer are formed by polyethylenimine (PEI) that has been modified with lipid molecules such as cholesterol and subsequent conjugation on to the polymer backbone. They share the properties of cationic liposomes and polycations. Lipopolymer becomes one of the attractive carriers as DNA can be readily packed as well as DNA can escape from the endosomes in the cytoplasm (Li and Szoka 2007).

## 12.6.2    Techniques of Formulation

### 12.6.2.1    Direct Mixing

Traditional lipoplexes are formed by direct mixing. The technique involves rapid mixing of preformed cationic liposomes with plasmid DNA. Sometimes polycation such as polylysine is used to precondense the DNA prior to the incorporation into cationic lipid complex resulting in the formation of condensed DNA surrounded by cationic bilayer. When protamine enriched with arginine residues is employed, the formed particles are known as lipid-protamine-DNA complex. The complex has potential for *in vivo* transfection (Martimprey et al. 2009).

### 12.6.2.2    Detergent Dialysis

The detergent dialysis method was initially adopted for the preparation of relatively stable cationic lipid-DNA particles by dissolving the DNA and cationic lipid mixture in a detergent solution followed by dialysis to remove the detergent. The resultant particles were found to have lesser *in vitro* transfection efficiency as compared to simply mixed lipoplexes. However, particles were found to be stable for a prolonged period and were found to be more active in the presence of serum-containing medium. Moreover, the technique did not have any positive impact on *in vivo* performance. Octyl glucoside (OG) is most widely used as detergent since it has high critical micelle concentration (CMC) as well as its nonionic nature prevents the interference of a charge–charge interaction between DNA and cationic lipid (Juliano et al. 2008).

When a PEG-lipid is included in the formulation, a stabilized plasmid-lipid particle (SPLP) is formed using detergent lysis method. The formation of SPLP involves two steps: (a) mixing of DNA with a cationic lipid, neutral lipid, and PEG-lipid in a detergent (OG) solutions and (b) removal of detergent from a solution/suspension formed in the above step. The particle size of SPLP generally varies between 60 and 100 nm. SPLP contains encapsulated DNA, surface-bound DNA, and free DNA.

### 12.6.2.3 Ethanol Dialysis

The detergent dialysis method has certain limitations, as it is very sensitive to ionic strength of the formulation buffer, the temperature of dialysis, and the detergent concentration. Therefore, those conditions should be properly optimized, otherwise, it may result in the formation of larger particles, as well as a decrease in encapsulation efficiency. Moreover, the high cost of OG and other detergents makes the process even more difficult to scale up. To overcome the drawbacks, ethanol can be used to form small unilamellar vesicle (diameter about 60 nm) by injecting ethanolic lipid mixture into water. Cationic liposomes formed by ethanolic injection possess the same transfection efficiency as compared to liposome obtained by the standard evaporation/sonication method. Ethanol has the ability to condense DNA into compact structures such as toroids, rod, fiber, in the presence of a critical concentration of multivalent cations (Fattal and Barratt 2009).

## 12.7 RECENT FINDINGS

Oligonucleotide-based therapeutics have raised new hope for targeted therapy. Though much preclinical data have already been generated, the quantum of clinical translation is still less due to various challenges. Very few ASOs actually reached the market. The liposomal formulation of c-raf ASO in the clinical development stage is possibly the first example of this kind (Zhang et al. 2009). Mipomersen, an ASO against ApoB100 intended for treating homozygous familial hypercholesterolemia, got FDA approval in 2013 and until now the only successful commercially available ASO. In recent years, the increasing numbers of clinical trials of ASOsprove the growing interest of this field after a short pause in the last decade. In this section, we will discuss some recent research findings of ASOs in targeted therapy.

The effect of the combined therapy bcl-2 ASO and dacarbazine in patients suffering from bcl-2 overexpressed advanced malignant melanoma was investigated. The proto-oncogene bcl-2 is associated with chemoresistance of malignant melanoma. During the dacarbazine and bcl-2 ASO combination therapy, ASO decreased bcl-2 protein expression and led to an increased tumor cell apoptosis in a mouse xenograft model (Jansen et al. 2000).

In a recent study, ASOs directed against microRNA-21 and microRNA-221 enhanced neoplastic cell death in pancreatic adenocarcinoma. The ASOs also sensitized the effects of gemcitabine exerting a synergistic effect in HS766T cells. ASOs atnanomolar concentration reduced pancreatic HS766T cell proliferation *in vitro* and apoptosis was found to enhance by 3–6-folds. When miR-21 and miR-221 were targeted by ASOs, tumor suppressor protein levels were increased (Jong-Kook et al. 2009).

Ghosh et al. (2014) studied the anticancer prospective of ASOs against IGF-II mRNA in rat hepatocarcinoma. Specific phosphorothioate ASOs against the coding exons of IGF-II gene were investigated. Hepatic lesions and tumors were absent in 40% of the treated rats (those who received two oligomers against exon-1 or -3). The rest 60% of animals were still having lesions but with reduced numbers of hepatic-altered foci. As compared to carcinogen control rats, significant reductions (64% and 53%) in the total lesion area were observed for ASOs against exon-1 and exon-3, respectively. The kind of treatment with antisense IGF-II ASOs revealed better integrity for hepatic cells along with modulation in various preneoplastic/neoplastic marker isoenzyme/enzyme levels (Ghosh et al. 2014). In another study, the ASO when directed against hTERT (Cantide) remarkably decreased tumor size in orthotopic primary hepatic lymphoma in mice. In the experimental animals, Cantide reduced tumor weight and serum LDH activity significantly in a dose-dependent manner as compared to the ASO control animals. Importantly, the survival rate was much prolonged in Cantide-treated tumor-bearing animals when compared to mock-treated mice (Yang et al. 2012). In another recent study, ASO targeting human telomerase RNA (hTR ASODN) in combination with radiation therapy showed preferential therapeutic effect in nasopharyngeal carcinoma as compared to conventional radiotherapy. The targeted ASO considerably decreased cell proliferation as well as reduced the length of telomere of CNE 2 human nasopharyngeal carcinoma cells. By this combination therapy (ASO and radiotherapy), the rate of apoptosis as well as the cleavage of caspase 9 were enhanced in the treated cells. The finding demonstrates the potential of hTR ASO as an adjuvant agent in the treatment of nasopharyngeal carcinoma with improved survival rates (Yu et al. 2015).

Further, ASO against activin receptor-like kinase 5 (ARK-5) was found to be potentially effective for treatment of Dupuytren's disease. In this hand disease, the extracellular matrix proteins are produced excessively and a compact fibrous tissue forms between the palm and fingers, due to which fine movement ability is disrupted permanently. In the *ex vivo* study, ARK-5 (also known as transforming growth factor-β)-targeted ASO significantly decreased the collagen protein expression and collagen structures were degraded/reorganized. The targeted ASOs induced exon skipping of ARK-5 precursor mRNA transcript and a functionally impaired protein formed (Karkampouna et al. 2014).

A study shows that the lethal effect of tumor necrosis factor (TNF) was prevented by ASO therapy. TNF is reported to exhibit notable antitumor property; however, often associated with toxic inflammatory effects. The treatment with TNF receptor targeted ASOs significantly reduced the protein expression levels in vital organs and protected the mice from acute toxicity of TNF (Hauwermeiren et al. 2014).

## 12.8 CONCLUSION

Despite the immense potential of ASOs as therapeutics against an array of genetic disorders, there are various key issues, which should be investigated and addressed properly in order to develop efficient antisense therapeutics. Although the research on oligonucleotides had started its journey three decades ago, but the progress and outcome of ASO-related research are limited. This clearly gives evidence in support of hurdles, which hinder the translation of ASO research from the laboratory to clinic. Smarter design, favorable pharmacokinetic profiles, improved target specificity, and target site-specific biological activity may potentiate ASOs for a better therapeutic outcome in the near future.

# REFERENCES

Aartsma-Rus, A., De winter, C., Janson, A. et al. 2005. Functional analysis of 114 exon-internal ions for targeted DMD exon skipping: Indication for steric hindrance of SR protein binding sites. *Oligonucleotides* 15:284–97.

Aartsma-Rus, A., van Vliet, L., Hirschi, M. et al. 2008. Guidelines for antisense oligonucleotide design and insight into splice-modulating mechanisms. *Mol Ther* 17:548–53.

Alam, S., Bromberg-White, J., Mclaughlin-Drubin, M. et al. 2005. Activity and therapeutic potential of ori-1001 antisense oligonucleotide on human papillomavirus replication utilizing a model of dysplastic human epithelium. *Anticancer Res* 25:765–78.

Alama, A., Barbieri, F., Cagnoli, M. et al. 1997. Antisense oligonucleotides as therapeutic agents. *Pharmacol Res* 36:171–8.

Alper, O.È.Z., Hacker, N.F., and Cho-Chung, Y.S. 1999. Protein kinase A-Ia subunit-directed antisense inhibition of ovarian cancer cell growth: Crosstalk with tyrosine kinase signaling pathway. *Oncogene* 18:4999–5004.

Altmann, K.H. and Dean, N.M., Fabbro, D. et al. 1996. Second generation of antisense oligonucleotides: From nuclease resistance to biological efficacy in animals. *Chimia* 50:168–76.

Amantan, A., Iversen, P.L. 2005. Phamacokinetics and biodistribution of phosphorodiamidate morpholino antisense oligomers. *Curr Opin Pharmacol* 5:550–5.

Andronescu, M., Zhang, Z., and Condon, A. 2005. Secondary structure prediction of interacting RNA molecules. *J Mol Biol* 345:987–1001.

Aoki, K., Yoshida, T., Sugimura, T. et al. 1995. Liposome-mediated *in vivo* gene transfer of antisense K-ras construct inhibits pancreatic tumor dissemination in the murine peritoneal cavity. *Cancer Res* 55:3810–6.

Arora, V., Knapp, D. C., Smith, B. L. et al. 2000. c-Myc antisense limits rat liver regeneration and indicates role for c- Myc in regulating cytochrome P-450 3A activity. *J Pharmacol Exp Ther* 292:921–8.

Bai, H., You, Y., Bo, X. et al. 2013. Antisense antivirals: Future oligonucleotides-based therapeutics for viral infectious diseases. In *Microbial Pathogens and Strategies for Combating Them: Science, Technology and Education*, ed. A. Méndez-Vilas, 1517–28. Spain: Formatex Research Centre.

Baker, B.F., Lot, S.S., Condon, T.P. et al. 1997. 2'-O-(2-methoxy)ethyl-modified anti-intercellular adhesion molecule 1 (ICAM-1) oligonucleotides selectively increase the ICAM-1 mRNA level and inhibit formation of the ICAM-1 translation initiation complex in human umbilical vein endothelial cells. *J Biol Chem* 272:11994–2000.

Balaji, K.C., Koul, H., Mitra, S. et al. 1997. Antiproliferative effects of C-Myc antisense oligonucleotide in prostate cancer cells: A novel therapy in prostate cancer. *Urology* 50:1007–15.

Ball, H.A., Van Scott, M.R., and Robinson, C.B. 2004. Sense and antisense: Therapeutic potential of oligonucleotides and interference RNA in asthma and allergic disorders. *Clin Rev Allergy Immunol* 27:207–17.

Bates, P., Kahlon, J.S., Thomas, J. et al. 1999. Antiproliferative activity of G-rich oligonucleotides correlates with protein binding. *J Biol Chem* 274:26369–77.

Behlke, M.A., Devor, E.J., and Goodchild, J., 2005. Designing antisense oligonucleotides. Available at: http://cdn.idtdna.com/support/technical/technicalbulletinpdf/designing_antisense_oligonucleotides.pdf. Accessed July 2016.

Bendigs, S., Salzer, U., Lipford, G. et al. 1999. Cpg-oligodeoxynucleotides co-Stimulate primary T cells in the absence of antigen-presenting cells. *Eur J Immunol* 29:1209–18.

Benimetskaya, L., Berton, M., Kolbanovsky, A. et al. 1997. Formation of a G-tetrad and higher order structures correlates with biological activity of the RelA (NF-kappaB p65) 'antisense' oligodeoxynucleotide. *Nucl Acids Res* 25:2648–56.

Benimetskaya, L., Takle, G., Vilenchik, M. et al. 1998. Cationic porphyrins: Novel delivery vehicles for antisense oligodeoxynucleotides. *Nucl Acids Res* 26:5310–7.

Bennett, C.F., Chiang, M.Y., Chan, H. et al. 1992. Cationic lipids enhance cellular uptake and activity of phosphorothioate antisense oligonucleotides. *Mol Pharmacol* 41:1023–33.

Bharti, A.C., Shukla, S., and Mahata, S. et al. 2009. Anti-human papillomavirus therapeutics: Facts and future. *Indian J Med Res* 130:296–310.

Blake, K., Murakami, A., Spitz, S. et al. 1985. Hybridization arrest of globin synthesis in rabbit reticulocyte lysates and cells by oligodeoxyribonucleoside methylphosphonates. *Biochemistry* 24:6139–45.

Bo, X. and Wang, S. 2005. Target finder: A software for antisense oligonucleotide target site selection based on MAST and secondary structures of target mRNA. *Bioinformatics* 21:1401–2.

Boffa, L.C., Morris, P.L., Carpaneto, E.M. et al. 1996. Invasion of the CAG triplet repeats by a complementary peptide nucleic acid inhibits transcription of the androgen receptor and TATA-binding protein genes and correlates with refolding of an active nucleosome containing a unique AR gene sequence. *J Biol Chem* 271:13228–33.

Bruice, T.W. and Lima, W.F. 1997. Control of complexity constraints on combinatorial screening for preferred oligonucleotide hybridization sites on structured RNA. *Biochemistry* 36:5004–19.

Buhr, C.A., Wagner, R.W., Grant, D. et al. 1996. Oligodeoxynucleotides containing C-7 propyne analogs of 7-deaza-2′- deoxyguanosine and 7-deaza-2′-deoxyadenosine. *Nucleic Acids Res* 24:2974–80.

Caffo, M., Caruso, G., Passalacqua, M. et al. 2013. Antisense oligonucleotides therapy in the treatment of cerebral gliomas: A review. *J Genet Syndr Gene Ther* 4:194.

Cai, J., Mao, Z., Hwang, J.J. et al. 1998. Differential expression of methionine adenosyltransferase genes influences therate of growth of human hepatocellular carcinoma cells. *Cancer Res* 58:1444–50.

Cartegni, L., Chew, S., and Krainer, A. 2002. Listening to silence and understanding nonsense: Exonic mutations that affect splicing. *Nat Rev Genet* 3:285–98.

Cazenave, C., Frank, P., Toulme, J.J. et al. 1994. Characterization and subcellular localization of ribonuclease H activities from Xenopus laevis oocytes. *J Biol Chem* 269:25185–92.

Chan, J.H.P., Lim, S., and Wong, W.S. 2006. Antisense oligonucleotides: From design to therapeutic application. *Clin Exp Pharmacol Physiol Clinical* 33:533–40.

Chen, J.K., Schultz, R.G., Lloyd, D.H. et al. 1995. Synthesis of oligodeoxyribonucleotide N3 → P5 phosphoramidates. *Nucleic Acids Res* 23:2661–8.

Cheng, J.Q., Ruggerit, B., Klein, W.M. et al. 1996. Amplification of AKT2 in human pancreatic cancer cells and inhibition of AKT2 expression and tumorigenicity by antisense RNA. *Proc Natl Acad Sci* 93:3636–41.

Chernicky, C.L., Yi, L., Tan, H. et al. 2000. Treatment of human breast cancer cells with antisense RNA to the type I insulin-like growth factor receptor inhibits cell growth, suppresses tumorigenesis, alters the metastatic potential, and prolongs survival *in vivo*. *Cancer Gene Ther* 7:384–95.

Chi, K.N., Eisenhauer, E., Fazli, L. et al. 2005. A phase I pharmacokinetic and pharmacodynamic study of OGX-011, a 2′-methoxyethyl antisense oligonucleotide to clusterin, in patients with localized prostate cancer. *J Natl Cancer Inst* 97:1287–96.

Chi, K.N., Gleave, M.E., Klasa, R. et al. 2001. A phase I dose-finding study of combined treatment with an antisense bcl-2 oligonucleotide (genasense) and mitoxantrone in patients with metastatichormone-refractory prostate cancer. *Clinical Cancer Research* 7:3920–7.

Cho-Chung, Y.S., Nesterova, M., Kondrashin, A. et al. 1997. Antisense-protein kinase A: A single-gene-based therapeutic approach. *Antisense Nucl Acid Drug Dev* 7:217–23.

Chou, S., Zhu, L., and Reid, B. 1994. The unusual structure of the human centromere (GGA)2 motif. *J Mol Biol* 244:259–68.

Ciardiello, F., Bianco, R., Damiano, V. et al. 2000. Antiangiogenic and antitumor activity of anti-epidermal growth factor receptor C225 monoclonal antibody in combination with vascular endothelial growth factor antisense oligonucleotide in Human GEO colon cancer cells. *Clin Cancer Res* 6:3739–47.

Cowsert, L., Fox, M., Zon, G. et al. 1993. *In vitro* evaluation of phosphorothioate oligonucleotides targeted to the E2 mRNA of papillomavirus: Potential treatment for genital warts. *Antimicrob Agents Chemother* 37:171–7.

Crooke, S.T. 2000. Progress in antisense technology: The end of the beginning. *Methods Enzymol* 313:3–45.

Crooke, S.T. and Geary, R.S. 2013. Clinical pharmacological properties of mipomersen (Kynamro), a second generation antisense inhibitor of apolipoprotein. *J Clin Pharmacol* 76:269–76.

Cutrona, G., Carpaneto, E.M., Ulivi, M. et al. 2000. Effects in live cells of a c-*myc* anti-gene PNA linked to a nuclear localization signal. *Nat Biotechnol* 18:300–3.

Dagle, J.M., Weeks, D.L., and Walder, J.A. 1991. Pathways of degradation and mechanism of action of antisense oligonucleotides inxenopus laevis embryos. *Antisense Res Dev* 1:11–20.

Dai, D.J., Lu, C.D., Lai, R.Y. et al. 2005. Survivin antisense compound inhibits proliferation and promotes apoptosis in liver cancer cells. *World J Gastroenterol* 11:193–9.

Damha, M., Noronha, A.C., Wilds, J. et al. 2001. Properties of arabinonucleic acids (ANA and 20′F-ANA): Implications for the design of antisense therapeutics that invoke RNase h cleavage of RNA. *Nucleos Nucleot Nucl* 20:429–40.

Das, T., Patra, F., and Mukherjee, B. 2010. Effect of antisense oligomer in controlling c-raf.1 overexpression during diethylnitrosamine-induced hepatocarcinogenesis in rat. *Cancer Chemother Pharmacol* 65:309–18.

Datta, R., Oki, E., Endo, K. et al. 2000. XIAP regulates DNA damage-induced apoptosis downstream of caspase-9 cleavage. *J Biol Chem* 275:31733–8.

Dean, N.M. and Bennett, C.F. 2003. Antisense oligonucleotide-based therapeutics for cancer. *Oncogene* 22:9087–96.

Dean, N.M., McKay, R., Condon, T.P. et al. Inhibition of protein kinase C-α expression in human A549 cells by antisense oligonucleotides inhibits induction of intercellular adhesion molecule 1 (ICAM-1) mRNA by phorbol esters. *J Biol Chem* 269:16416–24.

De Diesbach, P., Berens, C., and Kuli, N. 2000. Identification, purification and partial characterization of an oligonucleotide receptor in membranes of HepG2 cells. *Nucleic Acids Residence* 28:868–74.

Denham, D.W., Franz, M.G., Denham, W. et al. 1998. Directed antisense therapy confirms the role of protein kinase C–α in the tumorigenicity of pancreatic cancer. *Surgery* 124:218–24.

De Smet, M.D., Meen ken, C.J., and van den Horn, G.J. 1999. Fomivirsen—A phosphorothioate oligonucleotide for the treatment of CMV retinitis. *Ocul Immunol Inflamm* 7:189–98.

Deutekom, J.C., Janson, A.A., Ginjaar, I.B. et al. 2007. Local dystrophin restoration with antisense oligonucleotide PRO051. *N Engl J Med* 357:2677–86.

Deveraux, Q.L., Takahashi, R., Salvesen, G.S. et al. 1997. X-linked IAP is a direct inhibitor of cell-death proteases. *Nature (Lond.)* 388:300–4.

Dias, N., Dheur, S., Nielsen, P.E. et al. 1999. Antisense PNA tridecamers targeted to the coding region of Ha-*ras* mRNA arrest polypeptide chain elongation. J Mol Biol 294:403–16.

Dias, N. and Stein, C.A. 2002. Antisense oligonucleotides: Basic concepts and mechanisms. *Mol Cancer Ther* 1:347–55.

Ding, Y. and Lawrence, C.E. 2001. Statistical prediction of single-stranded regions in RNA secondary structure and application to predicting effective antisense target sites and beyond. *Nucl Acids Res* 29:1034–46.

Disney, M., Testa, S., and Turner, D. 2000. Targeting a pneumocystis carinii group I intron with methylphosphonate oligonucleotides: Backbone charge is not required for binding or reactivity. *Biochemistry* 39:6991–7000.

Ebinuma, H., Saito, H., Kosuga, M. et al. 2003. Reduction of c-myc expression by an antisense approach under Cre/loxP switching induces apoptosis in human liver cancer cells. *Gastroenterology* 124:202–16.

Eder, P., Walder, R., and Walder, J. 1993. Substrate specificity of human Rnase H1 and its role in excision repair of ribose residues misincorporated in DNA. *Biochimie* 75:123–6.

Eder, P.S., DeVine, R.J., Dagle, J.M. et al. 1991. Substrate specificity and kinetics of degradation of antisense oligonucleotides by a 3′ exonuclease in plasma. *Antisense Res De* 1:141–51.

Eder, P.S. and Walder, J.A. 1991. Ribonuclease H from K562 human erythroleukemia cells: Purification, characterization, and substrate specificity. *J Biol Chem* 266:6472–9.

Egholm, M., Buchardt, O., and Christensen, L. 1993. PNA hybridizes to complementary oligonucleotides obeying the Watson-Crick hydrogen-bonding rules. *Nature* 365: 566–8.

Farhood, H., Serbina, N., and Huang, L. 1995. The role of dioleoyl phosphatidylethanolamine in cationic liposome mediated gene transfer. *Biochem Biophys Acta* 1235:289–95.

Fattal, E. and Barratt, G. 2009. Nanotechnologies and controlled release systemsfor the delivery of antisense oligonucleotides andsmall interfering RNA. *British J Pharmacol* 157:179–94.

Fattal, E. and Bochot, A. 2008. State of the art and perspectives for the delivery of antisense oligonucleotides and siRNA by polymeric nanocarriers. *Int J Pharm* 364:237–48.

Fattal, E., Couvreur, P., and Dubernet, C. 2004. 'Smart' delivery of antisense oligonucleotides by anionic pH-sensitive liposomes. *Adv Drug Deliv Rev* 56:931–46.

Fei, J. and Zhang, Y. 2005. Prediction of VEGF mRNA antisense oligodeoxynucleotides by RNA structure software and their effects on HL60 and K562 cells. *Cell Biol Int* 29:737–41.

Felgner, J.H., Kumar, R., Sridhar, C.N. et al. 1994. Enhanced gene delivery and mechanism studies with a novel series of cationic lipid formulations. *J Biol Chem* 269:2550–61.

Flanagan, W.M., Kothavale, A., and Wagner, R.W. 1996. Effects of oligonucleotide length, mismatches and mRNA levels on C-5 propyne-modified antisense potency. *Nucleic Acids Res* 24:2936–41.

Fluiter, K., Frieden, M., Vreijling, J. et al. 2005. On the *in vitro* and *in vivo* properties of four locked nucleic acid nucleotides incorporated into an anti-H-Ras antisense oligonucleotide. *Chembiochemistry* 6:1104–9.

Fong, S., Itahana, Y., Sumida, T. et al. 2003. Id-1 as a molecular target in therapy for breast cancer cell invasion and metastasis. *PNAS* 100:13543–8.

Frank, P., Albert, S., Cazenave, C. et al. 1994. Purification and characterization of human ribonuclease HII. *Nucl Acids Res* 22:5247–54.

Galderisi, U., Cascino, A., and Giordano, A. 1999. Antisense oligonucleotides as therapeutic agents. *Journal of Cellular Physiology* 181:251–7.

Geary, R.S. 2009. Antisense oligonucleotide pharmacokinetics and metabolism. *Expert Opin Drug Metab Toxicol* 5:381–91.

Geary, R.S., Wancewicz, E., Matson, J. et al. 2009. Effect of dose and plasma concentration on liver uptake and pharmacologic activity of a 2′-methoxyethyl modified chimeric antisense oligonucleotide targeting PTEN. *Biochem Pharmacol* 78:284–91.

Geiger, T., Muller, M., Dean, N.M. et al. 1998. Antitumor activity of a PKC antisense oligonucleotide in combination with standard chemotherapeutic agents against various human tumors transplanted into nude mice. *Anticancer Drug Des* 13:35–45.

Ghosh, M.K., Patra, F., Ghosh, S. et al. 2014. Antisense oligonucleotides directed against insulin-like growth factor-II messenger ribonucleic acids delay the progress of rat hepatocarcinogenesis. *J Carcinog* 13:1–9.

Giles, R., Spiller, D., Grzybowski, J. et al. 1998. Selecting optimal oligonucleotide composition for maximal antisense effect following streptolysin o-mediated delivery into human leukaemia cells. *Nucl Acids Res* 26:1567–75. Oxford University Press (OUP).

Giles, R.V., Spiller, D.G., Clark, R.E. et al. 1999. Antisense morpholino oligonucleotide analog induces missplicing of C-myc mRNA. *Antisense Nucleic Acid Drug Dev* 9:213–20.

Gleave, M.E. and Monia, B.P. 2005. Antisense therapy for cancer. *Nat Rev Cancer* 5:468–79.

Goodchild, J., Carroll, E., and Greenberg, J. 1988. Inhibition of rabbit β-globin synthesis by complementary oligonucleotides: Identification of m-RNA sites sensitive to inhibition. *Arch Biochem Biophys* 263:401–9.

Gray, G.D., Basu, S., and Wickstrom, E. 1997. Transformed and immortalized cellular uptake of oligodeoxynucleoside phosphorothioates, 3′-alkylamino oligodeoxynucleotides, 2′-*O*-methyl oligoribonucleotides, oligodeoxynucleoside methylphosphonates, and peptide nucleic acids. *Biochem Pharmacol* 53:1465–76.

Griffey, R.H., Monia, B.P., Cummins, L.L. et al. 1996. 2′-O-aminopropyl ribonucleotides: A zwitterionic modification that enhances the exonuclease resistance and biological activity of antisense oligonucleotides. *J Med Chem* 39:5100–9.

Griffin, L., Toole, J., and Leung, L. 1993. The discovery and characterization of a novel nucleotide-based thrombin inhibitor. *Gene* 137:25–31.

Gritsko, T., Williams, A., Turkson, J. et al. 2006. Persistent activation of Stat3 signaling induces survivin gene expression and confers resistance to apoptosis in human breast cancer cells. *Clin Cancer Res* 12:2001–16.

Gryaznov, S., Skorski, T., Cucco, C. et al. 1996. Oligonucleotide N3 → P5 phosphoramidates as antisense agents. *Nucleic Acids Res* 24:1508–14.

Gryaznov, S.M., Lloyd, D.H., Schultz, J.K. et al. 1995. Oligonucleotide N3 → P5 phosphoramidates. *Natl Acad Sci* 92:5798–802.

Hanvey, J.C., Peffer, N.J, Bisi, J.E. et al. 1992. Antisense and antigene properties of peptide nucleic acids. *Science* 258:1481–5.

Hauwermeiren, F.V., Vandenbroucke, R.E., Grine, L. et al. 2014. Antisense oligonucleotides against TNFR1 prevent toxicity of TNF/IFNc treatment in mouse tumor models. *Int J Cancer* 135:742–50.

Hnik, P., Boyer, D.S., Grillone, L.R. et al. 2009. Antisense oligonucleotide therapy in diabetic retinopathy. *J Diabetes Sci Technol* 3:924–30.

Ho, S.P., Britton, D.H.O., Stone, B.A. et al. 1996. Potent antisense oligonucleotides to the human multidrug resistance-1mRNA are rationally selected by mapping RNA-accessible sites with oligonucleotide libraries. *Nucleic Acids Res* 24:1901–7.

Hotz, H.J., Hines, O.J., Masood, R. et al. 2004. VEGF antisense therapy inhibits tumor growth and improves survival in experimental pancreatic cancer. *Surgery* 137:192–9.

Hu, Y.P., Cherton-Horvat, G., Dragowska, V. et al. 2003. 2,3, Antisense oligonucleotides targeting XIAP induce apoptosis and enhance chemotherapeutic activity against human lung. cancer cells *in vitro* and *in vivo*. *Clin Cancer Res* 9:2826–36.

Hudziak, R.M., Barofsky, E.D., Barofsky, F. et al. 1996. Resistance of morpholino phosphorodiamidate oligomers to enzymatic degradation. *Antisense Nucleic AcidDrug Dev* 6:267–72.

Hung, G., Xiao, X., Peralta, R. et al. 2013. Characterization of target mRNA reduction through *in situ* RNA hybridization in multiple organ systems following systemic antisense treatment in animals. *Nucleic Acid Ther* 23:369–78.

Ichimura, E., Maeshima, A., Nakajima, T. et al. 1996. Expression of c-met/HGF receptor in human non-small cell lung carcinomas *in vitro* and *in vivo* and its prognostic significance. *Jpn J Cancer Res* 87:1063–9.

Ishikawa, K., Okumura, M., Katayanagi, K. et al. 1993. Crystal structure of ribonuclease H from thermus thermophilus HB8 refined at 2.8 Å resolution. *J Mol Biol* 230:529–42.

Iversen, P.L. 2001. Phosphorodiamidate morpholino oligomers: Favorable properties for sequence-specific gene inactivation. *Curr Opin Mol Ther* 3:235–8.

Iversen, P.L., Arora, V., Acker, A.J. et al. 2003. Efficacy of antisense morpholino oligomer targeted to c-*myc* in prostate cancer xenograft murine model and a phase-I safety study in humans. *Clin Cancer Res* 9:2510–9.

Jakob, T., Walker, P., Krieg, A. et al. 1999. Bacterial DNA and CpgandNdash; containing Oligodeoxynucleotides activate dutaneous dendritic cells and induce ILandNdash; 12 production: Implications for the augmentation of Th1 responses. *Int Arch Allergy Immunol* 118:457–61.

Jansen, B., Wacheck, V., Heere-Ress, E. et al. 2000. Chemosensitisation of malignant melanoma by BCL2 antisensetherapy. *Lancet* 356:1728–33.

Jarver, P. and Langel, U. 2004. The use of cell-penetrating peptides as a tool for gene regulation. *Drug Discov Today* 9:395–402.

Jhaveri, M.S., Rait, A.S., Chung, K.N. et al. 2004. Antisense oligonucleotides targeted to the human A folate receptor inhibit breast cancer cell growth and sensitizethe cells to doxorubicin treatment. *Mol Cancer Ther* 3:12.

Jiménez, J.L. and Durbin, R. 2006. [X]uniqMAP: Unique gene sequence regions in the human and mouse genomes. *BMC Genomics* 7:249.

Johansson, I., Oscarson, M., Yue Q.Y. et al. 1994. Genetic analysis of the Chinese cytochrome P4502D locus: Characterization of variant CYP2 D6 genes present in subjects with diminished capacity for debrisoquine hydroxylation. *MolPharmacol* 46:452–9.

Jong-Kook, P., Eun Joo, L., Christine, E., Thomas, D.S. et al. 2009. Antisense inhibition of microRNA-21 or -221 arrests cell cycle, induces apoptosis, and sensitizes the effects of gemcitabine in pancreatic adenocarcinoma. *Pancreas* 38:190–9.

Juliano, R., Alam, Md. R., Dixit, V. et al. 2008. Survey and summary mechanisms and strategies for effective delivery of antisense and siRNA oligonucleotides. *Nucleic Acids Research* 36:4158–71.

Juliano, R., Bauman, J., Kang, H. et al. 2009. Biological barriers to therapy with antisense and siRNA oligonucleotides. *Mol Pharmaceutics* 6:686–95.

Karkampouna, S., Kruithof, B., and Kloen, P. 2014. Novel *ex-vivo* culture method for the study of Dupuytren's disease: Effects of TGFβ Type 1 receptor modulation by antisense oligonucleotides. *Molecular Therapy-Nucleic Acids* 3:e142.

Karras, J.G., Maier, M.A., Lu, T. et al. 2001. Peptide nucleic acids are potent modulators of endogenous pre-mRNA splicing of the murine interleukin-5 receptor-α chain. *Biochemistry* 40:7853–9.

Kole, R. and Sazani, P. 2001. Antisense effects in the cell nucleus: Modification of splicing. *Curr Opin Mol Ther* 4:229–34.

Kole, R., Williams, T., and Cohen, L. 2004. RNA modulation, repair and remodeling by splice switching oligonucleotides. *Acta Biochim Pol* 51:373–8.

Koller, E., Vincent, T.M., Chappell, A. et al. 2011. Mechanisms of single-stranded phosphorothioatemodified antisense oligonucleotide accumulation in hepatocytes, *Nucleic Acids Res* 39:4795–807.

Koppelhus, U. and Nielsen, P.E. 2003. Cellular delivery of peptide nucleic acid (PNA). *Adv Drug Deliv Rev* 55:267–80.

Kordasiewicz, H.B., Stanek, L.M., Wancewicz, E.V. et al. 2012. Sustained therapeutic reversal of Huntington's disease by transient repression of huntingtin synthesis. *Neuron* 74:1031–44.

Koziolkiewicz, M., Gendaszewska, E., Maszewska, M. et al. 2001. The mononucleotide-dependent, nonantisense mechanism of action of phosphodiester and phosphorothioate oligonucleotides depends upon the activity of an ecto-5-nucleotidase. *Blood* 98:995–1002.

Koziolkiewicz, M., Krakowiak, A., Kwinkowski, M. et al. 1995. Stereodifferentiation—The effect of P chirality of oligo (nucleoside phosphorothioates) on the activity of bacterial Rnase H. *Nucl Acids Res* 23:5000–5.

Kretschmer-Kaxemi Far, R., Nedbal, W., and Sczakiel, G. 2001. Concepts to automate the theoretical design of effective antisense oligonucleotides. *Bioinformatics* 17:1058–61.

Krieg, A.M. and Stein, C.A. 1995. Phosphorothioate oligodeoxynucleotides: Antisense or antiprotein? *Antisense Res Dev* 5:241.

Krieg, A.M., Yi, A.K., and Hartmann, G. 1999. Mechanisms and therapeutic applications of immune stimulatory cpG DNA. *Pharmacol Ther* 84:113–20.

Kronenwett, R., Haas, R., and Sczakiel, G. 1996. Kinetic selectivity of complementary nucleic acids: Bcr-Abl-directed antisense RNA and ribozymes. *J Mol Biol* 259:632–44.

Kurreck, J., Wyszko, E., Gillen, C., Erdmann, V.A. et al. 2002. Design of antisense oligonucleotides stabilized by locked nucleic acid. *Nucleic Acids Res* 30:1911–8.

Kuwasaki, T., Hosono, K., Takai, K. et al. 1996. Hairpin antisense oligonucleotides containing 2′-methoxynucleosides with base-pairing in the stem region at the 3′-end: Penetration, localization, and Anti-HIV activity. *Biochem Biophys Res Commun* 228:623–31.

Larrouy, B., Blonski, C., Boiziau, C. et al. 1992. RNase H-mediated inhibition of translation by antisense oligodeoxy ribonucleotides: Use of backbone modification to improve specificity. *Gene* 121:189–94.

Lee, Y., Vassilakos, A., and Feng, N. 2003. GTI-2040, an antisense agent targeting the small subunit component (R2) of human ribonucleotide reductase, shows potent antitumor activity against a variety of tumors. *Cancer Res* 63:2802–11.

Levin, A.A. 1999. A review of the issues in the pharmacokinetics and toxicology of phosphorothioate antisense oligonucleotides. *Biochim Biophys Acta* 1489:69–84.

Levin, A.A., Yu, R.Z., and Geary, R.S. 2007. Basic principles of the pharmacokinetics of antisense oligonucleotide drugs. In *Antisense Drug Technology: Principles, Strategies, And Applications*, ed. S.T. Crooke, 183–216. Boca Raton, FL: CRC Press.

Li, W. and Szoka, C.F. 2007. Lipid-based nanoparticles for nucleic acid delivery. *Pharmaceutical Research* 24:438–48.

Li, W.C., Ye, S.L., Sun, R.X. et al. 2006. Inhibition of growth and metastasis of human hepatocellular carcinoma by antisense oligonucleotide targeting signal transducer and activator of transcription. *Clin Cancer Res* 12:7140–8.

Liang, X.H., Shen, W., Sun, H. et al. 2014. TCP1 complex proteins interact with phosphorothioate oligonucleotides and can co-localize in oligonucleotideinduced nuclear bodies in mammalian cells. *Nucleic Acids Res* 42:7819–32.

Ling, Y.D., Pu, L., and Pei, G. 1998. Antisense oligonucleotide to insulin-like growth factor II induces apoptosis in human ovarian cancer AO cell Line. *Cell Res* 8:159–65.

Lipford, G., Bauer, M., Blank, C. et al. 1997. Cpg-containing synthetic oligonucleotides promote B and cytotoxic T cell responses to protein antigen: A new class of vaccine adjuvants. *Eur J Immunol* 27:2340–4.

Loke, S.L., Stein, C.A., Zhang, X.H. et al. 1989. Characterization of oligonucleotide transport into living cells. *Proc Natl Acad Sci USA* 86:3474–8.

Lysik, M.A. and Wu-Pong, S. 2003. Innovations in oligonucleotide drug delivery. *J Pharm Sci* 92:1559–73.

Ma, D.D. and Wei, A.Q. 1996. Enhanced delivery of synthetic oligonucleotides to human leukaemic cells by liposomes and immunoliposomes. *Leuk Res* 20:925–30.

Mann, M. J. and Dzau, V.J. 2000. Therapeutic applications of transcription factor decoy oligonucleotides. *J. Clin. Invest.* 106:1071–75.

Martimprey, H., Vauthier, C., Malvy, C. et al. 2009. Polymer nanocarriers for the delivery of small fragments of nucleic acids: Oligonucleotides and siRNA. *Eur J Pharm Biopharm* 71:490–4.

Martin-Banderas, L., Holgado, M.A., Venero, J.L. et al. 2011. Nanostructures for drug delivery to the brain. *Curr Med Chem* 18:5303–21.

Mathews, D.H., Burkard, M.E., Freier, S.M. et al. 1999. Predicting oligonucleotide affinity to nucleic acid targets. *RNA* 5:1458–69.

Mathews, D.H., Disney, M.D., Childs, J.L. et al. 2004. Incorporating chemical modification constraints into a dynamic programming algorithm for prediction of RNA secondary structure. *Proc Natl Acad Sci USA* 101:7287–92.

Matveeva, O., Felden, B., Audlin, S. et al. 1997. A rapid *in vitro* method for obtaining RNA accessibility patterns for complementary DNA probes: Correlation with an intracellular pattern and known RNA structures. *Nucleic Acids Res* 25:5010–6.

Matveeva, O.V., Mathews, D.H., Tsodikov, A.D. et al. 2003. Thermodynamic criteria for high hit rate antisense oligonucleotide design. *Nucleic Acids Res* 31:4989–94.

Matveeva O.V., Tsodikov, A.D., Giddings, M. et al. 2000. Identification of sequence motifs in oligonucleotides whose presence is correlated with antisense activity. *Nucleic Acids Res* 28:2862–5.

McKay, R.A., Cummins, L.L., Graham, M.J. et al. 1996. Enhanced activity of an antisense oligonucleotide targeting murine protein kinase C-alpha by the incorporation of 2′-O-propyl modifications. *Nucleic Acids Res* 24:411–7.

McKay, R.A., Miraglia, L.J., Cummins, L.L. et al. 1999. Characterization of a potent and specific class of antisense oligonucleotide inhibitor of human protein kinase C-a expression. *J Biol Chem* 274:1715–22.

McMahon, B.M., Mays, D., Lipsky, J. et al. 2002. Pharmacokinetics and tissue distribution of a peptide nucleic acid after intravenous administration. *Antisense Nucleic Acid Drug Dev* 12:65–70.

Mendell, J.R., Rodino-Klapac, L.R., Sahenk, Z. et al. 2013. Eteplirsen for the treatment of Duchenne muscular dystrophy. *Ann Neurol* 74:637–47.

Mercatante, D., Bortner, C., Cidlowski, J., and Kole, R. 2001. Modification of alternative splicing of Bcl-X pre-Mrna in prostate and breast cancer cells: Analysis of apoptosis and cell death. *J Biol Chem* 276:16411–7.

Metelev, V., Lisziewicz, J., and Agrawal, S. 1994. Study of antisense oligonucleotide phosphorothioates containing segments of 29-O-methyloligoribonucleotides. *Bioorganic Med Chem Lett* 4:2929–34.

Miller, P.S., Yano, J., Yano, E. et al. 1979. Nonionic nucleic acid analogues. Synthesis and characterization of dideoxyribonucleoside methylphosphonates. *Biochemistry* 18:5134–43.

Miller, T.M., Pestronk, A., David, W. et al. 2013. An antisense oligonucleotide against SOD1 delivered intrathecally for patients with SOD1 familial amyotrophic lateral sclerosis: A phase 1, randomised, first-in-man study. *Lancet Neurol* 12:435–42.

Miyake, H., Chi, K.N., and Gleave, M.E. 2000. Antisense TRPM-2 oligodeoxynucleotides chemosensitize human androgen-independent PC-3 prostate cancer cells both *in vitro* and *in vivo*. *Clin Cancer Res* 6:1655–63.

Mologni, L., lecoutre, P., Nielsen, P.E. et al. 1998. Additive antisense effects of different Pnas on the *in vitro* translation of the PML/Raralpha gene. *Nucl Acids Res* 26:1934–8.

Monia, B.P., Johnston, J.F., Sasmor, H. et al. 1996. Nuclease resistance and antisense activity of modified oligonucleotides targeted to Ha-ras. *J Biol Chem* 271:14533–40.

Monia, B.P., Lesnik E.A., Gonzalez, C. et al. 1993. Evaluation of 2′-modified oligonucleotides containing 2′-deoxy gaps as antisense inhibitors of gene expression. *J Biol Chem* 268:14514–22.

Monteleone, G., Fantini, M.C., Onali, S. et al. 2012. Phase I clinical trial of smad7 knockdown using antisense oligonucleotide in patients with active Crohn's disease. *The American Society of Gene and Cell Therapy* 20:870–4.

Moriya, K.M. and Matsukura, K. 1996. *In vivo* inhibition of hepatitis B virus gene expression by antisense phosphotothioate oligonucleotides. *Biochem Biophys Res Commun* 218:217–23.

Moulds, C., Lewis, J.G., Froehler, B.C. et al. 1995. Site and mechanism of antisense inhibition by C-5 propyne oligonucleotides. *Biochemistry* 34:5044–53.

Mukherjee, B., Ghosh, S., Das, T. et al. 2005. Characterization of insulin-like-growth factor II (IGF II) mRNApositive hepatic altered foci and IGF II expression in hepatocellularcarcinoma during diethylnitrosamine-inducedhepatocarcinogenesis in rats. *J Carcinogenesis* 4:12.

Mulamba, G.B., Hu, A., Azad, R.F. et al. 1998. Human cytomegalovirus mutant with sequence-dependent resistance to the phosphorothioate oligonucleotide fomivirsen (isis 2922). *Antimicrob Agents Chemother* 42:971–3.

Murchie, A.I. and Lilley, D.M. 1994. Tetraplex folding of telomere sequences and the inclusion of adenine bases. *EMBO J* 13:993–1001.

Nakano, M., Aoki K., Matsumoto, N. et al. 2001. Suppression of colorectal cancer growth using an adenovirus vector expressingan antisense K-*ras* RNA. *Mol Ther* 3:491–9.

Nelson, M.H., Stein, D.A., Kroeker, A.D. et al. 2005. Arginine-rich peptide conjugation to morpholino oligomers: Effects on antisense activity and specificity. *Bioconjug Chem* 16:959–66.

Nielsen, P.E. 2004. PNA technology. *Mo. Biotechnol* 26:233–48.

Nielsen, P.E., Egholm, M., Berg, R.H. et al. 1991. Sequence selective recognition of DNA by strand displacement with a thyminesubstituted polyamide. *Science* 254:1497–500.

Olie, R.A., Simões-Wüst, A.P., Baumann, B. et al. 2000. A novel antisense oligonucleotide targeting survivin expression induces apoptosis and sensitizes lung cancer cells to chemotherapy. *Cancer Res* 60:2805–9.

Olivas, W.M. and Maher, L.J. 1995. Overcoming potassium-mediated triplex inhibition. *Nucleic Acids Res* 23:1936–41.

Passini, M.A., Bu, J., Richards, A.M. et al. 2011. Antisense oligonucleotides delivered to the mouse CNS ameliorate symptoms of severe spinal muscular atrophy. *Sci Transl Med* 3:72ra18.

Patzel, V., Steidl, U., Kronenwett, R. et al. 1999. A theoretical approach to select effective antisense oligodeoxyribonucleotides at high statistical probability. *Nucleic Acids Res* 27:4328–34.

Paulasova, P. and Pellestor, F. 2004. The peptide nucleic acid (PNAs): A new generation of probles for genetic and cytogenetic analyses. *Ann Genet* 47:349–58.

Phillips, F.M.C., Mullen, P., Monia, B.P. et al. 2001. Association of c-Raf expression with survival and its targeting with antisense oligonucleotides in ovarian cancer. *Brit J Cancer* 85:1753–8.

Pirollo, K.F., Rait, A., Sleer, L.S. et al. 2003. Antisense therapeutics: From theory to clinical practice. *Pharmacol Ther* 99:55–77.

Popadiuk, C.M., Xiong, J., Wells, M.G. et al. 2006. Antisense suppression of pygopus2 results in growth arrest of epithelial ovarian cancer. *Clin Cancer Res* 12:2216–23.

Prakash, T., Graham, M., Yu, J. et al. 2014. Targeted delivery of antisense oligonucleotides to hepatocytes using triantennary n-acetyl galactosamine improves potency 10-fold in mice. *Nucl Acids Res* 42:8796–807.

Puri, N. and Chattopadhyaya, J. 1999. How kinetically accessible is an RNA target for hybridization with an antisense oligo? A lesson from an RNA target which is as small as a 20mer. *Tetrahedron* 55:1505–16.

Qian, C.N., Guo, X., and Cao, B. 2002. Met protein expression level correlates with survival in patients with late-stage nasopharyngeal carcinoma. *Cancer Res* 62:589–96.

Rayburn, E.R. and Zhang, R. 2008. Antisense, RNAi, and gene silencing strategies for therapy: Mission possible or impossible? *Drug Discovery Today* 13:513–21.

Rigo, F., Chun, S.J., Norris, D.A. et al. 2012. Adult and juvenile monkey pharmacokinetics (PK) of a uniformly modified 2′-(2-Methoxyethyl) antisense oligonucleotide (ASO; ISIS-SMNRx) in development for treatment of spinal muscular atrophy (SMA). *AAPS Annual Meeting and Exposition*, Chicago 14–18.

Rigo, F., Chun, S.J., Norris, D.A. et al. 2014. Pharmacology of a central nervous oligonucle-otide in mice and nonhuman primates. *J Ther* 350:46–55.

Rittner, K., Burmester, C., and Sczakiel, G. 1993. *In vitro* selection of fast-hybridizing and effective antisense RNAs directed against the human immunodeficiency virus Type 1. *Nucl Acids Res* 21:1381–7.

Rudin, C.M., Marshall, J.L., Huang, C.H. et al. 2004. Delivery of a liposomal c-raf-1 anti-sense oligonucleotide by weekly bolus dosing in patients with advanced solid tumors: A phase I study. *Clin Cancer Res* 10:441–51.

Saini, S.S. and Klein, M.A. 2011. Targeting cyclin D1 in non-small cell lung cancer and meso-thelioma cells by antisense oligonucleotides. *Anticancer Res* 31:3683–90.

Saonere, J.A. 2011. Antisense therapy, a magic bullet for the treatment of various diseases: Present and future prospects. *Journal of Medical Genetics and Genomics* 3:77–83.

Sasaki, H., Sheng, Y.L., Kotsuji, F. et al. 2000. Down-regulation of X-linked inhibitor of apoptosis protein induces apoptosis in chemoresistant human ovarian cancer cells. *Cancer Res* 60:5659–66.

Schultz, R.G. and Gryaznov, S.M. 1996. Oligo-2′-fluoro-2′-deoxynucleotide N3′ → P5′ phos-phoramidates: Synthesis and properties. *Nucleic Acids Res* 24:2966–73.

Schultze, P., Macaya, R., and Feigon, J. 1994. Three-dimensional solution structure of the thrombin-binding DNA aptamer D(GGTTGGTGTGGTTGG). *J Mol Biol* 235:1532–47.

Sczakiel, G. 2000. Theoretical and experimental approaches to design effective antisense oligonucleotides. *Front Biosci* 5:d194–d201.

Shogan, B., Kruse, L., Mulamba, G.B. et al. 2006. Virucidal activity of a GT-Rich oligo-nucleotide against herpes simplex virus mediated by glycoprotein B. *J Virol* 80:4040–7.

Simões-Wüst, A.P., Schürpf, T., Hall, J. et al. 2002. Bcl-2/bcl-xL bispecific antisense treat-ment sensitizes breast carcinoma cells to doxorubicin, paclitaxel and cyclophospha-mide. *Breast Cancer Res Treat* 76:157–66.

Siva, K., Covello, G. and Denti, M.A. 2014. Exon-Skipping antisense oligonucleotides to cor-rect missplicing in neurogenetic diseases. *Nucleic Acid Therapeutics* 24:69–86.

Skorski, T., Perrotti, D., Nieborowska-Skorska, M. et al. 1997. Antileukemia effect of C-Myc N3′ → P5′ phosphoramidate antisense oligonucleotides *in vivo*. *Proc Natl Acad Sci* 94:3966–71.

Smith, R.A., Miller, T.M., Yamanaka, K. et al. 2006. Antisense oligonucleotide therapy for neurodegenerative disease. *J Clin Invest* 116:2290–6.

Southwell, A.L., Skotte, N.H., Bennett, C.F. et al. 2012. Antisense oligonucleotide therapeu-tics for inherited neurodegenerative diseases. *Trends Mol Med* 18:634–43.

Stabile, L.P., Lyker, J.S., Huang, L. et al. 2004. Inhibition of human non-small cell lung tumors by a c-Met antisense/U6 expression plasmid strategy. *Gene Ther* 11:325–35.

Summerton, J. and Weller, D. 1997. Morpholino antisense oligomers: Design, preparation, and properties. *Antisense Nucleic Acid Drug Dev* 7:187–95.

Sun, S., Zhang, X., Tough, D. et al. 1998. Type I Interferon-mediated stimulation of T cells by Cpg DNA. *J Exp Med* 188:2335–42.

Tamm, I., Wang, Y., Sausville, E. et al. 1998. IAP-family protein survivin inhibits caspase activity and apoptosis induced by Fas (CD95), Bax, caspases, and anticancer drugs. *Cancer Res* 58:5315–20.

Tanabe, K., Kim, R., Inoue, H. et al. 2003. Antisense Bcl-2 and Her-2 oligonucleotide treatment of breast cancer cells enhances their sensitivity to anticancer drugs. *Int J Oncol* 22:875–81.

Tian, H., Wittmack, E.K., and Jorgensen, T.J. 2000. p21WAF1/CIP1 Antisense therapy radio-sensitizes human colon cancer by converting growth arrest to apoptosisl. *Cancer Res* 60:679–84.

Tonkinson, J. and Stein, C. 1994. Patterns of intracellular compartmentalization, trafficking and acidification of 5×-fluorescein labeled phosphodiester and phosphorothioate oligo-deoxynucleotides in HL60 cells. *Nucl Acids Res* 22:4268–75.

Toth, P.P. 2013. Emerging LDL therapies: Mipomersen—antisense oligonucleotide therapy in the management of hypercholesterolemia. *J Clin Lipidol* 7:S6–S10.

Toulme, J.J. 1996. Ribonuclease H: From enzymes to antisense effects of oligonucleotides. In *DNA and RNA Cleavers and Chemotherapy of Cancer and Viral Diseases*, ed. B. Meunier, 271–288. Amsterdam, Netherlands: Kluwer Academic Publishers.

Tyler, B., Jansen, K., McCormick, D. et al. 1999. Peptide nucleic acids targeted to the neurotensin receptor and administered I.P. cross the blood-brain barrier and specifically reduce gene expression. *Proc Natl Acad Sci* 96:7053–8.

Vaerman, J.L., Moureau, P., Deldime, F. et al. 1997. Antisense oligodeoxyribonucleotides suppress hematologic cell growth through stepwise release of deoxyribonucleotides. *Blood* 90:331–9.

Vester, B. and Wengel, J. 2004. LNA (locked nucleic acid): High-affinity targeting of complementary RNA and DNA. *Biochemistry* 43:233–41.

Vickers, T.A., Griffith, M.C., Ramasamy, K. et al. 1995. Inhibition of NF κ B specific transcriptional activation by PNA strand invasion. *Nucl Acids Res* 23:3003–8.

Vollmer, J., Jepsen, J., Uhlmann, E. et al. 2004. Modulation of Cpg oligodeoxynucleotide-mediated immune stimulation by locked nucleic acid (LNA). *Oligonucleotides* 14:23–31.

Wagner, R., Matteucci, M., Grant, D. et al. 1996. Potent and selective inhibition of gene expression by an antisense heptanucleotide. *Nat Biotechnol* 14:840–4.

Wang, W., Chen, H., Sun, J. et al. 1998. A comparison of guanosine-quartet inhibitory effects versus cytidine homopolymer inhibitory effects on rat neointimal formation. Antisense nucleic acid. *Drug Dev* 8:227–36.

Warfield, K.L., Swenson, D.L., Olinger, G.G. et al. 2006. Gene-specific counter measures against Ebola virus based on antisense phosphorodiamidate morpholino oligomers. *Plos Pathogens* 2:5–13.

Webb, A., Cunningham, D., Cotter, F. et al. 1997. BCL-2 antisense therapy in patients with non-Hodgkin lymphoma. *Lancet* 349:1137–41.

Wilds, C. and Damha, M. 2000. 2′-Deoxy-2′-fluoro-beta-d-arabinonucleosides and oligonucleotides (2′F-ANA): Synthesis and physicochemical studies. *Nucl Acids Res* 28:3625–35.

Williamson, J., Raghuraman, M., and Cech, T. 1989. Monovalent cation-induced structure of telomeric DNA: The G-Quartet model. *Cell* 59:871–80.

Wilton, S., Fall, A., Harding, P. et al. 2007. Antisense oligonucleotide-induced exon skipping across the human dystrophin gene transcript. *Mol Ther* 15:1288–96.

Wolfe, M.S. 2014. Targeting mRNA for Alzheimer's and related dementias. *Scientifica* 2014:1–13.

Wu, H.P., Feng, G.S., Liang, H.M. et al. 2004. Vascular endothelial growth factor antisense oligodeoxynucleotides with lipiodol in arterial embolization of liver cancer in rats. *World J Gastroenterol* 10:813–8.

Wyatt, J., Vickers, T., Roberson, J. et al. 1994. Combinatorially selected guanosine-quartet structure is a potent inhibitor of human immunodeficiency virus envelope-mediated cell fusion. *Proc Natl Acad Sci* 91:1356–60.

Xin, H., Sha, X., Jiang, X. et al. 2012. The brain targeting mechanism of Angiopep-conjugated poly(ethyleneglycol)-co-poly(Iμ-caprolactone) nanoparticles. *Biomaterials* 33:1673–81.

Yang, B., Yu, R., Tuo, S. et al. 2012. Antisense oligonucleotide against hTERT (Cantide) inhibits tumor growth in an orthotopic primary hepatic lymphoma mouse model. *Plos One* 7:e41467.

Yang, H. 2010. Nanoparticle-mediated brain-specific drug delivery, imaging, and diagnosis. *Pharm Res* 27:1759–71.

Yang, S.P., Song, S.T., Tang, Z.M. et al. 2003. Optimization of antisense drug design against conservative local motif in stimulant secondary structures of HER-2 mRNA and QSAR analysis. *Acta Pharmacol Sin* 24:897–902.

Yasui, M., Yamamoto, H., Ngan, C.Y. et al. 2006. Antisense to cyclin D1 inhibits vascular endothelial growth factorstimulated growth of vascular endothelial cells: Implication of tumor vascularization. *Clin Cancer Res* 12:4720–9.

Yu, C. and Xu, Z. 2015. Antisense oligonucleotides targeting human telomerase mRNA increases the radiosensitivity of nasopharyngeal carcinoma cells. *Mol Med Rep* 11:2825–30.

Yu, R.Z., Geary, R.S., Monteith, D.K. et al. 2004. Tissue disposition of 2'-O-(2- mehtoxy) ethyl modified antisense oligonucleotides in monkeys. *J Pharm Sci* 93:48–59.

Yu, R.Z., Grundy, J.S., and Geary R.S. 2013. Clinical pharmacokinetics of second generation antisense oligonucleotides. *Expert Opin Drug Metab Toxicol* 9:169–82.

Yu R.Z., Kim, T.W., Hong, A. et al. 2007. Cross-species pharmacokinetic comparison from mouse to man of a second-generation antisense oligonucleotide, ISIS 301012, targeting human apolipoprotein B-100. *Drug Metab Dispos* 35:460–8.

Yu, R.Z., Lemonidis, K.M., Graham, M.J. et al. 2009. Cross-species comparison of *in vivo* PK/PD relationships for second-generation antisense oligonucleotides targeting apolipoprotein B-100. *Biochem Pharmacol* 77:910–9.

Zemany, L., Bhanot, S., Peroni, O.D. et al. 2015. Transthyretin antisense oligonucleotides lower circulating 1 RBP4 levels and improve insulin sensitivity in obese mice. *Diabetes* 64(5):1603–14.

Zhang, C., Newsome, J.T., Mewani, R. et al. 2009. Systemic delivery and pre-clinical evaluation of nanoparticles containing antisense oligonucleotides and siRNAs. *Methods Mol Biol* 480:65–83.

Zhang, H., Cook, J., Nickel, J. et al. 2000. Reduction of liver Fas expression by an antisense oligonucleotide protects mice from fulminant hepatitis. *Nat Biotechnol* 18:862–7.

Zuker M. 2003. Mfold web server for nucleic acid folding and hybridization prediction. *Nucl Acids Res* 31:3406–15.

# 13 Biodegradable Polymeric Carriers for Delivery of siRNA

*Sanika A. Rege, Sudip K. Das, and Nandita G. Das*

## CONTENTS

## 13.1 INTRODUCTION

The concept of gene therapy has now been around for several decades. The Human Genome Project brought hope of unique treatments for hitherto incurable diseases, cancers, genetic disorders, etc. The idea of introducing foreign genetic material into cells/tissues/organs to correct/treat hereditary or acquired diseases is straightforward, but the translation of this concept into clinical practice has been less than successful mainly because of two reasons: suboptimal efficacy or unacceptable toxicities, and the mode of drug delivery could be a contributor to either or both aforementioned issues. With the focus shifting from total correction to partial but effective correction with potentially lower toxicity, ribonucleic acid (RNA) interference, primarily aiming at the destruction of specific mRNA molecules, emerged as a fresh therapeutic approach in the late 1990s. Initially, researchers utilized double-stranded RNAs with more than 30 base pairs (bps) for specific gene silencing, but these were found to cause undesirable immune responses in mammalian cells. Using short 21 bp, siRNAs have now become a promising approach for the specific, posttranscriptional knockdown of disease-causing genes (Elbashir et al. 2001; Tuschl 2001). However, siRNAs possess unfavorable biopharmaceutical properties. Rapid degradation by serum nucleases, combined with low membrane penetration capabilities owing to their negative charge, lead to insufficient concentration in the cellular cytoplasm, which is the target location for RNA interference (RNAi) (Kurreck 2009). siRNAs do not readily pass cellular membranes due to their negative charge and are rapidly degraded by serum nucleases. Despite considerable progress in the field, improving the safety and effectiveness of nucleic acid carriers for their application in human therapy remains a major challenge. The primary limitation to clinical success, however, is the lack of efficient delivery systems that can deliver the siRNA to the target cell population.

Both viral and nonviral delivery vectors have been reported to aid in the delivery of siRNA (Guo et al. 2010). Viral vectors have high transfection efficiency, but are plagued with production, cost, and safety issues. Common concerns regarding the clinical usage of viral vectors are their immunogenicity, potential infectivity, and inflammation at the delivery site. For these reasons, nonviral vectors have gained an increasing interest in gene therapy research over the course of the past two decades (Li and Huang 2006; Schaffert and Wagner 2008). Nonviral vectors, in particular, offer many advantages over their viral counterparts, including improved safety, low immune response that enables repeated use, and ease of production (Zhang et al. 2007; Whitehead et al. 2009). There are two main chemical approaches used in nonviral vector delivery systems, in which siRNA could either be (i) conjugated to carrier molecules such as lipid, peptide, and polymer, or (ii) encapsulated into a cationic delivery system (Lee et al. 2007; Wolfrum et al. 2007).

The top three categories of nonviral vectors include liposomes, complexes of negatively charged plasmid with cationic polymers, and nanoparticles. Vesicle-like carriers such as liposomes typically exhibit relatively low encapsulation efficiency, poor storage stability, and rapid clearance from the blood. Therefore, nonviral systems based on cationic polymers containing several amine groups in their backbone have been used extensively as an alternate form of the gene carrier (Li and Huang 2000). Among all these approaches, much attention has been given to the design of polymeric nanoparticles because biodegradable nanoparticles can safely transport the genetic materials without exhibiting detrimental toxicity and immune responses, and can be produced on a large scale. An ideal polymeric carrier should form a stable complex with nucleic acids to maintain its stability in the biological solution, hide from the host surveillance systems, and deliver the therapeutic nucleic acid to the desired cell population by recognizing a specific characteristic on the cell surface. After being localized within the cells, the carrier should provide appropriate functionalities to escape from the endocytic pathways, deliver the complexes to the vicinity of the targets, perform decomplexation (or unpacking) in response to a certain intracellular environment, such as pH or redox potential, and actively transport the nucleic acids to the target (Jeong et al. 2007).

Cationic polymers have widely been investigated for siRNA delivery due to their great flexibility, ease of manufacturing, and modification possibilities (Howard 2009). Due to their positive charge, cationic polymers are able to complex with anionic nucleic acids to form polyplexes (Thomas and Klibanov 2003). However, many of these cationic polymers show considerable toxicity in both *in vitro* cultured cells and *in vivo* animal models. Typical examples include poly (ethylenimine) (PEI) (Urban-Klein et al. 2005), poly (L-lysine) (PLL) (Sato et al. 2007), chitosan (Howard et al.

2006), and poly (2-(dimethylamino) ethylmethacrylate) (pDMAEMA) (Varkouhi et al. 2011a). The N/P ratio, which is the molar ratio of amines in the polymer backbone to phosphates in the nucleic acid, is an important factor to be considered while formulating stable nanoparticles for siRNA delivery.

A recent study by Zheng et al. (2012) noted that high N/P ratios might not be optimal for efficient nanoparticle formation. Their study found that stable nanoparticles were formed at a high N/P ratio of 20 as well as at a low N/P ratio of 2. When comparing knockdown levels, N/P ratios of 2 and 20 mediated similar knockdown efficiency, but cytotoxicity was negligible at the low N/P ratio compared to around 25% at the higher N/P ratio of 20, thus demonstrating the potential utility of low N/P ratios for polycation-mediated siRNA delivery. Structures of commonly used biodegradable polymers for nucleic acid delivery are depicted in Figure 13.1.

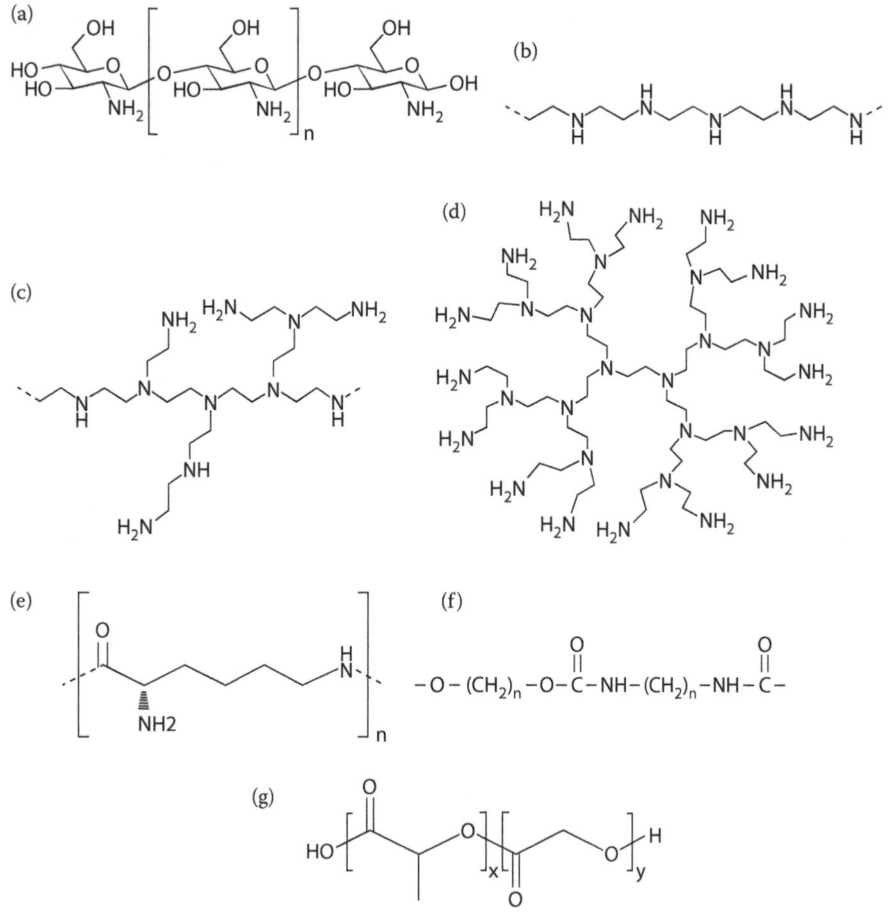

FIGURE 13.1 Structures of commonly used biodegradable polymers. (a) Chitosan. (b) Linear backbone of PEI. (c) Branched backbone of PEI. (d) Dendrimeric PEI generation 4. (e) Poly (L-lysine). (f) Polyurethane. (g) Poly (lactic-co-glycolic acid). x = number of units of lactic acid; y = number of units of glycolic acid.

## 13.2   BIODEGRADABLE POLYMERIC CARRIERS

### 13.2.1   Chitosan

Chitosan [(1 → 4) 2-amino-2-deoxy-β-D-glucan] is a linear polysaccharide composed of randomly distributed N-acetyl-D-glucosamine (GlcNAc) and β-(1, 4)-linked D-glucosamine (GlcN) (Rojanarata et al. 2008; Lai and Lin 2009). Chitosan is relatively reactive and can be produced in various forms such as powder, paste, film, fiber, etc. (Agnihotri et al. 2004). Commercially, chitosan is derived from chitin, the structural element in the exoskeleton of crustaceans (crabs, shrimp, etc.) and cell walls of fungi. Chitosan is formed via alkaline deacetylation of chitin, which is performed by decolorizing crab and shrimp shells with potassium permanganate followed by boiling in sodium hydroxide (Lai and Lin 2009; Mao et al. 2010). Chitosan is a weak base with a $pK_a$ value of the D-glucosamine residue of approximately 6.2–7.0, and is therefore insoluble at neutral and alkaline pH, but soluble in acid media such as acetic acid, citric acid, glutamic acid, aspartic acid, hydrochloric acid, and lactic acid (Rojanarata et al. 2008). Chitosan has been used in drug delivery as an absorption enhancer (Kumar et al. 2004) as well as a vector for gene delivery. Compared to many other natural polymers, chitosan has a net positive charge and is mucoadhesive (Agnihotri et al. 2004). It is known to be biocompatible with low toxicity, low immunogenicity, and degradable by enzymes (Rao and Sharma 1997; Varum et al. 1997; Kumar et al. 2004), and has therefore been proposed as a safer alternative to other nonviral vectors, such as cationic lipids and cationic polymers (Rojanarata et al. 2008). In addition, chitosan is capable of encapsulating siRNA to protect it from enzymatic degradation (Rudzinski and Aminabhavi 2010). Owing to the interaction of anionic siRNA and cationic chitosan, siRNA can be readily complexed and loaded into chitosan nanoparticles.

The proportion of GlcNAc and D-glucosamine determines the degree of deacetylation of the polymer. By affecting the number of protonatable amine groups, the degree of deacetylation fundamentally determines the critical polymer properties including solubility, hydrophobicity, and the ability to interact electrostatically with polyanions (Kiang et al. 2004; Huang et al. 2005). At acidic pH below its $pK_a$, primary amines of glucosamine-repeating units are protonated conferring positive charges on the chitosan backbone. These protonated amines provide the site for chitosan to bind to the negatively charged deoxyribonucleic acid (DNA) or RNA and self-assemble into particles by simple mixing (Techaarpornkul et al. 2010). A high degree of deacetylation of chitosan will increase its positive charge enabling greater binding with siRNA (Rao and Sharma 1997). The percentage of deacetylated primary amine groups along the chitosan chain determines the potential, while the positive charge density is responsible for the electrostatic interaction with negatively charged siRNA (Rao and Sharma 1997). The positive surface charge of chitosan allows it to interact with mucus, negatively charged mucosal surfaces, as well as the plasma membrane (Mansouri et al. 2004; Rojanarata et al. 2008). The electrostatic interaction of the amine groups of chitosan with the anionic mucus layer also enhances its mucoadhesive property (Rao and Sharma

1997). Moreover, the configuration of chitosan is fully displaced under acidic pH, and it is in this configuration that chitosan can trigger the opening of tight junctions, and hence enhances the paracellular transport of hydrophilic agents (Rojanarata et al. 2008).

Commercially available chitosan has an average molecular weight (MW) ranging between 3800 and 20,000 Da and is 66%–95% deacetylated (Lai and Lin 2009). Generally, chitosans with lower MWs and lower degrees of deacetylation exhibit greater solubility and faster degradation compared to their high-MW counterparts (Zhang and Neau 2001, 2002; Köping-Höggård et al. 2004; Ren et al. 2005). Chitosan can be degraded by enzymes that hydrolyze glucosamine–glucosamine, glucosamine-$N$-acetyl-glucosamine, and $N$-acetyl-glucosamine-$N$-acetyl-glucosamine linkages. Some of these enzymes include lysozyme, β-$N$-acetyl hexosaminidase, chitosanase, chitinase, and chitin deacetylase (Garcia-Alonso et al. 1990; Watanabe et al. 1999; Wiwat et al. 1999; Martinou et al. 2002; Andronopoulou and Vorgias 2003; Fu et al. 2003; Sanon et al. 2005; Akeboshi et al. 2007; Miyauchi et al. 2007). Lysozyme has been found to efficiently degrade chitosan; 50% acetylated chitosan had 66% loss in viscosity after a 4-h incubation *in vitro* at pH 5.5 (0.1 M phosphate buffer, 0.2 M NaCl, 37°C) (Onishi and Machida 1999). This degradation appears to be dependent on the degree of acetylation (more chitin like) resulting in the fast rate (Zhang and Neau 2002; Senel and McClure 2004). Other modifications such as covalent crosslinking and thiolation have been shown to alter degradation profiles (Kafedjiiski et al. 2007; Lu et al. 2009). It is somewhat unclear what the mechanism of degradation is when chitosan is injected intravenously. It is likely that the distribution, degradation, and elimination processes are strongly dependent on the MW and degree of acetylation of the chitosan. Possible sites of degradation, inferred due to the localization of chitosan, may be the liver and kidney. In one of the few studies reported, chitosan oligosaccharides were found to upregulate lysozyme activity in the blood of rabbits injected intravenously with 7.1–8.6 mg/kg chitosan (Hirano et al. 1991; Guo et al. 2006). After oral administration, some degradation of chitosan is expected to occur in the gastrointestinal tract. The digestion of chitosan has been found to be species dependent with hens being shown to be more efficient digesters (67%–98% degradation after oral ingestion) than rabbits (39%–83% degradation). In the same study, the digestion of $N$-stearoyl chitosan was negligible, indicating that the enzymatic degradation is dependent on chitosan's $-NH_2$ availability (Sarmento and Neves 2012). The studies discussed above show that chitosan degradation rate can be affected by the polymer's MW and degree of acetylation. Further, N-substitution may affect enzymatic degradation and this should be considered when new derivatives are suggested for systemic administration. It must be noted that the degradation mechanism of chitosan (and derivatives) is not fully understood when used *in vivo* but there may be adaptive mechanisms, which increase its degradation over time (Kean and Thanou 2010).

The biodistribution of chitosan is related to all aspects of the chitosan formulation from the MW and degree of deacetylation to the nanoparticle size. In the case of a nanoparticulate formulation, the kinetics and biodistribution will initially be controlled by the size and charge of the nanoparticles and not by the chitosan's composition. However, after particle decomposition to chitosan and the free drug, inside the cells or in the target tissue, free chitosan will distribute in the body and eliminate accordingly. The degradation profile of chitosan can be modified in such a way that it is possible to create formulations that exhibit quick release characteristics, or long-acting dosage forms, which are affected, by chitosan kinetics, metabolism, and excretion (Sarmento and Neves 2012). The effects of chitosan with differing MWs and the degree of deacetylation on Caco-2 cells, HT29-H cells, and *in situ* rat jejunum were studied. The toxicity of chitosan was found to be dependent on the degree of deacetylation and MW. At high degrees of deacetylation, the toxicity is related to the MW and the concentration, whereas at lower degrees of deacetylation, toxicity is less pronounced and less related to the MW (Schipper et al. 1996, 1999).

A comparison of the $IC_{50}$ values of chitosan and its derivatives indicate that most chitosans (and derivatives) are less toxic compared to the popular cationic polymers such as PEI ($LD_{50} < 20$ μg/mL against MCF7 and COS7 cells) (Kean et al. 2005). It appears that the toxicity of chitosan is related to the charge density of the molecule, with the toxicity increasing with the increasing charge density. There is a threshold level below which there are too few contact points between a molecule and the cell components to produce a significant toxic effect. *In vivo* toxicity, particularly after long-term administration, is of high importance for the design of drug delivery forms based on chitosan (Sarmento and Neves 2012). In a relatively long-term study (65 days), no detrimental effect on body weight was found when chitosan oligosaccharides were injected (7.1–8.6 mg/kg over 5 days) in rabbits (Kafedjiiski et al. 2007). Rao and Sharma (1997) reported no significant toxic effects of chitosan in acute toxicity tests in mice, and no eye or skin irritation in rabbits and guinea pigs, respectively. In the same study, it was also concluded that chitosan was not pyrogenic.

From a survey of the current studies, it could be inferred that, when used in the appropriate conditions, chitosan exhibits low toxicity in the μg/mL range (Sarmento and Neves 2012). However, regulatory agencies continue to encounter difficulties in approving all existing chitosans as generally recognized as safe (GRAS) materials. At this time, all chitosan GRAS applications are "At notifier's request, FDA ceased to evaluate the notice." Chitosan's chemical versatility and the variety of formulations confuse researchers and regulatory scientists. This is one reason why a thorough description of the specific chitosan used in a study should be included. A systematic study on biodistribution, *in vivo,* and *in vitro* toxicity using various chitosans and derivatives would provide data that could help correlate chitosan's structure and safety profile (Sarmento and Neves 2012).

To achieve an efficient siRNA delivery to overcome P-glycoprotein (P-gp) mediated multidrug resistance in cancer, Yhee et al. (2015) incorporated P-gp-targeted self-polymerized 5′-end thiol-modified siRNA (poly-siRNA) entrapped in tumor-targeted glycol chitosan nanoparticles. The Pgp-targeted poly-siRNA (psi-Pgp) and thiolated glycol chitosan polymers (tGC) formed stable nanoparticles (psi-Pgp-Tgc NPs), and the resulting nanoparticles successfully protected siRNA from enzymatic degradation (Figure 13.2).

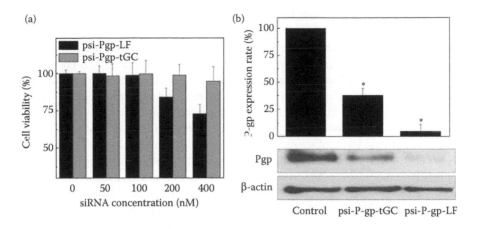

**FIGURE 13.2** (a) Cytotoxicity of psi-Pgp-tGC NPs and psi-Pgp-lipofectamine (LF). (b) Pgp downregulation in psi-Pgp-tGC NP-treated cells using Western blotting. LF 2000 was used as a positive control. (Reprinted from *J Control Release*, 198, Yhee, J.Y. et al., Cancer-targeted MDR-1 siRNA delivery using self-cross-linked glycol chitosan nanoparticles to overcome drug resistance, 1–9, Copyright 2015, with permission from Elsevier.)

### 13.2.2 POLYETHYLENIMINE

PEI is a commercially available synthetic cationic polyamine. It was first introduced by Boussif et al. (1995) and remains one of the most successful and widely studied gene delivery agents (Godbey et al. 1999b; Kunath et al. 2003; Lungwitz et al. 2005). It has been extensively used as a carrier for oligonucleotides, plasmids, and siRNA (Leng et al. 2009). Due to the high cationic charge density, PEI can compact DNA and siRNA into complexes that are effectively taken up in cells to make nucleic acid delivery and gene therapy possible (Godbey et al. 1999d; Urban-Klein et al. 2005; Yhee et al. 2015). Noncovalent complexation of synthetic siRNA with low-MW PEI stabilizes siRNA and enhances its intracellular delivery (Godbey et al. 1999d). The high positive charge density of PEI causes a strong electrostatic interaction with negatively charged polyanions such as pDNA, ODNs, and siRNA to form a small and complex (polyplex) that confers protection to the nucleic acids from degradation by nuclease activity. Every third atom of PEI is a protonable amino nitrogen atom, which makes the polymeric network an effective "proton sponge" at virtually any pH. Thus, PEI acts as an excellent transfection reagent that could facilitate endosomal escape after entering the cells as it acts as a "proton sponge" during acidification of the endosome (Godbey and Mikos 2001; Dailey et al. 2004). In addition to its ability to condense nucleic acids, the pH-buffering property of PEI disrupts endosomes, thereby enabling nucleic acids to reach the cytosol. Although plasmids must still reach the nucleus, siRNA only needs to reach the cytosol where the RNA-induced silencing complex (RISC) is formed to degrade mRNA (Boussif et al. 1995; Behr 1997; Ferrari et al. 1999; Godbey et al. 2000). Although the polymer itself is used as a delivery vehicle, PEI can also be attached to the nanoparticle surface through covalent and electrostatic interactions to achieve the same goal (McBain et al. 2007; Fuller et al. 2008; Park et al. 2008; Elbakry et al. 2009; Liong et al. 2009). Complexing the polymer with the nanoparticles has the potential advantage of facilitating DNA and siRNA delivery by a multifunctional platform that also allows imaging, targeting, and concurrent drug delivery. PEI–DNA or PEI–siRNA complex can be lyophilized in 5% glucose without the loss of transfection efficacy and produces efficient transfection *in vivo* after systemic and regional delivery (Werth et al. 2006; Thomas et al. 2007). For example, intrathecal injection of a branched PEI (B-PEI)–siRNA complex targeting the pain receptor NMDA-R2B decreased the mRNA and protein levels of the receptor and abolished formalin-induced pain in rats (Tan et al. 2005). Intraperitoneal administration of PEI–siRNA targeting the cerbB2/neu (HER-2) receptor resulted in siRNA delivery to the tumor, receptor downregulation, and significant reduction of tumor growth, whereas naked siRNA failed to produce comparable effects (Figure 13.3) (Urban-Klein et al. 2005).

PEI is highly soluble in water with a high capacity for positive charge formation (Kircheis et al. 2001a; Chirila et al. 2002; Panyam and Labhasetwar 2003) and is available in both linear and branched forms with MWs ranging from 1 to 1000 kDa (Godbey and Mikos 2001; Urban-Klein et al. 2005). The branched form of PEI contains primary, secondary, and tertiary amines whereby each of them has the potential to be protonated, and this has been used as the standard for gene delivery, due to its greater success in cell transfection compared to the linear form (Godbey et al. 1999b). B-PEI (800 kDa) conjugated to ligands such as transferrin has been shown to transfect well *in vitro* and *in vivo* (Kircheis et al. 1997, 1998; Ogris et al. 1998, 1999). Recently, the application of linear PEI or low molecular PEI as a siRNA

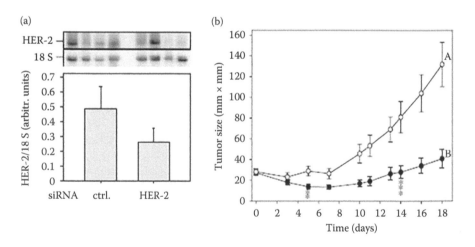

**FIGURE 13.3**   (a) After 2 weeks, northern blotting revealed an ~50% reduction of HER-2 mRNA levels in the treatment group. (b) Intraperitoneal (IP) injection of naked HER-2-specific siRNAs fails to exert an inhibitory effect on tumor growth (A, open circles) while PEI complexation of HER-2-siRNA (B, closed circles) results in a significantly reduced tumor growth. (Reprinted by permission from Springer Nature. *Gene Ther*, Urban-Klein, B. et al. RNAi-mediated gene-targeting through systemic application of polyethylenimine (PEI)-complexed siRNA *in vivo* 12:461–6. Copyright 2005.)

delivery vector has become hotly debated (Boussif et al. 1995; Hassani et al. 2005; Thomas et al. 2005; Niola et al. 2006). Among a variety of options in terms of MWs (0.8–800 kDa) and topological isomers (branched or linear structures), both linear high-molecular-weight PEI (L-HMW PEI) (e.g., 22 kDa) and branched high-molecular-weight PEI (B-HMW PEI) (e.g., 25 kDa) have demonstrated sufficient efficiency. Although both polymers have shown high gene delivery efficiency *in vitro* (Thomas et al. 2005), L-HMW PEI has shown much higher *in vivo* gene expression with less cytotoxicity in comparison to B-PEI (Goula et al. 1998a,b; Bragonzi et al. 1999; Zou et al. 2000; Brissault et al. 2006). However, linear PEI shows limited efficiency in siRNA delivery, although the reason for such a discrepancy between DNA and siRNA delivery efficiencies is unclear (Hassani et al. 2005; Richards Grayson et al. 2006).

Hydrophobic modifications of PEI have shown improved particle stability. Modification of 600-Da PEI with alkyl chains of 8–13 carbons led to the formation of stable nanoparticles, whereas unmodified PEI did not form particles under the same conditions. This result was further confirmed by gene-silencing studies, in which the alkyl-modified PEI showed greater than 80% gene knockdown, whereas unmodified PEI did not mediate any gene silencing. The hydrophobic modifications increased the toxicity significantly, especially at N/P ratios greater than 12, possibly owing to the enhanced ability for membrane disruption (Guo et al. 2013).

Werth et al. (2006) used gel permeation chromatography to fractionate a commercially available 25-kDa PEI, whereby they obtained a low-molecular-weight polyethylenimine (PEI F25-LMW) with superior transfection efficacy and low toxicity in various cell lines. Results showed that PEI F25-LMW was able to complex and fully protect siRNAs against nucleolytic degradation, and deliver siRNAs into cells where they displayed bioactivity. Upon lyophilization and reconstitution of PEI F25-LMW-based siRNA complexes, siRNAs were still able to efficiently induce RNAi. Treatment of PC-3 prostate carcinoma cells with fresh (or with lyophilized) complexes resulted in decreased cell proliferation in different assays due to the siRNA-mediated downregulation of vascular endothelial growth factor (VEGF) (Werth et al. 2006).

As is the case with other cationic polymers, PEI is associated with dose-dependent toxicity, which probably explains why it has not yet been used in clinical studies. Many factors affect the efficiency/cytotoxicity profile of PEI polyplexes such as MW, degree of branching, ionic strength of the solution, zeta potential, and particle size (Kircheis et al. 1999; Kunath et al. 2003). One study, for instance, showed that low MW (10 kDa) moderately branched polymer resulted in efficient delivery with low toxicity in comparison with the commercial high-MW PEI (Fischer et al. 1999; Godbey et al. 1999c). It is documented that, while low-MW PEI is not cytotoxic, these polymers are ineffective at transfecting nucleotides in contrast to the high-MW PEI. In this regard, it has been demonstrated that the size (MW), compactness, and chemical modification of PEI affect the efficacy and toxicity of this polymer (Florea et al. 2002; Neu et al. 2005). Godbey et al. (1999a) reported that there are mainly two types of cytotoxicities in the process of PEI-mediated cell transfection. One is an immediate toxicity associated with free PEI, while the other is a delayed toxicity associated with cellular processing of PEI/DNA complexes (Godbey et al. 2001).

When administered in the circulatory system, the free PEIs interact with negatively charged serum proteins (such as albumin) and red blood cells, precipitate in huge clusters, and adhere to the cell surface (Fischer et al. 1999). This effect could destabilize the plasma membrane and induce the immediate toxicity.

Although unmodified B-PEI nanoparticles have frequently been used for cell culture transfection experiments, a few studies have demonstrated the use of unmodified PEI nanoparticles in animal models. For example, Ge et al. (2004) found that systemically delivered PEI–siRNA nanoparticles inhibited influenza virus in mouse lungs. When 60 μg of siRNA were injected i.v., there was a 10-fold reduction in virus titers in the lungs and when 120 μg of siRNA were used, more than a 1000-fold reduction in lung virus titers was observed in some mice. Despite this apparent success, branched 25 kDa PEI in complex with nucleic acids is known to be toxic to the lungs and it is likely that some of the reduction in viral titers in the lungs may have been due to toxicity of the PEI polyplex itself. Moreover, the commonly used branch PEI in cell culture transfection experiments and/or siRNA delivery is generally known to be toxic to most cells. Thus, there is significant concern regarding the toxicity of nanoparticles formed from siRNA and unmodified PEI with high molecular masses (e.g., 25 kDa) and doses, and thus limits the clinical use of high MW unmodified PEI (Kircheis et al. 1999, 2001b). It has been reported that modifications of PEI, such as polyethylene glycol (PEG) grafting or biodegradable PEIs, may cause reduction of complex charges and much less cytotoxicity than PEI after applying them at high concentration to mouse fibroblasts (Petersen et al. 2002a,b).

### 13.2.3 POLY-L-LYSINE

PLL is a cationic polymer that has been commonly used as the gene carrier and is one of the earliest cationic polymers employed for gene transfer (Wu and Wu 1987; Yamagata et al. 2007). Until 2000, PLL was one of the most used cationic polymers for DNA delivery (Wu and Wu 1987; Wolfert et al. 1999). It is a linear polypeptide with the amino acid lysine as the repeat unit; thus, it possesses biodegradable characteristics that become very useful for *in vivo* applications. However, when entered into the circulatory system, PLL polyplexes were rapidly bound to plasma proteins and cleared from the circulation (Ziady et al. 1999; Ward et al. 2001), which could lead to lower transfection efficiency. Meyer et al. (2009) demonstrated that a siRNA-polymer conjugate consisting of PLL, PEG, and dimethylmaleic anhydride (DMMAn) showed high *in vitro* biocompatibility and efficient sequence specific gene silencing at around ≥25-nM siRNA, comparable to the corresponding electrostatic polyplexes (Figure 13.4).

The effect of the MW of PLL on the efficiency of gene delivery has been examined. It was shown that the longer the chain length of the L-lysine entity (hundreds of thousands of kDa), the more stable the complex was in blood, and the longer the gene expression lasted *in vitro* and *in vivo* (Wu and Wu 1987; Ward et al. 2001). Many PLL polymers with different MWs were tested and evaluated for gene transfer (Nishikawa et al. 1998; Mannisto et al. 2002; Ward et al. 2002). It has been shown that DNA condensation and transfection efficiency increased with high-MW PLL, which was also associated with undesirable high toxicity (Wolfert et al. 1999). Low-MW (<3 kDa) PLL cannot form stable complexes (Kwoh et al. 1999). Furthermore, it appears that

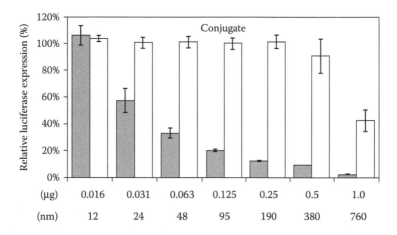

**FIGURE 13.4** siRNA gene-silencing efficiency of PEG–PLL–DMMAn–Mel-siRNA. Luciferase knockdown activity of the siRNA conjugate was compared with the knockdown activity of the counterpart containing the nonsilencing control of siRNA. The used amount of siRNA is indicated as (μg) per well and concentration (nM). Gray bars indicate the transfection with luciferase siRNA conjugates; white bars indicate the control of siRNA conjugates. (Reprinted with permission from Meyer, M. et al., Synthesis and biological evaluation of a bioresponsive and endosomolytic siRNA–polymer conjugate, *Mol Pharm*, 752–62. Copyright 2009 American Chemical Society.)

high-MW PLL is more suitable for gene delivery via systemic injection, with PLL 211 kDa/DNA complexes displaying levels in the blood up to 20-fold higher after 30 min compared to PLL 20 kDa/DNA complexes. Indeed, the complexes formed with low-MW PLL are fixed by the complement system *in vitro*, are less soluble *in vivo* (aggregation), and are thus rapidly removed by the Kupffer cells of the liver. Destabilization of these constructs in the blood is described as a possible mechanism for their removal from the blood circulation (Ward et al. 2001).

To obtain prolonged circulation in blood flow and targeted gene delivery *in vivo*, PEG modification of PLL and oligolysine has been performed. In the case of PEG-modified PLL, PLL forms a stable core-shell-type complex, that is, the condensed complex of PLL and the DNA chain forms the core, and the PEG coats the core to form the shell. Ligand modification of PEG-coated DNA complexes enables targeted gene delivery into target sites after systemic injection *in vivo* (Kwok et al. 1999, 2003; Collard et al. 2000; Harada-Shiba et al. 2002).

Ohsaki et al. (2002) and Okuda et al. (2004) reported that dendritic PLL of the sixth generation (KG6) showed high transfection efficiency, without significant toxicity or cell specificity. In another study by Yamagata et al. (2007), KG6 was advantageous for endosomal escape due to the proton sponge effect and for transcription of DNA, resulting from weak compaction of DNA by KG6; however, the total amount of DNA binding and KG6-mediated uptake into cells were lower than those with PLL. Like amphiphilic PEI, the creation of amphiphilic PLL, by linking both PEG and palmitoyl groups to the polymer, reduced toxicity without compromising the gene delivery efficiency (Brown et al. 2000). The PLL/DNA polyplexes are internalized in a way comparable to PEI/DNA complexes, but their transfection efficiency is weak due to a lack of amino groups allowing endosomolysis (Merdan et al. 2002). Meyer et al. (2009) used PLL as siRNA binding and protecting polycation to form a siRNA-polymer conjugate. The conjugate showed excellent *in vitro* knockdown activity, bioactivity, and biocompatibility. However, intravenous and intratumoral *in vivo* applications in mice showed unexpectedly high toxicity (Ziady et al. 1999). The advantages and disadvantages of different biodegradable polymers are provided in Table 13.1.

### 13.2.4 POLYURETHANE

Polyurethane (PUR) is a class of biodegradable polymers with urethane linkages in the backbone. It has been studied as a biomaterial in tissue engineering (Zhou and Yi 1999; Zhang et al. 2000; Guan et al. 2005). Recently, it has been demonstrated that cationic PUR can introduce DNA into cells (Yang et al. 2004). To study the crucial factors of polymers and complexes efficient for gene delivery, for example, size distribution of PUR/plasmid DNA (pDNA) complexes and structural effect of PUR, polymers with different structures of the PUR (e.g., tertiary amine groups in the side chain or backbone) have been prepared. Nelson et al. (2012) studied the application of PUR to deliver pH-responsive micellar NPs designed for the intracellular delivery of siRNA. They successfully incorporated siRNA nanocarriers within the PUR scaffold, which showed sustained, diffusion-controlled release of intact nanoparticles and maintenance of gene-silencing bioactivity of the released siRNA NPs. Modified PUR was developed by Tseng and Tang (2007) as a carrier for siRNA delivery. They demonstrated that PaEGU, an engineered polyester urethane (PEU) with amine and PEG units in the backbone, was able to associate with siRNA through electrostatic interaction that led to the self-assembly of polyplexes.

Porous, noncytotoxic, and biodegradable polyester PURs comprise a promising class of synthetic, injectable biomaterials that can provide both mechanical support and controlled drug release to regenerating tissues (Guelcher 2008). Further advantages of PURs are that they adhere to the tissue, do not stimulate inflammation (Li et al. 2009), and biodegrade into nontoxic by-products at rates that can be tuned based on the polyester triol and isocyanate precursor compositions (Li et al. 2009). Importantly, the use of lysine-derived polyisocyanates in the PUR scaffolds makes them more clinically translatable because they can be synthesized using a two-component foaming process that allows a short manipulation time for filling of any shape or size defect, followed by rapid curing *in situ* (Hafeman et al. 2008; Adolph et al. 2012).

### 13.2.5 POLY (DL-LACTIDE-*co*-GLYCOLIDE)

Poly(DL-lactide-*co*-glycolide) (PLGA) is an FDA-approved biodegradable polymer. PLGA nanoparticles have been used as a gene vector for siRNA and functional pDNA delivery in recent years. Compared with conventional carrier systems, the biodegradable polymeric nanoparticle system offers improved formulation stability, which is practically beneficial and may be used in future clinical studies of siRNA therapeutics (Yuan et al. 2006).

**TABLE 13.1**

**Advantages and Disadvantages of Biodegradable Polymers**

| Polymer | Advantages | Disadvantages |
|---|---|---|
| Chitosan | Abundant natural supply, safe, nontoxic and biocompatible and biodegradable in normal biological environment (Yamagata ct al. 2007). Control the release of encapsulated genes to a certain extent (Brown et al. 2000). Can enter the nucleus and deliver drugs/genes directly (Merdan et al. 2002). Able to overcome permeability barrier posed by the epithelium (Chandy and Sharma, 1990). Protects against enzymatic degradation (Zhang and Neau 2001). Possesses apoptotic properties of its own (Hasegawa et al. 2001). Mucoadhesive properties allow a cargo drug molecule increased interaction with membrane epithelium, thus leading to more efficient uptake (Rao and Sharma 1997; Lee et al. 2001). | Positively charged surface prevents interaction with other positively charged molecules. More studies have to be conducted regarding its drug/gene release profiles in view of gene therapy (Hasegawa et al. 2001; Tan et al. 2009). |
| Polyethyleneimine | High positive charge and proton buffering capacity forms strong complex with siRNA that helps in escaping lysosomal compartments and release the loaded gene (Varkouhi et al. 2011b; Cho et al. 2003). | Highly cytotoxic due to nonspecific interaction with negatively charged cell membranes (Illum et al. 1994; Tan et al. 2009). High molecular weight PEI is nonbiodegradable and cannot be cleared by the renal system after use. Leads to cumulative toxicity (Wang et al. 2015). |
| Poly-l-Lysine | Less toxic than PEI (Farrell et al. 2007). | No intrinsic proton buffering effect leading to low level of transgene expression (Tan et al. 2009). Highly cytotoxic due to nonspecific interaction with negatively charged cell membranes (Park et al. 2006). |
| Polyurethane | Biodegradable, nontoxic, do not stimulate inflammation (Cherng et al. 2013). | Low thermal stability and mechanical strength (Guelcher 2008). |
| Poly (DL-lactide-*co*-glycolide) | Nontoxic, biodegradable (Li et al. 2009). Improved formulation stability (Vasir and Labhasctwar 2007). | Non-modified PLGA shows inefficient endosomal escape that delays the release of encapsulated siRNA (Kwon 2011). |

Biodegradable PLGA nanoparticles were prepared to load siRNA oligonucleotides with the desired physicochemical properties. The green fluorescent protein (GFP) siRNA oligonucleotides were successfully loaded into PLGA nanoparticles and delivered to 293 T cells (Yuan et al. 2006, 2011). The siRNA can be encapsulated in the core of PLGA nanoparticles by the double-emulsion solvent evaporation method. An encapsulation efficiency of up to 57% was achieved by adjusting the inner water-phase volume, the PLGA concentration, the emulsification sonication time, and stabilization of the water–oil interface with serum albumin (Cun et al. 2010). Potential problems for PLGA nanoparticles are the efficient endosomal escape and timely release of encapsulated siRNA. Therefore, PLGA nanoparticles are sometimes modified to have more efficient endosomal escape and better release profile (Yuan et al. 2006). Small interfering RNA-loaded PLGA nanoparticles have been studied extensively for various types of cancer. PLGA nanoparticles were used to deliver the anticancer drug, paclitaxel, simultaneously with P-gp-targeted siRNA to overcome tumor drug resistance. The dual-agent combination nanoparticles were surface functionalized with biotin for active tumor targeting and showed significantly higher cytotoxicity *in vitro* compared to nanoparticles loaded with paclitaxel alone. *In vivo* studies in a mouse model of drug-resistant tumor demonstrated greater inhibition of tumor growth following the treatment with biotin-functionalized combination nanoparticles encapsulating both paclitaxel and P-gp-targeted siRNA at a paclitaxel dose that was ineffective in the absence of gene silencing (Yuan et al. 2006).

PLGA nanoparticles loaded with pDrive-sh AnxA2 p DNA were capable of sustained delivery of pDrive-sh AnxA2 pDNA vector for long-term siRNA-mediated downregulation of annexin A2. Intratumoral administration of pDrive-sh AnxA2-loaded nanoparticles to xenograft prostate tumors in nude mice demonstrated an overall decrease in tumor growth (Braden et al. 2010). Methyl-CpG binding domain protein 1 (MBD1) is a transcriptional regulator that binds methylated CpG islands of tumor suppressor genes and represses their transcription. MBD1–siRNA plasmid loaded into poloxamer-modified PLGA nanoparticles was successfully transfected into pancreatic tumor cells, and the MBD1 nanoparticle compound inhibited cell growth and induced apoptosis (Luo et al. 2009).

One concern with PLGA nanoparticles is that the polymer is anionic. To achieve positive charge, the surface of PLGA nanoparticles can be decorated with PEI using a cetyl derivative to improve surface functionalization and siRNA delivery. Submicrometer particles were produced by an emulsion–diffusion method using benzyl alcohol. The modified particles were able to bind and mediate siRNA delivery into the human osteosarcoma cell line U2OS and the murine macrophage cell line J774.1 (Andersen et al. 2010). Cationic PEI is sometimes used to convert the PLGA particles into positive charge. PEI was incorporated in the PLGA matrix to improve siRNA encapsulation and transfection efficiency in PLGA nanoparticles formulated using a double-emulsion solvent evaporation technique. The presence of PEI in the PLGA nanoparticle matrix increased siRNA encapsulation to about twofold and also resulted in twofold-higher cellular uptake of nanoparticles. Serum stability and lack of cytotoxicity further add to the potential of PLGA–PEI nanoparticles in gene silencing-based therapeutic applications (Patil and Panyam 2009).

Katas et al. (2009) developed PEI-incorporated PLGA nanoparticles by spontaneous emulsion–diffusion method resulting in particle size of around 100 nm, suitable for siRNA delivery. The incorporation of PEI into PLGA particles with the PLGA-to-PEI weight ratio 29:1 was found to produce spherical and positively charged nanoparticles. Particle size of ~100 nm was obtained when 5% (m/v) polyvinyl acetate (PVA) was used as a stabilizer. The PLGA–PEI nanoparticles were able to completely bind siRNA at N/P ratio 20:1 and to provide protection for siRNA against nuclease degradation. *In vitro* cell culture studies subsequently revealed that the PLGA–PEI nanoparticles with adsorbed siRNA could efficiently silence the targeted gene in mammalian cells, better than PEI alone, with acceptable cell viability. PLGA–PEI nanoparticles have been found to be superior to their cationizing parent compound, the PEI polymer.

To improve siRNA delivery for possible clinical applications, PLGA nanoparticles were also modified with biocompatible and biodegradable chitosan so that the nanoparticles possessed a positive surface charge, high siRNA loading, high transfection efficiency, and low toxicity. It was found that the nanoparticle diameter and positive zeta potential increase as the chitosan-coating concentration increases. Chitosan-modified PLGA nanoparticles showed an excellent siRNA-binding ability and effective protection of oligonucleotides from ribonuclease (RNase) degradation (Patil and Panyam 2009).

Small interfering RNA-loaded nanoparticles were successfully delivered into the HEK 293 T cell line, and the silencing of GFP expression was observed. In addition, the cytotoxicity assay revealed that nanoparticles had relatively low cytotoxicity (Yuan et al. 2010). Chitosan surface-modified PLGA nanospheres were also prepared by an emulsion solvent diffusion (ESD) method. The chitosan-modified PLGA nanoparticles showed much higher encapsulation efficiency compared to unmodified plain PLGA nanoparticles. Small interfering RNA-loaded chitosan-modified PLGA nanoparticles were taken up more effectively by the cells than plain PLGA nanoparticles. The gene-silencing efficiency of modified PLGA nanoparticles was higher and more prolonged than that of plain PLGA nanoparticles and naked siRNA (Tahara et al. 2010). Chitosan-modified PLGA nanoparticles loaded with TF-siRNA were used in a new external stent prepared by the hybrid ultrafine fibrous membrane as a potential therapy for vein graft disease. Because of the introduction of chitosan, which is a naturally hydrophilic polymer, the hybrid membranes showed good water absorption properties. It was found that the external stent prepared by chitosan–PLGA nanoparticles inhibits early neointima formation in rat vein grafts (Qiu et al. 2009).

Sustained release of siRNA has been achieved with biodegradable PLGA copolymer microspheres (Khan et al. 2004) possessing a matrix through which siRNA slowly diffuses. Instead of releasing siRNA all at once, these sustained-release vehicles protect their contents against serum nucleases and can provide a slow, continuous supply of pristine siRNA to cells, prolonging the RNAi effect. Clearly, a delivery system that enables precise control of the onset and duration of the RNAi effect would be an ideal platform for *in vivo* therapeutic applications of siRNA.

## 13.2.6  CATIONIC STAR POLYMERS

One of the groups of polymers, which have shown great promise for siRNA delivery, includes the cationic star polymers. Star polymers possess singular properties, originating from their multiarmed structures, leading to a unique solution behavior and enhanced cell uptake. Boyer et al. (2013) synthesized cationic star polymers using reversible addition–fragmentation transfer polymerization (RAFT) by polymerizing dimethylaminoethyl methacrylate (DMAEMA) followed by chain extension in the presence of two cross-linkers N,N-bis(acryloyl) cystamine and DMAEMA. The resulting star polymers were self-assembled with siRNA to form small uniform nanoparticle complexes. Treatment with star polymer–siRNA complexes resulted in an uptake of siRNA into pancreatic cancer cells and a significant decrease in target mRNA and protein levels (Figure 13.5).

## 13.2.7  ANIONIC POLYMERS

Schlegel et al. (2011) studied the effect of incorporating nontoxic biodegradable anionic polymers into siRNA lipoplexes. Table 13.2 lists the structures and characteristics of these anionic polymers.

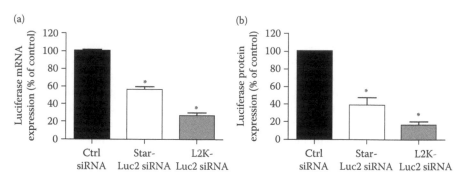

**FIGURE 13.5**  Gene-silencing efficiency of star polymer–siRNA complexes in pancreatic cancer cells. (a) The representative graph demonstrating firefly luciferase 2 (Luc2) mRNA levels in a pancreatic cancer cell line (MiaPaCa-2-Luc2) that stably expresses high levels of the Luc2 gene 24 h posttransfection with star polymer–Luc2 siRNA (100 nM) complexes. Cells treated with the nonsilencing control of siRNA (100 nM) complexed to the star polymer were used as controls (Ctrl siRNA). Cells treated with LF 2000–Luc 2 siRNA (100 nM) were used as a positive control (L2K-siRNA). Each value represents the mean ± standard deviation of duplicate wells from three independent experiments (*$p < 0.05$). All data were normalized to the housekeeping gene β2-microglobulin. (b) A representative graph showing firefly luciferase (Luc2) protein expression in MiaPaCa-2-Luc2 expressing cells 24 h posttransfection with star polymer–Luc2 siRNA (100 nM). Cells treated with the nonsilencing control of siRNA (100 nM) complexed to the star polymer were used as controls (Ctrl siRNA). Cells treated with LF 2000–Luc2 siRNA (100 nM) were used as a positive control (L2K-siRNA). Each value represents the mean ± standard deviation of duplicate wells from four independent experiments (*$p < 0.05$). All data were normalized to the total cellular protein. (Reprinted with permission from Boyer, C. et al., Effective delivery of siRNA into cancer cells and tumors using well-defined biodegradable cationic star polymers, *Mol Pharm*, 2435–44. Copyright 2013 American Chemical Society.)

**TABLE 13.2**
**Particle Size (Diameter in nm)**

| DSPE-PEG (%) | Size (nm; pdi) | | | | |
|---|---|---|---|---|---|
| | | Lipoplexes Charge Ratio | | | |
| | Liposome | 2 | 4 | 6 | 8 |
| 0 | 114 ± 2 (0.24) | p | p | 206 + 4 (0.27) | 141 ± 4 (0.22) |
| 1 | 152 ± 5 (0.27) | 204 ± 6 (0.33) | 142 ± 6 (0.36) | 127 ± 4 (0.42) | 138 ± 6 (0.44) |
| 3 | 85 ± 1 (0.41) | 103 ± 7 (0.46) | 147 ± 9 (0.33) | 103 ± 12 (0.59) | 103 ± 5 (0.51) |
| 5 | 115 ± 16 (0.38) | 75 ± 4 (0.47) | 67 ± 1 (0.22) | 69 ± 1 (0.22) | 68 ± 1 (0.24) |

*Source:* Reprinted from *J Control Release*, 152, Schlegel, A. et al., Anionic polymers for decreased toxicity and enhanced *in vivo* delivery of siRNA complexed with cationic liposomes, 393–401, Copyright 2011, with permission from Elsevier.

*Note:* Lipoplexes were formed at various charge ratios with siRNA pre associated with PG (1.5 k) (1/1,w/w) and cationic liposomes containing various amounts of DSPE-PEG (0, 1, 3, or 5 mol% of the total lipid). Measures have been performed in triplicates. pdi is given in brackets. p indicates that assayed particles were precipitated and no accurate measurement could be obtained.

It was concluded that the addition of these polymers led to decreased cellular toxicity and enhanced gene-silencing efficiency depending on the type of polymer used. Polyglutamate (PG) was identified as the most efficient polymer as it led to siRNA lipoplexes that are efficient at all charge ratios with satisfying colloidal stability and moderate cellular toxicity. The siRNA lipoplexes formed with PG were pegylated to enhance their stability toward aggregation in serum. The particle structure, stability, and gene-silencing efficiency of pegylated or unpegylated siRNA lipoplexes formed with PG are shown in Figure 13.6.

## 13.2.8 Recombinant Polymers

Unlike gene therapy where the gene must be delivered to the nucleus of the cell to initiate transcription, the site of action for siRNA is the cytoplasm. Therefore, for maximum efficacy, a delivery system must be designed that can differentiate between delivery to the cytoplasm and the nucleus. Most of the currently available delivery systems such as cationic lipids and polymers allow the siRNA to enter any cellular compartment nondiscriminatorily and do not necessarily localize the siRNA to the cytoplasm. Canine et al. (2011) designed a biopolymeric platform to selectively deliver siRNA of the cytoplasm and pDNA to the cell nucleus. This biopolymer consisted of M9 nuclear localization signal (NLS) as one of its components to enhance active translocation of the genetic material toward the cell nucleus. Figure 13.7 shows the schematics of the multidomain biopolymers. The biopolymers were complexed with pEGFP and GFP-siRNA and used to transfect SKOV-3 (HER2+) ovarian cancer cells.

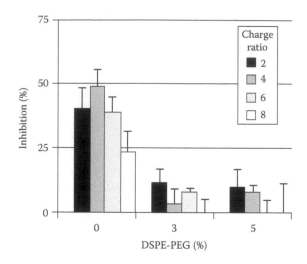

**FIGURE 13.6** Particle charge and gene-silencing efficiency of pegylated or unpegylated siRNA lipoplexes formed with polygutamate. Gene knockdown of luciferase activity in B16-Luc with unpegylated or pegylated (3% or 5% DSPE-PEG) siRNA lipoplexes formed with PG. (Reprinted from *J Control Release*, 152, Schlegel, A. et al., Anionic polymers for decreased toxicity and enhanced *in vivo* delivery of siRNA complexed with cationic liposomes, 393–401, Copyright 2011, with permission from Elsevier.)

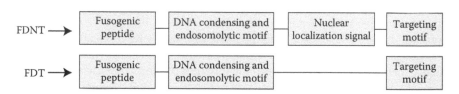

**FIGURE 13.7** Schematics of the multidomain biopolymers designed for siRNA delivery (FDT) and DNA delivery (FDNT). (Reprinted from *J Control Release*, 151, Canine, B.F. et al., Development of targeted recombinant polymers that can deliver siRNA to the cytoplasm and plasmid DNA to the cell nucleus, 95–101, Copyright 2011, with permission from Elsevier.)

It was observed that the biopolymer with NLS (FNDT) is suitable for pDNA delivery and the biopolymer without NLS (FDT) was suitable for siRNA delivery to the cytoplasm (Canine et al. 2011) (Figure 13.8).

## 13.3  CLINICAL TRIALS INVOLVING siRNA DELIVERY

The first direct evidence of an RNAi effect in Phase-I clinical trials was reported by Davis et al. (2010), when a siRNA was delivered using a nanoparticulate delivery system comprising of cyclodextrin cationic polymer, PEG, and a transferring ligand against the M2 subunit of ribonucleotide reductase (RRM2). These nanoparticles accumulated in the tumor tissue and mediated a significant reduction of mRNA level

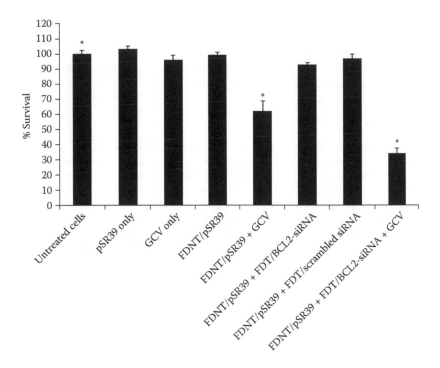

**FIGURE 13.8** Evaluation of cell-killing efficiency by biopolymers in complex with thera-peutic nucleic acids. Cells were transfected with two test groups (FDNT/pSR39 plus GCV, and FDNT/pSR39 plus GCV + FDT/BCL2-siRNA) and six control groups. The cell survival was assessed at day 8 with WST-1 cell toxicity assay. (Reprinted from *J Control Release*, 151, Canine, B.F. et al., Development of targeted recombinant polymers that can deliver siRNA to the cytoplasm and plasmid DNA to the cell nucleus, 95–101, Copyright 2011, with permission from Elsevier.)

for RRM2. In another study, Tabernero et al. (2013) studied the delivery of siRNA targeting VEGF and kinesin spindle protein using a lipoplex system for the treatment of patients with advanced cancer and liver metastases. Following an intravenous injection of nanoparticles at siRNA doses ranging from 0.4 to 1.5 mg/kg, the cleav-age of VEGF was detected at the target site, leading to lower mRNA levels and disease control in four patients, including complete regression in one patient with significant liver metastases.

## 13.4 CONCLUSION

Owing to the polyanionic nature of siRNA, cationic vectors are required to form a complex with it and efficiently transport it to its target. There are several barriers associated with the transport of siRNA to the target. These include catalytic enzymes, defense mechanisms, and other degradation processes. As seen from various stud-ies, polymeric nonviral vectors can be successfully used to target genes to specific cancer targets. Table 13.1 lists a brief summary of the advantages and disadvantages of various nonviral vectors discussed in this chapter. Compared to the viral vectors, which possess a considerable risk of toxicity, nonviral vectors can be promising for gene delivery and their chemistry, toxicity, stability, and biodegradability can be modified to make them most suitable for gene delivery.

## REFERENCES

Adolph, E.J., Hafeman, A.E., Davidson, J.M., Nanney, L.B., and Guelcher, S.A. 2012. Injectable polyurethane composite scaffolds delay wound contraction and support cellular infiltration and remodeling in rat excisional wounds. *J Biomed Mater Res A* 100:450–61.

Agnihotri, S.A., Mallikarjuna, N.N., and Aminabhavi, T.M. 2004. Recent advances on chitosan-based micro- and nanoparticles in drug delivery. *J Control Release* 100:5–28.

Akeboshi, H., Chiba, Y., Kasahara, Y. et al. 2007. Production of recombinant beta-hexosaminidase A, a potential enzyme for replacement therapy for Tay– Sachs and Sandhoff diseases, in the methylotrophic yeast *Ogataea minuta*. *Appl Environ Microbiol* 73:4805–12.

Andersen, M.O., Lichawska, A., Arpanaei, A. et al. 2010. Surface functionalisation of PLGA nanoparticles for gene silencing. *Biomaterials* 31:5671–7.

Andronopoulou, E. and Vorgias, C.E. 2003. Purification and characterization of a new hyperthermostable, allosamidin-insensitive and denaturation-resistant chitinase from the hyperthermophilic archaeon *Thermococcus chitonophagus*. *Extremophiles* 7:43–53.

Behr, J.P. 1997. The proton sponge: A trick to enter cells the viruses did not exploit. *Chimia Int J Chem* 51:34–6.

Boussif, O., Lezoualc'h, F., Zanta, M.A. et al. 1995. A versatile vector for gene and oligonucleotide transfer into cells in culture and *in vivo*: Polyethylenimine. *Proc Natl Acad Sci USA* 92:7297–301.

Boyer, C., Teo, J., Phillips, P. et al. 2013. Effective delivery of siRNA into cancer cells and tumors using well-defined biodegradable cationic star polymers. *Mol Pharm* 10:2435–44.

Braden, A.R., Kafka, M.T., Cunningham, L., Jones, H., and Vishwanatha, J.K. 2010. Polymeric nanoparticles for sustained down-regulation of annexin A2 inhibit prostate tumor growth. *J Nanosci Nanotechnol* 9:2856–65.

Bragonzi, A., Boletta, A., Biffi, A. et al. 1999. Comparison between cationic polymers and lipids in mediating systemic gene delivery to the lungs. *Gene Ther* 6:1995–2004.

Brissault, B., Leborgne, C., Guis, C., Danos, O., Cheradame, H., and Kichler, A. 2006. Linear topology confers *in vivo* gene transfer activity to polyethylenimines. *Bioconjug Chem* 17:759–65.

Brown, M.D., Schatzlein, A., Brownlie, A. et al. 2000. Preliminary characterization of novel amino acid based polymeric vesicles as gene and drug delivery agents. *Bioconjug Chem* 11:880–91.

Canine, B.F., Wang, Y., Ouyang, W., and Hatefi, A. 2011. Development of targeted recombinant polymers that can deliver siRNA to the cytoplasm and plasmid DNA to the cell nucleus. *J Control Release* 151:95–101.

Chandy, T. and Sharma, C.P. 1990. Chitosan as a biomaterial. *Biomater Artif Cells Artif Organs* 18:1–24.

Cherng, J.Y., Hou, T.Y., Shih, M.F., Talsma, H., Hennink, W.E. 2013. Polyurethane-based drug delivery systems. *Int J Pharm* 450(1):145–62.

Chirila, T.V., Rakoczy, P.E., Garrett, K.L., Lou, X., and Constable, I.J. 2002. The use of synthetic polymers for delivery of therapeutic antisense oligodeoxynucleotides. *Biomaterials* 23:321–42.

Cho, Y.W., Kim, J.D., and Park, K. 2003. Polycation gene delivery systems: Escape from endosomes to cytosol. *J Pharm Pharmacol* 55(6):721–34.

Collard, W.T., Yang, Y., Kwok, K.Y., Park, Y., and Rice, K.G. 2000. Biodistribution, metabolism, and *in vivo* gene expression of low molecular weight glycopeptide polyethylene glycol peptide DNA co-condensates. *J Pharm Sci* 89:499–512.

Cun, D., Foged, C., Yang, M. et al. 2010. Preparation and characterization of poly (DL-lactide-*co*-glycolide) nanoparticles for siRNA delivery. *Int J Pharm* 390:70–5.

Dailey, L.A., Kleemann, E., Merdan, T., Petersen, H., Schmehl, T., and Gessler, T. 2004. Modified polyethylenimines as non viral gene delivery systems for aerosol therapy: Effects of nebulization on cellular uptake and transfection efficiency. *J Control Release* 100:425–36.

Davis, M.E., Zuckerman, J.E., Choi, C.H. et al. 2010. Evidence of RNAi in humans from systemically administered siRNA via targeted nanoparticles. *Nature* 464:1067–70.

Elbakry, A., Zaky, A., Liebl, R., Rachel, R., Goepferich, A., and Breunig, M. 2009. Layer-by-layer assembled gold nanoparticles for siRNA delivery. *Nano Lett* 9:2059–64.

Elbashir, S.M., Harborth, J., Lendeckel, W., Yalcin, A., Weber, K., and Tuschl, T. 2001. Duplexes of 21-nucleotide RNAs mediate RNA interference in cultured mammalian cells. *Nature* 6836:494–8.

Farrell, L.L., Pepin, J., Kucharski, C., Lin, X., Xu, Z., and Uludag, H. 2007. A comparison of the effectiveness of cationic polymers poly-L-lysine (PLL) and polyethylenimine (PEI) for non-viral delivery of plasmid DNA to bone marrow stromal cells (BMSC). *Eur J Pharm Biopharm* 65(3):388–97.

Ferrari, S., Pettenazzo, A., Garbati, N., Zacchello, F., Behr, J.P., and Scarpa, M. 1999. Polyethylenimine shows properties of interest for cystic fibrosis gene therapy. *Biochim Biophys Acta* 1447:219–25.

Fischer, D., Bieber, T., Li, Y., Elsasser, H.P., and Kissel, T. 1999. A novel non-viral vector for DNA delivery based on low molecular weight, branched polyethylenimine: Effect of molecular weight on transfection efficiency and cytotoxicity. *Pharm Res* 16: 1273–9.

Florea, B.I., Meaney, C., Junginger, H.E., and Borchard, G. 2002. Transfection efficiency and toxicity of polyethylenimine in differentiated Calu-3 and nondifferentiated COS-1 cell cultures. *AAPS Pharm Sci* 4:E12.

Fu, J.Y., Wu, S.M., Chang, C.T., and Sung, H.Y. 2003. Characterization of three chitosanase isozymes isolated from a commercial crude porcine pepsin preparation. *J Agric Food Chem* 51:1042–8.

Fuller, J.E., Zugates, G.T., Ferreira, L.S. et al. 2008. Intracellular delivery of core_shell fluorescent silica nanoparticles. *Biomaterials* 29:1526–32.

Garcia-Alonso, J., Reglero, A., and Cabezas, J.A. 1990. Purification and properties of beta-*N*-acetylhexosaminidase A from pig brain. *Int J Biochem* 22: 645–51.

Ge, Q., Filip, L., Bai, A., Nguyen, T., Eisen, H.N., and Chen, J. 2004. Inhibition of influenza virus production in virus infected mice by RNA interference. *Proc Natl Acad Sci USA* 101:8676–81.

Godbey, W.T., Barry, M.A., Saggau, P., Wu, K.K., and Mikos, A.G. 2000. Poly(ethylenimine)-mediated transfection: A new paradigm for gene delivery. *J Biomed Mater Res* 51:321–8.

Godbey, W.T. and Mikos, A.G. 2001. Recent progress in gene delivery using non-viral transfer complexes. *J Control Release* 72:115–25.

Godbey, W.T., Wu, K.K., Hirasaki, G.J., and Mikos, A.G. 1999a. Improved packing of poly-(ethylenimine)/DNA complexes increases transfection efficiency. *Gene Ther* 6:1380–8.

Godbey, W.T., Wu, K.K., and Mikos, A.G. 1999b. Poly(ethylenimine) and its role in gene delivery. *J Control Release* 60:149–60.

Godbey, W.T., Wu, K.K., and Mikos, A.G. 1999c. Size matters: Molecular weight affects the efficiency of poly(ethylenimine) as a gene delivery vehicle. *J Biomed Mater Res* 45:268–75.

Godbey, W.T., Wu, K.K., and Mikos, A.G. 1999d. Tracking the intracellular path of poly(ethylenimine)/DNA complexes for gene delivery. *Proc Natl Acad Sci USA* 96:5177–81.

Godbey, W.T., Wu, K.K., and Mikos, A.G. 2001. Poly(ethylenimine)-mediated gene delivery affects endothelial cell function and viability. *Biomaterials* 22:471–80.

Goula, D., Benoist, C., Mantero, S., Merlo, G., Levi, G., and Demeneix, B.A. 1998a. Polyethylenimine-based intravenous delivery of transgenes to mouse lung. *Gene Ther* 5:1291–5.

Goula, D., Remy, J.S., Erbacher, P. et al. 1998b. Size, diffusibility and transfection performance of linear PEI/DNA complexes in the mouse central nervous system. *Gene Ther* 5:712–7.

Guan, J.J., Fujimoto, K.L., Sacks, M.S., and Wagner, W.R. 2005. Preparation and characterization of highly porous, biodegradable polyurethane scaffolds for soft tissue applications. *Biomaterials* 26:3961–71.

Guelcher, S.A. 2008. Biodegradable polyurethanes: Synthesis and applications in regenerative medicine. *Tissue Eng Part B Rev* 14:3–17.

Guo, G., Zhou, L., Chen, Z. et al. 2013. Alkane-modified low-molecular-weight polyethylenimine with enhanced gene silencing for siRNA delivery. *Int J Pharm* 450:44–52.

Guo, J., Fisher, K., Darcy, R., Cryan, J.F., and O'Driscoll, C. 2010. Therapeutic targeting in the silent era: Advances in non-viral siRNA delivery. *Mol Biosyst* 6:1143–61.

Guo, Z., Chen, R., Xing, R. et al. 2006. Novel derivatives of chitosan and their antifungal activities *in vitro*. *Carbohydr Res* 341:351–4.

Hafeman, A.E., Li, B., Yoshii, T., Zienkiewicz, K., Davidson, J.M., and Guelcher, S.A. 2008. Injectable biodegradable polyurethane scaffolds with release of platelet-derived growth factor for tissue repair and regeneration. *Pharm Res* 25:2387–99.

Harada-Shiba, M., Yamauchi, K., Harada, A., Takamisawa, I., Shimokado, K., and Kataoka, K. 2002. Polyion complex micelles as vectors in gene therapy—Pharmacokinetics and *in vivo* gene transfer. *Gene Ther* 9:407–14.

Hasegawa, M., Yagi, K., Iwakawa, S., and Hirai, M. 2001. Chitosan induces apoptosis via caspase-3 activation in bladder tumour cells. *Jpn J Cancer Res* 92:459–66.

Hassani, Z., Lemkine, G.F., Erbacher, P. et al. 2005. Lipid mediated siRNA delivery downregulates exogenous gene expression in the mouse brain at picomolar levels. *J Gene Med* 7:198–207.

Hirano, S., Iwata, M., Yamanaka, K., Tanaka, H., Toda, T., and Inui, H. 1991. Enhancement of serum lysozyme activity by injecting a mixture of chitosan oligosaccharides intravenously in rabbits. *Agric Biol Chem* 55:2623–5.

Howard, K.A. 2009. Delivery of RNA interference therapeutics using polycation-based nanoparticles. *Adv Drug Deliv Rev* 61:710–20.

Howard, K.A., Rahbek, U.L., Liu, X. et al. 2006. RNA interference *in vitro* and *in vivo* using a novel chitosan/siRNA nanoparticle system. *Mol Ther* 14:476–84.

Huang, M., Fong, C.W., Khor, E., and Lim, L.Y. 2005. Transfection efficiency of chitosan vectors: Effect of polymer molecular weight and degree of deacetylation. *J Control Release* 106:391–406.

Illum, L., Farraj, N.F., and Davis, S.S. 1994. Chitosan as a novel nasal delivery system for peptide drugs. *Pharm Res* 11:1186–9.

Jeong, J., Kim, S., and Park, T. 2007. Molecular design of functional polymers for gene therapy. *Prog Polym Sci* 32:1239–74.

Kafedjiiski, K., Foger, F., Hoyer, H., Bernkop-Schnurch, A., and Werle, M. 2007. Evaluation of *in vitro* enzymatic degradation of various thiomers and cross-linked thiomers. *Drug Dev Ind Pharm* 33:199–208.

Katas, H., Cevher, E., and Alpar, H.O. 2009. Preparation of polyethylenimine incorporated poly(D,L-lactide-coglycolide) nanoparticles by spontaneous emulsion diffusion method for small interfering RNA delivery. *Int J Pharm* 369:144–54.

Kean, T., Roth, S., and Thanou, M. 2005. Trimethylated chitosans as non-viral gene delivery vectors: Cytotoxicity and transfection efficiency. *J Control Release* 103: 643–53.

Kean, T. and Thanou, M. 2010. Biodegradation, biodistribution and toxicity of chitosan. *Adv Drug Deliv Rev* 62:3–11.

Khan, A., Benboubetra, M., Sayyed, P.Z. et al. 2004. Sustained polymeric delivery of gene silencing antisense ODNs, siRNA, DNAzymes and ribozymes: *In vitro* and *in vivo* studies. *J Drug Target* 12:393–404.

Kiang, T., Wen, J., Lim, H.W., and Leong, K.W. 2004. The effect of the degree of chitosan deacetylation on the efficiency of gene transfection. *Biomaterials* 25:5293–301.

Kircheis, R., Kichler, A., Wallner, G. et al. 1997. Coupling of cell-binding ligands to polyethylenimine for targeted gene delivery. *Gene Ther* 4:409–18.

Kircheis, R., Schuller, S., Brunner, S. et al. 1999. Polycation-based DNA complexes for tumor-targeted gene delivery *in vivo*. *J Gene Med* 1:111–20.

Kircheis, R., Wightman, L., Schreiber, A. et al. 2001a. Polyethylenimine/DNA complexes shielded by transferrin target gene expression to tumors after systemic application. *Gene Ther* 8:28–40.

Kircheis, R., Wightman, L., Wagner, E. 2001b. Design and gene delivery activity of modified polyethylenimines. *Adv Drug Deliv Rev* 53:341–58.

Köping-Höggård, M., Vårum, K.M., Issa, M. et al. 2004. Improved chitosan mediated gene delivery based on easily dissociated chitosan polyplexes of highly defined chitosan oligomers. *Gene Ther* 11:1441–52.

Kumar, M.N., Muzzarelli, R.A., Muzzarelli, C., Sashiwa, H., and Domb, A.J. 2004. Chitosan chemistry and pharmaceutical perspectives. *Chem Rev* 104:6017–84.

Kunath, K., von Harpe, A., Fischer, D. et al. 2003. Low-molecular-weight polyethylenimine as a non-viral vector for DNA delivery: Comparison of physicochemical properties, transfection efficiency and *in vivo* distribution with high-molecular-weight polyethylenimine. *J Control Release* 89:113–25.

Kurreck, J. 2009. RNA interference: From basic research to therapeutic applications. *Angew Chem Int Ed Engl* 48:1378–98.

Kwon, Y.J. 2011. Before and after endosomal escape: Roles of stimuli-converting siRNA/polymer interactions in determining gene silencing efficiency. *Acc Chem Res* 45(7):1077–88.

Kwoh, D.Y., Coffin, C.C., Lollo, C.P. et al. 1999. Stabilization of poly-L-lysine/DNA polyplexes for *in vivo* gene delivery to the liver. *Biochim Biophys Acta* 1444:171–90.

Kwok, K.Y., McKenzie, D.L., Evers, D.L., and Rice, K.G. 1999. Formulation of highly soluble poly(ethylene glycol)-peptide DNA condensates. *J Pharm Sci* 88:996–1003.

Kwok, K.Y., Park, Y., Yang, Y., McKenzie, D.L., Liu, Y., and Rice, K.G. 2003. *In vivo* gene transfer using sulfhydryl cross-linked PEG-peptide/glycopeptide DNA co-condensates. *J Pharm Sci* 92:1174–85.

Lai, W.F. and Lin, M.C. 2009. Nucleic acid delivery with chitosan and its derivatives. *J Control Release* 134:158–68.

Lee, M., Nah, J.W., Kwon, Y., Koh, J.J., Ko, K.S., and Kim, S.W. 2001. Water-soluble and low molecular weight chitosan-based plasmid DNA delivery. *Pharm Res* 18:427–31.

Lee, S.H., Kim, S., and Park, T.G. 2007. Intracellular siRNA delivery system using polyelectrolyte complex micelles prepared from VEGF siRNA–PEG conjugate and cationic fusogenic peptide. *Biochem Biophys Res Commun* 357:511–6.

Leng, Q., Woodle, M.C., Lu, P.Y., and Mixson, A.J. 2009. Advances in systemic siRNA delivery. *Drugs Future* 34:721.

Li, B., Davidson, J.M., and Guelcher, S.A. 2009. The effect of the local delivery of platelet derived growth factor from reactive two-component polyurethane scaffolds on the healing in rat skin excisional wounds. *Biomaterials* 30:3486–94.

Li, S. and Huang, L. 2000. Nonviral gene therapy: Promises and challenges. *Gene Ther* 7:31–4.

Li, S. and Huang, L. 2006. Gene therapy progress and prospects: Non-viral gene therapy by systemic delivery. *Gene Ther* 13:1313–9.

Liong, M., France, B., Bradley, K.A., and Zink, J.I. 2009. Antimicrobial activity of silver nanocrystals encapsulated in mesoporous silica nanoparticles. *Adv Mater* 21:1684–9.

Lu, G., Sheng, B., Wang, G. et al. 2009. Controlling the degradation of covalently cross-linked carboxymethyl chitosan utilizing bimodal molecular weight distribution. *J Biomater Appl* 23:435–51.

Lungwitz, U., Breunig, M., Blunk, T., and Go-pferich, A. 2005. Polyethylenimine based non-viral gene delivery systems. *Eur J Pharm Biopharm* 60:247–66.

Luo, G., Jin, C., Long, J., Fu, D., Yang, F., and Xu, J. 2009. RNA interference of MBD1 in BxPC-3 human pancreatic cancer cells delivered by PLGA–poloxamer nanoparticles. *Cancer Biol Ther* 8:594–8.

Mannisto, M., Vanderkerken, S., Toncheva, V. et al. 2002. Structure–activity relationships of poly(L-lysines): Effects of pegylation and molecular shape on physicochemical and biological properties in gene delivery. *J Control Release* 83:169–82.

Mansouri, S., Lavigne, P., Corsi, K., Benderdour, M., Beaumont, E., and Fernandes, J.C. 2004. Chitosan–DNA nanoparticles as non-viral vectors in gene therapy: Strategies to improve transfection efficacy. *Eur J Pharm Biopharm* 57:1–8.

Mao, S., Sun, W., and Kissel, T. 2010. Chitosan-based formulations for delivery of DNA and siRNA. *Adv Drug Deliv Rev* 62:12–27.

Martinou, A., Koutsioulis, D., and Bouriotis, V. 2002. Expression, purification, and characterization of a cobalt-activated chitin deacetylase (cda2p) from *Saccharomyces cerevisiae*. *Protein Expr Purif* 24:111–6.

McBain, S.C., Yiu, H.H.P., Haj, A.E., and Dobson, J. 2007. Polyethylenimine functionalized iron oxide nanoparticles as agents for DNA delivery and transfection. *J Mater Chem* 17:2561–5.

Merdan, T., Kopecek, J., and Kissel, T. 2002. Prospects for cationic polymers in gene and oligonucleotide therapy against cancer. *Adv Drug Deliv Rev* 54:715–58.

Meyer, M., Dohmen, C., Philipp, A., Kiener, D., Maiwald, G., and Scheu, C. 2009. Synthesis and biological evaluation of a bioresponsive and endosomolytic siRNA–polymer conjugate. *Mol Pharm* 6:752–62.

Miyauchi, K., Tsubuku, N., Matsumiya, M., and Mochizuki, A. 2007. Characterization of lysozymes from three shellfish. *Fish Sci* 73:1404–6.

Nelson, C.E., Gupta, M.K., Adolph, E.J., Shannon, J.M., Guelcher, S.A., and Duvall, C.L. 2012. Sustained local delivery of siRNA from an injectable scaffold. *Biomaterials* 33:1154–61.

Neu, M., Fischer, D., and Kissel, T. 2005. Recent advances in rational gene transfer vector design based on poly(ethylene imine) and its derivatives. *J Gene Med* 7:992–1009.

Niola, F., Evangelisti, C., Campagnolo, L. et al. 2006. A plasmid encoded VEGF siRNA reduces glioblastoma angiogenesis and its combination with interleukin-4 blocks tumor growth in a xenograft mouse model. *Cancer Biol Ther* 5:174–9.

Nishikawa, M., Takemura, S., Takakura, Y., and Hashida, M. 1998. Targeted delivery of plasmid DNA to hepatocytes *in vivo*: Optimization of the pharmacokinetics of plasmid DNA/galactosylated poly(L-lysine) complexes by controlling their physicochemical properties. *J Pharmacol Exp Ther* 287:408–15.

Ogris, M., Brunner, S., Schuller, S., Kiacheis, R., and Wagner, E. 1999. PEGylated DNA/transferrin–PEI complexes: Reduced interaction with blood components, extended circulation in blood and potential for systemic gene delivery. *Gene Ther* 6:595–605.

Ogris, M., Steinlein, P., Kursa, M., Mechtler, K., Kircheis, R., and Wagner, E. 1998. The size of DNA/transferrin–PEI complexes is an important factor for gene expression in cultured cells. *Gene Ther* 5:1425–33.

Ohsaki, M., Okuda, T., Wada, A., Hirayama, T., Niidome, T., and Aoyagi, H. 2002. *In vitro* gene transfection using dendritic poly(L-lysine). *Bioconjug Chem* 13:510–7.

Okuda, T., Sugiyama, A., Niidome, T., and Aoyagi, H. 2004. Characters of dendritic poly(L-lysine) analogues with the terminal lysines replaced with arginines and histidines as gene carriers *in vitro*. *Biomaterials* 25:537–44.

Onishi, H. and Machida, Y. 1999. Biodegradation and distribution of water-soluble chitosan in mice. *Biomaterials* 20:175–82.

Panyam, J. and Labhasetwar, V. 2003. Biodegradable nanoparticles for drug and gene delivery to cells and tissue. *Adv Drug Deliv Rev* 55:329–47.

Park, T.G, Jeong, J.H., and Kim, S.W. 2006. Current status of polymeric gene delivery systems. *Adv Drug Deliv Rev* 58(4):467–86.

Park, I.Y., Kim, I.Y., Yoo, M.K., Choi, Y.J., Cho, M.-H., and Cho, C.S. 2008. Mannosylated polyethylenimine coupled mesoporous silica nanoparticles for receptor-mediated gene delivery. *Int J Pharm* 359:280–7.

Patil, Y. and Panyam, J. 2009. Polymeric nanoparticles for siRNA delivery and gene silencing. *Int J Pharm* 367:195–203.

Petersen, H., Fechner, M.F., Martin, A.L. et al. 2002b. Polyethylenimine-graft-poly(ethylene glycol) copolymers: Influence of copolymer block structure on DNA complexation and biological activities as gene delivery system. *Bioconjug Chem* 13:845–54.

Petersen, H., Merdan, T., Kunath, K., Fischer, D., and Kissel, T. 2002a. Poly (ethylenimine-co-L-lactamide-co-succinamide): A biodegradable polyethylenimine derivative with an advantageous pH-dependent hydrolytic degradation for gene delivery. *Bioconjug Chem* 13:812–21.

Qiu, X.F., Dong, N.G., Sun, Z.Q., Hu, P., Su, W., and Shi, J.W. 2009. Controlled release of siRNA nanoparticles loaded in a novel external stent prepared by emulsion electrospinning attenuates neointima hyperplasia in vein grafts. *Zhonghua Yi Xue Za Zhi* 89:2938–42.

Rao, S.B. and Sharma, C.P. 1997. Use of chitosan as a biomaterial: Studies on its safety and hemostatic potential. *J Biomed Mater Res* 34:21–8.

Ren, D., Yi, H., Wang, W., and Ma, X. 2005. The enzymatic degradation and swelling properties of chitosan matrices with different degrees of N-acetylation. *Carbohydr Res* 340:2403–10.

Richards Grayson, A.C., Doody, A.M., and Putnam, D. 2006. Biophysical and structural characterization of polyethylenimine-mediated siRNA delivery *in vitro*. *Pharm Res* 23:1868–76.

Rojanarata, T., Opanasopit, P., Techaarpornkul, S., Ngawhirunpat, T., and Ruktanonchai, U. 2008. Chitosan–thiamine pyrophosphate as a novel carrier for siRNA delivery. *Pharm Res* 25:2807–14.

Rudzinski, W.E. and Aminabhavi, T.M. 2010. Chitosan as a carrier for targeted delivery of small interfering RNA. *Int J Pharm* 399:1–11.

Sanon, A., Tournaire-Arellano, C., El Hage, S.Y., Bories, C., Caujolle, R., and Loiseau, P.M. 2005. N-acetyl-beta-D-hexosaminidase from *Trichomonas vaginalis*: Substrate specificity and activity of inhibitors. *Biomed Pharmacother* 59:245–8.

Sarmento, B. and Neves, J.d. 2012. *Chitosan-Based Systems for Biopharmaceuticals Delivery, Targeting, and Polymer Therapeutics*. Chichester, West Sussex: John Wiley & Sons.

Sato, A., Choi, S.W., Hirai, M. et al. 2007. Polymer brush-stabilized polyplex for a siRNA carrier with long circulatory half-life. *J Control Release* 122:209–16.

Schaffert, D. and Wagner, E. 2008. Gene therapy progress and prospects: Synthetic polymer-based systems. *Gene Ther* 15:1131–8.

Schipper, N.G., Varum, K.M., and Artursson, P. 1996. Chitosans as absorption enhancers for poorly absorbable drugs. 1: Influence of molecular weight and degree of acetylation on drug transport across human intestinal epithelial (Caco-2) cells. *Pharm Res* 13:1686–92.

Schipper, N.G., Varum, K.M., Stenberg, P., Ocklind, G., Lennernas, H., and Artursson, P. 1999. Chitosans as absorption enhancers of poorly absorbable drugs. 3: Influence of mucus on absorption enhancement. *Eur J Pharm Sci* 8:335–43.

Schlegel, A., Largeau, C., Bigey, P. et al. 2011. Anionic polymers for decreased toxicity and enhanced *in vivo* delivery of siRNA complexed with cationic liposomes. *J Control Release* 152:393–401.

Senel, S. and McClure, S.J. 2004. Potential applications of chitosan in veterinary medicine. *Adv Drug Deliv Rev* 56: 1467–80.

Tabernero, J., Shapiro, G.I., LoRusso, P.M. et al. 2013. First in-human trial of an RNA interference therapeutic targeting VEGF and KSP in cancer patients with liver involvement. *Cancer Discov* 3:406–17.

Tahara, K., Yamamoto, H., Hirashima, N. et al. 2010. Chitosan-modified poly(D,L-lactide-*co*-glycolide) nanospheres for improving siRNA delivery and gene-silencing effects. *Eur J Pharm Biopharm* 74:421–6.

Tan, M.L., Choong, P.F., and Dass, C.R. 2009. Cancer, chitosan nanoparticles and catalytic nucleic acids. *J Pharm Pharmacol* 61:3–12.

Tan, P.H., Yang, L.C., Shih, H.C., Lan, K.C., and Cheng, J.T. 2005. Gene knockdown with intrathecal siRNA of NMDA receptor NR2B subunit reduces formalin-induced nociception in the rat. *Gene Ther* 12:59–66.

Techaarpornkul, S., Wongkupasert, S., Opanasopit, P., Apirakaramwong, A., Nunthanid, J., and Ruktanonchai, U. 2010. Chitosan-mediated siRNA delivery *in vitro*: Effect of polymer molecular weight, concentration and salt forms. *AAPS Pharm Sci Tech* 11:64–72.

Thomas, M. and Klibanov, A.M. 2003. Non-viral gene therapy: Polycation-mediated DNA delivery. *Appl Microbiol Biotechnol* 62:27–34.

Thomas, M., Lu, J.J., Chen, J., and Klibanov, A.M. 2007. Non-viral siRNA delivery to the lung. *Adv Drug Deliv Rev* 59:124–33.

Thomas, M., Lu, J.J., Ge, Q., Zhang, C., Chen, J., and Klibanov, A.M. 2005. Full deacylation of polyethylenimine dramatically boosts its gene delivery efficiency and specificity to mouse lung. *Proc Natl Acad Sci USA* 102:5679–84.

Tseng, S.J. and Tang, S.C. 2007. Development of poly(amino ester glycol urethane)/siRNA polyplexes for gene silencing. *Bioconjug Chem* 18:1383–90.

Tuschl, T. 2001. RNA interference and small interfering RNAs. *Chembiochem* 2:239–45.

Urban-Klein, B., Werth, S., Abuharbeid, S., Czubayko, F., and Aigner, A. 2005. RNAi-mediated gene-targeting through systemic application of polyethylenimine (PEI)-complexed siRNA *in vivo*. *Gene Ther* 12:461–6.

Varkouhi, A.K., Lammers, T., Schiffelers, R.M., van Steenbergen, M.J., Hennink, W.E., and Storm, G. 2011a. Gene silencing activity of siRNA polyplexes based on biodegradable polymers. *Eur J Pharm Biopharm* 77:450–7.

Varkouhi, A.K., Scholte, M., Storm, G., and Haisma, H.J. 2011b. Endosomal escape pathways for delivery of biologicals. *J Control Release* 51(3):220–8.

Varum, K.M., Myhr, M.M., Hjerde, R.J., and Smidsrod, O. 1997. *In vitro* degradation rates of partially N-acetylated chitosans in human serum. *Carbohydr Res* 299:99–101.

Vasir, J.K. and Labhasetwar, V. 2007. Biodegradable nanoparticles for cytosolic delivery of therapeutics. *Adv Drug Deliv Rev* 59(8):718–28.

Wang, X., Niu, D., Hu, C., and Li, P. 2015. Polyethyleneimine-based nanocarriers for gene delivery. *Curr Pharm Des* 21(42):6140–56.

Ward, C.M., Pechar, M., Oupicky, D., Ulbrich, K., and Seymour, L.W. 2002. Modification of PLL/DNA complexes with a multivalent hydrophilic polymer permits folate-mediated targeting *in vitro* and prolonged plasma circulation *in vivo*. *J Gene Med* 4:536–47.

Ward, C.M., Read, M.L., and Seymour, L.W. 2001. Systemic circulation of poly(L-lysine)/DNA vectors is influenced by polycation molecular weight and type of DNA: Differential circulation in mice and rats and the implications for human gene therapy. *Blood* 97:2221–9.

Watanabe, T., Kanai, R., Kawase, T. et al. 1999. Family 19 chitinases of streptomyces species: Characterization and distribution. *Microbiology* 145:3353–63.

Werth, S., Urban-Klein, B., Dai, L. et al. 2006. A low molecular weight fraction of polyethylenimine (PEI) displays increased transfection efficiency of DNA and siRNA in fresh or lyophilized complexes. *J Control Release* 112:257–70.

Whitehead, K.A., Langer, R., and Anderson, D.G. 2009. Knocking down barriers: Advances in siRNA delivery. *Nat Rev Drug Discov* 8:129–38.

Wiwat, C., Siwayaprahm, P., and Bhumiratana, A. 1999. Purification and characterization of chitinase from Bacillus circulans no. 4.1. *Curr Microbiol* 39:134–40.

Wolfert, M.A., Dash, P.R., Nazarova, O. et al. 1999. Polyelectrolyte vectors for gene delivery: Influence of cationic polymer on biophysical properties of complexes formed with DNA. *Bioconjug Chem* 10:993–1004.

Wolfrum, C., Shi, S., Jayaprakash, K.N. et al. 2007. Mechanisms and optimization of *in vivo* delivery of lipophilic siRNAs. *Nat Biotechnol* 25:1149–57.

Wu, G.Y. and Wu, C.H. 1987. Receptor-mediated *in vitro* gene transformation by a soluble DNA carrier system. *J Biol Chem* 262:4429–32.

Yamagata, M., Kawano, T., Shiba, K., Mori, T., Katayama, Y., and Niidome, T. 2007. Structural advantage of dendritic poly(L-lysine) for gene delivery into cells. *Bioorg Med Chem* 15:526–32.

Yang, T.F., Chin, W., Cherng, J., and Shau, M. 2004. Synthesis of novel biodegradable cationic polymer: *N,N* diethylethylene diamine polyurethane as a gene carrier. *Biomacromolecules* 5:1926–32.

Yhee, J.Y., Song, S., Lee, S.J. et al. 2015. Cancer-targeted MDR-1 siRNA delivery using self-cross-linked glycol chitosan nanoparticles to overcome drug resistance. *J Control Release* 198:1–9.

Yuan, X., Li, L., Rathinavelu, A. et al. 2006. SiRNA drug delivery by biodegradable polymeric nanoparticles. *J Nanosci Nanotechnol* 6:2821–8.

Yuan, X., Naquib, S., and Wu, Z. 2011. Recent advances of siRNA delivery by nanoparticles. *Expert Opin Drug Deliv* 8:521–36.

Yuan, X., Shah, B.A., Kotadia, N.K. et al. 2010. The development and mechanism studies of cationic chitosan-modified biodegradable PLGA nanoparticles for efficient siRNA drug delivery. *Pharm Res* 27:1285–95.

Zhang, H. and Neau, S.H. 2001. *In vitro* degradation of chitosan by a commercial enzyme preparation: Effect of molecular weight and degree of deacetylation. *Biomaterials* 22:1653–8.

Zhang, H. and Neau, S.H. 2002. *In vitro* degradation of chitosan by bacterial enzymes from rat cecal and colonic contents. *Biomaterials* 23:2761–6.

Zhang, J.Y., Beckman, E.J., Piesco, N.P., and Agarwal, S. 2000. A new peptide-based urethane polymer: Synthesis, biodegradation, and potential to support cell growth *in vitro*. *Biomaterials* 21:1247–58.

Zhang, S., Zhao, B., Jiang, H., Wang, B., and Ma, B. 2007. Cationic lipids and polymers mediated vectors for delivery of siRNA. *J Control Release* 123:1–10.

Zheng, M., Pavan, G.M., Neeb, M. et al. 2012. Targeting the blind spot of polycationic nanocarrier-based siRNA delivery. *ACS Nano* 6:9447–54.

Zhou, C.R. and Yi, Z.J. 1999. Blood-compatibility of polyurethane/liquid crystal composite membranes. *Biomaterials* 20:2093–9.

Ziady, A.G., Ferkol, T., Dawson, D.V., Perlmutter, D.H., and Davis, P.B. 1999. Chain length of the polylysine in receptor-targeted gene transfer complexes affects duration of reporter gene expression both *in vitro* and *in vivo*. *J Biol Chem* 274(8):4908–16.

Zou, S.M., Erbacher, P., Remy, J.S., and Behr, J.-P. 2000. Systemic linear polyethylenimine (L-PEI)-mediated gene delivery in the mouse. *J Gene Med* 2:128–34.

# 14 Organic–Inorganic Nanocomposites for Biomedical Applications

*Subham Banerjee and Animesh Ghosh*

## CONTENTS

## 14.1 INTRODUCTION

The suitable integration of numerous organic–inorganic nanocomposites with outstanding biomedical performance has attracted significant attention in the research community. In particular, these nanocomposites show great potential in biomedicine and biological research because of their multifunctional properties, improved biocompatibility, long-term stability under physiological conditions, fewer immunogenic responses, prolonged blood circulation times, and targeting ability. Thus, the suitability of polymer nanotechnology and/or bio-nanocomposites to these evolving biomedical and biotechnological applications is a fast-growing area, which will be highlighted in this chapter.

Similarly, bio-nanocomposites produce an outstanding subject area relating to more than one branch of knowledge that combines polymer science, nanotechnology, and biology. Novel bio-nanocomposites are influencing interdisciplinary areas, especially in the field of biomedical science. Usually, polymer nanocomposites are the outcome of the combined form of inorganic/organic fillers and polymers at the nanometric range. The amazing uniqueness of these novel materials comes from the huge assortment of fillers and biopolymers accessible to investigators. These include current biopolymers, but are not restricted to proteins, polyesters, polysaccharides, poly nucleic acids, and polypeptides, while diluents comprise hydroxyapatite, metal nanoparticles, and clays (Ruiz-Hitzky et al. 2005). The formation of molecular associations in the polymer matrix is governed by the intermolecular interaction between nanocomposite filler components at the nanometric range. Bio-nanocomposites provide a supreme dimension to these improved features in which they are biocompatible and bioerodable/biodegradable components. The biodegradable components may be defined as components that are progressively eroded/degraded, absorbed, and/or excreted from the body, under which the process of degradation is usually mediated by hydrolysis or metabolic processes (Daniels et al. 1990). Therefore, these nanocomposites are receiving huge attention in the biomedical field, for example, bone repair/replacement, tissue engineering, and dental and controlled drug delivery applications. In Table 14.1, we list the name and structure of some frequently used biopolymers in biomedical applications. Hence, existing openings for nanocomposites based on polymeric materials in the field of biomedical science generates a large number of potential applications and enormously diverse requisite necessities individually from such potential applications. This chapter therefore focuses on present significant efforts with special emphasis on major research experiments in the evolving practice of organic–inorganic polymer nanocomposites for significant biomedical applications.

The aim of this chapter is to give some interesting classical examples from biomedical fields in order to provide a prospective of the applications of nanocomposites to drug delivery, biotechnology, and biomedical application, a field that is rapidly expanding and will continue to do so for the foreseeable future.

## 14.2   BIOMEDICAL APPLICATIONS OF ORGANIC NANOCOMPOSITES

Polymer matrix organic nanocomposites have been dominating the selection for components in bioerodable drug delivery systems and/or release applications. Copolymers and homopolymers derived from lactic acid or lactide, glycolic acid or glycolide, and ε-caprolactone explore the gravity of this investigation. Such polymers are mostly degraded by acid/base-catalyzed hydrolysis or by enzymatic hydrolytic action (Edlund and Albertsson 2002). One of the most prevalent polyesters, namely, poly(L-lactic acid) (PLA) shows the potential application in the biomedical sector. The reason behind the utilization and applicability of PLA in the biomedical field is its capability to be copolymerized and mixed to get nanocomposite such as components with anticipated features. The accumulation of nanoparticles can deliver an obstruction to release the drug in a very systematic manner and more controlled

**TABLE 14.1**

**Name and Structure of Some Commonly Used Polymers for Potential Biomedical Applications**

| Name of Polymer | Molecular Structure of Polymer |
| --- | --- |
| Chitosan | |
| Collagen | |

*(Continued)*

**TABLE 14.1 (*Continued*)**
**Name and Structure of Some Commonly Used Polymers for Potential Biomedical Applications**

| Name of Polymer | Molecular Structure of Polymer |
|---|---|
| Gelatin | |
| Cellulose | |
| Poly(lactic) acid | |

(Continued)

**TABLE 14.1 (*Continued*)**
**Name and Structure of Some Commonly Used Polymers for Potential Biomedical Applications**

| Name of Polymer | Molecular Structure of Polymer |
|---|---|
| Poly($\varepsilon$-caprolactone) | |
| Poly(L-lysine) | |
| Poly(vinyl alcohol) | |

fashion, by reducing the degree of swelling (Zhang et al. 2007) and improved mechanical integrity (Haraguchi and Li 2006) of hydrogel-based nanocomposites.

### 14.2.1 TARGETED DELIVERY OF SMALL MOLECULES WITH SPECIAL EMPHASIS ON ANTICANCER DRUG DELIVERY

Organic nanoparticles have recently attracted growing interest in anticancer drug delivery because they have one important advantage over inorganic ones. The advantage is that many organic nanoparticles are biodegradable, which can overcome the risk of chronic toxicity of non-biodegradable nanoparticles to cells or tissues (Uhrich et al. 1999; Panyam and Labhasetwar 2003). Antineoplastic drugs having low molecular weight results in fast elimination from the body, showing low therapeutic index, necessitating the elevation of doses in cancer patients, and consequently expanding the occurrence of cellular toxicity and additional side effects (Rashdan et al. 2013). Furthermore, a single administration of anticancer drugs showed a lack of specificity and led to a noteworthy destruction of normal cells, thereby producing unwanted adverse events, including alopecia (loss of hair), suppression of bone marrow cells, and sloughing of gastrointestinal epithelial cells (Bharali and Mousa 2010; Rashdan et al. 2013). Thus, it is essential to realize that the addition of chemotherapeutic molecules into the core of nanoparticles cannot just possibly diminish their unfavorable cytotoxic impacts, but in several circumstances may enhance the drug accumulation in the tumor vasculature, which is commonly recognized as an enhanced permeation and retention effect. Abraxane® and Doxil® are the two classical examples of marketed anticancer formulations based on nanotechnology that obtained U.S. Food and Drug Administration (USFDA) approval for metastatic breast cancer and ovarian carcinoma, respectively (Bharali and Mousa 2010; Rashdan et al. 2013).

Some of the significant anticancer drug delivery applications of organic small molecule-based nanotherapeutics are therapeutically effective against various cancer therapies. These examples are listed in Table 14.2.

### 14.2.2 TARGETED DELIVERY OF LARGE MOLECULES WITH SPECIAL EMPHASIS ON GENE, DNA/RNA, AND PROTEIN DELIVERY

Safe and effective gene delivery systems are tremendously important in gene therapy for tackling various viral infections, cardiovascular diseases, and genetic disorders. Gene therapy is a process by which genetic material in the form of oligonucleotides or plasmids is introduced into specific target cells to recover or induce the expression of a normal protein to treat human disorders. In addition to these therapeutic applications, gene delivery is widely used with significant potential impact to biotechnology, basic research, and diagnostic applications (Ziauddin and Sabatini 2001). Gene delivery systems comprise of both viral and nonviral vectors. Viral vectors are the most effective, but there is an increasing concern about their application, which is limited by their limited DNA loading capacity, and in therapeutic application by their oncogenicity and immunogenicity. Nonviral vectors are more reproducible, unable to represent DNA size limits, and are safer and less costly than viral vectors (Hyndman et al. 2004; Tokunaga et al. 2004).

**TABLE 14.2**
**Polymer Nanoparticles for Anticancer Drug Delivery**

| Biopolymer | Anticancer Drug | Therapeutic Improvement | References |
|---|---|---|---|
| Poly(lactic-co-glycolic acid) | Paclitaxel | Greater tumor growth inhibition effect compared with Taxol® | Danhier et al. (2009) |
| | Paclitaxel | Better antitumor effect than free paclitaxel | Fonseca et al. (2002) |
| | 9-Nitrocamptothecin | Controlled release up to 160 h | Derakhshandeh et al. (2007) |
| | Docetaxel | A greater extent of intracellular uptake compared to nontargeted NPs | Esmaeili et al. (2008) |
| Poly(caprolactone) | Tamoxifen | Preferential tumor targeting and high drug accumulation within tumor than controls | Shenoy and Amiji (2005) |
| | Taxol | Nanoparticle intensely adhered to the mucosal layer of the intestine and reduced glycemia at fasting condition in a dose-dependent manner | Kim and Lee (2001) |
| | Docetaxel | Showed prominent antitumor efficacy and prolonged survival rate in B16 melanoma-bearing mice compared with Duopafei | Zheng et al. (2009) |
| | Vinblastine | Showed efficient cellular uptake and cancer cell mortality | Prabu et al. (2008) |
| Gelatin | Paclitaxel | Paclitaxel-loaded nanoparticles were active against human RT4 bladder transitional cancer cells | Lu et al. (2004) |
| Chitosan | Cyclosporin A (Cy A) | Achieved therapeutic drug concentration in external ocular tissues significantly higher than those obtained following instillation of a chitosan solution containing CyA and an aqueous CyA suspension | De Campos et al. (2001) |
| Poly (butyl-cyanoacrylate) | Doxorubicin | After coating with Tween 80, there is a 60-fold increase in concentration in the brain than with simple saline solution | Gulyaev et al. (1999) |
| | Florafur | A drug combination of uracil and tegafur. Uracil generates bulk therapeutic concentrations of 5-FU to remain in the tumor cells and destroy them, whereas tegafur is captured by tumor cells/tissues and converted to 5-FU, a drug that destroys tumor cells/tissues | Arias et al. (2007) |

Small interfering RNA (siRNA), an innovative group of bio-drugs, are employed for silencing an extensive variety of target genes to cure a wide range of diseases, especially cancers (de Fougerolles et al. 2007; Oh and Park 2009). The administration of bare siRNA to the targeted site is still a significant obstacle due to fast enzymatic breakdown, restricted translocation through cellular membranes, and inadequate endosomal release (Sonawane et al. 2003). Hence, the main challenge in siRNA delivery is to locate an appropriate carrier for gene delivery, for clinical applications, that should be less toxic and have high efficiency of transfection. One apparent strategy for RNA/DNA delivery is through the viral vectors (Toh et al. 2011). However, the issue of immunogenicity, carcinogenicity, and inflammation that are associated with viral vectors makes it necessary to go for the advancement of nonviral drug delivery systems established on cationic polymers, cell-penetrating peptides (CPPs), and lipids. Among these, cationic polymers are strongly favored for reasons of their being more stable and cost effective than CPPs and/or cationic liposomes and easier to operate (de Ilarduya et al. 2010; O'Rorke et al. 2010). Of the cationic polymers presently applied in siRNA-based delivery, polyethyleneimine (PEI), which is a protonable amino nitrogen, is an excellent core for the fabrication of many sophisticated devices (Chen et al. 2011). Some cationic polymer (both synthetic and naturally occurring)-based nanocarriers for DNA/RNA drug delivery systems are discussed herein.

The major limitation of nanoparticles for the delivery of DNA/RNA is largely their nonpositive charge that restricts the interaction between the nonpositive charged DNA. Through the cationic surface modification of DNA-encapsulated poly(lactic-*co*-glycolic acid) (PLGA), nanoparticles can escape these drawbacks and can freely attach and condense DNA. From among numerous polycationic materials, e.g., PEI, cetyltrimethylammonium bromide, chitosan, etc., the naturally occurring carbohydrate polymer chitosan is well known as a potential polymer for the transfection of gene and its expression due to its inherent biocompatibility, mucoadhesion, biodegradability, and enhanced permeability and retention (EPR) characteristics (Dodane et al. 1999). In a related investigation, chitosan-coated nanoparticles (cNPs) were observed to release DNA at pH 7, prerequisite to qualify for drug release, to enhance the permeation of the entrapped macromolecules in mucosal surfaces, and to expedite *in vivo* gene delivery and gene expression with improved efficiency and without producing any inflammation (Munier et al. 2005). Further studies on the effect of size and charge of cNPs led us to arrive at the conclusion that a small-sized nanoparticle fraction resulted in 27 times greater transfection in COS-7 cells and four times greater transfection in HEK-293 cells for the similar dose of bigger-size cNPs (Pitts and Corey 1998). Above all, the appropriateness of such nanoparticles, as gene carriers, is analyzed by examining their bonding with antisense oligonucleotides 2'-*O*-methyl-RNA (OMR), against human telomerase RNA template region, for the management of carcinoma in the lungs by inhibiting the enzyme telomerase (Pitts and Corey 1998).

### 14.2.3 MISCELLANEOUS BIOMEDICAL APPLICATIONS

A new imprint lithography technique on a crosslinked perfluoroether templating mold has been designed to develop a uniform micro- and nanoparticulate system harvested from organic polymers, including biodegradable and biocompatible polymers (Rolland et al. 2005). The addition of inorganic nanoparticles has applicability in a varied field of biomedical applications by adapting this technique. Poly($\alpha$-hydroxyl acids) such as poly(glycolic acid), poly(lactic acid), poly($\varepsilon$-caprolactone), and poly(lactic acid-*co*-glycolic acid) fulfill a major number of the scaffold's component necessities, and their large-scale utilization as a component for tissue scaffolding is intended for tissue engineering applications (Lo et al. 1995; Garmendia et al. 2013). Hydrophobic polymeric scaffolds are highly porous chemically, biologically inert, and comparatively feeble, and this restricts their potential application *in vivo* for bone tissue regeneration, specifically in the site of implantations (Garmendia et al. 2013).

## 14.3  BIOMEDICAL APPLICATIONS OF INORGANIC NANOCOMPOSITES

Organic nanoparticles have been widely investigated to date, with liposomes, polymersomes, polymer constructs, and micelles, all being employed for imaging or gene and drug delivery techniques (Bae et al. 2011; Koo et al. 2011). However, inorganic nanoparticles have gained huge attention in the past few years because of their distinctive features and size-dependent physiochemical aspects which are very hard to achieve for solid/liquid lipid materials and/or nanoparticles based on the polymeric matrix. Thus, inorganic nanoparticles, especially silver, gold, and hydroxyapatite-based polymer nanocomposites, have vast potential in modern biomedical applications. The following text describes these commonly used inorganic nanocomposites and details their various successes in the biomedical field.

### 14.3.1  Application as Antimicrobial Agent: Silver Nanocomposites

High fraction of surface atoms and high specific surface area of metal nanoparticles possess excellent antimicrobial properties in comparison to raw metal. Many researchers have investigated the antimicrobial properties of metal nanoparticles (Kumar et al. 2004; Jain and Pradeep 2005; Lok et al. 2006; Williams et al. 2006; Ruparelia et al. 2008; Sathishkumar et al. 2009; Akhavan and Ghaderi 2010; Chamakura et al. 2011; Rashdan et al. 2013). In the experiment with *E. coli*, it was found that the other metal nanoparticles, for example, Au, Pt, FeO, Si and their oxides, and Ni, have less antibacterial effects, while Ag nanoparticles have significant antimicrobial properties (Lok et al. 2006; Williams et al. 2006; Rashdan et al. 2013). Therefore, we focus our observation along these lines based on the recent significant research output of the antimicrobial properties/activities of Ag nanoparticles.

Sathishkumar et al. (2009) with the aid of green synthesis developed Ag nanoparticles and investigated their antimicrobial activity. They described the process for the biosynthesis of Ag nanoparticles from Ag precursors by employing *Cinnamomum zeylanicum* powder and bark extract and examined the antimicrobial efficacy.

Ruparelia et al. (2008) developed and examined the antimicrobial activities of Ag nanoparticles by using four different strains of *E. coli*. It was stated that minimum inhibitory concentration (MIC) ranged from 40 to 180 µg/mL and minimum bactericidal concentration (MBC) ranged from 60 to 220 µg/mL for different *E. coli* strains. Around 50% zone of inhibition was reported with Ag nanoparticles concentration of 100 mg/L within 18 h for the *E. coli* strain. A similar observation was made by Sathishkumar et al. (2009) with the same concentration of Ag nanoparticles with a 55.8% zone of inhibition.

Lok et al. (2006) explored the antibacterial mechanism of Ag nanoparticles with a possible mechanism of action based on the antibacterial effect of nano-Ag by proteomic examination. The Ag nanoparticle proteomic marks treated with *E. coli* strains were evaluated via accumulating envelope protein precursors. This reflects the fact that the developed Ag nanoparticles may trigger nano-Ag on the bacterial membrane (Lok et al. 2006).

Tran et al. (2010) investigated the antibacterial effects of nano-Ag against several Gram-positive (*Bacillus subtilis*, *Lactobacillus fermentum*, and *Staphylococcus aureus*) and Gram-negative (*Escherichia coli* and *Pseudomonas aeruginosa*) bacteria and yeast (*Candida albicans*). The antibacterial effects were influenced by the structure of the bacterial cell membrane, where Gram-negative bacteria were strongly inhibited compared to Gram-positive bacteria (Tran et al. 2010) as described by their investigation.

Sadeghi et al. (2010) again described the nano-Ag bactericidal activity against *E. coli* and *S. aureus*. To examine this, their group evaluated the MIC values against these two bacterial species by measuring the absorbance value of the bacterial culture solution comprising several dilutions after 24 h (Figure 14.1).

In recent times, Chamakura et al. (2011) demonstrated that the antimicrobial activity of the nanoparticles is largely dependent on the nanoparticle-to-*E. coli* ratio than on the Ag nanoparticle concentration.

Polymeric nanoparticles made up of silver, silver oxide, and silver salts offer excellent antimicrobial activity (Chen and Cooper 2000; Hung and Hsu 2007). Ag nanoparticles were revealed to have a considerably better antimicrobial effect than

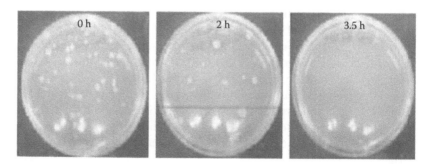

**FIGURE 14.1** Zone of inhibition of *E. coli* colonies after treatment with Ag nanoparticle having a concentration of 10 ppm with respect to various incubation hours such as 0, 2, and 3.5 h. (From Rashdan, S. et al. 2013. Nanoparticles for biomedical applications: Current status, trends and future challenges. In *Biomaterials and Medical Tribology: Research and Development*, ed. J.P. Davim, 1–132, Cambridge: Woodhead Publishing Limited. With permission.)

silver microparticles (Damm et al. 2008) due to the greater release profile of Ag ion. Ag nanoparticles with a particle size range between 5 and 50 nm at a concentration range between 0.1 and 1.0 on a percent weight basis in poly(methyl methacrylate) bone cement showed excellent antimicrobial features for joint arthroplasty (Alt et al. 2004) without influencing the cellular toxicity of Ag salts.

## 14.3.2 APPLICATION AS ANTICANCER AGENT: GOLD NANOCOMPOSITES

An area that has seen rapid development over the years is the application of Au nano-composites in the field of biological sciences. Specially, Au nanocomposites exhibit tremendous application in the biomedical sector due to their easy synthesis process, modification of surface, and improved optical properties with significant biocompat-ibility. From the biomedical applications points of view, Au nanoparticle surface modification is indispensable so as to locate them to particular diseased conditions (Fu et al. 2005; Rashdan et al. 2013).

Nano-Au can be directly combined with antimicrobial drugs/antibiotics via chem-ical bonding, or via physical absorption. The antibiotics-combined Au nanoparticles deliver excellent results for the therapeutic management of intracellular infections (Thirumurugan and Dhanaraju 2001; Chen et al. 2007; Saha et al. 2007; Gu et al. 2003; Rashdan et al. 2013). An example is methotrexate (MTX), an anticancer drug having the ability to inhibit cellular folate metabolism, tagged with 13-nm Au col-loids (Chen et al. 2007). After overnight incubation of the carboxylic groups of MTX conjugated with 13-nm colloidal Au nanoparticles, it has been stated that the con-jugated MTX concentration to Au nanoparticles is greater than pristine MTX. The anticancer activity of MTX tagged with nano-Au is almost seven times greater than pristine MTX in Lewis lung cancer cells (Figure 14.2).

Most delivery strategies, such as using cancer-targeting moieties conjugated to NPs for delivery into cancer tissue, are very similar to those used for magnetic and other types of NP. Indeed, gold has been used for many years to deliver molecules into cells. For delivery applications, AuNPs are employed for their small size, colloidal stability, ease of synthesis and conjugation, and their inert, biocompatible nature. Introduction into cells can either be forced, for example, with gene guns, or be achieved via cellular uptake. With regard to the gene gun technique, DNA is adsorbed onto the surface of AuNPs, which are then essentially shot into the cells (Sperling et al. 2008).

Gold nanomaterials are broadly employed to be a noninvasive surgery technique to kill cancer cells and remove diseased tissues (Kennedy et al. 2011). In addition to this, this photothermal effect can also be used to improve drug permeability and control drug release. In this section, we will focus on examples of using gold nano-materials to enhance drug release controllability, targeted delivery, and drug perme-ability followed by a brief summary of applying gold nanomaterial directly for drug delivery.

The application of GNPs in imaging includes dark-field imaging, which uses the light scattering of GNPs; this technique has been widely used for imaging of gold nanoparticles (GNPs) in tumoral cells (Loo et al. 2005). For instance, this technique has been used to image gold nanorods (GNRs) while they were being used to treat and diagnose head and neck cancer (Huang et al. 2006).

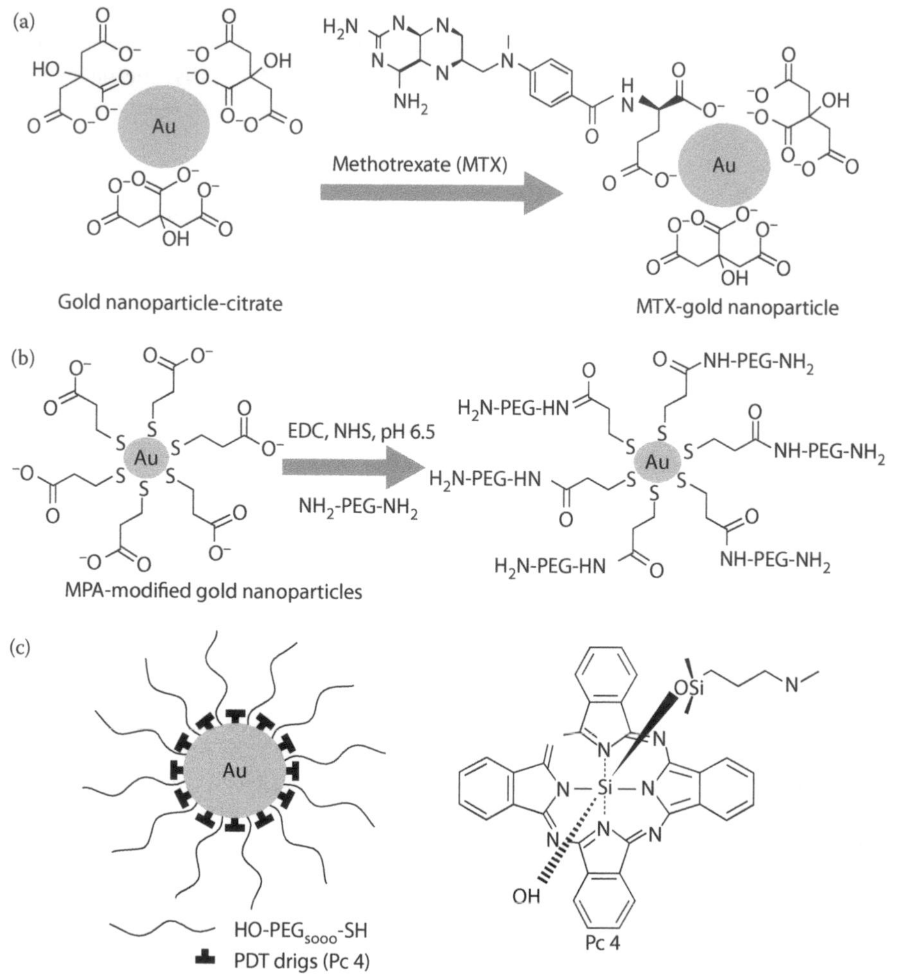

**FIGURE 14.2** Schematic diagram of MTX tagged with the spherical surface of nano-Au: (a) surface modification of nano-Au, (b) photodynamic drug therapy, and (c) tagged to PEGylated spherical nano-Au. (Reprinted from *J Control Release*, 149, Pissuwan, D., Niidome, T., and Cortie, M.B., The forthcoming applications of gold nanoparticles in drug and gene delivery systems, 65–71, Copyright 2011, with permission from Elsevier.)

The ease in attaching tumoral markers or drugs to GNPs has also been made use of to add specificity to the multitasking of GNP-based nanosystems. El-Sayed et al. carried out extensive work in this direction; for instance, they proved the ability of GNPs functionalized with a monoclonal antibody against epidermal growth factor receptor (anti-EGRF) to target only malignant epithelial cells (El-Sayed et al. 2006). Different labeling molecules such as platinum prodrugs, tamoxifen, folic acid, and transcription necrosis factor (a-TNF) have been used to functionalize GNPs to target different cancer cells such as lung and colon, breast (Dreaden et al. 2011), and MC-38 colon carcinomas, respectively (De et al. 2008).

### 14.3.3 APPLICATION IN TISSUE ENGINEERING FIELD

For tissue repair and rapid tissue regeneration, the field of tissue engineering has revealed marvelous potential in generating many biological substitutes (Mooney and Mikos 1999). Porous scaffold functions like a provisional base for sowing the cells *in vitro* and the generation of new tissues (Hua et al. 2002; Garmendia et al. 2013). It is to be noted that a perfect scaffold must be a three-dimensional highly porous network structure with excellent biocompatibility (Karageorgiou and Kaplan 2005; Rezwana et al. 2006; Garmendia et al. 2013) and suitable proliferation for cell adhesion cum differentiation as well (Cima et al. 1991; Garmendia et al. 2013). It should also permit the invasion of blood vessels for the supply of nutrients into the cells (Mikos et al. 1993; Garmendia et al. 2013). It should also offer the required tensile strength in order to uphold appropriate mechanical strength till new tissue generates (Uemura et al. 2003). Such tissue scaffolds should be biodegradable and biocompatible with a controlled rate of degradation and become ultimately obsolete when the development of new tissue begins. In such cases, the three-dimensional network structure is engaged by porous scaffolds that would be substituted by freshly generated tissue (Langer and Vacanti 1993; Holy et al. 1999; Garmendia et al. 2013).

Wei and Ma (2004) created a nano-hydroxyapatite (nHAp)/poly-L-lactide (PLLA) composite scaffold for the regeneration of bone tissues. They demonstrated that the elastic modulus elevated considerably when the amount of nHAp attained a PLLA composite of 30%. Furthermore, the incorporation of nHAp improved protein absorption, therefore, increasing the adhesion of cells.

Kothapalli et al. (2005) projected the HA/PLLA composite scaffold at the nano level. They state that those scaffolds are intended for applications in the biomedical tissue engineering field. Currently, Wang et al. (2010) experienced good *in vitro* behavior from porous nano-hydroxyapatite/polycaprolactone (nHAp/PCL) scaffolds. After the completion of a 7-day culture period, the bone marrow stromal cells (BMSCs) coalesced and generated big and flat cell layers and shields on nHA/PCL scaffolds. This outcome depicts that nHA/PCL scaffolds might be an alternative and significant approach in bone tissue engineering applications.

Many bioactive ceramics such as bicalcium phosphate, tricalcium phosphate, HAp, and glasses are applied to the tissue engineering field for bone regeneration (Vallet-Regi 2001). Among several synthetic ceramics, HAp is extensively used in the bone tissue engineering field because of its chemical resemblances to hard tissue inorganic components (Akao et al. 1981; Dewith et al. 1981). Many ceramic scaffolds that are bioactive in nature can either generate bone formation from neighboring tissue or can acts as a carrier-mediated matrix system for improved bone tissue regeneration via techniques such as the migration of cells, proliferation of cells, or by the cell differentiation process. The major drawback is that due to the brittleness of ceramics characters, they are not fit to bear over payloads (Paul and Sharma 2006). In order to avoid this difficult situation, the concept and application of polymer-ceramic-based nanocomposites have been proposed. Polymer-ceramic-based nanocomposites show similar natural bone composition in terms of both inorganic–organic compositions, the nanometric size range of the inorganic materials, especially bone-like apatite. To add high mechanical strength, nanostructured calcium phosphates are incorporated into this polymeric matrix system that mimics the natural strength of bone (Rho et al. 1998). Likewise, with these nanostructured matrix constituents, the osteoconductivity of tissue scaffolds and capability of bone joining are immensely improved where the adherence, *in situ* osteoblasts, and osteoprogenitor cells migration and their cell differentiation take place and lead to a newly developed bone matrix system (Zhang and Ma 1999; Ma et al. 2001; Garmendia et al. 2013).

Another area of extensive investigation deals with electrospinning approaches for manufacturing bioresorbable nanofiber scaffolds intended for applications in the field of tissue engineering. Such bioresorbable nanofiber scaffolds could be constructed in the form of nanocomposite matrix systems in order to permit resultant scaffolds for the progression of the cell by generating an exclusive composite matrix system. Construction of polymer bio-nanocomposites based on biomimetic methods has currently been an area of interest for investigators. Among various nanocomposite materials, HAp-polymer-based biocompatible nanocomposite materials are extensively utilized in osteoconductive replacement for the repair of bone and implantation (Hule and Pochan 2007). HAp, a chief component of hard tissue, displays unsatisfactory mechanical features if directly utilized. Therefore, polymer-based matrix composites are preferred to achieve satisfactory mechanical properties. Matrix biodegradation is preferred to permit the development of new bone at the site of repair. Frequently, biocompatible and biodegradable natural and/or synthetic polymers are used as essential components of the matrix system in their investigations. Gelatin derived from collagen (Kim et al. 2005) and poly-2-hydroxyethylmethacrylate/poly (3-caprolactone) (Huang et al. 2007) nanocomposites based on HAp are classical examples that have been examined for the repair of bone systems. Electrospinning technology is also a very popular method to develop nanofiber scaffolds from biodegradable polymer solutions for applications in the tissue engineering field.

### 14.3.4 Application as MRI Contrast Agent and Hyperthermia Drug Delivery: Magnetic Iron Oxide Nanocomposites

Another important classical example of inorganic nanocomposites for biomedical application is superparamagnetic iron oxide nanoparticles (SPIO-NPs). SPIO-NPs have a broad range of applications in the field of both biomedical and pharmaceutical sciences in terms of MRI contrast agents and drug delivery applications, respectively (Jain et al. 2005; Laurent et al. 2008). SPIO-NPs are excellent as MRI contrast agents in the imaging of cellular and molecular basic functional mechanisms and their processes of dysfunctionality (Modo and Bulte 2007). Biodegradable and biocompatible iron oxide mineral core are the basic components of SPIO-NPs. They are capable of uptake by the reticulo-endothelial system and follow the biochemical pathways of normal physiological iron metabolism process (Bulte and Kraitchman 2004). Additionally, the coating at the surface of SPIO-NPs permits the linkage between the chemical groups of functional moieties with the vectorizing molecules, which enable them to trigger specific organs of interest or any other diseased conditions (Thorek et al. 2006; Bae et al. 2009). Such SPIO-NPs can easily cross various physiological barriers being considered as an external particle by the mononuclear phagocyte system (MPS) (Owens and Peppas 2006) when they are administered into living biological systems. Thus, inside the living systems, they are generally considered as a foreign particle, a foreign molecule, or a foreign body, and for that particular reason, the MPS attempts to remove them from the systemic circulation by three basic mechanisms: first, the plasmatic proteins opsonization method; second, macrophages recognition; and finally, by the process of cellular phagocytosis.

As SPIO-NPs are comparatively large in size, their primary demerit in the application as an MRI contrast agent is rapid opsonization by the plasmatic proteins when administered intravenously. After administration of SPIO-NPs intravenously, it rapidly undergoes a huge capture by macrophages, especially by the hepatic Kupffer cells. The fate of their distribution range inside the liver is 80%–90%, inside the spleen is 5%–8%, and in the bone marrow varies from 1% to 2%. But the basic interesting fact is that ultrasmall superparamagnetic iron oxides (USPIOs) nanoparticles, which are relatively smaller in size compared to SPIO-NPs, undergo reduced cellular uptake by the hepatic Kupffer cells. Thus, they have a relatively longer biological half-life than various generic names of iron oxide contrast agents, for example, AMI–227 has 200 min, MION-46 has 180 min, SHU–555 has 10 min, and AMI–25 has 8 min. These USPIOs are predominantly operative as an MRI contrast agent for imaging lymph nodes because of their gradual uptake by the macrophages present in healthy lymphatic nodes (Pultrum et al. 2009). These USPIOs are capable of moving across the interstitium in a nonspecific vesicular manner to enter the lymphatic ganglion region at which they will be gathered by the reticulo-endothelial system. Such occurrence of this opsonization process is restricted when the surface of the NPs are relatively tiny, neutral, and water soluble (Yoo et al. 2008).

Again, owing to intense uptake by the macrophages present in the hepatic Kupffer cells, iron oxide NPs with tiny particle size are excellent in imaging tumors in the liver and in liver metastases as well. Additionally, SPIO-NPs with a particle size range between 7 and 30 nm are ideal for blood-pool imaging as it holds ultimate *in vivo* physicochemical and biological characteristics.

An approach for enhancing the particles' vascular remanence is adsorption with neutral and superabsorbent polymers such as poly(ethylene glycol) or poly(ethylene oxide) on their particle surface (Kohler et al. 2004). Apart from these polymers, other less effective polymers such as carboxymethyl cellulose, poly (vinyl alcohol), and dextran also cause interference in the process of protein phagocytosis. However, albumin protein are quite incapable of delay in phagocytic uptake by the macrophage present in the liver due to the presence of positive and nonpositive charges in carboxylate and amine moieties, respectively. Thus, in a huge number of cases, such delay in phagocytic uptake by the macrophage present in the liver is influenced by the adsorbed polymer nature and their concentration at the surface of the particles.

Therefore, the prime limitation of such MRI contrast agents is not to resolve and detect the actual image of the liver tumors in the MPS such as hepatocarcinomas, hemangiomas, and lymphomas or other drastic hepatic pathological conditions such as liver cirrhosis, hepatitis (inflammation in liver where macrophages play an active role), etc. However, by grafting the functional moieties with a specific ligand targeting their surface, there is a good way to avoid these severe limitations. One of the most modern approaches for this is the process of vectorization (Thorek and Tsourkas 2008) via attaching the surface of the NPs with various ligands such as oligosaccharides (Mornet et al. 2006); with various proteins such as lactoferrin, transferrin, insulin, secretin, peptides, and antibodies (Burtea et al. 2005; Kou et al. 2008); or with many other vitamins such as $B_9$, folic acid, etc., where the overexpression of tumors cell receptors is occasionally observed (Di Marco et al. 2007). Definite localization of the vectors by the receptors present in the cellular membrane leads to the capture of such NPs by the cells.

Since we are particularly focused on cellular and molecular MRI-related issues in this section, it is stated here that the concentration in the range between micromolar to nanomolar of the MRI contrast agent is capable of resolving a superior image at the surrounding areas of the tissue via elevating their aqueous protons' rate of relaxation and thereby restricting them to keep out from the image acquisition areas (Bulte and Kraitchman 2004; Liu and Frank 2009). Though, expending sufficient image acquirement factors, it is seen that the image contrast result of SPIO-NPs is much better than paramagnetic NPs. Hence, the lower intensity produced by contrast agents such as iron oxide-NPs during an MRI will always happen at a smaller relaxing ion concentration than the higher signal generated from Gd complexes (Liu and Frank 2009). This reduces the specificity in the detection limit during the image acquisition process in the perspective of molecular or cellular MRI tests using SPIO-NPs. The methods for visualizing iron oxide-labeled cells of positive contrast were recently developed (Liu and Frank 2009).

Apart from as an MRI contrast agent, some other interesting biological applications of SPIO-NPS are as follows:

- Specific and nonspecific cell labeling
- Hyperthermia
- Drug delivery

We have already discussed the necessary features and biological applications of SPIO-NPs in terms of specific and nonspecific cell labeling. Now, we proceed toward one of the most important and interesting therapeutic strategies of SPIO-NPs in the field of hyperthermia or magnetic drug delivery (MDD) applications (Nelson et al. 2008), targeting the principle of magnetic resonance that offers remarkable therapeutic prospects.

MDD approaches separately constitute an excellent notion for cell treatment because of their ability in combining with the active pharmaceutical ingredients (APIs) or drugs specifically and selectively to their targeted sites (Douziech-Eyrolles et al. 2007; Cengelli et al. 2009; Sajja et al. 2009). They have an ability to alter or differ as per the treatment needs and types considered, such as gene therapy or vectorization of the gene, chemotherapy with anticancer drugs, or CNS therapy for lesion repair. Such approaches help to minimize the excessive use of drugs and thereby reduce drug-related adverse events. In a tumor site, magnetic vectors are maintained in huge concentration with the application of an external magnetic field.

Owing to the magnetic properties of iron oxide, a major interesting therapeutic application of SPIO-NPs lies in the treatment of hyperthermia (Purushotham et al. 2009). Such approaches comprise of exposing the subject to the environment of a nonionizing electromagnetic wave to elevate the temperature of the surrounding tumor cells, because in comparison to healthy cells, these cells are far more sensitive to generate external energy; thus, a selective pathological section abolition can be attained. In order to exhibit a desired therapeutic effect at the site of the tumor, a temperature beyond 43°C must be reached (Renard et al. 2009). There are some interesting examples given for the therapeutic applications of SPIO-NPs in the treatment of hyperthermia.

Wust et al. (2006) displayed that using superparamagnetic nanocrystals is likely to get a considerably higher transformation of magnetic energy rate into heat energy as compared to ferromagnetic crystals. Such an approach was successfully used for the inhibition of the growth of tumors in the "mammary carcinoma" in experimental animal models such as mice.

As per the first clinical cases (Maier-Hauff et al. 2007), in certain clinical reports, the collective use of radiotherapy and thermotherapy could enhance the survival rate up to twofold after 2 years, in comparison with normal radiotherapeutic treatment. The chemotherapeutic drug-mediated resistance of cancer cells in certain cancer treatments is also a matter of great concern (Szakacs et al. 2006). Such cancers, stubborn to chemotherapy, are likely to proliferate, enhance the supply of blood , and ultimately encourage the stage of metastases regardless of the treatment.

Again, in molecular imaging, specificity, as well as furtivity, is the essential factor. The surface characteristics of NPs are subsequently decisive for their cellular interactions with each other. So far, the SPIO-NPs are coated with albumin or dextran in order to interact with the cells. However, their cell internalization remains quite feeble. For this, SPIO-NP-mediated cell tagging is presently studied systematically to visualize the imaging of cell movement and trafficking. There is an unmet need to establish an effective cell tagging protocol by these SPIO-NPs.

Arbab et al. (2006) showed an exciting cellular MRI imaging for nonphagocyte cells by SPIO-NPs. Many methodologies were adapted to achieve the desired cell capture. For instance, cationic transfecting agents such as poly (L-lysine) can be put to those NPs in order to enhance the magnetic labeling of the cells. Definitely, in charged conditions, the SPIO-NPs mediate an electrostatic interaction with the membranes of the cells for the development of endocytosis blisters (Chen et al. 2009).

Wilhelm and Gazeau (2009) again examined a large degree of cellular internalization that took place with the anionic maghemite NPs as compared to those NPs entrapped in dendrimer transfection agent (Jing et al. 2008) or united with the polypeptide chain (Wilhelm and Gazeau 2009), resulting from HIV virus protein, and also recognized its unique features of cellular penetration. A well-organized, stepwise labeling procedure helps in the detection of *in vivo* images by MRI cells conjugated with SPIO-NPs after parenteral administration (Heyn et al. 2006).

An additional policy to raise the growth of magnetic NPs adopted by cells is to generate a high degree of flexibility for the cell trafficking imaging that indicates the usage of micrometer-range particles size, as recommended earlier by Shapiro et al. (2005, 2006). Such micron-level particles can also be utilized as a transporter, especially for synthetic organic dyes to merge both optical and magnetic imaging (Sumner et al. 2007). The process of labeling can also be understood *in vivo* via intravenous administration of SPIO-NPs for examining the inflammatory response of the cells or tissues (Dunn et al. 2005), which is exciting for organ rejection study

(Wu et al. 2006). Generally, macrophages are considered as immune cells that can play an immense role in the study of the rejection of organs and their presence is mostly understood by biopsy. MRI-mediated macrophages imaging after tagging with SPIO-NPs performs a very encouraging carrier-mediated system for understanding of the organs' rejection mechanism in future transplants.

Contrast substances also do adjust themselves *in situ*. By adopting this concept, an MR reporter gene is founded on this context and it establishes an additional field in healthcare research (Gilad et al. 2007). It is essential to say that the reporter gene is that coding gene that is responsible for the generation of the contrast agent. The enzyme, namely, luciferase is usually utilized for investigating the activity of the other genes for delivery of the therapeutic gene. The MR reporter gene is very interesting due to its distinct optical imaging features; it offers excellent anatomical resolution. Inside the cell, the gathering of various metalloid compounds from the endogenous or exogenous origin (Batya et al. 2005; Genove et al. 2005) persuades a local alteration at the time of relaxation and permits MRI-based molecular imaging (Cohen et al. 2007).

Magnetic particle imaging (MPI), an innovative tomographic imaging method, has been recently established (Rahmer et al. 2009; Weizenecker et al. 2009). This technique is sufficient for imaging of magnetic tracers at high temporal and spatial resolutions. MPI possesses significant applications in the area of material and medicinal science.

## 14.3.5 Imaging-Guided Drug Delivery Using Upconverting Luminescent Nanoparticles

Today, lanthanide-doped UCNPs have appeared as greatly encouraging nanomaterials for a wide range of research applications in the area of biomedical science and engineering. For that, functional agents are designed for the application in cell labeling and cell tracking (Nam et al. 2011), small animal imaging (Nyk et al. 2008), delivery of drugs (Wang et al. 2011), and genetic material (Guo et al. 2011), photodynamic therapy (Shan et al. 2011), and photothermal therapy (Cheng et al. 2011a). All these applications are laid on the concrete and robust design to project and develop UCNPs-based image-guided nanoscale templates for molecular diagnostics and drug delivery applications. UCNPs hold manifold characteristics that allow them to design well, in organized materials for drug delivery application, various therapeutical utilizations, and concurrent imaging. Initially, UCNPs were selected for biomedical imaging because of their upconverting features, depicting a nonlinear optical method where the consecutive absorption of several low-energy photons with a broad wavelength affects the emission of light at shorter wavelengths more than the excited photons (Auzel 2004). Specifically, UCNPs are strongly excited by near-infrared (NIR) and they discharge higher-energy excited photons at ultraviolet (UV) to NIR range. These supreme photoluminescence features are usually absent for current exogenous and endogenous fluorophores, contributing several unique features to UCNPs for application in biomedical optical imaging. These UCNPs are particularly stable in the presence of light and are nonblinking in nature (Wu et al. 2009), making them extremely promising for both *in vitro* and *in vivo* therapeutics tracking.

Blue, green, and red visible emissive UCNPs have been applied for imaging multi-color cellular components, and imaging in thin tissues (Cheng et al. 2011b) and optically transparent organisms (such as *Caenorhabditis elegans*) (Zhou et al. 2011). In respect to the small animal imaging, the UCNPs with both NIR excitation and NIR emission are predominantly promising, as they transmit minimal scattering of light and superior tissue permeation for *in vivo* imaging as both excitation and emission wavelengths range within the biological NIR optical transmission window (700–1000 nm). Monodisperse UCNPs can be easily developed by varying their morphology in terms of smaller sizes and shapes, with a relatively broad area of the surface for well-organized drug conjugation and targeting ligands (Verma and Rotello 2005) as well. Moreover, UCNPs are highly biocompatible because they are entirely free of class A heavy metals (such as cadmium, mercury, and lead) and class B heavy metals (such as selenium, arsenic, and other miscellaneous toxicological substances). These may ultimately lead to promising substances for numerous biomedical applications, especially in small animal imaging and biosensing (Chatterjee et al. 2010; Fischer et al. 2011; Wang et al. 2011; Zhou et al. 2012).

Currently, UCNP-based drug delivery systems are categorized into three major groups: (1) hydrophobic pockets, (2) mesoporous silica shells, and (3) hollow meso-porous-coated spheres. In case of "hydrophobic pockets," water-insoluble APIs are entrapped into the cavity of such pockets at the surface of the UCNP. The water-insoluble polymeric side chains of these polymers are tagged at the surface ligands of UCNP through van der Waals interactions, creating the "hydrophobic pockets" where anti-neoplastic drug moieties could be bound. The amount of drug loading and the cumulative percent release of the antineoplastic drug were regulated via altering the pH of the system, with an experiential rise in the rate of drug dissociation in gastrointestinal conditions, which is considered as an outstanding candidate for sustained-release drug delivery system in abnormal cells. The same method was adopted with this and with iron oxide nanoparticles. These nanocomposites loaded with drugs qualified for simultaneous optical imaging and for targeted drug delivery systems (Xu et al. 2011).

In *mesoporous silica shell* method, drugs are encapsulated into mesoporous silica shell pores coated into the UCNPs surface. Such mesoporous structured silica shells possess a broad specific surface area and large pore volume, permitting them to incorporate a bulk quantity of APIs. The drug-loading content values were quantified by estimating the rate and extent of transfer in energy between the drug and matrix. The following rate of release of drug substances were measured by the recovery of the intensity of upconverting luminescence light emission (Hou et al. 2011).

In the final process, therapeutically active moieties are embedded onto a hollow sphere that is coated with the shells of UCNP. Such hollow spheres permit a considerable amount of theoretical drug loading while upholding upconverting-imaging ability. In this respect, a nanorattle system was newly established by an ion-exchange process (Xu et al. 2011). Such a system comprises water-soluble, $NaYF_4$ shells having an inner magnetic nanoparticle. Basically, it is a multifunctional mesoporous nanostructure with added upconverting luminescent and magnetic features, which offers high drug-encapsulation efficiency values. For such carriers, the adsorption and/or desorption isotherm measurement of nitrogen was engaged to estimate their internal volume for drug content. Such NPs are thought to localize inside the cells by either the nonspecific endocytosis method or the receptor-mediated endocytosis process. The surface coating of UCNPs can proceed to pH-related osmotic swelling and endosomal disruption at the consequent cytosolic drug delivery through the proton sponge effect.

Moreover, UCNPs having sole upconverting optical and chemical characteristics have been managed in gene therapy to deliver nucleic acids. Again, disease-related site-specific drug delivery could be attained by the receptor-mediated endocytosis pathway. This site-specific drug delivery can be achieved by conjugating a drug-delivery matrix with other ligands or molecules that precisely distinguish the surface-bound target cells receptor. These target-specific findings contain a ligand as an acceptor and antigen for antigen–antibody interactions. So far, such targeted imaging of the cell has been accomplished by covalently binding UCNPs with many biomolecules, such as folic acid (Hu et al. 2009), biotin (Sivakumar et al. 2006), antibodies (Wang et al. 2009a, b), and peptides (Zako et al. 2009).

## 14.4   CONCLUDING REMARKS

Many investigations are performed at nano-range for various applications in the area of biomedical research, especially in the treatment of cancer. We have talked about the evolving group of both organic–inorganic nanocomposite materials in light of different organic and inorganic polymeric nanpcomposites and nanofillers that are utilized broadly or show potential aspect in the research area of biomedical science. These materials are quite different from ceramic/inorganic reinforced nanocomposites for mechanical improvement and peptide-based nanomaterials where peptides act as either fillers or matrix, with the technology intended to make the biocompatible material. Increased attention for these nanocomposites are quite differ from their application-oriented strategy to understand a large number of structure—property—relations. Essential practical criteria comprise biodegradability, mechanical strength, morphology, biocompatibility, and a large group of different parameters, contingent upon end utilization.

## 14.5   FUTURE TRENDS

The advances are marvelous and the results are extremely encouraging, though a lot of research work is required regarding the treatment of different sorts of human ailments, with more emphasis in terms of toxicity and environmental issues. Again, the initial level of action of such nanocomposites is to interact between natural and synthetic polymers or matrix/fillers that can be tailored and can grab particular necessities. We do further expect that advanced research in these areas will be profitable in examining the configuration of innovative organic–inorganic nanocomposites intended for biomedical applications.

## ACKNOWLEDGMENTS

Appreciation goes to all the authors of papers, books, websites, and all published sources (listed in the references) that were used to prepare the contents of this chapter.

# REFERENCES

Akao, M., Aoki, H., and Kato, K. 1981. Mechanical properties of sintered hydroxyapatite for prosthetic applications. *J Mater Sci* 16:809–12.

Akhavan, O. and Ghaderi, E. 2010. Self-accumulated Ag nanoparticles on mesoporous $TiO_2$ thin film with high bactericidal activities. *Surf Coat Technol* 204:3676–83.

Alt, V., Bechert, T., Steinrucke, P. et al. 2004. An *in vitro* assessment of the antibacterial properties and cytotoxicity of nanoparticulate silver bone cement. *Biomaterials* 25:4383–91.

Arbab, A.S., Liu, W., and Frank, J.A. 2006. Cellular magnetic resonance imaging: Current status and future prospects. *Expert Rev Med Devices* 3:427–39.

Arias, J.L., Gallardo, V., Ruiz, M.A. et al. 2007. Ftorafur loading and controlled release from poly(ethyl-2-cyanoacrylate) and poly(butylcyanoacrylate) nanospheres. *Int J Pharm* 337:282–90.

Auzel, F. 2004. Upconversion and anti-stokes processes with f and d ions in solids. *Chem Rev* 104:139–73.

Bae, K.H., Chung, H.J., and Park, T.G. 2011. Nanomaterials for cancer therapy and imaging. *Mol Cells* 31:295–302.

Bae, S.J., Park, J.A., Lee, J.J. et al. 2009. Ultrasmall iron oxide nanoparticles: Synthesis, physicochemical, and magnetic properties. *Curr Appl Phys* 9:S19–S21.

Batya, C., Hagit, D., Gila, M. et al. 2005. Ferritin as an endogenous MRI reporter for noninvasive imaging of gene expression in C6 glioma tumors. *Neaplasia* 7:109–17.

Bharali, D.J. and Mousa, S.A. 2010. Emerging nanomedicines for early cancer detection and improved treatment: Current perspective and future promise. *Pharmacol Ther* 128:324–35.

Bulte, J.W.M. and Kraitchman, D.L. 2004. Iron oxide MR contrast agents for molecular and cellular imaging. *NMR Biomed* 17:484–499.

Burtea, C., Laurent, S., Vander, E.L. et al. 2005. Cellular magnetic-linked immunosorbent assay, a new application of cellular ELISA for MRI. *J Inorg Biochem* 99:1135–44.

Cengelli, F., Grzyb, J.A., Montoro, A. et al. 2009. Surface-functionalized ultrasmall superparamagnetic nanoparticles as magnetic delivery vectors for camptothecin. *Chem Med Chem* 4:988–97.

Chamakura, K., Perez-Ballestero, R., Luo, Z. et al. 2011. Comparison of bactericidal activities of silver nanoparticles with common chemical disinfectants. *Colloids Surf B* 84:88–96.

Chatterjee, D.K., Gnanasammandhan, M.K., and Zhang, Y. 2010. Small upconverting fluorescent nanoparticles for biomedical applications. *Small* 6:2781–95.

Chen, B., Liu, M., Zhang, L. et al. 2011. Polyethylenimine-functionalized graphene oxide as an efficient gene delivery vector. *J Mater Chem* 21:7736–41.

Chen, C.B., Chen, J.Y., and Lee, W.C. 2009. Fast transfection of mammalian cells using superparamagnetic nanoparticles under strong magnetic field. *J Nanosci Nanotechnol* 9:2651–59.

Chen, C.Z. and Cooper, S.L. 2000. Recent advances in antimicrobial dendrimers. *Adv Mater* 12:843–46.

Chen, Y.H., Tsai, C.Y., Huang, P.Y. et al. 2007. Methotrexate conjugated to gold nanoparticles inhibits tumor growth in a syngeneic lung tumor model. *Mol Pharm* 4:713–22.

Cheng, L., Yang, K., Li, Y.G. et al. 2011a. Facile preparation of multifunctional upconversion nanoprobes for multimodal imaging and dual-targeted photothermal therapy. *Angew Chem Int Ed* 50:7385–90.

Cheng, L.A., Yang, K., Shao, M.W. et al. 2011b. Multicolor *in vivo* imaging of upconversion nanoparticles with emissions tuned by luminescence resonance energy transfer. *J Phys Chem C* 115:2686–92.

Cima, L.G., Vacanti, J.P., Vacanti, C. et al. 1991. Tissue engineering by cell transplantation using degradable polymer substrates. *J Biomech Eng* 113:143–51.

Cohen, B., Ziv, K., Plaks, V. et al. 2007. MRI detection of transcriptional regulation of gene expression in transgenic mice. *Nature Med* 13:498–503.

Damm, C., Münstedt, H., and Rösch, A. 2008. The antimicrobial efficacy of polyamide 6/ silver-nano- and microcomposites. *Mater Chem Phys* 108:61–66.

Danhier, F., Lecouturier, N., and Vroman, B. 2009. Paclitaxel-loaded PEGylated PLGA-based nanoparticles: *In vitro* and *in vivo* evaluation. *J Control Release* 133:11–17.

Daniels, A.U., Chang, M.K.O., and Andriano, K.P. 1990. Mechanical properties of biodegradable polymers and composites proposed for internal fixation of bone. *J Appl Biomater* 1:57–58.

De Campos, A.M., Sanchez, A., and Alonso, M.J. 2001. Chitosan nanoparticles: A new vehicle for the improvement of the delivery of drugs to the ocular surface. Application to cyclosporin A. *Int J Pharm* 224:159–68.

de Fougerolles, A., Vornlocher, H.P., Maraganore, J. et al. 2007. Interfering with disease: A progress report on siRNA-based therapeutics. *Nat Rev Drug Discov* 6:443–53.

de Ilarduya, C.T., Sun, Y., and Düzgünes, N. 2010. Gene delivery by lipoplexes and polyplexes. *Eur J Pharm Sci* 40:159–70.

De, M., Ghosh, P.S., and Rotello, V.M. 2008. Applications of nanoparticles in biology. *Adv Mater* 20:4225–41.

Derakhshandeh, K., Erfan, M., and Dadashzadeh, S. 2007. Encapsulation of 9-nitrocamptothecin, a novel anticancer drug, in biodegradable nanoparticles: Factorial design, characterization and release kinetics. *Eur J Pharm Biopharm* 66:34–41.

Dewith, G., van Dijk, H.J.A., Hattu, N. et al. 1981. Preparation, microstructure and mechanical properties of dense polycrystalline hydroxyapatite. *J Mater Sci* 6:1592–98.

Di Marco, M., Sadun, C., Port, M. et al. 2007. Physicochemical characterization of ultrasmall superparamagnetic iron oxide particles (USPIO) for biomedical application as MRI contrast agents. *Int J Nanomedicine* 2:609–22.

Dodane, V., Khan, M.A., and Merwin, J.R. 1999. Effect of chitosan on epithelial permeability and structure. *Int J Pharm* 10:21–32.

Douziech-Eyrolles, L., Marchais, H., Hervé, K. et al. 2007. Nanovectors for anticancer agents based on superparamagnetic iron oxide nanoparticles. *Int J Nanomed* 2:541–50.

Dreaden, E.C., Mackey, M.A., Huang, X. et al. 2011. Beating cancer in multiple ways using nanogold. *Chem Soc Rev* 40:391–404.

Dunn, E.A., Weaver, L.C., and Dekaban, G.A. 2005. Cellular imaging of inflammation after experimental spinal cord injury. *Mol Imaging* 4:53–62.

Edlund, U. and Albertsson, A.C. 2002. Degradable polymer microspheres for controlled drug delivery. *Adv Polym Sci* 157:67–112.

El-Sayed, I.H., Huang, X.H., and El-Sayed, M.A. 2006. Selective laser photo-thermal therapy of epithelial carcinoma using anti-EGFR antibody conjugated gold nanoparticles. *Cancer Lett* 239:129–35.

Esmaeili, F., Ghahremani, M.H., Ostad, S.N. et al. 2008. Folate-receptor-targeted delivery of docetaxel nanoparticles prepared by PLGA–PEG–folate conjugate. *J. Drug Target* 16:415–423.

Fischer, L.H., Harms, G.S., and Wolfbeis, O.S. 2011. Upconverting nanoparticles for nanoscale thermometry. *Angew Chem Int Ed* 50:4546–51.

Fonseca, C., Simoes, S., and Gaspar, R. 2002. Paclitaxel-loaded PLGA nanoparticles: Preparation, physicochemical characterization and *in vitro* anti-tumoral activity. *J Control Release* 83:273–86.

Fu, W., Shenoy, D., Li, J. et al. 2005. Biomedical applications of gold nanoparticles functionalized using hetero-bifunctional poly(ethylene glycol) spacer. *Mater Res Soc Symp Proc* 845:AA5.4.1–6.

Garmendia, N., Olalde, B., and Obieta, I. 2013. Biomedical applications of ceramic nanocomposites. In *Ceramic Nanocomposites*, ed. R. Banerjee, I. Manna, 530–547, Cambridge: Woodhead Publishing Limited.

Genove, G., DeMarco, U., Xu, H. et al. 2005. A new transgene reporter for *in vivo* magnetic resonance imaging. *Nat Med* 11:450–54.

Gilad, A.A., Winnard, P.T., Jr. van Zijl, P.C.M. et al. 2007. Developing MR reporter genes: Promises and pitfalls. *NMR Biomed* 20:275–90.

Gu, H., Ho, P.L., Tong, L. et al. 2003. Presenting vancomycin on nanoparticles to enhance antimicrobial activities. *Nano Lett* 3:1261–63.

Gulyaev, A.E., Gelperina, S.E., Skidan, I.N. et al. 1999. Significant transport of doxorubicin into the brain with polysorbate 80-coated nanoparticles. *Pharm Res* 16:1564–69.

Guo, H.C., Idris, N.M., and Zhang, Y. 2011. LRET-based biodetection of DNA release in live cells using surface-modified upconverting fluorescent nanoparticles. *Langmuir* 27:2854–60.

Haraguchi, K. and Li, H.J. 2006. Mechanical properties and structure of polymer-clay nanocomposite gels with high clay content. *Macromolecules* 39:1898–905.

Heyn, C., Ronald, J.A., Mackenzie, L.T. et al. 2006. *In vivo* magnetic resonance imaging of single cells in mouse brain with optical validation. *Magn Reson Med* 55:23–29.

Holy, C.E., Dang, S.M., Davies, J.E. et al. 1999. In vitro degradation of a novel poly(lactide-*co*-glycolide) 75/25 foam. *Biomaterials* 20:1177–85.

Hou, Z.Y., Li, C.X., Ma, P.A. et al. 2011. Electrospinning preparation and drug-delivery properties of an up-conversion luminescent porous NaYF(4):Yb(3+), Er(3+)@silica fiber nanocomposite. *Adv Funct Mater* 21:2356–65.

Hu, H., Xiong, L., Zhou, J. et al. 2009. Multimodal-luminescence coreshell nanocomposites for targeted imaging of tumor cells. *Chemistry* 15:3577–84.

Hua, F.J., Kim, G.E., Lee, J.D. et al. 2002. Macroporous poly(L-lactide) scaffold. Preparation of a macroporous scaffold by liquid-liquid phase separation of a PLLA-dioxane-water system. *J Biomed Mater Res* 63:161–67.

Huang, J., Lin, Y.W., Fu, X.W. et al. 2007. Development of nano-sized hydroxyapatite reinforced composites for tissue engineering scaffolds. *J Mater Sci Mater Med* 18:2151–57.

Huang, X., El-Sayed, I.H., Qian, W. et al. 2006. Cancer cell imaging and photothermal therapy in the near-infrared region by using gold nanorods. *J Am Chem Soc* 128:2115–20.

Hule, R.A. and Pochan, D.J. 2007. Polymer nanocomposites for biomedical applications. *MRS Bull* 32:354–58.

Hung, H.S. and Hsu, S.H. 2007. Biological performances of poly(ether)urethane-silver nanocomposites. *Nanotechnology* 18:475101.

Hyndman, L., Lemoine, J.L., Huang, L. et al. 2004. HIV-1 Tat protein transduction domain peptide facilitates gene transfer in combination with cationic liposomes. *J Control Release* 99:435–44.

Jain, P. and Pradeep, T. 2005. Potential of silver nanoparticle coated polyurethane foam as an antibacterial water filter. *Biotechnol Bioeng* 90:59–63.

Jain, T.K., Morales, M.A., Sahoo, S.K. et al. 2005. Iron oxide nanoparticles for sustained delivery of anticancer agents. *Mol Pharm* 2:194–205.

Jing, Y., Mal, N., Williams, P.S. et al. 2008. Quantitative intracellular magnetic nanoparticle uptake measured by live cell magneto-phoresis. *FASEB J* 22:4239–47.

Karageorgiou, V. and Kaplan, D. 2005. Porosity of 3D biomaterial scaffolds and osteogenesis. *Biomaterials* 26:5474–91.

Kennedy, L.C., Bickford, L.R., Lewinski, N.A. et al. 2011. A new era for cancer treatment: Gold-nanoparticle-mediated thermal therapies. *Small* 7:169–83.

Kim, H.W., Kim, H.E., and Salih, V. 2005. Stimulation of osteoblast responses to biomimetic nanocomposites of gelatin-hydroxyapatite for tissue engineering scaffolds. *Biomaterials* 26:5221–30.

Kim, S.Y. and Lee, Y.M. 2001. Taxol-loaded block copolymer nanospheres composed of methoxy poly(ethylene glycol) and poly(epsilon-caprolactone) as novel anticancer drug carriers. *Biomaterials* 22:1697–704.

Kohler, N., Fryxell, G.E., and Zhang, M.A. 2004. Bifunctional poly(ethylene glycol) silane immobilized on metallic oxide-based nanoparticles for conjugation with cell targeting agents. *J Am Chem Soc* 126:7206–11.

Koo, H., Huh, M.S., Sun, I.C. et al. 2011. In vivo targeted delivery of nanoparticles for theranosis. *Acc Chem Res* 44:1018–28.

Kothapalli, C.R., Shaw, M.T., and Wei, M. 2005. Biodegradable HA-PLA 3D porous scaffolds: Effect of nano-sized filler content on scaffold properties. *Acta Biomat* 1:653–62.

Kou, G., Wang, S., Cheng, C. et al. 2008. Development of SM5-1-conjugated ultrasmall superparamagnetic iron oxide nanoparticles for hepatoma detection. *Biochem Biophys Res Commun* 374:192–97.

Kumar, V.S., Nagaraja, B.M., Shashikala, V. et al. 2004. Highly efficient Ag/C catalyst prepared by electro-chemical deposition method in controlling microorganisms in water. *J Mol Catal A*. 223:313–19.

Langer, R. and Vacanti, J.P. 1993. Tissue engineering. *Science* 260:920–26.

Laurent, S., Forge, D., Port, M. et al. 2008. Magnetic iron oxide nanoparticles: Synthesis, stabilization, vectorization, physico-chemical characterizations and biological applications. *Chem Rev* 108:2064–10.

Liu, W. and Frank, J.A. 2009. Detection and quantification of magnetically labeled cells by cellular MRI. *Eur J Radiol* 70:258–64.

Lo, H., Ponticiello, M.S., and Leong, K.W. 1995. Fabrication of controlled release biodegradable foams by phase separation. *Tissue Eng* 1:15–28.

Lok, C.N., Ho, C.M., Chen, R. et al. 2006. Proteomic analysis of the mode of antibacterial action of silver nanoparticles. *J Proteome Res* 5:916–24.

Loo, C., Lowery, A., Halas, N.J. et al. 2005. Immunotargeted nanoshells for integrated cancer imaging and therapy. *Nano Lett* 5:709–11.

Lu, Z., Yeh, T.K., Tsai, M. et al. 2004. Paclitaxel-loaded gelatin nanoparticles for intravesical bladder cancer therapy. *Clin Cancer Res* 10:7677–84.

Ma, P.X., Zhang, R., Xiao, G. et al. 2001. Engineering new bone tissue *in vitro* on highly porous poly($\alpha$-hydroxyl acids)/hydroxyapatite composite scaffolds. *J Biomed Mater Res* 54:284–93.

Maier-Hauff, K., Rothe, R., Scholz, R. et al. 2007. Intracranial thermotherapy using magnetic nanoparticles combined with external beam radiotherapy, results of a feasibility study on patients with glioblastoma multiforme. *J Neurooncol* 81:53–60.

Mikos, A.G., Sarakinos, G., Leite, S.M. et al. 1993. Laminated three-dimensional biodegradable foams for use in tissue engineering. *Biomaterials* 5:323–30.

Modo, M.M.J. and Bulte, J.W.M. 2007. *Molecular and Cellular MR Imaging*. London, UK: CRC Press.

Mooney, D.J. and Mikos, A.G. 1999. Growing new organs. *Sci Am* 280:60–65.

Mornet, S., Vasseur, S., Grasset, F. et al. 2006. Magnetic nanoparticle design for medical applications. *Progress Solid State Chem* 34: 237–47.

Munier, S., Messai, I., Delair, T. et al. 2005. Cationic PLA nanoparticles for DNA delivery: Comparison of three surface polycations for DNA: Binding, protection and transfection properties. *Colloids Surf B Biointerfaces* 43:163–73.

Nam, S.H., Bae, Y.M., Il Park, Y. et al. 2011. Long-term real-time tracking of lanthanide ion doped upconverting nanoparticles in living cells. *Angew Chem Int Ed* 50:6093–97.

Nelson, G.N., Roh, J.D., Mirensky, T.L. et al. 2008. Initial evaluation of the use of USPIO cell labeling and noninvasive MR monitoring of human tissue-engineered vascular grafts *in vivo*. *FASEB J* 22:3888–95.

Nyk, M., Kumar, R., Ohulchanskyy, T.Y. et al. 2008. High contrast *in vitro* and *in vivo* photoluminescence bioimaging using near infrared to near infrared up-conversion in $TM^{3+}$ and $Yb^{3+}$ doped fluoride nanophosphors. *Nano Lett* 8:3834–38.

O'Rorke, S., Keeney, M., and Pandit, A. 2010. Non-viral polyplexes: Scaffold mediated delivery for gene therapy. *Prog Polym Sci* 35:441–58.

Oh, Y.K. and Park, T.G. 2009. SiRNA delivery systems for cancer treatment. *Adv Drug Deliv Rev* 61:850–62.

Owens, D.E. and Peppas, N.A. 2006. Opsonization, biodistribution, and pharmacokinetics of polymeric nanoparticles. *Int J Pharm* 307:93–102.

Panyam, J. and Labhasetwar, V. 2003. Biodegradable nanoparticles for drug and gene delivery to cells and tissue. *Adv Drug Deliv Rev* 55:329–47.

Paul, W. and Sharma, C.P. 2006. Nanoceramic matrices: Biomedical applications. *Am J Biochem Biotech* 2:41–48.

Pissuwan, D., Niidome, T., and Cortie, M.B. 2011. The forthcoming applications of gold nanoparticles in drug and gene delivery systems. *J Control Release* 149:65–71.

Pitts, E.A. and Corey, D.R. 1998. Inhibition of human telomerase by 2'-*O*-methyl-RNA. *Proc Natl Acad Sci* 95:11549–54.

Prabu, P., Chaudhari, A.A., Dharmaraj, N., Khil, M.S., Park, S.Y., and Kim. H.Y. 2008. Preparation, characterization, in-vitro drug release and cellular uptake of poly(caprolactone) grafted dextran copolymeric nanoparticles loaded with anticancer drug. *J Biomed Mater Res A* 90:1128–36.

Pultrum, B.B., Van der Jagt, E.J., Van Westreenen, H.L. et al. 2009. Detection of lymph node metastases with ultrasmallsuperparamagnetic iron oxide (USPIO)-enhanced magnetic resonance imaging in oesophageal cancer. A feasibility study. *Cancer Imaging* 9:19–28.

Purushotham, S., Chang, P.E., Rumpel, H. et al. 2009. Thermoresponsive core-shell magnetic nanoparticles for combined modalities of cancer therapy. *Nanotechnology* 20:305101.

Rahmer, J., Weizenecker, J., Gleich, B. et al. 2009. Signal encoding in magnetic particle imaging: Properties of the system function. *BMC Med Imaging* 9:1–21.

Rashdan, S., Selva Roselin, L., Selvin, R., Lemine, O.M., and Bououdina, M. 2013. Nanoparticles for biomedical applications: Current status, trends and future challenges. In *Biomaterials and Medical Tribology: Research and Development*, ed. J.P. Davim, 1–132, Cambridge: Woodhead Publishing Limited.

Renard, P.E., Buchegger, F., Petri-Fink, A. et al. 2009. Local moderate magnetically induced hyperthermia using an implant formed *in situ* in a mouse tumor model. *Int J Hyperthermia* 25:229–39.

Rezwana, K., Chena, Q.Z., Blakera, J.J. et al. 2006. Biodegradable and bioactive porous polymer/inorganic composite scaffolds for bone tissue engineering. *Biomaterials* 27:3413–31.

Rho, J.Y., Kuhn-Spearing, L., and Zioupos, P. 1998. Mechanical properties and the hierarchical structure of bone. *Med Eng Phys* 20:92–112.

Rolland, J.P., Maynor, B.W., Euliss, L.E. et al. 2005. Direct fabrication and harvesting of monodisperse, shape-specific nanobiomaterials. *J Am Chem Soc* 127:10096–100.

Ruiz-Hitzky, E., Darder, M., and Aranda, P. 2005. Functional biopolymer nanocomposites based on layered solids. *J Mater Chem* 15:3650–62.

Ruparelia, J.P., Chatterjee, A.K., Duttagupta, S.P. et al. 2008. Strain specificity in antimicrobial activity of silver and copper nanoparticles. *Acta Biomater* 4: 707–16.

Sadeghi, B., Jamali, M., Kia, S. et al. 2010. Synthesis and characterization of silver nanoparticles for antibacterial activity. *Int J Nano Dimen* 1:119–24.

Saha, B., Bhattacharya, J., Mukherjee, A. et al. 2007. *In vitro* structural and functional evaluation of gold nanoparticles conjugated antibiotics. *Nanoscale Res Lett* 2:614–22.

Sajja, H.K., East, M.P., Mao, H. et al. 2009. Development of multifunctional nanoparticles for targeted drug delivery and noninvasive imaging of therapeutic effect. *Curr Drug Discov Technol* 6:43–51.

Sathishkumar, M., Sneha, K., Won, S.W. et al. 2009. Cinnamon zeylanicum bark extract and powder mediated green synthesis of nano-crystalline silver particles and its bactericidal activity. *Colloids Surf B* 73:332–38.

Shan, J.N., Budijono, S.J., Hu, G.H. et al. 2011. Prud'homme, Pegylated composite nanoparticles containing upconverting phosphors and meso-tetraphenyl porphine (TPP) for photodynamic therapy. *Adv Funct Mater* 21:2488–95.

Shapiro, E.M., Sharer, K., Skrtic, S. et al. 2006. *In vivo* detection of single cells by MRI. *Magn Reson Med.* 55:242–49.

Shapiro, E.M., Skrtic, S., Koretsky, A.P. et al. 2005. Cellular MRI using micron-sized iron oxide particles. *Magn Reson Med* 53:329–38.

Shenoy, D.B. and Amiji, M.M. 2005. Poly(ethylene oxide)-modified poly(ε-caprolactone) nanoparticles for targeted delivery of tamoxifen in breast cancer. *Int J Pharm* 293:261–70.

Sivakumar, S., Diamente, P.R., and van Veggel, F.C. 2006. Silica-coated $Ln^{3+}$doped $LaF_3$ nanoparticles as robust down- and upconverting biolabels, *Chemistry* 12:5878–5884.

Sonawane, N.D., Szoka, F.C., and Verkman, A.S. 2003. Chloride accumulation and swelling in endosomes enhances DNA transfer by polyamine-DNA polyplexes. *J Biol Chem* 278:44826–31.

Sperling, R.A., Rivera Gil, P., Zhang, F. et al. 2008. Biological applications of gold nanoparticles. *Chem Soc Rev* 37:1896–908.

Sumner, J.P., Conroy, R., Shapiro, E.M. et al. 2007. Delivery of fluorescent probes using iron oxide particles as carriers enables *in vivo* labeling of migrating neural precursors for magnetic resonance imaging and optical imaging. *J Biomed Optics* 12:051504–606.

Szakacs, G., Paterson, J.K., Ludwig, J.A. et al. 2006. Targeting multidrug resistance in cancer. *Nat Rev Drug Discov* 5:219–34.

Thirumurugan, G. and Dhanaraju, M.D. 2001. Novel biogenic metal nanoparticles for pharmaceutical applications. *Adv Sci Lett* 4:339–48.

Thorek, D.L. and Tsourkas, A. 2008. Size, charge and concentration dependent uptake of iron oxide particles by non-phagocytic cells. *Biomaterials* 29:3583–90.

Thorek, D.L.J., Chen, A.K., Czupryna, J. et al. 2006. Superparamagnetic iron oxide nanoparticle probes for molecular imaging. *Ann Biomed Eng* 34:23–38.

Toh, E.K.W., Chen, H.Y., Lo, Y.L. et al. 2011. Succinated chitosan as a gene carrier for improved chitosan solubility and gene transfection. *Nanomed Nanotechnol* 7:174–83.

Tokunaga, M., Hazemoto, N., and Yotsuyanagi, T. 2004. Effect of oligopeptides on gene expression: Comparison of DNA/peptide and DNA/peptide/liposome complexes. *Int J Pharm* 269:71–80.

Tran, H.V., Tran, L.D., Ba, C.T. et al. 2010. Synthesis, characterization, antibacterial and antiproliferative activities of mono-disperse chitosan-based silver nanoparticles. *Colloids Surf A* 360:32–40.

Uemura, T., Dong, J., Wang, Y. et al. 2003. Transplantation of cultured bone cells using combinations of scaffolds and culture techniques. *Biomaterials* 24:2277–86.

Uhrich, K., Cannizzaro, S., Langer, R. et al. 1999. Polymeric systems for controlled drug release. *Chem Rev* 99:3181–98.

Vallet-Regi, M. 2001. Ceramics for medical applications. *J Chem Soc Dalton Trans* 2001:97–108.

Verma, A. and Rotello, V.M. 2005. Surface recognition of biomacromolecules using nanoparticle receptors. *Chem Commun* 3:303–12.

Wang, C., Cheng, L.A., and Liu, Z.A. 2011. Drug delivery with upconversion nanoparticles for multi-functional targeted cancer cell imaging and therapy, *Biomaterials* 32:1110–20.

Wang, G.F., Peng, Q., and Li, Y.D. 2011. Lanthanide-doped nanocrystals: Synthesis, optical-magnetic properties, and applications. *Acc Chem Res* 44:322–32.

Wang, M. Hou, W. Mi, C.C. et al. 2009a. Immunoassay of goat antihuman immunoglobulin G antibody based on luminescence resonance energy transfer between near-infrared responsive NaYF4:Yb, Er upconversion fluorescent nanoparticles and gold nanoparticles. *Anal Chem* 81:8783–89.

Wang, M., Mi, C.C., Wang, W.X. et al. 2009b. Immunolabeling and NIR-excited fluorescent imaging of HeLa cells by using NaYF(4):Yb,Er upconversion nanoparticles. *ACS Nano* 3:1580–86.

Wang, Y., Liu, L., and Guo, S. 2010. Characterization of biodegradable and cytocompatible nano-hydroxyapatite/polycaprolactone porous scaffolds in degradation *in vitro*. *Polym Degrad Stabil* 95:207–13.

Wei, G. and Ma, P.X. 2004. Structure and properties of nano-hydroxyapatite/polymer composite scaffolds for bone tissue engineering. *Biomaterials* 25:4749–57.

Weizenecker, J., Gleich, B., Rahmer, J. et al. 2009. Three-dimensional real-time *in vivo* magnetic particle imaging. *Phys Med Biol* 54:L1–L10.

Wilhelm, C. and Gazeau, F. 2009. Magnetic nanoparticles: Internal probes and heaters within living cells. *J Magn Magn Mater* 321:671–74.

Williams, D.N., Ehrman, S.H., and Holoman, T.R.P. 2006. Evaluation of the microbial growth response to inorganic nanoparticles. *J. Nanobiotechnol* 4:1–8.

Wu, S.W., Han, G., Milliron, D.J. et al. 2009. Non-blinking and photostable upconverted luminescence from single lanthanide doped nanocrystals. *Proc Natl Acad Sci USA* 106:10917–21.

Wu, Y.L., Ye, Q., Foley, L.M. et al. 2006. *In situ* labeling of immune cells with iron oxide particles. An approach to detect organ rejection by cellular MRI. *PNAS* 103:1852–57.

Wust, P., Gneveckow, U., Johannsen, M. et al. 2006. Magnetic nanoparticles for interstitial thermotherapy- feasibility, tolerance and achieved temperatures. *Int J Hyperthermia* 22:673–85.

Xu, H., Cheng, L., Wang, C. et al. 2011. Polymer encapsulated upconversion nanoparticle/iron oxide nanocomposites for multimodal imaging and magnetic targeted drug delivery. *Biomaterials* 32:9364–73.

Yoo, H.J., Lee, J.M., Lee, M.W. et al. 2008. Hepatocellular carcinoma in cirrhotic liver: Double-contrast-enhanced, high-resolution 3.0 T-MR imaging with pathologic correlation. *Invest Radiol* 43:538–46.

Zako, T., Nagata, H., Terada, N. et al. 2009. Cyclic RGD peptide-labeled upconversion nanophosphors for tumor cell-targeted imaging. *Biochem Biophys Res Commun* 27:54–58.

Zhang, Q., Zha, L., Ma, J. et al. 2007. A novel route to the preparation of poly(N-isopropylacrylamide) microgels by using inorganic clay as a cross-Linker. *Macromol Rapid Commun* 28:116–20.

Zhang, R. and Ma, P.X. 1999. Poly($\alpha$-hydroxyl acids)/hydroxyapatite porous composites for bone-tissue engineering. I. Preparation and morphology. *J Biomed Mater Res* 44:446–55.

Zheng, D., Li, X., Xu, H. et al. 2009. Study on docetaxel-loaded nanoparticles with high antitumor efficacy against malignant melanoma. *Acta Biochim Biophys Sin* 41:578–87.

Zhou, J., Liu, Z., and Li, F. 2012. Upconversion nanophosphors for small-animal imaging. *Chem Soc Rev* 41:1323–49.

Zhou, J.C., Yang, Z.L., Dong, W. et al. 2011. Bioimaging and toxicity assessments of near-infrared upconversion luminescent NaYF4:Yb, Tm nanocrystals. *Biomaterials* 32:9059–67.

Ziauddin, J. and Sabatini, D.M. 2001. Microarrays of cells expressing defined cDNAs. *Nature* 411:107–10.

# 15 Carbon Nanotube-Induced Targeted Drug Delivery

*Anish Bhattacharya and Amit K. Chakraborty*

## CONTENTS

## 15.1 INTRODUCTION

The approaches, formulations, technologies, and systems for transporting a pharmaceutical compound in the body required to safely achieve its desired therapeutic effect is known as drug delivery. It is typically concerned with both quantity and duration of the presence of a drug. In general, drug delivery refers to the delivery of a drug's chemical formulation, but it may also refer to the delivery of medical devices or products of drug-device combination.

Technologies of drug delivery may include processes such as drug attachment to a carrier/vehicle, drug transportation, drug release followed by drug absorption, distribution and elimination in a manner that ensures improved efficacy and safety of the product, as well as the convenience and compliance of the patient. The most common routes of administration are noninvasive peroral (through the mouth), inhalation, transmucosal (nasal, buccal/sublingual, vaginal, ocular, and rectal) and topical (skin) routes. Many medications such as protein and peptide, antibody, gene and vaccine-based drugs may not be delivered using these routes because of their vulnerability to enzymatic degradation or due to inefficient absorption into the circulation system due to size and charge of the molecules, thereby limiting their therapeutic efficacy. For this reason, many protein and peptide drugs are delivered by injection or a nanoneedle array.

Many diseases such as various kinds of cancers, microbial diseases, and inflammatory diseases target only specific cells in our body, but the treatment procedures and drugs prescribed often affect a larger area than just the rogue cells. This can result in various unforeseen problems/side effects in the patient on whom the treatment is administered, and hence need remedy. Targeted drug delivery is a mechanism through which delivery structures for the drugs are created such that these can introduce the drug directly to the affected cells only without affecting other cells. In general, it refers to the methods of delivering drugs to a patient in a manner that increases its concentration in some parts of the body relative to others. The inherent advantage of this approach is the reduction in dose and side effects of the drug. For example, cisplatin (*cis*-diamminedichloroplatinum) has very effective cell killing properties but tends to reduce excretion and create depositions in tissues due to its bonding with plasma proteins (Borowiak-Palen et al. 2011). If delivery mechanisms are developed such that these compounds can be delivered to the rogue cells directly, then the side effects can be avoided. Another example is an antimicrobial drug known as Amphotericin B (AMB). As a polyene macrolide, AMB is considered one of the first-line agents to combat opportunistic systemic fungal infections in immune-compromised patients. Despite its broad activity spectrum, high effectiveness, and relatively low cost, the use of AMB is however limited by numerous undesirable side effects such as infusion reaction to nephrotoxicity due to its natural tendency to aggregate due to its poor water solubility (Czub and Baginski 2009).

As the carbon nanotubes (CNTs) have no affinity to normal or cancer cells, and hence in order to make them enter the cell membrane (normal or cancer cells), they need to be chemically functionalized. However, the question that remains is the mechanism in which the drug carrier finds its target and then enters it. One of the most common approaches to understanding the internalization process uses techniques such as epifluorescence, confocal microscopy, and flow cytometry studies (Pantarotto et al. 2004a; Kam and Dai 2006) to track the CNTs by labeling them with a fluorescent agent and then monitoring the uptake. This approach along with other studies suggests that the internalization of drugs occurs primarily in two mechanisms, *endocytosis-dependent pathway* and *endocytosis-independent pathway*, which includes diffusion, membrane fusion, or direct pore transport of the extracellular material into the cell (Fisher et al. 2012). The choice of the mechanism depends on several factors, such as the size, length, nature of functional groups, hydrophobicity, and the surface chemistry of CNTs (Fisher et al. 2012; Kostarelos et al. 2012).

In *endocytosis-dependent mechanism*, engulfing of extracellular materials (e.g., drug-loaded CNT) occurs within a segment of the cell membrane to form a saccule or a vesicle which is then pinched off intracellularly into the cytoplasm of the cell (Mu et al. 2009).The materials after coming in close proximity to the cellular membrane may directly interact with the membrane-embedded receptors or indirectly interact with them by means of the bilayer. It is an energy- and temperature-dependent transport process. The cellular requirements govern the selective incorporation of materials, whereas the contents are transported to early endosomes depending on the mode of internalization which can be either nonreceptor-mediated or receptor-mediated (Steinman et al. 1983).

In a *nonreceptor-mediated endocytosis* (Figure 15.1a), the drug-loaded CNT approaches a small portion of the plasma membrane and gets surrounded before pinching off intracellularly as an endocytic vesicle. The endosomes thus formed are eventually converted into the lysosomes and ultimately result in the drug release.

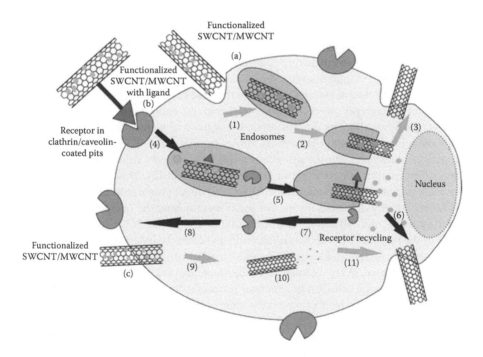

**FIGURE 15.1** Pathways for the penetration of CNTs into the cell. (a) Nonreceptor-mediated endocytosis: (1) membrane that surrounds the drug-loaded functionalized CNTs, (2) internalization of drug-loaded CNTs, and (3) release of drug; (b) receptor-mediated endocytosis: (4) membrane surrounds the CNT-receptor conjugate by forming endosomes followed by internalization, (5) release of drug, and (6, 7, 8) regeneration of receptor; (c) endocytosis-independent pathway: (9) direct penetration of drug-loaded functionalized CNT and (10) release of the drug. (Reprinted with permission from Copyright © 2014 Rastogi, V. et al. *Journal of Drug Delivery*, 1–23, 2014: Article ID 670815, DOI: http://dx.doi.org/10.1155/2014/670815.)

The internalization of single-walled carbon nanotubes (SWCNTs) occurs through the endocyte formation which was confirmed by an increase in the endosome numbers per cell with increased time of exposure to SWCNTs (Yaron et al. 2011). The authors further confirmed the strong energy-dependence of the endocytosis process by a dramatic reduction in SWCNT uptake into the hematoxylin-labeled HeLa cell incubated for 2 days at 4°C compared with that at 37°C.

In case of *receptor-mediated endocytosis* (Figure 15.1b), ligand-conjugated drug-loaded CNT binds to the complementary transmembrane receptor proteins before entering the cell as receptor–ligand complexes in the clathrin-coated vesicles. The ligand then dissociates from the receptor due to a drop in the pH in the endosomes (the vesicles carrying the extracellular particle) thus formed. When the receptors are released, the endosomes fuse with the lysosomes, and thus trigger the release of the drug particle by the action of lysozymes on the endosomes, and the free receptors thus formed are recycled to the plasma membrane for conjugating with another ligand-conjugated CNT (Alberts et al. 2002).

Receptor-mediated internalization is observed in pitted membrane regions which are lined either by the assistor proteins caveolin-1 or clathrin. Although each endocytic mechanism has a similar vesicle formation and the action of the pluripotent scission GTPase dynamin-2 is required, the intracellular fate of endosomes and their contents have to be distinctly regulated. Clathrin-dependent internalization follows a well-defined path where the mature endosomes are fused with the lysosomal vesicles resulting in the enzymatic destruction of a variety of molecules which include lipids, carbohydrates, nucleic acids, and viruses. In the case of clathrin-independent uptake, subcellular movements proceed through a series of endosomal compartments of increasing acidity allowing the hydrolytic breakdown of molecules.

In *endocytosis-independent pathway,* direct incorporation of CNT occurs through the plasma membrane into the cytoplasm much like a needle penetration, and hence often called *nanoneedle* mechanism (Kostarelos et al. 2012). This pathway includes processes such as diffusion, membrane fusion, and direct pore transport (Figure 15.1c). CNTs individually dispersed in aqueous solutions can enter the cytoplasm of cells by directly crossing the cell membrane. Such cellular uptake of CNT, which is not influenced by the presence of endocytosis inhibitors indicates the independence of the internalization process from endocytosis. The exact mechanism of how CNT enters the cells is poorly understood. What we know from the theoretical studies is that the internalization of CNTs into the cells is a spontaneous process mediated by the lipid membrane, and that the hydrophilic interactions and/or static charge interactions between the CNTs and the lipid membrane drive the translocation of the CNTs into the cell (Tran et al. 2009). Rastogi et al. (2014) reviewed this topic with detailed discussion and illustration (as shown in Figure 15.1) of various mechanisms of drug delivery using CNTs.

A nanomaterial being used as a transport module for another substance, such as a drug, is known as a nanocarrier. Carbon-based nanostructures, micelles, polymers, and liposomes and metal nanoparticles are commonly used carriers for drugs. Due to the unique characteristics demonstrated by CNT, researchers have shown great interest in this novel nanostructure of carbon as a suitable carrier for drug delivery. The focus of this chapter is to review the state of the art of targeted drug delivery using CNT as a drug carrier.

## 15.2 CNT AS DRUG CARRIER: WHY?

CNTs, discovered by S. Iijima in 1991 (Iijima 1991), are cylindrical-shaped hollow tubes consisting of repetitive arrangement of hexagonal units of sp2-hybridized carbon atoms. Typically, their diameter is in the range of a few nanometers and length up to tens of hundreds of microns. The interatomic separation in a C—C bond is about 1.4 Å. Due to many interesting physical properties, CNTs have been shown to have applications in almost every field of science and engineering. However, the main reasons for application of CNTs in targeted drug delivery are as follows:

- Small size and cylindrical shape allow CNTs to easily penetrate cells and tissues.
- Enhanced permeability and retention (EPR) effect associated with nanoscale materials.
- The ability to chemically functionalize with suitable compounds including drugs, and its biocompatibility.
- The ability to fill the hollow structure of the nanotubes with suitable drugs.
- Increased drug-loading capacity.
- The intrinsic spectroscopic properties of CNTs, such as Raman and photo-luminescence, allow for tracking and real-time monitoring of drug delivery efficacy *in vivo*.
- Capability of controlled drug release inside tumor cells.

Tasis et al. (2006) showed that CNTs can be used as drug-delivery vehicles in therapies for various diseases without causing toxicity to healthy tissue while allow-ing prolonged release of the drug. As SWCNTs provide potential advantages over metal nanoparticle systems, hence they have attracted considerable interest in recent years with regard to targeted drug delivery. For example, the high loading capacity, intrinsic stability, and structural flexibility of CNTs provide prolonged circulation time of the drug molecules it carries, thus increasing the drug's bioavailability as has been shown by several researchers (Wang et al. 2004; Singh et al. 2006; Worle-Knirsch et al. 2006; Liu et al. 2008). Further, the ability of SWCNTs to enter mam-malian cells has been shown by a number of research groups (Kam et al. 2004; Cai et al. 2007; Chin et al. 2007; Kostarelos et al. 2007). SWCNTs have also been tested for intracellular transport of nucleic acids, proteins, and drug molecules due to their promising properties (Kam et al. 2004; Pantarotto et al. 2004b; Cai et al. 2007; Kam and Dai 2005; Liu et al. 2007a,b,c; Chen et al. 2008; Prato and Kostarelos 2008). In view of the above, CNTs have attracted great interest from researchers as a vehicle to deliver drugs on target cells for therapeutic use.

## 15.3 DRUG ATTACHMENT TO CNTs

In spite of the potential advantages offered by CNTs for targeted drug delivery applications, their use has to be weighed against their biocompatibility. CNTs are known to be cytotoxic in nature rendering them unsafe for use in medical applications. Interestingly, toxicological profiles of pristine CNTs have shown that the cytotoxicity reduces drastically when CNTs are functionalized on the surface. SWCNTs wrapped with polymers have been shown to be compatible with water and, in particular, the physiological environments (Hasegawa et al. 2004; Numata et al., 2004; Yuan et al. 2006; Cai et al. 2007; Liu et al. 2007b; Abarrategi et al. 2008; Fu et al. 2008; Long et al. 2008; Zhang et al. 2008). Functionalization of SWCNTs with antibodies and low-molecular-weight targeting agents result in high efficiency of nanotube internalization in cells (Bottini et al. 2006; Shao et al. 2007; Cato et al. 2008; Kang et al. 2008Welsher et al. 2008; Zeineldin et al. 2009). Moreover, functionalization improves the loading capacity of CNT and allows incorporation of targeting agents and stealth molecules to evade the immune system (Heister et al. 2009; Mu et al. 2009). In view of the above, attachment of the drugs to CNT often involves functionalization of the CNT with some suitably chosen group and then binding the drug or in some cases direct functionalization with the drug itself. The drug can be loaded on the external surface of the CNT or inside the hollow tube.

### 15.3.1 ATTACHMENT BY FILLING OF CNTs

Ajayan and Iijima were the first to fill CNTs with metal. They showed that annealing of the CNTs in the presence of liquid lead (Pb) results in the opening of the capped tube ends and subsequent filling of the nanotubes with molten material through capillary action (Ajayan and Iijima 1993).

Although the mechanism of filling CNTs with liquids is not very well understood, the filling of CNT with a given material certainly requires the knowledge of the wetting and capillarity properties (surface tensions, viscosity etc.) of the solution and the type of CNT it needs to be filled in. In fact, a cut-off value for surface tension of 200 mN/m was also reported by some researchers, which is at the border of wetting and nonwetting liquids (Dujardin et al. 1994). The authors further established a rough relation between the inner diameter of CNT and their capillarity behavior. They suggested that inorganic salts of V, Co, and Pb having low surface tensions could fill CNTs having narrow diameter (inner diameters of 1–2 nm) while salts with higher values of surface tension could only fill CNTs having larger inner diameter (>4 nm).

It has been further observed that the metallic compounds and elements having higher surface tension tend to fill the CNT in oxidizing atmosphere apparently owing to lower surface tension values for oxides than the compounds/elements. Indirect filling of CNTs is also possible by the prior introduction of appropriate precursor in the CNT and subjecting them to some chemical reaction or physical interaction such as electron beam irradiation (Lukanov et al. 2011). The filling processes for CNTs can be broadly classified into *physical and wet chemical processes.*

### 15.3.1.1 Physical Filling of CNT from Molten Phase

Filling with molten materials was among the first procedures by which materials were tried to be encapsulated inside the CNT. This is a restrictive method used for filling of CNTs, that is, the materials satisfying certain conditions can only be used as filling material in this method. The conditions that need to be satisfied are:

1. The material should not decompose when melted.
2. The melting point of the material should be compatible with the melting point of the CNT.
3. In case of multiwalled carbon nanotube (MWCNT), the surface tension of the filler material should be in between 100 and 200 $N/cm^2$, so that the capillary action can take place.

The general procedure of filling of CNT by melted phases require a close mixing of CNT with the desired amount of filler material by gentle grinding and vacuum sealing the mixture in a silica ampoule (Lukanov et al. 2011). The ampoule is slowly heated to a temperature above the melting point of filler and then it is slowly cooled. Opening of the nanotube is not a necessary step in this process, although it has been shown that the nanotube opening is dependent on the chemical aggressiveness of the molten material toward carbon.

### 15.3.1.2 Wet Chemical Process

In this method, open CNTs are put in contact with a concentrated solution of the desired material, also known as precursor material. It can be done by either opening and filling the CNTs simultaneously (one step method) or opening the CNTs first and filling them subsequently (two-step method). The flexibility and experimental control over the process are the main advantages of this method. This is the desired route for filling of CNTs when the physical properties of filing material are not suitable for filling through melted or gaseous phase.

The main parameters involved in filling of CNT from solutions are as follows:

1. Diameter of the CNT
2. Physicochemical properties of the material to be inserted
3. Process of insertion

Pederson and Broughton (1992) showed that the insertion energy and the overlapping repulsion activation barrier decreased with increase in the tube size. Hence, they concluded that the encapsulation in SWCNT would be unfavorable compared to the encapsulation in MWCNT. In a study using MWCNT and DWCNT (double-walled carbon nanotube), it was observed that MWCNT having inner diameter from six to several hundreds of nanometers were easier to fill than the small-sized SWCNT having a diameter of 1.22 nm or DWCNT having a diameter of 1.5 nm (Tsang et al. 1994).

As most applications of CNT require smaller tube diameters, efforts have been made to fill SWCNTs and DWCNTs. SWCNTs were first filled with ruthenium chloride (Sloan et al. 1998). A high yield of filling was achieved by using ferromagnetic material like iron as the filling material. The advantages of this method for filling of CNT are high filling yields, simplicity, and versatility although the filler material requires exhibiting a certain value of surface tension at the filling temperature. The posteriori method of filling of CNT provides better control over the filling process than the *in situ* during the synthesis of CNT.

### 15.3.2 ATTACHMENT BY CHEMICAL FUNCTIONALIZATION OF CNT

Due to lack of solubility in aqueous medium and due to hydrophobic surfaces, several constraints have been attached to the use of CNT in biological and biomedical applications. CNTs are cytotoxic to certain mammalian cells, for example, pure MWCNTs can injure plasma membrane of human macrophages (Zhu and Li 2008). A process called functionalization in which the functional groups are attached on CNT surfaces renders the CNTs biocompatible mainly by changing them from hydrophobic to hydrophilic. Several methods exist for functionalization of the CNTs. These can be primarily classified into two types: physical and chemical. Physical functionalization involves processes that incur physical damage to the surfaces, such as ultrasonication, electron-irradiation, and ion-irradiation, and so on (Chakraborty et al. 2007).

Chemical functionalization uses chemical methods to alter the surface chemistry by attaching other atoms/molecules/radicals to the CNT surface. Chemical functionalization can be further classified into two main categories: *covalent functionalization,* and *noncovalent functionalization.*

#### 15.3.2.1 Covalent Functionalization

The oxidation of CNTs using strong acids (conc. $H_2SO_4$ or $HNO_3$) generates substitutable hydrophilic functional groups, such as COOH and OH, on the CNTs which then become ready to undergo subsequent chemical reactions with other functional groups such as amine, halide, and so on. The CNT end caps open up when they undergo the oxidation process due to which carboxylic groups are generated, which in turn improves the biocompatibility due to enhanced solubility. The CNTs become more hydrophilic when a highly negative charge due to carboxylation is developed. CNTs can be functionalized with active molecules through side wall covalent attachment of functional groups by the addition of carbine, nitrenes, nucleophilic cyclopropanation, and electrophiles (Hirsch 2002). Covalent functionalization of the CNTs with the therapeutically active molecule is governed by such oxidation. The covalent linkage of polyethylene glycol increases the sizes of CNTs due to which the rate of clearance through kidneys is reduced, thus increasing the circulation time in the plasma (Roman et al. 2011).

The authors of this article have reported the synthesis of SWCNTs covalently functionalized with tertiary phosphines (Suri et al. 2008). The debundling of SWCNTs was found to be significant as a result of modification with phosphines. We have also functionalized SWCNTs with groups, such as iodine and 4-nitrophenyl groups using simple chemical routes (Coleman et al. 2007; Chakraborty et al. 2009). Khabashesku et al. (2008) reported the side wall functionalization of SWCNTs through C—F bonds leading to the possibility of a wide range of further SWCNTs derivatizations, such as with amino acids, DNA, and polymer matrix. Another method which is worth mentioning is the one in which the oxidized CNTs in industrial scale were prepared (Tour et al. 2007).

### 15.3.2.2 Noncovalent Functionalization

In noncovalent functionalization, the interaction between molecules, such as van der Waals interactions, $\pi$–$\pi$ interactions, and so on, are used to attach biocompatible functional groups with CNT. In this method, the damage inflicted on the surface of the CNTs is minimal. The noncovalent functionalization is suggested to preserve the aromatic structure and electronic character as compared to pristine CNTs. To achieve noncovalent functionalization, weak bonds are used to add biopolymers, hydrophilic polymers, and surfactants to the walls of CNTs. For nanotube dispersion in aqueous media, a series of anionic, cationic, and nonionic surfactants have already been proposed. For example, sodium dodecyl sulfate (SDS) and benzylalkonium chloride aggregate to the nanotube side walls and ease the dissolution of the CNTs in water. The $\pi$–$\pi$ stacking interactions between the surfactant and nanotube walls strengthen the adhesion between them. We have shown that the noncovalent attachment of a commercial dispersing agent (surfactant) significantly improves the dispersion of MWCNTs in a high viscous epoxy resin matrix (Chakraborty and Coleman 2008). Studies have shown that the functionalization of CNT by anticancer drugs on the external wall enhances the therapeutic effect while keeping the cell-killing effect preserved, which is indeed a very positive sign for CNT-based drug delivery.

Apart from the above processes, there are some reports showing that CNTs can also be incorporated into other parent drug carriers, for instance hydrogels or electron spun fibers, as adjunct material to modulate the loading and release of coencapsulated drugs.

## 15.4 RELEASE OF DRUGS FROM CARBON NANOTUBES

The binding and release of drugs from CNT have been reported to be pH dependent. The drug release behavior of the drug rhodamine B (RB) encapsulated in CNT was found to be pH dependent (Zhang and Olin 2012). The pH value influenced the interaction affinities between RB and the CNTs, where the affinities were found to be dependent on the surface chemistry.

In another study, the pH of cell endosomes was exploited to release cisplatin by reducing the disulfide linkages, and it was observed that the efficiency of cisplatin enhanced when they were carried by CNT as it eliminated drug inactivation and improved its circulation (Feazell et al. 2007). Similar findings on drug release were reported in the studies conducted by Liu and colleagues (Liu et al. 2007b). In another interesting study, a near-infrared (NIR)-sensitive CS/PNIPAAm@CNT nanoparticle was synthesized for the intracellular delivery of doxorubicin (DOX) (Qin et al. 2015). MTT assays showed the minimal cytotoxicity of as prepared CS/PNIPAAm@CNT nanoparticles and significantly enhanced cell killing behavior upon NIR laser irradiation suggesting an NIR light-triggered drug release mechanism.

## 15.5 ANTICANCER DRUG DELIVERY USING CNT

### 15.5.1 BLADDER CANCER

Any cancer arising from the epithelial lining of the urinary bladder is termed bladder cancer. Rarely the bladder is involved in nonepithelial cancers, such as lymphoma or sarcoma, but these are not ordinarily included in the colloquial term *bladder cancer* (BCa). It is a disease in which the abnormal cells multiply without control in the bladder. The most common type of bladder cancer recapitulates the normal histology of the urothelium and is known as transitional cell carcinoma or more properly urothelial cell carcinoma. A study showed that MWCNTs with a wall diameter of 11–18 nm and a mean length of up to 1.5 μm could adhere to the urothelium of mouse bladders (Reiger et al. 2015). Further, the MWCNTs were observed to mediate a low-to-moderate impairment of cellular function in two different BCa cell lines. It was also observed that the CNTs with more defects and lower aspect ratios had more deleterious effects than the ones with fewer defects and higher aspect ratios. Functionalization and modification of the CNTs with lower defects can be proposed as a promising strategy for improving the mucoadhesive properties.

## 15.5.2   Blood Cancer

A cancer that begins in the bone marrow is termed leukemia. One of the four main types of leukemia is acute lymphoblastic leukemia which is a slow-growing blood cancer that starts in the bone marrow cells known as lymphocytes. The blood is invaded fairly quickly by the continuously reproducing lymphocytes. Recently, a research group (Taghdisi et al. 2011) developed a tertiary complex of Sgc8c aptamer, daunorubicin (DAU) and SWCNT which showed an enhanced drug delivery of DAU to the acute lymphoblastic lymphoma. Flow cytometric analysis showed that the tertiary complex was internalized effectively into the human T cell leukemia cell (MOLT-4 cells) but not to U266 myeloma cells.

## 15.5.3   Brain Cancer

The most aggressive brain cancers are anaplastic astrocytomas (survival period of 24 months) and glioblastomas (survival period of 9 months). The therapy of brain cancers causes substantial adverse effects on the nervous system (Huse and Holland, 2010). The blood–brain barrier (BBB) restricts the penetration of most of the drugs in the brain rendering them therapeutically less effective. Hence, the CNT-based targeting approaches have been studied for the treatment of brain cancer. Ou et al. (2009) conjugated the protein A with SWCNTs functionalized with phospholipid-bearing polyethylene glycol (PL-PEG). The final product was then coupled with the fluorescein-labeled integrin $\alpha_v\beta_3$ monoclonal antibody to form SWCNT-integrin $\alpha_v\beta_3$ mAb (SWCNT-PEG-mAb) as shown in Figure 15.2. By using confocal microscopy, it was found that the SWNT-PEG-mAb showed a much higher fluorescence signal on integrin $\alpha_v\beta_3$-positive U87MG cells and presented a high targeting efficiency with low cellular toxicity. For integrin $\alpha_v\beta_3$-negative MCF-7 cells, no obvious fluorescence was observed indicating low targeting efficiency of the functionalized SWCNTs. Hence, it can be concluded that the specific targeting of integrin $\alpha_v\beta_3$-positive U87MG cells was caused by the specific recognition of integrin $\alpha_v\beta_3$ on the cellular membrane by the $\alpha_v\beta_3$ mAb.

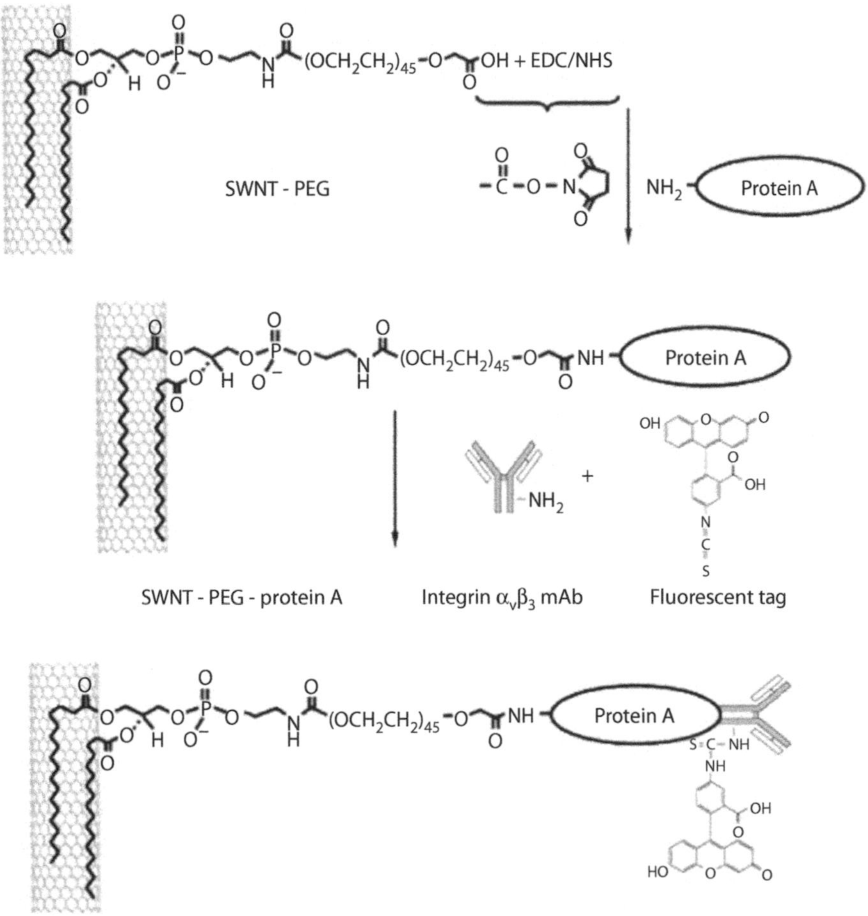

**FIGURE 15.2** An illustration of the procedures of producing SWCNT-PEG-mAb. (Ou, Z. et al., 2009. Functional single-walled carbon nanotubes based on an integrin $\alpha_v\beta_3$ monoclonal antibody for highly efficient cancer cell targeting. *Nanotechnology* 20(10): 105102, ©IOP Publishing. Reproduced with permission.)

It is difficult to distribute oxidized MWCNTs in the brain as they tend to accumulate in the tumor when conjugated with specific ligands. Anticancer drugs can be loaded efficiently on them due to their ultrahigh surface area. A dual targeting PEGylated MWCNTs was developed by Ren et al. which was then loaded with a targeting ligand angiopep2 (ANG) and DOX, respectively (Ren et al. 2012). The low-density lipoprotein receptor-related protein receptor which is overexpressed on the BBB and C6 glioma cells were targeted in this study. The *in vitro* intracellular tracking and *in vivo* fluorescence imaging showed that the ideal dual targeting of the developed system was attained by the high transcytosis capacity and parenchymal accumulation through angiopep-2. Hence, it is evident that this can be considered a material of choice to cross the BBB and to specifically recognize their lipoprotein receptors present on the glioma cells for directing the site-specific release of the drug. The C6 cytotoxicity, hematology analysis, and cluster of differentiation 68 (CD68) immunohistochemical analysis show that better antiglioma effect with good biocompatibility and low toxicity can be achieved for O-MWCNTs-PEGANG as compared with those of free DOX.

In a study by Moore et al. (2013), it was shown that the toxicity of the CNT was reduced and its therapeutic efficacy was increased by the critical aspect of hydrophobic and hydrophilic polymers. In the work, the hydrophobic drugs were mediated and encapsulated with the help of polylactide (PLA) layer, and the pharmacokinetic properties and aqueous solubility were found to improve due to the presence of PEG layer. Block copolymers composed of PLA-PEG were reported to reduce short-term and long-term toxicity, sustain drug release of paclitaxel (PTX), and prevent aggregation. The toxicity also decreased significantly as compared to the results obtained when pristine CNTs were used. Compared to the noncoated CNTs, *in vivo* studies showed no long-term inflammatory response with CNT coated with PLA-PEG (CLP) and the surface coating significantly lowered the acute toxicity by doubling the maximum tolerated dose in mice.

### 15.5.4 BREAST CANCER

Overexpression of human epidermal growth factor receptor 2 (HER2), also known by the names c-erbB-2 or HER2/neu, has been accounted as approximately 20%–25% responsible for invasive breast cancer. As the understanding of the role of HER2 in tumor angiogenesis, metastasis, and proliferation has increased, novel special treatment strategies for this HER2-positive subtype of breast cancer have been validated and have been increasingly used in clinical practice. One of the major treatment strategies is blocking the signal pathway of HER2/neu, that is, targeted therapy. In a study by Liu et al., a dose of 20–50 mg/kg of taxol was found to be the tolerance limit for BALB/c mice (Liu et al. 2008). A dose of 5 mg/kg of SWCNT-PTX was observed to be successful in suppressing the tumor growth once every 6 days suggesting drug delivery via SWCNT with low side effects. The water-soluble SWCNT-PTX was reported to be cremophor free. To deliver 5 mg/kg dose of PTX, only ~4 mg/kg of SWCNT was required compared to ~420 mg/kg cremophor in taxol for the same dose of PTX.

The efficiency of MWCNT to deliver the gene to the tumor cell was investigated by Pan et al. (2009). In the study, the fabricated MWCNTs were modified with polyamidoamine dendrimer that was further conjugated with fluorescein isothiocyanate (FITC)-labeled antisense c-Myc oligonucleotides (asODN). The modified MWCNTs were then incubated with human breast cancer cell line MCF-7 and MDA-MB-435 cells. The cellular uptake of asODN-d-MWCNTs was then observed with fluorescence developed by the FITC within 15 min. The cell growth was inhibited by the composites in both time- and dose-dependent means. Downregulation of the expression of c-Myc gene (overexpression of this gene amplifies the expression of HER2) and c-Myc protein was also observed in this study.

The NIR/pH-responsive drug delivery using SWCNTs was investigated by Wang et al. (2013). In this study, the functionalized SWCNTs were linked with NGR (Asn-Gly-Arg) tumor-targeting peptide with polymerized polymeric polyethyleneimine

(PEI) via the maleimide group of DSPE-PEG2000-maleimide and sulfhydryl group of cysteine. The delivery system displayed sustained release effect and slow release of DOX, which in turn reduced the systemic toxicity. The release of DOX was found to be dependent on NIR laser irradiation or pH of the environment such that under 808 nm NIR laser irradiation or a lower value of pH would accelerate the release of DOX. Hence, the delivery system can be seen as a promising candidate, owing to the hyperthermia sensitizer effect of DOX.

In another study, a steroid-macromolecular bioconjugate based on PEG-linked 17β-estradiol ($E_2$) was appended to MWCNTs which were intrinsically able to penetrate the cells (Das et al. 2013). In the study, DOX was taken as a model anticancer agent and it was observed that the combination of DOX with this delivery system facilitated nuclear targeting through an estrogen receptor as well as deciphered an *in vivo* synergistic anticancer response. The antitumor efficacy of the delivery system was found to be more than saline, drug-deprived $E_2$-PEG-MWCNTs, free DOX, and DOX with m-PEG-MWCNTs. Unlike DOX, the delivery system did not produce any severe cardiotoxicity, and the levels of hepatotoxicity and nephrotoxicity were also nonperceivable. Hence, the delivery system can be used for cancer cell-selective intranuclear drug delivery.

In another study, the *in vitro* and *in vivo* efficacy of DOX-loaded SWCNTs as theranostic nanoprobes was investigated in a murine breast cancer model (Faraj et al. 2014). For this investigation, iron-tagged SWCNTs was synthesized in conjugation with endoglin/CD105 antibody with or without DOX. The biocompatibility was assessed *in vitro* in luciferase (Luc2)-expressing 4T1 (4T1-Luc2) murine breast cancer cells. The results displayed enhanced therapeutic efficacy of DOX delivered through antibody-conjugated SWCNT. The drug delivery system also acted as a sensitive imaging biomarker as diffusion-weighted MRI was possible for diffusion coefficient measurements.

In an interesting work, SWCNTs were shown to have explosive properties such that these can act as potent therapeutic nanobomb agents for killing breast cancer cells (Panchapakesan et al. 2005). In this study, water molecules were first adsorbed on the SWCNTs which were then exposed to laser light of 800 nm at light intensities of approximately 50–200 mW/cm$^2$. It was observed that the thermal energy was sufficient to cause the water molecules to evaporate, which in turn built extreme pressure in SWCNT causing them to explode in the human BT-474 breast cancer cells suspended in saline solution buffered with phosphate and rendering the cells to death. The presence of bubbles around the dead cells accounted for the boiling effect caused by SWCNT explosions.

The drug delivering properties of SWCNT were also investigated with PTX as the choice of drug (Shao et al. 2015). In this work, a novel SWCNT-based drug delivery system conjugated with human serum albumin (HSA) nanoparticles was synthesized for loading of PTX antitumor agent. High intracellular delivery efficiency with cell uptake rate of 80% was observed in MCF-7 breast cancer cells. The drug loading on this drug delivery system was achieved through high binding affinity of PTX to HSA proteins. Greater growth inhibition was observed when SWCNT-HSA was used to load PTX instead of HSA. Hence, it can be suggested that the SWCNT-based antitumor agent developed in this study is functional and effective.

## 15.5.5  Cervical Cancer

Nearly 70% of cervical cancers are caused by oncogenic human papillomavirus (HPV) types 16 and 18. As the oncogenic antigens E6 and E7 are overexpressed on tumor cells in HPV-associated cancers, they are the most suitable target for developing antigen-specific immunotherapy for the control of cervical cancer.

In a recent study, MWCNTs were investigated for the delivery of DOX into cancer cells overexpressing CD44 receptors (Cao et al. 2015). MWCNTs were modified with PEI and then they were conjugated with fluorescein isothiocyanate (FITC) and hyaluronic acid (HA). The above complex with a drug loading percentage of 72% has been shown to be water-soluble and stable. The drug release was observed to be higher in acidic pH of 5.8 than in physiological conditions which have a pH of 7.4. The carrier material was shown to be biocompatible in the tested concentration range. The complex was found to be capable of specifically targeting cancer cells overexpressing CD44 receptors and exert growth-inhibition effects.

For imaging tumor cells, a novel approach of utilizing natural biocompatible polymer chitosan was developed by Wu and co-workers (2009). In this study, SWCNT was modified by chitosan fluorescein isothiocyanate (FITC-CHIT). This product was then conjugated with folic acid (FA) to construct the functional FITC-CHIT-SWCNT-FA conjugate, as folic acid receptors are overexpressed in most cancer cells. These novel functionalized SWCNTs were soluble and stable in phosphate buffer saline and were found to be easily transportable inside the human cervical carcinoma HeLa cells. FITC-CHIT-SWCNT-FA can be used as potential devices for targeting the drug into the tumor cells and also for imaging them as we get the combination of intrinsic properties of CNTs, versatility of chitosan, and folic acid. In the study of Qin et al., a thermos/pH-sensitive nanocarrier was used for NIR light remote-controlled drug delivery (Qin et al. 2015). For this study, an amphiphilic chitosan derivative-coated SWCNT was encapsulated in a thermos/pH-sensitive nanogel (CS/PNIPAAm@CNT). HeLa cells were chosen as the model tumor cells and DOX was chosen as the model drug. The study suggested that the multifunctional DOX-loaded CS/PNIPAAm@CNT nanocomposite is a promising therapeutic nanocarrier for intracellular drug delivery with potential for targeted cancer therapy.

## 15.5.6  Colon Cancer

Colon cancer is generally viewed as a homogeneous entity developing through multiple genetic and epigenetic abnormalities, such as defective DNA mismatch repair (dMMR) and the CpG island methylator phenotype (CIMP).

In a study, the suppression of c-Myc was investigated when the downregulation of ATP-binding cassette transporters in human colon carcinoma cells was done through MWCNTs (Wang et al. 2015). It was observed that MWCNTs reduced the transport activity and expression of ABC transporters including ABCB1/Pgp and ABCC4/MRP4 in human colon adenocarcinoma Caco-2 cells. Proto-oncogene c-Myc, which directly regulates ABC gene expression, also decreased in MWCNT-treated cells. In turn, the forced overexpression of c-Myc reversed the inhibitory effects on ABCB1 and ABCB4 expression. Hence, it was implied that MWCNTs may act on c-Myc

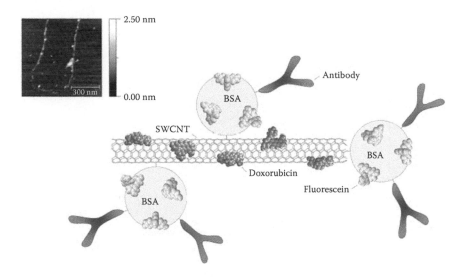

**FIGURE 15.3** Schematic illustration of the DOX–fluorescein–BSA–antibody-SWCNT complexes. Insert: AFM image of DOX–fluorescein–BSA–SWCNT complexes (without antibodies). (Reprinted from *Carbon*, 47, Heister, E. et al., Triple functionalization of single-walled carbon nanotubes with doxorubicin, a monoclonal antibody, and a fluorescent marker for targeted cancer therapy, 2152–2160, Copyright 2009, with permission from Elsevier.)

without oxidative stress to downregulate ABC transporter expression. CNT's novel cellular effects can be used to develop drug delivery systems to overcome the ABC transporter-mediated cancer chemoresistance.

Heister et al. (2009) fabricated triple functionalized SWCNTs with an anticancer drug (DOX), a mAb, and a fluorescent marker (fluorescein) at the noncompetitive binding sites on the SWCNTs for targeting the cancer cells (Figure 15.3). The authors chose noncovalent functionalization for attachment of DOX to allow for its release after cellular uptake, whereas covalent functionalization to attach fluorescein and carcinoembryonic antigen (CEA) antibodies to oxidized SWCNTs. The latter was achieved by using the hydrophilic protein bovine serum albumin (BSA) as a multifunctional linker. Confocal laser microscopy revealed the BSA-antibody-specific receptor-mediated uptake of SWCNTs by the human colon adenocarcinoma cell WiDr with subsequent targeting of DOX intracellularly to the nucleus.

### 15.5.7 Lymph Node Metastasis

One of the strongest indicators in prognoses of distant metastasis and survival in cancers is the presence of lymph node invasion. Invasion into a vascular or a lymphatic system is believed to be a key step of dissemination of tumor cells in the process of cancer metastasis development. After acquiring the abilities of intravasation and survival in an unfavorable vascular environment, the tumor cells circulate around the whole body parts to form new tumors at the secondary sites. Lymph node metastasis can act as an accurate predictor of recurrence and death of patients with cutaneous melanoma.

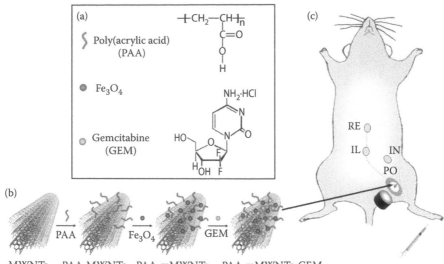

**FIGURE 15.4** Magneticlymphatic drug delivery system. (a) Molecular structures of poly(acrylic acid) and gemcitabine (GEM). (b) Schematic synthetic route of magnetic multiwalled carbon nanotubes (mMWNTs) and illustration of chemical reactions used to attach gemcitabine onto mMWNTs. (c) Schematic drawing of magnetic lymphatic targeted chemotherapy in mice. mMWNTs-GEM were subcutaneously injected into a mouse that had cancer lymph node metastasis via the left rear footpad, and were taken up into lymphatic vessels and retained in the targeted lymph node under the magnetic field. For clarity, different parts are drawn at arbitrary scales. PO, popliteal lymph node; IN, inguinal lymph node; IL, para-iliac lymph node; RE, renal hilar lymph nodes. (Reprinted from *Eur J Cancer*, 47, Yang, F. et al., Magnetic functionalised carbon nanotubes as drug vehicles for cancer lymph node metastasis treatment, 1873–1882, Copyright 2011, with permission from Elsevier.)

The potential therapeutic effect of gemcitabine (GEM)-loaded magnetic MWCNTs (mMWCNTs) with that of GEM-loaded magnetic-carbon particles was compared by Yang et al. (2011), in both *in vitro* and *in vivo* studies (Figure 15.4). The observation in this study showed the high antitumor activity of both the systems in human pancreatic cancer BxPC-3 cells compared with the free drug. The superparamagnetic behavior of mMWCNTs-GEM causes their magnetic moments to align along the applied field leading to a net magnetization, which in turn greatly affects the interaction of mMWCNTs-GEM with the cellular membrane, and thus they provide better results as compared to mACs-GEM in successful inhibition of lymph node metastasis when subcutaneous administration is performed under the impact of magnetic field.

## 15.5.8 Prostate Cancer

In prostate cancer, the affected cells propagate before the cancer becomes clinical. Prostate cancer antigen 3 (PCA3) is the principal molecule associated with prostate cancer. This molecule has been classified as the prostate cancer-specific gene and

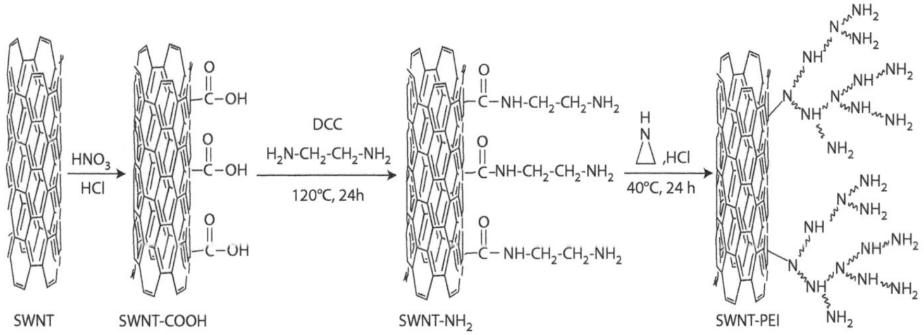

**FIGURE 15.5** The approach for preparing SWCNT-PEI. (Reprinted from *Biomaterials*, 34, Wang, L. et al., Synergistic anticancer effect of RNAi and photothermal therapy mediated by functionalized single-walled carbon nanotubes, 262–274, Copyright 2013, with permission from Elsevier.)

is usually found on chromosome 9q21-22. All prostate tumor specimens and PCA metastasis have a high overexpression of PCA3.

MWCNTs conjugated with prostate stem cell antigen antibody were investigated for ultrasound imaging and drug delivery (Wu et al. 2014). For the investigation, MWCNTs were cut short, grafted with PEI and conjugated covalently with FITC and prostate stem cell antigen (PSCA) mAb (Figure 15.5). The toxicity data revealed that the prepared conjugate had a good biocompatibility. The conjugate was confirmed to be capable of specifically targeting the cancer cells which were overexpressing PSCA. It can also be used as an ultrasound contrast agent. The testing on PC-3 tumor-bearing mice suggested that CNT-PEI(FITC)-mAb can deliver the drug to the target and suppress tumor growth. Hence, the prepared conjugate can be used for simultaneous ultrasound imaging and drug delivery. By using SWCNTs chemically functionalized with PEI and bound by DSPE-PEG 2000 maleimide for further conjugation with Asn-Gly-Arg peptide, Wang et al. (2013) developed a novel targeting SiRNA delivery system. This novel system was found to efficiently cross the human prostate cancer PCA3 cell membrane and induce severe apoptosis and suppression of the proliferating cells. The antitumor activity decreased when a combination of NIR photothermal therapy and RNA interference (RNAi) was used without any toxicity affecting other organs.

### 15.5.9 OTHER CANCERS

Among other types of cancers which are relatively rare are cancers that originate in the kidney, liver, and so on. Obviously, the CNTs are not well studied as drug carrier for the treatment of these types of cancers. The few existing literature reports are discussed below.

The cancer that starts in the kidney cells is known as kidney cancer. Renal cell carcinoma (RCC) and transitional cell carcinoma (TCC) are the two most common types of kidney cancers named after the type of cell from which they originate. The interaction between human embryo kidney HEK-293 cells and SWCNTs was investigated in a study by Cui et al. (2005). It was found that SWCNTs can inhibit the proliferation of HEK-293 cells, induce the cell apoptosis, and decrease cell adhesive ability in a time- and dose-dependent manner. SWCNTs induce changes in the cell cycle which could be attributed to the decrease in the number of cells in the S-phase due to upregulated expression of P16 which inhibits the cyclin-dependent kinase activity of CdK2, CdK4, and CdKr, and therefore prevents the cells from entering an S-phase and subsequently arresting the cell cycle in the G1 phase.

A cancer that originates in the liver is termed liver cancer or hepatic cancer. The tumors in the liver are either discovered on medical imaging equipment or present themselves symptomatically as an abdominal mass, pain, yellow skin, nausea, or liver dysfunction. Often this cancer is linked with the overexpression of c-Myc gene. To suppress the expression of c-Myc gene and c-Myc protein in the tumor-bearing cell, polyamidoamine dendrimer-modified MWCNTs (dMWCNTs) were fabricated for the efficient delivery of antisense c-Myc oligonucleotide (asODN) into liver cancer cell line HepG2 cells (Pan et al. 2009). The composites of asODN-dMWCNTs were incubated with HepG2 cells and laser confocal microscopy confirmed that they enter the tumor cells within 15 min. These composites inhibited the cell growth in a time- and dose-dependent manner, and downregulated the expression of the c-Myc gene and c-Myc protein. These composites exhibited maximal transfection efficiencies and inhibition effects on tumor cells when compared to CNT-NH$_2$-asODN and dendrimer (asODN) alone. In another report, the effective killing of hepatocellular carcinoma SMMC-7221 cell lines and depressed growth of liver cancer was achieved along with decrease in *in vivo* toxicity as compared to free DOX by the chitosan and folic acid-modified SWCNTs (Figure 15.6) (Ji et al. 2012).

For convenience of the reader, we have summarized in Table 15.1 the most important results reported to date on the use of various drugs in anticancer treatment in which CNTs have been used as the carrier for targeted drug delivery.

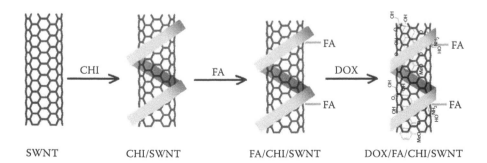

**FIGURE 15.6** Scheme for preparation of DOX/FA/CHI/SWCNT. (Reprinted from *J Colloid Interface Sci*, 365, Ji, Z. et al., Targeted therapy of SMMC-7721 liver cancer in vitro and in vivo with carbon nanotubes based drug delivery system, 143–149, Copyright 2012, with permission from Elsevier.)

**TABLE 15.1**

**Summary of the Most Significant Work on CNT-Based Targeted Drug Delivery for the Treatment of Cancer**

| Name of the Drug | Targeted Disease | Drug Carrier Used | Targeted Cell Line | Main Findings | Reference |
|---|---|---|---|---|---|
| 5-fluorouracil (5-FU) | Multiple cancer types | GN-CNT-$Fe_3O_4$ nanoparticle hybrid | HepG2 cell line | The anticancer drug 5-FU was loaded onto the GN-CNT-$Fe_3O_4$ hybrid in capacities as high as 0.27 mg mg$^{-1}$, and the release of the drug from this hybrid is found higher under acidic conditions than under neutral conditions. Furthermore, the GN-CNT-$Fe_3O_4$ hybrid was proven to have excellent biocompatibility by the cell uptake studies and MTT assay | Fan et al. (2013) |
| Doxorubicin/ Daunorubicin | Blood cancer | SWCNTs functionalized with Sgc8c aptamer | Human T cell, leukemia cell MOLT-4, and U266 myeloma cells | The tertiary complex of Dau-aptamer-SWCNTs was effectively internalized in MOLT-4 cells but not in U266 cells. The pH of the environment played an important role in the release of Dau. The system showed potential to reduce cytotoxic effects of Dau by selective delivery and controlled release of the drug | Taghdisi et al. (2011) |
| | Brain cancer | PEGylated oxidized MWCNTs-modified with angiopep-2 | C6 glioma cells | Angiopep-2-modified PEGylated MWCNTs showed better antiglioma activity, good compatibility, and low toxicity | Ren et al. (2012) |
| | | MWCNTs functionalized with Poly(acrylic acid) and folic acid | U87 human glioblastoma cells | The cytotoxicity toward U87 human glioblastoma cells was enhanced when dual targeting of DOX from magnetic MWCNTs was done instead of free DOX | Lu et al. (2012) |
| | Breast cancer | Functionalized SWCNTs with polyethyleneimine linked to NGR (Asn-Gly-Arg) attached with DOX | MCF-7 human breast cancer cell line | A SWCNT-PEI/DOX/NGR intelligent delivery system was developed, which showed the characteristics of NIR/pH responsiveness, sustained release, and tumor targeting. DOX delivered by SWCNT-PEI/DOX/NGR presented significant tumor growth inhibition *in vitro* and *in vivo* | Wang et al. (2013) |

*(Continued)*

**TABLE 15.1 (Continued)**
**Summary of the Most Significant Work on CNT-Based Targeted Drug Delivery for the Treatment of Cancer**

| Name of the Drug | Targeted Disease | Drug Carrier Used | Targeted Cell Line | Main Findings | Reference |
|---|---|---|---|---|---|
| | | Estradiol–PEG-appended MWCNTs | A549 cell line, MCF 7 cell line | Conjugation of estradiol with MWCNTs not only augmented the nuclear targeting index of the carrier-bound drug but also improved its therapeutic efficacy both *in vitro* and *in vivo* | Das et al. (2013) |
| | | SWCNTs tagged with iron and conjugated with endoglin/CD105 | Luciferase (Luc2)- expressing 4T1 (4T1-Luc2) murine breast cancer cells | The antibody-conjugated magnetic SWCNTs showed enhanced therapeutic efficacy of DOX. Further, diffusion-weighted MRI was found to be a sensitive imaging biomarker for assessment of treatment-induced changes | Faraj et al. (2014) |
| | Cervical cancer | SWCNTs functionalized with polysaccharides (sodium alginate [ALG] and chitosan [CHI] | HeLa cells | When DOX was released from the modified nanotubes on the cancer cells, the nuclear DNA was damaged and the cell proliferation was inhibited | Zhang et al. (2009) |
| | | MWCNTs functionalized with folic acid & iron (FA-MWCNT@Fe) | HeLa cells | High load capacity of DOX on FA-MWCNT@Fe and prolonged release property was observed. FA-MWCNT@Fe also showed sixfold higher delivery efficiency of DOX than free DOX | Li et al. (2011) |
| | Colon cancer | SWCNTs functionalized with bovine serum albumin antibody, fluorescein | WiDr Human colon adeno-carcinoma cell | The intake of the DOX-attached functionalized SWCNT by cancer cells were efficiently accompanied by intracellular release of DOX | Heister et al. (2009) |
| | Liver cancer | SWCNTs functionalized with chitosan and folic acid | Hepatocellular carcinoma SMMC -7721 cell lines | The DOX attached with chitosan-folic acid-modified SWCNTs killed cancer cells more efficiently and displayed less toxicity than free DOX | Ji et al. (2012) |

*(Continued)*

**TABLE 15.1 (Continued)**
**Summary of the Most Significant Work on CNT-Based Targeted Drug Delivery for the Treatment of Cancer**

| Name of the Drug | Targeted Disease | Drug Carrier Used | Targeted Cell Line | Main Findings | Reference |
|---|---|---|---|---|---|
| | Prostate cancer | MWCNT grafted with polyethyleneimine (PEI) conjugated to FITC and prostate stem cell antigen monoclonal antibody (mAb) | PC-3 cells overexpressing PSCA, MCF-7 cells low-expressing PSCA | The CNT-PEI(FITC)-mAb showed good biocompatibility. The presence of the antibody enhanced the cellular uptake capability of the material by PSCA-overexpressed cancer cells. CNT-PEI(FITC)-mAb/DOX can selectively accumulate in the malignant tumor tissues and inhibit the tumor growth | Wu et al. (2014) |
| | Multiple cancer types | Quantum dots conjugated with $Fe_3O_4$-Filled CNT ($Fe_3O_4$@CNT-HQDs) | HeLa cell lines, HEK293 human kidney cell lines | $Fe_3O_4$@CNT-HQDs showed fluorescence allowing simultaneous imaging/mapping of cancer cells, rendering double-targeted drug delivery. Attractive fluorescent marker, high drug loading efficiency, low cytotoxicity, and high sensitivity for effective diagnostic and therapeutic applications. Efficient DOX release into the cell nucleus causing cell death | Moore et al. (2013) |
| | | Branched gold–CNT hybrid | Human lung cancer cell line A549 | DOX was loaded in high quantity on both inner and outer surfaces of the oxidized CNTs which showed efficient transport and delivery of DOX inside the cells | Minati et al. (2012) |
| | | MWCNTs functionalized with polyamidoamine-polyethyleneglycol polyamidoamine (PAMAM-PEG-PAMAM) ABA type linear dendritic polymers | Mouse tissue connective fibroblast adhesive cell line (L929) | CNTs functionalized with PAMAM-PEG-PAMAM linear dendritic copolymers are able to load anticancer drugs and kill cancer cells efficiently even in low concentrations | Adeli et al. (2013) |

*(Continued)*

**TABLE 15.1 (*Continued*)**

**Summary of the Most Significant Work on CNT-Based Targeted Drug Delivery for the Treatment of Cancer**

| Name of the Drug | Targeted Disease | Drug Carrier Used | Targeted Cell Line | Main Findings | Reference |
|---|---|---|---|---|---|
| | | Lysine-modified SWCNTs | Human hepatic carcinoma cell line (SMMC-7721) *in vitro* and sarcomaia 180-bearing mice *in vivo* | The release of DOX from DOX-Lys/ SWCNT- thermo-sensitive liposomes depended strongly on the temperature. Studies on the therapeutic effects showed higher antitumor efficacy in cultured SMMC-7721 cells *in vitro* and in S180-bearing mice *in vivo* with minimal side effects. | Zhu et al. (2014) |
| FITC-labeled antisense c-Myc oligonucleotides | Breast and liver cancer | MWCNTs functionalized with polyamidoamine dendrimer(d) | MCF-7 and MDA-MB-435 human breast cancer cell line, HEP-G2 human hepatoma cells | The MWCNTs readily entered the tumor cell within 15 min of incubation. The cell growth was inhibited and the expression of c-Myc gene and c-Myc protein was downregulated | Pan et al. (2009) |
| Gemcitabine | Lymph node metastasis | Magnetically functionalized MWCNTs | BxPC-3 pancreatic cancer cells | Subcutaneous administration of GEM-loaded mMWCNTs resulted in successful regression and inhibition of lymph node metastasis under the action of magnetic field, in particular, in the high dose cases | Cui et al. 2005 |
| Irinotecan | Colon cancer | Folate–chitosan conjugate of CNT followed by microencapsulation with Eudragit S-100 copolymer | HT 29 cells | Controlled release of irinotecan was observed at a pH of 7.4 through the dissolution of the coating layer of Eudragit-100. Further, the CNTs not loaded with irinotecan showed no cell proliferation | Zhou et al. (2014) |

*(Continued)*

**TABLE 15.1 (Continued)**

**Summary of the Most Significant Work on CNT-Based Targeted Drug Delivery for the Treatment of Cancer**

| Name of the Drug | Targeted Disease | Drug Carrier Used | Targeted Cell Line | Main Findings | Reference |
|---|---|---|---|---|---|
| Mitoxantrone (MTO) | Breast cancer | MWCNT functionalized with carboxylic acid | MDA 231 breast cancer cell line, NIH3T3 nonneoplastic fibroblast cell line | MWCNT–MTO cytotoxic effects were time and dose dependent, and greatly improved with respect to the soluble drug at long incubation times. Internalization is more visible at shorter incubation times for NIH3T3 cells, whereas for MDA 231 cells the internalization is prominent at longer incubation times. Efficient delivery of MTO to both cell types, without distinction between cancer and nonneoplastic cells | Risi et al. (2014) |
| Paclitaxel (PTX) | Brain cancer | Poly(lactide)-co-poly(ethylene glycol)-coated MWCNTs | U-87 glioblastoma cells | Significantly, less toxic in both *in vitro* and *in vivo* when compared to pristine CNTs. Moreover, PTX was released to the cells for a prolonged period of time | Song et al. (2013) |
| | Breast cancer | SWCNTs functionalized with polyethyleneglycol | 4T1 murine breast cancer cell line | Higher efficacy in suppressing tumor growth was achieved without causing obvious toxic effects to normal organs by the use of SWCNT-PTX | Liu et al. (2008) |
| | Breast cancer | SWCNT conjugated with human serum albumin (HSA) | MCF-7 breast cancer cells | SWCNT-HAS-formulated PTX displayed stronger antitumor effect than HSA nanoparticle-formulated PTX. The increased cancer cell inhibition effect was linked with SWCNTmediated efficient cell internalization | Shao et al. (2015) |
| Silibinin (SB) | Multiple cancer types | MWCNTs grafted with –COOH groups to which silibinin was conjugated by a cross-linker agent | HepG2 cell lines, A549 cell lines | Release of the drug from the nanocarrier was observed to occur in a sustained and pH-dependent manner. Proliferation assay against cancer cells indicated that SB-MWCNTs expressed cytotoxicity at lower dosages in comparison with free drug. | Tan et al. (2014) |
| SiRNA and DNA | Brain cancer | MWCNTs | V2 microglia and GL261 glioma cell | Efficient uptake of MWCNTs was observed in both BV2 and GL261 cells without any noticeable cytotoxicity | Kateb et al. (2007) |

*(Continued)*

**TABLE 15.1 (*Continued*)**
**Summary of the Most Significant Work on CNT-Based Targeted Drug Delivery for the Treatment of Cancer**

| Name of the Drug | Targeted Disease | Drug Carrier Used | Targeted Cell Line | Main Findings | Reference |
|---|---|---|---|---|---|
| Small interfering RNA (SiRNA) | Breast cancer | SWCNTs functionalized with DSPE-PEG-Amine carrying siRNA, and mouse double minute 2 homolog (MDM2) | B-Cap-37 breast carcinoma cells | Significant efficiency was shown in carrying SiRNA and SiRNA-MDM2 complexes in B-Cap-37 cells by f-SWCNTs causing inhibition of proliferating cells. It was obvious that MDM2 can serve as a novel therapeutic target by an effective carrier system of DSPE-PEG-amine-functionalized SWNTs | Chen et al. (2012) |
| | Prostate cancer | SWCNTs functionalized with polyethylenimine (PEI), and tumor targeting NGR (Cys-Asn-Gly-Arg-Cys) peptide | Human prostate carcinoma (GIV) cell PC-3 cells | Severe apoptosis was induced by SWCNT-PEI-SiRNA-NGR and the proliferation of PC-3 cells was suppressed. The delivery system exhibited higher antitumor activity due to more accumulation in tumor without obvious toxicity in the main organs | Wang et al. (2013) |
| Multiple drugs: Hematoxylin-eosin (H&E) | Bladder cancer | Four different MWCNTs with separate lengths, diameters, defects and aspect ratios | Urinary bladders explanted from mice and human BCa cell line UM-UC-3 | MWCNTs with wall diameter of 11 nm or 18 nm and mean length of upto 1500 nm are able to adhere to the urothelium of mouse bladders. The CNTs mediated a low-to-moderate impairment of cellular function in two different BCa cell lines. CNTs with more defects and lower aspect ratios were more deleterious than the ones with fewer defects and higher aspect ratios | Reiger et al. (2015) |
| No specific drugs | Brain cancer | SWCNTs conjugated with phospholipid-polyethylene glycol, protein A, fluorescein-labeled integrin $\alpha_v\beta_3$ mAb(SWNT-PEG-mAb) | Integrin $\alpha_v\beta_3$-positive human U-87MG cell, integrin $\alpha_v\beta_3$-negative MCF-7 | *In vitro* study revealed that SWNT-PEG-mAb presented a high targeting efficiency on integrin $\alpha_v\beta_3$-positive U87MG cells with low cellular toxicity, while for integrin $\alpha_v\beta_3$-negative MCF-7 cells, the system had a low targeting efficiency indicating that the high targeting to U87MG cells was due to the specific integrin targeting of the monoclonal antibody | Ou et al. (2009) |

*(Continued)*

**TABLE 15.1 (*Continued*)**
**Summary of the Most Significant Work on CNT-Based Targeted Drug Delivery for the Treatment of Cancer**

| Name of the Drug | Targeted Disease | Drug Carrier Used | Targeted Cell Line | Main Findings | Reference |
|---|---|---|---|---|---|
| | | SWCNTs functionalized with CD133 monoclonal antibody (mAb) | CD133(+) cancer stem-like cells (CSCs) in glioblastoma (GBM) | GBM-CD133$^+$ cells were selectively targeted and eradicated, whereas CD133$^-$ cells remained viable after incubation with SWCNT-CD133-mAb. GBM-CD133$^+$ cells pretreated with SWCNT-CD133mAb and irradiated by near-infrared laser for 2 days did not exhibit sustainability of cancer stem-like cells feature for tumor growth | Wang et al., (2011) |
| | Breast cancer | Carboxylated SWNTs conjugated with anti-HER2 chicken IgY antibody | HER2-expressing SK-BR-3 cells and HER2-negative MCF-7 cells | The SWCNT complex specifically targeted HER2-expressing SK-BR-3 cells but not receptor-negative MCF-7 cells showing their potential for both detection and selective photothermal ablation of receptor-positive breast cancer cells without the need of internalization by the cells | Xiao et al. (2009) |
| | | SWCNTs containing adsorbed water | Human BT-474 breast cancer cells | Water molecules adsorbed in SWCNT sheets or loosely adsorbed on top of cells got heated to more than 100°C upon exposure to laser light of 800 nm at light intensities of approx 50–200 mW/cm². Thermal energy generated from the laser and the subsequent confinement of thermal energy in SWCNT caused the water molecules to evaporate and develop extreme pressures in SWCNT causing them to explode leading to killing of human BT474 breast cancer cells | Panchapakesan et al. (2005) |

*(Continued)*

**TABLE 15.1 (Continued)**
**Summary of the Most Significant Work on CNT-Based Targeted Drug Delivery for the Treatment of Cancer**

| Name of the Drug | Targeted Disease | Drug Carrier Used | Targeted Cell Line | Main Findings | Reference |
|---|---|---|---|---|---|
| | | SWCNTs functionalized with PEG and poly (maleic anhydride-alt-1-octadecene) ($PMHC_{18}$) | Balb/c mice bearing 4T1 tumors | The blood circulation and biodistribution of the PEG-$PMHC18$-coated SWCNTs in mice after intravenous injection revealed ultra-long blood circulation half-lives with high uptake in the tumor, along with accumulation in the skin dermis. A surface coating which affords SWNTs a blood half-life of 12–13 h appeared to be optimal to balance the tumor-to-normal organ (T/N) uptake ratios of CNTs in major organs | Liu et al. (2011) |
| | Cervical cancer | SWCNTs functionalized with chitosan and folic acid, FITC | Human cervical carcinoma HeLa cells | Conjugates provide new options for targeted drug delivery and tumor cell sensing because of the combined intrinsic properties of CNTs and the versatility of chitosan | Wu et al. (2009) |
| | Kidney cancer | SWCNTs | Human kidney embryo cell HEK-293 cells | The proliferation of HEK293 cells was inhibited by inducing cell apoptosis and decreasing the adhesive ability of the cell. Also, HEK293 cells mobilized active responses by isolation of SWCNTs attached cells from the cell mass with the help of small *isolation* proteins. These proteins can act as target molecules | Yang et al. (2011) |
| | Multiple cancer types | Drug-loaded liposomes covalently attached to CNT forming a CNT-liposome conjugate (CLC) | Diploid primary human fibroblast-adherent cell line derived from fetal lung tissue IMR90 (CCL-186) and HEK 294 | Large amount of drug can be delivered into the cells by the CLC system, thus preventing potential adverse systemic effects of CNT when administered at high doses. The calcein uptake proves this point. Targeting may be achieved by covalent attachment of targeting (to desired cells) agents to thus enhancing delivery of drugs to diseased cells and reducing drug interaction with healthy tissue | Karchemski et al. (2012) |

## 15.6 FACTORS LIMITING THE APPLICATION OF CNT IN TARGETED DRUG DELIVERY

The amazing prospects put forward by CNTs are loomed by certain factors limiting its application. The presence of impurities, large surface area (leads to protein opsonization), nonuniform morphology and structure, insolubility, hydrophobicity, and tendency of CNTs to bundle together are some obstacles encountered in their medical applications.

The toxicity of CNTs is another key obstacle. The nanoparticles enter the body physically and usually only few nanoparticles enter the body; however, the nanoparticles are persistent inside the body due to their limited metabolisms. As their removal is slow, chronic cumulative health effects have been observed due to that. The mechanisms underlying toxicity of CNT include oxidative stress, production of chemokines, cytokines and inflammatory responses, DNA damage and mutation (errors in chromosome number as well as disruption of the mitotic spindle), malignant transformation, interstitial fibrosis, and the formation of granulomas (Liu et al. 2012).

Warheit et al. (2004) introduced SWCNTs directly into the lungs of rats to study the carcinogenicity of CNTs. It was observed that exposure to high concentration of SWCNTs leads to the development of granulomas in rodents and a concentration of 0.5 mg/m$^3$ and 2.5 mg/m$^3$ of MWCNTs induced microgranulomas with inflammation in the alveoli. Kanno et al. (2008) demonstrated that MWCNTs could induce multiple mesothelioma with severe peritoneal adhesion when it was administered intraperitoneally to p53 heterozygous mice. The reason for this could be the structural similarity as well as the persistency of CNT in the organism with asbestos. In another study, Takanashi et al. (2012) reported that subcutaneously implanting the MWCNTs into the rasH2 mice did not develop neoplasm.

Among the many pathways of toxicity, interference with fibrous mechanisms and cytoskeleton, cell signaling, and membrane perturbations are some of the effects caused by exposure to CNTs (Rodriguez-Yanez et al. 2013). In another work, the length of MWCNTs were reported to exert effects on the biomembranes; when the distribution of MWCNTs (3–14 μm length) in RAW264 cells were observed under a light microscope, MWCNTs increased the permeability defects of the plasma membrane lipid bilayer while the toxicity of shorter (1.5 μm) MWCNTs was significantly less (Liu et al. 2012). A study on the interference of CNTs in cytoskeleton revealed that the exposure of cultured human epidermal keratinocytes (HaCaT) to SWCNTs induced oxidative stress and resulted in the loss of cell viability, indicating that the dermal exposure to CNTs may lead to altered skin conditions (Shvedova et al. 2003). Thus, widespread application of CNTs as drug carrier in targeted drug delivery requires that the above issues arising from the toxicity of CNTs need to be addressed properly such that the appropriate safeguards can be observed against the possible side effects.

## 15.7 CONCLUSION

The use of CNTs in targeted drug delivery has been widely investigated with encouraging results. The enhanced properties of CNT, such as high aspect ratio and surface area, the ability to attach additional molecules for targeting or imaging purposes, elevated cellular internalization and preferential tumor accumulation have made CNTs versatile and effective drug carriers for many small-molecule drugs.

Generally, there are three approaches by which CNTs can be loaded with drugs and be used for drug delivery: (i) some drugs can be encapsulated within the interior cavity of CNTs via chemically wet procedures or through physical procedures (using capillary effect), (ii) drugs that possess conjugated aromatic ring systems are able to physically adsorb onto the surface of CNTs via noncovalent $\pi$–$\pi$ and hydrophobic interactions, and (iii) drugs can be chemically functionalized either permanently or via cleavable linkers onto the surface of CNTs.

In addition to the above three, CNTs can also be incorporated into other parent drug carriers, for instance hydrogels or electron spun fibers, as adjunct material to modulate the loading and release of coencapsulated drugs. In order to further enhance the therapeutic efficacy and diminish the inherent cytotoxicity of CNTs, other molecules are used to functionalize CNTs. Polymers such as PEG and chitosan are popularly used to impart water dispersibility to the CNTs and offer additional sites for conjugation of other molecules. Some CNT–drug constructs are tethered to targeting molecules, be it small chemical molecules, complex biological entities, or even magnetic NP to selectively target certain cells or tissues.

## 15.8   FUTURE WORK

CNTs show great potential as an effective carrier for targeted drug delivery for many therapeutic applications including treatment of cancer. However, despite the various successes and promising potential CNT have demonstrated in targeted drug delivery, there still exist a few concerns regarding their clinical applications. One of the major concerns is the issue of CNT toxicity which is still a topic of debate as the long-term effects of CNTs are yet to be established. As we have seen in this review that the toxicity differs with surface functionalization and physical properties of CNTs, comprehensive studies on the long-term impacts of CNTs and their functionalized derivatives to be used as carrier for drug delivery on humans are necessary before these nanostructures can be considered as safe for drug delivery applications. The existing literature is mostly on short-term effects and that too most of them do not study the effects on humans, and hence conclusions drawn from these literatures on the toxicity of CNTs are not beyond doubt. Moreover, the varieties of CNTs in terms of their size, length, number of walls, and the presence of defects and functional groups pose further difficulty in providing batch-to-batch consistency of fabricated CNT–drug complexes. Further studies are needed on the adequate and controlled delivery and release of drugs by CNTs to the targeted cells, as for any therapeutic application the dose of drugs plays a very significant role. In order for CNT-induced targeted drug delivery to be commercially successful, future researchers should be directed to meet the following needs such as (i) better control on the size, diameter, and properties of CNTs, (ii) more knowledge on the long-term impacts (cytotoxicity) of various types of functionalized CNTs on humans, and (iii) better understanding on the mechanisms of drug delivery and release in the target cells from CNT-drug carrier, which in turn could help in obtaining better control on the dosage of drugs.

## ACKNOWLEDGMENT

The authors acknowledge the MHRD, Govt. of India for the partial support in the form of a "Centre of Excellence in Advanced Materials" grant under its Technical Education Quality Improvement program (TEQIP-II).

## REFERENCES

Abarrategi, A., Gutierrez, M.C., Morenovicente, C. et al. 2008. Multiwall carbon nanotube scaffolds for tissue engineering purposes. *Biomaterials* 29:94–102.

Adeli, M., Ashiri, M., Chegeni, B.K. et al. 2013. Tumor targeted drug delivery systems based on supramolecular interactions between iron oxide–carbon nanotubes PAMAM–PEG–PAMAM linear dendritic copolymers. *J Iranian Chem Soc* 10:701–08.

Ajayan, P.M. and Iijima, S. 1993. Capillarity-induced filling of carbon nanotubes. *Nature* 361:333–34.

Alberts, B., Johnson, A., Lewis, J. et al. 2002. *Transport into the Cell from the Plasma Membrane: Endocytosis, Molecular Biology of the Cell*, 4th edition. New York: Garland Science.

Borowiak-Palen, E., Tripisciano, C., Rummeli, M. et al. 2011. Filling of carbon nanotubes: Containers for magnetic probes and drug delivery. In *Carbon Nanotubes for Biomedical Applications*, eds. R. Klingeler, R.B. Sim, 67–82. Germany: Springer.

Bottini, M., Cerignoli, F., Dawson, M.I. et al. 2006. Full-length single-walled carbon nanotubes decorated with streptavidin-conjugated quantum dots as multivalent intracellular fluorescent nanoprobes. *Biomacromolecules* 7: 2259–63.

Cai, D., Doughty, C.A., Potocky, T.B. et al. 2007. Carbon nanotube-mediated delivery of nucleic acids does not result in non-specific activation of B lymphocytes. *Nanotechnology* 18:365101.

Cao, X., Tao, L., Wen, S. et al. 2015. Hyaluronic acid-modified multiwalled carbon nanotubes for targeted delivery of doxorubicin into cancer cells. *Carbohydr Res* 405:70–77.

Cato, M.H., D'annibale, F., Mills, D.M. et al. 2008. Cell-type specific and cytoplasmic targeting of PEGylated carbon nanotube-based nanoassemblies. *J Nanosci. Nanotechnol* 8:2259–69.

Chakraborty, A.K. and Coleman, K.S. 2008. Poly(ethylene)glycol/single-walled carbon nanotube composites. *J Nanosci Nanotechnol* 8:4013.

Chakraborty, A.K., Dhanak, V.R., and Coleman, K.S. 2009. Electronic fine structure of 4-nitrophenyl functionalized SWCNTs. *Nanotechnology* 20:155704.

Chakraborty, A.K., Woolley, R.A.J., Butenko, Y.V. et al. 2007. Ion irradiation induced defects in carbon nanotubes: A photoelectron spectroscopy study. *Carbon* 45:2744.

Chen, H., Ma, X., Li, Z. et al. 2012. Functionalization of singlewalled carbon nanotubes enables efficient intracellular delivery of siRNA targeting MDM2 to inhibit breast cancer cells growth. *Biomed Pharmacother* 66:334–38.

Chen, J.Y., Chen, S.Y., Zhao, X.R. et al. 2008. Functionalized single-walled carbon nanotubes as rationally designed vehicles for tumor-targeted drug delivery. *J Am Chem Soc* 130:16778–85.

Chin, S.F., Baughman, R.H., Dalton, A.B. et al. 2007. Amphiphilic helical peptide enhances the uptake of single walled carbon nanotubes by living cells. *Exp Biol Med (Maywood)* 232:1236–44.

Coleman, K.S., Chakraborty, A.K., Bailey, S.R. et al. 2007. Iodination of single walled carbon nanotubes. *Chem Mater* 19:1076.

Cui, D., Tian, F., Ozkan, C.S. et al. 2005. Effect of single wall carbon nanotubes on human HEK293 cells. *Toxicol Lett* 155:73–85.

Czub, J. and Baginski, M. 2009. Amphotericin B and its new derivatives mode of action. *Curr Drug Metab* 10:459–69.

Das, M., Singh, R.P., Datir, S.R. et al. 2013. Intranuclear drug delivery and effective *in vivo* cancer therapy via estradiol–PEG-appended multiwalled carbon nanotubes. *Mol Pharm* 10:3404–16.

Dujardin, E., Ebbesen, T.W., and Krishnan, A. 1994. Wetting of Single Carbon nanotubes. *Adv Mater* 10: 1472–75.

Fan, X., Jiao, G., Gao, L. et al. 2013. The preparation and drug delivery of a graphene–carbon nanotube–$Fe_3O_4$ nanoparticle hybrid. *J Mater Chem* B 1:2658–64.

Faraj, A.A., Shaik, A.P., and Shaik, A.S. 2014. Magnetic SWCNTs as efficient drug delivery nanocarriers in breast cancer murine model: non-invasive monitoring using diffusion-weighted magnetic resonance imaging as imaging sensitive biomarker. *Int J Nanomedicine* 10:157–68.

Feazell, R.P., Nakayama-Ratchford, N., Dai, H. et al. 2007. Soluble single walled Carbon nanotubes as long boat delivery systems from Platinum (IV) anti-cancer drug design. *J Am Chem Soc* 129:8438–39.

Fisher, C., Rider, A.E., Han, Z. J. et al. 2012. Applications and nanotoxicity of carbon nanotubes and graphene in biomedicine. *J Nanomaterials* 2012:15185–203.

Fu, Q., Weinberg, G., and Su, D-S. 2008. Selective filling of carbon nanotubes with metals by selective washing. *New Carbon Mater* 23:17–20.

Hasegawa, T., Fujisawa, T., Numata, M. et al. 2004. Single-walled carbon nanotubes acquire a specific lectin-affinity through supramolecular wrapping with lactose-appended schizophyllan. Chem Commun 2004: 2150–51.

Heister, E., Neves, V., Tilmaciu, C. et al. 2009. Triple functionalisation of single-walled carbon nanotubes with doxorubicin, a monoclonal antibody, and a fluorescent marker for targeted cancer therapy. *Carbon* 47:2152–60.

Hirsch, A. 2002. Functionalization of Single-Walled Carbon nanotubes. *Angew Chem Int Ed* 41:1853–53.

Huse, J.T. and Holland, E.C. 2010. Targeting brain cancer: Advances in the molecular pathology of malignant glioma and medulloblastoma. *Nat Rev Cancer* 10:319–31.

Iijima, S. 1991. Helical microtubules of graphitic carbon. *Nature* 354:56–58.

Ji, Z., Lin, G., Lu, Q. et al. 2012. Targeted therapy of SMMC-7721 liver cancer *in vitro* and *in vivo* with carbon nanotubes based drug delivery system. *J Colloid Interface Sci* 365:143–49.

Kam, N.W.S. and Dai, H.J. 2005. Carbon nanotubes as intracellular protein transporters: generality and biological functionality. *J Am Chem Soc* 127:6021–26.

Kam, N.W.S. and Dai, H. 2006. Single walled carbon nanotubes for transport and delivery of biological cargos. *Phys Stat Sol B* 243: 3561–66.

Kam, N.W.S., Jessop, T.C., Wender, P.A. et al. 2004. Nanotube molecular transporters: Internalization of carbon nanotube-protein conjugates into mammalian cells. *J. Am Chem Soc* 126:6850–51.

Kang, B., Yu, D.C., Chang, S.Q. et al. 2008. Intracellular uptake, trafficking and subcellular distribution of folate conjugated single walled carbon nanotubes within living cells. *Nanotechnology* 19:375103.

Kanno, J., Takagi, A., Hirose, A. et al. 2008. Induction of mesothelioma in p53+/− mouse by intraperitoneal application of multi-wall carbon nanotube. *J Toxicol Sci* 33:105–16.

Karchemski, F., Zucker, D., Barenholz, Y. et al. 2012. Carbon nanotubes-liposomes conjugate as a platform for drug delivery into cells. *J Control Release* 160:339–45.

Kateb, B., Van Handel, M., Zhang, L. et al. 2007. Internalization of MWCNTs by microglia: Possible application in immunotherapy of brain tumors. *Neuroimage* 37:S9–S17.

Khabashesku, V.N., Margrave, M.L., Stevens, J.L. et al. 2008. Side wall functionalization of single-wall carbon nanotubes through CN bending forming substitution of fluoronanotubes. US Patent 7452519.

Kostarelos, K., Bianco, A., Lacerda, L. et al. 2012. Translocation mechanisms of chemically functionalised carbon nanotubes across plasma membranes, *Biomaterials* 33:3334–43.

Kostarelos, K., Lacerda, L., Pastorin, G. et al. 2007. Cellular uptake of functionalized carbon nanotubes is independent of functional group and cell type. *Nat Nanotechnol* 2:108–13.

Li, R., Wu, R., Zhao, L. et al. 2011. Folate and iron difunctionalized multiwall carbon nanotubes as dual-targeted drug nanocarrier to cancer cells. *Carbon* 49:1797–05.

Liu, D., Wang, L. Wang, Z. et al. 2012. Different cellular response mechanisms contribute to the length-dependent cytotoxicity of multi-walled carbon nanotubes. *Nanoscale Res Lett* 7:361.

Liu, X., Tao, H., Yang, K. et al. 2011. Optimization of surface chemistry on single-walled carbon nanotubes for *in vivo* photothermal ablation of tumors. *Biomaterials* 32:144–51.

Liu, Y., Zhao, Y., Sun, B. et al. 2013. Understanding the toxicity of carbon nanotubes. *Acc Chem Res* 46:702–13.

Liu, Z., Cai, L.H.E., Nakayama, N. et al. 2007a. In vivo biodistribution and highly efficient tumor targeting of carbon nanotubes in mice. *Nat Nanotechnol* 2:47–52.

Liu, Z., Chen, K., Davis, C. et al. 2008. Drug delivery with carbon nanotubes for *in vivo* cancer treatment. *Cancer Res* 68:6652–60.

Liu, Z., Sun, X.M., Nakayama-Ratchford, N. et al. 2007b. Supramolecular chemistry on water-soluble carbon nanotubes for drug loading and deliver. *ACS Nano* 1:50–56.

Liu, Z., Winters, M., Holodniy, M. et al. 2007c. RNA delivery into human T cells and primary cells with carbon nanotube transporters. *Angew Chem Int Ed* 46:2023–27.

Long, D.W., Wu, G.Z., and Zhu, G.L. 2008. Noncovalently modified carbon nanotubes with carboxymethylated chitosan: A controllable donor-acceptor nanohybrid. *Int J Mol Sci* 9:120–30.

Lu, Y.-J., Wei, K.-C., Ma, C.-C. et al. 2012. Dual targeted delivery of doxorubicin to cancer cells using folate-conjugated magnetic multi-walled carbon nanotubes. *Colloids Surf B Biointerfaces* 89:1–9.

Lukanov, P., Tilmaciu, C.-M., Galibert, A.M. et al. 2011. Filling of carbon nanotubes with compounds in solid or melted phase. In *Carbon Nanotubes for Biomedical Applications*, Eds. R. Klingeler, R.B. Sim, 41–65, Germany: Springer.

Minati, L., Antonini, V., Dalla Serra. et al. 2012. Multifunctional branched gold–carbon nanotube hybrid for cell imaging and drug delivery. *Langmuir* 28:15900–06.

Moore, T.L., Pitzer, J.E., Podila, R. et al. 2013. Multifunctional polymer-coated carbon nanotubes for safe drug delivery. *Part Part Syst Char* 30:365–73.

Mu, Q., Broughton, D.L., and Yan, B. 2009. Endosomal leakage and nuclear translocation of multiwalled carbon nanotubes: Developing a model for cell uptake. *Nano Lett* 9:4370–75.

Mu, Q.X., Broughton, D.L., and Yan, B. 2009. Endosomal leakage and nuclear translocation of multiwalled carbon nanotubes: Developing a model for cell uptake. *Nano Lett* 9:4370–75.

Numata, M., Asai, M., Kaneko, K. et al. 2004. Curdlan and schizophyllan (beta- 1,3-glucans) can entrap single-wall carbon nanotubes in their helical superstructure. *Chem Lett* 33:232–33.

Ou, Z., Wu, B., Xing, D. et al. 2009. Functional single-walled carbon nanotubes based on an integrin $\alpha_v\beta_3$ monoclonal antibody for highly efficient cancer cell targeting. *Nanotechnology* 20(10):105102.

Pan, B., Cui, D., Xu, P. et al. 2009, .Synthesis and characterization of polyamidoamine dendrimer-coated multi-walled carbon nanotubes and their application in gene delivery systems. *Nanotechnology* 20:125101.

Panchapakesan, B., Lu, S., Sivakumar, K. et al. 2005. Single-wall carbon nanotube nanobomb agents for killing breast cancer cells. *Nanobiotechnology* 1:133–39.

Pantarotto, D., Briand, J.-P., Prato, M. et al. 2004a. Translocation of bioactive peptides across cell membranes by carbon nanotubes. *Chem Commun* 1:16–17.

Pantarotto, D., Singh, R., Mccarthy, D. et al. 2004b. Functionalized carbon nanotubes for plasmid DNA gene delivery. *Angew Chem Int Ed* 43:5242–46.

Pederson, M.R. and Broughton, J.Q. 1992. Nanocapillarity in fullerene tubules. *Phys Rev Lett* 69:2689–92.

Prato, M. and Kostarelos, K.A.B. 2008. Functionalized carbon nanotubes in drug design and discovery. *Acc Chem Res* 41:60–68.

Qin, Y., Chen, J., Bi, Y. et al. 2015. Near-infrared light remote-controlled intracellular anti-cancer drug delivery using thermo/pH sensitive, nanovehicle. *Acta Biomaterialia* 17:201–09.

Rastogi, V., Yadav, P., Bhattacharya, S.S. et al. 2014. Carbon nanotubes: An emerging drug carrier for targeting cancer cells. *J Drug Deliv* 2014:1–23.

Reiger, C., Kunhardt, D., Kaufmann, A. et al. 2015. Characterization of different carbon nanotubes for the development of a mucoadhesive drug delivery system for intravesical treatment of bladder cancer. *Int J Pharm* 479:357–63.

Ren, J., Shen, S., Wang, D. et al. 2012. The targeted delivery of anticancer drugs to brain glioma by PEGylated oxidized multiwalled carbon nanotubes modified with angio-pep-2. *Biomaterials* 33:3324–33.

Risi, G., Bloise, N., Merli, D. et al. 2014. In vitro study of multiwall carbon nanotubes (MWCNTs) with adsorbed mitoxantrone (MTO) as a drug delivery system to treat breast cancer. *RSC Adv* 4:18683–93.

Rodriguez-Yanez, Y., Munoz, B., and Albores, A. 2013. Mechanisms of toxicity by carbon nanotubes. *Toxicol Mech Methods* 23:178–95.

Roman, J.A., Niedzielko, T.L., Haddon, R.C. et al. 2011. Single-walled carbon nanotubes chemically functionalized with polyethylene glycol promote tissue repair in a rat model of spinal cord injury. *J Neurotrauma* 28:2349–62.

Shao, N., Lu, S., Wickstrom, E. et al. 2007. Integrated molecular targeting of IGF1R and HER2 surface receptors and destruction of breast cancer cells using single wall carbon nanotubes. *Nanotechnology* 18:315101.

Shao, W., Paul, A., Rodes, L. et al. 2015. A new carbon nanotube based breast cancer drug delivery system: Preparation and *in vitro* analysis using paclitaxel. *Cell Biochem Biophys* 71:1405–14.

Shvedova, A.A., Castranova, V., Kisin, E.R. et al. 2003. Exposure to carbon nanotube material: Assessment of nanotube cytotoxicity using human keratinocyte cells. *J Toxicol Environ Health A* 66:1909–26.

Singh, R., Pantarotto, D., Lacerda, L. et al. 2006. Tissue biodistribution and blood clearance rates of intravenously administered carbon nanotube radiotracers. *Proc Natl Acad Sci. USA* 103:3357–62.

Sloan, J.J., Hammer, J., Sibley, M.Z. et al. 1998. The opening and filling of single walled carbon nanotubes (SWNTs). *Chem Commun* 1998:347–48.

Song, H., Su, C., Cui, W. et al. 2013. Folic acid-chitosan conjugate nanoparticles for improving tumor-targeted drug delivery. *Biomed Res International* 2013:723518–23.

Steinman, R.M., Mellman, I.S., Muller, W.A. et al. 1983. Endocytosis and the recycling of plasma membrane. *J Cell Biol* 96:1–27.

Suri, A., Chakraborty, A.K., and Coleman, K.S. 2008. A facile, solvent-free, non-covalent, and nondisruptive route to functionalize single-wall carbon nanotubes by tertiary phosphenes. *Chem Mater* 20:1705.

Taghdisi, S.M., Lavaee, P., Ramezani, M. et al. 2011. Reversible targeting and controlled release delivery of daunorubicin to cancer cells by aptamer-wrapped carbon nanotubes. *Eur J Pharm Biopharm* 77:200–06.

Takanashi, S., Hara, K., Aoki, K. et al. 2012. Carcinogenicity evaluation for the application of carbon nanotubes as biomaterials in rasH2 mice. *Sci Rep* 2:498.

Tan, J.M., Karthivashan, G., Arulselvan, P. et al. 2014. Characterization and *in vitro* sustained release of silibinin from pH responsive carbon nanotube-based drug delivery system. *J Nanomaterials* 2014:439873–82.

Tasis, D., Tagmatarchis, N., Bianco, A. et al. 2006. Chemistry of carbon nanotubes. *Chem Rev* 106:1105–36.

Tour, J.M., Hudson, J.L., Dyke, C.R. et al. 2007. Functionalization of carbon nanotubes in acidic media., US Patent 0280876.

Tran, P.A., Zhang, L., and Webster, T.J. 2009. Carbon nanofibers and carbon nanotubes in regenerative medicine. *Adv Drug Deliv Rev* 61:1097–14.

Tsang, S.C., Chen, Y.K., Harris, P.J.F. et al. 1994. A sample chemical method of opening and filling carbon nanotubes. *Nature* 372:159–61.

Wang, C.-H., Chiou, S.-H., Chou, C.-P. et al. 2011. Photothermolysis of glioblastoma stem-like cells targeted by carbon nanotubes conjugated with CD133 monoclonal antibody. *Nanomedicine* 7:69–79.

Wang, H.F., Wang, J., Deng, X.Y. et al. 2004. Biodistribution of carbon single-wall carbon nanotubes in mice. *J Nanosci Nanotechnol* 4:1019–24.

Wang, L., Shi, J., Jia, X. et al. 2013. NIR-/pH-responsive drug delivery of functionalized single-walled carbon nanotubes for potential application in cancer chemo-photothermal therapy. *Pharm Res* 30:2757–71.

Wang, L., Shi, J., Zhang, H. et al. 2013. Synergistic anticancer effect of RNAi and photothermal therapy mediated by functionalized single-walled carbon nanotubes. *Biomaterials* 34:262–74.

Wang, Z., Xu, Y., Meng, X. et al. 2015. Suppression of c-Myc is involved in multi-walled carbon nanotubes' down-regulation of ATP-binding cassette transporters in human colon adenocarcinoma cells. *Toxicol Appl Pharmacol* 282:42–51.

Warheit, D.B., Laurence, B.R., Reed, K.L. et al. 2004. Comparative pulmonary toxicity assessment of single-wall carbon nanotubes in rats. *Toxicol Sci* 77:117–25.

Welsher, K., Liu, Z., Daranciang, D. et al. 2008. Selective probing and imaging of cells with single walled carbon nanotubes as near-infrared fluorescent molecules. *Nano Lett* 8:586–90.

Worle-Knirsch, J.M., Pulskamp, K., and Krug, H.F. 2006. Oops they did it again! Carbon nanotubes hoax scientists in viability assays. *Nano Lett* 6:1261–68.

Wu, B., Ou, Z., and Xing, D. 2009. Functional single-walled carbon nanotubes/chitosan conjugate for tumor cells targeting. In *8th International Conference on Photonics and Imaging in Biology and Medicine (PIBM '09)*, vol. 7519 of Proceedings of SPIE, Wuhan, China, August, 75190K1–75190K8.

Wu, H., Shi, H., Zhang, H. et al. 2014. Prostate stem cell antigen antibody-conjugated multiwalled carbon nanotubes for targeted ultrasound imaging and drug delivery. *Biomaterials* 35:5369–80.

Xiao, Y., Gao, X., Taratula, O. et al. 2009. Anti-HER2 IgY antibody-functionalized single-walled carbon nanotubes for detection and selective destruction of breast cancer cells. *BMC Cancer* 9: 351.

Yang, F., Jin, C., Yang, D. et al. 2011. Magnetic functionalised carbon nanotubes as drug vehicles for cancer lymph node metastasis treatment. *Eur J Cancer* 47:1873–82.

Yaron, P.N., Holt, B.D., Short, P.A. et al. 2011. Single wall carbon nanotubes enter cells by endocytosis and not membrane penetration. *Journal of Nanobiotechnology* 9:45–59.

Yuan, W.Z., Sun, J.Z., Dong, Y.Q. et al. 2006. Wrapping carbon nanotubes in pyrene-containing poly(phenylacetylene) chains: Solubility, stability, light emission, and surface photovoltaic properties. *Macromolecules* 39:8011–20.

Zeineldin, R., Al-Haik, M., and Hudson, L.G. 2009. Role of polyethyleneglycol integrity in specific receptor targeting of carbon nanotubes to cancer cells. *Nano Lett* 9:751–57.

Zhang, R.Y. and Olin, H. 2012. Carbon nanotubes as drug carriers: Real time drug release investigation. *Mater Sci Eng C* 32:1247–52.

Zhang, X., Meng, L., Lu, Q. et al. 2009. Targeted delivery and controlled release of doxorubicin to cancer cells using modified single wall carbon nanotubes. *Biomaterials* 30:6041–47.

Zhang, X.K., Wang, X.F., Lu, Q.H. et al. 2008. Influence of carbon nanotube scaffolds on human cervical carcinoma HeLa cell viability and focal adhesion kinase expression. *Carbon* 46:453–60.

Zhou, M., Peng, Z., Liao, S. et al. 2014. Design of microencapsulated carbon nanotube-based microspheres and its application in colon targeted drug delivery. *Drug Deliv* 21:101–09.

Zhu, X., Xie, Y., Zhang, Y. et al. 2014. Thermo-sensitive liposomes loaded with doxorubicin and lysine modified single-walled carbon nanotubes as tumor-targeting drug delivery system. *J Biomater Appl* 29:769–79.

Zhu, Y. and Li, W.X. 2008. Cytotoxicity of carbon nanotubes. *Sci China Ser B Chem* 51:1021–29.

# 16 Functionalized Cyclodextrin

## A Versatile Supramolecular System for Drug Delivery

*Subrata Jana and Sougata Jana*

## CONTENTS

## 16.1   INTRODUCTION

The prime objective of drug delivery systems is to deliver the necessary amount of drug to the targeted site for a necessary period of time, both efficiently and precisely (Juliano 1980; Anderson and Kim 1984; Buri and Gumma 1985; Wong and Choi 2015).

To achieve this goal, suitable carrier materials are used that can prevent undesired biochemical reactions in other parts of the body except the targeted region. Hence, different types of high-performance biomaterials focusing on effective drug delivery application are being constantly developed. Biomaterials are the hot target for this purpose over synthetic polymeric materials due to nontoxicity and biodegradability in the biological setup. In the recent past, different natural polysaccharides such as sodium alginate, tamarind seed polysaccharides (TSP), etc. have been used as the delivery vehicle for drug molecules. Natural polymers are very attractive for use in the field of drug delivery due to their easy availability, high swelling capacity, low extraction cost, overall cost effectiveness, biodegradability and biocompatibility, nontoxicity, and sometimes biorecognition properties.

Cyclodextrins (CDs) are one such biomaterial that have been used for such a role due to their ability to form a complex with a wide variety of guests, specifically drug molecules. These properties of CD vary depending upon the cavity size of CDs. The major advantages of CDs are to enhance the solubility, stability, and bioavailability of drug molecules. Though natural CDs are not very soluble, their different synthetic modified forms perform the drug delivery role very effectively. Efficient assembly in host–guest interactions is crucial to supramolecular nanotechnology-based drug delivery systems. CDs have improved the biocompatibility of nanodelivery systems that make CDs versatile in design and respective applications as functional delivery systems. Among all CDs (Table 16.1), α-CD and β-CD are commonly used for the construction of supramolecular structures due to their high inclusion ability (Hu et al. 2014).

## 16.2   STRUCTURAL FEATURES OF CD

CDs are important carriers for drug molecules as these can modulate the physical, chemical, and biological properties of guest molecules through the formation of inclusion complexes. The supramolecular complex that is formed by the complexation of CDs and drug molecules will also effectively deliver in the targeted region. Carbohydrates, such as cellulose, starch, and sucrose, are probably the most abundant organic substances in nature. Humans have processed carbohydrates from ancient times through fermentation, applying enzymatic degradation. It is now well established that these processes lead to the formation of mixtures of monosaccharides, disaccharides, and various oligosaccharides, such as linear and branched dextrins and that, under certain conditions, small amounts of cyclic dextrins or CDs are also formed during these degradation processes. From these degradation processes, mainly α-, β-, and γ-CDs are prepared. The α-, β-, and γ-CDs are widely used natural CDs, consisting of six, seven, and eight D-glucopyranose residues, respectively, linked by R-1,4 glycosidic bonds into a macrocycle (Figure 16.1). So, cavity size along with inclusion phenomenon will vary depending on the number of glucopyranose residues.

## TABLE 16.1
## Common Abbreviations of Cyclodextrins as Reported in the Literature

| Cyclodextrin (CD) | Abbreviation |
|---|---|
| α-Cyclodextrin | α-CD |
| β-Cyclodextrin | β-CD |
| γ-Cyclodextrin | γ-CD |
| Heptakis-β-cyclodextrin | Heptakis-β-CD |
| Hydroxyethyl-β-CD | HE-β-CD |
| Hydroxypropyl-α-CD | HP-α-CD |
| Hydroxypropyl-β-CD | HP-β-CD |
| 2-Hydroxypropyl-β-cyclodextrin | 2-HP-β-CD |
| Monochlorotriazinyl-β-cyclodextrin | MCT-β-CD/β-CDMCT |
| Sulfobutylether-β-CD | SB-β-CD |
| Methyl-β-CD | M-β-CD/Me-β-CD |
| Dimethyl-β-CD | DM-β-CD (DIMEB) |
| Randomly dimethylated-β-CD | RDM-β-CD |
| Randomly methylated-β-CD | RM-β-CD/RAME-β-CD (RAMEB) |
| Carboxymethyl-β-CD | CM-β-CD |
| Carboxymethyl ethyl-β-CD | CME-β-CD |
| Diethyl-β-CD | DE-β-CD |
| O-Methylated-β-cyclodextrin | OM-β-CD |
| Tri-O-methyl-β-CD | TRIMEB |
| Tri-O-ethyl-β-CD | TE-β-CD |
| Tri-O-butyryl-β-CD | TB-β-CD |
| Tri-O-valeryl-β-CD | TV-β-CD |
| Di-O-hexanoyl-β-CD | DH-β-CD |
| Glucosyl-β-CD | G1-β-CD |
| Maltosyl-β-CD | G2-β-CD |
| 2-Hydroxy-3-trimethyl-ammoniopropyl-β-CD | HTMAPCD |
| γ-Cyclodextrin-2,3-di-O-hexanoyl cyclomaltooctaose | γ-CDC6 |
| (Sulfobutylether)$_{7m}$-β-cyclodextrin | (SBE)7m-β-CD |
| Permethyl-β-cyclodextrin | PM-β-CD |

The outer surface of the CD is hydrophilic in nature due to the presence of a primary hydroxyl group, whereas the inner core is hydrophobic in nature due to the presence of an inward-directed ethereal oxygen atom of the ring. The guest molecules usually fit in the inner hydrophobic core depending on the size of the cavity as well as that of the guest molecules (Bender and Komiyama 1978; Saenger 1980; Szejtli 1982). The most common pharmaceutical application of CDs is to enhance the solubility, stability, and bioavailability of drug molecules (Szejtli 1983; Duchene 1987; Szejtli 1988). However, natural CDs have relatively low solubility, both in water and organic solvents, which thus limits their uses in pharmaceutical formulations. The chemical structure of CDs and hydrophobic as well as hydrophilic areas are shown in Figure 16.2. Some characteristics of different CDs are given in Table 16.2.

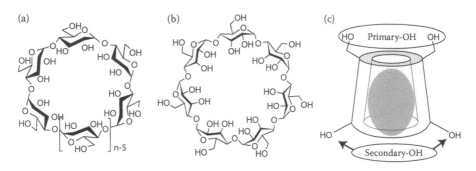

**FIGURE 16.1** Cyclodextrins: α-cyclodextrin ($n = 6$), β-cyclodextrin ($n = 7$), and γ-cyclodextrin ($n = 8$) (a); β-cyclodextrin with primary and secondary hydroxyl groups (b); three-dimensional model of cyclodextrin with primary and secondary hydroxyl exterior (c).

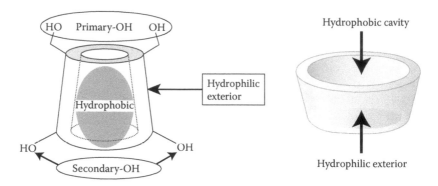

**FIGURE 16.2** Cyclodextrin with hydrophobic inner and hydrophilic exterior.

**TABLE 16.2**
**Some Characteristics of α-, β-, and γ-CD**

| Type of CD Units | Number of Glucose | Cavity Diameter (Å) | External Diameter (Å) | Molecular Weight | Solubility |
|---|---|---|---|---|---|
| α-CD | 6 | 4.7–5.3 | 14.6 ± 0.4 | 972 | 14.5 |
| β-CD | 7 | 6.0–6.5 | 15.4 ± 0.4 | 1135 | 1.85 |
| γ-CD | 8 | 7.5–8.3 | 17.5 ± 0.4 | 1297 | 23.2 |

*Source:* Adapted from Loftsson, T., Brewester, M. 1996. *J Pharm Sci* 85:1017–25.; Endo, T., Nagase, H., Ueda, H. et al. 1997. *Chem Pharm Bull* 45:532–36.

## 16.3  SYNTHETIC MODIFICATION OF CD

### 16.3.1  FUNCTIONALIZATION AND ALKYLATION

Recently, various kinds of CD derivatives have been prepared to improve physio-chemical properties and host–guest interaction of natural CDs (Pitha et al. 1983;

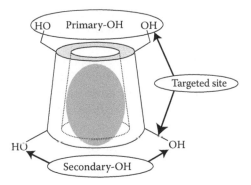

**FIGURE 16.3** Cyclodextrin with targeted hydroxyl group for synthetic modification.

Uekama 1985; Uekama and Otagiri 1987; Uekama and Hirayama 1988; Duchene 1991; Brewster et al. 1988; Khan et al. 1998; Uekama et al. 1998; Szejtli 1992). Thus, the modified CD derivatives have immense applications as efficient drug carriers for a wide range of drug molecules (Uekama et al. 1993; Frömming and Szejtli 1994; Albers and Muller 1995; Atwood et al. 1996; Loftsson and Brewester 1996; Rajewski and Stella 1996; Connors 1997; Irie and Uekama 1997; Stella and Rajewski 1997; Thompson 1997).

CD molecules contain mainly two types of hydroxyl groups in the external hydrophilic surface. Among these two types, primary hydroxyl groups are more potent to modify synthetically (Figure 16.3).

CD can easily conjugate to different proteins, fluorophores, etc. Besides this, CD can also be tagged with hydrogel networks to prepare stimuli-responsive hydrogel-based biomaterial for pharmaceutical applications through click chemistry (Cai et al. 2012).

### 16.3.2 Mono-Modification at the C6-Position

6-Tosyl β-CD are important precursors for the synthesis of a variety of functionalized modified CDs because a nucleophile can attack the electrophilic carbon at the 6-position to produce the desired compounds (Scheme 16.1). The reaction of β-CD with tosyl chloride in aqueous alkaline medium produces 6-mono-tosylated β-CD in good yield (Takahashi et al. 1984).

Other than tosylation, modifications of C6 of CD are very rare. But a single-step quantitative-yield synthesis of CD monoaldehydes has been reported. The CD was dissolved in an organic solvent followed by the addition of Dess–Martin perodinane and the mixture was stirred for 1 h at room temperature. The reaction mixture was cooled down and acetone was added to obtain the precipitate of the

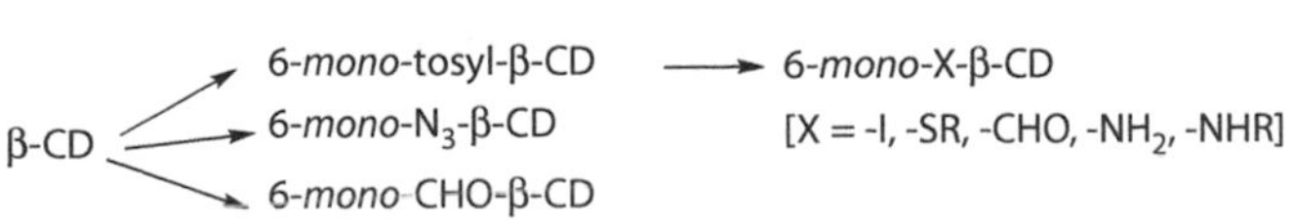

$$
\beta\text{-CD}
\begin{cases}
6\text{-}mono\text{-tosyl-}\beta\text{-CD} \longrightarrow 6\text{-}mono\text{-X-}\beta\text{-CD} \\
6\text{-}mono\text{-N}_3\text{-}\beta\text{-CD} \qquad [\text{X} = \text{-I, -SR, -CHO, -NH}_2\text{, -NHR}] \\
6\text{-}mono\text{-CHO-}\beta\text{-CD}
\end{cases}
$$

**SCHEME 16.1** Mono-modification at the C6-position of β-cyclodextrin (β-CD).

crude, which is isolated by filtration (Cornwell et al. 1995). The other protocol for the synthesis of this type of compound is the oxidation method using IBX (1-hydroxy-1,2-benziodoxol-3(1$H$)-one 1-oxide) as the oxidant in dimethyl sulfoxide (DMSO). The incorporation of β-CD into chitosan was reported by the mono-oxidation of β-CD followed by a reductive coupling reaction (Jimenez et al. 2003). The direct azidation of CDs with sodium azide in the presence of triphenylphosphine-carbon tetrabromide has also been reported (Jimenez-Blanco et al. 1997).

### 16.3.3   Mono-Modification at Any One of the C2-, C3-, or C6-Positions

To carry out mono-modification at C2 and C3, three types of reaction of 2,3-anhydro-β-CDs, namely, nucleophilic ring-opening, reduction to 2-enopyranose, and reduction to 3-deoxypyranose, were investigated and reported in the secondary face of β-CD, keeping regio- and stereoselectivity. The nucleophilic ring-opening reaction was carried out using various nucleophiles at both 2,3-mannoepoxy and 2,3-alloepoxy-β-CDs, producing 2- and 3-modified CD derivatives (Scheme 16.2). The 3-position is more sterically accessible than the 2-position.

Thiourea also reacts with the CD epoxides, producing thiirane and olefin species instead of any ring-opening products. By modulating the reaction conditions, CD olefin, diene, and triene are synthesized in moderate to good yields (Yuan et al. 2003).

A method for selective monoalkylation of the C2- or C3-mono-hydroxy groups of β-CDs is being reported. Besides β-CDs, dimers are also produce by this reaction by linking in either a 2–2′ or 3–3′ fashion at their secondary faces (Chiu et al. 2000).

The carboxyl group of predonisolone-21-hemisuccinate is connected to the C2 and C3 hydroxyl groups of α-, β-, and γ-CDs using carbonyldiimidazole as a coupling agent (Yano et al. 2001).

### 16.3.4   Per-Modification

Per-$O$-alkylated CDs are synthesized by the reaction of an excess of an alkylhalide with CD in the presence of a base. CD superimposed with one D-glucose is represented in Figure 16.4. Subunit per-$O$-methylation of CD has been obtained using sodium hydroxide, potassium hydroxide (Ciucanu and Kerek 1984), or sodium hydride (Nowotny et al. 1989) as the base with methyl iodide as the alkylating agent in DMSO as the solvent. Heptakis (2,3,6-tri-$O$-methyl)-β-CD is now commercially available. However hexakis (2,3,6-tri-$O$-methyl)-α-CD and heptakis (2,3,6-tri-$O$-methyl)-β-CD are soluble in both aqueous and organic media. The alkylation at

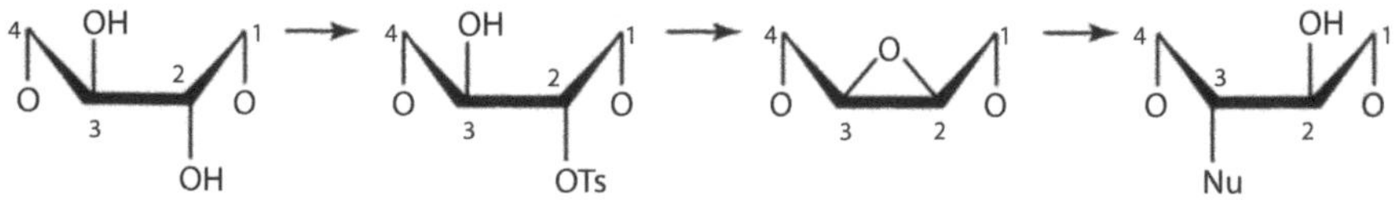

**SCHEME 16.2**  Nucleophilic ring-opening of 2,3-mannoepoxy to 3-deoxypyranose. (Adapted from Chiu, S.H., Myles, D.C., Garrel, R.L. et al. 2000. *J Org Chem* 65:2792–96.)

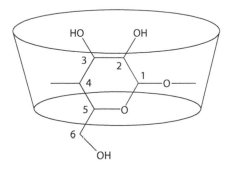

**FIGURE 16.4** Representation of cyclodextrin superimposed with one D-glucose subunit.

2- and 6-positions of the CD is achieved with an excess of 1-bromopentane in DMSO in the presence of sodium hydroxide. The less reactive 3-position was being alkylated by the reaction of the 2,6-di-*O*-pentyl CD with excess 1-bromopentane in tetrahydrofuran (THF) in the presence of sodium hydride (Konig et al. 1988a,b) or by the addition of more sodium hydroxide and 1-bromopentane with continuing reflux for five days. Per-substituted 2,6-di-*O*-pentyl- α-, β-, or γ-CDs are prepared by alkylation using 1-bromopentane as the alkylating agent in the presence of sodium hydroxide as the base in DMSO (Berthod et al. 1990).

Using the same protocol, per-substituted (2,6-di-*O*-methyl)-β-CD was also synthesized by the reaction of 1-bromopentane and sodium hydride to alkylate at the 3-position (Bicchi et al. 1992). Besides, other positions are alkylated by selective blockage of certain positions followed by alkylation and deprotection based on the reactivity of the 2-, 3-, and 6-hydroxy groups in the presence of suitable bases. By exploiting the strategy of protection and deprotection as well as reactivity and suitability of the bases, different substituted CDs have been synthesized. When sodium hydride (seven equivalent) is used as a base, it selectively promotes the reaction at the 2-position of CD to synthesize heptakis (2-*O*-methyl)-β-CD in the presence of methyl iodide as the alkylating agent (Rong and D'Souza 1990).

The heptakis (2,3-di-*O*-pentyl-6-*O*-methyl)-β-CD has been prepared by protection, alkylation followed by deprotection, and further alkylation strategy. The first step involved the protection of the 6-position by using *tert*-butyldimethylsilyl chloride (TBDMS) and imidazole in dimethylformamide (DMF) (Konig et al. 1990). Now, TBDMS protected CD reacts with *n*-pentyl iodide in the presence of sodium hydride in DMF followed by deprotection of TBDMS using tetrabutylammonium fluoride (TBAF) in THF and further alkylation at the 6-position by reaction with methyl iodide and sodium hydride in THF.

Heptakis (2-*O*-methyl-3,6-di-*O*-pentyl)-β-CD has been synthesized by using a similar procedure as mentioned. The 6- and 2-positions were first protected using TBDMS and allyl bromide, respectively. Then deprotection of the 6-position followed by alkylation allowed the preparation of the desired compound. The methylation at 2-position was performed by using methyl iodide and sodium hydride (Konig et al. 1991). By using different established protocols, the heptakis (2,3-di-*O*-methyl)- or heptakis (2,6-di-*O*-methyl)-β-CD derivatives are prepared and a further mono-oct-7-enyl derivative can be obtained using one equivalent of sodium hydride and 8-bromo-1-octene in DMF. The remaining hydroxy groups were then methylated using the methylation technique (Cousin et al. 2003).

### 16.3.5 CONJUGATION

The delivery vehicle for targeted drug delivery systems has been synthesized from the conjugation of CD derivatives with different molecules. One such compound is prepared by the monoconjugation of C6-amino-β-CD with mannosyl-coated dendritic branches followed by an interactive thiourea-forming using convergent strategy (Baussanne et al. 2000). The C6-mono-glucose-branched CDs were synthesized from amino-β-CD and arbutin. The glucose residues acted as reactive glycosyl acceptors for the transglycosylation by the enzyme to produce sialo-complex-type oligosaccharide-branched CDs (Yamanoi et al. 2005). Mono-6-alkylamino-β-CDs are prepared from corresponding 6-β-CD following substitution reaction (Petter et al. 1990). Lactose-containing CDs are prepared and conjugated with hydrophobic polymers in aqueous medium to form lectins, binding dynamic multivalent lactosides. The lactose-containing α- and β-CDs were synthesized by the coupling C6-monoamino CDs with lactosyl propionic acid using HBTU-BF$_4$ to activate the carboxylic acid (Nelson and Stoddart 2003).

The synthesis of monosubstituted β-CD with different sugar moieties such as β-D-glucose, β-D-galactose, α-D-mannose, and β-L- and β-D-fucose is carried by the synthesis of stable amide linkage via a C9 spacer chain by the coupling of the NCS (N-chlorosuccinimide) sugar derivatives to monomethyl nonanedicarboxylate (Parrot-Lopez et al. 1993). The mono-C6-amino-β-CDs and other C6-mono-alkylene-diamino-β-CDs are used to modify bovine pancreatic trypsin using EDC (1-ethyl-3-(3-dimethylaminopropyl)carbodiimide) as the coupling agent (Nelissen et al. 2002; Aoki et al. 2004). By this chemical conjugation, thermal stability and catalytic properties are enhanced (Fernandez ct al. 2003). Branched β-CDs containing mannosyl mimetic derivatives allow the inclusion of other guest substrates even in the presence of self-inclusion of water (Yockot et al. 2003). The ionic interactions between C6 and C3 mono-positively charged CDs and anionic guests are also exploited for the synthesis of supramolecular drug delivery systems (Alvarez-Parrilla et al. 2003).

The heterocyclic moieties containing β-CDs were prepared by the condensation of indol-3-yl butyric acid with the oligo(aminoethylamino)-β-CDs using *N,N′*-dicyclohexylcarbodiimide (DCC) (Liu et al. 2002).

The monoazacoronand of C6-mono-substituted α-CD and β-CD were synthesized by the acylation of 6-aminohexylamino-CDs (Lock et al. 2004). The saccharide coupled with CDs enhances the mobility of galactose and recognition by lectin. For this purpose, different synthetic compounds are prepared by linking β-CD and the sugar moiety having spacer arms of 3, 4, 5, 6, and 9 carbons (Kassab et al. 1997). The targeted delivery system was also synthesized by the coupling of C6-amino-β-CD with mannosyl-coated dendritic branches (Baussanne et al. 2000).

The site-directed drug delivery system is exceptionally important for the delivery of drugs in a specific area of the body. To achieve this, CDs are conjugated with protein or peptides with variable length and desired sequence. CD–peptide conjugates are promising compounds for targeting host/guest complexes or covalently bound cytotoxic drugs to specific tumor cells. The size of the CD is very crucial for the designing of CD–peptide conjugates for effective delivery. Generally, larger CD-based carriers spoil the recognition process at the molecular level. The C-terminal tetrapeptide amide of gastrointestinal hormone was connected to mono-(6-succinylamino-6-deoxy)-β-CD or heptakis-(6-succinylamino-6-deoxy)-β-CD to study the effect of the bulky cyclic carbohydrate moiety on recognition of the peptides by the G-protein-coupled CCK-B/gastrin receptor and on their signal transduction potencies (Schaschke et al. 1998).

Another site-directed drug carrier system was synthesized by the conjugation of β-CD with an endo-epoxysuccinyl peptide inhibitor MeO–Gly–Gly–Leu–(2S,3S)–tEPS–Leu–Pro–OH as an appropriate sequence for cathepsin B (Schaschke et al. 2000). Many tumor cell lines are known to secrete cathepsin B and/or to contain membrane-associated cathepsin B, which is thought to be involved in invasion and metastasis. This peptide selectively inhibits cathepsin B among the cysteine proteinases by the specific salt-bridge interaction of the C-terminal carboxylate function with histidine 110 and 111 of the occluding loop and spans the whole substrate binding groove at the S subsite with the propeptide portion.

### 16.3.6 POLYMERIC MODIFICATION

Different polymeric substrates are attached with CD for the preparation of a modified polymer, which has a better drug entrapment efficiency along with an effective delivery system (Heidel and Schluep 2012; Liu et al. 2005). Here, the general protocol is to prepare a system that has a dual role in forming a better platform as a nanoparticle, microparticle, etc. and better drug entrapment. In this type of CD-tagged polymeric system, polymeric parts create the platform and CD or a functionalized CD moiety either covalently linked or noncovalently recognized with drug molecules. In this way, prepared nanoparticles are being widely used as potential therapeutics for different applications in medicine and have been shown to significantly improve the circulation, biodistribution, efficacy, and safety profiles of multiple classes of drugs. The drug molecules are released in the targeted site by the effect of pH or different biochemical as well as enzymatic reactions.

The β-CD-grafted poly(acrylic acid) (PCDAA) polymer was synthesized by radical polymerization, and then embedded on the surface of polycaprolactone–poly(ethylene glycol) (PCL–PEG–PCL) nanoparticles through host–guest interaction and hydrogen bonding between oxygen of PEG and carboxyl group of PCDAA. These PCDAA embedded PCL–PEG–PCL (PCDAA@PCL–PEG–PCL) nanoparticles showed a spherical core–shell structure and their size was less than 200 nm with high drug loading capacity for pentoxifylline (PTX) compared to the original PCL–PEG–PCL nanoparticles. This happened because some amount of drug was encapsulated in the hydrophobic cavities of β-CD. These drug-loaded nanoparticles showed significant cytotoxicity against HepG2 cells compared to blank nanoparticles (Ahmed et al. 2015).

The polymeric conjugation between chitosan and CD has been recently reported along with its inclusion properties where 6-deoxy-6-(4-oxobutyramido)-β-CD and 6-oxo-β-CD gave two β-CD-linked chitosan derivatives with C4 (4-butylamido) and C0 linking arms, respectively, by reductive alkylation of the amino group of chitosan with β-CD aldehyde derivatives. The degree of substitution (D.S.) of both C4-β-CD and C0-β-CD linked chitosan was controlled by the ratio of starting materials. It is revealed that the length of the linking arms between CD and chitosan is influenced by their inclusion property (Buranaboripan et al. 2014).

## 16.4   CD-BASED DELIVERY SYSTEM

Different carriers for drug delivery systems have been reported so far. In all reports, CD molecules are linked with different polymeric substances to prepare different transportation media, among which liposomes, microspheres, microcapsules, nanoparticles, CD-grafted cellulosic fabric, hydrogels, nanosponges, beads, nanogels/nanoassemblies, and CD-containing polymers are very important (Gidwani and Vyas 2015). The various CD-based delivery systems are summarized in Table 16.3.

### 16.4.1   CD-BASED LIPOSOMES

Liposomes are phospholipid vesicles composed of lipid bilayers enclosing one or more aqueous compartments. CDs have the ability to increase drug solubility and entrapment efficiency at the aqueous compartment of liposomes. Liposomes have been widely used as safe and effective carriers for both hydrophilic and lipophilic drugs. Generally, lipophilic drugs incorporated in the membrane bilayers can be released quickly, which limits the effectiveness of this drug delivery system. But the coupling of both delivery systems containing encapsulating CD/drug inclusion complex into liposomes is very effective (Gharib et al. 2015) compared to liposomes prepared by polymeric compounds. These protocols have also been used for the improved delivery of flurbiprofen (Zhang et al. 2015; Maestrelli et al. 2005a), ketoprofen (Maestrelli et al. 2005b), 9-nitrocamptothecin (Chen et al. 2015), and itraconazole (ITZ) (Alomrani et al. 2014).

### 16.4.2   CD-BASED NANOPARTICLES

Nanoparticles are solid colloidal particles composed of natural, synthetic, or semi-synthetic polymers with size range from 10 to 1000 nm. CD enhances the drug loading capacity of nanoparticles. The release of camptothecin (CAM) has been extended to 12 days with amphiphilic β-CD nanoparticles and 48 h with polymeric nanoparticles, showing the massive enhancement of drug release of CD-based nanoparticles over conventional polymeric nanoparticles (Cirpanli et al. 2009). The anticancer efficacy of amphiphilic CD nanoparticles was higher than that of PLGA/PCL nanoparticles loaded with camptothecin (CPT) and its solution in DMSO. Agueros et al. (2009) reported that paclitaxel encapsulation in hydroxypropyl-β-cyclodextrin (HP-β-CD) and poly(anhydride) nanoparticles was higher and showed threefold increase in loading capacity. The nanoparticles of the amphiphilic CD heptakis (2-O-oligo-(ethyleneoxide)-6-hexadecylthio-)-β-CD (SC16OH) entrapping docetaxel (Doc) achieved prolonged drug release. Biodegradable β-CD-based nanoparticles

**TABLE 16.3**

**Novel Drug Delivery Systems Incorporating Cyclodextrin–Drug Complex**

| Cyclodextrin Type | Drugs | Delivery System | Objective/Purpose/Work Done | Reference |
|---|---|---|---|---|
| β-CD, HP-β-CD, SBE-β-CD | Flurbiprofen | Liposomes | HIV therapeutics | Zhang et al. (2015) |
| HP-β-CD | 9-Nitrocamptothecin | Liposomes | Strongest cytotoxicity to tumor cells | Chen et al. (2015) |
| HP-β-CD | Itraconazole | Liposomes | Transdermal drug delivery | Alomrani et al. (2014) |
| β-CD, HP-β-CD | Ketoprofen | Liposomes | Transdermal drug delivery | Maestrelli et al. (2005b) |
| HP-β-CD, DM-β-CD, OM-β-CD | Gonadorelin, leuprolide acetate, recombinant human growth hormone, lysozyme | Injectable microspheres | Protein stability and sustained release | Lee et al. (2007) |
| β-CD, HP-β-CD | Niclosamide | Dendrimers | Solubility enhancement and controlled release | Devarakond et al. (2005) |
| β-CD | Poly(propylene glycol) bisamine | Hydrogels | Biomedical materials for tissue engineering and drug carriers with controlled release | Yu et al. (2007) |
| β-CD | Dexamethasone, flurbiprofen, doxorubicin hydrochloride | Nanosponges | Nanoparticulate system as drug carriers | Loukas et al. (1998) |
| 2-HP-β-CD | Glutathione | Microparticles | Oral sustained-release delivery systems for tripeptide with reduced peptide degradation | Trapani et al. (2007) |
| HP-α-CD, HP-β-CD | Triclosan, furosemide | Nanoparticles | Transmucosal delivery of hydrophobic compounds | Maestrelli et al. (2006) |
| α-CD, β-CD, γ-CD | Insulin | Pegylated insulin/CD polypseudorotaxanes | Polypseudorotaxanes as sustained release system | Higashi et al. (2007) |

*(Continued)*

**TABLE 16.3 (Continued)**
**Novel Drug Delivery Systems Incorporating Cyclodextrin–Drug Complex**

| Cyclodextrin Type | Drugs | Delivery System | Objective/Purpose/Work Done | Reference |
|---|---|---|---|---|
| β-CD, M-β-CD, HP-β-CD, SB-β-CD | Estradiol | Hydrogels | Hydrogels as controlled release delivery system | Rodriguez-Tenreiro et al. (2007) |
| γ-CDC6 | Progesterone | Nanospheres | Feasibility of preparing nanospheres | Lemos-Senna et al. (1998) |
| HP-β-CD | Nifedipine | Microspheres | Solubility enhancement | Skalko et al. (1996) |
| HP-β-CD | Hydrocortisone | Microspheres | Release and stability were investigated | Filipovic-Grcic et al. (2000) |
| 2-HP-β-CD | Insulin | Nanoparticles | Oral insulin delivery | Sajeesh and Sharma (2006) |
| HP-β-CD | Carvedilol | Buccal tablets | Bioadhesive sustained-release buccal delivery | Cappello et al. (2006) |
| HP-β-CD | Insulin | Large porous particles | Dry powders for the sustained release for pulmonary delivery | Ungaro et al. (2006) |
| β-CD hydrate | Amlodipine | Liposomes | Stability against photodegradation | Ragnoa et al. (2003) |
| HP-β-CD | Methoxydibenzoylmethane | Lipospheres | Photostability | Scalia et al. (2006b) |
| HP-β-CD | Insulin | Microspheres | Protein release kinetics | Rosa et al. (2005) |
| β-CDMCT | Octyl methoxycinnamate | Cellulose fabric | Incorporation of sunscreen agent into cyclodextrin bound to cloth fiber and evaluation of its sun protective capacity | Scalia et al. (2006a) |
| Heptakis-β-CD | TPPS | Nanoparticles | Photodynamic activity | Sortino et al. (2006) |
| HP-β-CD | Saquinavir | Nanoparticles | Improve oral delivery | Boudad et al. (2001) |
| β-CD, 2-HP-β-CD | Hydrocortisone, progesterone | Solid lipid nanoparticles | To modulate the release kinetics | Cavalli et al. (1999) |

*(Continued)*

**TABLE 16.3 (*Continued*)**
**Novel Drug Delivery Systems Incorporating Cyclodextrin–Drug Complex**

| Cyclodextrin Type | Drugs | Delivery System | Objective/Purpose/Work Done | Reference |
|---|---|---|---|---|
| Bis-CD | Bovine serum albumin | Nanoparticles | As protein delivery system | Gao et al. (2006) |
| HP-β-CD | Bovine serum albumin | Microspheres | To investigate the conformational stability of protein | Kang and Singh (2003) |
| α-, β-, γ-CD | Gabexate mesylate | Bioadhesive microspheres | Bioadhesive nasal delivery system | Fundueanua et al. (2004) |
| β-CDC6 | Tamoxifen citrate | Nanoparticles, nanospheres, nanocapsules | Developed nanoparticulate drug delivery systems | Memisoglu-Bilensoya et al. (2005) |
| HP-β-CD | Itraconazole | Vaginal cream | Developed mucoadhesive vaginal cream | Francois et al. (2003) |
| α-, β-, γ-CD | Indomethacin, furosemide, naproxen | Nanoparticles | Developed nanoparticles as delivery systems and solubility enhancement | Wongmekiat et al. (2006) |
| β-CD, HP-β-CD | Nifedipine | Suppositories | Improved the release property | Nishimura et al. (2006) |
| β-CD | Amikacin | Microparticles microspheres for pulmonary drug delivery | Pulmonary drug delivery | Skiba et al. (2005) |
| HP-β-CD, γ-CD, RM-β-CD | Methacholine | Nebulized aerosols | Pulmonary administration | Evrarda et al. (2004) |
| (SBE)$_{7m}$-β-CD | Chlorpromazine hydrochloride | Osmotic pump tablets | Controlled release of poorly water-soluble drugs | Okimoto et al. (1999) |
| α-CD | Isotretinoin | Oil beads | Oral bioavailability of lipophilic drugs | Trichard et al. (2007) |

*(Continued)*

**TABLE 16.3 (*Continued*)**
**Novel Drug Delivery Systems Incorporating Cyclodextrin–Drug Complex**

| Cyclodextrin Type | Drugs | Delivery System | Objective/Purpose/Work Done | Reference |
|---|---|---|---|---|
| MCT-β-CD | Miconazole | Fabric with antibacterial abilities | Incorporation of the antibacterial agent into cyclodextrin on to covalently bonded cloth fibers | Wang and Cai (2008) |
| SBE7-β-CD | Carbamazepine | Beads | Sustained release and solubility enhancement | Smith et al. (2005) |
| β-CD | Retinoic acid | Hydrogel topical formulation | Improve efficacy and tolerability of retinoic acid | Anadolu et al. (2004) |
| HP-β-CD | Rh-interferon α-2a | Lipidic implants | Controlled protein release | Herrmann et al. (2007) |
| α-CD | Droepiandrosterone | Matrix tablets | Sustained release | Mora et al. (2007) |
| β-CD, HP-β-CD, Me-β-CD | Flurbiprofen | Fast-dissolving tablets | Solubility enhancement | Maestrelli et al. (2005a) |
| β-CD | Naproxen, ibuprofen | Water-soluble epichlorohydrin-β-cyclodextrin polymer | To modulate the kinetic release and solubility enhancement | Martin et al. (2006) |
| β-CD, Me-β-CD | Piroxicam | Gels | Development of topical dosage form to overcome side effects connected with the oral use | Jug et al. (2005) |
| α-CD, β-CD, HP-β-CD, RAME-β-CD | Melarsoprol | Oral form and parenteral aqueous solution | Solubility enhancement and to improve tolerability and safety | Gibaud et al. (2005) |
| HP-β-CD, PM-β-CD | Bupranolol | Solution/suspension | Transdermal penetration enhancement | Babu and Pandit (2004) |
| β-CD | Diclofenac | Solution | Permeation enhancement studies using silicone as a model membrane | Dias et al. (2003) |

containing tertiary amine groups (CD-p-AE) have been developed through the Michael addition of acrylated CD macromer and 1,4-butanediol diacrylate with *N,N*-dimethylethyldiamine (Gil et al. 2012). These nanoparticles are chemically nontoxic with highly permeable efficiency to the *in vitro* blood brain barrier (BBB) without disturbing the integrity of the BBB and could sustain the release of doxorubicin for at least 4 weeks.

### 16.4.3 CD-Grafted Polymeric Nanocarriers

This type of delivery system is produced by grafting natural CDs with polymers and then applied for complexation with drugs. These complexes are further loaded into nanocarriers. The grafting is carried out by the conjugating several units of CDs on polymer through chemical tools, to enhance the binding ability of guest molecules. These polymers containing several CD units improve the entrapment for drug molecules within the nanostructure. Recently, Zhang et al. (2011) designed a β-CD functionalized hyperbranched polyglycerol (HPG-β-CD) of paclitaxel to obtain high drug loading capacity and solubility in aqueous medium. The prepared nanoparticles had shown good biocompatibility and proved to be a promising delivery system for hydrophobic drugs. Zeng et al. (2013) also prepared hollow nanospheres of CAM complexed with β-CD-graftPAsp (β-CD grafted with polyaspartic acid). Complexation led to improvement in aqueous solubility and stability of CPT. The nanoassemblies (nanospheres) of CPT-β-CD-graft-PAsp led to passive targeting of drug, sustained release, and decreased cytotoxicity (Zeng et al. 2013). Thus, CD-grafted polymeric nanocarriers showed promising multifunctional properties for effective delivery of anticancer drugs.

### 16.4.4 CD-Based Nanosponges

Nanosponges are a class of tiny sponges with cavities a few nanometers wide. These types of compounds have excellent encapsulation properties for a wide range of guests, specifically drug molecules (Subramanian et al. 2012). These particles are capable of carrying both lipophilic and hydrophilic substances and improving the solubility of poorly water-soluble molecules (Trotta et al. 2007). The nanosponge is spherical, about the size of a virus, with a backbone of biodegradable polyester. The spherical substrates are synthesized by the small molecule crosslinking of the polymeric polyester compounds. The crosslink zone of the polyester has many pockets to store drug molecules. So, during administration in the targeted area of the body, these drug-conjugated polyesters easily break down to release the drug molecules (Tambe et al. 2015). The major benefit is that encapsulated drugs are transported through aqueous media. The nanosponges are solid particles with spherical morphology with very high solubilizing effect. The size of this type of carrier can be modulated depending upon reaction condition. It also depend on the size and chemical nature of the polymer as well as crosslinker, which are used for the synthesis of drug carrier. Subsequently, the drug loading capacity of the nanosponge varied with the nature of the cavity. The CD:crosslinker ratio improves the drug loading capacity of nanosponge manifold, which can also modulate the drug release profile

(Cavalli et al. 2006). Shende et al. (2015) developed β-CD-based nanosponges for the inclusion complexes of meloxicam to increase their solubility and stability as well as to prolong the release of the drug. The interaction of the meloxicam- and CD-based nanosponges was proved by different experimental techniques. The *in vitro* and *in vivo* drug release studies revealed the controlled release of meloxicam that sustained its anti-inflammatory and analgesic effects for 24 h.

Swaminathan et al. (2010) prepared CD-based nanosponges containing CAM for enhancement of the shelf life and duration of the drug release. CAM is administered less frequently for therapeutic use due to its poor solubility and instability of the lactone ring as well as serious side effects (Swaminathan et al. 2010). Upon encapsulation of CAM in CD-based nanosponges, the drug release lasted for 24 h. The cytotoxicity study revealed that the formulations containing CAM were more cytotoxic than pure CAM.

Mognetti et al. (2012) reported the preparation of paclitaxel-loaded β-CD nanosponges. It is formed as a water-soluble colloidal system. The *in vitro* release studies revealed that the release was completed within 2 h without an initial burst effect. The delivery of paclitaxel via nanosponges enhanced the amount of paclitaxel entering cancer cells and lowered the paclitaxel IC50 (Mognetti et al. 2012).

Gold nanoparticles containing β-CD polyurethane nanosponges with 1,6-hexamethylene diisocyanate as crosslinker for catalytic purposes have been reported by Vasconcelos et al. (2016). The Au concentrations modulate the formation $Au_n$ clusters in the nanosponge and showed higher catalytic effect toward the reduction of 4-nitrophenol.

## 16.4.5   CD-Based Hydrogels

Hydrogels are one of the very important carriers for drug and other biologically important substrates. Hydrogels are receiving much attention from the scientific community for numerous applications in medical devices and scaffolds for tissue regeneration and substitution. These hydrogels may be used for the preparation of a variety of dosage forms as well as the possibility to administer poorly water-soluble drugs by a different delivery mode. Thus, the crosslinked CDs as well as CDs containing polymeric systems are very important for the preparation of hydrogels by complexing drugs and drug-like molecules. These carriers are tools for efficient delivery at the targeted site (Liu et al. 2004a; Kanjickal et al. 2005).

The CD tagged in polymeric materials promotes the ability for better complexation of drug molecules in aqueous media compared to free CD. When these solutions of drug–CD complexes are administered and diluted in physiological fluids, the release of the drug starts instantaneously. By contrast, the CD units are covalently attached to the polymeric backbone to produce multifunctional networks in CD-tagged hydrogels, which have better ability to entrap drug molecules compared to normal CD-based hydrogels. In this way, the release of drug from the polymeric CD enhanced manifold (Liu et al. 2004a,b; Siemoneit et al. 2006; Rodriguez-Tenreiro et al. 2007).

The copolymerization technique between CD and acrylic or vinyl monomers produces different hydrogels with improved delivery efficiency. HP-β-CD hydrogels using diglycidyl ethers as crosslinking agents can be produced directly in order to avoid drawbacks during the chemical modification of CDs. Higashi et al. reported that the pegylated insulin forms polypseudorotaxanes with α- and γ-CDs as PEG does. The produced polypseudorotaxanes are less soluble in water and the rate of release from the pegylated drug is controlled by regulating the threading and dethreading rates of the polypseudorotaxanes by modulating the administration conditions such as the amount of injection medium and concentration of CDs in the medium. So, polypseudorotaxane is another important tool for sustained drug delivery techniques for pegylated proteins and peptides (Higashi et al. 2007).

A dual pH- and temperature-responsive hydrogel system has been developed and successfully used for the delivery of atorvastatin in 2-methylacrylic acid-modified β-CD, which is copolymerized with 2-methylacrylic acid and $N,N'$-methylene diacrylamide. It works in the media and reduces the swelling rate of hydrogels with low (pH ≤ 3.84) and high (pH ≥ 10.34) pH values whereas swelling of the hydrogel is increased with increasing temperature from 30°C to 45°C. This hydrogel system improved the solubility of atorvastatin and the drug release rate. The release of atorvastatin was as high as 90.5% in alkaline (pH 8.0) buffer solution (Yang et al. 2016).

The other instance of preparation of β-CD-based hydrogels has been reported using nontoxic reagent sodium trimetaphosphate (STMP) in basic aqueous media with dextran. Here, hydrogels contain dextran chains, phosphate groups, and β-CD units. Methylene blue and benzophenone compounds entrap in this hydrogel system through different interaction mechanisms as electrostatic and inclusion complex interactions (Wintgens et al. 2015).

## 16.5 APPLICATION OF CD IN DRUG DELIVERY

### 16.5.1 ORAL DELIVERY

The oral route is the most convenient pathway for drug administration due to easy administration, high patient compliance, and comfort. Insulin is the most effective and durable drug in the treatment of advanced-stage diabetes Noninvasive insulin delivery systems today remain a major challenge in the area of pharmaceutical research and development. The oral delivery of insulin undergoes a first hepatic pass, and will produce a similar effect as pancreas-secreted insulin by inhibiting hepatic gluconeogenesis and suppressing hepatic glucose production. In these background, Sajeesh and Sharma (2006) developed β-CD–insulin (HP-β-CD–I) complex-encapsulated polymethacrylic acid–chitosan–polyether (polyethylene glycol–polypropylene glycol copolymer) (PMCP) nanoparticles. Methacrylic acid in the presence of chitosan and polyether in a solvent/surfactant-free medium were used for the preparation of nanoparticles by free radical polymerization. The particle size determined by dynamic light scattering (DLS) method and the size range of the nanoparticle was 500–800 nm. The noncovalent inclusion complex with insulin (HP-β-CD–I) was analyzed by Fourier transform infrared (FTIR) and fluorescence spectroscopic techniques. The enzyme-linked immunosorbent assay (ELISA) study was used for the encapsulation of insulin in the nanoparticles. The in vitro release profile of HP-β-CD–I nanoparticles was evaluated at acidic/alkaline pH. The result stated that CD complexed insulin nanoparticles acted as a good candidate for oral insulin delivery.

### 16.5.2 Colon Drug Delivery

In recent years, colon drug delivery systems have gained increasing importance, not only for more effective treatment of local pathologies, such as Chron's diseases, ulcerative colitis, and colorectal cancer, but also for the systemic therapy of both conventional and labile molecules. Thus, colon-specific delivery is advantageous in terms of the oral bioavailability improvement of drugs, reduction in the total administered dose, and a decrease in side effects. Mennini et al. (2012) formulated chitosan-Ca-alginate microspheres for colon delivery of celecoxib-HP-β-CD-PVP (polyvinylpyrrolidone) complex by the quality by design (QbD) approach. The outcomes of the experiment concluded that the dual use of drug–CD complex and chitosan-Ca-alginate microsphere carrier could be effective for colon-targeted delivery of celecoxib.

### 16.5.3 Transdermal Delivery

Transdermal drug delivery systems are prepared to deliver drugs through the skin at a predetermined rate escaping the first-pass effect by the liver and the high risk of adverse gastrointestinal effects. The skin barrier is most difficult for transdermal drug delivery. The rate-limiting step of transdermal drug delivery is the skin, stratum corneum. Liposomal formulations are used as a transdermal delivery due to their versatility and clinical efficacy. Maestrelli et al. (2006) investigated liposomes encapsulating ketoprofen–CD complexes (ketoprofen and HP-β-CD for transdermal delivery. Formulations (liposomes) were prepared with different techniques, such as thin layer evaporation, freezing and thawing, and extrusion through microporous membrane. The size and morphology of the different types of liposomes were investigated by light scattering analysis, confocal laser scanning microscopy, and transmission electron microscopy. CD complexation increases drug solubilization and entrapment into the aqueous liposomal phase.

### 16.5.4 Buccal Delivery

The buccal delivery of drugs has various advantages such as high blood levels circumventing first-pass metabolism and avoiding degradation in the gastrointestinal tract by enzymes and bacteria. Cappello et al. (2006) developed CD-containing poly(ethyleneoxide) (PEO) tablets for the delivery of poorly soluble drugs (carvedilol), where PEO was the bioadhesive sustained-release platform and HP-β-CD the modulator of drug release. The buccal delivery of carvedilol was performed by permeation experiments on pig excised mucosa.

Jug et al. (2010) designed and evaluated the binary products of bupivacaine hydrochloride (BVP.HCl), an amide-type local anesthetic, with parent β-CD (β-CD) and its soluble β-CD–epichlorohydrin polymer (EPI-β-CD), a novel mucoadhesive formulation for buccal delivery. The products were characterized by differential scanning calorimetry (DSC), x-ray powder diffractometry (XRPD), FTIR, and environmental scanning electron microscopy (ESEM). The results showed that complexation of BVP.HCl with β-CD and EPI-β-CD was an effective tool for properly tailoring the dissolution properties of the drug and thus could be favorably used as a buccal drug delivery system.

### 16.5.5 OCULAR DRUG DELIVERY

Nanotechnology is now massively exploited for ocular drug delivery, mainly because of the association of an active molecule to a nanocarrier that allows the molecule to intimately interact with specific ocular structures, overcome ocular barriers, and prolong its residence in the target tissue. This technology was used for the promising solution of formulating various poorly water-soluble drugs in the form of eye drops. The development of efficient ocular delivery nanosystems remains a major challenge for pharmaceutical research scientists in achieving a sustained therapeutic effect. For this reason, Mahmoud et al. (2011) prepared chitosan nanoparticles using sulfobutylether-β-CD (SBE-β-CD) as a polyanionic crosslinker and investigated the potential of these nanostructures as an ocular delivery system for econazole nitrate (ECO). The results revealed that the prepared nanoparticles were predominantly spherical in shape having average particle diameter from 90 to 673 nm with positive zeta potential (22–33 mV) and drug content values (13%–45%). The release profile of nanoparticles followed zero-order kinetics. The optimized nanoparticles released approximately 50% of the original amount over an 8 h. *Albino* rabbits were used for ocular drug delivery study of optimized nanoparticles.

### 16.5.6 INTRANASAL DELIVERY

In recent years, nasal drug delivery systems are attracting more attention as a convenient and reliable method for the systemic administration of drugs. The various advantages of the nasal route include fast onset of therapeutic action by rapid absorption, avoidance of liver or gastrointestinal metabolism, higher bioavailability allowing lower doses, and enhanced patient compliance by self-medication. Cho et al. (2010) prepared microparticles (MPs) of granisetron (GRN) in combination with HP-β-CD and sodium carboxymethylcellulose (CMC-Na) by the simple freeze-drying method for intranasal delivery. The drug–excipients interaction were evaluated by FTIR, powder x-ray diffraction (PXRD) analysis and DSC studies. The *in vitro* release of drug from MPs was determined in phosphate buffered saline (pH 6.4) at 37°C. Cytotoxicity of the MPs and *in vitro* permeation studies were performed in primary human nasal epithelial (HNE) cells and their monolayer system cultured by the air–liquid interface (ALI) method. The data stated that new complex (GRN, HP-β-CD, and CMC-Na) could be used as an alternative to oral and intravenous administration of GRN for intranasal delivery.

### 16.5.7 PULMONARY DELIVERY

Pulmonary delivery of antifungal drugs as aerosols are important for high drug concentrations targeted directly at the site of infection while minimizing potential systemic toxicity. Yang et al. (2010) fabricated and evaluated 2-HP-β-CD solubilized

ITZ solution (HP-β-CD-ITZ) suitable for pulmonary delivery by nebulization, and compared the pharmacokinetics of inhaled nebulized aerosols of HP-β-CD-ITZ versus a colloidal dispersion of ITZ nanoparticulate formulation (URF-ITZ).

### 16.5.8 ANTICANCER AND ANTI-HIV DRUG DELIVERY

Recently, a series of CD-based glyconjugates, including glycoclusters and glyco-polymers, were synthesized and reported, which showed their high-affinity binding to the human transmembrane lectin DC-SIGN (dendritic cell-specific intercellular adhesion molecule-3-grabbing nonintegrin) and act as an inhibitor to prevent the binding of HIV envelop protein gp120 to DC-SIGN at nanomolar concentration. The block glycopolymers showed high loading capacity of hydrophobic anticancer and anti-HIV drugs, which indicates promising application in HIV therapeutics and efficient drug delivery (Zhang et al. 2014).

The delivery system containing hydrophilic and hydrophobic terminal of CDs has been prepared by exploiting *click* chemistry and successfully used as *tumor-triggered targeting* for the treatment of cancer. A peptide containing the Arg–Gly–Asp (RGD) sequence was introduced to self-assembled noncovalently connected micelles (NCCMs) through host–guest interactions. In this case, pegylated technology was employed via benzoic-imine bonds to protect the ligands in normal tissues and body fluid. Here, the release of drug only happens when drug containing supra-molecular systems reach the tumor site (Quan et al. 2010). The drug targeting events are shown in Figure 16.5.

A drug delivery system is prepared by graft polymerization technique of chito-san-coated magnetic nanoparticles (CS-MNP) with acrylic acid and grafted with ethylenediamine derivative of β-CD (β-CD). This can be used for the sustained and controlled release of the anticancer drug, curcumin. The carrier was efficiently complexed with curcumin due to hydrogen bonding and van der Waal's interactions at an optimum pH of 5.0. The cytotoxicity studies revealed that CUR-loaded-(CS-MNP)-g-poly(AA) works well as compared to (CS-MNP)-g-poly(AA) in MCF-7 cells (Anirudhan et al. 2016).

## 16.6    CONCLUSION

This chapter highlights the basic information focusing on a wide audience range. It outlines how well CDs can satisfy the requirements of a wide range of researchers. It is primarily used as an encapsulating agent in its natural condition. But today its synthetic modification is extremely important for chemists, whereas the product is tremendously useful for pharmaceutical, polymer, and material scientists for bio-medical applications. Presently, synthetically modified CDs are being used for drug delivery based on different carrier systems. These starch derivatives interact with drug/drug-like substrates via dynamic complex formations using different covalent bonding as well as noncovalent interactions to form a purposeful supramolecular assembly to achieve efficient and precise drug delivery. In this way, a wide range of drug molecules, specifically poorly water-soluble drug candidates, are being

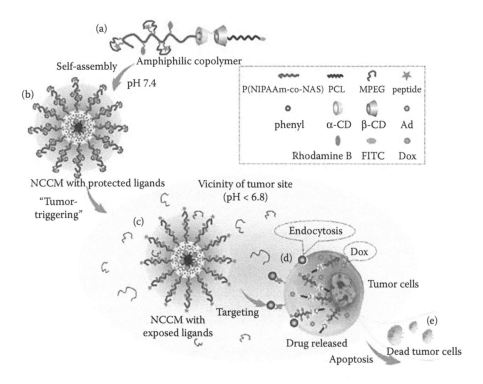

**FIGURE 16.5** Delivery of anticancer drug: Diblock copolymer connected by α-β cyclo-dextrin dimer (a); Noncovalently connected micelle (NCCM) with protected target ligands at pH = 7.4 (b); Tumor-triggered deshielding to switch on the targeting property through removal of PEG segments at pH < 6.8 (c); Endocytosis of NCCM and drug release after destruction of the shell-core structure of NCCM at $T >$ LCST (d); Apoptosis of tumor cells (e). (Reprinted with permission from Quan, C.-Y., Chen, J.-X., Wang, H.-Y. et al. 2010. Core-shell nanosized assemblies mediated by the α-, β-cyclodextrin dimer with a tumor-triggered targeting property. *ACS Nano* 4:4211–19. Copyright 2010 American Chemical Society.)

delivered to targeted sites. Overall, this chapter briefly describes the different carrier systems and their potential applications in a wide range of delivery mechanisms in different sites with prolonged and sustained release.

## REFERENCES

Agueros, M., Ruiz-Gaton, L., Vauthier C. et al. 2009. Combined hydroxypropyl-β-cyclodextrin and poly(anhydride) nanoparticles improve the oral permeability of pacli-taxel. *Eur J Pharm Sci* 38:405–13.

Ahmed, A., Wang, H., Yu, H. et al. 2015. Surface engineered cyclodextrin embedded poly-meric nanoparticles through host–guest interaction used for drug delivery. *Chem Eng Sci* 125:121–28.

Albers, E. and Muller, B.W. 1995. Cyclodextrin derivatives in pharmaceuticals. *CRC Crit Rev Ther Drug Carrier Syst* 12:311–37.

Alomrani, A.H., Shazly, G.A., Amara, A.A. et al. 2014. Itraconazole-hydroxypropyl-β-cyclodextrin loaded deformable liposomes: *In vitro* skin penetration studies and antifungal efficacy using *Candida albicans* as model. *Colloids Surf B Biointerfaces* 121:74–81.

Alvarez-Parrilla, E., Cabrer, P.R., De La Rosa, L.A. et al. 2003. "Three-in-one" complexes formed by anionic guests and monosubstituted cationic alkyldiamino β-cyclodextrin derivatives. *Supramol Chem* 15:207–11.

Anadolu, R.Y., Sen, T., Tarimci, N. et al. 2004. Improved efficacy and tolerability of retinoic acid in acne vulgaris: A new topical formulation with cyclodextrin complex. *JEADV* 18:416–21.

Anderson, J.M. and Kim, S.W. 1984. *Recent Advances in Drug Delivery Systems*. New York: Plenum Press.

Anirudhan, T.S., Divya, P.L., and Nima, J. 2016. Synthesis and characterization of novel drug delivery system using modified chitosan based hydrogel grafted with cyclodextrin. *Chem Engg J* 284:1259–69.

Aoki, N., Arai, R., and Hattori, K. 2004. Improved synthesis of chitosan-bearing bold beta-cyclodextrin and its adsorption behavior towards bisphenol A and 4-nonylphenol. *J Inclusion Phenom* 50:115–20.

Atwood, J.L., Davies, J.E.D., MacNicol, D.D., and Vogtle, F. 1996. *Comprehensive Supramolecular Chemistry*. Oxford, UK: Pergamon.

Babu, R.J. and Pandit, J.K. 2004. Effect of cyclodextrins on the complexation and transdermal delivery of bupranolol through rat skin. *Int J Pharm* 271:155–65.

Baussanne, I., Benito, J.M., Mellet, C.O., Fernández, J.M.G., Law, H., and Defaye, J. 2000. Synthesis and comparative lectin-binding affinity of mannosyl-coated β-cyclodextrin-dendrimer constructs. *Chem Commun* 16:1489–90.

Bender, M.L. and Komiyama, M. 1978. *Cyclodextrin Chemistry*. Berlin: Springer-Verlag.

Berthod, A., Li, W.Y., and Armstrong, D.W. 1990. Chiral recognition of racemic sugars by polar and nonpolar cyclodextrin-derivative gas chromatography. *Carbohydr Res* 201:175–84.

Bicchi, C., Artuffo, G., D'Amato, A., Manzin, V., Galli, A., and Galli, M. 1992. Cyclodextrin derivatives in the GC separation of racemic mixtures of volatile compounds. Part V: Heptakis-2,6-dimethyl-3-pentyl-β-cyclodextrins. *J High Resolut Chromatogr* 1992:710–14.

Boudad, H., Legrand, P., Lebas, G. et al. 2001. Combined hydroxypropyl-beta-cyclodextrin and poly (alkylcyanoacrylate) nanoparticles intended for oral administration of saquinavir. *Int J Pharm* 218:113–24.

Brewster, M.E., Estes, K., Loftsson, T. et al. 1988. Improved delivery through biological membranes. XXXL: Solubilization and stabilization of an estradiol chemical delivery system by modified beta-cyclodextrins. *J Pharm Sci* 77:981–85.

Buranaboripan, W., Lang, W., Motomura, E. et al. 2014. Preparation and characterization of polymeric host molecules, β-cyclodextrin-linked chitosan derivatives having different linkers. *Int J Bio Macromol* 69:27–34.

Buri, P. and Gumma, A. 1985. *Drug Targeting*. Amsterdam: Elsevier.

Cai, T., Yang, W.J., Zhang, Z. et al. 2012. Preparation of stimuli-responsive hydrogel networks with threaded β-cyclodextrin end-capped chains via combination of controlled radical polymerization and click chemistry. *Soft Matter* 8:5612–20.

Cappello, B., De Rosa, G., Giannini, L. et al. 2006. Cyclodextrin-containing poly(ethyleneoxide) tablets for the delivery of poorly soluble drugs: Potential as buccal delivery system. *Int J Pharm* 319:63–70.

Cavalli, R., Peira, E., Caputo, O. et al. 1999. Solid lipid nanoparticles as carriers of hydrocortisone and progesterone complexes with betacyclodextrins. *Int J Pharm* 182:59–69.

Cavalli, R., Trotta, F. and Tumiatti, W. 2006. Cyclodextrin-based nanosponges for drug delivery. *J Incl Phenom Macrocycl Chem* 56:209–13.

Chen, J., Lu, S., Gu, W. et al. 2015. Characterization of 9-nitrocamptothecin-in-cyclodextrin-in-liposomes modified with transferrin for the treating of tumor. *Int J Pharm* 490:219–28.

Chiu, S.H., Myles, D.C., Garrel, R.L. et al. 2000. Novel ether-linked secondary face-to-face 2 – 2′ and 3 – 3′ β-cyclodextrin dimers. *J Org Chem* 65:2792–96.

Cho, H.-J., Balakrishnan, P., Shim, W.-S. et al. 2010. Characterization and *in vitro* evaluation of freeze-dried microparticles composed of granisetron-cyclodextrin complex and carboxymethylcellulose for intranasal delivery. *Int J Pharm* 400:59–65.

Cirpanli, Y., Bilensoy, E., Dogan, A.L. et al. 2009. Comparative evaluation of polymeric and amphiphilic cyclodextrin nanoparticles for effective camptothecin delivery. *Eur J Pharm Biopharm* 73:82–9.

Ciucanu, I. and Kerek, F. 1984. A simple and rapid method for the permethylation of carbohydrates. *Carbohydr Res* 131:209–17.

Connors, K.A. 1997. The stability of cyclodextrin complexes in solution. *Chem Rev* 97:1325–58.

Cornwell, M.J., Huff, J.B., and Bieniarz, C. 1995. A one-step synthesis of cyclodextrin monoaldehydes. *Tetrahedron Lett* 36:8371–74.

Cousin, H., Trapp, O., Peulon-Agasse, V. et al. 2003. Synthesis, NMR spectroscopic characterization and polysiloxane-based immobilization of the three regioisomeric monooctenylpermethyl-β-cyclodextrins and their application in enantioselective GC. *Eur J Org Chem* 2003:3273–87.

Devarakond, B., Hill, R.A., Liebenberg, W. et al. 2005. Comparison of the aqueous solubilization of practically insoluble niclosamide by polyamidoamine (PAMAM) dendrimers and cyclodextrins. *Int J Pharm* 304:193–209.

Dias, M.M.R., Raghavan, S.L., Pellett, M.A. et al. 2003. The effect of β-cyclodextrins on the permeation of diclofenac from supersaturated solutions. *Int J Pharm* 263:173–81.

Duchene, D. 1987. *Cyclodextrins and Their Industrial Uses*. Paris: Editions de Sante.

Duchene, D. 1991. *New Trends in Cyclodextrins and Derivatives*. Paris: Editions de Sante.

Endo, T., Nagase, H., Ueda, H. et al. 1997. Isolation, purification, and characterization of cyclomaltodecaose (curly epsilon-cyclodextrin), cyclomaltoundecaose (zeta-cyclodextrin) and cyclomaltotridecaose (é-cyclodextrin). *Chem Pharm Bull* 45:532–36.

Evrarda, B., Bertholeta, P., Guedersc, M. et al. 2004. Cyclodextrins as a potential carrier in drug nebulization. *J Control Release* 96:403–10.

Fernandez, M., Fragoso, A., Cao, R. et al. 2003. Improved functional properties of trypsin modified by monosubstituted amino-β-cyclodextrins. *J Mol Catalysis B: Enzymatic* 21:133–41.

Filipovic-Grcic, J., Voinovich, D., Moneghini, M. et al. 2000. Chitosan microspheres with hydrocortisone and hydrocortisone hydroxypropyl-β-cyclodextrin inclusion complex. *Eur J Pharm Sci* 9:373–79.

Francois, M., Snoeckx, E., Putteman, P. et al. 2003. A mucoadhesive, cyclodextrin-based vaginal cream formulation of itraconazole. *AAPS PharmSci* 5:50–54.

Frömming, K.-H. and Szejtli, J. 1994. *Cyclodextrins in Pharmacy*. Dordrecht, The Netherlands: Kluwer.

Fundueanua, G., Constantinb, M., Dalpiaza, A. et al. 2004. Preparation and characterization of starch/cyclodextrin bioadhesive microspheres as platform for nasal administration of Gabexate Mesylate (Foys) in allergic rhinitis treatment. *Biomaterials* 25:159–70.

Gao, H., Yang, Y., Fan, Y. et al. 2006. Conjugates of poly(DL-lactic acid) with ethylenediamino or diethylenetriamino bridged bis(β-cyclodextrin)s and their nanoparticles as protein delivery systems. *J Control Release* 112:301–11.

Gharib, R., Greige-Gerges, H., Fourmentin, S. et al. 2015. Liposomes incorporating cyclodextrin–drug inclusion complexes: Current state of knowledge. *Carbohydr Polym* 129:175–86.

Gibaud, S., Zirar, S.B., Mutzenhardt, P. et al. 2005. Melarsoprol–cyclodextrins inclusion complexes. *Int J Pharm* 306:107–21.

Gidwani, B. and Vyas, A. 2015. A comprehensive review on cyclodextrin-based carriers for delivery of chemotherapeutic cytotoxic anticancer drugs. *BioMed Res Int* 2015:1–15.

Gil, E.S., Wu, L., Xu, L. et al. 2012. β-Cyclodextrin-poly(β-amino ester) nanoparticles for sustained drug delivery across the blood–brain barrier. *Biomacromolecules* 13:3533–41.

Heidel, J.D. and Schluep, T. 2012. Cyclodextrin-containing polymers: Versatile platforms of drug delivery materials. *J Drug Deliv* 2012:1–17.

Herrmann, S., Winter, G., Mohl, S. et al. 2007. Mechanisms controlling protein release from lipidic implants: Effects of PEG addition. *J Control Release* 118:161–68.

Higashi, T., Hirayama, F., Arima, H. et al. 2007. Polypseudorotaxanes of pegylated insulin with cyclodextrins: Application to sustained release system. *Bioorg Med Chem Lett* 17:1871–74.

Hu, Q.-D., Tang, G.-P., and Chu, P. K. 2014. Cyclodextrin-based host–guest supramolecular nanoparticles for delivery: From design to applications. *Acc Chem Res* 47:2017–25.

Irie, T. and Uekama, K. 1997. Pharmaceutical applications of cyclodextrins. III. Toxicological issues and safety evaluation. *J Pharm Sci* 86:147–62.

Jimenez V., Belmar J., and Alderete J.B. 2003. Determination of the association constant of 6-thiopurine and chitosan grafted β-cyclodextrin. *J Inclusion Phenom* 47:71–75.

Jimenez-Blanco, J.L., Garcia-Fernandez, J.M., Gadelle, A. et al. 1997. A mild one-step selective conversion of primary hydroxyl groups into azides in mono- and oligo-saccharides. *Carbohydr Res* 303:367–72.

Jug, M., Becirevic-Lacan, M., Kwokal, A. et al. 2005. Influence of cyclodextrin complexation on piroxicam gel formulations. *Acta Pharm* 55:223–36.

Jug, M., Maestrelli, F., and Bragagni, M. 2010. Preparation and solid-state characterization of bupivacaine hydrochloride cyclodextrin complexes aimed for buccal delivery. *J Pharm Biomed Anal* 52:9–18.

Juliano, R. L. 1980. *Drug Delivery System.* New York: Oxford University Press.

Kang, F. and Singh, J. 2003. Conformational stability of a model protein (bovine serum albumin) during primary emulsification process of PLGA microspheres synthesis. *Int J Pharm* 260:149–56.

Kanjickal, D., Lopina, S., Evancho-Chapman, M.M. et al. 2005. Improving delivery of hydrophobic drugs from hydrogels through cyclodextrins. *J Biomed Mater Res* 74A:454–60.

Kassab, R., Felix, C., Parrot-Lopez, H. et al. 1997. Synthesis of cyclodextrin derivatives carrying bio-recognisable saccharide antennae. *Tetrahedron Lett* 38:7555–58.

Khan, A.R., Forgo, P., Stine, K.J. et al. 1998. Methods for selective modifications of cyclodextrins. *Chem Rev* 98:1977–96.

Konig, W.A., Icheln, D., Runge, T. et al. 1990. Cyclodextrins as chiral stationary phases in capillary gas chromatography. Part VII: Cyclodextrins with an inverse substitution pattern-synthesis and enantioselectivity. *J High Resolut Chromatogr* 13:702–07.

Konig, W.A., Icheln, D., Runge, T. et al. 1991. Gas chromatographic enantiomer separation of agrochemicals using modified cyclodextrins. *J High Resolut Chromatogr* 14:530–36.

Konig, W.A., Lutz, S., and Wenz, G. 1988a. Modified cyclodextrins—Novel highly enantioselective stationary phases for gas chromatography. *Angew Chem Int Ed Engl* 27:979–80.

Konig, W.A., Lutz, S., Mischnick-Lubbecke, P. et al. 1988b. Modified cyclodextrins—A new generation of chiral stationary phase for capillary gas chromatography. *Starch* 40:472–76.

Lee, E.S., Kwon, M.J., Lee, H. et al. 2007. Stabilization of protein encapsulated in poly(lactide-*co*-glycolide) microspheres by novel viscous S/W/O/W method. *Int J Pharm* 331:27–37.

Lemos-Senna, E., Wouessidjewe, D., Lesieur, S. et al. 1998. Preparation of amphiphilic cyclodextrin nanospheres using the emulsification solvent evaporation method. Influence of the surfactant on preparation and hydrophobic drug loading. *Int J Pharm* 170:119–28.

Liu, Y., You, C.C., He, S., Chen, G.-S., and Zhao, Y.-L. 2002. Synthesis of novel indolyl modified β-cyclodextrins and their molecular recognition behavior controlled by the solution's pH value. *J Chem Soc Perkin Trans 2* 2002:463–69.

Liu, Y.Y., Fan, X.D., Hu, H. et al. 2004a. Release of chlorambucil from poly(N-isopropylacryl-amide) hydrogels with beta-cyclodextrin moieties. *Macromol Biosci* 4:729–36.

Liu, Y.Y., Fan, X.D., Kang, T. et al. 2004b. A cyclodextrin microgel for controlled release driven by inclusion effects. *Macromol Rapid Commun* 25:1912–16.

Liu, Y.Y., Fan, X.D., and Zhao, Y.B. 2005. Synthesis and characterization of a poly(N iso-propylacrylamide) with β-cyclodextrin as pendant groups. *J Polym Sci A Polym Chem* 43:3516–24.

Lock, J.S., May, B.L., Clements, P., Lincoln, S.F., and Easton, C.J. 2004. Intra- and intermo-lecular complexation in C(6) monoazacoronand substituted cyclodextrins. *Org Biomol Chem* 2:1381–86.

Loftsson, T. and Brewester, M. 1996. Pharmaceutical applications of cyclodextrins. 1. Drug solubilization and stabilization. *J Pharm Sci* 85:1017–25.

Loukas, Y.L., Vraka, V., and Gregoriadis, G. 1998. Drugs, in cyclodextrins, in liposomes: A novel approach to the chemical stability of drugs sensitive to hydrolysis. *Int J Pharm* 162:137–42.

Maestrelli, F., Corti, G., Mura, P. et al. 2005a. Development of fast-dissolving tablets of flur-biprofen–cyclodextrin complexes. *Drug Dev Ind Pharm* 31:697–707.

Maestrelli, F., Gonzalez-Rodrıguez, M.L., Rabasco, A.M. et al. 2005b. Preparation and char-acterisation of liposomes encapsulating ketoprofen–cyclodextrin complexes for trans-dermal drug delivery. *Int J Pharm* 298:55–67.

Maestrelli, F., Garcia-Fuentes, M., Mura, P. et al. 2006. A new drug nanocarrier consisting of chitosan and hydoxypropylcyclodextrin. *Eur J Pharm Biopharm* 63:79–86.

Mahmoud, A.A., El-Feky, G.S., Kamel, R. et al. 2011. Chitosan/sulfobutylether-β-cyclodextrin nanoparticles as a potential approach for ocular drug delivery. *Int J Pharm* 413:229–36.

Martin, R., Nchez, I.S., Cao, R. et al. 2006. Solubility and kinetic release studies of naproxen and ibuprofen in soluble epichlorohydrin-β-cyclodextrin polymer. *Supramol Chem* 18:627–31.

Memisoglu-Bilensoya, E., Vurala, T.I., Bochotb, A. et al. 2005. Tamoxifen citrate loaded amphiphilic h-cyclodextrin nanoparticles: *In vitro* characterization and cytotoxicity. *J Control Release* 104:489–96.

Mennini, N., Furlanetto, S., Cirri, M. et al. 2012. Quality by design approach for developing chitosan-Ca-alginate microspheres for colon delivery of celecoxib-hydroxypropyl-β-cyclodextrin-PVP complex. *Eur J Pharm Biopharm* 80:67–75.

Mognetti, B., Barberis, A., Marino, S. et al. 2012. *In vitro* enhancement of anticancer activity of paclitaxel by a Cremophor free cyclodextrin-based nanosponge formulation. *J Incl Phenom Macrocycl Chem* 74:201–10.

Mora, P.C., Cirri, M., and Mura, P. 2007. Development of a sustained-release matrix tab-let formulation of DHEA as ternary complex with a-cyclodextrin and glycine. *J Incl Phenom Macrocycl Chem* 57:699–704.

Nelissen, H.F.M., Feiters, M.C., and Nolte, R.J.M. 2002. Synthesis and self-inclusion of bipyridine-spaced cyclodextrin dimers. *J Org Chem* 67:5901–06.

Nelson, A. and Stoddart, J.F. 2003. Dynamic multivalent lactosides displayed on cyclodextrin beads dangling from polymer strings. *Org Lett* 5:3783–86.

Nishimura, K., Hidaka, R., Hirayama, F. et al. 2006. Improvement of dispersion and release properties of nifedipine in suppositories by complexation with 2-hydroxypropyl-β-cyclodextrin. *J Incl Phenom Macrocycl Chem* 56:85–88.

Nowotny, H.-P., Schmalzing, D., Wistuba, D. et al. 1989. Extending the scope of enantio-mer separation on diluted methylated β-cyclodextrin derivatives by high resolution gas chromatography. *J High Resolut Chromatogr* 12:383–93.

Okimoto, K., Ohike, A., Ibuki, R. et al. 1999. Factors affecting membrane-controlled drug release for an osmotic pump tablet (OPT) utilizing (SBE)-β-CD as both a 7 m solubi-lizer and osmotic agent. *J Control Release* 60:311–19.

Parrot-Lopez, H., Leray, E., and Coleman, A.W. 1993. New β-cyclodextrin derivatives possessing biologically active saccharide antennae. *Supramol Chem* 3:37–42.

Petter, R.C., Salek, J.S., Sikorski, C.T. et al. 1990. Cooperative binding by aggregated mono-6-(alkylamino)-beta-cyclodextrins. *J Am Chem Soc* 112:3860–68.

Pitha, J., Szente, L., and Szejtli, J. 1983. Molecular encapsulation of drugs by cyclodextrins and congeners. In *Controlled Drug Delivery*, ed. S.D. Bruck, 125, Boca Raton, FL: CRC Press.

Quan, C.-Y., Chen, J.-X., Wang, H.-Y. et al. 2010. Core-shell nanosized assemblies mediated by the α-, β-cyclodextrin dimer with a tumor-triggered targeting property. *ACS Nano* 4:4211–19.

Ragnoa, G., Cione, E., Garofalo, A. et al. 2003. Design and monitoring of photostability systems for amlodipine dosage forms. *Int J Pharm* 265:125–32.

Rajewski, R.A. and Stella, V.J. 1996. Pharmaceutical applications of cyclodextrins. 2. In vivo drug delivery. *J Pharm Sci* 85:1142–69.

Rodriguez-Tenreiro, C., Alvarez-Lorenzo, C., Rodriguez-Perez, A. et al. 2007a. Estradiol sustained release from high affinity cyclodextrin hydrogels. *Eur J Pharm Biopharm* 66:55–62.

Rodriguez-Tenreiro, C., Alvarez-Lorenzo, C., Rodriguez-Perez, A. et al. 2007b. Estradiol sustained release from high affinity cyclodextrin hydrogels. *Eur J Pharm Biopharm* 66:55–62.

Rong, D., D'Souza, V.T. 1990. A convenient method for functionalization of the 2-position of cyclodextrins. *Tetrahedron Lett* 31:4275–78.

Rosa, G.D., Larobina, D., La Rotonda, M.I. et al. 2005. How cyclodextrin incorporation affects the properties of protein-loaded PLGA-based microspheres: The case of insulin/hydroxypropyl-β-cyclodextrin system. *J Control Release* 102:71–83.

Saenger, W. 1980. Cyclodextrin inclusion compounds in research and industry. *Angew Chem Int Ed Engl* 19:344–62.

Sajeesh, S. and Sharma, C.P. 2006. Cyclodextrin–insulin complex encapsulated polymethacrylic acid based nanoparticles for oral insulin delivery. *Int J Pharm* 325:147–54.

Scalia, S., Tursilli, R., Bianchi, A. et al. 2006a. Incorporation of the sunscreen agent, octyl methoxycinnamate in a cellulosic fabric grafted with β-cyclodextrin. *Int J Pharm* 308:155–59.

Scalia, S., Tursilli, R., Sala, N. et al. 2006b. Encapsulation in lipospheres of the complex between butyl methoxydibenzoylmethane and hydroxypropyl-β-cyclodextrin. *Int J Pharm* 320:79–85.

Schaschke, N., Assfalg-Machleidt, I., Machleidt, W. et al. 2000. β-Cyclodextrin/epoxysuccinyl peptide conjugates: A new drug targeting system for tumor cells. *Bioorg Med Chem Lett* 10:677–80.

Schaschke, N., Fiori, S., Weyher, E. et al. 1998. Cyclodextrin as carrier of peptide hormones. Conformational and biological properties of β-cyclodextrin/gastrin constructs. *J Am Chem Soc* 120:7030–38.

Shende, P.K., Gaud, R.S., Bakal, R. et al. 2015. Effect of inclusion complexation of meloxicam with β-cyclodextrin and β-cyclodextrin-based nanosponges on solubility, *in-vitro* release and stability studies. *Colloids Surf B Biointerfaces* 136:105–10.

Siemoneit, U., Schmitt, C., Alvarez-Lorenzo, C. et al. 2006. Acrylic/cyclodextrin hydrogels with enhanced drug loading and sustained release capability. *Int J Pharm* 312:66–74.

Skalko, N., Brandl, M., Ladan, M.B. et al. 1996. Liposomes with nifedipine and nifedipine–cyclodextrin complex: Calorimetrical. *Eur J Pharm Sci* 4:359–66.

Skiba, M., Bounoure, F., Barbot, C. et al. 2005. Development of cyclodextrins microspheres for pulmonary drug delivery. *J Pharm Pharm Sci* 8:409–18.

Smith, J.S., MacRaea, R.J., and Snowden, M.J. 2005. Effect of SBE7-β-cyclodextrin complexation on carbamazepine release from sustained release beads. *Eur J Pharm Biopharm* 60:73–80.

Sortino, S., Mazzagliab, A., Scolaroc, L.M. et al. 2006. Nanoparticles of cationic amphiphilic cyclodextrins entangling anionic porphyrins as carrier-sensitizer system in photodynamic cancer therapy. *Biomaterials* 27:4256–65.

Stella, V.J. and Rajewski, R.A. 1997. Cyclodextrins: Their future in drug formulation and delivery. *Pharm Res* 14:556–67.

Subramanian, S., Singireddy, A., Krishnamoorthy, K. et al. 2012. Nanosponges: A novel class of drug delivery system—Review. *J Pharm Pharm Sci* 15:103–11.

Swaminathan, S., Pastero, L., Serpe, L. et al. 2010. Cyclodextrin-based nanosponges encapsulating camptothecin: Physicochemical characterization, stability and cytotoxicity. *Eur J Pharm Biopharm* 74:193–201.

Szejtli, J. 1982. *Cyclodextrins and Their Inclusion Complexes*. Budapest: Akademiai Kiado.

Szejtli, J. 1983. Dimethyl-β-cyclodextrin as parenteral drug carrier. *J Inclusion Phenom* 1:135–50.

Szejtli, J. 1988. *Cyclodextrin Technology*. Dordrecht, The Netherlands: Kluwer.

Szejtli, J. 1992. The properties and potential uses of cyclodextrin derivatives. *J Inclusion Phenom Mol Recognit Chem* 14:25–36.

Takahashi, K., Hattori, K., and Toda, F. 1984. Monotosylated α- and β-cyclodextrins prepared in an alkaline aqueous solution. *Tetrahedron Lett* 25:3331–34.

Tambe, R.S., Battase, P.W., Arane, P.M., Palve, S.A., Talele, S.G., and Chaudhari, G. 2015. Review on nanosponges: As a targeted drug delivery system. *Am J Pharm Tech Res* 5:215–24.

Thompson, D.O. 1997. Cyclodextrin-enabling excipients: Their present and future use in pharmaceuticals. *CRC Crit Rev Ther Drug Carrier Syst* 14:1–104.

Trapani, A., Laquintana, V., Denora, N. et al. 2007. Eudragit RS 100 microparticles containing 2-hydroxypropyl-β-cyclodextrin and glutathione: Physicochemical characterization, drug release and transport studies. *Eur J Pharm Sci* 30:64–74.

Trichard, L., Fattal, E., Besnard, M. et al. 2007. α-Cyclodextrin/oil beads as a new carrier for improving the oral bioavailability of lipophilic drugs. *J Control Release* 122:47–53.

Trotta, F., Cavalli, R., Tumiatti, W. et al. 2007. Ultrasound-assisted synthesis of cyclodextrin-based nanosponges. Patent no. EP 1 786 841 B1.

Uekama, K. 1985. Pharmaceutical applications of methylated cyclodextrin derivatives. *Pharm Int* 6:61–65.

Uekama, K. and Hirayama, F. 1988. *Handbook of Amylases and Related Enzymes*. Oxford: The Amylase Research Society of Japan, Pergamon Press.

Uekama, K., Hirayama, F., and Irie, T. 1993. In *Drug Targeting Delivery*, ed. A.G. Boer, 411. Amsterdam: Harwood Publishers.

Uekama, K., Hirayama, F., and Irie, T. 1998. Cyclodextrin drug carrier systems. *Chem Rev* 98:2045–76.

Uekama, K. and Otagiri, M. 1987. Cyclodextrins in drug carrier system. *CRC Crit Rev Ther Drug Carrier Syst* 3:1–40.

Ungaro, F., De Rosa, G., Miro, A. et al. 2006. Cyclodextrins in the production of large porous particles: Development of dry powders for the sustained release of insulin to the lungs. *Eur J Pharm Sci* 28:423–32.

Vasconcelos, D.A., Kubota, T., Santos, D.C. et al. 2016. Preparation of $Au_n$ quantum clusters with catalytic activity in β-cyclodextrin polyurethane nanosponges. *Carbohydr Polym* 136:54–62.

Wang, J. and Cai, Z. 2008. Incorporation of the antibacterial agent, miconazole nitrate into a cellulosic fabric grafted with β-cyclodextrin. *Carbohydr Polym* 72:695–700.

Wintgens, V., Lorthioir, C., Dubot, P. et al. 2015. Cyclodextrin/dextran based hydrogels prepared by cross-linking with sodium trimetaphosphate. *Carbohydr Polym* 132:80–88.

Wong, P.T. and Choi, S.K. 2015. Mechanisms of drug release in nanotherapeutic delivery systems. *Chem Rev* 115:3388–32.

Wongmekiat, A., Yoshimatsu, S., Tozuka, Y. et al. 2006. Investigation of drug nanoparticle formation by co-grinding with cyclodextrins: Studies for indomethacin, furosemide and naproxen. *J Incl Phenom Macrocycl Chem* 56:29–32.

Yamanoi, T., Yoshida, N., Oda, Y. et al. 2005. Synthesis of *mono*-glucose-branched cyclodextrins with a high inclusion ability for doxorubicin and their efficient glycosylation using *Mucor hiemalis endo*-β-*N*-acetylglucosaminidase. *Bioorg Med Lett* 15:1009–13.

Yang, W., Chow, K.T., Lang, B. et al. 2010. In vitro characterization and pharmacokinetics in mice following pulmonary delivery of itraconazole as cyclodextrin solubilized solution. *Eur J Pharm Sci* 39:336–47.

Yang, K., Wan, S., Chen, B. et al. 2016. Dual pH and temperature responsive hydrogels based on β-cyclodextrin derivatives for atorvastatin delivery. *Carbohydr Polym* 136:300–06.

Yano, H., Hirayama, F., Arima, H. et al. 2001. Preparation of prednisolone-appended α-, β- and γ-cyclodextrins: Substitution at secondary hydroxyl groups and *in vitro* hydrolysis behavior. *J Pharm Sci* 90:493–503.

Yockot, D., Moreau, V., Demailly, G. et al. 2003. Synthesis and characterization of mannosyl mimetic derivatives based on a β-cyclodextrin core. *Org Biomol Chem* 1:1810–18.

Yu, H., Wei, H., Hou, D. et al. 2007. Composite hydrogels filled with inclusion complexes made from β-cyclodextrins with poly(propylene glycol) bisamine. *Curr Appl Phys* 7:116–19.

Yuan, D.Q., Tahara, T, Chen, W.H. et al. 2003. Functionalization of cyclodextrins via reactions of 2,3-anhydrocyclodextrins. *J Org Chem* 68:9456–66.

Zeng, J., Hunag, H., and Liu, S. 2013. Hollow nanospheres fabricated from β-cyclodextrin-grafted, α, β-poly (aspartic acid) as the carrier of camptothecin. *Colloids Surf B Biointerfaces* 105:120–27.

Zhang, L., Zhang, Q., Wang, X. et al. 2015. Drug-in-cyclodextrin-in-liposomes: A novel drug delivery system for flurbiprofen. *Int J Pharm* 492:40–45.

Zhang, Q., Su, L., Collins, J. et al. 2014. Dendritic cell lectin-targeting sentinel-like unimolecular glycoconjugates to release an anti-HIV drug. *J Am Chem Soc* 136:4325–32.

Zhang, X., Zhang, X., and Wuetal, Z. 2011. A hydrotropic β-cyclodextrin grafted hyperbranched polyglycerol co-polymer for hydrophobic drug delivery. *Acta Biomaterialia* 7:585–92.

# 17 Nanopolymer Scaffolds as Novel Carriers for Cells and Drugs

*Aum Solanki, Bhargav Hirapara,*
*Emily Johnson, and Yashwant Pathak*

## CONTENTS

## 17.1 INTRODUCTION

Tissue engineering is a growing field of research that shows great promise in the treatment of many human disorders. Infections and other diseases can often lead to the degeneration of certain tissue to a degree at which surgery and other techniques are rendered ineffective. Such degeneration is the cause of the high demand of organ transplants, which are highly problematic because of the increasing demand and decreasing prevalence of organ donations worldwide (Escudero and Otero 2015). For this reason, the development of techniques that allow for the treatment of such failures is necessary, and a promising avenue for this treatment is tissue engineering using polymer-based scaffolds. Such regeneration can be achieved through the use of biomaterials to

1. Implant cells into organisms.
2. Deliver tissue-growth-inducing substances such as growth factors.
3. Place cells within different matrices (Langer and Peppas 2003).

Scaffolds are known to be incredibly useful in this field, as they are able to mimic the extracellular matrix in which cells of the desired tissue can be implanted and grown (Freed et al. 1994). Polymers can be developed and shaped into architectures that make the scaffold applicable for the desired tissue (Dhandayuthapani et al. 2011). Figure 17.1 illustrates the general process of scaffold optimization.

Optimization of such scaffolds would result in those that promote cell-to-cell adhesion, promote extracellular matrix accumulation, permit transport of gases,

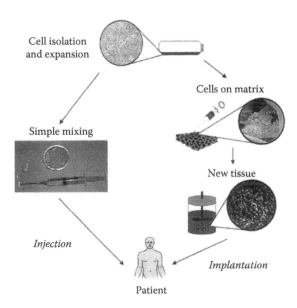

**FIGURE 17.1** The possible processes for tissue regeneration. (Reprinted with permission from Lee, K.Y. and Mooney, D.J. Hydrogels for tissue engineering. *Chem Rev* 101(7):1869–1880. Copyright 2001 American Chemical Society.)

nutrients, and regulatory factors, biodegrade at a controllable rate that allows tissue regeneration, and minimizes any toxic side effects *in vivo* (Langer and Tirrell 2004). Because of the selective ability of tissue engineering scaffolds, they have the potential to treat and cure a wide variety of disorders. The optimal selection of biomaterial, cell type, and growth factors can allow for regeneration for many different types of tissue. This includes the regeneration of lost bone tissue, irreparable heart valves, and even the treatment of permanently scarred skin tissue.

As stated, scaffold selection and optimization is a very important factor in the efficacy of the treatment. Different polymers have very different properties that have a high impact on scaffold efficacy. These include mechanical properties such as porosity and surface area, biocompatibility with targeted tissue, and biodegradability. As such, there is a wide variety of biomaterials available for researchers to choose. These materials can range from naturally occurring extracellular matrix (ECM) polymers such as collagen to selectively synthesized polymers such as polylactic acid (PLA) and poly(lactide-*co*-glycolide) (PLGA). This chapter will discuss the advantages and limitations of these biomaterials.

In addition to the scaffold composition, an important factor in the efficacy of the scaffold is the fabrication of the scaffold. The method used can greatly impact many of the scaffold properties, with each method having its own advantages and limitations. This chapter will elaborate on the most commonly used fabrication techniques.

## 17.2   METHODS OF FABRICATION

Polymer scaffolds are nanoscale structures that must be fabricated with high precision in order to ensure proper replication of ECM properties. Because of this, the proper method of fabrication must be used so that the particular scaffolding material used may function as desired.

### 17.2.1   ELECTROSPINNING

This is a very prevalent method used in the field of tissue engineering, as it creates structures similar to the fibrous proteins of the ECM, regardless of their chemical compositions (Goldberg et al. 2007). The method uses electric fields to collect polymers to produce a fiber mesh with very small diameters (several hundred nanometers). This simplicity makes this method of fabrication very popular in the field. Most of the polymers used are synthetic polymers such as PCL, PLA, and PLGA, but natural molecules such as collagen and fibrinogen have been synthesized with this method as well. The fibers that are synthesized using this method have strong mechanical properties that maximize interactions between the polymer and the cells because of the high porosity, surface area, and spatial interconnectivity.

As stated, the majority of electrospun scaffolds are composed of synthetic polymers, which explains the issue of biocompatibility largely associated with electrospinning. This issue can be rectified through the incorporation of natural materials on the surface of the fiber or the creation of a synthetic/natural hybrid scaffold. Kwon et al. created such hybrid scaffolds by co-electrospinning poly(L-lactide-*co*-ε-caprolactone) (PLCL) with type-I collagen. The study revealed that greater concentrations of collagen within the scaffold resulted in increasing loss of mechanical strength. This loss of strength was significant, as the PLCL/50%wt collagen scaffolds were able to withstand much smaller amounts of stress when compared to pure PLCL scaffolds. However, collagen incorporation did result in the desired cellular properties, as cellular adhesion was promoted by the ligands present on the collagen (Kwon and Matsuda 2005).

Electrospinning natural polymers into scaffolds is difficult because natural proteins such as collagen can be denatured or disrupted when used in this process. However, studies that use natural polymers for electrospinning have been conducted with certain modifications in order to enhance mechanical properties and biocompatibility. Mechanical properties, specifically, can be enhanced by the implementation of chemical cross-linkers with the natural polymer (Goldberg et al. 2007). Rho et al. (2006) were able to formulate scaffolds using electrospun collagen. This process resulted in fibers with strong mechanical properties that even held up in aqueous solution. However, when comparing the cell adhesion properties of electrospun collagen fibers to electrospun collagen fibers coated with type-I collagen or laminin, it was revealed that the electrospinning greatly reduced the fibers' biocompatibility; a problem that was fixed by type-I collagen or laminin coating. This illustrates the ability of this fabrication method to create strong scaffolds even with natural polymers, but also reveals the disruptive effects the process can have on the biocompatibility of the scaffolds. To improve this biocompatibility, scaffold coating appears to be an appropriate and effective modification.

Another limitation of electrospinning is its difficulty to form a complex 3D structure (Goldberg et al. 2007). This is a very serious limitation, especially *in vivo*, because cell penetration within the scaffold is very important, as it allows for further proliferation. Ekaputra et al. (2008) produced scaffolds through different types of electrospinning in order to produce a more complex and implantable 3D structure. Their studies revealed that the use of collagen along with medical-grade PCL resulted in greater cellular penetration. Figure 17.2 illustrates the methods employed for their fabrication, while Figure 17.3 provides some illustrations of the morphology of some of their fabricated scaffolds.

## 17.2.2 Knitting

Recent developments in scaffold formulation have utilized techniques that originate in textile manufacturing. These include weaving, knitting, and braiding, which are more complex than electrospinning, as they involve molding of fibers followed by assembly of the smaller subunits (Wang et al. 2011; Tamayol et al. 2013). These techniques are useful in the formulation of the 3D scaffolds for tissue engineering because they allow for the manipulation and control of the pore size and composition (Mikučionienė et al. 2010). This type of manipulation is very useful because optimization of pore size and composition is one of the most important factors affecting the cellular attachment properties of the scaffold.

Knitting is a widely used technique in scaffold fabrication. Through this technique, fibers of the desired polymer are interwoven in a series of connected loops that produces thin structures that have strong mechanical properties (Tamayol et al.

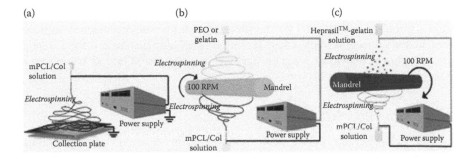

**FIGURE 17.2** Three different electrospinning setups were used: conventional flat plate collection (a), two-capillary coelectrospinning system (b), and two-capillary electrospinning–electrospraying system (c). In all three systems, medical-grade poly(ε-caprolactone)/collagen (mPCL/Col) was used as the main fiber material. PEO or gelatin was used as water-soluble fiber material for selective leaching approach (b). Heprasil hydrogel was embedded in the mPCL/Col mesh using a simultaneous electrospraying–electrospinning setup (c). To ensure homogeneous collection of PEO, gelatin, and Heprasil, a rotating mandrel collector was used at a speed of 100 RPM. (Reprinted with permission from Ekaputra, A.K. et al. Combining electrospun scaffolds with electrosprayed hydrogels leads to three-dimensional cellularization of hybrid constructs. *Biomacromolecules* 9(8):2097–2103. Copyright 2008 American Chemical Society.)

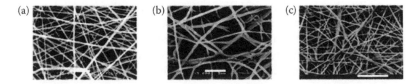

**FIGURE 17.3** Electron microscope images of the scaffolds produced. (a) Conventional poly(ε-caprolactone)/collagen coelectrospun scaffolds, (b) poly(ε-caprolactone) coelectrospun with collagen containing poly(ethylene oxide), and (c) poly(ε-caprolactone) coelectrospun with collagen containing gelatin. (Reprinted with permission from Ekaputra, A.K. et al. Combining electrospun scaffolds with electrosprayed hydrogels leads to three-dimensional cellularization of hybrid constructs. *Biomacromolecules* 9(8):2097–2103. Copyright 2008 American Chemical Society.)

2013). They are generally known to have high porosity, which promotes cellular penetration through the scaffold. Like electrospinning, the most common polymers used for knitting are synthetic polymers because of their mechanical strength and ease of use. Similarly, these synthetic polymers are often co-knitted or coated with natural polymers in order to enhance the biocompatible properties of the scaffold (Wang et al. 2011).

Knitted meshes also have applications when used in conjunction with natural polymer-containing scaffolds in order to optimize their mechanical properties. Chen et al. (2005) created hybrid scaffolds by incorporation of collagen microsponges onto knitted PLGA scaffolds. The collagen decreased the mechanical strength of the scaffold but greatly increased biocompatibility. Results revealed that cellular concentration on pure PLGA scaffolds was not as uniformly distributed as in the hybrid scaffold, which allowed for the successful skin tissue regeneration in mice. There are many possible applications for knitted scaffolds such as soft tissue regeneration, ligament and tendon repair, cartilage repair, and skin and blood vessel regeneration (Wang et al. 2011).

### 17.2.3 MOLECULAR SELF-ASSEMBLY

Molecular self-assembly, unlike electrospinning, forms scaffolds by emulating the ECM's assembly process. It incorporates ECM components within the desired scaffold and uses the intermolecular interactions between these components to form a larger, 3D structure that resembles the ECM. This method utilizes amphiphilic peptides present in physiological conditions. The hydrophobic and hydrophilic residues interact in ways that promote the self-assembly of the molecule into the desired conformation. Hartgerink et al. (2001) were able to develop such a scaffold that can be cross-linked, which can result in a conformational alignment that reflects the alignment of collagen fibrils in bone tissue. Modification of these scaffolds can be done through altering the amino acid composition. This would alter the intermolecular interactions, which would have a substantial effect on the final shape of the scaffold.

There are several issues associated with this method of fabrication. It has limited applicability because of its inability to be readily scaled up (Goldberg et al. 2007). Furthermore, the rate of degradation has not been characterized properly for scaffolds created through this method. This degradation can impact interaction with enzymes, as the self-assembly and degradation are uncontrolled processes. This dynamic nature results in mechanically weak scaffolds.

### 17.3 TYPES OF POLYMERIC SCAFFOLDS

Polymeric scaffolds differentiate themselves from other types of biomaterials because of their high surface-to-volume ratio, high porosity, small pore size, biodegradation, biocompatibility, and chemical versatility (Dhandayuthapani et al. 2011). This versatility is due to the fact that the chemical properties of polymers are determined by the combination of factors such as the composition, structure, and arrangement of the polymer's macromolecules. There are three types of polymers used in tissue engineering: naturally occurring, synthetic biodegradable, and synthetic non-biodegradable.

### 17.3.1 NATURAL POLYMERS

The advantage of using this type of polymer is that they have a greater ability to have interactions with and enhance the performance of cells in that particular biological system. They can be in the form of proteins (collagen, elastin, etc.), polysaccharides (cellulose, amylose, etc.), or nucleotides (DNA, RNA) (Yannas 2004). These were the first polymers clinically used as biomaterials for tissue engineering (Dhandayuthapani et al. 2011).

Natural polymers have been widely used to treat wounds. These wound dressings are useful due to their compatibility, biodegradability, and similarity to the macromolecules in the human body (Mogoşanu and Grumezescu 2014). Some of the most common used in wounds include polysaccharides and proteins (heparin and collagen, respectively). These are important for the migration, cell proliferation, and differentiation of the skin. This later leads to re-epithelialization, which is essential for positive wound repair (Wasiak et al. 2013).

### 17.3.1.1 Polysaccharides

Polysaccharides show great promise as cell carriers because of their high biocompatibility, availability, and their ease of fabrication (Bačáková et al. 2014). Polysaccharides are secreted in large quantities by many organisms, which makes them easy to obtain. Furthermore, they can be fabricated and customized because of their ability to form diverse structures. This customization could occur through the specific choice of the monosaccharide composition and the bonding structure (branched/unbranched). This customization is important because of its impact on the biocompatibility of the scaffold. Further addition of other polymers, synthetic or natural, has the ability to enhance the scaffold's mechanical properties. These choices should be made with regard to the type of tissue that the scaffold is meant to regenerate.

The composition of the scaffold determines the immune response it causes. This immune response results in the increased presence of selectin and immunoglobulin in the tissue surrounding the scaffold. A severe immune response to the scaffold may render the scaffold completely ineffective. Weidenbecher et al. (2007), while attempting to regenerate cartilage tissue using hyaluronic acid-based scaffolds, experienced this problem, as the scaffold triggered a severe immune response, causing complete degradation of the scaffold. Because of this degradation, the engineered cartilage graft failed *in vivo*. Factors that cause such an immune response include presence of contaminating DNA, mitogens, endotoxins, and cross-linking agents (Bačáková et al. 2014). Polysaccharides with low immune toxicity include chitosan, gellan gum, cellulose, dextran, and starch.

### 17.3.1.2 Proteins

There is a great applicability of proteins as polymers for scaffold fabrication because they make up many of the major components of the ECM. They hold an advantage over polysaccharides because they provide binding sites for cellular adhesion, whereas polysaccharides require additional modification to adhere and proliferate (Li et al. 2015).

Collagen is the most prevalent protein in humans and it is a very commonly used polymer for scaffold fabrication. Its biocompatibility with various tissues such as skin, bone, and vascular tissue makes it an ideal carrier for cells. Because of its high concentration in the body, obtaining and isolating collagen is relatively easy, as it is available in many different connective tissues of the body. There are many different collagens that can be used for scaffold formation, and the choice of the scaffold is made based upon the type of tissue being treated. Type-I collagen, specifically, has a fibrillar structure (fibrils of 50–100 nm) that allows for attachment and proliferation. For vascular tissue engineering, type-I collagen is highly favored because it is the major protein within the ECM of the walls of blood vessels (Pang and Greisler 2010). Weinberg and Bell were the first to engineer a blood vessel, and they did this using type-I collagen obtained from smooth muscle cells (Weinberg and Bell 1986). Their findings showed that collagen, when used alone, has low mechanical strength that must be remedied in order to be an effective biomaterial for tissue engineering. This low mechanical strength, coupled with the rapid rate of biodegradation, can be remedied by using cross-linkers and other modifications that increase the mechanical strength of the scaffold (Yamada et al. 2014).

Fibrin is another protein that is being widely used for tissue engineering applications. This protein has many benefits such as a controllable degradation rate and nontoxic degradation products that distinguish it from other polymers, namely, collagen (Li et al. 2015). The fibrin gel can be injected to the desired site where it will cure because it mimics the final step of blood coagulation, forming fibrin clots. This ability to cure *in situ* makes it a desirable compound to be used in tissue repair, as well as tissue engineering.

### 17.3.2 SYNTHETIC BIODEGRADABLE POLYMERS

Synthetic polymers are different from natural polymers in that they give researchers the ability to design and optimize the polymers' properties (degradation rate, elastic modulus, and tensile strength) to their specific needs (Gunatillake et al. 2006). They are also often cheaper than natural polymers, as they can be synthesized in large quantities that are more uniform and last for longer periods of time. Synthetic biodegradable polymers are the predominant variety for tissue engineering. This is due to the above-discussed properties as well as their flexible processing abilities and ability to avoid immunological responses. The flexible processing abilities include mimicking the properties of extracellular matrix (nano- and micro-features) (Guo and Ma 2014).

#### 17.3.2.1 Structural Polymers

The most common of these polymers are aliphatic polyesters such as polylactide (Cleary et al. 1992), polyglycolide (PGA), and PLGA (Ma 2004). These polymers undergo controlled degradation through hydrolysis of their backbones' ester groups. This degradation is controlled by the structure, composition, and molecular weight of the polymer (Guo and Ma 2014). PGA is one of the first of these polymers to be used. It is a highly crystalline, rapidly degrading polymer that is widely used in tissue engineering. PLA is a much more mechanically stable polymer than PGA because of its greater hydrophobicity. Poly($\varepsilon$-caprolactone) is another popular aliphatic polyester used for tissue engineering. This polymer degrades very slowly compared to the others discussed previously, causing it to be more prevalent in long-term implants and drug delivery systems and not in short-term tissue repair/regeneration. Other classes of structural polymers include polyanhydrides, polyphosphazenes, and polyurethanes.

Synthetic polymers are being increasingly used for orthopedics, particularly for implants that are not permanent (Middleton and Tipton 2000). The main materials for such polymers are PLA and PGA. Their synthesis, degradation, and composition are essential for success in the human body. Autografts, allografts, and nondegradable polymers are some materials that are used for bone replacement (Middleton and Tipton 2000). These polymers are not only similar in characteristic to the human body but also have the ability to induce tissue growth and mold cell function (Hofmann 1992).

### 17.3.2.2 Functional Polymers

As mentioned in the previous section concerning fabrication methods, polymers such as PGA, PLA, and PLGA are predominantly structural polymers that interact minimally with the *in vivo* environment therein. There is a separate class of synthetic polymers that interact with the cellular environment in different ways. Certain polymers contain electrical and optical properties that allow them to have an effect on certain cellular activities. These activities may include cell-to-cell adhesion, cell differentiation, and cell proliferation. This ability to control the activities of the cellular environment around the scaffold makes them a very appealing candidate for tissue engineering. They are limited, however, by their mechanical properties, making composites of these conducting polymers and other synthetic biodegradable polymers an important area of study. Guo and coworkers have conducted many such studies and combined such conducting polymers with aliphatic polyesters (PLA, PCL) to create an architecturally diverse group of molecules. These include linear, star-shaped, hyperbranched, and cross-linked hybrids (Guo et al. 2010a, b, 2011a–e).

Another major group of functional synthetic polymers is those that are synthesized in a way that allows them to interact with enzymes in the cellular environment. Interaction with enzymes is desirable because it can act as a means to control the degradation rate of that particular scaffold. In order to infuse this enzymatic activity, poly(ethylene glycol) (PEG) is added to the scaffold. This addition has shown to be effective in inducing degradation by specific ECM enzymes (Guo and Ma 2014). Such replication of the ECM is what is desired in tissue engineering scaffolds, making this a promising group of scaffolds.

### 17.3.3 SYNTHETIC NON-BIODEGRADABLE POLYMERS

These biomaterials can form by the reaction between physiological fluids and bioactive ceramics and glasses (Dhandayuthapani et al. 2011). They lack potential in clinical use due to low biodegradability. This can be improved by blending natural polymers with these synthetic materials so that biocompatibility as well as biodegradability can be greatly increased (Ciardelli et al. 2005).

Ozawa et al. (2002) conducted experiments with polymers using the non-biodegradable polymer polytetrafluoroethylene (PTFE) as the control. This experiment found that the PTFE patch did not change size or shape. The only quality of the PTFE patch that changed was its thickness. This increased depth was due to the fibroblast and collagen on either side of the patch.

Further experimentation has been done using non-biodegradable polymers and patients with diabetes mellitus. Wiemer et al. (2015) tested the efficacy of a DES (third-generation drug-eluting stent) that had a biodegradable polymer. The biodegradable polymer was contained within the population of diabetes mellitus. The results of the study concluded that candidates with diabetes mellitus had excellent outcomes while candidates with insulin-dependent diabetes mellitus had an adverse reaction (Stein et al. 1995). This comes down to the treatment being specialized for what type of diabetes mellitus is present (Gollop et al. 2013).

## 17.4    TYPES OF TISSUE TARGETED

As mentioned before, polymers can be developed and shaped into architectures that make the scaffold applicable for the desired tissue. These distinctions can be made through choosing the types of polymers and fabrication techniques to be used. There are many different scaffolds that can be used for different types of tissues.

### 17.4.1    BONE TISSUE ENGINEERING

There is a great need for the development for novel methods for the regeneration of bone tissue. This is because traditional methods such as autografts, allografts, xenografts, and other substitutes are associated with many adverse effects (Goldberg 1992; Costantino and Friedman 1994). For example, the dependency of bone grafts on diffusion causes them to be limited by the size of the defect and host bed viability (Burg et al. 2000). Autografting techniques are also limited by the scarcity of donor tissue, while allografting is associated with risk of infection. Bone tissue engineering shows promise in treating many skeletal disorders while also avoiding the negative properties associated with traditional treatments.

The ability of highly porous poly(L-lactic acid) (PLLA) (Ma et al. 2001) and PLLA/hydroxyapatite (HAP) (Dhandayuthapani et al. 2011) scaffolds was studied to generate bone tissue *in vitro*. Osteoblast cells were seeded on these scaffolds and results showed that they were able to penetrate deeper and uniformly spread in the PLLA/HAP scaffold than in the PLLA scaffolds. This uniform distribution allowed for the tissue to grow continuously in the PLLA/HAP scaffolds whereas it only grew near the surface in the PLLA scaffolds. These results show the importance of osteoconduction, which is the ability to grow capillaries and cells from host into three-dimensional (3D) structure that forms bone (Burg et al. 2000). HAP is able to enhance the osteoconductive ability of the scaffold, thereby allowing the tissue to grow in a much more uniform manner. This osteoconductivity of HAP is thought to stem from its similarity to the mineral composition of the bone (LeGeros 2002). Wei et al. were then able to incorporate nano-size HAP particles into porous PLLA scaffolds and found that the nano-hydroxyapatite (NHAP)/polymer composite scaffolds were superior for bone tissue engineering (Wei and Ma 2004).

Polycaprolactone (PCL) is another polymer that has been focused on for bone tissue engineering because of its sustained biodegradability, elasticity, and low inflammatory response (Zhang et al. 2014). Zhang et al. developed a method to test the effect that HAP had on the cellular behavior of human fetal osteoblasts. This was done by preparing HAP/PCL composites with different weight ratios and examining the effect these had on alkaline phosphatase activity. It was found that the 1:4 weight ratio of HAP/PCL showed the highest level of ALP mineralization, thereby being optimal best composition for bone tissue regeneration.

Alothman et al. (2013) studied the effects that gamma radiation and accelerated aging had on such polymer/inorganic particle composite scaffolds. This was done by irradiating high density polyethelene (HDPE)/HAP scaffolds and accelerating aging through incubation in saline solution at 80° for 4 weeks. Results showed that radiation improved the properties of the composites whereas the aging process did significant damage to the composites.

These examples show how composites of highly porous polymers and inorganic materials such as HAP can be incredibly useful in the field of bone tissue engineering because of their high osteoconductivity and mechanical strength.

### 17.4.2 SKIN TISSUE ENGINEERING

The regeneration of skin tissue is a highly studied area of tissue engineering. Injuries such as burns may leave permanent scars on skin, making tissue engineering a desirable method of treatment. Early attempts at skin tissue engineering attempted to use sheets of cultured keratinocytes (KCs) (O'Connor et al. 1981). This method was unsuccessful because the sheets did not have the mechanical strength for proper grafting. This mechanical strength can be provided by the use of polymer scaffolds. Properties that epidermal scaffolds should have include

1. Mechanical strength that allows transfer of the graft from culture dish to wound
2. Favorable, biocompatible environment for KC growth

Many such scaffolds have been used that range in composition from purely natural-to-natural-synthetic blends. Natural polymer scaffolds include those made of collagen (Powell and Boyce 2009; Kempf et al. 2011) cellulose, alginate, and even glycosaminoglycans (Bačáková et al. 2014). Possible hybrid scaffold combinations include gelatin and PCL (Chong et al. 2007; Duan et al. 2013), PCL and PLGA, chitosan and gelatin in hydrogel form (Franco et al. 2011) polyglycolic acid (PGA) and gelatin, and many others.

Kempf et al. (2011) studied the ability of large pore bovine collagen to act as a scaffold for human KCs and fibroblast cell scaffolds. Results showed that in long-term co-cultures of the fibroblasts and KC, fibroblasts were able to penetrate into and spread throughout the scaffold, while the KCs were able to form stratified layers on top of the scaffold, therefore, forming a human skin equivalent structure.

Gelatin has been shown to improve the biodegradability and biocompatibility of polymer scaffolds. Feng et al. (2012) developed gelatin/PCL composite scaffolds (50:50 composition) with small amounts of acetic acid that improved the mechanical properties of these electrospun scaffolds. The biocompatible properties of these scaffolds were then studied by seeding them with human KCs, for which results were also positive (Duan et al. 2013). *In vivo* testing of this model showed the ability of the scaffold to heal the injury in 14 days, showing that this is a very suitable scaffold for epidermis engineering. These studies show how natural and natural/synthetic composite scaffolds can be used to develop treatments for various epidermal injuries.

### 17.4.3 CARDIOVASCULAR TISSUE ENGINEERING

Cardiovascular tissue engineering is an incredibly import field of study, as cardiovascular failures are known to be some of the leading causes of death around the world.

Heart valve diseases are one of the most prevalent cardiovascular disorders that cause its high mortality. Possible treatment includes prosthetic valve implantation with the use of a mechanical prosthetic valve or a bioprosthetic heart valve (Zhou et al. 2013). Mechanical prosthetics have potential complications such as thrombus formation, endocarditis, or even hemorrhage (Filova et al. 2009). Bioprosthetic heart valves may result in immune reactions that complicate surgical procedures even more while endangering the patient. Therefore, tissue-engineered heart valves show great promise in treating heart valve disorders while avoiding the complications related to prosthetic heart valves.

Possible methods to tissue engineer heart valves include the acellular matrix xenograft, bioresorbably scaffold, collagen-based constructs, elastin-based constructs, fibronectin-based constructs, and fibrin-based constructs (Vesely 2005; Filova et al. 2009). Type-I collagen scaffolds, when prepared from decellularized porcine pericardium treated with collagen-stabilizing agent, penta-galloyl glucose, were able to form *in vivo* with great mechanical properties and cellular penetration, while avoiding calcification (Tedder et al. 2008). Such scaffolds may be of great use in cardiovascular disease treatment.

Tissue therapy for organs is limited in the cardiovascular field. Hinderer et al. (2015) tested replacement therapies including the use of generation methods (myocardium, heart valves, and blood vessels). These methods involve the heart and the myocardial tissue. These have been further tested by Cebotari and coworkers. Their findings included using artificial and biological scaffolds for alternatives for reconstruction. This reconstruction includes repairment, replacement, and regeneration of the damaged tissue (Cebotari et al. 2010). Akhyari et al. (2011) also tested the use of cardiac tissue engineering. The limitations found included long-term durability and the long-term need for anticoagulation therapy. These issues for tissue engineering are present in not just cardiac engineering but all tissue engineering.

### 17.4.4 Neural Tissue Engineering

Axonal regeneration in the nervous system, especially the peripheral, can occur spontaneously by the body. However, the functionality of these regenerated axons is very poor compared to that of normally functioning neurons. Here again, collagen shows potential as a scaffolding material (Chen et al. 2003; Michelini et al. 2006; Führmann et al. 2010). 3D collagen scaffolds were seeded with human neural progenitor-derived astrocytes (Führmann et al. 2010). The porous structure of the scaffold allowed for substantial intermixing of the astrocytes with the co-cultured fibroblasts. This enhanced cell interaction revealed that the astrocytes promoted significant neurite outgrowth. Other possible scaffolding materials include hyaluronan (Seckel et al. 1995; Horn et al. 2007; Zhang et al. 2008; Khaing et al. 2011; Khaing and Schmidt 2012), fibrin (Bhang et al. 2007; Johnson et al. 2010), and agarose (Stokols and Tuszynski 2006).

Neural tissue engineering has also been used for the treatment of spinal cord injury. This has been researched by Liu et al. (2014). The imperative qualities for spinal cord treatment with tissue engineering include a suitable cell source and the scaffold. Liu and associates used mouse embryonic fibroblasts to directly reprogram into stem cells, neural (iNSCs). The qualities tested of the iNSCs included cell growth, survival, and proliferation. The result was that iNSCs had similar features of biological neural stem cells, wild type, and had a variety of neural stem cell marker genes. Consequently, the iNSCs survived and were able to undergo differentiation. Further investigation of spinal injury has been researched by Kang et al. (2011).

## 17.5 POLYMER SCAFFOLDS AS CARRIERS FOR DRUGS

Cell carrying polymers are integral to the field of tissue engineering. In addition to this, polymer scaffolds can also act as carriers for drugs that can aid the tissue regeneration process. In recent years, nanopolymer scaffolds have shown great promise as possible drug delivery systems for the treatment of various disorders. Scaffolds are injected into the body to deliver a particular drug for targeted disease treatment with high efficiency, high load capacity, and sustained release. There are multiple types of polymeric scaffolds available for drug delivery. These include: (1) nanofibrous matrix, (2) a typical 3D porous matrix, (3) porous microsphere, and (4) thermo-sensitive sol–gel transition hydrogel. As stated in the previous section, fabrication of a scaffold can be from natural or synthetic polymers. Possible synthetic polymers include polyvinyl alcohol and PGA and possible natural polymer include collagens, gelatin, fibrins, proteins, and alginate. Several different methods or techniques can be used to fabricate scaffolds, including supercritical fluid technology, leaching, thermally induced phase separation, freeze drying, sol–gel, melt molding, powder compaction, and rapid prototyping (Garg et al. 2012). Recent discoveries indicate that incorporation of drug into polymer scaffolds can also prevent infection after surgery.

Drug delivery scaffolds are 3D porous structures capable of highly efficient and sustained drug delivery (Garg et al. 2012). Drug delivery scaffolds should contain these following features:

1. Ability to release the drug at desired rate
2. Equal drug distribution throughout the scaffold
3. Stable chemical structure, biological activity, and physical dimensions over prolonged periods of time (Romagnoli et al. 2013)
4. High loading capacity to allow for prolonged release after insertion
5. Low binding affinity to drug to allow for effective release

### 17.5.1 MATERIAL USED FOR DRUG CARRYING SCAFFOLDS

There are many different types of biomaterials used to manufacture both synthetic and natural polymers scaffold for drug delivery. These are similar to the materials used for cell carrying scaffolds. Tables 17.1 and 17.2 list some of the advantages and disadvantages of the most commonly used synthetic and natural polymers. Table 17.1 highlights the characteristics of synthetic polymers, while Table 17.2 highlights the characteristics of natural polymers.

### 17.5.2 APPLICATIONS OF DRUG DELIVERY SCAFFOLDS

As mentioned, drugs may be added to scaffolds in order to accelerate tissue regeneration or the scaffold may act as a drug delivery system for prolonged release of the particular drug.

**TABLE 17.1**

**Advantages and Disadvantages of Using Synthetic Polymers for Scaffold Fabrication**

| Polymer | Advantages | Disadvantages | Reference |
|---|---|---|---|
| Poly(lactic acid) | Functions accurately without harming other tissue<br>Great ability for controlled biodegradation | Inflammatory response possibility<br>Compacting strength | Sokolsky-Papkov et al. (2007) |
| PAA | Highly hydrophilic<br>Nontoxic<br>Versatile | High degradation | Romagnoli et al. (2013) |
| N-Succinimidyl tartarate mono-amine | Biocompatible<br>Easier to control surface composition | Hard to process | |
| Polydepsipeptide | High level of function control<br>Controlled interaction with cells | Low stability over long periods of time | Romagnoli et al. (2013) |
| Polyphosphazene | Effective function without affecting other tissues | Low mechanical strength | |

## 17.5.2.1 Drug Carriers that Enhance Tissue Regeneration

Tissue regeneration is a very complex field of study that has grown rapidly over the past three decades. The earlier focus in this field surrounded the development of scaffolds and the ability of these scaffolds to act as carriers for cells that can proliferate into a regenerated tissue. To achieve this goal, the scaffolds were fabricated in such a manner that they replicated the properties of the ECM. However, this may not be a sufficient treatment in certain cases, which is why the ability of scaffolds to act as carriers for drugs is gaining more traction in the field. The ability to carry growth factors for accelerated cell proliferation and drugs for the treatment of certain diseases greatly increases the functionality of the scaffolds.

A possible cause for lost/damaged tissue that needs to be replaced is microbial infection. Furthermore, the possibility of complications following device implantation is high and the ability of the immune system to fight these infections is often a determining factor of the efficacy of the particular implant (Zilberman and Elsner 2008). Before tissue regeneration can take place, reduction of the pathogen concentration in the area is required, which can be accomplished through the use of antibiotics with the scaffold. Polymer scaffolds are the ideal delivery method for these antibiotics because of their ability to provide a targeted and controlled release of therapeutic concentrations of the drug, while also avoiding toxic effects caused by systemic administration.

Bone tissue engineering implants widely experience such issues. The most commonly used antibiotic for bone trauma is gentamicin. This is because of its broad-spectrum characteristics (Mouriño and Boccaccini 2009). This antibiotic has been

**TABLE 17.2**
**Advantages and Disadvantages of Using Natural Polymers for Scaffold Fabrication**

| Polymer | Advantages | Disadvantages | Reference |
|---|---|---|---|
| Gelatin | Low antigenicity<br>Easier to process into a variety of shapes | Poor mechanical properties | Liu et al. (1998) |
| Collagen | Low antigenicity<br>High biocompatibility<br>High mechanical strength | High costs<br>Difficult to process | Mikos et al. (1993) |
| Soy | Low costs<br>Renewable<br>Environment friendly | Low usage in research | Baker et al. (2008) |
| Elastin | Controlled degradation | Insoluble at high temperatures | |
| Chitosan | Low costs<br>Nontoxic | | |
| Starch | Low costs<br>Renewable | Difficult to process | Nam and Park (1999) |
| Alginate | Resistance to acidic condition<br>Effective function without harm to other tissue | Drug loss due to leaching | Lahner et al. (2014) |
| Hyaluronic acid | Effective function without harm to other tissue<br>Easy to produce in large quantities | Weak mechanical properties | Nam and Park (1999) |
| Dextran | Effective function without harm to other tissue<br>Available in many molecular weights<br>Low aggregation | High costs | Martins et al. (2008) |
| Carrageenans | Flexible | High melting points | |
| Gellan gum | Resistant to high temperature<br>Resistant to low pH | | Garg et al. (2012) |

reported to be nephrotoxic and ototoxic when administered systemically, which is an issue that is avoided by scaffold-based delivery systems, as they allow for effective local administration (Smith et al. 1980). Its implementation in a PLGA composite scaffold to combat *Staphylococcus aureus* infections on open fractures was studied in an ovine medial femoral condyle model (McLaren et al. 2014). Results revealed a controlled release of gentamicin with effective infection prevention for at least 7 days. Gentamicin was then combined with other antibiotics, which revealed that such a combined administration released the antibiotics for 19–21 days *in vitro*. This controlled release was sufficient in allowing the growth of new bone tissue on the scaffold for a 13-week period, showing that successful antimicrobial activity took place.

In many forms of scaffolds, the matrix is usually prepared by using biodegradable polymers and inclusions, in the form of fibers, of HA and TCP to improve the mechanical strength and bioactivity (Roether et al. 2002). Furthermore, over the past few decades, dozens of release dosage forms have been developed for drug delivery. However, one common problem is the existence of a large burst over a narrow time period during the early stages of release (Nie et al. 2008). As a strategy to solve that issue, research has been conducted with fibrous PLGA/nanoparticles HA (Cunha-Vaz 1997) composite scaffold as a better release dosage form because of its favorable properties and morphology (Wei et al. 2006). Compacted fibrous scaffold, compared with microsphere, can give the cell stable 3D growth environment and provide a good support to the new generated bone (Nie et al. 2008).

### 17.5.2.2   Application in Cancer Treatment

Cancer is a leading cause of death since the beginning of the twenty-first century. There are hundreds of techniques that attempt to overcome cancer. Nanopolymer scaffolds show promise as carriers for drug to target, particularly cancer cells while not affecting normal cells. Recent research has shown that the use of polymer scaffold-based drug delivery systems can increase efficiency while decreasing the side effects of anticancer drugs. Polymeric scaffold were tested as carrier vehicles for the anticancer drug doxorubicin (Chun et al. 2009). A polymeric scaffold with L-isoleucine ethyl ester (IleOEt), glycine glycine allyl ester (GlyGlyOAll), and α-amino-ω-methoxy-poly(ethylene glycol) substituents was synthesized and subsequently conjugated with doxorubicin (DOX) through the pendant carboxylic acid groups after removing the allyl protecting groups on glycine. These materials were shown to be injectable as a solution and precipitate into a gel material upon heating, which is suitable for targeted drug delivery applications as it maintains the drug in the desired location, especially tumor sites (Baillargeon and Mequanint 2014). The material was tested *in vitro* for degradation properties, drug (doxorubicin) release profile, and anti-tumor activity. Release profile and degradation studies were performed in PBS at 37°C over a period of 30 days. The mass loss after 30 days was approximately 60% of the original mass and the molecular weight decline was slightly less than 40% of the original. The doxorubicin release profile demonstrated a sustained release of the drug, which is ideal for most drug delivery techniques and polymeric scaffold founded as suitable drug carrier for cancer treatment (Chun et al. 2009).

### 17.5.2.3   Application in Treating Spinal Cord Injury

There are many challenges in treating spinal cord injury because of its complex structure. A recent study has found a way to treat spinal cord injury most efficiently via polymeric scaffold drug delivery system to deliver drug efficiently at the damaged site. The current practice in treating spinal cord injury involves large dose and sometimes results in side effects. Recently developed drug delivery system for the treatment of spinal cord injury showed extreme efficiency by targeting a specific part of the body and limiting side effects and loss of drugs in spinal cord injury treatment with faster recovery (Bracken et al. 1997).

## 17.6 CONCLUSIONS AND FURTHER RESEARCH

As stated, polymeric scaffolds are an essential component in the field of tissue engineering. They act as sites for cellular attachment and proliferation by mimicking the properties of the ECM of the particular tissue being targeted. Many different types of polymers are in use for this purpose. Synthetic biodegradable polymers are the most widely used but 3D self-assembled scaffolds show a great potential due to their ability to replicate the properties of the ECM. In order to aid the tissue repair process, these scaffolds are able to act as carriers for drugs and growth factors. They show great potential as drug delivery systems because of their ability to deliver sufficient therapeutic concentrations for sustained periods of time. These advantages make polymer scaffolds exciting technologies that could greatly improve human health in the future.

## REFERENCES

Akhyari, P., Minol, P., Assmann, A., Barth, M., Kamiya, H., and Lichtenberg, A. 2011. Tissue engineering of heart valves. *Der Chirurg* 82:311–18.

Alothman, O.Y., Almajhdi, F.N., and Fouad, H. 2013. Effect of gamma radiation and accelerated aging on the mechanical and thermal behavior of HDPE/HA nano-composites for bone tissue regeneration. *Biomed Eng Online* 12:95–109.

Bačáková, L., Novotná, K., and Pařízek, M. 2014. Polysaccharides as cell carriers for tissue engineering: The use of cellulose in vascular wall reconstruction. *Physiol Res* 63:29–47.

Baillargeon, A.L. and Mequanint, K. 2014. Biodegradable polyphosphazene biomaterials for tissue engineering and delivery of therapeutics. *BioMed Res Int* 2014:1–16.

Baker, B.M., Gee, A.O., Metter, R.B. et al. 2008. The potential to improve cell infiltration in composite fiber-aligned electrospun scaffolds by the selective removal of sacrificial fibers. *Biomaterials* 29:2348–58.

Bhang, S.H., Lee, Y.E., Cho, S.-W. et al. 2007. Basic fibroblast growth factor promotes bone marrow stromal cell transplantation-mediated neural regeneration in traumatic brain injury. *Biochem Biophys Res Commun* 359:40–45.

Bracken, M.B., Shepard, M.J., Holford, T.R. et al. 1997. Administration of methylprednisolone for 24 or 48 hours or tirilazad mesylate for 48 hours in the treatment of acute spinal cord injury: Results of the Third National Acute Spinal Cord Injury Randomized Controlled Trial. *JAMA* 277:1597–604.

Burg, K.J., Porter, S., and Kellam, J.F. 2000. Biomaterial developments for bone tissue engineering. *Biomaterials* 21:2347–59.

Cebotari, S., Tudorache, I., Schilling, T., and Haverich, A. 2010. Heart valve and myocardial tissue engineering. *Herz* 35:334–41.

Chen, G., Sato, T., Ohgushi, H., Ushida, T., Tateishi, T., and Tanaka, J. 2005. Culturing of skin fibroblasts in a thin PLGA–collagen hybrid mesh. *Biomaterials* 26:2559–66.

Chen, S.S., Revoltella, R.P., Papini, S. et al. 2003. Multilineage differentiation of rhesus monkey embryonic stem cells in three-dimensional culture systems. *Stem Cells* 21:281–95.

Chong, E., Phan, T., Lim, I. et al. 2007. Evaluation of electrospun PCL/gelatin nanofibrous scaffold for wound healing and layered dermal reconstitution. *Acta Biomater* 3:321–30.

Chun, C., Lee, S.M., Kim, C.W. et al. 2009. Doxorubicin–polyphosphazene conjugate hydrogels for locally controlled delivery of cancer therapeutics. *Biomaterials* 30:4752–62.

Ciardelli, G., Chiono, V., Vozzi, G. et al. 2005. Blends of poly-(ε-caprolactone) and polysaccharides in tissue engineering applications. *Biomacromolecules* 6:1961–76.

Cleary, P., Schlievert, P., Handley, J. et al. 1992. Clonal basis for resurgence of serious *Streptococcus pyogenes* disease in the 1980s. *Lancet* 339:518–21.

Costantino, P.D. and Friedman, C.D. 1994. Synthetic bone graft substitutes. *Otolaryngol Clin North Am* 27:1037–74.

Cunha-Vaz, J.G. 1997. The blood-ocular barriers: Past, present, and future. *Documenta Ophthalmol* 93:149–57.

Dhandayuthapani, B., Yoshida, Y., Maekawa, T., and Kumar, D.S. 2011. Polymeric scaffolds in tissue engineering application: A review. *Int J Polym Sci* 2011:1–19.

Duan, H., Feng, B., Guo, X. et al. 2013. Engineering of epidermis skin grafts using electrospun nanofibrous gelatin/polycaprolactone membranes. *Int J Nanomed* 8:2077–84.

Ekaputra, A.K., Prestwich, G.D., Cool, S.M., and Hutmacher, D.W. 2008. Combining electrospun scaffolds with electrosprayed hydrogels leads to three-dimensional cellularization of hybrid constructs. *Biomacromolecules* 9:2097–103.

Escudero, D. and Otero, J. 2015. Intensive care medicine and organ donation: Exploring the last frontiers? *Medicina Intensiva* 39:373–81.

Feng, B., Tu, H., Yuan, H., Peng, H., and Zhang, Y. 2012. Acetic-acid-mediated miscibility toward electrospinning homogeneous composite nanofibers of GT/PCL. *Biomacromolecules* 13:3917–25.

Filova, E., Straka, F., Mirejovský, T., Mašín, J., and Bačáková, L. 2009. Tissue-engineered heart valves. *Physiol Res* 58:141–58.

Franco, R.A., Nguyen, T.H., and Lee, B.-T. 2011. Preparation and characterization of electrospun PCL/PLGA membranes and chitosan/gelatin hydrogels for skin bioengineering applications. *J Mater Sci Mater Med* 22:2207–18.

Freed, L.E., Vunjak-Novakovic, G., Biron, R.J. et al. 1994. Biodegradable polymer scaffolds for tissue engineering. *Biotechnology (N Y)* 12:689–93.

Führmann, T., Hillen, L.M., Montzka, K., Wöltje, M., and Brook, G.A. 2010. Cell–cell interactions of human neural progenitor-derived astrocytes within a microstructured 3D-scaffold. *Biomaterials* 31:7705–15.

Garg, T., Singh, O., Arora, S., and Murthy, R. 2012. Scaffold: A novel carrier for cell and drug delivery. *Crit Rev Ther Drug Carrier Syst* 29:1–63.

Goldberg, M., Langer, R., and Jia, X. 2007. Nanostructured materials for applications in drug delivery and tissue engineering. *J Biomater Sci Polym Ed* 18:241–68.

Goldberg, V.M. 1992. Natural history of autografts and allografts. In *Bone Implant Grafting*, ed. M.W.J. Older, 9–12, London: Springer.

Gollop, N.D., Henderson, D.B., and Flather, M.D. 2013. Comparison of drug-eluting and bare-metal stents in patients with diabetes undergoing primary percutaneous coronary intervention: What is the evidence? *Interact Cardiovasc Thorac Surg* 18:112–16.

Gunatillake, P., Mayadunne, R., and Adhikari, R. 2006. Recent developments in biodegradable synthetic polymers. *Biotechnol Annu Rev* 12:301–47.

Guo, B., Finne-Wistrand, A., and Albertsson, A.-C. 2010a. Enhanced electrical conductivity by macromolecular architecture: Hyperbranched electroactive and degradable block copolymers based on poly (ε-caprolactone) and aniline pentamer. *Macromolecules* 43:4472–80.

Guo, B., Finne-Wistrand, A., and Albertsson, A.-C. 2010b. Molecular architecture of electroactive and biodegradable copolymers composed of polylactide and carboxyl-capped aniline trimer. *Biomacromolecules* 11:855–63.

Guo, B., Finne-Wistrand, A., and Albertsson, A.-C. 2011a. Degradable and electroactive hydrogels with tunable electrical conductivity and swelling behavior. *Chem Mater* 23:1254–62.

Guo, B., Finne-Wistrand, A., and Albertsson, A.-C. 2011b. Facile synthesis of degradable and electrically conductive polysaccharide hydrogels. *Biomacromolecules* 12:2601–09.

Guo, B., Finne-Wistrand, A., and Albertsson, A.-C. 2011c. Simple route to size-tunable degradable and electroactive nanoparticles from the self-assembly of conducting coil-rod–coil triblock copolymers. *Chem Mater* 23:4045–55.

Guo, B., Finne-Wistrand, A., and Albertsson, A.-C. 2011d. Universal two-step approach to degradable and electroactive block copolymers and networks from combined ring-opening polymerization and post-functionalization via oxidative coupling reactions. *Macromolecules* 44:5227–36.

Guo, B., Finne-Wistrand, A., and Albertsson, A.-C. 2011e. Versatile functionalization of polyester hydrogels with electroactive aniline oligomers. *J Polym Sci A Polym Chem* 49:2097–105.

Guo, B. and Ma, P.X. 2014. Synthetic biodegradable functional polymers for tissue engineering: A brief review. *Sci China Chem* 57:490–500.

Hartgerink, J.D., Beniash, E., and Stupp, S.I. 2001. Self-assembly and mineralization of peptide-amphiphile nanofibers. *Science* 294:1684–88.

Hinderer, S., Brauchle, E., and Schenke-Layland, K. 2015. Generation and assessment of functional biomaterial scaffolds for applications in cardiovascular tissue engineering and regenerative medicine. *Adv Healthc Mater* 4:2326–41.

Hofmann, G. 1992. Biodegradable implants in orthopaedic surgery—A review on the state-of-the-art. *Clin Mater* 10:75–80.

Horn, E.M., Beaumont, M., Shu, X.Z. et al. 2007. Influence of cross-linked hyaluronic acid hydrogels on neurite outgrowth and recovery from spinal cord injury. *J Neurosurg Spine* 6:133–40.

Johnson, P.J., Tatara, A., McCreedy, D.A., Shiu, A., and Sakiyama-Elbert, S.E. 2010. Tissue-engineered fibrin scaffolds containing neural progenitors enhance functional recovery in a subacute model of SCI. *Soft Matter* 6:5127–37.

Kang, K.N., Lee, J.Y., Kim, D.Y. et al. 2011. Regeneration of completely transected spinal cord using scaffold of poly (D,L-lactide-*co*-glycolide)/small intestinal submucosa seeded with rat bone marrow stem cells. *Tissue Eng A* 17:2143–52.

Kempf, M., Miyamura, Y., Liu, P.-Y. et al. 2011. A denatured collagen microfiber scaffold seeded with human fibroblasts and keratinocytes for skin grafting. *Biomaterials* 32:4782–92.

Khaing, Z.Z., Milman, B.D., Vanscoy, J.E., Seidlits, S.K., Grill, R.J., and Schmidt, C.E. 2011. High molecular weight hyaluronic acid limits astrocyte activation and scar formation after spinal cord injury. *J Neural Eng* 8:046033.

Khaing, Z.Z. and Schmidt, C.E. 2012. Advances in natural biomaterials for nerve tissue repair. *Neurosci Lett* 519:103–14.

Kwon, I.K. and Matsuda, T. 2005. Co-electrospun nanofiber fabrics of poly (L-lactide-*co*-ε-caprolactone) with type I collagen or heparin. *Biomacromolecules* 6:2096–105.

Lahner, M., Kalwa, L., Olbring, R., Mohr, C., Göpfert, L., and Seidl, T. 2014. Biomimetic structured surfaces increase primary adhesion capacity of cartilage implants. *Technol Health Care* 23:205–13.

Langer, R. and Peppas, N.A. 2003. Advances in biomaterials, drug delivery, and bionanotechnology. *AIChE J* 49:2990–3006.

Langer, R. and Tirrell, D.A. 2004. Designing materials for biology and medicine. *Nature* 428:487–92.

LeGeros, R.Z. 2002. Properties of osteoconductive biomaterials: Calcium phosphates. *Clin Orthopaed Relat Res* 395:81–98.

Li, Y., Meng, H., Liu, Y., and Lee, B.P. 2015. Fibrin gel as an injectable biodegradable scaffold and cell carrier for tissue engineering. *Sci World J* 2015:1–10.

Liu, C., Huang, Y., Pang, M. et al. 2014. Tissue-engineered regeneration of completely transected spinal cord using induced neural stem cells and gelatin-electrospun poly (lactide-*co*-glycolide)/polyethylene glycol scaffolds. *PloS One* 10:e0117709.

Liu, Q., de Wijn, J.R., and van Blitterswijk, C.A. 1998. Composite biomaterials with chemical bonding between hydroxyapatite filler particles and PEG/PBT copolymer matrix. *J Biomed Mater Res* 40:490–97.

Ma, P.X. 2004. Scaffolds for tissue fabrication. *Mater Today* 7:30–40.

Ma, P.X., Zhang, R., Xiao, G., and Franceschi, R. 2001. Engineering new bone tissue in vitro on highly porous poly(α-hydroxyl acids)/hydroxyapatite composite scaffolds. *J Biomed Mater Res* 54:284–93.

Martins, A.M., Pham, Q.P., Malafaya, P.B. et al. 2008. The role of lipase and α-amylase in the degradation of starch/poly (ε-Caprolactone) fiber meshes and the osteogenic differentiation of cultured marrow stromal cells. *Tissue Eng A* 15:295–305.

McLaren, J., White, L., Cox, H. et al. 2014. A biodegradable antibiotic-impregnated scaffold to prevent osteomyelitis in a contaminated in vivo bone defect model. *Eur Cell Mater* 27:332–49.

Michelini, M., Franceschini, V., Sihui Chen, S. et al. 2006. Primate embryonic stem cells create their own niche while differentiating in three-dimensional culture systems. *Cell Proliferation* 39:217–29.

Middleton, J.C. and Tipton, A.J. 2000. Synthetic biodegradable polymers as orthopedic devices. *Biomaterials* 21:2335–46.

Mikos, A.G., Sarakinos, G., Leite, S.M., Vacant, J.P., and Langer, R. 1993. Laminated three-dimensional biodegradable foams for use in tissue engineering. *Biomaterials* 14:323–30.

Mikučionienė, D., Čiukas, R., and Mickevičienė, A. 2010. The influence of knitting structure on mechanical properties of weft knitted fabrics. *Mater Sci (Medžiagotyra)* 16:221–25.

Mogoşanu, G.D. and Grumezescu, A.M. 2014. Natural and synthetic polymers for wounds and burns dressing. *Int J Pharm* 463:127–36.

Mouriño, V. and Boccaccini, A.R. 2009. Bone tissue engineering therapeutics: controlled drug delivery in three-dimensional scaffolds. *J R Soc Interface* 7:209–27.

Nam, Y.S. and Park, T.G. 1999. Biodegradable polymeric microcellular foams by modified thermally induced phase separation method. *Biomaterials* 20(19):1783–1790.

Nie, H., Soh, B.W., Fu, Y.C., and Wang, C.H. 2008. Three-dimensional fibrous PLGA/HAp composite scaffold for BMP-2 delivery. *Biotechnol Bioeng* 99:223–34.

O'Connor, N., Mulliken, J., Banks-Schlegel, S., Kehinde, O., and Green, H. 1981. Grafting of burns with cultured epithelium prepared from autologous epidermal cells. *Lancet* 317:75–78.

Ozawa, T., Mickle, D.A., Weisel, R.D. et al. 2002. Histologic changes of nonbiodegradable and biodegradable biomaterials used to repair right ventricular heart defects in rats. *J Thoracic Cardiovasc Surg* 124:1157–64.

Pang, Y. and Greisler, H.P. 2010. Using a type I collagen based system to understand cell-scaffold interactions and to deliver chimeric collagen binding growth factors for vascular tissue engineering. *J Investig Med* 58:845–48.

Powell, H.M. and Boyce, S.T. 2009. Engineered human skin fabricated using electrospun collagen–PCL blends: Morphogenesis and mechanical properties. *Tissue Eng A* 15:2177–87.

Rho, K.S., Jeong, L., Lee, G. et al. 2006. Electrospinning of collagen nanofibers: Effects on the behavior of normal human keratinocytes and early-stage wound healing. *Biomaterials* 27:1452–61.

Roether, J., Boccaccini, A.R., Hench, L., Maquet, V., Gautier, S., and Jérôme, R. 2002. Development and in vitro characterisation of novel bioresorbable and bioactive composite materials based on polylactide foams and Bioglass® for tissue engineering applications. *Biomaterials* 23:3871–78.

Romagnoli, C., D'Asta, F., and Brandi, M.L. 2013. Drug delivery using composite scaffolds in the context of bone tissue engineering. *Clin Cases Miner Bone Metab* 10:155–61.

Seckel, B.R., Jones, D., Hekimian, K., Wang, K.K., Chakalis, D., and Costas, P. 1995. Hyaluronic acid through a new injectable nerve guide delivery system enhances peripheral nerve regeneration in the rat. *J Neurosci Res* 40:318–24.

Smith, C.R., Lipsky, J.J., Laskin, O.L. et al. 1980. Double-blind comparison of the nephrotoxicity and auditory toxicity of gentamicin and tobramycin. *New Engl J Med* 302:1106–09.

Sokolsky-Papkov, M., Agashi, K., Olaye, A., Shakesheff, K., and Domb, A.J. 2007. Polymer carriers for drug delivery in tissue engineering. *Adv Drug Deliv Rev* 59:187–206.

Stein, B., Weintraub, W.S., Gebhart, S.S. et al. 1995. Influence of diabetes mellitus on early and late outcome after percutaneous transluminal coronary angioplasty. *Circulation* 91:979–89.

Stokols, S. and Tuszynski, M.H. 2006. Freeze-dried agarose scaffolds with uniaxial channels stimulate and guide linear axonal growth following spinal cord injury. *Biomaterials* 27:443–51.

Tamayol, A., Akbari, M., Annabi, N., Paul, A., Khademhosseini, A., and Juncker, D. 2013. Fiber-based tissue engineering: Progress, challenges, and opportunities. *Biotechnol Adv* 31:669–87.

Tedder, M.E., Liao, J., Weed, B. et al. 2008. Stabilized collagen scaffolds for heart valve tissue engineering. *Tissue Eng A* 15:1257–68.

Vesely, I. 2005. Heart valve tissue engineering. *Circ Res* 97:743–55.

Wang, X., Han, C., Hu, X. et al. 2011. Applications of knitted mesh fabrication techniques to scaffolds for tissue engineering and regenerative medicine. *J Mech Behav Biomed Mater* 4:922–32.

Wasiak, J., Cleland, H., Campbell, F., and Spinks, A. 2013. Dressings for superficial and partial thickness burns. *Cochrane Database Syst Rev* 4:CD002106.

Wei, G., Jin, Q., Giannobile, W.V., and Ma, P.X. 2006. Nano-fibrous scaffold for controlled delivery of recombinant human PDGF-BB. *J Control Release* 112:103–10.

Wei, G., Ma, P.X. 2004. Structure and properties of nano-hydroxyapatite/polymer composite scaffolds for bone tissue engineering. *Biomaterials* 25:4749–57.

Weidenbecher, M., Henderson, J.H, Tucker, H.M., Baskin, J.Z., Awadallah, A., and Dennis, J.E. 2007. Hyaluronan-based scaffolds to tissue-engineer cartilage implants for laryngotracheal reconstruction. *Laryngoscope* 117:1745–49.

Weinberg, C.B. and Bell, E. 1986. A blood vessel model constructed from collagen and cultured vascular cells. *Science* 231:397–400.

Wiemer, M., Danzi, G.B., West, N. et al. 2015. Drug-eluting stents with biodegradable polymer for the treatment of patients with diabetes mellitus: clinical outcome at 2 years in a large population of patients. *Medical Devices (Auckland, NZ)* 8:153–60.

Yamada, S., Yamamoto, K., Ikeda, T., Yanagiguchi, K., and Hayashi, Y. 2014. Potency of fish collagen as a scaffold for regenerative medicine. *BioMed Res Int* 2014:1–8.

Yannas, I.V. 2004. Classes of materials used in medicine. In *Biomaterials Science—An Introduction to Materials in Medicine*, eds. B. Ratner, A. Hoffman, F. Schoen, J. Lemons, 127–36. San Diego: Elsevier Academic Press.

Zhang, H., Wei, Y.T., Tsang, K.S. et al. 2008. Implantation of neural stem cells embedded in hyaluronic acid and collagen composite conduit promotes regeneration in a rabbit facial nerve injury model. *J Transl Med* 6:67–77.

Zhang, X., W. Chang, P. Lee et al. 2014. Polymer-ceramic spiral structured scaffolds for bone tissue engineering: Effect of hydroxyapatite composition on human fetal osteoblasts. *PloS One* 9:e85871.

Zhou, J., Hu, S., Ding, J., Xu, J., Shi, J., and Dong, N. 2013. Tissue engineering of heart valves: PEGylation of decellularized porcine aortic valve as a scaffold for in vitro recellularization. *Biomed Eng Online* 12:87.

Zilberman, M. and Elsner, J.J. 2008. Antibiotic-eluting medical devices for various applications. *J Control Release* 130:202–15.

# 18 Bio-Targets for Polyionic Glucan Derivatives and Their Nano-Therapeutic Systems

*Manabendra Dhua, Kalyan Kumar Sen, and Sabyasachi Maiti*

## CONTENTS

## 18.1 INTRODUCTION

Glucans are D-glucose polymers having unique physical properties and are produced by a wide variety of microorganisms. Glucans are high-molecular-weight (MW) microbial polysaccharides (Manjanna et al. 2009) and can be divided into three groups: exocellular, cell wall, and intercellular. The exocellular polysaccharides (EPS) are easy to isolate from cell culture medium because these are constantly diffused outside the cells. Other types of polysaccharides are integral parts of the cell wall or capsular products, and therefore difficult to separate from cell biomass.

Ionic or nonionic glucans are primarily linear polysaccharides of 3–7 different monosaccharides to which side chains of varying length and complexity are attached at regular intervals. Monosaccharides, such as pentoses and hexoses, amino sugars, or uronic acids are arranged in groups of 10 or less to form repeating units (gellan or xanthan) (Purama et al. 2009). In a homopolysaccharide repeating unit, sugar monomers are bound to form either linear chains (pullulan, levan, curdlan, or bacterial cellulose) or ramified chains (dextran). Despite close similarity of building blocks such as pyranose or furanose ring structures, they differ structurally and functionally.

Over the past few years, exopolysaccharides are being used in the food and pharmaceutical industries. These are also being investigated as drug delivery carriers for a variety of drug candidates. The bacterial species *Leuconostoc*, *Sphingomonas*, and *Alcaligenes* are the key sources for dextran, gellan, and curdlan, respectively. It is worth mentioning that the fermentation conditions, such as pH, temperature, oxygen concentration, and agitation and composition of the culture medium, are controlling factors for EPS production (Kazak et al. 2010; Nicolaus et al. 2010). Besides, the type of strain may also influence the chemical structure, monomer composition, and physicochemical and rheological properties of the final product. For example, the mesophilic strains produce maximum levels of EPS (Degeest et al. 2001). Microbial sources are preferred for the production of bioactive polysaccharides within days or weeks as opposed to plants. Further, the effect of geographical or seasonal variation on EPS production as is observed for plant polysaccharides can be minimized (Donot et al. 2012).

This diverse class of polysaccharides, especially glucans and their derivatives, has been screened for antiviral, antitumor, and anticoagulant activities. They have recently become a popular area of research for finding new functional entities with comparable bioactivity and for designing new drug carriers.

Most of the natural polysaccharides are nontoxic, biodegradable, and biocompatible (Beneke et al. 2009). Owing to the presence of a large number of hydrophilic groups such as hydroxyl, carboxyl, and amino groups, glucans are amenable for chemical modification (Raman et al. 2005). The new biological activities and physicochemical properties can be promoted with the use of new reaction schemes (d'Ayala et al. 2008; Laurienzo 2010). Moreover, their biological activity is currently being recognized for human applications (Rinaudo 2008). The activities such as antitumor (Lu et al. 2010; Chen et al. 2013), antibacterial (Harada et al. 1988), antiviral (González et al. 1987; Talarico et al. 2005), anti-inflammatory (Blondin et al. 1994; Joseph et al. 2011), and antioxidant (Vamanu 2012) have been reported for some of the polyionic glucans. However, the poor solubility, high viscosity, and unstable physicochemical properties must be overcome to make these bioactive glucans clinically relevant. The structural modification of polysaccharides is an interesting option of modifying and improving physicochemical and bioactive properties. Research reports on the biological activities of modified polysaccharides are limited (Yoshida et al. 1995; Fernandes et al. 2014).

The structural modifications of polysaccharides have expanded the area of application in the field of drug delivery science and technology. Even their structures can be manipulated to obtain biologically active polysaccharides. However, reports on the toxicity and biodegradability of modified glucans are limited in the literature. Modified bioactive polysaccharides can be used to augment the therapeutic efficacy of an associated drug. Thus, a great deal of attention is required to unveil the real potential the polysaccharide might possess. Considering the available works, it seems to be particularly interesting to modify glucan polysaccharides and explore their biological activities and drug targeting potentials.

In recent years, numerous efforts have been put in developing nanometric glucan-based drug delivery systems with enhanced biopharmaceutical properties, target selectivity, decreased toxicity, and, therefore, improved therapeutic profiles compared to traditional formulations.

A basic knowledge of the structural features of glucan polymers is essential to do justice with the material during derivatization. Therefore, the source and chemical structures have been a matter of concern at the onset of discussion of this chapter.

Pullulan is a water-soluble neutral glucan commercially produced by the fungi *Aureobasidium pullulans*. In this polymer, maltotriose repeating units are linked by $\alpha$-1 $\rightarrow$ 6 glycosidic bonds (Kimoto et al. 1997). Dextrans are $\alpha$-D-glucans produced by several kinds of *Leuconostoc*, *Streptococcus*, and *Lactobacillus* bacterial strains. The soluble dextrans with more than 95% $\alpha$-1 $\rightarrow$ 6 linked D-glucosyl units are used for most industrial applications. Curdlan is a linear polysaccharide composed of $\beta$-1 $\rightarrow$ 3-linked D-glucose units produced by a strain of *Alcaligenes faecalis* (Gao et al. 2008). Its biological applications, particularly in the medicine and food industries, are limited due its inherent low water solubility. The derivatization reaction can solve the problem. The sulfate, carboxymethyl ($-CH_2COO-$), and phosphoryl derivatives of curdlan can have better biological activities. The chemical structures of various glucans are given in Figure 18.1.

This chapter describes the potential bioactivity of glucans, such as curdlan, pullulan, and dextrans, and relationships with the structural features. Moreover, nanotherapeutic systems have also been described with updated information on drug targeting.

## 18.2  POTENTIAL BIOACTIVITY OF POLYIONIC GLUCANS AND STRUCTURAL RELATIONSHIP

Evidence indicates that the bioactivity of glucans is intimately connected with the structure and physicochemical properties of these polymers. Virtually all natural products of glucan isolates are a distribution of glucan polymers with varying polymer length, MW, and in most cases, difference in branching frequency, length of side chain and solution conformation (random coil or helix), chemical composition, and potential derivatization. This level of structural complexity and variability presents a number of challenges to the investigators attempting to understand the chemistry and biology of glucans.

A common procedure in the development of new drugs is the evaluation of structure–activity relationships. There are numerous studies on the chemical modifications of glucan polysaccharide derivatives and their biological activities in the literature. Most of them exhibit antitumor, immunomodulatory, or antiviral or anticoagulatory properties. The structure–activity relationship of some of these polysaccharides is discussed here. The probable sites of chemical modification have been indicated in Figure 18.2.

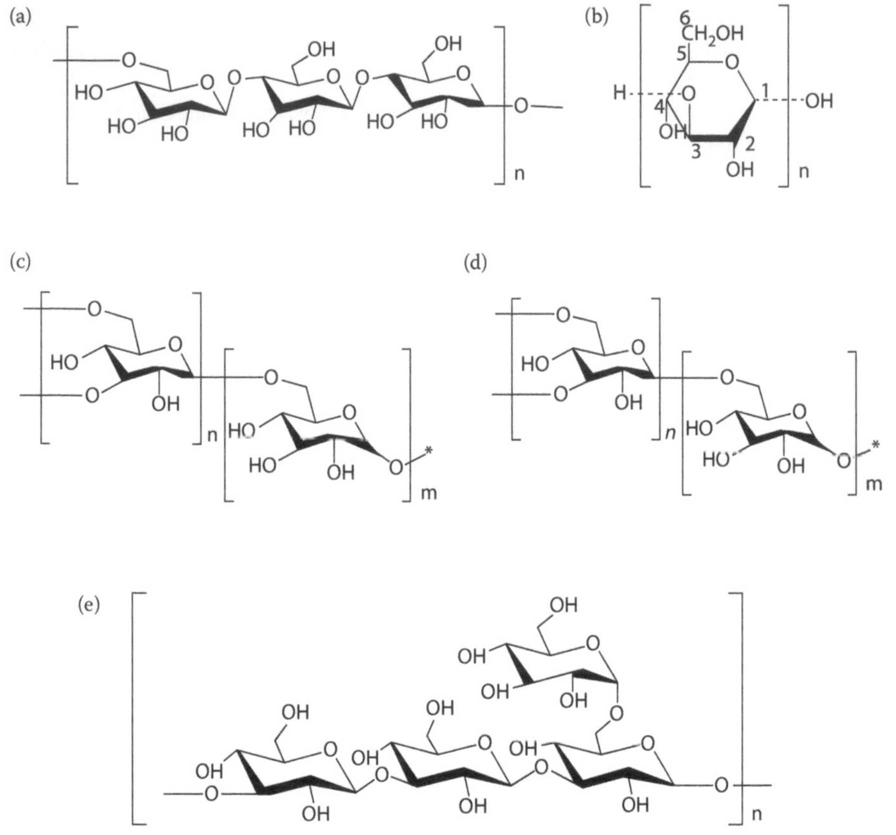

**FIGURE 18.1** Chemical structures of various glucans: (a) pullulan, (b) curdlan, (c) dextran, (d) lentinan, and (e) scleroglucan.

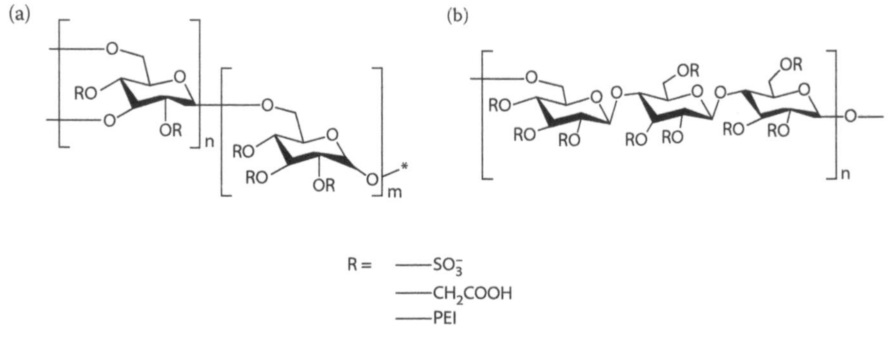

$$R = \quad\text{—}SO_3^-$$
$$\text{—}CH_2COOH$$
$$\text{—}PEI$$

**FIGURE 18.2** Probable sites of sulfation, carboxymethylation, and cationic molecule conjugation such as polyethyleneamine (PEI) of (a) dextran and (b) pullulan.

## 18.2.1 ANTITUMOR ACTIVITY

Most of the research have focused on potential antitumor properties of glucan polysaccharides such as schizophyllan, lentinan, sinofilan, and curdlan. Coadministration of schizophyllan with antineoplastic drugs could prolong the lifespan of patients suffering from lung, gastric, colorectal, and breast cancer as well as stage II or III cervical cancers (Chihara et al. 1987; Ikekawa 2001; Wasser 2002). Simultaneous chemotherapy or irradiation improved their effectiveness. The antitumor effects are mediated through natural killer (NK) cell activation and T-cell-mediated cytotoxicity, increased production of cytokines—interferons (IFNs), interleukines (ILs), and tumor necrosis factor-$\alpha$ (TNF-$\alpha$)—activation of peripheral mononuclear cells (PMNCs), and stimulation of phagocytosis by neutrophils (Falch et al. 2000; Giavasis et al. 2005). Sinofilan, a scleroglucan, can stimulate phagocytic cells, and monocyte, neutrophil, and platelet hemopoietic activity (Giavasis et al. 2005). The unmodified curdlan does not provoke immune responses; however, its ether, sulfate, sulfoethyl, and sulfoalkyl derivatives do exert antitumor activity against sarcoma 180 (Bohn and BeMiller 1995).

Sulfoethyl, sulfopropyl, and sulfobutyl derivatives of the $(1 \rightarrow 3)$-$\beta$-D-glucan curdlan were very active against the sarcoma 180 tumor but the $(1 \rightarrow 3/1 \rightarrow 4)$-$\beta$-D-glucan lichenan derivatives remained inactive (Demleitner et al. 1992a, b). This revealed that $(1 \rightarrow 3)$ glycosidic linkage was essential for antitumor activity and the influence of chain length of the substituent was insignificant. Sasaki et al. (1978, 1979) synthesized carboxymethyl derivative of $(1 \rightarrow 3)$-$\beta$-D-glucan, obtained from *Alcaligenes faecalis* var. *myxogenes* to modulate the antitumor activity of native glucan. Beta-D-glucan did not lose antitumor activity after carboxymethylation. The degree of carboxymethylation had a prime role in determining antitumor activity. A carboxymethyl substitution of 0.47 and 0.65 demonstrated strong antitumor activity at a dose of 5 mg/kg, relatively lower than parent glucan. It is interesting to note that excessive substitution increased water solubility with simultaneous changes in molecular orientation and conformational structure, eventually causing the loss of antitumor activity. In a preclinical study, dextran sulfate (DS) decreased pulmonary metastasis in rat models significantly at a dose of 300 mg/kg intraperitoneal (IP), prior to intravenous (IV) injection of lung carcinoma cells (Suemasu and Ishikawa 1970).

## 18.2.2 ANTIVIRAL ACTIVITY

Lentinan is active against the influenza virus. It can prevent proliferation of the polio virus and increase host resistance to the AIDS virus. The antiviral action is supposed to occur by induction of IFNs (Markova et al. 2002).

At an early stage of the infection, scleroglucan is active against the herpes simplex virus (HSV) and rubella virus because bioactive glucan binds to host cell membrane glycoproteins, and hinders interactions between virus and host cell plasma (Mastromarino et al. 1997). Nucleoside analogs, such as azidothymidine (AZT), inhibit viral reverse transcriptase of the human immunodeficiency virus (HIV) and terminate DNA chain synthesis from viral RNA inside the infected cells. However, these nucleosides are associated with bone marrow toxicity. Owing to the multiplicity of AIDS viruses, the production of a viral vaccine is difficult. Therefore, anti-AIDS compounds having a different blocking mechanism are of great interest.

Yoshida et al. (1994a, b) have synthesized sulfated polysaccharides that demonstrate high anti-AIDS virus activity. The mechanism of action of sulfated polysaccharides is different from that of nucleosides. The sulfated derivatives are supposed to bind HIV virions and prevent them from penetrating into target cells. The complex HIV-1 replication cycle suggests that several stages of replication might represent potential targets for anti-HIV-1 therapy. Recently, considerable attention has been paid to the sulfated polysaccharides that are potent inhibitors of HIV-1 replication *in vitro* (Baba et al. 1990).

The mechanism of anti-HIV-1 activity of sulfated polysaccharides is contradictory. A number of mechanisms have been proposed, including a direct antiviral action (Bagasra and Lischner 1988), inhibition of virion binding to the CD4 receptor (Lederman et al. 1989), binding to the V3 loop glycoprotein and inhibition of postreceptor binding events (Callahan et al. 1991), and interaction with an 18-kDa cellular protein of unknown function (McClure et al. 1992). The discrepancies may arise possibly due to the following reasons. The differences in chemical structures and MW of various sulfated polysaccharides may determine the binding capacity to their viral and/or cellular ligands and show toxicity for the host cells. The plastic-bound recombinant CD4 and/or gp120 used in solid-phase assays may alter conformational structures compared to their cellular counterparts and may show contradictory results. The tumor cell lines may differ in their susceptibility to infection and sensitivity to the compound. Therefore, a contradictory array of results is not unexpected.

The anti-HIV-1 and immunosuppressive effects have been found for curdlan sulfate (CRDS) and DS. They can provide complete protection to the MT-4 cells against HIV-1 cytopathogenicity at a concentration of 3.3 and 25 µg/mL, respectively (Baba et al. 1988; Kaneko et al. 1990). Clinical studies revealed that DS administered intravenously to HIV-1-infected patients had anticoagulant activity and caused thrombocytopenia (Flexner et al. 1991). In view of the unsuccessful clinical trial with DS, a better understanding of the mechanism of anti-HIV-1 activity of sulfated polysaccharides together with the evaluation of their toxic effect on the activation of peripheral blood mononuclear cells could provide useful information for the synthesis of more effective and less toxic therapeutic drugs. CRDS was investigated in detail because of its strong anti-HIV-1 activity *in vitro*, manifested by the inhibition of cell-free HIV-I infection and syncytium formation, absence of antigenicity, and 10-fold lower anticoagulant activity than that of heparin in guinea pigs and rats (Kaneko et al. 1990).

Jagodzinski et al. (1994) analyzed the effects of CRDS on HIV-1 infection of SupT-1 cells and peripheral blood mononuclear cells. Perhaps, CRDS impedes virus attachment to the host cell. There is also the possibility of virus encapsulation by the glucan after cellular internalization. However, CRDS was devoid of any effects on virions. CRDS was required at the time of exposure of cells to HIV-1 and for at least 24 h during infection to achieve complete protection. The kinetic analysis of HIV-1 infection in the presence of CRDS alone and together with a reverse transcription inhibitor, dideoxyinosine, indicated that CRDS delayed the early events in viral replication, which preceded the proviral transcription. The treatment of cellular gp120 glycoprotein with CRDS, followed by 0.5β mAb (Matsushita et al. 1988) or F105 mAb (Posner et al. 1991) inhibited the binding of antibodies to their respective epitopes within the V3 loop, which mediates postreceptor binding steps in virus entry (Freed et al. 1991). The induced changes in some components of CD4 binding region of gp120 were responsible for the inhibitory effect of CRDS on HIV-1 infection and syncytium formation. The analysis suggested that both the continuous epitopes on the V3 loop and the discontinuous CD4 binding site of gp120 represent a target for CRDS. The target site of sulfated glucans in HIV replicating cycles is illustrated in Figure 18.3.

Another derivative, CRDS, completely inhibited AIDS virus infection of T lymphocytes at a dose of 3.3 μg/mL *in vitro*. The derivative accumulated in the liver, kidneys, lymph nodes, and bone marrow within 1 h, and resisted degradation for 10 days (Kaneko et al. 1990). To sustain the concentration of CRDS in plasma, L-glucose and L-mannose sugars were conjugated to curdlan to design L-glycosyl-branched curdlan (Yoshida et al. 1994b). However, the retention time remained unaltered, indicating negligible effect of L-glucose. The anticoagulant activity of the sulfated glucan was lower (10 units/mg) than that of standard DS (20.6 units/mg).

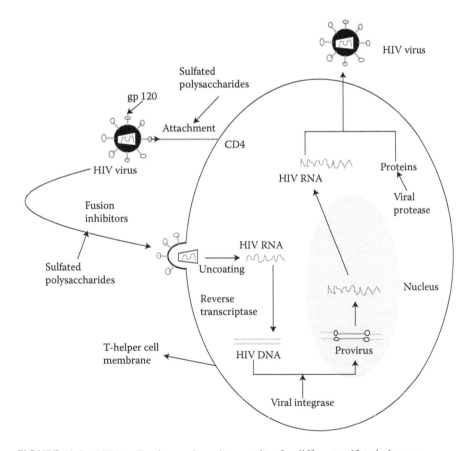

**FIGURE 18.3** HIV replicating cycle and target sites for different sulfated glucans.

Thus, CRDS could be an anti-AIDS drug through the inhibition of virus entry in the early stages of infection. The sulfated compound accelerated dose-dependent CD4 lymphocyte production in HIV-infected patients without any evidence of clinical side effects (Gordon et al. 1994).

Yoshida et al. (1995) evaluated the antiviral activity of CRDS in HIV-$1_{HTLV-IIIB}$ or HIV-$2_{ROD}$ strains-infected human T-lymphotropic virus type-I positive human T cell line (MT-4 cell). HIV-infected and HIV-uninfected MT-4 cells ($3 \times 10^5$ cells/mL) were cocultured with various concentrations of CRDS for 5 days. CRDS completely inhibited the HIV-1 and HIV-2 infections on MT-4 cells at an $EC_{50}$ value of 0.4 and 0.6 µg/mL, respectively. MT-4 cells, which escaped viral infections by the inhibitory effects of CRDS, proliferated at a rate similar to uninfected cells. The derivative was noncytotoxic against HIV-uninfected MT-4 cells up to 1 mg/mL, because MT-4 cells grew normally. The CRDS obtained by the piperidine-$N$-sulfonic acid (PSA) method had high anti-AIDS virus activity and low cytotoxicity against both HIV strains. The derivative, obtained by sulfur trioxide ($SO_3$) in pyridine (SPC) or the chlorosulfonic acid (CSA) method, exhibited slightly higher cytotoxic effects than that prepared by the PSA method. The presence of inorganic sulfate impurities in CRDSs could be responsible for the difference.

To establish the relation between antiviral activity and structural properties, CRDSs of different degree of sulfation and molecular mass were synthesized by the SPC or CSA method. Products with molecular masses $>1.0 \times 10^4$ and degrees of sulfation $>1.0$ had potent anti-AIDS virus activity against both HIV-1 and HIV-2 strains. The inhibitory effect against HIV-2 was not substantially higher at low degree of sulfation of around 1.0 than HIV-1. The infectious strength of HIV-2 strain might have been weaker than that of the HIV-1 strain. CRDSs obtained by PSA method exhibited lower anticoagulant activity (10 units/mg) than those prepared by the SPC (36 units/mg) and CSA (22–39 units/mg) methods. The SPC- or CSA-modified CRDSs demonstrated higher anticoagulation activities due to their helical structure. On the contrary, the rigid and rod-like structures were responsible for low anticoagulant activity of PSA-modified CRDS. Therefore, CRDSs with low anticoagulant activities, obtained by the PSA method, were preferable anti-AIDS drugs.

Notable anti-HIV activity was reported by Osawa et al. (1993) for sulfopropyl derivative of curdlan. The anti-HIV activity was expressed as $EC_{50}$ and SI (selectivity index, $CC_{50}/EC_{50}$, where $CC_{50}$ is 50% cytotoxic concentration of polysaccharide solution). The lower $EC_{50}$ and higher SI indicated safety and drug efficacy. The action of sulfopropyl curdlan on HIV ($EC_{50}$ 13.4 µg/mL and SI >75) was weaker than standard CRDS ($EC_{50}$ 0.018 µg/mL and SI >55600) at a degree of sulfation of 0.41. The trimethylene spacer perhaps weakened anti-HIV activity; however, the activity became higher at higher substitution ranging from 2.46 to 3.18.

All serotypes of dengue virus (DENV-1 to DENV-4) cause diseases in humans, ranging from mild dengue fever to dengue shock syndrome, which can be fatal (Rigau-Perez et al. 1998). The primary infection by the virus induces immunity against the infecting serotype. However, a secondary infection with a different serotype may enhance the risk of developing hemorrhagic fever or shock syndrome (Heinz and Stiasny 2012). The multistep DENV infection cycle presents potential drug targets during entry, viral membrane fusion, translation, assembly, and maturation.

X-ray crystallographic and electron microscopic analyses of the DENV structure and its component proteins have led to the identification of new drug targets (Modis et al. 2003; Rey 2003; Mondotte et al. 2007). Consequently, E-protein has emerged as a promising target for the inhibition of virus entry into cells (Li et al. 2008; Zhou et al. 2008; Kampmann et al. 2009). The class II viral fusion protein, DENV-2 E, consists of three domains. Owing to domain rearrangements, the protein can assume different conformations in the mature, immature, and postfusion stages (Modis et al. 2004). A hydrophobic pocket near the hinge region between domains II and I of the E-protein can accommodate the $N$-octyl-$\beta$-D-glucoside molecule. This finding has provided an impetus to the development of drugs that can target the viral entry process.

An *in silico* blind docking study identified CRDS as a probable inhibitor of the DENV E protein (Ichiyama et al. 2013). The binding site of the CRDS molecule resided near the fusion loop at the interface of domain II and domain III of the E protein dimer. They identified that this polysaccharide can block DENV at both the binding and fusion steps very efficiently. CRDS showed a favorable selectivity index against all serotypes of DENV. Since the compound has already been tested in humans without serious side effects, it offers the possibility of clinical application. One of the major concerns toward the possible clinical use of sulfated polysaccharides is their extent of anticoagulant activity. CRDS is largely devoid of anticoagulant potential compared to another polyanionic substance, heparin (Kaneko et al. 1990).

Of all the serotypes, DENV-1 appeared to be less susceptible to CRDS. The extent of the anti-DENV effect of CRDS was dependent on the serotypes of the virus (Talarico et al. 2005). CRDS inhibited DENV-2 replication at a concentration starting as low as 0.1 µg/mL in LLC-MK2 cells in a dose-dependent manner up to 400 µg/mL. CRDS showed an $EC_{50}$ of 7 µg/mL and a $CC_{50}$ of more than 10 mg/mL with LLC-MK2 cells. Thus, a selectivity index >1428 indicated potent and selective antiviral action of CRDS. CRDS showed much higher anti-DENV activity as compared to its two analogs, both with higher extent of sulfation, but lower MW than CRDS. Hence, the MW could dictate the anti-DENV activity of CRDS.

The results of clinical trials on HIV-infected patients were disappointing in terms of efficacy. The continuous administration of CRDS was suggested for the suppression of viral effects. The coagulation effect must be monitored by dose adjustment associated during long-term use of CRDS. Extra care is required for monitoring the time of blood coagulation since DENV infection is notoriously associated with hemorrhage. According to Ichiyama et al. (2013), CRDS interferes with the viral binding and membrane fusion steps, early and critical events of virus replication in DENV infection.

The sulfated polysaccharide, DS, was a potent and selective inhibitor of HIV-1 *in vitro* (Baba et al. 1988). DS (MW 5000) guaranteed protection to MT-4 cells against HIV-1-induced cytopathogenicity at 25 µg/mL dose ($EC_{50\%}$ ~9.1 µg/mL). The compound was safe for the host cells up to 625 µg/mL. The low-molecular-weight DSs (5000, 8000) exerted significant inhibitory effect on HIV-1 without marked anticoagulant effect. DS acted as HIV-1 reverse transcriptase inhibitor at concentrations in excess of that inhibited HIV-1 replication. It was most effective against HIV-1 replication at an early stage because DS could inhibit virus adsorption to the host cells.

The dextrans derivatized with varying percentages of carboxymethyl, benzyl-amide, and sulfonate groups showed activity against retroviruses and herpes viruses, probably due to the inhibition of virus binding to the cells (Neyts et al. 1995).

In the late 1980s, dextran was tested for HIV treatment. The sulfated derivative of dextran (MW 8000) blocked the binding of the virus to T-lymphocytes and interfered with HIV-1 replication *in vitro* at concentrations clinically attainable in humans. The sulfated molecules suppressed the replication of HIV-2 *in vitro* (Mitsuya et al. 1988). Clinical trial revealed that the oral DS was well tolerated without apprecia-ble changes in CD4 lymphocyte numbers and reduction in $\beta$-2 microglobulin levels (Abrams et al. 1989).

The anti-HIV effect of DS was reported to be strain dependent such that two clini-cal isolates of HIV-1 (transmembrane, TM; single peptide, SP) were less susceptible to DS compared to two cloned isolates (HIV-1 World Medical Fund, WMF and HIV-2 ROD isolate) and prototype laboratory strain (HIV-IIIB). High-MW DS ($5 \times 10^5$) was antagonistic (in HIV-1 TM and SP isolate) when DS was given in combination with dideoxynucleosides (Busso and Resnick 1990). As was reported by Nakashima et al. (1989), the anti-HIV-1 activity was dependent on the sulfate content of DS.

### 18.2.3 Antioxidant

Oxidation phenomena are concerns for many human diseases, such as diabetes mel-litus, arteriosclerosis, Alzheimer's disease (AD), and cancers. Therefore, natural antioxidants represent one of the most useful nutraceuticals and functional foods for health protection and disease prevention. The regioselective conversion of the C6 primary hydroxyl groups of curdlan into carboxylate groups with negligible depolymerization, and aldehyde formation led to considerable antioxidant activity (Yan et al. 2014). The carboxyl contents of hydrophilic oxidized curdlans moved from 2.07 mmoL/g to 4.87 mmoL/g. The antioxidant effect of the highly substituted derivative was attributed to the ability of donating hydrogen (Wang et al. 2010). However, the antioxidant curdlans must be evaluated in biological systems to prove and understand the health effects and potential value.

### 18.2.4 Anticoagulant Activity

In addition to the degree of sulfation, the actual location of the sulfate groups on heparin chains was vital for anticoagulant activity. Keeping this in mind, Franz and Alban (1995) devoted their study to establish the influence of pattern of sulfate group substitution on the anticoagulant activity of curdlan. According to them, sulfation of C6 hydroxyl groups was not essential for the anticoagulant effect. However, the uni-form distribution of sulfate ester groups on the glucose units improved anticoagulant activities considerably. The activity of the derivatives with higher MW and DS were comparable.

The structurally well-defined $\beta$-1 $\rightarrow$ 3-glucan sulfates, such as CRDS, exhibited structure-dependent potent anticoagulant and antithrombotic activities. The mode of action of these heparinoids differed from that of heparin. The anticoagulant effect of CRDS was nonspecific due to their anionic character, but they might interfere specifically with the coagulation process at several sites depending on their individual structure. The derivatives reduced the thrombin formation by principally inhibiting intrinsic FXa generation. They could eliminate already generated thrombin by accelerating the HCII-mediated thrombin inactivation, whereas their AT-mediated antithrombin activity was found lower than that of heparin. Further, they prevented thrombin-mediated fibrin polymerization by directly interfering with the thrombin action on fibrinogen as well as by binding to fibrinogen. Lastly, they activated the contact system as a consequence of the potential fibrinolytic effect. The marked anticoagulant activity of CRDS in the APTT (40 U/mg) and TT (25 U/mg) suggested that the derivatives interfered with the intrinsic step of coagulation as well as with the last step, i.e., the thrombin-mediated fibrin formation (Alban and Franz 2000).

Furthermore, they provided a valuable guide that the primary hydroxyl-linked sulfo groups are needed to qualify for the anticoagulant effect (Alban and Franz 1994). The above prerequisite plus glycosidic branching could determine the activity of β-1,3 sulfoglucan not only *in vitro* but also *in vivo*. Although the *in vivo* antithrombotic activity of the sulfoglucans seemed comparable to heparin, the coagulation assays indicated that these compounds had lower activity. This hinted the involvement of additional mechanisms, other than anticoagulation toward their antithrombotic activity. CRDS could serve anticoagulant functions through various mechanisms that involved (a) direct inhibition of the thrombin–fibrinogen interactions, (b) activation of plasma inhibitors (predominantly HCII) and thrombin inhibition, and (c) complexation with fibrinogen. CRDS could also activate the contact system as indicated by its ability to stimulate kalikrein generation, though this property was least significant with regard to anticoagulant activity *in vivo*.

Carboxymethyl dextran benzylamide sulfonate/sulfates (CMDBS) are synthetic polysaccharides with anticoagulant activity (Figure 18.4). Raucourt et al. (1998) compared the anticoagulant activity of eight different highly substituted CMDBS and one CM-DS.

CMDBS directly inhibited thrombin formation and accelerated the rate of thrombin inhibition by heparin cofactor II (HCII). The anticoagulant and catalytic activities of CMDBS correlated to their MW and sulfate content. The interaction of the dextran derivatives with thrombin does not involve the active site of the enzyme. All CMDBS had higher affinity for thrombin than heparin. Nevertheless, the same was lower for CMDBS than DS. This suggested that the benzylamide and sulfate groups potentiated the interaction between the dextran derivatives and thrombin.

Logeart-Avramoglou and Jozefonvicz (1999) postulated that the CMDBS compounds were capable of ensuring specific interactions with biological constituents, depending on the overall polymer composition. It appeared that dextran derivatives could mimic heparin action with regard to their interactions with antithrombin and serine proteases, involved in blood coagulation. In addition to their numerous biological properties and biospecificity, the functionalized dextrans were relatively simple to manufacture, free from donor contaminant, and thus were attractive materials for a variety of clinical applications.

Large-MW dextran (~1000 kDa) was partially degraded to low-MW dextran of 10–100 kDa size (Alban 1997). The unsulfated dextran had low efficacy as an anticoagulant. DS could exhibit heparin-like activity. Further, a very narrow therapeutic

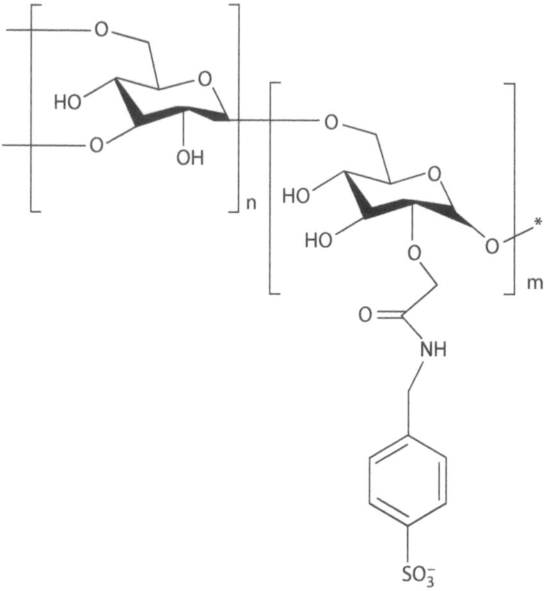

**FIGURE 18.4** Structural features of carboxymethyl dextran benzylamide sulfonate/sulfates (CMDBS) having anticoagulant potential.

index of DS is an undesirable therapeutic application (Nagasawa et al. 1972). A direct relationship existed between MW and toxicity in that the molecules <10 kDa showed little or no toxicity. The dextran required a relatively high degree of sulfonation (~1.3) for maximum activity (Zhang et al. 2005).

The type of glycosidic linkages may create differences in chemical reactivities of the polymers because of distinct distribution of hydroxyl groups within the pyranose rings. In pullulan, nine OH groups could be approximated per repeating unit for functional modifications, though the relative reactivities of these groups differed based on the solvent polarity and the reagents. Carboxymethyl substitution in an alcohol–water mixture was less pronounced at the C-6 hydroxyl group of pullulan and preceded at the C-2 position (Glinel et al. 2000). The introduction of negative charges in the form of carboxymethyl groups prolonged the retention of the drug carrier within the body (Yamaoka et al. 1993).

In the case of pullulan, the degree of sulfate substitution was higher than that of dextran under identical conditions due to the presence of a large number of primary hydroxyl groups in its structure. Mähner et al. (2001) obtained sulfated derivatives of dextran and pullulan via SPC method. The reactivity of the polysaccharide C-atoms for dextran was C-3 > C-2 > C-4, while for pullulan it was C-6 > C-3 > C-2 > C-4. This corroborated with the reports of Alban et al. (2002) who found that the sulfation occurred in the order of C-6 > C-2 > C-3 > C-4, irrespective of MW and degree of sulfation in pullulan. The anticoagulant activity of sulfated pullulan was almost identical to that of heparin. However, the action relied upon the MW, degree of sulfation, and distribution of the sulfated groups on the various positions of the glucose residue.

## 18.2.5 General Considerations

A good relationship between chemical structure and physical properties and the function of polysaccharides was of prime importance in understanding the mechanism of action and relevant bio-targets. Nevertheless, the relationship is controversial, and the results concerning structure–activity were highly variable for several polysaccharides.

Many researchers suggested that the β-(1 → 3) linkage of the glucan chain was vital for the expression of antitumor activity. In general, the glucans with (1 → 6) linkages exhibit comparatively low antitumor activity (Bohn and BeMiller 1995). This was reasonable to some extent owing to β-(1 → 3)-glucan receptors in macrophages and macrophage activation. The immunoactive glucans with β-(1 → 6)-linked backbone ensured that the β-(1 → 3)-linked main chain was not always a prerequisite for the antitumor activity of glucans (Osawa et al. 1993; Wasser, 2002).

The degree of branching (DB) might also affect the biological function of polysaccharides. The glucans with low, medium, or high DB might possess different immunostimulatory properties. Perhaps, a moderate branching (0.20–0.33) was desirable for the antitumor activity of glucans (Bohn and BeMiller 1995). Increased branching led to weaker interchain associations and favored the formation of single helices. The structural differences resulting from different DBs may relate to the variable biological activity of the polysaccharides of different origin and with different DBs (Jamas et al. 1991).

Two modified β-D-glucans were prepared by Cleary et al. (1999) from *Saccharomyces cerevisiae*. The glucans with more β-(1,6) side chains (20% by weight of the total molecule) led to higher stimulation of macrophages than the low-branch glucans (5%). This was ascribed to the ability of highly branched glucans to cross-link glucan receptors in macrophages.

The effect of MW on the biological function of the polysaccharides was also controversial. High MW usually promoted the antitumor and immunomodulating activity of glucans. It was argued that MW was more important than chemical structure in microbial glucans because they usually act via nonspecific immune mechanism (Bohn and BeMiller 1995). The MW, in turn, could affect the ordered conformation of glucans. The structural conformation and arrangements were crucial in eliciting the biological function for many microbial polysaccharides. Suzuki et al. (1992) elucidated the impact of ordered arrangements on the immunological characteristics of microbial glucans such as curdlan and schizophyllan. The triple helix activated the alternative pathways of complement (APC) in the immune system more strongly than did the single helix. On the other hand, the single helical conformation was more efficient in stimulating the classical pathways of complement (CPS). The activation of CPS by the single helix occurred via binding of the glucan to immunoglobulin in serum. Of course, the structural conformation seemed to influence the way by which glucans were recognized by each system.

Other factors that influence the performance of bioactive polysaccharide were the solubility and chemical substitution of the original biopolymer (sulfate, phosphate, carboxymethyl, or hydroxyl groups). The degree of branching/substitution enhanced water solubility to a limited extent. The solubility was determined by the number and type of hydrophilic groups positioned on the outer surface of the helix. The antitumor activity of many glucans was promoted for highly soluble derivatives. Carboxymethyl glucans with DS ranging from 0.4 to 1.15 were water soluble, while the samples with DS <0.4 were insoluble or partly soluble. The biological activity of carboxymethyl glucans was optimal in the DS range of 0.6–0.8. The immuno-modulating effect of carboxymethyl glucans gradually diminished with a DS value >1.0. A low DS value was appropriate in achieving the desired solubility of sulfoe-thyl *S. cerevisiae* glucans. The glucans having DS 0.3 were totally soluble when sulfoethyl groups were introduced, thus highlighting the importance of the ionic substituents (Šandula et al. 1999).

Many studies demonstrated that ionic substitution could enhance the biological activity of glucans. CRDS and sulfated schizophyllans inhibited the growth of the HIV virus, depending on their sulfur content. In fact, the sulfur content in schizo-phyllans mostly decided HIV efficacy rather than MW or chemical nature of the polysaccharide component (Jagodzinski et al. 1994).

As P-selectin is a promising target for anti-inflammatory agents, Hopfner et al. (2003) and Fritzsche et al. (2006) examined sulfated glucans such as curdlan or pul-lulan to analyze their inhibitory capacity by detecting the rolling of U937 cells on P-selectin layers. The charge density was a determinant in P-selectin binding. Highly sulfated samples were excellent as P-selectin inhibitors. Regardless of MW, the gly-cosidic linkage endowed certain importance on P-selectin binding. The study advo-cated that semisynthetic glucan sulfates with optimal structures could avidly bind P-selectin and thus showed promise as anti-inflammatory drugs to replace heparin (weaker P-selectin inhibitor) for certain applications.

## 18.3 NANO-SYSTEMS OF GLUCAN DERIVATIVES FOR DRUG TARGETING

Targeted nanodrug delivery technology represents one of the most rapidly advanc-ing areas of science due to some obvious advantages, including improved effi-cacy, reduced toxicity, improved patient compliance, and cost-effective therapeutic treatment.

The fundamental characteristics of the delivery systems are the ability to incor-porate drugs without damaging them, long *in vivo* stability, tunable release kinetics, and targeting to specific organs and tissues. *In vivo* stability of the system is strictly connected with external surface properties, and *stealth* surfaces were identified, as those not recognized as foreign bodies by the immune system. One strategy consists of the obtainment of a water-like shield, promoted by a strongly hydrophilic coating. Polyethylene glycol (PEG) polymers are widely used as coatings due to their high biocompatibility and antiopsonizing effect (Gref et al. 1994). Normally, particles that are small enough to escape the capillary bed of lungs are quickly scavenged by the reticular endothelial system (RES). This sequestration represents the major barrier to targeting cells or tissues elsewhere in the body. On intravenous administration, parti-cles are quickly coated by specific blood components (opsonins) and then recognized and captured by RES. The antiopsonizing action of polymers such as PEG consists of the adsorption of a surface water layer to form a hydrophilic barrier. PEG-based shields can be obtained by (1) physical coating of preformed nanoparticles (NPs) with PEG copolymers (Hawley et al. 1997); (2) PEG residues chemically linked to functional groups on the preformed NP surface (Dunn et al. 1994); and (3) thermodynamic assembly of block amphiphilic copolymers in core–shell nanostructures. In these cases, the hydrophilic domain constituted by PEG blocks results in an intrinsic shield action (Kim et al. 1998).

NPs are solid colloidal particles in which biologically active molecules can be trapped, dissolved, and/or encapsulated. The size and surface characteristics of NPs, such as charge, hydrophilic–hydrophobic balance, and presence of site-specific components dramatically influence their body distribution and targeting attitude (Kreuter 1994). Biodegradable matrices can provide a further control of drug release rates, by joining the typical diffusive mechanism with tunable polymer degradation.

Biodegradable polymers represent a valuable option for the formulation of nano-drug delivery systems, since they address the severe problem of the removal of the releasing device after drug depletion. In addition, most biodegradable polymers originate from natural monomeric precursors, thus improving the biocompatibility of the initial polymeric material and of its degradation products. In addition, their innate bioactivity can be combined with drugs to potentiate biological functions. Owing to their hydrophilic character, they can be used to prevent opsonizing effect by RES. In the subsequent section, the drug targeting potential of polyionic glucan NPs developed will be discussed.

## 18.3.1 CURDLAN DERIVATIVES

The tumor activity of carboxymethyl curdlan (CM–curdlan) has been appreciated earlier (Sasaki et al. 1978; 1979; Saitô et al. 1991). However, the mechanism of action remains unclear. Ohya et al. (1994) stated that immunoenhancement activity of the muramyl dipeptide analog (GADP)/CM–curdlan conjugate could be as a consequence of polymeric character of GADP and immunoactivity of polysaccharide. Na et al. (2000) reported self-assembled lactobionic acid (LBA)–CM–curdlan–sulfonylurea (SU) hydrogel NPs (<300 nm) that exerted immunomodulation activities via ligand–receptor interactions in HepG2 cells. The controlled drug release property suggested their potential as drug carrier for liver cancer cells.

A bile acid component, deoxycholic acid (DOCA), was conjugated to CM–curdlan (MW 81,000) by Gao et al. (2010). The epirubicin (EPB)-loaded conjugate NPs (EDNs) of 327.4–511.5 nm size showed considerable cytotoxicity in MCF-7 cells due to their enhanced cellular uptake. However, the cytotoxic effects of the conjugate were not appreciable *in vivo* even at a concentration of 100 µg/mL. DOCA conjugation enhanced EPB uptake in tumor cells with concomitant reduction in the kidney and heart, compared to the free drug in tumor-bearing mice models. EDNs retarded 70% growth of the tumor, whereas the inhibition was 58.9% for free EPB. The subacute toxicity of conjugate revealed normal functioning of liver and kidney based on the levels of aminotransferases, creatinine, and blood urea nitrogen (BUN) in all groups and supported the safety profile of the drug–carrier conjugates. The higher area under curve (AUC) of EDNs in plasma and tumor than free EPB was attributed to the reduced renal excretion and enhanced permeability and retention (EPR) effects, respectively (Duncan 1999). According to Brigger et al. (2002), the cut-off size for the pores in tumor vessels ranged from 380 to 780 nm. The direct observation of tumor vasculature reflected a tumor-dependent pore cut-off size, ranging from 0.2 to 2 μm (Hashizume et al. 2000), which may permit the access of drug-loaded NPs to malignant tumor cells. Therefore, EDNs (~350 nm) could accumulate in the tumor by EPR effect. EPB NPs reduced the inherent cardiac toxicity of EPB. Overall, the design of EDNs supported the therapeutic improvement of EPB.

Li et al. (2010) tested cholesterol–carboxymethyl curdlan nanoconjugates for the delivery of the same anticancer molecule. The EPB-loaded NPs were more cytotoxic for human cervical carcinoma (HeLa) cells than free EPB. The mean residence time (4-fold), half-life (4.31-fold), and $AUC_{0-\infty}$ (6.69-fold) in the rat model increased considerably for EPB-loaded NPs than for free EPB. The drug level was found to be significantly higher than free EPB. *In vivo* antitumor efficacy of EPB-loaded curdlan NPs was also greater than free EPB and thus exhibited their potential as anticancer drug carriers.

The application of NPs to deliver immunostimulatory signals to cells could be interesting. Realizing the immunostimulating property, curdlan was conjugated to poly(lactic-*co*-glycolic acid) (PLGA) via carbodiimide chemistry. The copolymer NPs were subsequently developed with the hope that immunostimulatory activity and noncytotoxic activity would also persist in the NPs (Tukulula et al. 2015). Immunostimulatory activity of the conjugate (C-PLGA) was characterized in THP-1-derived macrophages. C-PLGA NPs enhanced phosphorylated ERK production in macrophages, which was indicative of cell stimulation. A slow permeation of rifampicin across Caco-2 cells was noticed. $Ca^{2+}$-dependent uptake by the macrophages was assumed for the NPs. Therefore, C-PLGA NPs were found promising in terms of macrophage stimulation and delivery of drug to the infected tissues or organs.

## 18.3.2 Dextran Derivatives

It is a very interesting polysaccharide for chemical substitution of hydroxyl groups with hydrophilic or hydrophobic groups per monomer unit. Some groups modified this glucan for further development of drug-loaded nanostructures. Suzuki et al. (1977) introduced palmitoyl and phosphate groups into dextran (MW 38.000) for the syntheses of *O*-palmitoyldextran phosphates, *O*-palmitoyldextrans, and dextran phosphate. The derivative containing 7.3% phosphate and 47.8% palmitoyl groups had a remarkable growth-inhibitory effect against sarcoma 180 ascites tumors in mice when administered alone at a dose of 1 mg/kg for 5 days. On the contrary, a second derivative having 16% phosphate and 0.6% palmitoyl groups exhibited only a borderline inhibition. However, the latter compound had a synergistic effect with mitomycin C, showing 83% recession in the same tumor. All *O*-palmitoyldextrans and dextran phosphates were practically inactive against ascites tumor. This indicated that both fatty acid and phosphate groups are essential to manifest antitumor activity. Masunaga et al. (1996) and Ohya et al. (2009) designed Dach-Pt (*cis*-dichloro (cyclohexane-*trans*-*l*-1,2-diamine)platinum(II)) prodrug with reduced side effects. This water-soluble complex was conjugated to oxidized dextran and dicarboxymethyl–dextran through coordination bonding. The cytotoxicity of this complex against p388D$_1$ *lymphocytic leukemia* cells *in vitro* was similar to free Dach-Pt. However, the cytotoxic effects of free complex decreased by incubation in serum. This suggested that dextran complexation improved serum stability without loss of cytotoxic activity. The oxidized dextran conjugate showed effective tumor inhibition in colon 26 tumor-bearing mice. The dicarboxymethyl–dextran conjugation amplified antitumor activity compared to cisplatin alone in mice bearing colon cancer cells (Ichinose et al. 2000). Lam et al. (1999, 2000) showed that the antitumor activity of doxorubicin (DOX)-DS conjugate was superior to free DOX in multidrug-resistant cancer cell lines. DOX removal from P-glycoprotein overexpressing multidrug-resistant cells decreased after dextran conjugation and antitumor activity enhanced. DOX–DS conjugate was examined in a multidrug-resistance subline of human carcinoma to overcome DOX resistance; however, the same results persisted (Sheldon et al. 1989).

The esterification of hydroxyl group of 10 kDa dextran (Dex) with stearic acid (SA) resulted in the synthesis of stearate-*g*-dextran (Dex-SA) amphiphilic copolymer (Du et al. 2010). Dex-SA micelles conferred excellent internalization ability, and delivered DOX into tumor cells. Dex-SA/DOX micelles maintained the cytotoxicity of commercial DOX injection against drug-sensitive tumor cells. Moreover, the micelles caused the reversal of the activity against DOX-resistant cells. Compared to commercial injection, Dex-SA/DOX micelles were effective in the suppression of tumor growth and preclinical toxicity. DOX-loaded bovine serum albumin (BSA)–dextran–chitosan NPs largely decreased the toxicity of DOX and increased the survival rate of hepatoma H22 tumor-bearing mice (Qi et al. 2010). Lin et al. (2012) studied the *in vitro* effect of chitosan-carboxymethyl–dextran NP (CDNP) containing cefmetazole, 5-fluorouracil (5-FU) on mouse B16 melanoma cell proliferation. The CDNP elicited dose-dependent inhibitory effects on B16 tumor cell proliferation at a dose of 25–100 μg/mL. CDNP98 (98% degree of deacetylation of chitosan) was more effective than CDNP78 (78% degree of deacetylation of chitosan). Despite poor oral bioavailability, bisphosphonates (BPs) have interesting antitumor effects. Migianu-Griffoni et al. (2014) reported new carboxymethyl–dextran bioconjugates to overcome bioavailability and related problems. Its efficiency as a vector was biologically proven on mammalian carcinoma models in mice. At an alendronate concentration of 10 μM, the conjugates caused more than 80% carcinoma inhibition.

Thambi et al. (2014) synthesized bioreducible and amphiphilic carboxymethyl–dextran (CMD) conjugates (CMD–SS–LCAs) via chemical modification of CMD with lithocholic acid (LCA). The fluorescence signal of DOX, quenched in the bioreducible NPs, was highly recovered in the presence of glutathione (GSH), a tripeptide capable of reducing disulfide bonds in the intracellular compartments. DOX-loaded bioreducible NPs exhibited higher toxicity to SCC7 cancer cells than DOX-loaded NPs without disulfide linkages. The bioreducible NPs effectively delivered DOX into the nuclei of SCC7 cells. Following systemic administration into tumor-bearing mice, Cy5.5-labeled CMD–SS–LCAs selectively accumulated at tumor sites and exhibited better antitumor efficacy than reduction-insensitive control NPs.

### 18.3.3 PULLULAN CONJUGATES

The most extensively studied glucan polymer for drug targeting is pullulan. Several workers investigated the drug targeting potential of pullulan and its derivatives. Herein, we describe the key findings of their investigations.

The inherent liver affinity of pullulan prompted workers in preparing drug–polymer composites for therapy of hepatitis (Xi et al. 1996; Tabata et al. 1999). In chronic human liver disease, IFN-β is widely used to eliminate hepatitis C virus. However, its unsatisfactory clinical efficacy necessitates the targeting IFN-β to the liver with lowered side effects. Xi et al. (1996) showed that the liver accumulation of IFN was higher after conjugation with cyanurated pullulan, which retained ~7–9% of the IFN bioactivity. IV injection of the IFN–pullulan conjugate markedly enhanced the activity of 2′,5′-oligoadenylate synthetase (2–5AS) enzyme in the mice liver. The enzyme activity sustained for 3 days, but the same was lost within the first day after free IFN injection. Undoubtedly, IFN–pullulan conjugates exerted better antiviral activity in the liver.

Suginoshita et al. (2002) reported a similar finding with diethylenetriaminepentaacetic acid (DTPA)–pullulan–zinc coordinate conjugates with natural human IFN-β. Intravenous injection of the conjugate significantly induced an antiviral enzyme, 2-5AS, specifically in the liver as compared to free natural IFN-β. Moreover, the duration of the enzyme induction was longer than that by free natural IFN-β, exhibiting their potential in clinical applications.

The conjugation with pullulan via spacers suppressed the loss of activity of INF (Tabata et al. 1999). The pullulan covering physically prevented the interaction of INF molecules with cell surface receptors and caused lowering of antiviral action of IFN. The low molecular size of pullulan and the spacer incorporation reduced the steric hindrance of pullulan and enhanced the drug interaction with the receptors.

Some workers confirmed the receptor-mediated uptake of pullulan in liver cells (Kaneo et al. 2001; Tanaka et al. 2004). The hepatic uptake of fluorescein-labeled pullulan, FP-60, was dose-dependent in rats. The coadministration of asialofetuin and arabinogalactan markedly reduced the hepatic uptake. FP-60 circulated in blood for a long time due to the saturation of hepatic uptake at high doses. Since nonparenchymal cells occupy ~3% of the total rat liver, the majority of FP-60 accumulated in the parenchymal cells. Hence, pullulan was thought to endocytose via asialoglycoprotein (ASGPR) receptors in liver parenchymal cells. The stereochemistry of pyranose rings and mutual orientation of the polymer hydroxyl groups governs specific glucan–receptor binding (Iobst and Drickamer 1996). The conformational transitions from $C_{1a}$ to boat conformers, as found in α-(1,4)-linked glucose residues of pullulan, may determine the affinity of glucan toward ASGPR receptors (Yamaoka et al. 1993).

The carboxymethylation of pullulan (CMP) imparts negative charge, high MW (Nogusa et al. 1995), enhanced vascular permeability (Takakura and Hashida 1996), and pH sensitivity (Dulong et al. 2007). CMP could form hydrazone bonds with hydrophobic drugs and led to a higher amount of drug delivery to tumor cells in comparison to the drug alone. CMP released a higher amount of the drug *in vitro* at pH 5.0 than at pH 7.4 due to pH sensitivity (Lu et al. 2008). Recently, Li et al. (2014) targeted mouse fibroblast cells, human liver cancer cells, and human cervical carcinoma cells by conjugation of DOX with CMP at a pH range of 4.5–7.4. Higher liver targeting ability was achieved in the range of pH 4.5–6.5. CMP (>70 kDa)– drug conjugation through peptide spacers showed less systemic toxicity than DOX (Nogusa et al. 1995). Overall, the properties of drug, macromolecule, and spacer affected the desired drug release rate to the targeted organ/tissues (Nogusa et al. 2000). CMP–DOX conjugates bound via appropriate dipeptide spacers were more potent in exhibiting antitumor activity than DOX.

CMP and sialyl Lewis X conjugate was prepared by Horie et al. (1999). E-selectin, a cell-binding molecule, plays a vital role in receptor-mediated site-specific drug delivery by coupling with ligand incorporating sialyl Lewis X. It has been stated that the sialyl Lewis X–CMP complex is not susceptible to filtration at the glomerulus and can accumulate directly in inflammatory lesions (Takakura and Hashida 1996). Sialyl Lewis X–CMP complex at inflammatory sites upregulate the E-selectin expression and can be used in the treatment of a wide range of human diseases, such as atherosclerosis, ischemia, graft rejection, asthma, and tumor (Rohde et al. 1992; Bevilacqua et al. 1994). Cytokines (IL-1β and TNF-α) helps in the expression of E-selectin on the inflammatory sites (McEver 1992).

The conversion of polysaccharide into a reactive derivative is usually required to make a drug–polymer conjugate. These derivatives help in proper attachment and appropriate functionality of drugs or bioactive compounds with polysaccharides (Barker et al. 1971). However, the reports argued that derivatization could reduce or nullify the liver affinity of pullulan (Nogusa et al. 2000a).

Gene therapy could be a cure for various inherited disorders/diseases. However, the use of viral vectors can be risky mainly due to immunogenicity of viruses (Simon et al. 1993). Therefore, cationic polysaccharide derivatives are progressively being investigated as possible nonviral vectors for gene delivery (Lee and Kim 2002). Polyethylenimine (PEI) is an efficient cationic nonviral gene transfer vector. However, the toxicity and incompatibility with blood components are the major drawbacks (Lv et al. 2006; Masago et al. 2007). Further, the cationic surface charge of nonviral vectors may cause nonspecific interactions with cellular blood components, proteins, platelet activation, etc. PEGylation can improve blood compatibility, but reduces transfection efficiency (Liang et al. 2009). Hence, new cationic polymers must be sought to overcome these problems. Owing to its nontoxic, nonimmunogenic, nonmutagenic, and noncarcinogenic nature, pullulan has drawn attention of research workers to explore its potential for targeted drug delivery and gene delivery vectors (Kimoto et al. 1997; Na and Bae 2002).

Other groups reported cationic pullulan vectors for gene delivery. For example, Jo et al. (2006) established *in vivo* transfection efficiency of spermine pullulan. Juan et al. (2007) demonstrated the efficiency of a 3D DEAE–pullulan matrix for gene transfer directly to the tissue. The cationic pullulan derivatives formed nanocomplexes (<200 nm) with calf thymus DNA (Rekha and Sharma 2009). The derivative with lower cationic substitution of glycidyl trimethyl amine was frequently blood compatible. Despite derivatization, cationic pullulan retained liver binding affinity in amice model and show good transfection efficiency on HepG2 cells.

Pullulan derivatives can also be used as carriers for DNA plasmid or genes to various organs or cancer cells in the form of NPs or nanogels. Gupta and Gupta

(2004) modified pullulan to fabricate hydrogel NPs with hydrophilic core. The NPs encapsulated pBUDLacZ plasmid for targeted delivery without cytotoxic events. The ligand was attached to the surface of NPs for selective gene delivery. The particles provided protection to the DNA or gene from DNase degradation and also showed high transfection potential.

Owing to high charge density, polyethyleneimine (PEI)–pullulan was complexed with siRNA for treating almost incurable diseases—tumors, viral infections, etc. (Kim and Rossi 2008; Zhang et al. 2009). The injection of siRNA–PEI–pullulan conjugate through the tail vein resulted in increased mortality of mice (Kang et al. 2010). The cellular uptake, transfection efficiency to HepG2 cells, blood compatibility, and stability were higher for PEI–pullulan (Rekha and Sharma 2011). Thus, the carboxyl group of folate coupled with the amino group of PEI when branched on pullulan led to higher gene transfection and a silencing effect in the liver.

Out of three different cationic vectors, DTPA, triethylenetetramine, and spermine, the IV administration of $Zn^{2+}$-coordinated DTPA–pullulan–plasmid DNA exhibited a higher level of gene expression in liver cells over 12 days than that of free plasmid DNA (Hosseinkhani et al. 2002). The level of gene expression was dependent on the degree of chelation, plasmid DNA dose, and mixing ratio with DTPA residue in the conjugate. The level of gene expression following arabinogalactan and galactosylated albumin injection was significantly greater than mannosylated albumin, indicating DNA transfection in the hepatocytes. Thus, $Zn^{2+}$-coordinated pullulan conjugation could be a promising way of targeting plasmid DNA to the liver for gene expression and prolonged duration of gene expression.

According to Park et al. (2012), the cytotoxicity of pullulan-$g$-poly (1-lysine) NPs was negligible and was thus safe for gene delivery. The chemical structure of pullulan-$g$-poly (1-lysine) is shown in Figure 18.5.

Thomsen et al. (2011) conjugated plasmid DNA with pullulan–spermine complex and transfected rat and human brain endothelial cells. Pullulan–spermine complex helped in easy transfection of plasmid DNA, and consisted of cDNA encoding for the human growth hormone 1 (hGH1). They noted that the pullulan–spermine system could help in DNA and protein delivery to the brain endothelial cells. Kanatani et al. (2006) targeted T24 cells of human bladder cancer by conjugating DNA with pullulan–spermine complex. The tricomplex caused higher gene transfection and cellular uptake with acceptable cytotoxicity. This complex was also tested for neuronal gene

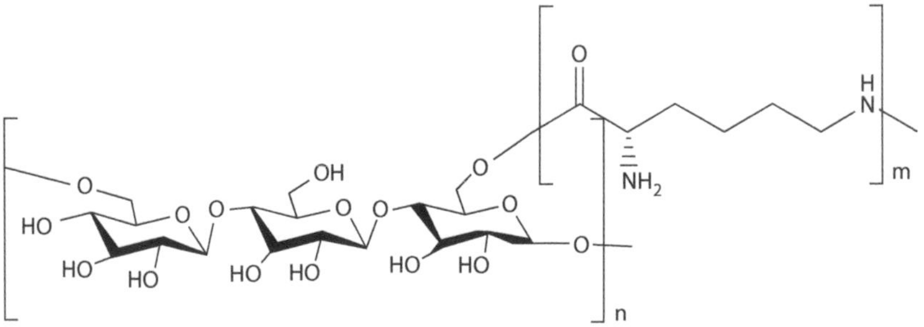

**FIGURE 18.5**    Chemical structure of pullulan-$g$-poly (1-lysine).

delivery (Thakor et al. 2009). Pullulan–spermine can also transfer the Notch intracellular domain gene to bone marrow stromal cells (BMSCs). The BMSCs are consisted of dopamine-producing neuronal cells, which secrete dopamine via reverse transfection technique for the treatment of Parkinson's disease (Nagane et al. 2009).

Several CMP–DOX micellar conjugates have higher affinity toward tumor cells than free DOX (Nogusa et al. 1995, 2000b). The length of the spacer connecting the carboxyl group of CMP and the amino group of DOX controlled the release of drug from CMP carriers (Nogusa et al. 2000).

Nanomicellar conjugates (70 nm) of sulfodimethoxine-pullulan acetate also exhibited tumor cell penetrability (Na and Bae 2002). The particle surface was decorated with hydroxyl and carboxy groups of pullulan acetate and sulfodimethoxine moieties. Depending on the ionization of carboxyl groups, the surface charge of particles differed over the pH range of 6.5–7.2. Ionization of the sulfodimethoxine moieties provoked repulsion between particles, and was thus stable above pH 7.4. The deionization of sulfodimethoxine groups at pH <7.2 accelerated drug release from the composites. Cholesteryl pullulan (CHP) bearing aminolactose, a cell recognition element, was a promising carrier for anticancer drugs (Yamamoto et al. 2000). Further, CHP–oncoprotein complex was efficient in cancer therapy (Gu et al. 1998).

Nogusa et al. (2000) studied the effect of phenylalanine–glycine spacer in CMP–FG–DOX carriers on the antitumor activity of DOX and compared it with that of DOX. The structure of the conjugate is depicted in Figure 18.6. The drug–carrier conjugate offered higher antitumor efficacy against Lewis lung carcinoma than DOX. Complete tumor regression and long-term tumor-free survival were noticed after multiple i.v. injections of 10 mg/kg dose. These effects of the conjugate were superior to DOX in M5076 murine reticulosarcoma model with lower systemic toxicity. The physical mixture of CMP with DOX did not attenuate the antitumor activity of DOX, showing the importance of covalent conjugation for improved antitumor activity. However, no improvement in antitumor efficacy was noted in the P388 leukemia nonsolid tumor model.

Scomparin et al. (2011) evaluated two new polymer therapeutic conjugates for tumor cell targeting. The design of the conjugates involved multiple components and steps. Briefly, the oxidized pullulan was functionalized with cysteamine and diaminepolyethylene glycol (PEG $(NH_2)_2$) by reductive conjugation. The thiol groups of cysteamine were then conjugated to DOX via a pH-sensitive hydrazone spacer in the first conjugate. The folic acid was attached to the pending PEG-NH$_2$ functions to obtain second polymer therapeutics.

After passive accumulation of the nonfolated conjugate in the tumor, the drug was selectively released into the cell by endosomal carrier disruption. The receptor-mediated endocytosis of the folated conjugate caused higher drug accumulation in cancer cells by a clathrin-independent mechanism (Anderson et al. 1992). Folic acid residues along the polysaccharide backbone can establish a cooperative interaction between the conjugate and folate receptors located on the cell membrane and enhance cellular uptake by caveolar mechanism. Both the conjugates self-assembled into NPs (100–150 nm). The particle size was less than the cut-off size for passive tumor disposition by EPR (Campbell 2006). Only ~20% of the conjugated DOX was released in 3 days from both derivatives. The hydrazone bond (stable at pH 7.4)

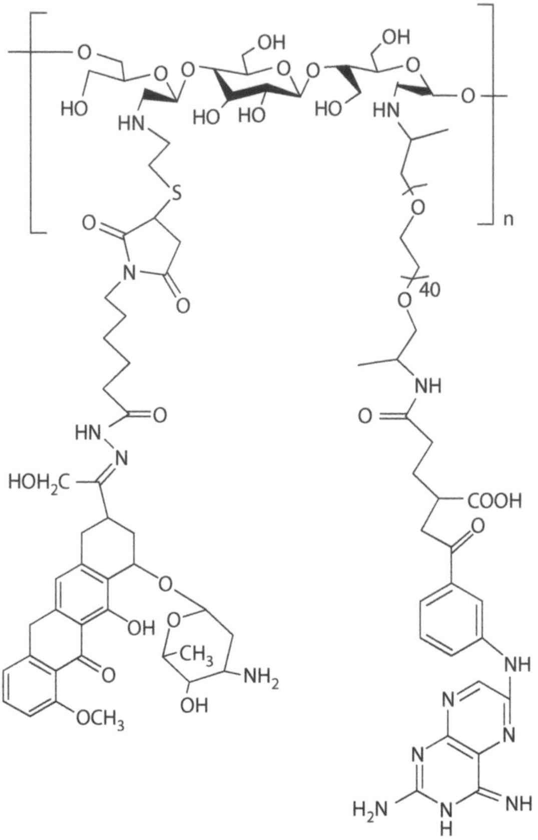

**FIGURE 18.6** Structure of the CMP–DOX conjugates with phenylalanine–glycine spacer.

prevented systemic toxicity of DOX and nonspecific organ/tissue diffusion. The complete drug release at pH 5.5 (lysosomal pH) in about 40 h ensured DOX release into tumor cells.

The folate conjugate was rapidly internalized by the HeLa cells as compared to the nonfolated derivative. Both conjugates showed higher $IC_{50}$ values than bare DOX with KB and MCF7 cells. The nonfolated bioconjugate displayed slightly better cytotoxicity against the MCF7 cell line, which does not overexpress the folate receptor. The small size, unconjugated amino groups, and the supramolecular structure of the nonfolated conjugate conveyed an overall positive charge to the carrier and thus favored its nonspecific cell uptake. However, the glucan conjugation reduced the activity of DOX in either MCF7 or KB cells. The folated conjugate reduced the efficacy of prodrug in MCF7 cells. However, this trend reversed in KB cells. Unexpectedly, the folate derivative was only slightly more active than the nonfolated conjugate toward folate receptor overexpressing KB cells. This indicated that folic acid conjugation was marginally effective in conveying targeting properties to the drug carrier. The glucan conjugation prolonged the circulation time of DOX *in vivo*.

Even 4 h after intravenous administration of the folated conjugate to Balb/c mice, about 40% of the administered drug was present in the bloodstream. In the case of nonconjugated DOX, 80% of the drug cleared within 0.5 h. Thus, the DOX–pullulan conjugates were suitable for passive tumor targeting. The folated conjugates had limited effects on selective cellular uptake (Scomparin et al. 2011).

The biotin–polymer conjugate are uptaken by cancer cells and may exhibit higher drug distribution in malignant tumors tissues than in the normal (Heo et al. 2012; Xiong et al. 2011). According to the literature, biotin receptors overexpress in numerous tumor cells (Li et al. 2013). In this regard, the report of Na et al. (2003) makes it easier to understand the real facts. They examined self-aggregated NPs of biotin-*g*-pullulan acetate (BPA) for targeting of the anticancer drug. The BPA NPs strongly adsorbed to the HepG2 cells, while the PA NPs did not interact significantly with the cells. Therefore, biotin was supposed to act as the active tumor targeting ligand for various anticancer drugs with low cytotoxicity. A low level of antigenicity and immunogenicity enhanced the intracellular uptake of drug further in cancer cells (Bu et al. 2013; Kim et al. 2012).

Various groups demonstrated that PEG conjugation could improve the blood compatibility but reduce the transfection efficiency of nonviral vectors (Mishra et al. 2004; Germershaus et al. 2008). PEI is one of the most successful and efficient nonviral gene transfer vectors. However, its toxicity and incompatibility with blood components are problematic (Lv et al. 2006; Masago et al. 2007). Hence, Rekha and Sharma (2011) developed a new cationic vector using pullulan and PEI (25 kDa) for gene delivery applications.

The pullulan conjugation drastically reduced the cytotoxicity and blood component interactions in comparison to PEI-based DNA nanocomplexes. The smallest size of polymer–DNA complexes ($69 \pm 0.99$ nm) was obtained at the N/P charge ratio of 4.0. The cytotoxicity was dependent on the polymer:vector ratio. At a ratio of 1:1 (PPE3), the cationic pullulan conjugate was most cytotoxic and least at a ratio of 1:0.25 (PPE1) for HepG2 cell lines. The cells were more than >80% viable for PPE1. This was <50% for PPE3 and PEI. The DNA complexation reduced the cytotoxicity in comparison to the polymer alone.

DNA transfection efficiency of PPE1 in HepG2 cells was comparable to PEI. The hydration of pullulan prevents nonspecific interactions between pullulan and blood cells/components and therefore, can improve its blood compatibility. Gonçalves et al. (2004) highlighted that PEI/DNAplexes were internalized via clathrin-dependent and clathrin-independent pathways. Clathrin-independent uptake occurs via phagocytosis or macropinocytosis. The inhibition of the clathrin-mediated pathway completely bared the uptake of nanoplexes. This was contradictory to that reported by Rekha and Sharma (2011). They observed higher uptake and transfection efficiency in the presence of chlorpromazine, a clathrin-mediated endocytosis inhibitor. A similar uptake mechanism was observed by Gabrielson and Pack (2009). They found a threefold increase in the delivery efficiency of PEI/DNA in the presence of chlorpromazine, perhaps due to increased uptake of polyplexes via nonspecific pathways. However, PPE1 facilitated higher intake due to specific ASGPR-mediated internalization because caveosomes were absent in HepG2 cells. In the presence of the ASGPR-mediated endocytosis inhibitor ASF, only partial inhibition of the nanoplexes uptake was noted possibly via other pathways. Because the particles of <200 nm size are internalized via clathrin-mediated endocytosis, DNA nanoplexes (~70 nm) were thus susceptible to clathrin-mediated endocytosis. The buffering capacity of all pullulan–PEI conjugates was similar to that of PEI and contributed to the transfection efficiency. Dextran conjugates, reported by Murthy et al. (1996), caused significant reduction of the transfection efficiency of PEI. The steric hindrances imparted by dextran might have decreased the protonation degree of PEI and reduced the cationic charge density on the vector. Overall, the buffering effect in conjunction with the subsequent osmotic effect, flexibility, and swelling of the polymer led to vesicle disruption and consequently released polyplexes into the cytosol (Rejman et al. 2005). The high flexibility debarred pullulan from imparting steric hindrance effect, resulting in reasonable transfection.

The interaction between the therapeutic gene and nano-delivery system must be optimal in that the electrostatically bound DNA remains stable in the extracellular environment and simultaneously releases DNA in the intracellular environment (Layek and Singh 2012).

Protamine, an arginine-rich cationic nuclear protein (MW ~ 4000 Da), is essential for sperm head condensation and DNA stabilization in spermatogenesis. The role of protamine in delivering the sperm DNA into the nucleus of the egg following fertilization is familiar to us. This function would ensure the transport of plasmid DNA from the cytoplasm into the nucleus (Sorgi et al. 1997). Therefore, polycationic protamine could be introduced for the design of novel gene delivery systems. It was shown that pullulan could be bound to cationic vectors for their efficient transportation in the targets (Kimoto et al. 1997). Because of questionable DNA aggregation, Priya et al. (2014) used protamine to cationize pullulan, improve DNA complexation, and its transfection ability.

The condensation of DNA with pullulan protamine (PPA) resulted in positively charged NPs (<100 nm) and protected from DNase degradation and plasma protein interactions. Owing to the hydrophilic nature of pullulan, PPA showed excellent hemocompatibility and improved cell viability. Moreover, the conjugates were more readily endocytosed and transfected efficiently, probably due to the unique nuclear localization signals of protamine even at 400 µg/mL polymer concentration. This property can be utilized for *in vivo* gene transfection, if a large amount of cationic polymers is needed.

Since low cytotoxicity and high transfection efficiency are essential for all gene delivery systems, the cell viability of cationic polymer was evaluated on C6 cells using different concentrations of PPA. Usually, the cytotoxicity is mediated through ionic interactions between anionic domains on cell surface and cationic vectors. Therefore, the aggregation and accumulation of polyplexes at the cell surface could severely impair membrane function, leading to cell death. However, the cell viability of the glucan derivative was ~70% at 200–400 µg/mL concentration range. The glucan conjugation might have reduced positive charge on the vector for anionic membrane interaction and consequently, the cells became more viable. The polymer conjugation did not affect the efficiency of protamine in delivering the DNA into the nucleus. The exogenous chlorpromazine, filipin, and amiloride can inhibit the clathrin-, caveolae-, and macropinocytosis-mediated endocytosis, respectively.

Therefore, they used these three inhibitors to confirm the mechanism of intracellular uptake of the nanoplexes. In the presence of all inhibitors, the uptake of the nano-plexes was reduced. Then, it was said that the multiple endocytotic pathways were involved in the uptake process. The reduction was much higher with the caveolae-mediated endocytic pathway. High p53 gene transfection efficiency was ensured at a polymer/DNA ratio of 25:1, where the cell death was greater than 80% after 48 h of incubation.

As reported by Boridy et al. (2009), the stable, monodisperse hydrogel NPs of CHP (20–30 nm) can interact with soluble proteins via hydrophobic bonding and may help in the amelioration of neurological disorders. The structure of CHP is given in Figure 18.7.

Both neutral and positively charged CHP NPs interacted with 42 amino acid vari-ant of β-amyloid (Ab$_{1-42}$) monomers and oligomers. However, the positively charged derivatives (CHPNH$_2$) exhibited toxicity to primary cortical cultures. The dual bind-ing of monomeric and oligomeric Ab$_{1-42}$ to CHP reduced Ab$_{1-42}$ toxicity significantly in primary cortical and microglial cells. Hence, the design of CHP nanogels may be a complementary approach to antibody immunotherapy in neurological disorders, such as AD.

The intra- and extracellular accumulation of protein aggregates promotes the formation of plaques. These are now considered as reservoirs for toxic oligomeric species of the β-amyloid peptide (Ab) (Haass and Selkoe 2007). For instance, Ab oligomers have been found to associate with senile plaques near excitatory synapses in a mouse model of AD (Koffie et al. 2009). The endogenous molecular chaperones are the proteins that assist in covalent folding or unfolding of other macromolecular structures. They are believed to regulate the aggregation and elimination of toxic misfolded species, thereby preventing cell damage, especially in susceptible postmi-totic cells, such as neurons. Hence, it strongly appeals that nanogels can prevent Ab fibrillation *in vitro* (Ikeda et al. 2006).

Only a few studies have been devoted to reveal the responsiveness of neural cells to nanomaterials, in particular, their effects on astrocyte and microglia activation (Maysinger 2007). Astrocytes are the principal macroglial cell types of the cen-tral nervous system (CNS). Their activation regulates neural cell responses to stress and brain injuries. The star-shaped neural cells can have both positive and adverse

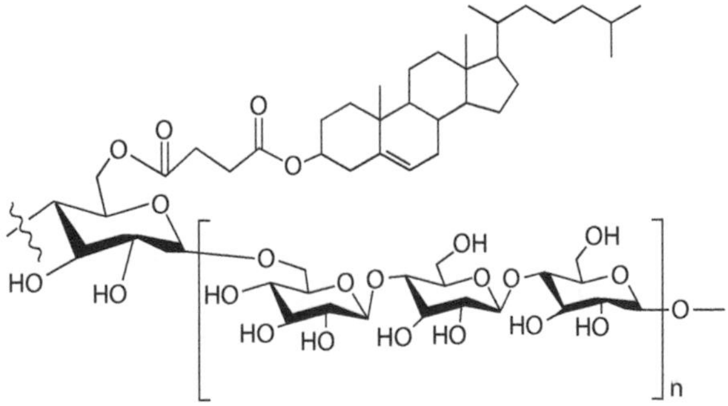

FIGURE 18.7   Modified cholesteryl pullulan structures.

effects on neuron health depending on their activation status (Emsley et al. 2004). Notably, their transition from the resting state to the activated state is associated with an upregulation of an intermediate filament, specific to astrocytes known as glial fibrillary acidic protein (GFAP). Microglia is the resident macrophage population of the CNS (Hanisch and Kettenmann 2007; Napoli and Neumann 2009), which responds robustly and rapidly to bio- or artificial particles (Nimmerjahn et al. 2005). However, a suitable nanomaterial for treating brain disorders is still lacking. Toxicity and specificity are two major concerns for nanomaterials (Lewinski et al. 2008). No doubt, CHP nanogels can favorably interact with Ab monomers and oligomers and enhance cell viability in microglia and primary cortical cultures. However, this attempt remains in its infancy.

## 18.4 FUTURE PERSPECTIVES

The well-documented advantages of glucan biopolymers have enabled global research workers to fabricate various nanodrug delivery systems. In particular, their chemical derivatives have been extensively investigated in this area. The glucan derivatives have been found to have various biological activities. Though the biopolymer derivatives are thought to behave as inert drug delivery carriers, recent findings demonstrated that the carrier might influence biological activities. However, the research workers may take the advantages of this property in combining the therapeutic efficacy of similar type of therapeutic entities. Furthermore, the hydrophilic character of the biopolymers may be utilized to devise nanocarriers to make them long circulating and prevent opsonization. This has drawn the attention of many research groups to improve the effectiveness of nanomedicines. Nonetheless, a number of hurdles have to be overcome. Being natural, the polymers are highly variable with regard to MW and sugar units. Since the biological activities are mostly affected, the reproducible production of polysaccharides needs to be addressed for further progress of polysaccharide-based nanomedicines. Furthermore, glucan polysaccharides are often contaminated with endotoxins and pathogens. The active contaminants may counteract the desired effect of the polysaccharide, and hence, purification to an unprecedented level is suggested. The mechanism of potential bioactivities of polysaccharides still remains to be identified. It is not surprising to notice contradictory bioactivities in the same polysaccharide. A subtle difference in MW, degree of branching, or the arrangement of monomers can cause significant differences in their biological effects. The biological effects and fates of excess polysaccharides and degradation products are not entirely clear yet. The optimization of NP properties is another concern in specific drug targeting, while minimizing nonspecific tissue residence. Thus, a better and thorough understanding of these issues is a prerequisite for successful therapeutic application of the glucan biopolymers as well as their nanoconjugations. However, the ultimate performance of the nanosystems described herein must be established in clinical trials. The toxicological information on polysaccharide-based NPs is scarce in the literature. Hence, the toxicity, pharmacokinetics, and biodistribution of the biopolymer derivatives are important considerations in the field of glucan polymer-based drug delivery applications.

# REFERENCES

Abrams, D.I., Kuno, S., Wong, R. et al. 1989. Oral dextran sulfate (UA 001) in the treatment of acquired immunodeficiency syndrome (AIDS) and AIDS-related complex. *Ann Intern Med* 110:183–88.

Alban, S. 1997. Carbohydrates with anticoagulant and antithrombotic properties. In *Carbohydrates in Drug Design*, ed. Z.J. Witezak, and K.A. Nieforth, 209. New York: Marcel Dekker.

Alban, S. and Franz, G. 1994. Gas liquid choromatography-mass spectrometry analysis of anticoagulant active curdlan sulfate. *Semin Thromb Hemost* 20:152–58.

Alban, S. and Franz, G. 2000. Characterization of the anticoagulant actions of a semisynthetic curdlan sulfate. *Thromb Res* 99:377–88.

Alban, S., Schauerte, A., and Franz, G. 2002. Anticoagulant sulfated polysaccharides: Part I. Synthesis and structure–activity relationships of new pullulan sulfates. *Carbohydr Polym* 47:267–74.

Anderson, R.G., Kamen, B.A., Rothberg, K.G., and Lacey, S.W. 1992. Potocytosis: Sequestration and transport of small molecules by caveolae. *Science* 255:410–11.

Baba, M., Pauwels, R., Balzarini, J., Arnout, J., Desmyter, J., and DeClercq, E. 1988. Mechanism of inhibitory effect of dextran sulfate and heparin on replication of human immunodeficiency virus in vitro. *Proc Natl Acad Sci USA* 85:6132–36.

Baba, M., Schols, D., Pauwels, R., Nakashima, H., and De Clercq, E. 1990. Sulfated polysaccharides as potent inhibitors of HIV-induced syncytium formation: A new strategy towards AIDS chemotherapy. *J Acquir Immune Defic Syndr* 3:493–99.

Bagasra, O. and Lischner H.W. 1988. Activity of dextran sulfate and other polyanionic polysaccharides against human immunodeficiency virus. *J Infect Dis* 158:1084–88.

Barker, S.A., Tun, H.C., Doss, S.H., Gray, G.J., and Kennedy, J.F. 1971. Preparation of cellulose carbonate. *Carbohydr Res* 17:471–74.

Beneke, C., Viljoen, A., and Hamman, J. 2009. Polymeric plant-derived excipients in drug delivery. *Molecules* 14:2602–20.

Bevilacqua, M.P., Nelson, R.M., Mannori, G., and Cecconi, O. 1994. Endothelial–leukocyte adhesion molecules in human disease. *Annu Rev Med* 45:361–78.

Blondin, C., Fischer, E., Boisson-Vidal, C., Kazatchkine, M.D., and Jozefonvicz, J. 1994. Inhibition of complement activation by natural sulfated polysaccharides (fucans) from brown seaweed. *Mol Immunol* 31:247–53.

Bohn, J.A., and BeMiller, J.N. 1995. $(1 \rightarrow 3)$-$\beta$-D-Glucans as biological response modifiers: A review of structure-functional activity relationships. *Carbohydr Polym* 28:3–14.

Boridy, S., Takahashi, H., Akiyoshi, K. et al. 2009. The binding of pullulan modified cholesteryl nanogels to A$\beta$ oligomers and their suppression of cytotoxicity. *Biomaterials* 30:5583–91.

Brigger, I., Dubernet, C., and Couvreur, P. 2002. Nanoparticles in cancer therapy and diagnosis. *Adv Drug Deliv Rev* 54:631–51.

Bu, L., Gan, L.C., Guo, X.Q. et al. 2013. Transresveratrol loaded chitosan nanoparticles modified with biotin and avidin to target hepatic carcinoma. *Int J Pharm* 452:355–62.

Busso, M.E. and Resnick, L. 1990. Anti-human immunodeficiency virus effects of dextran sulfate are strain dependent and synergistic or antagonistic when dextran sulfate is given in combination with dideoxynucleosides. *Antimicrob Agents Chemother* 34:1991–95.

Callahan, L.N., Phelan, M., Mallinson, M., and Norcross, M.A. 1991. Dextran sulfate blocks antibody binding to the principal neutralizing domain of human immunodeficiency virus type-1 without interfering with gp120-CD4 interactions. *J Virol* 65:1543–50.

Campbell, R.B. 2006. Tumor physiology and delivery of nanopharmaceuticals. *Anticancer Agents Med Chem* 6:503–12.

Chen, Y., Hu, M., Wang, C. et al. 2013. Characterization and *in vitro* antitumor activity of polysaccharides from the mycelium of *Sarcodon aspratus*. *Int J Biol Macromol* 52:52–58.

Chihara, G., Hamuro, J., Maeda, Y.Y. et al. 1987. Antitumor metastasis inhibitory activities of lentinan as an immunomodulator: An overview. *Cancer Detect Prev Suppl* 1:423–43.

Cleary, J.A., Kelly, G.E., and Husband, A.J. 1999. The effect of molecular weight and β-1,6-linkages on priming of macrophage function in mice by (1,3)-β-D-glucan. *Immunol Cell Biol* 77:395–403.

d'Ayala, G.G., Malinconico, M. and Laurienzo, P. 2008. Marine derived polysaccharides for biomedical applications: Chemical modification approaches. *Molecules* 13:2069–106.

Degeest, B., Vaningelgem, F., and De Vuyst, L. 2001. Microbial physiology, fermentation kinetics, and process engineering of heteropolysaccharide production by lactic acid bacteria. *Int Dairy J* 11:747–57.

Demleitner, S., Kraus, J., and Franz, G. 1992a. Synthesis and antitumour of curdlan and lichenan. *Carbohydr Res* 226:247–52.

Demleitner, S., Kraus, J., and Franz, G. 1992b. Synthesis and antitumour activity of derivatives of curdlan and lichenan branched at C-6. *Carbohydr Res* 226:239–46.

Donot, F., Fontana, A., Baccou, J.C., and Schorr-Galindo, S. 2012. Microbial exopolysaccharides: Main examples of synthesis, excretion, genetics and extraction. *Carbohyd Polym* 87:951–62.

Du, Y.Z., Weng, Q., Yuan, H., and Hu, F.Q. 2010. Synthesis and antitumor activity of stearate-g-dextran micelles for intracellular doxorubicin delivery. *ACS Nano* 11:6894–902.

Dulong, V., Mocanu, G., and LeCerf, D. 2007. A novel amphiphilic pH-sensitive hydro-gel based on pullulan. *Colloid Polym Sci* 285:1085–91.

Duncan, R. 1999. Polymer conjugates for tumour targeting and intracytoplasmic delivery. The EPR effect as a common gateway. *Pharm Sci Technol Today* 2:441–49.

Dunn, S.E., Brindley, A., Davis, S.S., Davies, M. C., and Illum L. 1994. Polystyrene-poly (ethylene glycol) (PS-PEG2000) particles as model systems for site specific drug delivery. 2. The effect of PEG surface density on the in vitro cell interaction and in vivo biodistribution. *Pharm Res* 11:1016–22.

Emsley, J.G., Arlotta, P., and Macklis, J.D. 2004. Star-cross'd neurons: Astroglial effects on neural repair in the adult mammalian CNS. *Trends Neurosci* 27:238–40.

Falch, B.H., Espevik, T., Ryan, L., and Stokke, B.T. 2000. The cytokine stimulating activity of (1 → 3)-β-D-glucans is dependent on the triple helix conformation, *Carbohydr Res* 329:587–96.

Fernandes, S.C.M., Sadoccom, P., Causio, J., Silvestre, A.J.D., Mondragon, I., and Freire, C.S.R. 2014. Antimicrobial pullulan derivative prepared by grafting with 3-aminopropyltrimethoxysilane: Characterization and ability to form transparent films. *Food Hydrocolloids* 35:247–52.

Flexner, C., Barditch-Crovo, P.A., Kornhauser, D.M. et al. 1991. Pharmacokinetics, toxicity, and activity of intravenous dextran sulfate in human immunodeficiency virus infection. *Antimicrob Agents Chemother* 35:2544–50.

Franz, G. and Alban, S. 1995. Structure-activity relationship of antithrombotic polysaccharide derivatives. *Int J Biol Macromol* 17:311–14.

Freed, E.O., Myers, D.J., and Risser, R. 1991. Identification of the principal neutralizing determinant of human immunodeficiency virus −1 as a fusion domain. *J Virol* 65:190–94.

Fritzsche, J., Alban, S., Ludwig, R.J. et al. 2006. The influence of various structural parameters of semisynthetic sulfated polysaccharides on the P-selectin inhibitory capacity. *Biochem Pharmacol* 72:474–85.

Gabrielson, N.P. and Pack, D.W. 2009. Efficient polyethylenimine-mediated gene delivery proceeds via a caveolar pathway in HeLa cells. *J Control Release* 136:54–61.

Gao, F., Li, L., Zhang. H. et al. 2010. Deoxycholic acid modified-carboxymethyl curdlan conjugate as a novel carrier of epirubicin: In vitro and in vivo studies. *Int J Pharm* 392:254–60.

Gao, F.-P., Zhang, H.-Z., Liu, L.-R. et al. 2008. Preparation and physicochemical characteristics of self-assembled nanoparticles of deoxycholic acid modified carboxymethyl
curdlan conjugates. *Carbohydr Polym* 71:606–13.

Germershaus, O., Mao, S., Sitterberg, J., Bakowsky, U., and Kissel, T. 2008. Gene delivery using chitosan, trimethyl chitosan or polyethyleneglycol–graft–trimethyl chitosan
block copolymers: Establishment of structure–activity relationships in vitro. *J Control
Release* 125:145–54.

Giavasis, I., Harvey, L.M., and McNeil, B. 2005. Scleroglucan polysaccharides. In *Biopolymers
Online*, ed. A. Steinbuchel, Weinheim: Wiley-VCH Verlag GmbH.

Glinel, K., Sauvage, J.P., Oulyadi, H., and Huguest, J. 2000. Determination of substituents distribution in carboxymethylpullulans by NMR spectroscopy. *Carbohydr Res*
328:343–54.

Gonçalves, C., Mennesson, E., Fuchs, R., Gorvel, J.-P., Midoux, P., and Pichon, C. 2004.
Macropinocytosis of polyplexes and recycling of plasmid via the clathrin-dependent
pathway impair the transfection efficiency of human hepatocarcinoma cells. *Mol Ther*
10:373–85.

González, M.E., Alarcón, B., and Carrasco, L. 1987. Polysaccharides as antiviral agents:
Antiviral activity of carrageenan. *Antimicrob Agents Chemother* 31:1388–93.

Gordon, M., Guralnik, M., Kaneko, Y., Mimura, T., Baker, M., and Lang, W. 1994. A phase
I study of curdlan sulfate—An HIV inhibitor. Tolerance, pharmacokinetics and effects
on coagulation and on CD4 lymphocytes. *J Med* 25:163–80.

Gref, R., Minamitake, Y., Peracchia, M.T., Trubetskoy, V., Torchilin, V., and Langer, R. 1994.
Biodegradable long-circulating polymeric nanospheres. *Science* 263:1600–03.

Gu, X.G., Schmitt, M., Hiasa, A. et al. 1998. A novel hydrophobized polysaccharide/onco-
protein complex vaccine induces in vitro and in vivo cellular and humoral immune
responses against HER2-expressing murine sarcomas. *Cancer Res* 58:3385–90.

Gupta, M. and Gupta, A.J. 2004. Hydrogel pullulan nanoparticles encapsulating pBUDLacZ
plasmid as an efficient gene delivery carrier. *J Control Release* 99:157–66.

Haass, C. and Selkoe, D.J. 2007. Soluble protein oligomers in neurodegeneration: Lessons
from the Alzheimer's amyloid beta-peptide. *Nat Rev Mol Cell Biol* 8:101–12.

Hanisch, U.K. and Kettenmann, H. 2007. Microglia: Active sensor and versatile effector cells
in the normal and pathologic brain. *Nat Neurosci* 10:1387–94.

Harada, H., Sakagami, H., Konno, K. et al. 1988. Induction of antimicrobial activity by antitumor substances from pine cone extract of *Pinus parviflora* sieb. et Zucc. *Anticancer
Res* 8:581–88.

Hashizume, H., Baluk, P., Morikawa, S. et al. 2000. Openings between defective endothelial
cells explain tumor vessel leakiness. *Am J Pathol* 156: 1363–80.

Hawley A. E., Illum L., and Davies S.S. 1997. Preparation of biodegradable, surface engineered PLGA nanospheres with enhanced lymphatic drainage and lymph node uptake.
*Pharm Res* 14: 657–61.

Heinz, F.X. and Stiasny, K. 2012. Flaviviruses and flavivirus vaccines. *Vaccine* 30:4301–06.

Heo, D.N., Yang, D.H., Moon, H.J. et al. 2012. Gold nanoparticles surface-functionalized
with paclitaxel drug and biotin receptor as theranostic agents for cancer therapy.
*Biomaterials* 33:856–66.

Hopfner, M., Alban, S., Schumacher, G., Rothe, U., and Bendas, G. 2003. Selectin-blocking
semisynthetic sulfated polysaccharides as promising anti-inflammatory agents. *Pharm
Pharmacol* 55:697–706.

Horie, K., Sakagami, M., Kuramochi, K., Hanasaki, K., Hamana, H., and Ito, T. 1999.
Enhanced accumulation of sialyl Lewis X-carboxymethylpullulan conjugate in acute
inflammatory lesion. *Pharm Res* 16:314–19.

Hosscinkhani, H., Aoyama, T., Ogawa, O., and Tabata, Y. 2002. Liver targeting of plasmid DNA
by pullulan conjugation based on metal coordination. *J Control Release* 83:287–302.

Ichinose, K., Tomiyama, N., Nakashima. M. et al. 2000. Antitumor activity of dextran derivatives immobilizing platinum complex (II). *Anticancer Drugs* 11:33–38.

Ichiyama, K., Reddy, S.B.G., Zhang, L.F. et al. 2013. Sulfated polysaccharide, curdlan sulfate, efficiently prevents entry/fusion and restricts antibody-dependent enhancement of dengue virus infection in vitro: A possible candidate for clinical application. *PLoS Negl Trop Dis* 7:e2188.

Ikeda, K., Okada, T., Sawada, S., Akiyoshi, K., and Matsuzaki, K. 2006. Inhibition of the formation of amyloid beta-protein fibrils using biocompatible nanogels as artificial chaperones. *FEBS Lett* 580:6587–95.

Ikekawa, T. 2001. Beneficial effects of edible and medicinal mushrooms in health care. *Int J Med Mushrooms* 3:291–99.

Iobst, S.T. and Drickamer, K. 1996. Selective sugar binding to the carbohydrate recognition domains of the rat hepatic and macrophage asialoglycoprotein receptors. *J Biol Chem* 12:6686–93.

Jagodzinski, P., Wiaderkiewicz, R., and Kursawski, G. 1994. Mechanism of the inhibitory effect of curdlan sulphate on HIV-1 infection *in vitro*. *Virol* 202:735–45.

Jamas, S., Easson, D.D., Ostroff, G.R., and Onderdonk, A.B. 1991. PGG-glucans: A novel class of macrophage-activating immunomodulators. *ACS Symp Ser* 469:44–51.

Jo, J., Yamamoto, M., Matsumoto, K., Nakamura, T., and Tabata, Y. 2006. Liver targeting of plasmid DNA with a cationized pullulan for tumor suppression. *J Nanosci Nanotechnol* 6:2853–59.

Joseph, S., Sabulal, B., George, V., Antony, K., and Janardhanan, K. 2011. Antitumor and anti-inflammatory activities of polysaccharides isolated from *Ganoderma lucidum*. *Acta Pharmaceutica* 61:335–42.

Juan, A.S., Hlawaty, H., Chaubet, F., Letourneur, D., and Feldman, L.J. 2007. Cationized pullulan 3D matrices as new materials for gene transfer. *J Biomed Mater Res A* 82:354–62.

Kampmann, T., Yennamalli, R., Campbell, P. et al. 2009. In silico screening of small molecule libraries using the dengue virus envelope E protein has identified compounds with antiviral activity against multiple flaviviruses. *Antiviral Res* 84:234–41.

Kanatani, I., Ikai, T., Okazaki, A. et al. 2006. Efficient gene transfer by pullulan-spermine occurs through both clathrin-and raft/caveolae-dependent mechanisms. *J Control Release* 116:75–82.

Kaneko, Y., Yoshida, O., Nakagawa, R. et al. 1990. Inhibition of HIV-1 infectivity with curdlan sulfate in vitro. *Biochem Pharmacol* 39:793–97.

Kaneo, Y., Tanaka, T., Nakano, T., and Yamaguchi, Y. 2001. Evidence for receptor-mediated hepatic uptake of pullulan in rats. *J Control Release* 70:365–73.

Kang, J.-H., Tachibana, Y., Kamata, W., Mahara, A., Harada-Shiba, M., and Yamaoka, T. 2010. Liver targeted siRNA delivery by polyethylenimine (PEI)-pullulan carrier. *Bioorg Med Chem* 18:3946–50.

Kazak, H., Toksoy, Ö.E., and Dekker, R.F.H. 2010. Extremophiles as sources of exopolysaccharides. In *Handbook of Carbohydrate Polymers: Development, Properties and Applications*, ed. R. Ito, and Y. Matsuo, 237–77, New York: Nova Science Publishers.

Kim, D.H. and Rossi, J.J. 2008. RNAi mechanisms and applications. *Biotechniques* 44:613–16.

Kim, J.H., Li, Y., Kim, M.S., Kang, S.W., Jeong, J.H., and Lee, D.S. 2012. Synthesis and evaluation of biotin-conjugated pII-responsive polymeric micelles as drug carriers. *Int J Pharm* 427:435–42.

Kim, S.Y., Shin, I.G., and Lee, Y.M. 1998. Preparation and characterization of biodegradable nanospheres composed of methoxy poly(ethylene glycol) and DL-lactide block copolymer as novel drug carriers. *J Control Release* 56:197–208.

Kimoto, T., Shibuya, T., and Shiobara, S. 1997. Safety studies of a novel starch, pullulan: Chronic toxicity in rats and bacterial mutagenicity. *Food Chem Toxicol* 35:323–29.

Koffie, R.M., Meyer-Luehmann, M., Hashimoto, T. et al. 2009. Oligomeric amyloid beta associates with postsynaptic densities and correlates with excitatory synapse loss near senile plaques. *Proc Natl Acad Sci USA* 106:4012–17.

Kreuter, J. 1994. Nanoparticles. In *Colloidal Drug Delivery Systems*, ed. J. Swarbrick, 31–71. New York: Marcel Dekker Inc.

Lam, W., Chan, H., Yang, M., Cheng, S., and Fong, W. 1999. Synergism of energy starvation and dextran-conjugated doxorubicin in the killing of multidrug-resistant KB carcinoma cells. *Anticancer Drugs* 10:171–78.

Lam, W., Leung, C.H., Chan, H.L., and Fong, W.F. 2000. Toxicity and DNA binding of dextran-doxorubicin conjugates in multidrug-resistant KB-V1 cells: Optimization of dextran size. *Anticancer Drugs* 11:377–84.

Laurienzo, P. 2010. Marine polysaccharides in pharmaceutical applications: An overview. *Marine Drugs* 8:2435–65.

Layek, B. and Singh, J. 2012. N-Hexanoyl, N-octanoyl and N-decanoyl chitosan: Binding affinity, cell uptake, and transfection. *Carbohydr Polym* 89:403–10.

Lederman, S., Gulick, R., Chess, L. et al. 1989. Dextran sulfate and heparin interact with CD4 molecules to inhibit the binding of coat protein (gp120) of HIV. *J Immunol* 143:1149–54.

Lee, M. and Kim, S.W. 2002. Polymeric gene carriers. *Pharmaceut. News* 9:407–15.

Lewinski, N., Colvin, V., and Drezek, R. 2008. Cytotoxicity of nanoparticles. *Small* 4:26–49.

Li, H., Bian, S., Huang, Y., Liang, J., Fan, Y., and Zhang, X. 2014. High drug loading pH-sensitive pullulan–DOX conjugate nanoparticles for hepatic targeting. *J Biomed Mater Res Part A* 102A:150–59.

Li, L., Gao, F.P., Tang H.B. et al. 2010. Self-assembled nanoparticles of cholesterol-conjugated carboxymethyl curdlan as a novel carrier of epirubicin. *Nanotechnology* 21:265601.

Li, M., Lam, J.W.Y., Mahtab, F. et al. 2013. Biotin decorated fluorescent silica nanoparticles with aggregation-induced emission characteristics: Fabrication, cytotoxicity and biological applications. *J Mater Chem B* 1:676–84.

Li, Z., Khaliq, M., Zhou, Z. et al. 2008. Design, synthesis, and biological evaluation of antiviral agents targeting flavivirus envelope proteins. *J Med Chem* 51:4660–71.

Liang, B., He, M.L., Chan C.Y. et al. 2009. The use of folate-PEG grafted-hybranched-PEI nonviral vector for the inhibition of glioma growth in the rat. *Biomaterials* 30:4014–20.

Lin, Y.S., Radzi, R., Morimoto, M., Saimoto, H., Okamoto, Y., and Minami, S. 2012. Characterization of chitosan-carboxymethyl dextran nanoparticles as a drug carrier and as a stimulator of mouse splenocytes. *J Biomater Sci Polym Ed* 23:1401–20.

Logeart-Avramoglou, D. and Jozefonvicz, J. 1999. Carboxymethyl benzylamide sulfonate dextrans (CMDBS), a family of biospecific polymers endowed with numerous biological properties: A review. *J Biomed Mater Res* 48:578–90.

Lu, D., Wen, X., Liang, J., Gu, Z., Zhang, X., and Fan, Y. 2008. A pH-sensitive nanodrug delivery system derived from pullulan/doxorubicin conjugate. *J Biomed Mater Res Part B* 89B:177–83.

Lu, Y.-Z., Geng, G.-X., Li, Q.-W., Li, J., Liu, F.-Z., and Ha, Z.-S. 2010. Antitumor activity of polysaccharides isolated from *Patrinia heterophylla*. *Pharm Biol* 48:1012–17.

Lv, H., Zhang, S., Wang, B., Cui, S., and Yan, J. 2006. Toxicity of cationic lipids and cationic polymers in gene delivery. *J Control Release* 114:100–09.

Mähner, C., Lechner, M.D., and Nordmeier, E. 2001. Synthesis and characterisation of dextran and pullulan sulphate. *Carbohydr Res* 331:203–08.

Manjanna, K.M., Shivakumar, B., and Pramodkumar, T.M. 2009. Natural exopolysaccharides as novel excipients in drug delivery: A review. *Arch Appl Sci Res* 1:230–53.

Markova, N., Kussovski, V., Radoucheva, T., Dilova, K., and Georgieva, N. 2002. Effects of intraperitoneal and intranasal application of lentinan on cellular response in rats. *Int Immunopharmacol* 2:1641–45.

Masago, K., Itaka, K., Nishiyama, N., Chung, U., and Kataoka, K. 2007. Gene delivery with biocompatible cationic polymer: Pharmacogenomic analysis on cell bioactivity. *Biomaterials* 28:5169–75.

Mastromarino, P., Petruzziello, R., Macchia, S., Rieti, S., Nicoletti, R., and Orsi, N. 1997. Antiviral activity of natural and semisynthetic polysaccharides on early steps of rubella virus infection. *J Antimicrob Chemother* 39:339–45.

Masunaga, T., Baba, T., and Ouchi, T. 1996. Synthesis and cytotoxic activity of dextran-immobilizing platinum (II) complex through chelate-type coordination bond. *J Macromol Sci Part A Pure Appl Chem* 33:1005–16.

Matsushita, S., Robert-Guroff, M., Rusche, J. et al. 1988. Characterization of a human deficiency virus neutralizing monoclonal antibody and mapping of neutralizing epitope. *J Virol* 62:2107–14.

Maysinger, D. 2007. Nanoparticles and cells: good companions and doomed partnerships. *Org Biomol Chem* 5:2335–42.

McClure, M.O., Moore, J.P., Blanc, D.F. et al. 1992. Investigation into the mechanism by which sulfated polysaccharides inhibit HIV infection in vitro. *AIDS Res Hum Retroviruses* 8:19–26.

McEver, R.P. 1992. Leukocyte–endothelial cell interactions. *Curr Opin Cell Biol* 4:840–49.

Migianu-Griffoni, E., Chebbi, I., Kachbi, S. et al. 2014. Synthesis and biological evaluation of new bisphosphonate-dextran conjugates targeting breast primary tumor. *Bioconjug Chem* 25:224–30.

Mishra, S., Webster, P., and Davis, M.E. 2004. PEGylation significantly affects cellular uptake and intracellular trafficking of non-viral gene delivery particles. *Eur J Cell Biol* 83:97–111.

Mitsuya, H., Looney, D.J., Kuno, S., Ueno, R., Wong-Staal, F., and Broder, S. 1988. Dextran sulfate suppression of viruses in the HIV family inhibition of virion binding to $CD4^+$ cells. *Science* 240:646–49.

Modis, Y., Ogata, S., Clements, D., and Harrison, S.C. 2003. A ligand-binding pocket in the dengue virus envelope glycoprotein. *Proc Natl Acad Sci USA* 100:6986–91.

Modis, Y., Ogata, S., Clements, D., and Harrison, S.C. 2004. Structure of the dengue virus envelope protein after membrane fusion. *Nature* 427:313–19.

Mondotte, J.A., Lozach, P., Amara, A., and Gamarnik, A.V. 2007. Essential role of dengue virus envelope protein n glycosylation at asparagine-67 during viral propagation. *J Virol* 81:7136–48.

Murthy, N., Robichaud, J.R., Tirrell, D.A., Stayton, P.S., and Hoffman, A.S. 1996. The design and synthesis of polymers for eukaryotic membrane disruption. *J Control Release* 61:137–43.

Na, K. and Bae, Y.H. 2002. Self-assembled hydrogel nanoparticles responsive to tumor extra-cellular pH from pullulan derivative/sulfonamide conjugate: Characterization, aggregation, and adriamycin release *in vitro*. *Pharm Res* 19:681–87.

Na, K., Bum, L.T., Park, K.H., Shin, E.K., Lee, Y.B., and Choi, H.K. 2003. Self-assembled nanoparticles of hydrophobically-modified polysaccharide bearing vitamin H as a targeted anti-cancer drug delivery system. *Eur J Pharm Sci* 18:165–73.

Na, K., Park, K.H., Kim, S.W., and Bae, Y.H. 2000. Self-assembled hydrogel nanoparticles from curdlan derivatives: Characterization, anti-cancer drug release and interaction with a hepatoma cell line (HepG2). *J Control Release* 69:225–36.

Nagane, K., Kitada, M., Wakao, S., Dezawa, M., and Tabata, Y. 2009. Practical induction system for dopamine-producing cells from bone marrow stromal cells using spermine–pullulan-mediated reverse transfection method. *Tissue Eng Part A* 15:1655–65.

Nagasawa, K., Harada, H., Hayashi, S., and Misawa, T. 1972. Sulfation of dextran with piperidine-*N*-sulfonic acid. *Carbohydr Res* 21:420–26.

Nakashima, H., Yoshida, O., Baba, M., De Clercq, E., and Yamamoto, N. 1989. Anti-HIV activity of dextran sulphate as determined under different experimental conditions. *Antiviral Res* 11:233–46.

Napoli, I. and Neumann, H. 2009. Microglial clearance function in health and disease. *Neuroscience* 158:1030–38.

Neyts, J., Reymen, D., and Letourneur, D. 1995. Differential antiviral activity of derivatized dextrans. *Biochem Pharmacol* 50:743–51.

Nicolaus, B., Kambourova, M., and Toksoy Ö.E. 2010. Exopolysaccharides from extremophiles: From fundamentals to biotechnology. *Environ Technol* 31:1145–58.

Nimmerjahn, A., Kirchhoff, F., and Helmchen, F. 2005. Resting microglial cells are highly dynamic surveillants of brain parenchyma *in vivo. Science* 308:1314–18.

Nogusa, H., Yamamoto, K., Yano, T., Kajiki, M., Hamana, H., and Okuno, S. 2000. Distribution characteristics of carboxymethylpullulan–peptide–doxorubicin conjugates in tumor-bearing rats: Different sequence of peptide spacers and doxorubicin contents. *Biol Pharm Bull* 23:621–26.

Nogusa, H., Yano, T., Okuno, S., Hamana, H., and Inoue, K. 1995. Synthesis of carboxymethylpullulan–peptide–doxorubicin conjugates and their properties. *Chem Pharm Bull* 43:1931–36.

Nogusa, H., Hamana, H., Uchida, N., Meakawa, R., and Yoshioka, T. 2000b. Improved in vivo antitumor efficacy and reduced systemic toxicity of carboxymethylpullulan-peptide-doxorubicin conjugates. *Jpn J Cancer Res* 91:1333–38.

Nogusa, H.T., Yano, T., Kashima, N., Yamamoto, K., Okuno, S., and Hamana, H. 2000a. Structure-activity relationships of carboxymethylpullulan-peptide-doxorubicin conjugates—Systematic modification of peptide spacers. *Bioorg Med Chem Lett* 10:227–30.

Ohya, Y., Masunaga, T., Ouchi, T. et al. 2009. Antitumor drug delivery by dextran derivatives immobilizing platinum complex (II) through coordination bond. In *Tailored Polymeric Materials for Controlled Delivery Systems*, ed. I. McCulloch, and S.W. Shalaby, 266–78. American Chemical Society Symposium Series.

Ohya, Y., Nishimoto, T., Murata, J., and Ouchi, T. 1994. Immunological enhancement activity of muramyl dipeptide analogue/CM-curdlan conjugate. *Carbohydr Polym* 23:47–54.

Osawa, Z., Morota, T., Hatanaka, K. et al. 1993. Synthesis of sulphated derivatives of curdlan and their anti-HIV activity. *Carbohydr Polym* 21:283–88.

Park, J.S., Park, J.-K., Nam, J.-P. et al. 2012. Preparation of pullulan-*g*-poly (l-lysine) and it's evaluation as a gene carrier. *Macromol Res* 20:667–72.

Posner, M.R., Hideshima, T., Cannon, T., Mukherjee, M., Mayer, K.H., and Byrn, R.A. 1991. An IgG human monoclonal antibody which reacts with HIV-1 gp120, inhibits virus binding to cells and neutralizes infection. *J Immunol* 146:4325–32.

Priya, S.S., Rekha, M.R., and Sharma, C.P. 2014. Pullulan–protamine as efficient haemocompatible gene delivery vector: Synthesis and in vitro characterization. *Carbohydr Polym* 102:207–15.

Purama, R.K., Goswami, P., Khan, A.T., and Goyal, A. 2009. Structural analysis and properties of dextran produced by *Leuconostoc mesenteroides* NRRL B-640. *Carbohydr Polym* 76:30–35.

Qi, J., Yao, P., He, F., Yu, C., and Huang, C. 2010. Nanoparticles with dextran/chitosan shell and BSA/chitosan core—Doxorubicin loading and delivery. *Int J Pharm* 393:176–84.

Raman, R., Sasisekharan, V., and Sasisekharan, R. 2005. Structural insights into biological roles of protein-glycosaminoglycan interactions. *Chem Biol* 12:267–77.

Raucourt, E.de., Mauray, S., Chaubet, F., Maiga-Revel, O., Jozefowicz, M., and Fischer, A. M. 1998. Anticoagulant activity of dextran derivatives. *J Biomed Mater Res* 41:49–57.

Rejman, J., Bragonzi, A., and Conese, M. 2005. Role of clathrin- and caveolae mediated endocytosis in gene transfer mediated by lipo-polyplexes. *Mol Ther* 12:468–74.

Rekha, M.R., and Sharma, C.P. 2011. Hemocompatible pullulan–polyethyleneimine conjugates for liver cell gene delivery: In vitro evaluation of cellular uptake, intracellular trafficking and transfection efficiency. *Acta Biomater* 7:370–79.

Rekha, M.R., and Sharma, C.P. 2009. Blood compatibility and in vitro transfection studies on cationically modified pullulan for liver cell targeted gene delivery. *Biomaterials* 30:6655–64.

Rey, F.A. 2003. Dengue virus envelope glycoprotein structure: New insight into its interactions during viral entry. *Proc Natl Acad Sci USA* 100:6899–901.

Rigau-Perez, J., Clark, G., Gubler, D. et al. 1998. Dengue and Dengue haemorrhagic fever. *Lancet* 352:971–77.

Rinaudo, M. 2008. Main properties and current applications of some polysaccharides as biomaterials. *Polym Int* 57:397–30.

Rohde, D., Schuluter-Wigger, W., Mielke, V., von den Driesch, P., von Gaudecker, B., and Sterry, W. 1992. Infiltration of both T cells and neutrophils in the skin is accompanied by the expression of endothelial leukocyte adhesion molecule-1(ELAM-1): An immunohistochemical and ultrastructural study. *J Invest Dermatol* 98:794–99.

Saitô, H., Yoshioka, Y., and Uehara, N. 1991. Relationship between conformation and biological response for (1–3)-β-D-glucans in the activation of coagulation factor G from limulus amebocyte lysate and host-mediated antitumor activity. Demonstration of single helix conformation as a stimulant. *Carbohydr Res* 217:181–90.

Šandula, J., Kogan, G., Kačuráková, M., and Machova, E. 1999. Microbial (1 → 3)-β-D glucans, their preparation, physicochemical characterization and immunomodulatory activity. *Carbohydr Polym* 38:247–53.

Sasaki, T., Abiko, N., Nitta, K., Takasuka, N., and Sugino, Y. 1979. Antitumor activity of carboxymethylglucans obtained by carboxymethylation of (1,3)-β-D-glucan from *Alcaligenes faecalis* var. *myxogenes* IFO13140. *Eur J Cancer* 15:211–15.

Sasaki, T., Abiko, N., Sugino, Y., and Nitta, K. 1978. Dependence on chain length of antitumor activity of (1 leads to 3)-beta-D-glucan from *Alcaligenes faecalis* var. *myxogenes*, IFO 13140, and its acid-degraded products. *Cancer Res* 38:379–83.

Scomparin, A., Salmosa, S., Bersani, S., Satchi-Fainaro, R., and Caliceti, P. 2011. Novel folated and non-folated pullulan bioconjugates for anticancer drug delivery. *Eur J Pharm Sci* 42:547–58.

Sheldon, K., Marks, A., and Baumal, R. 1989. Sensitivity of multidrug resistant KB-C1 cells to an antibody-dextran-adriamycin conjugate. *Anticancer Res* 9:637–41.

Simon, R.H., Engelhardt, J.F., Yang, Y. et al. 1993. Adenovirus-mediated gene transfer of the CFTR gene to the lung of nonhuman primates: A toxicity study. *Hum Gene Ther* 4:771–80.

Sorgi, F.L., Bhattacharya, S., and Huang, L. 1997. Protamine sulfate enhances lipid-mediated gene transfer. *Gene Ther* 4:961–68.

Suemasu, K. and Ishikawa, S. 1970. Inhibitive effect of heparin and dextran sulfate on experimental pulmonary metastasis. *Gan* 61:125–30.

Suginoshita, Y., Tabata, Y., Matsumura, T. et al. 2002. Liver targeting of human interferon-β with pullulan based on metal coordination. *J Control Release* 83:75–88.

Suzuki, M., Mikami, T., Matsumoto, T., and Suzuki, S. 1977. Preparation and antitumor activity of *O*-palmitoyldextran phosphates, *O*-palmitoyldextrans, and dextran phosphate. *Carbohydr Res* 53:223–29.

Suzuki, T., Ohno, N., Saito, K., and Yadomae, T. 1992. Activation of the complement system by (1,3)-beta-D-glucans having different degrees of branching and different ultrastructures. *J Pharmacobiodyn* 15:277–85.

Tabata, Y., Xi, K., Uno, K., and Ikada, Y. 1999. Chemical conjugation of interferon with pullulan and its antiviral activity. *STP Pharm Sci* 9:101–05.

Takakura, Y. and Hashida, M. 1996. Macromolecule carrier systems for targeted drug delivery: Pharmacokinetic consideration on for biodistribution. *Pharm Res* 13:820–31.

Talarico, L.B., Pujol, C.A., Zibetti, R.G. et al. 2005. The antiviral activity of sulfated polysaccharides against dengue virus is dependent on virus serotype and host cell. *Antiviral Res* 66:103–10.

Tanaka, T., Fujishima, Y., Hanano, S., and Kaneo, Y. 2004. Intracellular disposition of polysaccharides in rat liver parenchymal and nonparenchymal cells. *Int J Pharm* 286:9–17.

Thakor, D.K., Teng, Y.D., and Tabata, Y. 2009. Neuronal gene delivery by negatively charged pullulan–spermine/DNA anioplexes. *Biomaterials* 30:1815–26.

Thambi, T., You, D.G., Han, H.S. et al. 2014. Bioreducible carboxymethyl dextran nanoparticles for tumor-targeted drug delivery. *Adv Healthc Mater* 3:1829–38.

Thomsen, L.B., Lichota, J., Kim, K.S., and Moos, T. 2011. Gene delivery by pullulan derivatives in brain capillary endothelial cells for protein secretion. *J Control Release* 151:45–50.

Tukulula, M., Hayeshi, R., Fonteh, P. et al. 2015. Curdlan-conjugated PLGA nanoparticles possess macrophage stimulant activity and drug delivery capabilities. *Pharm Res* 32:2713–26.

Vamanu, E. 2012. Biological activities of the polysaccharides produced in submerged culture of two edible pleurotus ostreatus mushrooms. *J Biomed Biotechnol* 2012: 1–8.

Wang, J., Guo, H., Zhang, J. et al. 2010. Sulfated modification, characterization and structure-antioxidant relationships of *Artemisia sphaerocephala* polysaccharides. *Carbohydr Polym* 81:897–905.

Wasser, S.P. 2002. Medicinal mushrooms as a source of antitumor and immunomodulating polysaccharides. *Appl Microbiol Biotechnol* 60:258–74.

Xi, K., Tabata, Y., Uno, K. et al. 1996. Liver targeting of interferon through pullulan conjugation. *Pharm Res* 13:1846–50.

Xiong, X.Y., Gong, Y.C., Li, Z.L., Li, Y.P., and Guo, L. 2011. Active targeting behaviors of biotinylated pluronic/poly(lactic acid) nanoparticles in vitro through three-step biotin-avidin interaction. *J Biomater Sci Polym Ed* 22:1607–19.

Yamamoto, M., Ichinose, K., Ishii, N. et al. 2000. Utility of liposomes coated with polysaccharide bearing 1-amino-lactose as targeting chemotherapy for AH66 hepatoma cells. *Oncol Rep* 7:107–11.

Yamaoka, T., Tabata, Y., and Ikada, Y. 1993. Body distribution profile of polysaccharides after intravenous administration. *Drug Deliv* 1:75–82.

Yan, J.K., Ma, H.L., Cai, P.F. et al. 2014. Structural characteristics and antioxidant activities of different families of 4-acetamido-TEMPO-oxidised curdlan. *Food Chem* 143:530–35.

Yoshida, T., Wu, C., Song, L. et al. 1994a. Synthesis of cellulose-type polyriboses and their branched sulfates with anti-AIDS virus activity by selective ring-opening copolymerization of 1,4-anhydro-alpha-D-ribopyranose derivatives. *Macromolecules* 27:4422–28.

Yoshida, T., Yasuda, Y., Mimura, T. et al. 1995. Synthesis of curdlan sulfates having inhibitory effects in vitro against AIDS viruses HIV-1 and HIV-2. *Carbohydr Res* 276:425–36.

Yoshida, T., Yasuda, Y., Uryu, T. et al. 1994b. Synthesis and in vitro inhibitory effect of l-glycosyl-branched curdlan sulfates on AIDS virus infection. *Macromolecules* 27:6272–76.

Zhang, F., Yoder, P.G., and Linhardt, R.J. 2005. Synthetic and natural polysaccharides with anticoagulant properties. In *Polysaccharides: Structural Diversity and Functional Versatility*, ed. S. Dumitriu, 773–794. New York: Marcel Dekker.

Zhang, N., Tan, C., Cai, P., Zhang, P., Zhao, Y., and Jiang, Y. 2009. RNA interference in mammalian cells by siRNAs modified with morpholino nucleoside analogues. *Bioorg Med Chem* 17:2441–46.

Zhou, Z., Khaliq, M., Suk, J.E. et al. 2008. Antiviral compounds discovered by virtual screening of small-molecule libraries against dengue virus E-protein. *ACS Chem Biol* 3:765–75.

# 19 Novel Carriers for Targeted Delivery of Herbal Medicines

*Sankhadip Bose*

## CONTENTS

## 19.1 INTRODUCTION

Herbal and ayurvedic medicines are the oldest medicines in human society. From the beginning of civilization, herbal medicines were used to treat human diseases as much as possible (DeSmet 1997). The World Health Organization (WHO) has defined herbal medicines as finished, labeled medicinal products that contain active ingredients, aerial or underground parts of plants or other plant material or combinations. For primary health care, 80% of the total population of this world use herbal drugs (Atmakuri and Dathi 2010). Herbal drugs can be formulated by different techniques such as extraction, fractionation, and purification; and the structure of the marker compounds can be determined with the use of whole plant or specific plant parts (Cott 1995). Natural drugs do hold some merits over traditional medicines such as lower risk of side effects, widespread availability, low cost, and efficacy for lifestyle diseases for prolonged periods of time (Kumar and Rai 2012). Evidences on current drug therapies tell us that they simply suppress the symptoms and ignore the underlying disease processes. In contrast, many natural products appear to address the cause of many diseases and give superior clinical outcomes (Devi et al. 2010). Nevertheless, the different limitations such as stability in acidic medium and metabolism in the liver decrease the drug levels below therapeutic concentrations in the blood and show less therapeutic effects (Goyal et al. 2011). The alkaloids, flavonoids, tannins, and glycosides are polar and absorbed in very little amount due to their relatively large molecular size. This limits herbal drugs in crossing biological membranes, thereby reducing their absorption via the passive diffusion mechanism. Consequently, the bio-availability and therapeutic index of plant actives decrease (Giriraj 2011).

Hence, considerable attention has been given to the development of novel carriers for the delivery of herbal drugs. Ideally, novel carriers must fulfill two basic requirements: (a) they should deliver the drug at a rate in accordance with the needs of the body, over the period of treatment; and (b) they should channel the active constituents of herbal drugs to the desired site of action (Saraf 2010). The incorporation of herbal drugs into novel delivery carriers may reduce repeated administration to overcome noncompliance, and help enhance the therapeutic value by reducing toxicity and increasing bioavailability (Musthaba et al. 2009).

Novel drug delivery systems (NDDS) aim to sustain drug action at a predetermined rate, and minimize side effects by maintaining an effective concentration of the drug in the body. Various drug delivery technologies such as phytosomes, ethosomes, transfersomes, nanoparticles, herbal transdermal patches, and micro- and nano-emulsion have been developed for the delivery of herbal actives/extracts. Gradually, these are becoming popular in achieving better therapeutic response (Biju et al. 2006; Yadav et al. 2014).

The stability of herbal products in the gastric environment is poor and the first pass metabolism is higher, which creates a hindrance in their frequent use over synthetic molecules. However, the vesicular systems could help in targeted delivery of the desired constituents (Terreno et al. 2008).

The drug delivery has two parameters such as rate and extent, which increase the efficacy of the therapy, when they have been given as a supplemented drug with oriented drug delivery.

The most important potentials of herbal NDDS are

1. It enhances the solubility of the constituents.
2. It minimizes the associated toxic effects.
3. It improves the pharmacological actions.
4. The constituents can also be enhanced due to the presence of lipoidal content in the tissue macrophage uptake (He et al. 2008).

In recent days, new carriers are gaining popularity for the delivery of herbal medicines due to their obvious therapeutic benefits. The chapter deals with various novel technologies such as phytosomes, ethosomes, transfersomes, nanoparticles, and micro- and nano-emulsion for the delivery of plant extracts. Herbal drug targeting approaches are also discussed herein. All information available on novel carriers for targeted delivery of herbal medicines has been collected and accumulated as much as possible to construct this chapter.

## 19.2 NOVEL CARRIERS FOR HERBAL DRUGS

### 19.2.1 Phytosomes

Phytosomes are lipid-compatible molecular complexes, *phyto* meaning plant and *some* meaning cell-like (Amin and Bhat 2012). Most of the bioactive constituents of phytomedicines are water-soluble molecules such as phenolics, glycosides, and flavonoids. However, their effectiveness is limited because of poor absorption when taken orally or when applied topically as they cannot penetrate the lipoidal membrane barrier (Kumari et al. 2011). A patented technology formulated by a leading manufacturer of drugs and nutraceuticals called *phytosome*, where plant extracts or water-soluble phytoconstituents are incorporated into phospholipids and the lipid-compatible molecular complexes, improves their absorption and bioavailability (Jain et al. 2010). In phytosome technology, a little cell is produced to protect the valuable components of the herbal extracts from damage by different bacteria such as gut and digestive secretions. By this technology, the transition from a hydrophilic environment into the hydrophobic environment of the cell membrane is possible, thereby making their entry into the cell easier before reaching the bloodstream (Patil et al. 2012).

These are bilayer delivery systems that can deliver drugs of high molecular weight or polarity (e.g., flavonoids), where conventional delivery technology cannot generate the intended therapeutic benefits. Phytosomes could enhance the bioavailability of such chemical entities by overcoming this problem. In phytosome technology, drug molecules bind to phospholipids (phosphatidylcholine) in a stoichiometric ratio (1:1 or 1:2) and form lipid complexes (Maiti et al. 2007). As it involves bond formation, vesicles are more stable than liposomes. They can entrap higher amount of drugs, can overcome stability-related issues, and enhance cutaneous absorption. The use of phospholipids made them suitable for therapeutic delivery owing to their natural origin. Like liposomes, they can generate both lipophilic as well as hydrophilic domains because of which they can entrap lipophilic as well as hydrophilic drugs (Pathan and Bhandari 2011). Various herbal drugs have been formulated as phytosomes to make the therapy more effective (Table 19.1).

Since ancient times, the use of preparations of plants or their parts has been widespread in most of the world's population (Cott 1995). During the last century, chemical and pharmacological studies have been performed on different plant extracts in order to explore their chemical composition and confirm the indications of traditional medicine. It is often observed that the separation and purification of the various components of an extract lead to a partial loss of specific activity for the purified component. Over the past century, phytochemical and phytopharmacological sciences established the compositions, biological activities, and health-promoting benefits of numerous plant products. Most of the biologically active constituents of plants are large-molecular-weight polar or water-soluble molecules, thus limiting their absorption by passive diffusion (Manach et al. 2004). Phytosomes have been found to improve pharmacokinetic and pharmacological parameters, which in turn can be used for the effective treatment of acute and chronic liver diseases of toxic metabolic or infective origin or degenerative nature (Mascarella et al. 1993). Phytosomes are

**TABLE 19.1**
**Herbal Formulations as Phytosomes, Ethosomes, and Transfersomes**

| Herbal Formulations | Biological Source | Category | Application | Clinical Use | Active Ingredients |
|---|---|---|---|---|---|
| Herbal phytosome | Silibium marianum | Flavonoids | Increase in absorption by 4.6-fold | Hepatoprotective, antioxidants | Silybin |
| | Vitis vinifera | Proanthocyanidins | Increase in antioxidant property | Antioxidant, anticancer | Catechin, epicatechin |
| | Curcuma longa | Polyphenols | Increase in bioavailability | Antioxidant, anti-inflammatory, anticancer | Diferuloylmethane curcumin, demethoxycurcumin, bisdemethoxycurcumin |
| Herbal ethosomes | Glycyrrhiza glabra | Triterpenoid, saponin glycosides | Improved anti-inflammatory activity and sustained release action | Treatment of dermatitis, eczema, and psoriasis | Ammonium glycyrrhizinate |
| | Cannabis sativa | Resins | Improved patient compliance and increased skin permeation | Treatment of rheumatoid arthritis | Tetrahydrocannabidiol (THC) |
| | Tryptterigyum wilfordii | Diterpene oxide | Increase in percutaneous permeability | Anti-inflammatory, antitumor | Triptolide |
| Herbal transfersomes | Capsicum annum | Resins | Increased skin penetration | Treatment of rheumatism | Capsaicin |
| | Curcuma longa | Resins | Increased skin permeability | Anti-inflammatory | Curcumin |
| | Catharanthus roseus | Indole alkaloids | Increase in permeability | Anticancer | Vincristine |

obtained by reacting soy phospholipids with the selected botanical derivatives in an opportune solvent. On the basis of their physicochemical and spectroscopic characteristics, these complexes can be considered novel entities (Bombardelli et al. 1994).

### 19.2.1.1 Phytosome Technology

Polyphenols such as flavonoids and terpenoids of plant extracts are well suited for proper binding to phosphatidylcholine. A stoichiometric amount of the phospholipid with the standardized extract in a nonpolar solvent results in a *phytosome* (Bombardelli et al. 1989). Phosphatidylcholine has two different functions, the phosphatidyl moiety being the lipophilic one and the choline moiety being hydrophilic. Specifically, the choline head of the phosphatidylcholine molecules bind to these compounds and at the same time, the lipid-soluble phosphatidyl portion comprises the body and tail, which then envelops the choline-bound material. Therefore, the phytoconstituents produce a complex with phospholipids called a phyto-phospholipid complex. Molecules are attached through chemical bonds to the polar choline head of the phospholipids (Bombardelli 1991). Precise chemical analysis indicates that the unit phytosome is usually a flavonoid molecule linked with at least one phosphatidylcholine molecule. Therefore, a little microsphere or cell is produced. The morphological structure of phytosomes is represented in Figure 19.1. The phytosome technology produces a little cell, whereby the plant extract or its active constituent is protected from destruction by gastric secretions and gut bacteria owing to the gastroprotective property of phosphatidylcholine (Bombardelli and Spelta 1991).

### 19.2.1.2 Preparation of Phytosomes

Phytosomes are novel complexes, constructed by reacting 1 mol of natural or synthetic phospholipid such as phosphatidylcholine, phosphatidylethanolamine, or phosphatidylserine with 1 mol of constituent, for example, flavonolignans, either alone or in natural mixture in aprotic solvent such as dioxane or acetone. The complex

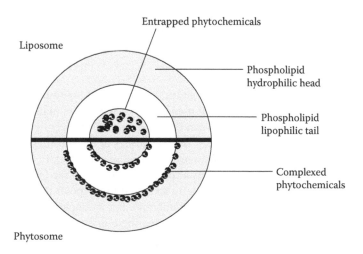

**FIGURE 19.1**  Construction of liposomes and phytosomes.

is isolated by precipitation with nonsolvents such as aliphatic hydrocarbons or by lyophilization or spray drying. In its structure, the ratio between these two moieties lies in the range of 0.5–2.0 moles. The preferable ratio of phospholipid to flavonoids is 1:1 (Jose and Bombardelli 1987). During phytosome preparation, phospholipids are selected from the group consisting of soy lecithin, from bovine or swine brain or dermis, phosphatidylcholine, phosphatidylethanolamine, phosphatidylserine, in which the acyl group may be the same or different and mostly derived from palmitic, stearic, oleic, and linoleic acid. Flavonoids are selected from the group consisting of quercetin, kaempferol, quercretin-3, rhamnoglucoside, quercetin-3-rhamnoside, hyperoside, vitexine, diosmine, 3-rhamnoside, (+) catechin, (−) epicatechin, apigenin-7-glucoside, luteolin, luteolin glucoside, ginkgonetine, isoginkgonetine, and bilobetine. Generally, liposomal complexes operate in aqueous media or buffer solution but phytosomes operate with the solvent having a reduced dielectric constant. Starting materials such as flavonoids are insoluble in chloroform, ethyl ether, or benzene. They become extremely soluble in these solvents after forming phytosomes. This change in chemical and physical properties is due to the formation of a true stable complex (Sharma and Sikarwar 2005).

### 19.2.1.3   Advantages of Phytosomes

The advantages of the phytosomes are as follows:

1. It increases the absorption of lipid-insoluble polar phytoconstituents through oral as well as topical route that show better bioavailability.
2. There is improved absorption of active constituent(s) leading to reduction of dose.
3. Besides acting as carriers, phosphatidylcholine used in the preparation of phytosomes also acts as a hepatoprotective, thus giving the synergistic effect when hepatoprotective substances are employed.
4. Chemical bonds are formed between phosphatidylcholine molecule and phytoconstituent, thus improving the stability of phytosomes.
5. Added nutritional benefit of phospholipids.

### 19.2.1.4   Properties of Phytosomes

*19.2.1.4.1   Chemical*

Phytosomes are a complex between a natural product and natural phospholipids. Based on spectroscopic data, the main phospholipid–substrate interaction is due to the formation of hydrogen bonds between the polar head of phospholipids (i.e., phosphate and ammonium groups) and polar functionalities of the substrate. Phytosomes assume a micellar shape forming liposome-like structures in water. In liposomes, the active principle is dissolved in the internal pocket or floated in the layer membrane, while in phytosomes, the active principle is anchored to the polar head of phospholipids, becoming an integral part of the membrane. For example, in the case of the catechin–distearoylphosphatidylcholine complex, there is the formation of H-bonds between the phenolic hydroxyls of the flavone moiety and the phosphate ion on the phosphatidylcholine side.

### 19.2.1.4.2 Biological

Phytosomes are the most modern system of herbal medicine that are better absorbed and consequently produce better results than conventional herbal extracts. The studies on experimental animals and human subjects demonstrated that the bioavailability of phytosomes is greater than noncomplexed botanical derivatives (Franco and Bombardelli 1998).

### 19.2.1.4.3 Characterization

The behavior of phytosomes in both physical and biological systems is governed by various factors such as physical size, membrane permeability, percentage of entrapped solutes, chemical composition, as well as the quantity and purity of starting materials. Therefore, phytosomes are characterized with respect to their shape, size, size distribution, drug entrapment efficiency, drug release behaviors, and chemical composition (Jain 2001).

## 19.2.2 ETHOSOMES

Owing to obstruction of the skin, the transdermal administration of drugs is generally limited. Vesicular systems are one of the most controversial methods for transdermal delivery of active substances (Gupta et al. 2012).

In recent years, ethosomes have become innovative vesicular systems in the field of pharmaceutical technology and drug delivery. Ethosomes are phospholipid vesicles intended to administer the drug via the transdermal route. These carriers can permeate intact through the human skin due to its high deformability. It has been demonstrated that the physicochemical characteristics of ethosomes allow this vesicular carrier to transport active substances more efficaciously through the stratum corneum into the deeper layers of the skin than conventional liposomes. Ethosomes can entrap drug molecules with diverse physicochemical characteristics, that is, hydrophilic, lipophilic, or amphiphilic (Patel and Bhargava 2012). They are soft, malleable noninvasive vesicles for enhanced delivery of active agents. The size range of ethosomes may vary from 10 to 1000 nm (Pratima and Shailee 2012). They are lipid-based elastic vesicles containing phospholipids, alcohol (ethanol and isopropyl alcohol) in relatively high concentration, and water. High concentration of ethanol (20%–45%) enhances topical drug delivery and prolongs the physical stability of ethosomes (Table 19.1) with respect to liposomes (Patel et al. 2012a). Ethanol being a chemical permeation enhancer disrupts the membrane barrier and thereby enhances the solubility. Moreover, it makes the vesicle flexible without altering the associated stability component.

## 19.2.3 TRANSFERSOMES

Transfersomes were discovered by Gregor Cevc in 1991. The name means *carrying body*, and is derived from the *Latin* word *transferre*, meaning "to carry across," and the *Greek* word *soma*, meaning a "body." It is an artificial vesicular system, which resembles natural cell vesicles, and thus is found suitable for targeted and controlled drug delivery (Kulkarni et al. 2011). Transfersomes are a complex that are adaptable and stress-responsive.

These are ultradeformable vesicles, which possess an aqueous core surrounded by a complex lipid bilayer. As local composition and shape of the bilayer depend on each other, the vesicles are both self-regulating and self-optimizing. This enables the transfersomes to cross various transport barriers efficiently, and makes them effective as a carrier for noninvasive, sustained release of therapeutic agents (Walve et al. 2011). These aggregates with the ultraflexible membrane are able to deliver the drug reproducibly either into or through the skin, depending on the choice of administration or application, with high efficiency (Bhardwaj et al. 2010). These vesicular transfersomes are more elastic than the standard liposomes and thus are well suited for skin penetration. Transfersomes can overcome the skin penetration difficulty by squeezing themselves along the intracellular sealing lipid of the stratum corneum. Flexibility of transferosome membrane is achieved by mixing suitable surface-active components in proper ratios (Girhepunje and Pal 2012). Transfersomes do so by utilizing the hydration and osmotic pressure of the skin. Transfersomes are used to deliver herbal constituents into the upper layers of the skin while for the deeper layers, they are usually given in the form of a cream or gel; being noninvasive, they have better patient compliance. There are a considerable number of examples where the efficacy of herbal constituent-mediated treatment has improved after adopting this delivery system (Table 19.1).

## 19.2.4 LIPOSOMES

These are spherical, colloidal, double-layer vesicles made up of biodegradable phospholipids.

The main structural units, phospholipids, are amphipathic molecules having bipolarity in their structure. Upon hydration with aqueous media, they adopt a characteristic spherical shape and construct an aqueous core within them. The polar heads of phospholipids orient themselves toward the aqueous media, while the hydrophobic tails constitute the interior part of the membrane. Thus, there are both hydrophilic as well as hydrophobic regions within the liposome structures. This property makes them versatile carriers for the delivery of both hydrophilic and hydrophobic drugs entrapped in them. Lipophilic drugs occupy the lipoidal domain within the bilayer membrane, while hydrophilic drugs occupy the aqueous core. The mean size of liposome varies from 0.05 to 5.0 μm. Various advantages associated with liposomes are given below:

1. Increasing solubility
2. Increasing bioavailability
3. Programmed targeting
4. Prolongation of action
5. Lipoidal nature increases the tissue macrophagial uptake of the entrapped constituents
6. Increasing stability
7. Enhanced absorption (Wen et al. 2010)

A significant progress in research works has been noted in the field of liposomal drug delivery and their wonderful applications have inspired research workers to get more benefits from this drug delivery system.

## 19.2.5 NANOPARTICLES

Nanotechnology is the science of matter and materials that deals with particle size in nanometers. The word *nano* is derived from the Latin word meaning dwarf (1 nm = $10^{-9}$ m). Pharmaceutical nanotechnology embraces the applications of nanoscience to pharmacy as nanomaterial, and as devices such as drug delivery, diagnostic imaging, and biosensor materials.

Pharmaceutical nanotechnology has provided more fine-tuned diagnosis and focused treatment of disease at the molecular level. It helps in detecting the antigen associated with cancer, diabetes mellitus, and neurodegenerative diseases as well as the microorganisms and viruses associated with infections. Nanoparticles are defined as particulate dispersions or solid particles with size in the range of 10–1000 nm. The drug is dissolved, entrapped, encapsulated, or attached to a nanoparticle matrix. Depending upon the method of preparation, either nanospheres or nanocapsules can be obtained. Nanocapsules are systems in which the drug is confined to a cavity surrounded by a unique polymer membrane, while nanospheres are matrix systems, in which the drug is uniformly dispersed. The major goals in designing nanoparticles as a delivery system are to control particle size, surface properties, and release of pharmacologically active agents in order to achieve the site-specific action of the drug at the therapeutically optimal rate and dose regimen (Maravajhala et al. 2012). Nanoparticles offer some specific advantages such as they help to increase the stability of drugs/proteins and possess useful controlled release properties. They can be modified to achieve both active and passive targeting and high drug loading, and can be administered by various routes such as parenteral, nasal, intraocular, and oral routes (Manmode et al. 2009).

Nanoparticulate matrices are made up of synthetic or natural polymers. Synthetic biodegradable polymers, which are generally recognized as safe (GRAS) by FDA, are used for human consumption, for example, polylactic acid, poly-α-cyanoacrylate alkyl esters, polyvinyl alcohols, and glycolic acid polymers. The natural polymers, used as a matrix in the design of nanoparticles, belong to two important categories: (a) polysaccharides (chitosan, cellulose, dextran, alginate, and their derivatives) and (b) proteins (gelatin, albumin, and several types of proteins from vegetative origin).

From the nanosphere, the drug release is controlled by two parameters, namely, dissolution and diffusion. The particles can show a burst release as well as surface erosion mechanism. Various methods have been reported for their preparation such as solvent diffusion methods, warm microemulsion formation method, sonication method, etc. The following are the advantages of nanoparticles:

1. Increase shelf life of product
2. Offer possibility of tailoring surface characteristics
3. Increase solubility
4. Minimize adverse reactions associated with dose

5. Target specific area of the body
6. Both hydrophilic and lipophilic drugs can be incorporated (Chang et al. 2011)

## 19.2.6 EMULSION-BASED SYSTEMS

Emulsion is a biphasic liquid system, in which an oil phase is dispersed in an aqueous phase in the form of globules. The globules are stabilized in the dispersion media with the aid of surfactants, which act on the interface between two phases and thereby minimize the interfacial energy so as to stabilize the system and prevent coalescence (Sun and Ouyang 2007; Lindenstorm et al. 2010). Surfactants can be cationic, anionic, and nonionic. The surfactants, which are regarded as safe by FDA, are only used for the preparation of consumable emulsions. Based on globule size, an emulsion system is classified into four categories, namely, ordinary emulsion, microemulsion, nanoemulsion, and sub-microemulsion (lipid emulsion) (Sun and Ouyang 2007; Cui et al. 2009). The size of emulsion globules is of the following order: ordinary (0.1–100 µm) > submicron emulsion (100–600 nm) > microemulsion (10–100 nm) (Sun and Ouyang 2007). Emulsion-based systems have various advantages, which make them suitable as carrier systems for herbal drug delivery ranging from specific locus targeting to sustained release, enhanced macrophagial uptake to minimize stability issues, increased permeability of herbal constituents, etc.

The microemulsion concept was introduced in 1940 where a monophasic solution was prepared after titration of a milky emulsion with alcohol (Lawrence and Rees 2000). Small emulsion-like structures appeared under the electron microscope and the term *microemulsion* was coined (Patel et al. 2012b). Microemulsion is a homogeneous, transparent, thermodynamically stable dispersion of water and oil, stabilized by a surfactant, usually in combination with a co-surfactant (Rao et al. 2009). It is a promising delivery system for sustained release after percutaneous, peroral, topical, transdermal, ocular, and parenteral administration of medicaments. There are several advantages such as increased absorption of drugs, modulation of the drug release kinetics, and decreased toxicity (Grampurohit et al. 2011). In such a system, the interface fluctuates continuously and spontaneously. Physically, they are divided into oil-in-water (O/W), water-in-oil (W/O), and bicontinuous microemulsions. Recently, there has been a considerable interest for microemulsion formulation, for the delivery of hydrophilic as well as lipophilic drug as drug carriers because of its improved drug solubilization capacity, long shelf life, ease of preparation, and improvement of bioavailability. A microemulsion system generally consists of four different components: a lipophilic phase, a hydrophilic phase, surfactant, and co-surfactant (Table 19.2) (Talegaonkar et al. 2008).

Nanoemulsions are submicron-sized emulsions that are under extensive investigation as drug carriers for improving the delivery of therapeutic agents. A nanoemulsion is a heterogeneous system and consists of two immiscible phases; one phase is the oil phase and the other is the aqueous phase, while the droplet is of submicron size range of 5–200 nm (Thakur et al. 2012). Nanoemulsions/submicron emulsions (SMEs)/miniemulsions/ultrafine emulsions are thermodynamically stable transparent (translucent) dispersions of oil and water stabilized by an interfacial film of surfactant and co-surfactant molecules having a droplet size of less than 100 nm. A nanoemulsion,

**TABLE 19.2**

**Herbal Formulations as Nanoparticles, Nanoemulsion, and Microemulsions**

| Type of Carriers | Biological Source | Category | Application | Use | Active Ingredients |
|---|---|---|---|---|---|
| Herbal nanoparticles | *Ginkgo biloba* | Flavonoids | Improved cerebral blood flow | Brain function activation | – |
| | *Berberis vulgaris* | Isoquinolin alkaloids | Sustained drug release | Anticancer | Berberine |
| | *Glycyrrhiza glabra* | Saponin glycosides | Improved bioavailability | Anti-inflammatory, antihepatotoxic, antiviral | Glycyrrhizic acid |
| Herbal nanoemulsion | *Silibum marianum* | Flavonolig-nans | Increase in solubility and therapeutic activity | Hepatoprotective | Silymarin |
| | *Berberis vulgaris* | Isoquinoline alkaloids | Improve residence time and absorption | Anticancer | Berberine |
| | *Curcuma zedooria* | Resins | Improved aqueous dispersibility, stability, and oral bioavailability | Hepatoprotection, anticancer, antibacterial | β-Elemene |
| Herbal microemulsion | *Ligusticum wallichii* | Alkaloids | Increased permeation rate | Inhibit platelet aggregation and lower blood levels | Ligustrazin phosphate |
| | *Tripterygium wilfordii* | Diterpene oxide | Enhance the penetration by increase hydration | Anti-inflammatory | Triplolide |
| | *Pilocarpus jaborandi* | Amino alkaloid | Improved ocular retention, reduced systemic side effects | Treatment of glaucoma | Pilocarpine |

which is categorized as multiphase colloidal dispersion, is generally characterized by its stability and clarity and is formed readily and sometimes spontaneously, generally without high-energy input. In many cases, a co-surfactant or co-solvent is used in addition to the surfactant, the oil phase, and the water phase (Patel and Joshi 2012). Nanoemulsions are made from surfactants approved for human consumption and common food substances that are GRAS by the FDA (Lovelyn and Attama 2011). High shear generation by microfluid or ultrasonic approach is generally used to reduce the droplet size to nanoscale. The transformation between oil-in-water (O/W) or water-in- oil (W/O), and bicontinuous types of emulsion can be achieved by varying the components of the emulsions (Bhatt and Madhav 2011).

## 19.3  TARGETING APPROACHES

Targeted drug delivery is a method of delivering medication to a patient in a manner that increases the concentration of the medication in some parts of the body relative to others. The delivery system design is a highly integrated approach and requires various disciplines, such as chemistry, biology, and engineering, to join forces to optimize this system. The goal of a targeted drug delivery system is to prolong, localize, and target the diseased tissues. This helps maintain the required plasma and tissue drug levels in the body, thereby preventing any damage to the healthy tissue via the drug. Targeted drug delivery systems have been developed to optimize regenerative techniques. Many common diseases such as diabetes, cardiovascular disease, and cancer are caused by several factors, such as physiological, pathological, environmental, and lifestyle. The main effort in the past was aimed at developing highly specific molecules acting on the targets which has now become successful.

Multitarget therapy is a new concept that tries to treat diseases with a multidrug combination in a more causally directed manner. It was recognized very early by physicians practicing phytotherapy that a greater efficacy could be achieved with the application of a combination of plant extracts than with a single use of a highly dosed drug. They noticed that this therapy concept at the same time has the advantage of reducing or eliminating side effects due to the lower doses of the single compounds or drug components within the extract mixtures. Hence, the current phytotherapy of the West, similar to the traditional medicine of China, India, Africa, and South America, uses phytopreparations, which are composed of several herbal drug extracts. The underlying concept of drug medication derives from the assumption that a complex multifactorial pathophysiology (multicausality) can be managed more effectively through the use of a correspondingly composed multidrug mixture than with a single drug. This concept aligns well with the experimental results of modern molecular biology, according to which optimal effects are achievable only with a medication directed simultaneously against the various causes of diseases and the already-existing cellular damages caused by the illness. In this context, it is not very surprising that also in chemotherapy, which for a long time advocated only mono-drug therapy; a gradual trend can be seen away from the mono-substance dogma toward multidrug application. Today, a series of illnesses such as cancer, AIDS, or hypertension are treated successfully with synthetic drug combinations containing 3–5 single individual components. Meanwhile, the advantages and superiorities of the drug combinations over single drug components have been assessed for chemotherapeutics and phytopreparations in several controlled clinical trials. The task for phytotherapy is to prove the advantages of the multidrug- and multitarget therapy in pharmacological and clinical studies as evidenced for the multiextract preparation

Iberogasts (Wagner 2006). For many years, the therapeutic superiority of a plant drug combination over a mono extract only had the support of practical experience. The first approach toward rationalizing this multidrug therapeutic concept was made by Berenbaum (1989) who described the results of synergy effects using two mathematical equations in which the effect of a drug combination is compared with that of its components. According to the first equation, "a total effect of a combination is greater than expected from the sum of the effects of the single components" that is, E (da, db) > E (da) + E (db). The second equation states that "synergy is deemed present if the effect of a combination is greater than that of each of the individual agents" (i.e., E (da, db) > E (da) and E (da, db) > E (db). (E ¼ observed effect and da and db are the doses of agents a and b) (Williamson and Evans 2000). How can the suggested synergy effect of a mixture containing two substances be determined? The method of choice or the determination of synergy effect of a mixture containing two substances is the isobole method, which is independent of the mechanism of action. An isobole is an *iso-effect* curve, in which a combination of constituents (da, db) is represented on a graph, the axes of which are the dose axes of the single agents (da and db) (Wagner 2006).

The drug targeting can be achieved by active and passive targeting approaches. The approaches can be best illustrated by considering the drug targeting mechanism of cancer cells. For example, current cancer treatments include surgical intervention, radiation, and chemotherapeutic drugs, which often also kill healthy cells and cause toxicity in the patient. It would therefore be desirable to develop chemotherapeutics that can either passively or actively target cancerous cells. Passive targeting exploits the characteristic features of tumor biology that allow nanocarriers to accumulate in the tumor by enhanced permeability and retention (EPR) effect.

Passively targeting nanocarriers first reached clinical trials in the mid-1980s, and the first products, based on liposomes and polymer–protein conjugates, were marketed in the mid-1990s. Later, therapeutic nanocarriers based on this strategy were approved for wider use and methods of further enhancing targeting of drugs to cancer cells were investigated. Active approaches achieve this by conjugating nanocarriers containing chemotherapeutics with molecules that bind to overexpressed antigens or receptors on the target cells. Recent reviews provide perspective on the use of nanotechnology as a fundamental tool in cancer research and nanomedicine (Ferrari 2005; Duncan 2006). Here, the potential of nanocarriers and molecules that can selectively target tumors has been focused on, and the challenges in translating some of the basic research to the clinic have been highlighted.

Although passive targeting approaches form the basis of clinical therapy, they suffer from several limitations. Ubiquitously targeting cells within a tumor is not always feasible because some drugs cannot diffuse efficiently and the random nature of the approach makes it difficult to control the process. This lack of control may induce multiple-drug resistance (MDR), a situation where chemotherapy treatments fail in patients owing to the resistance of cancer cells toward one or more drugs. MDR occurs because transporter proteins that expel drugs from cells are overexpressed on the surface of cancer cells (Gottesman et al. 2002; Peer and Margalit 2006). Expelling drugs inevitably lowers the therapeutic effect and cancer cells soon develop resistance to a variety of drugs. The passive strategy is further limited because certain tumors do not exhibit the EPR effect, and the permeability of vessels may not be the same throughout a single tumor (Jain 1994).

One way to overcome these limitations is to program the nanocarriers so they actively bind to specific cells after extravasation. This binding may be achieved by attaching targeting agents such as ligand molecules that bind to specific receptors on the cell surface—to the surface of the nanocarrier by a variety of conjugation chemistries (Torchilin 2005). Nanocarriers will recognize and bind to target cells through ligand–receptor interactions, and bound carriers are internalized before the drug is released inside the cell (Allen 2002; Pastan et al. 2006; Peer at al. 2007). In general, when using a targeting agent to deliver nanocarriers to cancer cells, it is imperative that the agent binds with high selectivity to molecules that are uniquely expressed on the cell surface. To maximize specificity, a surface marker (antigen or receptor) should be overexpressed on target cells relative to normal cells.

For example, a more significant therapeutic outcome was achieved when immuno-liposomes targeted to human blood cancer (B-cell lymphoma) were labeled with an internalizing anti-CD19 ligand rather than a noninternalizing anti-CD20 ligand (Sapra and Allen 2002). In contrast, targeting nanocarriers to noninternalizing receptors may sometimes be advantageous in solid tumors owing to the bystander effect, where cells lacking the target receptor can be killed through drug release at the surface of the neighboring cells, where carriers can bind (Allen 1994). It is generally known that higher binding affinity increases targeting efficacy. However, for solid tumors, there is evidence that high binding affinity can decrease penetration of nanocarriers due to a "binding-site barrier," where the nanocarrier binds to its target so strongly that penetration into the tissue is prevented (Adams et al. 2001; Allen 2002).

## 19.4 CLINICAL TRIALS OF HERBAL MEDICINES

Clinical trials of herbal medicines may have two types of objectives. One is to validate the safety and efficacy that is claimed for a traditional herbal medicine. The other is to develop new herbal medicines or examine a new indication for an existing herbal medicine or a change of dose formulation, or route of administration. In some cases, trials may be designed to test the clinical activity of a purified or semipurified compound derived from herbal medicines. Phase I study may not be necessary for herbal formulations but the need for testing its toxicity in animals has been considerably reduced. Toxicity studies may not be needed for phase II trial unless reports suggest them and a larger multicentric phase III trial is subsequently planned based on results of phase II study. These trials also have to be approved by the appropriate scientific and ethical committees of the concerned institutes. Many standardizations and clinical trials of herbal drugs have been completed and several are also in the pipeline. However, the clinical trials of different site-targeted modern dosage forms of herbal drugs are still awaited for their proper standardization and clinical trials. Although several modern formulations of herbal medicine have been recently prepared by scientists and their standardizations are also complete, clinical trials for those formulations are still in the process of beginning. In the near future, several modern herbal dosage forms will be available in the market with proper clinical trials.

## 19.5 FUTURE PERSPECTIVES

Therapeutic targeting is a domain that can be utilized to increase the therapeutic efficacy of the delivery system. Phytomedicinal research is now directed toward targeted delivery of herbal constituents. In multitarget therapies, phytotherapy will be more important and new phytodrug combinations will come up to combat diseases that are now being treated by chemotherapy. Many other NDDS can be utilized to enhance the efficacy of herbal medicines. For example, the mucoadhesive drug delivery system can also be utilized to enhance the efficacy of the therapy. The unit dosage form or multiparticulate system locates the drug around the absorption window of the drug molecule, which may lead to the enhancement of bioavailability. Floating drug delivery can also be used for drugs having their absorption window in the upper GI tract. However, its utilization is limited since most herbal drugs are unstable at gastric pH. Niosomes can also be used to deliver herbal drugs. They are cheaper than liposomes, and due to the use of nonionic surfactants, the associated toxicity of the carrier system is less with respect to ionic surfactants. Niosomes do not have issues related to oxidation, etc. as associated with liposomes because liposomes contain lipids, which contain double bonds in their structure. They are prone to free radical chain–mediated oxidation reactions. These types of NDDS are still awaiting utilization for the delivery of herbal molecules.

## REFERENCES

Adams, G.P., Schier, R., McCall, A.M. et al. 2001. High affinity restricts the localization and tumor penetration of single-chain Fv antibody molecules. *Cancer Res* 61:4750–55.

Allen, T.M. 1994. Long-circulating (sterically stabilized) liposomes for targeted drug-delivery. *Trends Pharmacol Sci* 15:215–20.

Allen, T.M. 2002. Ligand-targeted therapeutics in anticancer therapy. *Nat Rev Cancer* 2:750–63.

Amin, T. and Bhat, S.V. 2012. A review on phytosome technology as a novel approach to improve the bioavailability of nutraceuticals. *Int J Adv Res Technol* 1:1–15.

Atmakuri, L.R. and Dathi, S. 2010. Current trends in herbal medicines. *J Pharm Res* 3:109–13.

Berenbaum, M. 1989. What is synergy? *Pharmacol Rev* 41:93–141.

Bhardwaj, V., Shukla, V., Singh, A., Malviya, R., and Sharma, P.K. 2010. Transfersomes ultra flexible vesicles for transdermal delivery. *Int J Pharm Sci Res* 1:12–20.

Bhatt, P. and Madhav, S. 2011. A detailed review on nanoemulsion drug delivery system. *Int J Pharm Sci Res* 2:2482–89.

Biju, S.S., Talegaonkar, S., Mishra, P.R., and Khar, R.K. 2006. Vesicular system: An overview. *Indian J Pharm Sci* 68:141–53.

Bombardelli, E. 1991. Phytosome: New cosmetic delivery system. *Boll Chim Farm* 130:431–38.

Bombardelli, E., Cristoni, A., and Morazzoni, P. 1994. Phytosomes in functional cosmetics. *Fitoterapia* 95:387–01.

Bombardelli, E., Curri, S.B., Della, R.L., Del, N.P., Tubaro, A., and Gariboldi, P. 1989. Complexes between phospholipids and vegetal derivatives of biological interest. *Fitoterapia* 60:1–9.

Bombardelli, E. and Spelta, M. 1991. Phospholipid-polyphenol complexes: A new concept in skin care ingredients. *Cosm Toil* 106:69–76.

Chang, C.H., Huang, W.Y., Lai, C.H. et al. 2011. Development of novel nanoparticles shelled with heparin for berberine delivery to treat *Helicobacter pylori*. *Acta Biomater* 7:593–03.

Cott, J. 1995. Natural product formulation available in Europe for psychotropic indications. *Psychopharmacol Bull* 31:745–51.

Cui, J., Yu, B., Zhao, Yu. et al. 2009. Enhancement of oral absorption of curcumin by self microemulsifying drug delivery systems. *Int J Pharm* 371:148–55.

DeSmet, P.G.A.M. 1997. The role of plant derived drugs and herbal medicines in health care. *Drugs* 54:801–40.

Devi, V.K., Jain, N., and Valli, K.S. 2010. Importance of novel drug delivery systems in herbal medicines. *Pharmacog Rev* 4:27–31.

Duncan, R. 2006. Polymer conjugates as anticancer nanomedicines. *Nat Rev Cancer* 6:688–701.

Ferrari, M. 2005. Cancer nanotechnology: Opportunities and challenges. *Nat Rev Cancer* 5:161–71.

Franco, P.G. and Bombardelli, E. 1998. Complex compounds of bioflavonoids with phospholipids, their preparation, uses and pharmaceutical and cosmetic compositions containing them. U.S. Patent No. EPO 275005.

Girhepunje, K. and Pal, R. 2012. Potential role of transfersomes in transdermal drug delivery. *World J Pharm Res* 1:21–38.

Giriraj, K. 2011. Herbal drug delivery systems: An emerging area in herbal drug research. *J Chronother Drug Deliv* 2:113–19.

Gottesman, M.M., Fojo, T., and Bates, S.E. 2002. Multidrug resistance in cancer: Role of ATP-dependent transporters. *Nat Rev Cancer* 2:48–58.

Goyal, A., Kumar, S., Nagpal, M., Singh, I., and Arora, S. 2011. Potential of novel drug delivery systems for herbal drugs. *Indian J Pharm Edu Res* 45:225–35.

Grampurohit, N., Ravikumar, P., and Mallya, R. 2011. Microemulsions for topical use: A review. *Indian J Pharm Edu Res* 45:100–07.

Gupta, N.B., Loona, S., and Khan, M.U. 2012. Ethosomes as elastic vesicles in transdermal drug delivery: An overview. *Int J Pharm Sci Res* 3:682–87.

He, Z.F., Liu, D.Y., Zeng, S., and Ye, J.T. 2008. Study on preparation of ampelopsin liposomes. *China Journal of Chinese Mat Med* 33:27–30.

Jain, N., Gupta, B.P., Thakur, N., Jain, R., Banweer, J., Jain, D., and Jain, S. 2010. Phytosome: A novel drug delivery system for herbal medicine. *Int J Pharm Sci Drug Res* 2:224–28.

Jain, N.K. 2001. *Controlled and Novel Drug Delivery*. New Delhi: CBS Publisher.

Jain, R.K. 1994. Barriers to drug-delivery in solid tumors. *Sci Am* 271:58–65.

Jose, M.M. and Bombardelli, E. 1987. Pharmaceutical compositions containing flavanolignans and phospholipid active principles. U.S. Patent EPO209037.

Kulkarni, P.R., Yadav, J.D., Vaidya, A.K., and Gandhi, P.P. 2011. Transfersomes: An emerging tool for transdermal drug delivery. *Int J Pharm Sci Res* 2:735–41.

Kumar, K. and Rai, A.K. 2012. Miraculous therapeutic effect of herbal drug using novel drug delivery system. *Int Res J Pharm* 3:27–30.

Kumari, P., Singh, N., Cheriyan, B.P., and Neelam. 2011. Phytosome: A novel approach for phytomedicine. *Int J Inst Pharm Life Sci* 1:89–100.

Lawrence, M.J. and Rees, G.D. 2000. Microemulsion-based media as novel drug delivery systems. *Adv Drug Deliv Rev* 45:89–121.

Lindenstorm, T., Andersen, P., and Marie, A.E. 2010. Determining adjuvant activity on T-cell function *in vivo*: The cells. In *Vaccines Adjuvant Methods and Protocols*, eds. L. Gwyn, P. Davis, New Jersey: Humana Press.

Lovelyn, C. and Attama, A.A. 2011. Current state of nanoemulsions in drug delivery. *J Biomater Nano Biotechnol* 2:626–39.

Maiti, K., Mukherjee, K., Gantait, A., Saha, B.P., and Mukherjee, P.K. 2007. Curcumin-phospholipid complex: Preparation, therapeutic evaluation and pharmacokinetic study in rats. *Int J Pharm* 330:155–63.

Manach, C., Scalbert, A., and Morand, C. 2004. Polyphenols: Food sources and bioavailability. *Am J Clin Nutr* 79:727–47.

Manmode, A.S., Sakarka, D.M., and Mahajan, N.M. 2009. Nanoparticles—Tremendous therapeutic potential: A review. *Int J Pharm Tech Res* 1:1020–27.

Maravajhala, V., Papishetty, S., and Bandlapalli, S. 2012. Nanotechnology in development of drug delivery system. *Int J Pharm Sci Res* 3:84–96.

Mascarella, S., Giusti, A., Marra, F. et al. 1993. Therapeutic and antilipoperoxidant effects of silybin-phosphatidylcholine complex in chronic liver disease: Preliminary results. *Curr Ther Res* 53:98–102.

Musthaba, S.M., Baboota, S., Ahmed, S., Ahuja, A., and Ali, J. 2009. Status of novel drug delivery technology for phytotherapeutics. *Expert Opin Drug Deliv* 6:625–37.

Pastan, I., Hassan, R., FitzGerald, D.J., and Kreitman, R.J. 2006. Immunotoxin therapy of cancer. *Nat Rev Cancer* 6:559–65.

Patel, D. and Bhargava, P. 2012. Ethosomes—A phyto drug delivery system. *Adv Res Pharm Biologicals* 2:1–8.

Patel, D.R., Patel, N.M., Patel, M.R., and Patel, K.R. 2012b. Microemulsions: A novel drug carrier system. *Int J Drug Formul Res* 2:41–56.

Patel, N.S., Patel, S.N., Patel, K.R., and Patel, N.M. 2012a. A vesicular transdermal delivery system for enhance drug permeation—Ethosomes & transfersomes. *Internationale Pharmaceutica Sciencia* 2:24–32.

Patel, R.P. and Joshi, J.R. 2012. An overview on nanoemulsion: A novel approach. *Indian J Pharm Sci Res* 3:4640–50.

Pathan, R. and Bhandari, U. 2011. Preparation and characterization of embelin- phospholipid complex as effective drug delivery tool. *J Incl Phenom Macrocycl Chem* 9:139–47.

Patil, M.S., Patil, S.B., Chittam, K.P., and Wagh, R.D. 2012. Phytosomes: Novel approach in herbal medicines. *Asian J Pharm Sci Res* 2:1–9.

Peer, D. and Margalit, R. 2006. Fluoxetine and reversal of multidrug resistance. *Cancer Lett* 237:180–87.

Peer, D., Zhu, P., Carman, C. V., Lieberman, J., and Shimaoka, M. 2007. Selective gene silencing in activated leukocytes by targeting siRNAs to the integrin lymphocyte function-associated antigen-1. *Proc Natl Acad Sci* 104:4095–100.

Pratima, N.A. and Shailee, T. 2012. Ethosomes: A novel tool for transdermal drug delivery. *IJRPS* 2:1–20.

Rao, Y.S., Deepthi, K.S., and Chowdary, K.P.R. 2009. Microemulsions: A novel drug carrier system. *Int J Drug Deliv Technol* 1:39–41.

Sapra, P. and Allen, T.M. 2002. Internalizing antibodies are necessary for improved therapeutic efficacy of antibody-targeted liposomal drugs. *Cancer Res* 62:7190–94.

Saraf, A.S. 2010. Applications of novel drug delivery system for herbal formulations. *Fitoterapia* 81:680–89.

Sharma, S. and Sikarwar, M. 2005. Phytosome: A review. *Planta Indica* 1:1–3.

Sun, H.W. and Ouyang, W.Q. 2007. The preparation of neem oil microemulsion (*Azadirachta indica*) and the comparison of acaricidal time between neem oil microemulsion and other formulation *in vitro*. *J Shanghai Jiao Tong Univ (Agric Sci)*.

Talegaonkar, S., Azeem, A., Ahmad, F.J., Khar, R.K., Pathan, S.A., and Khan, Z.I. 2008. Microemulsions: A novel approach to enhanced drug delivery. *Recent Pat Drug Deliv Formul* 2:238–57.

Terreno, E., Delli, C.D., Cabella, C., et al. 2008. Paramagnetic liposomes as innovative contrast agents for magnetic resonance (MR) molecular imaging applications. *Chem Biodivers* 5:1901–12.

Thakur, N., Garg, G., Sharma, P.K., and Kumar, N. 2012. Nanoemulsions: A review on various pharmaceutical application. *Global J Pharmacol* 6:222–25.

Torchilin, V.P. 2005. Recent advances with liposomes as pharmaceutical carriers. *Nat Rev Drug Discov* 4:145–60.

Wagner, H. 2006. Multitarget therapy—The future of treatment for more than just functional dyspepsia. *Phytomedicine* 13SV:122–9.

Walve, J.R., Bakliwal, S.R., Rane, B.R., and Pawar, S.P. 2011. Transfersomes: A surrogated carrier for transdermal drug delivery system. *Int J Appl Biol Pharm Technol* 2:204–13.

Wen, Z., Liu, B., Zheng, Z., You, X., Pu, Y., and Li, Q. 2010. Preparation of liposomes entrapping essential oil from *Atractylodes macrocephala* Koidz by modified RESS technique. *Chem Eng Res Design* 88:1102–07.

Williamson, E.M. and Evans, F.J. 2000. Cannabinoids in clinical practice. *Drugs* 60:1303–14.

Yadav, M., Vidhi, J.B., Gaurav, D., and Keyur, S. 2014. Novel techniques in herbal drug delivery systems. *Int J Pharm Sci Rev Res* 28:83–89.

# 20 Toxicological Concerns Related to Nanoscale Drug Delivery Systems

*Sabyasachi Maiti, Sanmoy Karmakar, and Kumar Anand*

## CONTENTS

## 20.1 INTRODUCTION

Nanotechnology is an emerging science of precise manipulation of atomic or molecular structures of materials at the nanometer level with unique properties and applications (Miyazaki and Islam 2007; Bakand et al. 2012). Over the past few years, there has been an increasing toxicological concern regarding the safety of the developed nanosystems. Assuming biocompatibility and biodegradability of most of the materials, the toxicity issues caused by them when formulated into nanoparticles (NPs) are usually neglected by the scientific community. Only a few studies approach the toxicity of the nanosystems. However, nanomaterials safety data are limited to arrive at an overall picture of material-specific risks. The NP characteristics such as size, charge, and surface properties could influence their pharmacokinetics after oral administration.

In health-care delivery, the drug availability is a secondary reason rather than the therapeutic effect. For a therapeutic system to be successful, it is desirable that the required drug should reach the site of action without any undesirable interactions and be available for a sufficient period of time. On the contrary, there are many drug candidates that show low bioefficiency due to different factors such as undesirable interactions, immature degradation, and insufficient capability of tissue penetration. To overcome these problems and monitor therapeutic actions, novel drug delivery systems are being introduced to discharge the drug molecules at the desired site for sufficient duration.

Despite numerous benefits, the potential dangers of NP exposure cannot be ignored. The nanotoxicological responses of NPs are primarily observed on adult healthy animals. Therefore, their effects on susceptible populations are not well known. The perturbations of physiological structures and functions in susceptible populations may exhibit unusual pharmacokinetic profiles of the NPs.

The safety concern regarding the exposure of NPs in humans may restrict the wider application of these promising nanomaterials. NPs may enter the human body via respiratory pathways, digestive tract, intravenous (IV) injection, implantation, and other routes (Arora et al. 2012; Araújo et al. 2015). After absorption, the NPs are carried to distal organs by the bloodstream and the lymphatic system (Oberdörster et al. 2005). During this process, they interact with biological molecules and perturb physiological systems. Some ingested or absorbed NPs are eliminated and the rest remain in the body for a long time. The unexpected invasion of the physiological systems by the NPs disturbs normal cell signaling, cell and organ functions, and may even cause pathological disorders.

Nanotechnology offers some obvious benefits, including better treatment efficacy, specific localization, reduction of dosage regimens, and dose-related side effects (Sahoo and Labhasetwar 2003; Svenson and Tomalia 2005). Other nanostructures are also assumed for use in diagnostics known as nanodevices. However, the nanomaterials have a large surface area to volume ratio, which enables them to alter biological properties as compared to the parent molecule (Williams 2008). In the present scenario, a number of nanoscale drug delivery applications have been attempted for the treatment of cancer, central nervous system (CNS) disorders, and so on. Liposomes, nanoshells, nanotubes, dendrimers, NPs, nanospheres, aquasomes, and solid lipid NPs (SLNs) are among the different nanostructures, commonly known as nanocarriers (Moghimi et al. 2005). Nanomaterials can be generated from different parent materials in different shapes such as spheres, rods, wires, and tubes (Liu 2006). Apart from these differences, similar toxicological profiles are expected.

Further, the polymers are usually required for the fabrication of drug delivery systems. The polymers can help in achieving desired pharmacotherapy by stabilizing the proposed medication during production. Further, the structural manipulation of the polymers is required in order to fabricate different forms, namely, films, microspheres, monoliths, NPs, and polymeric prodrugs (Amsden and Cheng 1995;

Kim et al. 2009), and control the release of drug at the target organ/tissues (Vilar et al. 2012). However, the fabrication strategy depends on the intended route of administration, pharmacokinetics, and the drug efficacy. The design of polymeric carriers intended for delivery via different routes may raise some questions regarding the fate of carrier materials, that is, accumulation or elimination, in the body after completion of drug release. Indeed, the biodegradability of polymers is one of the most important regulatory issues since this property of the polymers determines their consequent removal from the body.

In this section, several nanosystems are described with special reference to their toxicological concerns. A number of NPs based on different types of materials such as polymers, lipids, carbon, and metals are reviewed. The different nanosystems may precipitate toxicity by perturbing different physiological systems as revealed by animal studies. Hence, the factors that contribute toxicity to the nanomaterials are also discussed herein.

## 20.2 FACTORS AFFECTING TOXICITY OF NANOCARRIERS

A generalization of nanomaterial-related toxicity is difficult due to a large difference in physicochemical properties between the materials and their products. The biokinetics are found to be affected by different physicochemical properties of the nanocarriers that encompass particle size, morphology, surface area, chemical reactivity, surface charge, and state of aggregation (Lockman et al. 2004; Radomski et al. 2005; Hardman 2006; Jiang et al. 2008; Sonavane et al. 2008). Undoubtedly, it is of utmost priority to figure out the different forms of toxicological phenomena and understand the effects caused by the novel nanocarriers, nanodevices, or by occupational exposure of the nanostructures. The biological effects may be beneficial or harmful as supported by very limited data. The physicochemical parameters must be cautiously manipulated before the design of various nanostructures using polymeric materials. Research on the potential health risks on exposure to NPs lags behind the rapid development of nanotechnology. In general, the biological impacts and toxicity of NPs are functions of multiple parameters, and therefore the various characteristics must be addressed for evaluating toxicity of the nanomaterials.

### 20.2.1 PARTICLE MORPHOLOGY

The biodistribution, biological fate, toxicity, as well as drug-targeting capacity depend on size and size distribution of NPs. De Jong et al. (2008) conducted an experiment with IV gold NPs (AuNPs; 10, 50, 100, and 250 nm) to investigate the impact of particle size on the biodistribution in mice. Irrespective of sizes, the majority of the Au was present in liver and spleen after 24 h. A clear distinction was evident between the distribution of the 10-nm particles and the larger particles. The smaller particles accumulated in almost all vital organs, whereas the larger ones were only detectable in blood, liver, and spleen. The results demonstrated size-dependent tissue distribution of AuNPs. The smaller size NPs showed the most widespread organ distribution. The discrepancy in tissue distribution pattern of different sized particles was likely to induce damage of varying degrees to tissues or organs.

Usually, the adsorption of opsonin increases the recognition of foreign materials for phagocytosis and cause rapid clearance of the circulating NPs. The polymer particles of hydrophilic surface (<100 nm) show prolonged circulation by delaying opsonization (Alexis et al. 2008; Bertrand and Leroux 2012). The particle size may have an impact on the entry mechanism into target cells and phagocytes. NPs that are <200 nm are internalized via clathrin-coated pits, whereas 500-nm particles are internalized via caveolae-meditated endocytosis (Rejman et al. 2004).

The NPs smaller than serum albumin (~40–50 kDa or a diameter of ≤4–6 nm) are eliminated primarily through the kidneys. The particles or aggregates (>10 μm) are passively entrapped within the lung capillaries. The particles greater than 3 μm size are transiently entrapped and are subsequently moved from lung to the liver (Deshmukh et al. 2012). Particles that lie in the range of 3–6 μm accumulate in the liver and spleen. Thus, the particle size determines the deposition sites in tissues. It is hypothesized that bulk materials (>1 μm) that are relatively inert may become toxic when their size is reduced to the nanoscale level. This could be attributed to greater biodistribution at high surface/volume ratio, and the ability of nanomaterials to traverse cell barriers. The surface of materials interacts with other nanoscale biological molecules such as deoxyribonucleic acid (DNA), proteins, and cell membranes (Xia et al. 2009). The interaction of molecular oxygen and electron donor or acceptor groups on the particle surface generates either superoxide or hydrogen peroxide. Both species can oxidize other compounds through an electron transfer mechanism (Semete et al. 2010a,b). The propagation of reactive oxygen species (ROS) is associated with nanometric size of the materials, and therefore constitutes a mechanism of generating potential toxicity (Nel et al. 2006).

PEGylated AuNPs (13 nm in size) demonstrated long circulating half-lives for ~1 week. They accumulated within the liver and spleen over the course of one week. The sequestration of AuNPs within lysosomes of Kupffer cells and spleen macrophages resulted in acute hepatic inflammation and apoptosis in mice (Cho et al. 2009, 2010).

In its extended conformation, polyethylene glycol (PEG) provides steric hindrance to the adsorption of serum proteins on the surface of NPs, thus delaying phagocytosis by macrophages, and rendering them long circulating property (Owens and Peppas 2006).

In addition to particle size, Chithrani et al. (2006) investigated the impacts of morphology on cellular intake of AuNPs. In spite of their similar dimensions (74 and 14 nm), the uptake of nanorods was slower than the spherical particles in HeLa cells. The ellipsoid particles are more readily engulfed by macrophages than that of spherical particles (Sharma et al. 2010). However, NPs with high aspect ratios (tubular shape vs. spherical) resist uptake by macrophages because of high curvature angles (Champion and Mitragotri 2006, 2009). Short-rod (aspect ratio = 1.5) mesoporous silica NPs are easily trapped in the liver, whereas long-rod (aspect ratio = 5) silica NPs distribute in the spleen (Huang et al. 2011). Thus, the particles with smaller aspect ratios exhibit more rapid clearance.

In addition to overall shape, the smoothness/roughness of the particle surface also affects the opsonization of the particle and its subsequent uptake by the mononuclear phagocyte system (MPS; Bertrand and Leroux 2012). Particle shape also affects potential toxicities. Titanium dioxide ($TiO_2$), in its fiber structure >15 μm, provokes an inflammatory response in alveolar macrophages. Due to alteration of shapes, it becomes difficult for the phagocytic cells to process the NPs, resulting in toxicity by lysosomal disruption (Hamilton et al. 2009).

The surface charge directly affects the interaction of NPs with biological surfaces, cell membranes, and proteins. Charged liposomes (positive or negative) undergo greater opsonization than neutral vesicles do and show greater accumulation in the MPS (Chonn et al. 1991). In mice, undesirable liver uptake has been observed for PEG-oligocholic acid-based micellar NPs with highly positive or highly negative surfaces, whereas liver uptake was low for slightly negatively charged NPs. The NPs had greater accumulation in ovarian tumors (Xiao et al. 2011).

Owing to negative charge on cell surface, the positively charged particles may cause higher nonspecific cellular internalization and relatively shorter circulation half-life. The particles with positive charges are more likely to accumulate within macrophages. The introduction of negative charges into the dextran molecule prolonged its circulation in blood. The derivatization with cationic diethylaminoethyl groups reduced its half-life. The polycationic dextran deposited in the liver more readily than the polyanionic and original dextran macromolecules. Approximately, 10% substitution of dextran with the diethylaminoethyl group was sufficient to enhance the accumulation of dextran in the liver and spleen (Yamaoka et al. 1995).

Conversely, the negatively charged/neutral materials experienced lower nonspecific uptake owing to steric/electrostatic repulsion (Alexis et al. 2008) and resisted the cytotoxic effects. A strong electrostatic barrier sometimes overrides the size or shape factors in exhibiting toxicity (El Badawy et al. 2011).

## 20.2.2 ROUTE OF EXPOSURE

The oral route is most popular among the others for the delivery of drugs due to various advantages like better patient compliance, ease of self-medication, pharmacoeconomic suitability, and painless delivery, which all combine to make this route suitable for chronic therapy (Das and Chaudhury 2011). Apart from these advantages, the oral delivery poses some problems, including the drug interaction with gastrointestinal tract (GIT) content, poor intestinal permeability, and intestinal transit. This can further be complicated by the low solubility and instability of small and large drug molecules. These problems can be resolved by adopting different nanotechnological aspects, where the metallic and polymeric nanocarriers are quite capable of crossing different barriers and enhancing bioavailability of the drug candidate.

Owing to the nature of mucus layer/secretions and turnover, it creates a major barrier for the penetration of NPs across the intestinal tract (Ensign et al. 2012). The mucoadhesive or mucolytic properties of the modified nanostructures could be helpful in crossing the mucosal barriers (Li et al. 2013; Araújo et al. 2014). The mucolytic NPs disrupt the natural mucus barriers, exposing the intestinal surface. This effect enhances the uptake of NPs and also the bacterial attachment and translocation, which may lead to infections (Albanese et al. 1994). Moreover, the cell surface is exposed to the harsh conditions of intestinal tract, leading to their further damage in the absence of mucus layer.

Regardless of the various advantages, the different nanostructures have mild-to-moderate toxicological effects on different tissues and cells. Most of the polymeric nanocarriers for oral administration contain surfactants for better absorption. In addition, these nanocarriers are often coated with different hydrophilic materials to serve the same purpose. It has been reported that the chronic use of nanostructures containing surfactants can cause disruption to the intestinal epithelium, which further enhances the entry of microorganisms resulting in various pathological changes. The coated biomaterials in the nanostructure can cause structural reorganization of the tight junctions (TJs), leading to disruption of epithelial integrity (Yeh et al. 2011; Sonaje et al. 2012).

The oral toxicity of the NPs may be local or systemic. The local toxicity involves direct interaction of the NPs with the intestinal cells by virtue of their size and charge. In systemic toxicity, all the characteristic features of NPs that influence their translocation and interaction with different tissues must be considered.

Most of the time, the toxic potentials of NPs are neglected when biocompatible and biodegradable materials are used to produce the NPs. The NPs composed of biodegradable materials can also precipitate cellular toxicity due to the intracellular changes caused by their accumulation inside the cells. Moreover, the material properties may change completely upon some chemical modifications. Besides the toxicity of the materials, their degradation products are also another concern.

Furthermore, the reagents used in the production of NPs should be less toxic and the final product must comply with the Food and Drug Administration (FDA) limits (Arora et al. 2012). The long-term use of absorption enhancers can lead to the damage of the intestinal epithelium with the possibility of promoting the passage of pathogens and toxins through the GIT (Fonte et al. 2013). Some biomaterials induce structural reorganization in the TJs or chelate the calcium causing the disruption of TJs (Werle et al. 2009; Yeh et al. 2011), thus enhancing the drug absorption. Often, the toxicity is associated with the materials that are part of the NPs. Nevertheless, the pharmacokinetic properties of a drug or excipient may change considerably following incorporation into nanoparticulate system (Chiu et al. 2009; Baldrick 2010).

Intravenous and subcutaneous injections of nanomaterial-based carriers deliver exogenous NPs into the body for better distribution. However, wider spreading may cause toxicity and undesirable interaction with biological macromolecules. Injected nanomaterials <100 nm are efficiently transported via interstitial flow to the draining lymphatics and lymph nodes. Meanwhile, they reach most of the organs based on their size and surface characteristics. Besides injection, other routes of exposure like nasal and dermal are also common.

It has been shown that metallic NPs <10 nm can penetrate the epidermal layers (Baroli et al. 2007). NPs may pass through the stratum corneum of damaged skin and may induce lung inflammation by stimulating pulmonary epithelial cells to generate proinflammatory cytokines (Nel et al. 2006).

A major challenge in drug delivery is to improve selective targeting and safe strategies, but major caution should be made in a special group of patients like pregnant women, infants, and aged people. For example, studies have shown that NPs can easily cross the placental barrier and induce pregnancy complications (Wick et al. 2010).

Most of the information about kinetics of materials comes from tests of materials in the normal size, and unsurprisingly, there is a lack of data about kinetics of nanosized materials that may have a major role in toxicity (Pourmand and Abdollahi

2012). Therefore, a data bank on biological effects, toxicity, biokinetics, as well as structure and molecular size can assist scientists to predict the toxicity of nanomaterials. The biokinetics of NPs is illustrated in Figure 20.1.

### 20.2.3 Polymer Characteristics

The polymers offer versatility in both structure and functions due to a wide variety of monomers available, and contribute to the advances of nanodrug delivery systems. Loading capacity as well as controlled delivery of the drug solely depends on the type of polymer used. Out of two kinds of polymers, that is, biodegradable and nonbiodegradable, mainly biodegradable polymers are used in drug delivery. Thus, the chances of toxicological manifestations related to polymeric nanocarriers in drug delivery are lessened. Regardless of this belief, the safety concerns related to polymeric nanocarriers are very important and require further attention.

Of all the polymers, chitosan (CS) is the most widely studied natural polymer for oral drug delivery application. Due to the nontoxic and biocompatible nature, it is approved by the FDA for wound dressing (Baldrick 2010).

Despite extensive investigation with CS, it did not get approval from the FDA for use in any product for the drug delivery, and as a consequence, very few companies are using this material for drug delivery applications (Kean and Thanou 2010).

Toxicity data are still needed to answer some safety concerns in order to include CS as an excipient in new drug formulations. CS is not absorbed by the GIT and is unlikely to show biodistribution, while CS oligosaccharides and its derivatives such as trimethyl chitosan (TMC) are absorbed to some extent (Chae et al. 2005; Zheng et al. 2007). It has been shown that *in vitro* Caco-2 cell or *in vivo* oral absorption of CS derivatives in rats depends on their molecular weight (MW; Chae et al. 2005). Low-MW oligomers (3.8 kDa, 88.4% deacetylation degree [DD]) show relatively higher absorption than high-MW CS (230 kDa, 84.9% DD), which remains almost unabsorbed.

The MW and DD also influence the toxicity of CS. *In vitro* studies have shown that at high DD, the CS toxicity is related to the MW and the polymer concentration; however, at lower DD, the toxicity is less marked and is merely associated with the MW (Schipper et al. 1996, 1999; Agrawal et al. 2014).

Arai et al. (1968) reported an $LD_{50}$ value of 16 g/kg for CS following oral administration to mice. No side effects were reported up to a dose of 4.5 g/day in humans. However, regular intake for 12 weeks produced mild nausea and constipation (Gades and Stern 2003; Baldrick 2010).

The toxicity data regarding the CS derivatives also exist in the literature. Yin et al. (2009) reported the toxicity of TMC–cysteine conjugate (500 kDa) solution. However, the NPs did not produce toxicity. Zheng et al. (2007) noted that TMC NPs could cause light diarrhea at high doses, which can be relieved by discontinuing the administration.

The toxicity of decanoic acid-*g*-oligo CS NPs has also been assessed in rats (Du et al. 2014). The histopathology studies did not exhibit significant differences between the experimental and the control groups. The villi structure of the intestinal epithelium was normal without the presence of inflammatory cells.

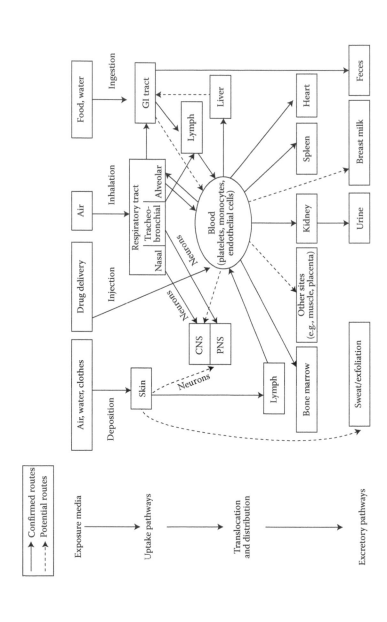

**FIGURE 20.1** Biokinetics of nanoparticles. PNS means peripheral nervous system. Although many uptake and translocation routes have been demonstrated, others still are hypothetical and need to be investigated. Translocation rates are largely unknown, as are accumulation and retention in critical target sites and their underlying mechanisms. These, as well as potential adverse effects, largely depend on physicochemical characteristics of the surface and core of nanosize particles (NSPs). Both qualitative and quantitative changes in NSP biokinetics in a diseased or compromised organism also need to be considered. (Reproduced from *Environmental Health Perspectives*, and the material was reproduced from the link: http://www.ncbi. nlm.nih.gov/pmc/articles/PMC1257642/.)

Like the natural polymers, the toxicity of synthetic polymers is not addressed in most of the studies. Poly(lactic-co-glycolic acid) (PLGA), a synthetic polymer is used to produce oral controlled release NPs for its biodegradable and biocompatible nature. The safety of PLGA as a drug delivery polymer is supported by a number of studies. In an *in vitro* study on lung epithelial cell line, Yang et al. (2012) concluded that the PLGA polymer is safe for use either alone or in combination with CS. The effect of native CS on cell death was likely more pronounced because it carries highly positive charge. The reduced cytotoxicity of the NPs was probably due to partial neutralization or encapsulation of the positive surface charge of CS by negatively charged low-MW heparin (LMWH) or by PLGA coating. In addition, the low cytotoxicity of CS-PLGA LMWH NPs was supposed to be influenced by the biocompatible and biodegradable nature of PLGA and CS polymers. PLGA NPs (200 nm) presented no toxicity either alone or coated with CS in Caco-2 and HT29 intestinal cell lines (Araújo et al. 2014).

Moreover, Semete et al. (2010a) evaluated the cytokines' expression of CS and PEG-coated PLGA NPs within 24 h of oral and peritoneal administration in Balb/C mice. The expression of proinflammatory cytokines interleukin (IL)-2, IL-6, IL-12p70, and tumor necrosis factor-$\alpha$ (TNF-$\alpha$) in the plasma and the peritoneal lavages persisted at low concentrations.

The oral toxicity of PLGA NPs stabilized with didodecyldimethylammonium bromide was studied in rats (Bhardwaj et al. 2009). Methylthiazolyldiphenyl-tetrazolium (MTT) and lactate dehydrogenase assays suggested that the cationic surfactant was safe in the cell cultures at concentrations <33 $\mu$m. PLGA NPs prepared with this stabilizer were found to be nontoxic on cell lines.

The biodistribution of the PLGA NPs was studied for 7 days after oral administration (Semete et al. 2010b). The cell viability was >75% for PLGA particles, but significantly reduced for zinc oxide particles. *In vivo* toxicity was assessed via histopathological evaluation, and no specific anatomical pathological changes or tissue damage were seen in the tissues of Balb/C mice. The results showed that about 40% of the particles were localized in the liver, 26% in the kidney, and 13% in the brain.

Jain et al. (2011) showed that tamoxifen-encapsulated PLGA NPs (165.58 ± 3.81 nm) could significantly reduce hepatotoxicity than its solution form. The liver section of rats treated with PLGA NPs presented normal histopathology in contrast to tamoxifen solution that presented edema and swelling of hepatocytes, necrosis, hyperplasia of Kupffer cells, and apoptosis.

Polylactic acid (PLA)/cholate NPs were nontoxic at a dose of 75 mg/kg after IV administration to rats (Plard and Bazile 1999). At higher doses, 220 and 440 mg/kg, mortality and marked clinical signs were observed with dose-related hematological changes. Methyl-PEGylated PLA NPs did not show any incidents of lethality and clinical complications even at dose of 440 mg/kg. PEGylation improved the safety profile of PLA/cholate NPs as compared to non-PEGylated NPs. They reasoned that the steric repulsion by the highly dense methyl-PEG chains on the NPs' surface prevented the coagulation cascade and associated toxicity.

Mura et al. (2011) tested lung toxicity of PLGA NPs on human bronchial Calu-3 cells. The positively charged, negatively charged, and neutral NPs were prepared by coating their surface with CS, poloxamer, or poly(vinyl alcohol), respectively.

Regardless of the surface charge, the cytotoxicity of the NPs was very limited, with no signs of inflammatory response.

However, in case of nondegradable block copolymers, there is risk of accumulation in the MPS or other tissues due to lack of elimination. Biodistribution, movement of materials through tissues, phagocytosis, opsonization, and endocytosis of nanosized materials are all likely to have an impact on potential toxicity, which in turn depends on the particle surface charge (Garnett and Kallinteri 2006).

PEG is generally regarded as safe with $LD_{50} > 10$ mg/kg. It has long been used as an excipient in pharmaceutical formulations intended for parenteral, oral, ocular, rectal, and topical use. A little toxicity associated with the exposures of 10 mg/kg for PEG up to 10 kDa are deemed acceptable. However, there are a few long-term toxicological data on PEGs > 10 kDa that are commonly used in the design of NPs. Most of the nonimmunogenic effect of PEG is due to the decrease in opsonin adsorption to the particles, thereby reducing phagocytosis by macrophages of the MPS (Fruijtier-Polloth 2005; Webster et al. 2009).

The hydrophilic natural polymers are currently being investigated as drug delivery carriers due to their nontoxic, biodegradable, and biocompatible nature. In a majority of cases, various polysaccharide derivatives are synthesized for drug delivery applications. The modified polymers are also being screened as potential therapeutic agents. Despite the desirable properties of the native polymers, the same are questionable for their derivatives. Hence, there is also need to consider toxicology of the polysaccharide derivatives, and their NPs since they may cause various interactions with fluids, cells, and tissues, starting at the portal of entry and then via a range of possible pathways toward target organs. At the target organ, the NPs may trigger mediators that may activate inflammatory or immunological responses (Donaldson et al. 2004).

## 20.2.4 TYPES OF NANOCARRIERS

### 20.2.4.1 Dendrimers

Dendrimers are micelle-like NPs that are composed of a hydrophobic core and a hydrophilic shell, constituted by polymeric branches (Svenson and Tomalia 2005). The inherent toxicity of dendrimers in biological system creates a barrier toward extensive pharmaceutical application (Jain et al. 2010). The cationic dendrimer surface interacts with negatively charged biological membranes *in vivo* and exhibits via nanohole generation, thinning, and erosion of membrane. The toxicity in biological system mainly includes hemolytic toxicity, cytotoxicity, and hematological toxicity. The toxicity can be minimized by designing biocompatible dendrimers and masking the peripheral charge by surface modification via PEGylation, acetylation, carbohydrate, and peptide conjugation, or by introducing negative charge such as half-generation dendrimers. Neutral and negatively charged dendrimers do not interact with biological environment, and hence are compatible for clinical applications.

The polymer, poly(amidoamine) or PAA, is a class of dendrimer, which is made up of repetitively branched subunits of amide and amine functionality. The relative ease/low cost of synthesis of PAA dendrimers, along with their biocompatibility, structural control, and functionalizability have made PAA viable candidates for application in drug delivery (Lee et al. 2005). Initial studies on PAA toxicity showed that PAA was less toxic than related dendrimers of minimal cytotoxicity (Haensler and Szoka Jr 1993; Fischer et al. 2003).

More recently, a series of studies by Mukherjee et al. (2010a,b) have shed some light on the mechanism of PAA cytotoxicity, providing evidence that the dendrimers cause harm to the cell's mitochondria and eventually leading to cell death. It has also been shown that PAA dendrimers cause rupturing of red blood cells (RBCs), or hemolysis (Malik et al. 2000).

Thiagarajan et al. (2013) reported that cationic dendrimers are more toxic than the anionic ones that are tolerated at 10 times higher doses. Moreover, larger dendrimers are more toxic, causing hemobilia and splenomegaly. However, the masking of cationic residues with noncharged groups can improve their safety and uptake by the epithelial cells (Wiwattanapatapee et al. 2000).

To date, a few in-depth studies on the *in vivo* behavior of PAA dendrimers have been carried out. The functionalization of PAA has a dramatic effect on their ability to diffuse in the CNS tissue *in vivo* and penetrate living neurons as shown by intra-parenchymal or intraventricular injections in animals. The G4-C12 PAA dendrimer can induce dramatic apoptotic cell death of neurons *in vitro* at a concentration of 100 nM. On the contrary, G4 PAA does not induce apoptotic cell death of neural cells in the submicromolar range of concentration and induces low microglia activation in brain tissue after a week (Albertazzi et al. 2013).

### 20.2.4.2 Lipid-Based Nanostructures

A number of lipid-based nanocarriers have been designed such as nanostructured lipid carriers (NLCs) and SLNs for the purpose of oral drug delivery. Both systems consist of many components like oil, surfactants, cosurfactants, and cosolvents. These components, especially the surfactants and cosurfactants, can precipitate toxic effects because a large amount of emulsifier is required for their preparation. The materials used for the formulation of SLNs include different triglycerides such as tricaprin, trilaurin, trimyristin, tripalmitin, tristearin, and hard fats such as different grades of Witepsol, Softisan, glyceryl monostearate, glyceryl behenate, stearic acid, palmitic acid, and so on. Regardless of these lipids, the emulsifiers such as soybean lecithin, egg lecithin, and phosphatidylcholine are also used. Emulsifiers are the most important component of SLNs/NLCs and maintain hydrophilic–lipophilic balance (HLB) with lipids to give stability to the formulation. In some cases, when any of the system components got unbalanced, it may cause toxicity. For example, use of Tween 80, a commonly used emulsifier having HLB value very high, can result in loosening of TJs of intestinal epithelium cells (Buyukozturk et al. 2010).

*In vivo* toxicology of SLNs by Cho et al. (2014) did not reveal any damage to the intestinal epithelium such as villi fusion, occasional epithelial cell shedding, and congestion of the mucosal capillary with blood and focal trauma even 8 h after oral administration.

According to Buyukozturk et al. (2010), the oil structure, surfactant HLB values, and surfactant to oil ratio are important considerations for the safety of the emulsion-based formulations. In case, some of these parameters are imbalanced, toxic effects may occur.

NPs can coexist with surfactants, but this coexistence of NPs and surfactants is likely to give rise to joint toxic effect on biological systems and environment (Wang et al. 2014). Indeed, research results have demonstrated that surfactants embedded into membranes as interstitial ingredients brought about alterations in bilayer structure, as well as dissolving capacity (Schreier et al. 2000). In addition, surfactants absorbed on the surface of NPs may cause surface charges, disparity, and toxicity (Lovern and Klaper 2006; Baalousha 2009), otherwise, they are more likely to decrease toxicity effects due to inhibiting interactions between NPs and bacteria by means of steric hindrance and charge repulsion (Zhang et al. 2007). Thus, safety studies of all the materials and lipid-based nanostructures are essential prerequisite before their clinical applications.

### 20.2.4.3   Carbon Nanotubes

Carbon nanotubes (CNTs) are cylindrical structures formed by rolling of single layer (single-walled CNT [SWCNTs]) or multiple layers (multiple-walled CNT [MWCNTs]) of graphene sheets with diameters of 1–2 nm and lengths of 0.05–1 μm (Foldvari and Bagonluri 2008). The allotrope of carbon consists of 60 carbon atoms joined together to form a cage-like structure. C60 is soluble in aromatic solvents (e.g., toluene or benzene), but insoluble in water and alcohol. However, C60 can be functionalized with $-OH$, $-COOH$, or $-NH_2$ to increase its hydrophilicity.

The cylindrical structures are capped at the ends by carbon networks. CNTs are being explored as drug nanocarriers due to their high surface area, conductivity, high tensile strength, and potential higher absorption capabilities (Beg et al. 2011). The hollow monolithic structure of CNT allows the incorporation of drug molecules for controlled and site-specific delivery (Heister et al. 2009). Moreover, the outer surface of CNTs can be functionalized to enhance their biocompatibility and biodegradability (Beg et al. 2011). There have been several toxicological studies after oral administration of CNTs. However, the study reports are contradictory to each other. Some workers reported acute toxicity and genotoxicity with CNTs, while the others reported no toxic influence of the CNTs. It was previously shown that CNT had some immunological reactions of CNTs. Later on, the effects were ascribed to the metallic impurities and contaminants present in the CNTs (Pulskamp et al. 2007).

The high purity and well-dispersed sample of SWCNTs ($3.0 \pm 1.1$ nm, length <1.2 μm) did not exhibit any genotoxic effects in both *in vitro* and *in vivo* experiments at a dose of 60 and 200 mg/kg BW (Naya et al. 2011). Single-dose genotoxicity study in Fischer 344 rats revealed that the nanotubes (0.9–1.7 nm) elevated the levels of 8-oxo-7,8-dihydro-2′-deoxyguanosine in lungs and liver at doses of 0.064 and 0.64 mg/kg BW. The nanotubes caused oxidative damage to DNA in liver and lung cells after oral administration (Folkmann et al. 2009). A single bolus dose of ultrashort and full-length SWCNTs (diameter = 1 nm  length = 2 μm–20 nm; 1000 mg/kg BW) in Swiss mice did not produce causalities and abnormalities. No acute oral toxicity was observed regardless of the length, surface area, and surface interactions (Kolosnjaj-Tabi et al. 2010).

Studies on carbon nanomaterials have indicated the potential neurotoxic effects after inhalation or systemic exposure. Oberdörster and coworkers (2004) showed that inhalation of elemental $^{13}C$ particles (36 nm) following whole-body exposure for a period of 6 h led to a significant and persistent increase in the accumulation of $^{13}C$ NPs in the rat's olfactory bulb, and the NP concentration gradually increased. However, different shapes of carbon nanomaterials may elicit different neuronal toxicity.

More specifically, pure graphene exhibits less toxicity than highly purified SWCNTs in a concentration-dependent manner after 24-h exposure of PC12 cells, involving the apoptosis pathway (Zhang et al. 2010b).

CNTs with surface coating of PEG are less toxic on mitochondrial function and membrane integrity than uncoated CNTs. A study has shown that oxidative stress is involved in this toxic pathway, with surface coating playing an important role (Zhang et al. 2011). It has been reported that 14-nm carbon black particles may translocate to the olfactory bulb through olfactory neurons, resulting in the activation of microglial cells, which induces proinflammatory cytokines and chemokines, suggesting an inflammatory response (Shwe et al. 2006). Further *in vivo* studies are needed to understand the effect of surface coating on the biocompatibility of these carbon-based nanomaterials prior to use in humans.

### 20.2.4.4   Metal NPs

The most widely studied metal NPs include AuNPs, and superparamagnetic iron oxides' ($Fe_2O_3$ or $Fe_3O_4$) NPs (SPIONs). However, the use of these NPs as oral delivery systems is very limited due to crisis in toxicological studies (Li and Chen 2011).

A report by Hillyer and Albrecht (2001) indicated that 4-nm AuNPs could cross the GIT more readily, resulting in higher accumulation in kidney, liver, spleen, lungs, and brain of mice compared to the particles of 10–58 nm size. Pokharkar et al. (2009) did not find any changes in clinical signs, body weight, food consumption rate, hematological parameters, organ weights, and histopathological observation for CS-coated AuNPs in rats after 28 days of oral administration. Moreover, the $LD_{50}$ was >2000 mg/kg BW. Zhang et al. (2010a) compared the toxicity of different oral doses (137.5–2200 μg/kg) of AuNPs of 13.5 nm size. The particles were almost nontoxic at lower doses. However, a reduced RBC count was noticed at higher doses, with higher accumulation in spleen. Thus, the factors such as size, surface coating, and the dose are among important considerations in developing oral AuNPs formulations.

The SPIONs are composed of $Fe_3O_4$ (magnetite) or $Fe_2O_3$ (maghemite) core. They are specifically used for brain imaging or brain-targeted drug/gene delivery due to their ability to cross the blood–brain barrier (BBB; Kong et al. 2012). Despite their desirable traits, the *in vivo* and *in vitro* toxicity data are of great concern before clinical application. SPIONs can interfere with gene expression, actin modulation, cell cycle regulation, and signaling pathways, and may lead to excessive ROS generation and disruption of iron homeostasis (Singh et al. 2010).

According to Wang et al. (2009), the transport of submicron level $Fe_3O_4$ NPs to the brain via the olfactory nerve pathway may cause oxidative stress-related damage in brain. They also demonstrated size-dependent effect on iron deposition in different brain regions after single intranasal exposure of 21-nm and 280-nm $Fe_2O_3$ NPs in mice (Wang et al. 2008). The iron content in olfactory bulb, hippocampus, cerebral cortex, and cerebellum to the brainstem significantly increased after administration of smaller particles. However, the iron deposition was significant only in olfactory bulb and hippocampus for larger particles. Even after 30 days, the iron content in these regions was lower than that in mice treated with 21-nm $Fe_2O_3$ NPs. The brain iron accumulation is associated with oxidative stress induced by the formation of the highly reactive *OH via the Fenton reaction (Kim et al. 2000; Castellani et al. 2007).

The generation of ROS is a well-established paradigm to explain the toxic effects of NPs. Wu et al. (2013) focused on the neurotoxicity of iron oxide NPs in the rat brain *in vivo*. Overall, the number of studies regarding the toxicity of metallic NPs is very limited. Most of the studies focused on the biodistribution of the NPs. However, the study regarding interaction of NPs with the tissues is lacking. Hence, the safety of metallic NPs must be ensured.

## 20.3   EXPERIMENTAL NANOTOXICITY ON DIFFERENT PHYSIOLOGICAL SYSTEMS

NPs enter the human body through various routes, including respiratory tract, GIT, skin contact, IV injection, and implantation. Following absorption, the NPs are carried to distal organs by the bloodstream and the lymphatic system. The possible nanotoxicity to physiological systems is represented in Figure 20.2.

### 20.3.1   CIRCULATORY SYSTEM

NPs are transported to distal organs through the blood. During translocation, NPs alter fluid dynamics of blood, affect vascular walls, and adhere to the blood vessel surfaces due to nonspecific van der Waals, electrostatic, and steric interactions (Decuzzi et al. 2005).

This trend may relate to physical properties of the NPs like size and shape. Oblate-shaped NPs adhered to the surface of blood vessels greater than spherical NPs of the same volume (Decuzzi and Ferrari 2006). In blood, the original properties of the NPs are changed by proteins that form a protein corona on their surface (Demir et al. 2011). This protein corona influences *in vivo* behavior of NPs such as cell uptake and biocompatibility. Protein adsorption also helps in better dispersion of NPs and causes a higher cellular accumulation of NPs.

NPs are harmful to the circulatory system also. After inhalation, the NPs may stimulate the generation of oxidative stress in the lungs of animal models and lead to the release of proinflammatory mediators and coagulation factors, which are then transmitted to the circulation, leading to cardiovascular lesions, including platelet aggregation, thrombosis, and cardiovascular malfunction (Donaldson et al. 2001).

NPs can induce circulation toxicity after entrance via inhalation as follows. The macrophages located in the alveolar epithelium release cytokines after NPs' uptake. These cytokines migrate across endothelium of the blood vessel and stimulate cardiovascular lesions. Moreover, the NPs migrated across the interstitium are picked up by endothelium macrophages. Consequently, cytokines are released into blood and aggravate cardiovascular lesions. The particles escaping interstitium and endothelium uptake are taken up by blood cells such as platelets and stimulate cardiovascular lesions. The events are described in Figure 20.3.

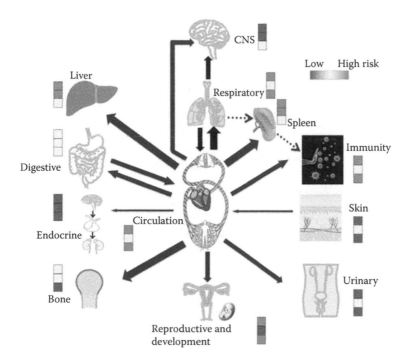

**FIGURE 20.2** A working model indicating possible nanotoxicity to physiological systems. In the three-frame bar, three frames (from upper to lower) represent the probability of nanoparticle accumulation in organs or systems, self-repair capability (including the inclination of nanoparticle degradation to facilitate the excretion) of the system, and the observed toxicity from the available literature. The nanoparticle accumulation, self-repair capability, and observed toxicity are intermediate between low and high levels for the digestive system. These effects appear in high-low-high order for reproductive system, respectively. For the rest of the organs or systems, open boxes indicate intermediate level of effects. The closed boxes indicate high level of effects for liver, spleen, circulation, immunity, and respiratory systems and low level of effects for endocrine, bone, CNS, skin, and urinary systems. The scale bar indicates low (left), intermediate (middle), and high (right) levels. These scales are only based on available data and are not conclusive because of differences in dose, nanoparticle preparation, and animal models. Arrows show the direction of nanoparticle translocation. The width of the lines indicates the readiness of nanoparticle translocation. Dashed lines show the reported cross-system effects. (Zhang, Y. et al. 2014. Perturbation of physiological systems by nanoparticles. *Chem Soc Rev* 43:3762–809. Reproduced by permission of The Royal Society of Chemistry.)

In blood, NPs also activate some coagulation pathways. MWCNTs with different habits like pristine, carboxylated, and amidated damage endothelial cell of blood vessel and trigger coagulation *in vivo*. *In vitro*, they exhibit obvious procoagulant activity with activated partial thromboplastin time (aPTT) assays (Burke et al. 2011). MWCNTs may activate both intrinsic and extrinsic pathways of coagulation via factor IX- and factor XII-dependent ways and stimulate thrombosis. Carbon NPs (MWCNTs, SWCNTs, and mixtures thereof) (Radomski et al. 2005) can enhance platelet aggregation and contribute to the vascular thrombosis. Though the thrombus

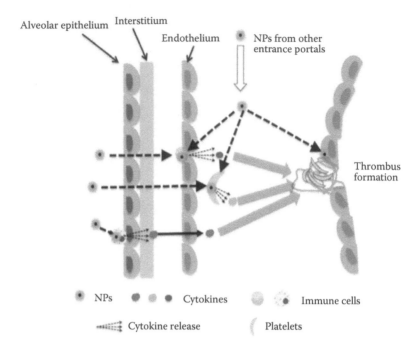

**FIGURE 20.3** Nanoparticle-induced circulation toxicity after inhalation. (Zhang, Y. et al. 2014. Perturbation of physiological systems by nanoparticles. *Chem Soc Rev* 43:3762–809. Reproduced by permission of The Royal Society of Chemistry.)

sizes are dose dependent (Hoet et al. 2004; Vermylen et al. 2005), these effects may cause significant risks in populations for atherothrombosis.

### 20.3.2 Bone Marrow

Under certain pathological conditions, the liver, thymus, and spleen may resume hematopoietic functions, causing their pathological enlargement. This may impair production of blood cells by affecting the hematopoietic stem cell functions, immune functioning, and lead to diseases like leucopenia, thrombocytopenia, neutropenia, and anemia.

Like the liver and spleen, bone marrow is one of the primary organs of the reticuloendothelial system where the production and maturation of most blood cells occur. Thus, the access of NPs to bone marrow is quite expected. After oral administration into mice, polystyrene NPs were detected in bone marrow. Polystyrene microparticles in the size range 50 nm–3 μ were fed by gavage to female Sprague Dawley rats daily for 10 days at a dose of 1.25 mg/kg. The extent of absorption of 50-nm particles under the conditions of these experiments was 34% and of the 100-nm particles was 26%. Particles larger than 100 nm did not reach the bone marrow (Jani et al. 1990). Hence, NPs are currently being investigated for targeting bone marrow for drug delivery.

The adverse hematopoietic effects of particles also depend on the route of administration. After 4 weeks, the inhalation of magnetic NPs decreased the mean corpuscular volume and hemoglobin content, two indicators of impaired erythrocyte function. Inhaled NPs also decreased the platelets production, increased white blood cell (WBC) count in the bone marrow, and induced extramedullary hematopoiesis in the mouse spleen, which was indicative of pathological conditions such as anemia (Kwon et al. 2009).

### 20.3.3 Reproductive Systems

The reproductive nanotoxicity takes into account the adverse effects on germ cells, physiological structure and function, fertility, and their effects on the offspring. The AuNPs (9 nm) penetrated the heads and tails of healthy male human sperm cells and caused 25% of sperm cells to become immotile at a concentration of 44 mg/mL (Wiwanitkit et al. 2009).

Repeated IV injection of water-soluble MWCNTs into male mice caused reversible testicular damage without affecting fertility (Bai et al. 2010). Nanotubes accumulated in the testes generated oxidative stress and lowered the thickness of seminiferous epithelium at day 15, but recovered after 60 and 90 days. The quantity, quality, and integrity of the sperm and the levels of sex hormone remained undisturbed throughout the entire study period.

After tail vein injection to pregnant Sprague Dawley rats, [$^{14}$C]C60 NPs (~0.3 mg/kg BW) cross the placenta and is transmitted to offspring via the dam's milk and subsequently drained into blood (Sumner et al. 2010). Some colloidal Au particles (5 and 30 nm) are transferred to the fetus 1 h after IV injection at gestational day 19. Small AuNPs exhibited a slightly higher transfer rate than 30-nm NPs (Takahashi and Matsuoka 1981). The NPs may be transferred from placenta to the fetus, where they may exhibit potential developmental toxicity. In one study, pregnant Slc mice were injected intraperitoneally (IP) with C60 NPs on gestational day 10 and the embryos were examined 18 h after injection (Tsuchiya et al. 1996).

At a dose of 50 mg/kg, the NPs distributed into the yolk sac and embryos, and half of the embryos deformed in the head and tail regions. At a dose of 25 mg/kg, abnormal embryos were less frequent; however, all embryos died at a dose of 137 mg/kg. It was speculated that C60 NPs caused severe dysfunction of the yolk sac and embryonic morphogenesis.

### 20.3.4 Gastrointestinal Tract

Upon oral administration, NPs have only transient contact with the oral cavity, pharynx, and esophagus. A majority of them are accumulated in the stomach and intestines and the unabsorbed fraction is quickly eliminated thorough feces. Due to protective mucous layer and the tight epithelial junctions, the rate of absorption of NPs from GIT is much lower than other routes. Under certain pathological conditions, however, the integrity or function of one or more GI layers is compromised and the layers become permeable, causing disorders such as inflammatory bowel disease.

The NPs retained in the GIT may adversely affect its structure and function. Recently, a pH-responsive NP system shelled with CS has been found to effectively increase the oral absorption of insulin and produce a hypoglycemic effect, presumably due to the CS-mediated TJ opening (Sonaje et al. 2011).

Using *in vitro* model of the intestinal epithelium and *in vivo* chicken intestinal loop model, acute and chronic oral exposures to polystyrene NPs were studied (Mahler et al. 2012). Intestinal cells showed increased iron transport at high doses due to disruption of cell membrane by the NPs. Chickens acutely exposed to carboxylated particles of 50-nm sizes had lower iron absorption than unexposed or chronically exposed birds. Chronic exposure possibly caused remodeling of the intestinal villi and increased the surface area available for iron absorption. In addition to potential impact of NPs on nutrient absorption, this report emphasized the complexity of interactions between NPs and GI tract.

### 20.3.5 URINARY NANOTOXICITY

Since NPs readily accumulate in kidney in addition to the reticuloendothelial system, their urinary toxicity is a prioritized concern. Further, the kidney is an important organ for the elimination of NPs (Li and Huang 2008).

Larger NPs are primarily localized in the liver and spleen. However, small particles (~5–10 nm) may pass glomerular barriers (glomerular endothelial cells

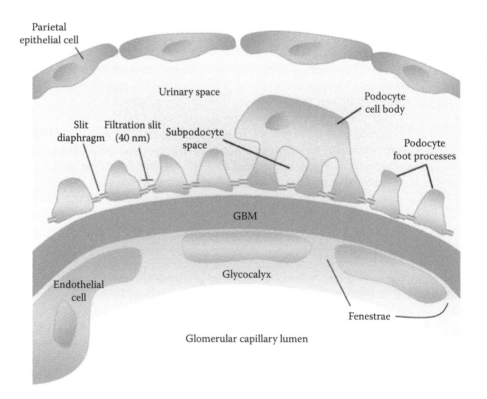

**FIGURE 20.4**  Glomerular capillary wall. (Reprinted from *Am J Kidney Dis*, 58, Jefferson, J.A. et al., Podocyte disorders: Core curriculum, 666–77, Copyright 2011, with permission from Elsevier.)

fenestrate, 80–100-nm wide pores, glomerular basement membrane, and podocytes) and excrete rapidly in urine (Longmire et al. 2008; Figure 20.4).

Larger NPs that diffuse through the glomerular endothelial cell fenestrate, 80–100-nm wide pores are further prevented by the glomerular basement membrane and podocyte foot processes. The basement membrane, together with the podocyte foot processes, imposes an apparent cutoff size of ~10 nm or MW of 30–50 kDa. However, some SWCNTs can penetrate the physical barriers and excrete in mice urine. SWCNTs (0.8–1.2 nm in diameters and 100–500 nm in length) are cleared intact by glomerular filtration, with partial tubular reabsorption and transient translocation into proximal tubular cell nuclei ($t_{1/2}$ ~ 6 min) after IV injection. The threshold MW for the glomerular filtration of polymers lies in the range of 30–50 kDa and depends on charge, molecular conformation, and deformation ability. Because of high aspect ratio ($d$ ~ 1 nm, $100 \leq L \leq 500$ nm), negative charge, and high MW (150–750 kDa), the construct largely exceeds structural sizes of the glomerular pores (at least in the longitudinal dimension). The renal elimination of ~65% of the recovered construct was observed with ~15% of the construct undergoing passive reabsorption within the tubules at 20 min postinjection. This can be regarded as an exceptional case, probably due to their needle-like shape (Ruggiero et al. 2010).

Studies indicated that kidney is relatively insensitive to the adverse effects of NPs. IP and IV injection of $N$-octyl-$O$-sulfate chitosan (NOSC) NPs into mice led to systemic toxicity, without any histopathological changes in the kidneys (Zhang et al. 2008). The $LD_{50}$ values of NOSC were found to be 102.59 and 130.53 mg/kg, respectively, after IV and IP administration. Almost 75% of the dose of tritium-labeled NOSC (13.44 mg/kg) was excreted in urine over 7 days. NOSC was predominantly excreted through urine, rather than bile or feces.

### 20.3.6 CENTRAL NERVOUS SYSTEM

The BBB and blood–cerebrospinal fluid barrier afford protection to the microenvironment of human CNS from hazardous xenobiotics, however, make CNS delivery of therapeutics difficult. Small particles could be advantageous as therapeutic carriers for the treatment of CNS diseases (Bharali et al. 2005) and raises concerns regarding their possible unwanted toxic effects.

The NPs can enter the CNS at least by three distinct ways. Firstly, NPs can penetrate the BBB without damaging its integrity. PEG-grafted CS copolymer (Veiseh et al. 2009) and silica-coated magnetic NPs (Kim et al. 2006) penetrated the BBB without affecting its functions. This mode of penetration is ideal for the therapy of CNS diseases. A biodistribution study suggested that only 0.3% of AuNPs (10 nm) were distributed in rat brain after 24 h, but no Au particles with diameters of 50, 100, or 250 nm were detected after injection into their tail vein (De Jong et al. 2008). Thus, the ability to cross the BBB probably depends on particle size.

The disruption of the BBB integrity is the second option for NPs' penetration. Direct disruption of the cell membrane caused by NPs will allow their entry into the brain. Breakdown of the BBB enables the passage of various serum components, including proteins and other toxic substances into brain microfluid environment. Polysorbate 80-coated poly($n$-butylcyanoacrylate) (PBCA) NPs are able to cross the

BBB *in vitro* and *in vivo* (Rempe et al. 2011). The disruption of the barrier by polysorbate 80-coated PBCA NPs became reversible after 4 h. Instead of incorporating therapeutic agents into the NP, the drugs may cross the BBB with simultaneous administration of the PBCA NPs.

Lastly, the NPs can translocate to the brain via olfactory nerve pathway, bypassing the BBB. The accumulation of NPs in the cerebral compartment generates oxidative stress and inflammation, causing damage to brain nerve cells. CNS damage may be more severe than other tissues due to weak antioxidant and the self-regenerative ability of neurons. After penetration, the NPs induce morphological changes of nerve cells in cerebral cortex, hippocampus, cerebellum, thalamus, hypothalamus, and brainstem, and cause damage to myelinated fibers as well as the degeneration of nerve cells (Sharma 2007).

### 20.3.7 HEPATOTOXICITY

Liver is the major organ for accumulation of NPs. In liver, both hepatocytes and Kupffer cells selectively take up surface-modified NPs. NPs can be excreted from liver via biliary pathway. Eleven days after exposure, approximately 5% of total hydroxylated SWCNTs administered IP were excreted in feces (Wang et al. 2004).

It possesses self-protecting capability due to its antioxidant system and various metabolizing enzymes. However, the long-term retention of NPs may increase the risk of hepatotoxicity (Yang et al. 2008). The prolonged retention of $TiO_2$ and CNT NPs caused injury to hepatocytes as was evident by histopathologic examination and abnormal serum levels of liver function indicators such as aspartate aminotransferase and alanine aminotransferase (Liang et al. 2008).

The hepatocytes cytoplasmic degeneration and nuclear destruction suggest that AuNPs interact with the proteins and enzymes of hepatic tissues, interferes with antioxidant defense mechanism, and leads to ROS generation. This in turn induces stress in the hepatocytes resulting in atrophy and necrosis (Abdelhalim and Jarrar 2012).

Injection of PEGylated AuNPs (15 nm) have been found to cause severe hepatic cell damage, acute inflammation, higher apoptosis, and ROS production in the livers of mice, which were on methionine- and choline-deficient (MCD) diet for 4 weeks. AuNPs demonstrated toxicity in a stressed liver environment by stimulating inflammatory response and accelerating stress-induced apoptosis (Hwang et al. 2012). Other effects of NPs on bile secretion, glucose and fatty acids synthesis, and blood iron content are largely unknown.

## 20.4 CONCLUSION

Despite a significant advancement on research works on nanoparticulate drug delivery systems, the number of marketed products is minimal. One possible reason could be the lack of toxicity and safety information related to different nanoparticulate systems that are needed to surpass the regulatory requirements.

Each nanosystem is unique based on material characteristics, and thus requires a case-specific toxicity study. A basic and conceptual understanding of the interactions of the nanosystems with the biological systems is needed in order to have safe and effective nanosystems for improved drug delivery applications. Overall, the information regarding the toxicology of the NPs is still very limited, which makes it difficult to draw any conclusions regarding the safety and efficacy of nanoparticulate drug delivery systems.

There is an urgent need to understand the potential toxicities of nanomaterials, which would provide useful information to develop safer and more efficient nano-formulations. The safety profiles of the materials used in the nanosystems cannot be directly translated to the final NPs. The size, charge, and surface chemistries of the NPs also influence the biokinetics and toxicity of the systems. Thus, detailed toxicity-safety profiles could help in fulfilling the stringent requirements by the regulatory authorities and will give faster acceptance.

A collaborative research between formulation development scientists and toxicologists is an hour of need to realize the benefits of nanotechnology in human health.

## REFERENCES

Abdelhalim, M.A.K. and Jarrar, B.M. 2012. Histological alterations in the liver of rats induced by different gold nanoparticle sizes, doses and exposure duration. *J Nanobiotechnol* 10:5–13.

Agrawal, U., Sharma, R., Gupta, M. et al. 2014. Is nanotechnology a boon for oral drug delivery? *Drug Discov Today* 19:1530–46.

Albanese, C.T., Cardona, M., Smith, S.D. et al. 1994. Role of intestinal mucus in transepithelial passage of bacteria across the intact ileum in vitro. *Surgery* 116:76–82.

Albertazzi, L., Gherardini, L., Brondi, M. et al. 2013. In vivo distribution and toxicity of pamam dendrimers in the central nervous system depend on their surface chemistry. *Mol Pharm* 10:249–60.

Alexis, F., Pridgen, E., Molnar, L.K. et al. 2008. Factors affecting the clearance and biodistribution of polymeric nanoparticles. *Mol Pharm* 5:505–15.

Amsden, B. and Cheng, Y.L. 1995. A generic protein system based on osmotically rupturable monoliths. *J Control Release* 33:99–105.

Arai, K., Kinumaki, T., and Fujita, T. 1968. Toxicity of chitosan. *Bull Tokai Reg Fish Res Lab* 56:89–94.

Araújo, F., Shrestha, N., Granja, P.L. et al. 2015. Safety and toxicity concerns of orally delivered nanoparticles as drug carriers. *Expert Opin Drug Metab Toxicol* 11:381–93.

Araújo, F., Shrestha, N., Shahbazi, M.A. et al. 2014. The impact of nanoparticles on the mucosal translocation and transport of GLP-1 across the intestinal epithelium. *Biomaterials* 35:9199–207.

Arora, S., Rajwade, J.M., and Paknikar, K.M. 2012. Nanotoxicology and in vitro studies: The need of the hour. *Toxicol Appl Pharmacol* 258:151–65.

Baalousha, M. 2009. Aggregation and disaggregation of iron oxide nanoparticles: Influence of particle concentration, pH and natural organic matter. *J Sci Total Env* 407:2093–01.

Bai, Y., Zhang, Y., Zhang, J. et al. 2010. Repeated carbon nanotube administrations in male mice cause reversible testis damage without affecting fertility. *Nat Nanotechnol* 5:683–9.

Bakand, S., Hayes, A., and Dechsakulthorn, F. 2012. Nanoparticles: A review of particle toxicology following inhalation exposure. *Inhal Toxicol* 24:125–35.

Baldrick, P. 2010. The safety of chitosan as a pharmaceutical excipient. *Regul Toxicol Pharmacol* 56:290–9.

Baroli, B., Ennas, M.G., Loffredo, F. et al. 2007. Penetration of metallic nanoparticles in human full-thickness skin. *J Investig Dermatol* 127:1701–12.

Beg, S., Rizwan, M., Sheikh, A.M. et al. 2011. Advancement in carbon nanotubes: Basics, biomedical applications and toxicity. *J Pharm Pharmacol* 63:141–63.

Bertrand, N. and Leroux, J.C. 2012. The journey of a drug-carrier in the body: An anatomo-physiological perspective. *J Control Release* 161:152–63.

Bharali, D.J., Klejbor, I., Stachowiak, E.K. et al. 2005. Organically modified silica nanoparticles: A nonviral vector for in vivo gene delivery and expression in the brain. *Proc Natl Acad Sci U S A* 102:11539–44.

Bhardwaj, V., Ankola, D.D., Gupta, S.C. et al. 2009. PLGA nanoparticles stabilized with cationic surfactant: Safety studies and application in oral delivery of paclitaxel to treat chemical-induced breast cancer in rat. *Pharm Res* 26:2495–503.

Burke, A.R., Singh, R.N., Carroll, D.L. et al. 2011. Determinants of the thrombogenic potential of multiwalled carbon nanotubes. *Biomaterials* 32:5970–8.

Buyukozturk, F., Benneyan, J.C., and Carrier, R.L. 2010. Impact of emulsion-based drug delivery systems on intestinal permeability and drug release kinetics. *J Control Release* 142:22–30.

Castellani, R.J., Moreira, P.I., Liu, G. et al. 2007. Iron: The redox-active center of oxidative stress in Alzheimer disease. *Neurochem Res* 32:1640–5.

Chae, S.Y., Jang, M.K., and Nah, J.W. 2005. Influence of molecular weight on oral absorption of water soluble chitosans. *J Control Release* 102:383–94.

Champion, J.A. and Mitragotri, S. 2006. Role of target geometry in phagocytosis. *Proc Natl Acad Sci U S A* 103:4930–4.

Champion, J.A. and Mitragotri, S. 2009. Shape induced inhibition of phagocytosis of polymer particles. *Pharm Res* 26:244–9.

Chithrani, B.D., Ghazani, A.A., and Chan, W.C. 2006. Determining the size and shape dependence of gold nanoparticle uptake into mammalian cells. *Nano Lett* 6:662–8.

Chiu, G.N., Wong, M.Y., Ling, L.U. et al. 2009. Lipid-based nanoparticulate systems for the delivery of anti-cancer drug cocktails: Implications on pharmacokinetics and drug toxicities. *Curr Drug Metab* 10:861–74.

Cho, W.S., Cho, M., Jeong, J. et al. 2010. Size-dependent tissue kinetics of PEG-coated gold nanoparticles. *Toxicol Appl Pharmacol* 245:116–23.

Cho, H.J., Park, J.W., Yoon, I.S. et al. 2014. Surface-modified solid lipid nanoparticles for oral delivery of docetaxel: Enhanced intestinal absorption and lymphatic uptake. *Int J Nanomedicine* 9:495–504.

Cho, W.S., Cho, M., Jeong, J. et al. 2009. Acute toxicity and pharmacokinetics of 13 nm-sized PEG-coated gold nanoparticles. *Toxicol Appl Pharmacol* 236:16–24.

Chonn, A., Cullis, P.R., and Devine, D.V. 1991. The role of surface charge in the activation of the classical and alternative pathways of complement by liposomes. *J Immunol* 146:4234–41.

Das, S. and Chaudhury, A. 2011. Recent advances in lipid nanoparticle formulations with solid matrix for oral drug delivery. *AAPS PharmSciTech* 12:62–76.

De Jong, W.H., Hagens, W.I., Krystek, P. et al. 2008. Particle size-dependent organ distribution of gold nanoparticles after intravenous administration. *Biomaterials* 29:1912–9.

Decuzzi, P. and Ferrari, M. 2006. The adhesive strength of non-spherical particles mediated by specific interactions. *Biomaterials* 27:5307–14.

Decuzzi, P., Lee, S., Bhushan, B. et al. 2005. A theoretical model for the margination of particles within blood vessels. *Ann Biomed Eng* 33:179–90.

Demir, E., Vales, G., Kaya, B. et al. 2011. Genotoxic analysis of silver nanoparticles in Drosophila. *Nanotoxicology* 5:417–24.

Deshmukh, M., Kutscher, H.L., Gao, D. et al. 2012. Biodistribution and renal clearance of biocompatible lung targeted poly(ethylene glycol) (PEG) nanogel aggregates. *J Control Release* 164:65–73.

Donaldson, K., Stone, V., Seaton, A. et al. 2001. Ambient particle inhalation and the cardiovascular system: Potential mechanisms. *Environ Health Perspect* 109:523–7.

Donaldson, K., Stone, V., Tran, C.L. et al. 2004. Nanotoxicology. *Occup Environ Med* 61:727–8.

Du, X., Zhang, J., Zhang, Y. et al. 2014. Decanoic acid grafted oligochitosan nanoparticles as a carrier for insulin transport in the gastrointestinal tract. *Carbohydr Polym* 111:433–41.

El Badawy, A.M., Silva, R.G., Morris, B. et al. 2011. Surface charge-dependent toxicity of silver nanoparticles. *Environ Sci Technol,* 45:283–7.

Ensign, L.M., Cone, R., and Hanes, J. 2012. Oral drug delivery with polymeric nanoparticles: The gastrointestinal mucus barriers. *Adv Drug Deliv Rev* 64:557–70.

Fischer, D., Li, Y., Ahlemeyer, B. et al. 2003. In vitro cytotoxicity testing of polycations: Influence of polymer structure on cell viability and hemolysis. *Biomaterials* 24:1121–31.

Foldvari, M. and Bagonluri, M. 2008. Carbon nanotubes as functional excipients for nano-medicines: I. Pharmaceutical properties. *Nanomedicine* 4:173–82.

Folkmann, J.K., Risom, L., Jacobsen, N.R. et al. 2009. Oxidatively damaged DNA in rats exposed by oral gavage to c60 fullerenes and single-walled carbon nanotubes. *Environ Health Perspect* 117:703–8.

Fonte, P., Araújo, F., Reis, S. et al. 2013. Oral insulin delivery: How far are we? *J Diabetes Sci Technol* 7:520–31.

Fruijtier-Polloth, C. 2005. Safety assessment on polyethylene glycols (PEGs) and their deriva-tives as used in cosmetic products. *Toxicology* 214:1–38.

Gades, M.D. and Stern, J.S. 2003. Chitosan supplementation and fecal fat excretion in men. *Obes Res* 11:683–8.

Garnett, M.C. and Kallinteri, P. 2006. Nanomedicines and nanotoxicology: Some physiologi-cal principles. *Occup Med* 56:307–11.

Haensler, J. and Szoka Jr F.C. 1993. Polyamidoamine cascade polymers mediate efficient transfection of cells in culture. *Bioconjugate Chemistry* 4:372–9.

Hamilton, R.F., Wu, N., Porter, D. et al. 2009. Particle length-dependent titanium dioxide nanomaterials toxicity and bioactivity. *Part Fibre Toxicol* 6:35–45.

Hardman, R. 2006. A toxicologic review of quantum dots: Toxicity depends on physico-chemical and environmental factors. *Environ Health Perspect* 114:165–72.

Heister, E., Neves, V., Tîlmaciu, C. et al. 2009. Triple functionalisation of single-walled car-bon nanotubes with doxorubicin, a monoclonal antibody, and a fluorescent marker for targeted cancer therapy. *Carbon* 47:2152–60.

Hillyer, J.F. and Albrecht, R.M. 2001. Gastrointestinal persorption and tissue distribution of differently sized colloidal gold nanoparticles. *J Pharm Sci* 90:1927–36.

Hoet, P.H., Brüske-Hohlfeld, I., and Salata, O.V. 2004. Nanoparticles—Known and unknown health risks. *J Nanobiotechnol* 2:12–27.

Huang, X., Li, L., Liu, T. et al. 2011. The shape effect of mesoporous silica nanoparticles on biodistribution, clearance, and biocompatibility *in vivo. ACS Nano* 5:5390–9.

Hwang, J.H., Kim, S.J., Kim, Y.H. et al. 2012. Susceptibility to gold nanoparticle-induced hepatotoxicity is enhanced in a mouse model of nonalcoholic steatohepatitis. *Toxicology* 294:27–35.

Jain, A.K., Swarnakar, N.K., Godugu, C. et al. 2011. The effect of the oral administration of polymeric nanoparticles on the efficacy and toxicity of tamoxifen. *Biomaterials* 32:503–15.

Jain, K., Kesharwani, P., Gupta, U. et al. 2010. Dendrimer toxicity: Let's meet the challenge. *Int J Pharm* 394:122–42.

Jani, P., Halbert, G.W., Langridge, J. et al. 1990. Nanoparticle uptake by the rat gastrointesti-nal mucosa: Quantitation and particle size dependency. *J Pharm Pharmacol* 42:821–6.

Jefferson, J.A., Nelson, P.J., Najafian, B. et al. 2011. Podocyte disorders: Core curriculum 2011. *Am J Kidney Dis* 58:666–77.

Jiang, W., Betty, Y.S.K., James, T.R. et al. 2008. Nanoparticle-mediated cellular response is size-dependent. *Nat Nanotechnol* 3:145–50.

Kean, T. and Thanou, M. 2010. Biodegradation, biodistribution and toxicity of chitosan. *Adv Drug Deliv Rev* 62:3–11.

Kim, J.S., Yoon, T.J., Yu, K.N. et al. 2006. Toxicity and tissue distribution of magnetic nanoparticles in mice. *Toxicol Sci* 89:338–47.

Kim, N.H., Park, S.J., Jin, J.K. et al. 2000. Increased ferric iron content and iron-induced oxidative stress in the brains of scrapieinfected mice. *Brain Res* 884:98–103.

Kim, S., Kim, J.H., Jeon, Q. et al. 2009. Engineered polymers for advanced drug delivery. *Eur J Pharm Biopharm* 71:420–30.

Kolosnjaj-Tabi, J., Hartman, K.B., Boudjemaa, S. et al. 2010. In vivo behavior of large doses of ultrashort and full-length single-walled carbon nanotubes after oral and intraperitoneal administration to swiss mice. *ACS Nano* 4:1481–92.

Kong, S.D., Lee, J., Ramachandran, S. et al. 2012. Magnetic targeting of nanoparticles across the intact bloodebrain barrier. *J Control Release* 164:49–57.

Kwon, J.T., Kim, D.S., Arash M.T. et al. 2009. Inhaled fluorescent magnetic nanoparticles induced extramedullary hematopoiesis in the spleen of mice. *J Occup Health* 51:423–31.

Lee, C.C., MacKay, J.A., Fréchet, J.M.J. et al. 2005. Designing dendrimers for biological applications. *Nat Biotechnol* 23:1517–26.

Li, S. and Huang, L. 2008. Pharmacokinetics and biodistribution of nanoparticles. *Mol Pharmaceutics* 5:496–504.

Li, X., Guo, S., Zhu, C. et al. 2013. Intestinal mucosa permeability following oral insulin delivery using core shell corona nanolipoparticles. *Biomaterials* 34:9678–87.

Li, Y.F. and Chen, C. 2011. Fate and toxicity of metallic and metal-containing nanoparticles for biomedical applications. *Small* 7:2965–80.

Liang, X.J., Chen, C., Zhao, Y. et al. 2008. Biopharmaceutics and therapeutic potential of engineered nanomaterials. *Curr Drug Metab* 9:697–709.

Liu, W.T. 2006. Nanoparticles and their biological and environmental applications. *J Biosci Bioeng* 102:1–7.

Lockman, P.R., Koziara, J.M., Mumper, R.J. et al. 2004. Nanoparticle surface charges alter blood–brain barrier integrity and permeability. *J Drug Target* 12:635–41.

Longmire, M., Choyke, P.L., and Kobayashi, H. 2008. Clearance properties of nano-sized particles and molecules as imaging agents: Considerations and caveats. *Nanomedicine* 3:703–17.

Lovern, S.B. and Klaper, R. 2006. Daphnia magna mortality when exposed to titanium dioxide and fullerene (C60) nanoparticles. *J Env Toxicol Chem* 25:1132–7.

Mahler, G.J., Esch, M.B., Tako, E. et al. 2012. Oral exposure to polystyrene nanoparticles affects iron absorption. *Nat Nanotechnol* 7:264–71.

Malik, N., Wiwattanapatapee, R., Klopsch, R. et al. 2000. Dendrimers. *J Control Release* 65:133–48.

Miyazaki, K. and Islam, N. 2007. Nanotechnology systems of innovation—An analysis of industry and academia research activities. *Technovation* 27:661–71.

Moghimi, S.M., Hunter, A.C., and Murray, J.C. 2005. Nanomedicine: Current status and future prospects. *FASEB J* 19:311–30.

Mukherjee, S.P., Davoren, M., and Byrne, H.J. 2010a. In vitro mammalian cytotoxicological study of PAMAM dendrimers—Towards quantitative structure activity relationships. *Toxicol in Vitro* 24:169–77.

Mukherjee, S.P., Lyng, F., Garcia, A. et al. 2010b. Mechanistic studies of in vitro cytotoxicity of poly(amidoamine) dendrimers in mammalian cells. *Toxicol Appl Pharmacol* 248:259–68.

Mura, S., Hillaireau, H., Nicolas, J. et al. 2011. Influence of surface charge on the potential toxicity of PLGA nanoparticles towards Calu-3 cells. *Int J Nanomed* 6:2591–605.

Naya, M., Kobayashi, N., Mizuno, K. et al. 2011. Evaluation of the genotoxic potential of single-wall carbon nanotubes by using a battery of *in vitro* and *in vivo* genotoxicity assays. *Regul Toxicol Pharmacol* 61:192–8.

Nel, A., Xia, T., Madler, L. et al. 2006. Toxic potential of materials at the nanolevel. *Science* 311:622–27.

Oberdörster, G., Oberdörster, E., and Oberdörster J. 2005. Nanotoxicology: An emerging discipline evolving from studies of ultrafine particles. *Environ Health Perspect* 113:823–39.

Oberdörster, G., Sharp, Z., Atudorei, V. et al. 2004. Translocation of inhaled ultrafine particles to the brain. *Inhal Toxicol* 16:437–45.

Owens, D.E. 3rd. and Peppas, N.A. 2006. Opsonization, biodistribution, and pharmacokinetics of polymeric nanoparticles. *Int J Pharm* 307:93–102.

Plard, J.P. and Bazile, D. 1999. Comparison of the safety profiles of PLA50 and Me. PEG-PLA50 nanoparticles after single dose intravenous administration to rat. *Colloids Surf B Biointerfaces* 16:173–83.

Pokharkar, V., Dhar, S., Bhumkar, D. et al. 2009. Acute and subacute toxicity studies of chitosan reduced gold nanoparticles: A novel carrier for therapeutic agents. *J Biomed Nanotechnol* 5:233–9.

Pourmand, A. and Abdollahi, M. 2012. Current opinion on nanotoxicology. *Daru* 20:95–7.

Pulskamp, K., Diabate, S., and Krug, H.F. 2007. Carbon nanotubes show no sign of acute toxicity but induce intracellular reactive oxygen species in dependence on contaminants. *Toxicol Lett* 168:58–74.

Radomski, A., Jurasz, P., Alonso-Escolano, D. et al. 2005. Nanoparticle-induced platelet aggregation and vascular thrombosis. *Br J Pharmacol* 146:882–93.

Rejman, J., Oberle, V., Zuhorn, I.S. et al. 2004. Size-dependent internalization of particles via the pathways of clathrin- and caveolae-mediated endocytosis. *Biochem J* 377:159–69.

Rempe, R., Cramer, S., Hüwel, S. et al. 2011. Transport of poly(n-butylcyano-acrylate) nanoparticles across the blood-brain barrier in vitro and their influence on barrier integrity. *Biochem Biophys Res Commun* 406:64–9.

Ruggiero, A., Villa, C.H., Bander, E. et al. 2010. Paradoxical glomerular filtration of carbon nanotubes. *Proc Natl Acad Sci U S A* 107:12369–74.

Sahoo, S.K. and Labhasetwar, V. 2003. Nanotech approaches to drug delivery and imaging. *Drug Discov Today* 8:1112–20.

Schipper, N.G., Varum, K.M., and Artursson, P. 1996. Chitosans as absorption enhancers for poorly absorbable drugs. 1: Influence of molecular weight and degree of acetylation on drug transport across human intestinal epithelial (caco-2) cells. *Pharm Res* 13:1686–92.

Schipper, N.G., Varum, K.M., Stenberg, P. et al. 1999. Chitosans as absorption enhancers of poorly absorbable drugs. 3: Influence of mucus on absorption enhancement. *Eur J Pharm Sci* 8:335–43.

Schreier, S., Malheiros, S.V.P., and de Paula, E. 2000. Surface active drugs: Self-association and interaction with membranes and surfactants. Physicochemical and biological aspects. *Biochim Biophys Acta* 1508:210–34.

Semete, B., Booysen, L.I., Kalombo, L. et al. 2010a. In vivo uptake and acute immune response to orally administered chitosan and PEG coated PLGA nanoparticles. *Toxicol Appl Pharmacol* 249:158–65.

Semete, B., Booysen, L., Lemmer, Y. et al. 2010b. In vivo evaluation of the biodistribution and safety of PLGA nanoparticles as drug delivery systems. *Nanomedicine* 6:662–71.

Sharma, G., Valenta, D.T., Altman, Y. et al. 2010. Polymer particle shape independently influences binding and internalization by macrophages. *J Control Release* 147:408–12.

Sharma, H.S. 2007. Nanoneuroscience: Emerging concepts on nanoneurotoxicity and nanoneuroprotection. *Nanomedicine* 2:753–8.

Shwe, T.T.W., Yamamoto, S., Ahmed, S. et al. 2006. Brain cytokine and chemokine mRNA expression in mice induced by intranasal instillation with ultrafine carbon black. *Toxicol Lett* 163:153–60.

Singh, N., Jenkins, G.J., Asadi, R. et al. 2010. Potential toxicity of superparamagnetic iron oxide nanoparticles (SPION). *Nano Rev* 1:5358.

Sonaje, K., Chuang, E.Y., Lin, K.J. et al. 2012. Opening of epithelial tight junctions and enhancement of paracellular permeation by chitosan: Microscopic, ultrastructural, and computed-tomographic observations. *Mol Pharm* 9:1271–9.

Sonaje, K., Lin, K.J., Tseng, M.T. et al. 2011. Effects of chitosan-nanoparticle-mediated tight junction opening on the oral absorption of endotoxins. *Biomaterials* 32:8712–21.

Sonavane, G., Tomoda, K., and Makino, K. 2008. Biodistribution of colloidal gold nanoparticles after intravenous administration: Effect of particle size. *Colloids Surf B Biointerfaces* 66:274–80.

Sumner, S.C.J., Fennell, T.R., Snyder, R.W. et al. 2010. Distribution of carbon-14 labeled C60 ([$^{14}$C]C60) in the pregnant and in the lactating dam and the effect of C60 exposure on the biochemical profile of urine. *J Appl Toxicol* 30:354–60.

Svenson, S. and Tomalia, D.A. 2005. Dendrimers in biomedical applications-reflections on the field. *Adv Drug Deliv Rev* 57:2106–29.

Takahashi, S. and Matsuoka, O. 1981. Cross placental transfer of 198Au-colloid in near term rats. *J Radiat Res* 22:242–9.

Thiagarajan, G., Greish, K., and Ghandehari, H. 2013. Charge affects the oral toxicity of poly (amidoamine) dendrimers. *Eur J Pharm Biopharm* 84:330–4.

Tsuchiya, T., Oguri, I., Yamakoshi Y.N. et al. 1996. Novel harmful effects of [60]fullerene on mouse embryos in vitro and in vivo. *FEBS Lett* 393:139–45.

Veiseh, O., Sun, C., Fang, C. et al. 2009. Specific targeting of brain tumors with an optical/ magnetic resonance imaging nanoprobe across the blood-brain barrier. *Cancer Res* 69:6200–7.

Vermylen, J., Nemmar, A., Nemery, B. et al. 2005. Ambient air pollution and acute myocardial infarction. *J Thromb Haemost* 3:1955–61.

Vilar, G., Tulla-Puche, J., and Albericio, F. 2012. Polymers and drug delivery systems. *Curr Drug Deliv* 9:367–94.

Wang, B., Feng, W., Zhu, M. et al. 2009. Neurotoxicity of low-dose repeatedly intranasal instillation of nano- and submicronsized ferric oxide particles in mice. *J Nanopart Res* 11:41–53.

Wang, B., Wang, Y., Feng, W. et al. 2008. Trace metal disturbance in mice brain after intranasal exposure of nano- and submicron-sized Fe$_2$O$_3$ particles. *Chem Anal* 53:927–42.

Wang, D., Lin, Z., Yao, Z. et al. 2014. Surfactants present complex joint effects on the toxicities of metal oxide nanoparticles. *J Chemosphere* 108:70–5.

Wang, H., Wang, J., Deng, X. et al. 2004. Biodistribution of carbon single-wall carbon nanotubes in mice. *J Nanosci Nanotechnol* 4:1019–24.

Webster, R., Elliott, V., Park, BK. et al. 2009. PEG and PEG conjugates toxicity: Towards an understanding of the toxicity of PEG and its relevance to PEGylated biologicals. In *PEGylated Protein Drugs: Basic Science and Clinical Applications*, ed. F.M. Veronese, 127–46. Basel, Switzerland: Birkhäuser.

Werle, M., Takeuchi, H., and Bernkop-Schnurch, A. 2009. Modified chitosans for oral drug delivery. *J Pharm Sci* 98:1643–56.

Wick, P., Malek, A., Manser, P. et al. 2010. Barrier capacity of human placenta for nanosized materials. *Environ Health Perspect* 18:432–6.

Williams, D. 2008. The relationship between biomaterials and nanotechnology. *Biomaterials* 29:1737–8.

Wiwanitkit, V., Sereemaspun, A., and Rojanathanes, R. 2009. Effect of gold nanoparticles on spermatozoa: The first world report. *Fertil Steril* 91:e7–e8.

Wiwattanapatapee, R., Carreño-Gómez, B., Malik, N. et al. 2000. Anionic PAMAM dendrimers rapidly cross adult rat intestine in vitro: A potential oral delivery system? *Pharm Res* 17:991–8.

Wu, J., Ding, T., and Sun, J. 2013. Neurotoxic potential of iron oxide nanoparticles in the rat brain striatum and hippocampus. *Neurotoxicology* 34:243–53.

Xia, T., Li, N., and Nel, A.E. 2009. Potential health impact of nanoparticles. *Annu Rev Public Health* 30:137–50.

Xiao, K., Li, Y., Luo, J. et al. 2011. The effect of surface charge on *in vivo* biodistribution of PEG-oligocholic acid based micellar nanoparticles. *Biomaterials* 32:3435–46.

Yamaoka, T., Kuroda, M., Tabata, Y. et al. 1995. Body distribution of dextran derivatives with electric charges after intravenous administration. *Int J Pharm* 113:149–57.

Yang, S.T., Wang, X., Jia, G. et al. 2008. Long-term accumulation and low toxicity of single-walled carbon nanotubes in intravenously exposed mice. *Toxicol Lett* 181:182–9.

Yang, T., Nyiawung, D., Silber, A. et al. 2012. Comparative studies on chitosan and polylactic-*co*-glycolic acid incorporated nanoparticles of low molecular weight heparin. *AAPS PharmSciTech* 13:1309–18.

Yeh, T.H., Hsu, L.W., Tseng, M.T. et al. 2011. Mechanism and consequence of chitosanmediated reversible epithelial tight junction opening. *Biomaterials* 32:6164–73.

Yin, L., Ding, J., He, C. et al. 2009. Drug permeability and mucoadhesion properties of thiolated trimethyl chitosan nanoparticles in oral insulin delivery. *Biomaterials* 30:5691–700.

Zhang, C., Qu, G., Sun, Y. et al. 2008. Biological evaluation of *N*-octyl-*O*-sulfate chitosan as a new nano-carrier of intravenous drugs. *Eur J Pharm Sci* 33:415–23.

Zhang, L.W., Zeng, L., Barron, A.R. et al. 2007. Biological interactions of functionalized single-wall carbon nanotubes in human epidermal keratinocytes. *Int J Toxicol* 26:103–13.

Zhang, X.D., Wu, H.Y., Wu, D. et al. 2010a. Toxicologic effects of gold nanoparticles in vivo by different administration routes. *Int J Nanomedicine* 5:771–81.

Zhang, Y., Ali, S.F., Dervishi, E. et al. 2010b. Cytotoxicity effects of graphene and single-wall carbon nanotubes in neural phaeochromocytoma-derived PC12 cells. *ACS Nano* 4:3181–6.

Zhang, Y., Bai, Y., Jia, J. et al. 2014. Perturbation of physiological systems by nanoparticles. *Chem Soc Rev* 43:3762–809.

Zhang, Y., Xu, Y., Li, Z. et al. 2011. Mechanistic toxicity evaluation of uncoated and PEGylated single-walled carbon nanotubes in neuronal PC12 cells. *ACS Nano* 5:7020–33.

Zheng, F., Shi, X.W., Yang, G.F. et al. 2007. Chitosan nanoparticle as gene therapy vector via gastrointestinal mucosa administration: Results of an *in vitro* and *in vivo* study. *Life Sci* 80:388–96.

# Index

## A

Milton Keynes UK
Ingram Content Group UK Ltd.
UKHW030901141024
449569UK00025B/1280